Y0-BZZ-948

CAM DESIGN
and
Manufacturing Handbook

Robert L. Norton P. E.

Worcester Polytechnic Institute

http://www.me.wpi.edu/Norton/

and

Norton Associates Engineering

http://www.designofmachinery.com

Industrial Press

New York

Library of Congress Cataloging-in-Publication Data

Norton, Robert L.

Cam Design and Manufacturing Handbook / Robert L. Norton; with contributions by Ronald G. Mosier.

p. cm

Includes index.

ISBN: 0-8311-3122-5

1. Cams—Design and construction. 2. Cams. Machinery, Kinematics of. 3. Machinery, Dynamics of. I. Title.

TJ206 .N67 2002

621'.8'38—dc21 2001039755

This book is dedicated to my former students,
and to all the folks at The Gillette Company,
from whom I have learned many things.

All text, drawings, and equations in this book were prepared and typeset electronically, by the author, on a *Macintosh*® computer using *Freehand*®, *MathType*®, and *Pagemaker*® desktop publishing software. The body text was set in Times Roman, and headings set in Avant Garde. Printing press plates were made directly from the author's disks.

Cover design by Janet Romano. Cover and title page photos courtesy of The Ferguson Company, St. Louis, MO.

Industrial Press, Inc.
200 Madison Avenue
New York, NY 10016-4078

Copyright © 2002 by Industrial Press Inc., New York. Printed in the United States of America. All rights reserved. This book, or any parts thereof, may not be reproduced, stored in a retrieval system, or transmitted in any form without the permission of the publisher.

10 9 8 7 6 5 4 3 2 1

FOREWORD

For many years, Professor Robert L. Norton has been known not only for his teaching skills, but also for his ability to tie his academic work very closely to practical applications in industry. He has always made time to work in the field, building and maintaining strong relationships with practicing engineers throughout industry. His teachings and practice range from fundamentals to state-of-the-art developments.

Professor Norton's newest book, *Cam Design and Manufacturing Handbook*, is the most comprehensive work we know of on a subject that is so important to machine design. From cover to cover, the book provides valuable information for both students and practicing engineers. It is a complete work, ranging from the basics of cam design and manufacture to advanced topics such as spline functions and vibration analysis.

At Gillette and throughout industry, engineers are constantly challenged to create mechanisms that perform faster and more efficiently than ever before. Sources of acoustic noise must be eliminated or minimized. Machines must run without failure and they must run for long periods of time between routine maintenance. Designing and building equipment that satisfy all of these criteria is essential to achieving high levels of productivity, product quality, and competitiveness. All of these challenges can be addressed through proper attention to cam design.

By working closely with Professor Norton to learn and apply the principles covered in this book, we have been able to manufacture new products with unprecedented levels of productivity, quality, and reliability. We have also been able to improve the performance of existing equipment.

I am confident that students and practicing engineers alike will find this book to be an essential text and reference.

Thomas J. Lyden
Group Director, Manufacturing, Blades & Razors
The Gillette Company
Boston, Massachusetts

PREFACE

Cam-follower systems are an extremely important and ubiquitous component in all kinds of machinery. It is difficult to find examples of machinery that do not use one or more cams in their design. Cams are the first choice of many designers for motion control where high precision, repeatability, and long life are required.* All automotive engines depend on cams for their proper valve function. Most automated production machinery uses cams extensively.

The design and manufacture of cams has changed dramatically in recent years. The development and proliferation of computers in engineering design and of numerical control in manufacturing have completely changed the process of cam design and manufacturing, and very much for the better. Until about the late 1960's cams were designed only by manual graphical layout techniques, manufactured in low quantities by manually controlled machining methods and in high quantities by analog duplication of a hand-dressed master cam. The subtleties of the effects of higher derivatives of the cam's chosen mathematical function were often ignored, due either to ignorance of their importance, or the inability to accurately determine their effects given the lack of computational facilities available at the time.

Currently, it is virtually universal and also very economical to use computer-aided engineering and design techniques to create cam geometry, including proper consideration of the effects of higher derivatives, and also to make the cam with high precision using continuous numerically controlled milling, grinding, or electrical discharge machining (EDM) equipment. A significant number of fundamental research results on the subject of cam design and manufacture have been published in recent years. This book is intended to provide a definitive reference for the design and manufacturing of cam-follower systems by bringing up-to-date cam design technology and cam research together between a single set of covers for the benefit of the design and manufacturing engineering community.

The book takes the subject from an introductory level through advanced topics needed to properly design, model, analyze, specify, and manufacture cam-follower systems. Beginning with a description of "how not to design a cam" in order to point out pitfalls that may not be obvious to the beginner, the proper way to design a cam for multiple and single-dwell situations is developed in detail. All the acceptable (and some unacceptable) classical cam functions are described and their mathematics defined for the common double-dwell application. Polynomial functions are introduced and used for both double- and single-dwell examples. Problems with polynomial cams are defined in detail and ways to design around these problems are discussed. Spline functions are introduced as a class of cam motion functions that can solve the most difficult cam design problems. Many examples are developed to show how splines, especially B-splines, can solve otherwise intractable cam design problems.

The issues of cam pressure angle and radius of curvature are fully addressed for various types of cams and followers: radial, barrel, translating, and oscillating, roller and

* Some machines use pneumatic devices for motion control rather than cams. The former have poor dynamics and accuracy compared to cam-driven machines.

flat-faced. The dynamics of the cam-follower system are introduced along with techniques for modeling the follower system as lumped parameters. Both the inverse dynamic (kinetostatic) and forward dynamic solutions are developed for a multiplicity of models of various degrees of freedom. The extensive literature on these topics is referenced in the bibliography. Residual vibrations in the follower train are addressed along with a number of cam functions that can reduce the level of vibration. Polydyne and splinedyne cams are defined and methods for their calculation described.

Calculations for the cam contour of radial and barrel cams with translating and oscillating roller or flat followers are defined. Cutter compensation algorithms and cam surface generation are defined for all common cam-follower configurations. Conjugate cam calculation is defined as well. Cam materials and manufacturing techniques are described and recommendations made.

Stress analysis of the cam-follower joint is presented in detail along with methods to determine the failure modes of typical cam/follower materials in surface contact under time-varying loads. Lubrication of the cam-follower interface is also addressed as is wear.

Methods for the experimental measurement of acceleration, velocity and displacement of cam-follower systems are described, and examples of such measurements taken on operating machinery are shown. Case studies from automotive and automated manufacturing machinery are presented.

Accompanying the book on CD-ROM is a limited-time trial demonstration copy of the Professional Version of program DYNACAM for WINDOWS V 7.0, written by the author. This program will solve most of the equations described in the book and allows (in its fully licensed version) the design, dynamic modeling, analysis, and generation of follower center, cam surface, and cutter coordinate data for any cam. It also defines conjugates for any cam design. Also included are limited-time trial demonstration versions of programs FOURBAR, SIXBAR, and SLIDER that allow the design and analysis of cam-driven linkages.

The author would like to express his sincere appreciation to Dr. Ronald G. Mosier who wrote Chapter 5 on spline functions and checked many of the equations in other chapters. Also, Dennis Klipp of Klipp Engineering, Waterville, ME, Paul Hollis of Tyco Electronics Corporation, Harrisburg, PA, R. Alan Jordan of Delta Engineering, Muncie IN, and Dr. Thomas A. Dresner, Mountain View, CA provided welcome and helpful comments on the book during its development. Many other people reviewed sections of the book or supplied data, illustrations, and information used in the book. I would like to especially thank Gregory Aviza, Al Duchemin, Charles Gillis, Robert Gordon, Joel Karsberg, Thomas Lyden, Corey Maynard, Edwin Ryan, Edward Swanson, and John Washington, all of the Gillette Company, Boston, MA, and Arthur Borgeson of Borg Engineering, Hanson, MA. Finally, the author thanks his editors at Industrial Press, John Carleo and Janet Romano, for making this the most pleasant and productive book development process yet experienced.

Every effort has been made to ensure that the material in this book is technically correct. If any errors remain, the author takes full responsibility, and will greatly appreciate their being pointed out to him. Please contact him by email at norton@wpi.edu or norton@ designofmachinery.com if you discover any errors in the text or in the accompanying programs. Information on book errata and program updates can be found at http://www.designofmachinery.com.

Robert L. Norton
Mattapoisett, Mass.
August, 2001

Contents

ABOUT THE AUTHORS

ROBERT L NORTON, editor and principal author, has over forty years experience in the practice and teaching of mechanical engineering. He has undergraduate degrees in Mechanical Engineering and in Industrial Technology from Northeastern University in Boston, and an M.S. in Engineering Design from Tufts University in Medford, Mass.

He first encountered and designed cams for camera mechanisms in 1960 at Polaroid Corporation and has been fascinated by them ever since. He has spent many years designing and analyzing cam-follower systems at Polaroid, Gillette, and many other companies through his consulting practice.

He has taught kinematics, dynamics, stress analysis, and machine design to mechanical engineering students for more than thirty years at Northeastern, Tufts, and at Worcester Polytechnic Institute, Worcester, Mass. where he is currently Professor of Mechanical Engineering, head of the design program in the Mechanical Engineering Department, and Director of The Gillette Project Center at WPI.

His cam design program, DYNACAM , is used around the world by many companies large and small and his textbooks on kinematics and machine design are widely used in four languages worldwide. He has published many technical papers in the literature and holds 13 U.S. patents. He is a member of SAE and a Fellow of the ASME. He hopes one day to really understand the mathematics of splines.

email: norton@wpi.edu web: http://www.designofmachinery.com

RONALD G. MOSIER, contributor of Chapter 5 on spline functions, has an M.S. in Physics from Carnegie-Mellon University, Pittsburgh, Pennsylvania and a Ph.D. in Mathematics from Wayne State University, Detroit, Michigan. He worked for over thirty years as an Applied Mathematician at Chrysler Corp., now DaimlerChrysler, AG. The only part of his career that did not pertain to mechanical engineering was shortly after the Berlin Wall was built, when the U.S. Army felt that he was needed to teach guided missile electronics at the Redstone Arsenal in Huntsville, Alabama. He intends to someday learn something about electronics.

Chapter 1

INTRODUCTION

1.0 CAM-FOLLOWER SYSTEMS

Cam-follower systems are everywhere. They are used in a wide variety of devices and machines. It is actually difficult to get through a normal day in an industrialized society without using one or even many of these systems, or using products that were made by cam-driven machines. If you shave, the razor you use was probably made with a cam-driven machine. When you turn the timer knob on the dishwasher or washing machine, you are setting cams that will slowly rotate to activate the wash cycle's events. When you drive to work, cams get you there. Perhaps the most common application for cams is valve actuation in internal combustion (IC) engines—the typical IC engine has two or more cams per cylinder. Many sewing machines use cams to obtain patterned stitching. If you go to a health club, you use cams to actuate many of the weight training machines. Figure 1-1 shows some examples of common cams.

This book will present practical information as well as the mathematical foundation needed to properly design and manufacture cams for a variety of applications, principally those that involve high speed operation and the need for accuracy, precision, and repeatability. Cams of this variety are used extensively in vehicles and in machinery of all types. Automated production machinery, such as screw machines, spring winders, and assembly machines, all rely heavily on cam-follower systems for their operation. The cam-follower is an extremely useful mechanical device, without which the machine designer's tasks would be much more difficult to accomplish.*

1.1 FUNDAMENTALS

Figure 1-2 shows two simple examples of cams and followers. The cam is a specially shaped piece of metal or other material arranged to move the follower in a controlled fashion. The follower's motion may be either rotation or translation. Figure 1-2a shows a rotating cam driving an oscillating (rotating or swinging) follower. Figure 1-2b shows a rotating cam driving a translating (sliding) follower. In these examples, a spring is used to maintain contact between cam and follower. This is referred to as a *force closed* cam joint, meaning that an external force is needed to keep them together. Just as you cannot push on a rope, you cannot pull on a force-closed cam joint.

* As Neklutin said,[4] *Cams can be rightfully considered as a universally useful mechanism. They have decided advantages over all other mechanisms where a stroke starts from a dwell and ends at a dwell, especially for intermittent motion.*

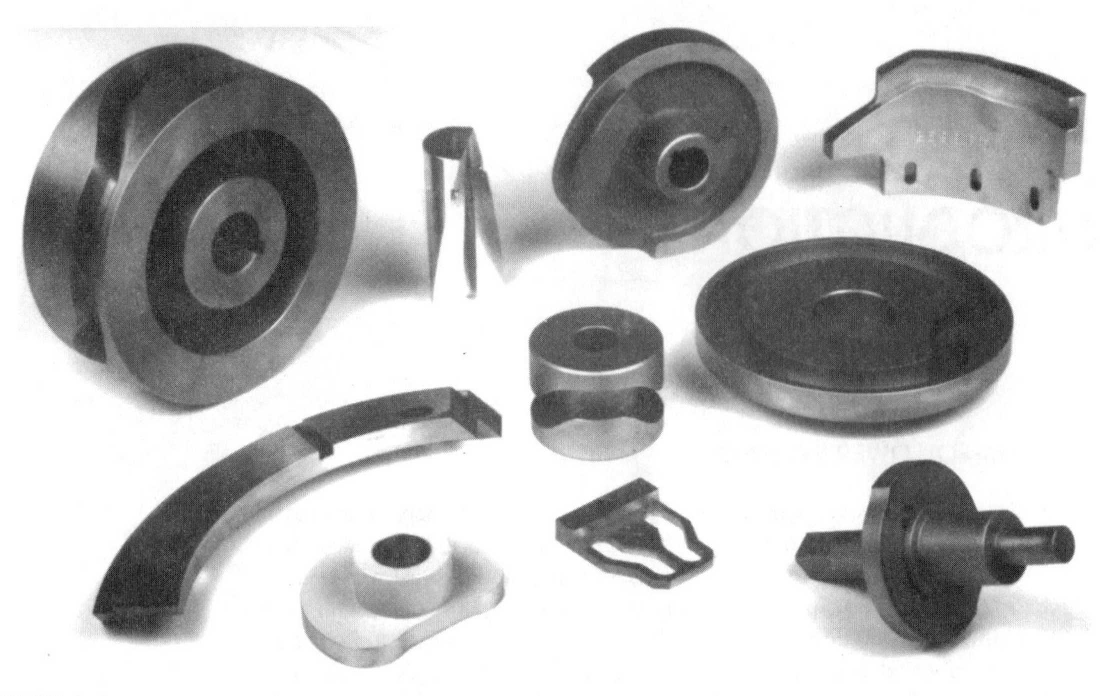

FIGURE 1-1

Examples of cams (Courtesy of Matrix Tool & Machine, Inc, Mentor, OH)

Figure 1-3 shows an alternate arrangement to connect the follower to the cam that does not need a spring. A track or groove in the cam traps the roller follower and now can both push and "pull"—actually it just pushes in both directions. This is called a *form closed* joint as the cam is "formed" around the follower, capturing it by geometry. This type of cam, when used for valve actuation in engines, is also known as **desmodromic**,

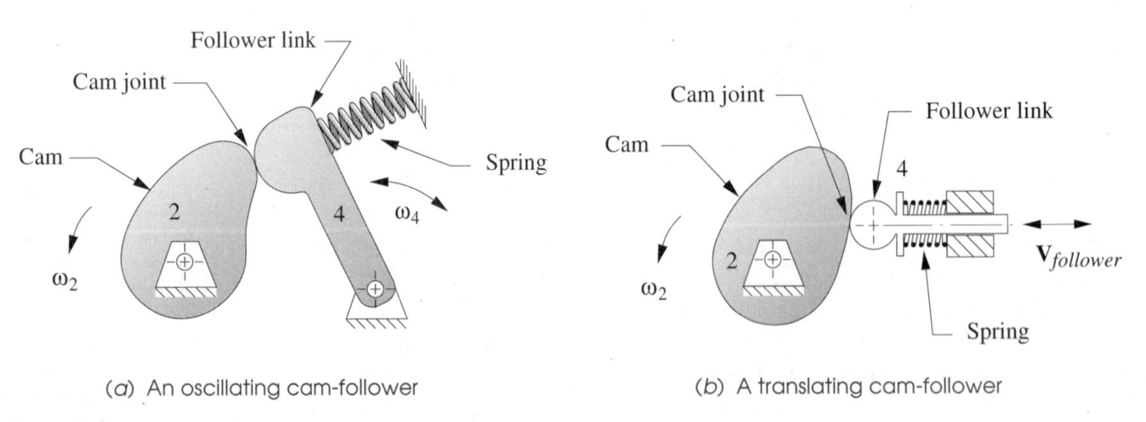

(*a*) An oscillating cam-follower

(*b*) A translating cam-follower

FIGURE 1-2

Force-closed cam-follower systems

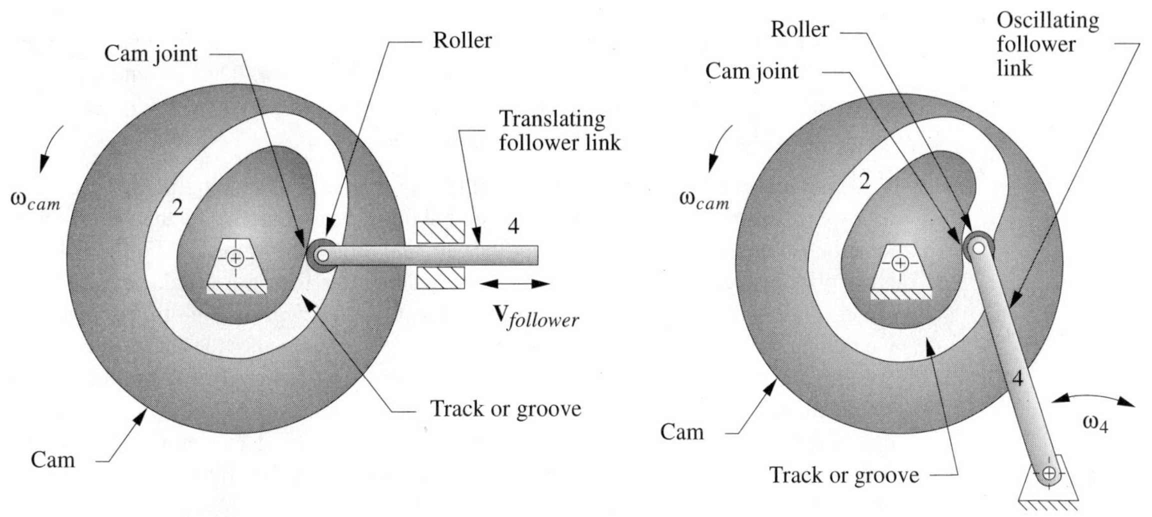

(*a*) Form-closed cam with translating follower link

(*b*) Form-closed cam with oscillating follower link

FIGURE 1-3

Form-closed cam-follower systems

from the French word *desmodromique* meaning *to force to follow a contour*.[1] Both form- and force-closed cams are used extensively in machinery. Their pros and cons will be explored throughout the book.

Kinematically, cam-follower systems are fourbar linkages with one degree of freedom.[2] Figure 1-4a shows the cam-follower of Figure 1-2a; Figure 1-4b then shows the effective linkage that is (for one instantaneous position) kinematically equivalent to the

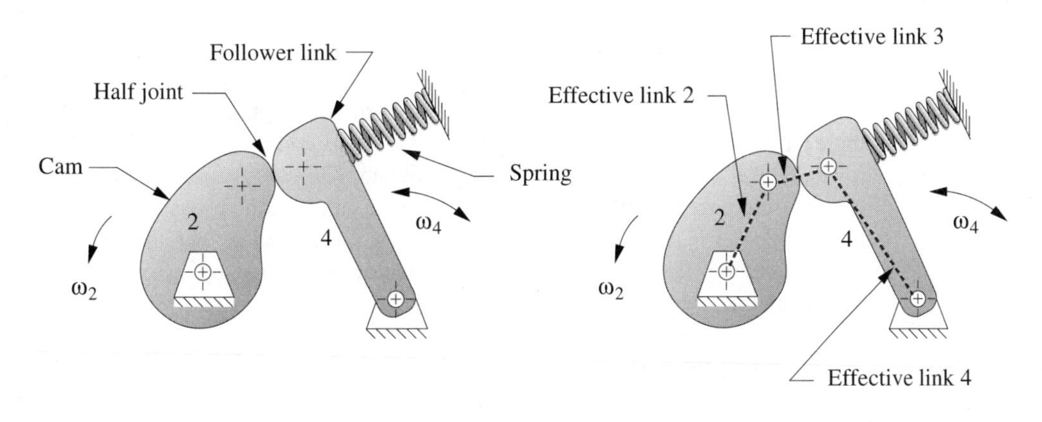

(*a*) An oscillating cam-follower

(*b*) Its effective fourbar linkage equivalent

FIGURE 1-4

Effective linkage of a cam-follower

cam-follower shown in Figure 1-4a. The cam and oscillating arm follower of Figure 1-4a is equivalent to a particular pin-jointed *fourbar crank-rocker* linkage that will change at each position. It is shown in Figure 1-4b for only one position. The cam and translating follower of Figure 1-2b is also equivalent to a particular pin-jointed *fourbar slider-crank* linkage that changes its geometry at each position.

This "equivalent linkage" then has different geometry (link lengths) for each cam-follower position. The lengths of the effective links are determined by the instantaneous locations of the centers of curvature of the cam and follower as shown in Figure 1-4. This is the principal advantage of the cam-follower compared to a "pure" linkage, i.e., it is, in effect, a variable-length linkage that provides a greater degree of motion control than would one with fixed link lengths. It is this difference that makes the cam-follower such a flexible and useful *function generator*. We can specify virtually any output function we desire and create a curved surface on the cam to generate a good approximation to that function in the motion of the follower. The velocities and accelerations of the cam-follower system can be found by analyzing the behavior of the effective linkage for any position. A proof of this can be found in McPhate.[2]

However, in engineering, all advantages come with concomitant disadvantages and the cam-follower system has many such trade-offs. For example, compared to linkages, cams are more compact and easier to design for a specific output function, but they are much more difficult and expensive to make than a linkage. Both positive and negative aspects of cam-follower systems will be explored in the ensuing chapters. For more information on linkage design and kinematics, see Norton.[3]

1.2 TERMINOLOGY

Cam-follower systems can be classified in several ways: by *type of follower motion*, by *type of joint closure*, by *type of follower*, by *type of cam*, by *type of motion constraints*, or by *type of motion program*.

Type of Follower Motion

Figure 1-2a (p. 2) shows a system with an oscillating (rotating or swinging) follower. All three terms are used interchangeably. Figure 1-2b (p. 2) shows a translating (sliding) follower. The choice between these two types of cam-follower is usually determined by the type of output motion desired. If true rectilinear translation is required, then the translating follower is needed. If a pure rotation output is needed, then the oscillator is the obvious choice. There are advantages to each of these approaches separate from their motion characteristics. These will be discussed in Chapters 7 and 18.

Type of Joint Closure

Force and form closure were introduced earlier. **Force closure**, as shown in Figure 1-2, *requires an external force to be applied to the joint* in order to keep the two links, cam and follower, physically in contact. This force is usually provided by a spring or sometimes by an air cylinder. This force, defined as positive in a direction that closes the joint,

Portions of this chapter were adapted from R. L. Norton, *Design of Machinery* 2ed. McGraw-Hill, 2001, with permission.

cannot be allowed to become negative. If it does, the links lose contact because a *force-closed joint can only push, not pull*. **Form closure**, as shown in Figure 1-3, *closes the joint by geometry*. No external force is required. There are really two cam surfaces in this arrangement, one surface on each side of the follower. Each surface pushes, in its turn, to drive the follower in both directions.

Figure 1-3a and b shows track or groove cams that capture a single follower in the groove and both push and pull on the follower link. Figure 1-5 shows another variety of form-closed cam-follower arrangement, called **conjugate cams**. There are two cams fixed on a common shaft that are mathematical conjugates of one another. Two roller followers, attached to a common arm, are each pushed in opposite directions by one conjugate cam. The conjugate nature of the two cam surfaces provides that the distance across the two cam surfaces between rollers remains essentially constant. When form-closed (or conjugate) cams are used in automobile or motorcycle engine valve trains, they are called **desmodromic** cams.[1] The advantages and disadvantages to both force- and form-closed arrangements will be discussed in later chapters.

Type of Follower

Follower, in this context, refers only to that part of the follower link which contacts the cam. Figure 1-6 shows three common arrangements, **flat-faced**, **mushroom** (curved), and **roller**. The roller follower has the advantage of lower (rolling) friction than the sliding contact of the other two, but can be more expensive. **Flat-faced followers** can package smaller than roller followers for some cam designs; they are often favored for that reason as well as cost for some automotive valve trains. Many modern automotive engine valve trains now use roller followers for their lower friction.

Roller followers are commonly used in production machinery where their ease of replacement and availability from bearing manufacturers' stock in any quantities are ad-

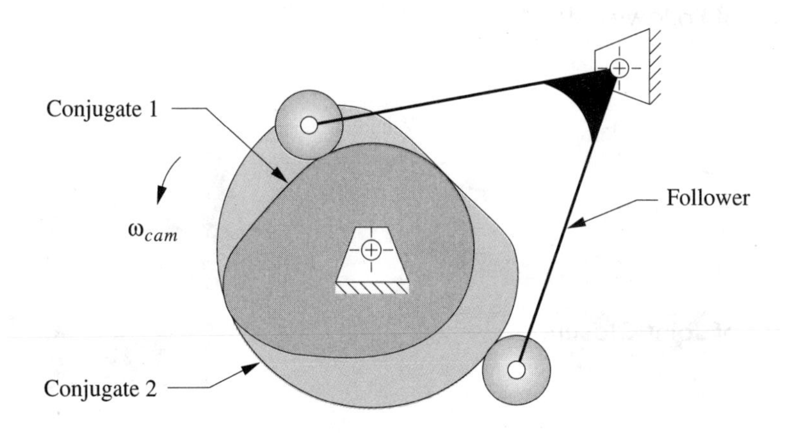

FIGURE 1-5

A conjugate cam pair

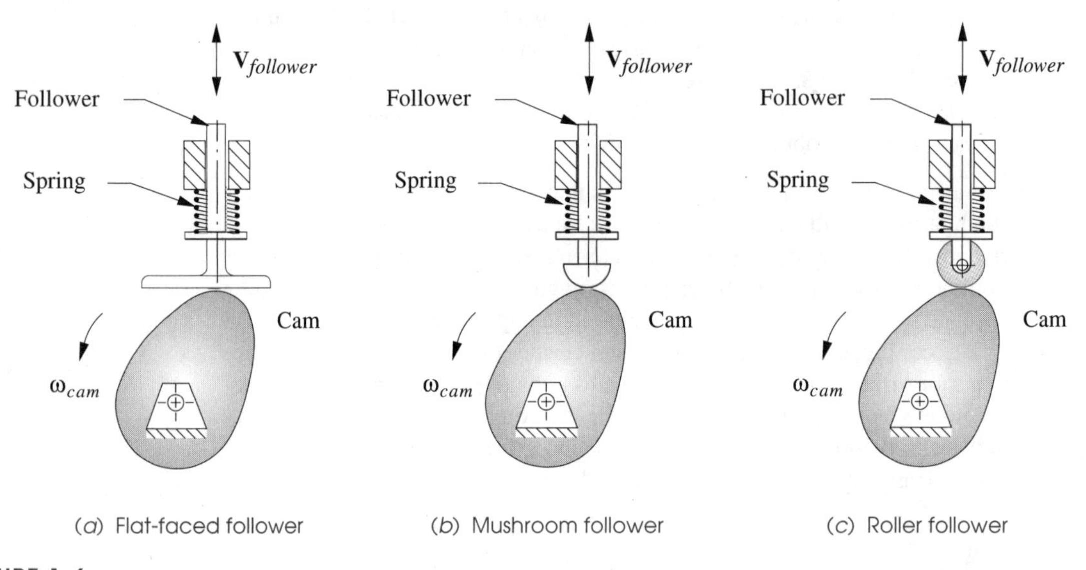

FIGURE 1-6

Three common types of cam followers

vantages. Grooved or track cams require roller followers. Roller followers are essentially ball or roller bearings with customized mounting details. Figure 1-7 shows two common types of commercial roller followers. Flat-faced or **mushroom followers** are usually custom designed and manufactured for each application. For high-volume applications such as automobile engines, the quantities are high enough to warrant a custom-designed follower.

Type of Cam

The direction of the follower's motion relative to the axis of rotation of the cam determines whether it is a **radial** or **axial** cam. All of the cams shown in Figures 1-2 to 1-6

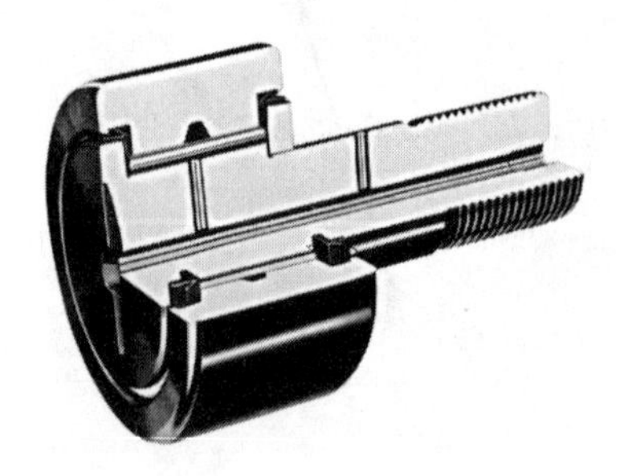

FIGURE 1-7

Commercial roller followers (Courtesy of McGill Manufacturing Co., South Bend, IN)

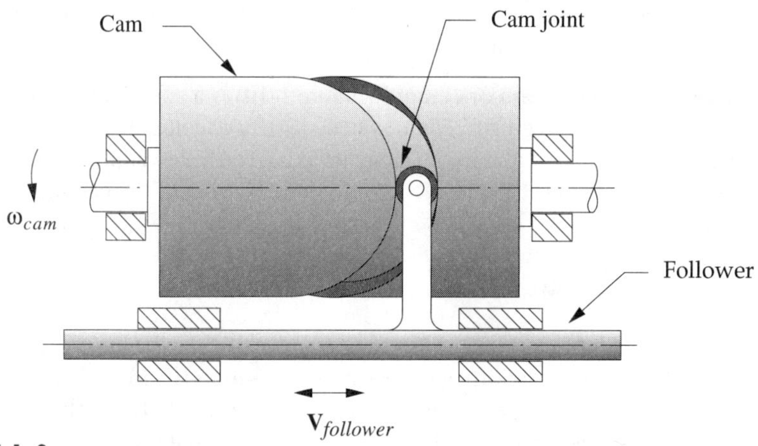

FIGURE 1-8

Axial, cylindrical, or barrel cam with form-closed, translating follower

are radial cams because the follower motion is generally in a radial direction. Open (force-closed) **radial cams** are also called **plate cams**.

Figure 1-8 shows an **axial cam** whose follower moves parallel to the axis of cam rotation. This arrangement is also called a **face** cam if open (force-closed) and a **cylindrical** or **barrel** cam if grooved or ribbed (form-closed).*

Figure 1-9 shows a selection of cams of various types. Clockwise from the lower left, they are: an open axial or face cam (force-closed); an axial grooved (track) cam

FIGURE 1-9

Commercial cams of various types (Courtesy of The Ferguson Co. St. Louis, MO)

* We will refer to this type of cam as a barrel cam throughout the book.

(form-closed) with external gear; an open radial or plate cam (force-closed); a ribbed axial cam (form-closed); and an axial grooved (barrel) cam.

A **three-dimensional cam** or **camoid** (Figure 1-10) is a combination of radial and axial cams. It is a two-degree-of-freedom system. The two inputs are rotation of the cam about its axis and translation of the cam along its axis. The follower motion is a function of both inputs. The follower tracks along a different portion of the cam depending on the axial input.

Type of Motion Constraints

There are two general categories of motion constraint, **critical extreme position** (CEP—also called endpoint specification) and **critical path motion** (CPM). **Critical extreme position** refers to the case in which the design specifications define only the start and finish positions of the follower (i.e., the extreme positions) but do not specify any constraints on the path motion between those extreme positions.* This case is the easier one to design as one has complete freedom to choose the cam functions that control the motion between the extremes. **Critical path motion** is a more constrained problem than CEP because the path motion and/or one or more of its derivatives are defined over all or part of the interval of motion. This requires the generation of a particular function to match the given constraints. It may only be possible to create an approximation of the specified function and still maintain suitable dynamic behavior.

Type of Motion Program

The motion programs **rise-fall** (RF), **rise-fall-dwell** (RFD), and **rise-dwell-fall-dwell** (RDFD)[4] all refer mainly to the CEP case of motion constraint. In effect they define how many dwells are present in the full cycle of motion, either none (RF), one (RFD), or more than one (RDFD). **Dwells,** defined as *no output motion for a specified period of input motion*, are an important feature of cam-follower systems.

FIGURE 1-10

A three-dimensional cam or camoid (Courtesy of Matrix Tool & Machine, Inc, Mentor, OH)

* In some applications it may be desireable to also maximize the area under the curve between the endpoints. An example is the displacement function for an automotive valve motion where area under the curve relates to air flow past the open valve. Such a constraint will affect the choice of follower motion function. Such considerations will be addressed in later chapters after spline functions are introduced.

A cam-follower is the mechanism of choice whenever a dwell is required as it is very easy to create precise dwells in cam-follower mechanisms. Pure linkages can, at best, only provide an approximate dwell. Single- or double-dwell linkages tend to be quite large for their output motion and are somewhat difficult to design. In general, cam-follower systems tend to be more compact than linkages for the same output motion. For the RF case (no dwell), a pin-jointed fourbar crank-rocker linkage is often a superior solution and should be considered before designing a cam-follower solution. See Norton.[3]

A cam-follower is an obvious choice for the RFD and RDFD cases. However, each of these two cases has its own set of constraints on the behavior of the cam functions at the interfaces between the segments that control the rise, fall, and dwell(s). The kinematic constraints that drive the choice of cam profile functions are different for the RFD and RDFD cases. Adding more dwells than two does not change the character of the kinematic constraints from that of the RDFD case. So, a multi-dwell cam is kinematically similar to a double-dwell cam, but both are different than a single-dwell cam in terms of the type of motion program needed. In general, we must match the **boundary conditions** (BC) of the follower displacement functions and their derivatives at all interfaces between the segments of the cam. This topic will be discussed in the next chapter.

1.3 APPLICATIONS

Figure 1-11 shows a cam-follower system used in an automotive valve actuation application. This is an overhead camshaft engine. The camshaft operates against an oscillating follower arm that, in turn, opens the valve. The cam joint is force-closed by the valve spring. The camshaft is typically driven from the engine's crankshaft by gears, chain, or

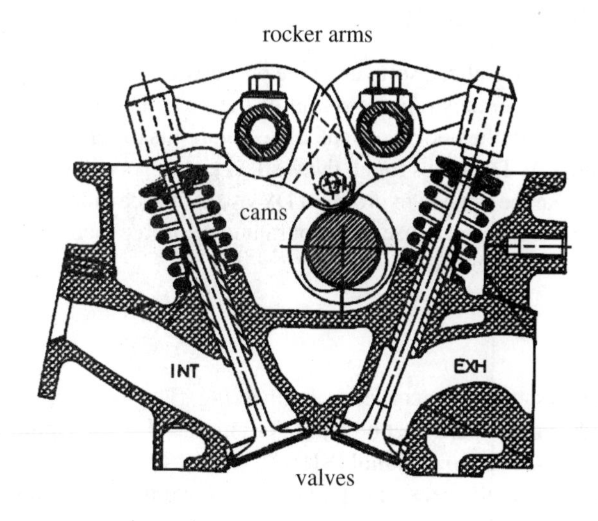

FIGURE 1-11

An overhead camshaft automotive valve train. (Courtesy of DaimlerChrysler Corp., Auburn Hills, MI)

toothed belt at a 1:2 reduction. Maximum camshaft speeds in these applications can range from about 2 500 rpm in large automobile engines to over 10 000 rpm in motorcycle racing engines.

Figure 1-12 shows a typical automated assembly machine cam-follower train. Two cams are shown, each of which drives a linkage that actuates tooling in one of several assembly stations along a conveyor line. The tooling will insert a part, crimp a fastener, or do some other operation on the product that is being automatically assembled as it is carried along on the conveyor. A machine of this type may have several dozen of these cam-follower trains arrayed along one or more large camshafts that run the length of the machine (ten meters or more). They may have a mix of force- and form-closed followers. Those shown in the figure are force-closed with air cylinders acting as springs. Maximum camshaft speeds in these applications typically range from about 100 to 1000 rpm.

1.4 TIMING DIAGRAMS

When a machine such as an internal combustion (IC) engine, or an assembly machine that requires several operations, is being designed, one of the first tasks is to define its timing diagram. The timing diagram specifies the relative phasing of all the related events within the machine's cycle. For an IC engine, the timing diagram defines the duration of each cylinder's valve openings and their phasing relative to some rotational reference (or fiduciary) such as top dead center (TDC). For an assembly machine, it defines the start and end points of all motions relevant to the assembly task and relates their phasing to a rotational reference. The cam designs will all flow from this timing diagram. Figure 1-13 shows a timing diagram for a subset of the tooling motions on a typical automated assembly machine.

1.5 CAM DESIGN SOFTWARE

Cam design requires specialized software to be properly done. This book contains a CD-ROM that contains a demonstration version of the program DYNACAM for Windows. Most of the examples in the book were calculated with this program. Instructions for its use are in Appendix A. Data files for all the worked examples in the book are also on the CD-ROM. These can be opened in program DYNACAM to see more details of their solution. Readers are encouraged to use this program to reinforce their understanding of the concepts presented in the book.

1.6 UNITS

There are several systems of units used in engineering. The most common in the United States are the **U.S. foot-pound-second (fps) system**, the **U.S. inch-pound-second (ips) system**, and the **System International (SI)**. All systems are created from the choice of three of the quantities in the general expression of Newton's second law:

$$F = \frac{ml}{t^2} \tag{1.1a}$$

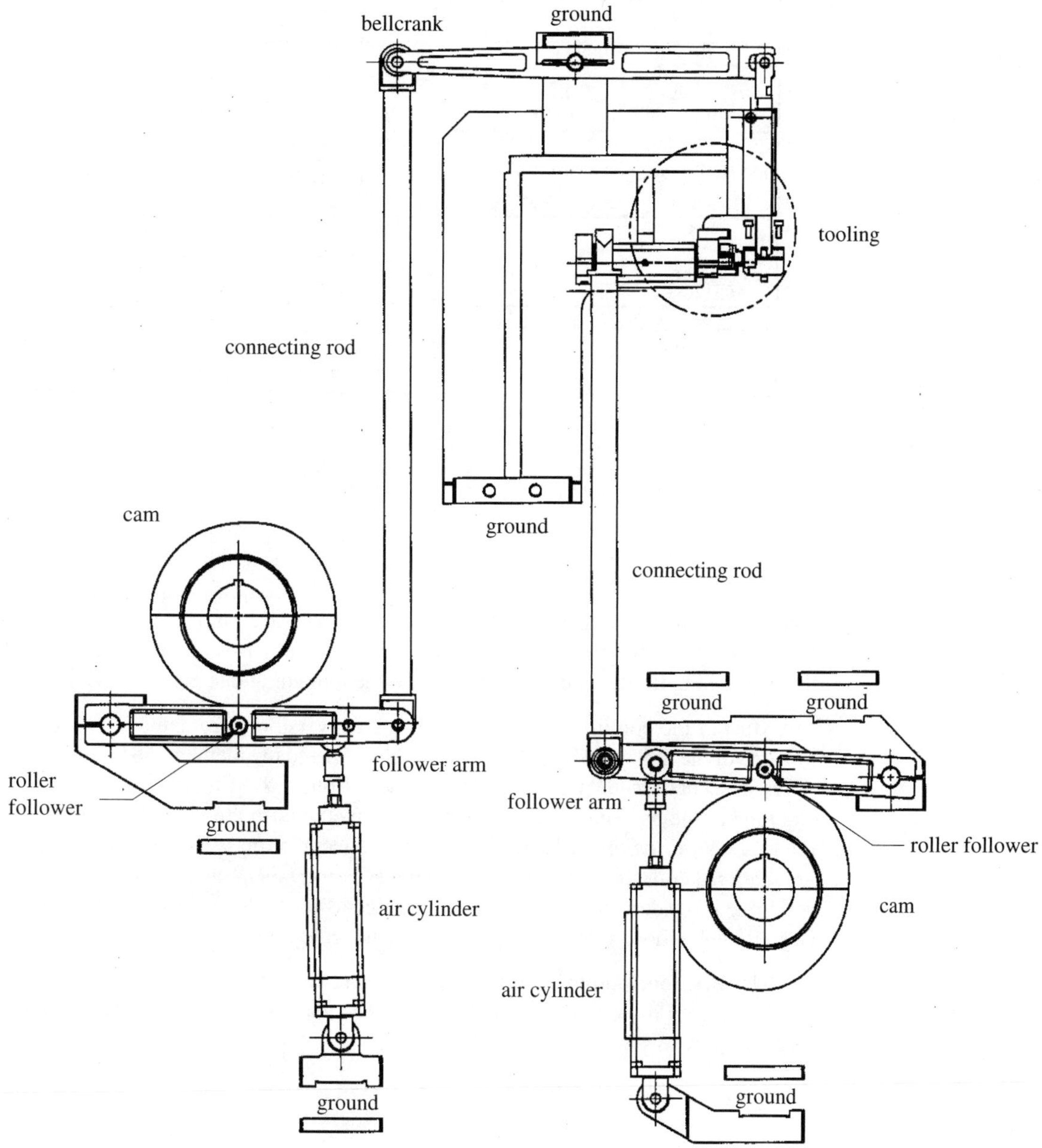

FIGURE 1-12

Cam-follower mechanisms for one station of an automated assembly machine (Courtesy of the Gillette Co., Boston, MA)

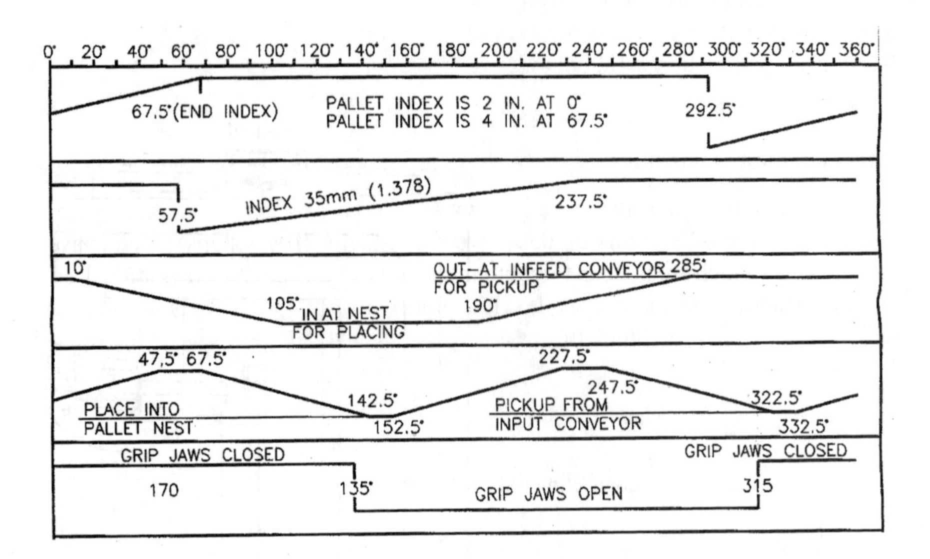

FIGURE 1-13

Partial timing diagram for an automated assembly machine (Courtesy of the Gillette Co., Boston, MA)

where F is force, m is mass, l is length, and t is time. The units for any three of these variables can be chosen and the other is then derived in terms of the chosen units. The three chosen units are called *base units*, and the remaining one is then a *derived unit.*

Most of the confusion that surrounds the conversion of computations between either one of the U.S. systems and the SI system is due to the fact that the SI system uses a different set of base units than the U.S. systems. Both U.S. systems choose ***force***, *length,* and *time* as the base units. Mass is then a derived unit in the U.S. systems, which are referred to as *gravitational systems* because the value of mass is dependent on the local gravitational constant. The SI system chooses ***mass***, *length,* and *time* as the base units and force is the derived unit. SI is then referred to as an *absolute system* since the mass is a base unit whose value is not dependent on local gravity.

The **U.S. foot-pound-second (fps)** system requires that all lengths be measured in feet (ft), forces in pounds (lb), and time in seconds (sec). Mass is then derived from Newton's law as

$$m = \frac{Ft^2}{l} \tag{1.1b}$$

and the units are:

Pounds seconds squared per **foot** (lb-sec^2/ft) = **slugs**

The **U.S. inch-pound-second (ips)** system requires that all lengths be measured in inches (in), forces in pounds (lb), and time in seconds (sec). Mass is still derived from Newton's law, equation 1.1b, but the units are now:

Pounds seconds squared per **inch** (lb-sec^2/in) = **blobs**

This mass unit is not slugs! It is worth twelve slugs or one blob.*

Weight is defined as the force exerted on an object by gravity. Probably the most common units error is to mix up these two unit systems (**fps** and **ips**) when converting weight units (which are pounds force) to mass units. Note that the gravitational acceleration constant (g) on earth at sea level is approximately 32.2 **feet** per second squared which is equivalent to 386 **inches** per second squared. The relationship between mass and weight is:

Mass = weight / gravitational acceleration

$$m = \frac{W}{g} \tag{1.2}$$

It should be obvious that, if you measure all your lengths in **inches** and then use g = 32.2 **feet**/sec^2 to compute mass, you will have an error of a *factor of 12* in your results. This is a serious error, large enough to crash the airplane you designed. Even worse off is one who neglects to convert weight to mass *at all* in his calculations. He will have an error of either 32.2 or 386 in his results. This is enough to sink the ship!

To add even further to the confusion about units is the common use of the unit of **pounds mass** (lb$_m$). This unit is often used in fluid dynamics and thermodynamics; it comes about through the use of a slightly different form of Newton's equation:

$$F = \frac{ma}{g_c} \tag{1.3}$$

where m = mass in lb$_m$, a = acceleration and g_c = the gravitational constant.

The value of the **mass** of an object measured in **pounds mass** (lb$_m$) is *numerically equal* to its **weight** in **pounds force** (lb$_f$). However, the student *must remember to divide* the value of m in lb$_m$ by g_c when substituting into this form of Newton's equation. Thus the lb$_m$ will be divided either by 32.2 or by 386 when calculating the dynamic force. The result will be the same as when the mass is expressed in either slugs or blobs in the $F = ma$ form of the equation. Remember that in round numbers at sea level on earth:

1 lb$_m$ = 1 lb$_f$ 1 slug = 32.2 lb$_f$ 1 blob = 386 lb$_f$

The **SI** system requires that lengths be measured in meters (m), mass in kilograms (kg), and time in seconds (sec). This is sometimes also referred to as the **mks** system. Force is derived from Newton's law, equation 1.1b and the units are:

kilogram-meters per second^2 (kg-m/sec^2) = newtons

Thus in the SI system, there are distinct names for mass and force which helps alleviate confusion. When converting between SI and U.S. systems, be alert to the fact that mass converts from kilograms (kg) to either slugs (sl) or blobs (bl), and force converts

* It is unfortunate that the mass unit in the **ips** system has never officially been given a name such as the term ***slug*** used for mass in the **fps** system. The author boldly suggests (with tongue only slightly in cheek) that this unit of mass in the **ips** system be called a *blob* (bl) to distinguish it more clearly from the ***slug*** (sl), and to help avoid some of the common units errors listed above.

Twelve slugs = one blob.

Blob does not sound any sillier than slug, is easy to remember, implies mass, and has a convenient abbreviation (bl) which is an anagram for the abbreviation for pound (lb). Besides, if you have ever seen a garden slug, you know it looks just like a "*little blob.*"

1

Table 1-1 Variables and Units
Base Units in Boldface – Abbreviations in ( )

Variable	Symbol	ips unit	fps unit	SI unit
Force	F	**pounds (lb)**	**pounds (lb)**	newtons (N)
Length	l	**inches (in)**	**feet (ft)**	**meters (m)**
Time	t	**seconds (sec)**	**seconds (sec)**	**seconds (sec)**
Mass	m	lb–sec^2/in (bl)	lb–sec^2/ft (sl)	**kilograms (kg)**
Weight	W	pounds (lb)	pounds (lb)	newtons (N)
Velocity	v	in/sec	ft/sec	m/sec
Acceleration	a	in/sec^2	ft/sec^2	m/sec^2
Jerk	j	in/sec^3	ft/sec^3	m/sec^3
Angle	θ	degrees (deg)	degrees (deg)	degrees (deg)
Angle	θ	radians (rad)	radians (rad)	radians (rad)
Angular velocity	ω	rad/sec	rad/sec	rad/sec
Angular acceleration	α	rad/sec^2	rad/sec^2	rad/sec^2
Angular jerk	φ	rad/sec^3	rad/sec^3	rad/sec^3
Torque	T	lb–in	lb–ft	N–m
Mass moment of inertia	I	lb–in–sec^2	lb–ft–sec^2	N–m–sec^2
Energy	E	in–lb	ft–lb	joules
Power	P	in–lb / sec	ft–lb / sec	watts
Volume	V	in^3	ft^3	m^3
Weight density	γ	lb/in^3	lb/ft^3	N/m^3
Mass density	ρ	bl/in^3	sl/ft^3	kg/m^3

from newtons (N) to pounds (lb). The gravitational constant (g) in the SI system is approximately 9.81 m/sec^2.

The principal system of units used in this book will be the U.S. **ips** system. Most machine design in the United States is still done in this system. Table 1-1 shows some of the variables used in this text and their units. Table 1-2 shows conversion factors between the U.S. and SI systems.

The reader is cautioned to always check the units in any equation written for a problem solution. If properly written, an equation should cancel all units across the equal sign. If it does not, then you can be *absolutely sure it is* ***incorrect.*** Unfortunately, a unit balance in an equation does not guarantee that it is correct, as many other errors are possible.

Table 1-2 Selected Units Conversion Factors

Multiply this	by	this	to get	this
acceleration				
in/sec^2	x	0.0254	=	m/sec^2
ft/sec^2	x	12	=	in/sec^2
angles				
radian	x	57.2958	=	deg
area				
in^2	x	645.16	=	mm^2
ft^2	x	144	=	in^2
area moment of inertia				
in^4	x	416 231	=	mm^4
in^4	x	4.162E–07	=	m^4
m^4	x	1.0E+12	=	mm^4
m^4	x	1.0E+08	=	cm^4
ft^4	x	20 736	=	in^4
density				
lb/in^3	x	27.6805	=	g/cc
g/cc	x	0.001	=	g/mm^3
lb/in^3	x	1 728	=	lb/ft^3
kg/m^3	x	1.0E–06	=	g/mm^3
force				
lb	x	4.448	=	N
N	x	1.0E+05	=	dyne
ton (short)	x	2 000	=	lb
length				
in	x	25.4	=	mm
ft	x	12	=	in
mass				
blob	x	386.4	=	lb
slug	x	32.2	=	lb
blob	x	12	=	slug
kg	x	2.205	=	lb
kg	x	9.8083	=	N
kg	x	1 000	=	g
mass moment of inertia				
lb-in-sec^2	x	0.1138	=	N-m-sec^2
moments and energy				
in-lb	x	0.1138	=	N-m
ft-lb	x	12	=	in-lb
N-m	x	8.7873	=	in-lb
N-m	x	0.7323	=	ft-lb
power				
HP	x	550	=	ft-lb/sec
HP	x	33 000	=	ft-lb/min
HP	x	6 600	=	in-lb/sec
HP	x	745.7	=	watts
N-m/sec	x	8.7873	=	in-lb/sec
pressure and stress				
psi	x	6 894.8	=	Pa
psi	x	6.895E-3	=	MPa
psi	x	144	=	psf
kpsi	x	1 000	=	psi
N/m^2	x	1	=	Pa
N/mm^2	x	1	=	MPa
spring rate				
lb/in	x	175.126	=	N/m
lb/ft	x	0.08333	=	lb/in
stress intensity				
MPa-m$^{0.5}$	x	0.909	=	ksi-in$^{0.5}$
velocity				
in/sec	x	0.0254	=	m/sec
ft/sec	x	12	=	in/sec
rad/sec	x	9.5493	=	rpm
volume				
in^3	x	16 387.2	=	mm^3
ft^3	x	1 728	=	in^3
cm^3	x	0.061023	=	in^3
m^3	x	1.0E+9	=	mm^3

1.7 REFERENCES

1 **Renstrom, R. C.** (1976). "Desmodromics: Ultimate Valve Gear?". *Motorcycle World*, May, 1976.

2 **McPhate, A. J., and L. R. Daniel**. (1962). "A Kinematic Analysis of Fourbar Equivalent Mechanisms for Plane Motion Direct Contact Mechanisms." *Proc. of Seventh Conference on Mechanisms*, Purdue University, pp. 61-65.

3 **Norton, R. L.** (2001). *Design of Machinery,* 2ed. McGraw-Hill: New York.

4 **Neklutin, C, N**. (1952). "Designing Cams for Controlled Inertia and Vibration". *Machine Design*, June, 1952, pp. 143-160.

Chapter 2

UNACCEPTABLE CAM CURVES

2.0 INTRODUCTION

The author can remember instances of machinists (many years ago) saying something like "Make a cam? No problem, I just lay it out with bluing using circle arcs and straight lines, cut it out on the band saw, sand it smooth with the belt sander and stick it in the machine." This may have worked (and apparently did) when machine speeds were relatively low, but it definitely will not work with current cam speeds that are typically hundreds to thousands of rpm, nor in cases requiring great precision.

This chapter will first show how one should NOT design a cam for a multi-dwell application that involves moderate to high speed operation. What does that mean? What is high speed? Well, it is relative. If the mass of the follower train is very small and its stiffness high, then one can get away with poor dynamic design at higher speeds than one can if the follower train has large mass and/or low stiffness. Newton's second law determines all in this context. Finally, the key to properly designing a cam for good dynamic behavior will be discussed.

2.1 *S V A J* DIAGRAMS

The first task faced by the cam designer, when presented with a timing diagram, is to select the mathematical functions to be used to define the motions of the followers. The easiest approach to this process is to "linearize" the cam, i.e., "unwrap it" from its circular shape* and consider it as a function plotted on Cartesian axes. We plot the displacement function s; its first derivative, velocity v; its second derivative, acceleration a; and its third derivative, jerk j, all on aligned axes as a function of camshaft angle θ as shown in Figure 2-1. Note that we can consider the independent variable in these plots to be either time t or shaft angle θ, as we typically know the angular velocity ω of the camshaft (assumed to be essentially constant) and can easily convert from angle to time and vice versa.

* Assuming it is a radial cam

2

TABLE 2-1 Notation Used in Chapters 2 Through 5

t = time, sec

θ = camshaft angle, degrees (deg) or radians (rad)

ω = camshaft angular velocity, rad/sec

β = total angle of any segment, rise, fall, or dwell, deg or rad

h = total lift (rise or fall) of any one segment, length units

s or S = follower displacement, length units

$v = ds/d\theta$ = follower velocity, length/rad

$V = dS/dt$ = follower velocity, length/sec

$a = dv/d\theta$ = follower acceleration, length/rad^2

$A = dV/dt$ = follower acceleration, length/sec^2

$j = da/d\theta$ =follower jerk, length/rad^3

$J = dA/dt$ = follower jerk, length/sec^3

$s\ v\ a\ j$ refer to the group of diagrams, length units versus radians

$S\ V\ A\ J$ refer to the group of diagrams, length units versus time

Portions of this chapter were adapted from R. L. Norton, *Design of Machinery* 2ed. McGraw-Hill, 2001, with permission.

$$\theta = \omega t \tag{2.1}$$

Figure 2-1a shows the specifications for a four-dwell cam that has eight segments, RDFDRDFD. Figure 2-1b shows the $s\ v\ a\ j$ curves for the whole cam over 360 degrees of camshaft rotation. A cam design begins with a definition of the required cam func-

Segment Number	Function Used	Start Angle	End Angle	Delta Angle
1	Constant velocity rise	0	60	60
2	Dwell	60	90	30
3	Constant acceleration fall	90	150	60
4	Dwell	150	180	30
5	Cubic displacement rise	180	240	60
6	Dwell	240	270	30
7	Simple harmonic fall	270	330	60
8	Dwell	330	360	30

(a) Cam program specifications

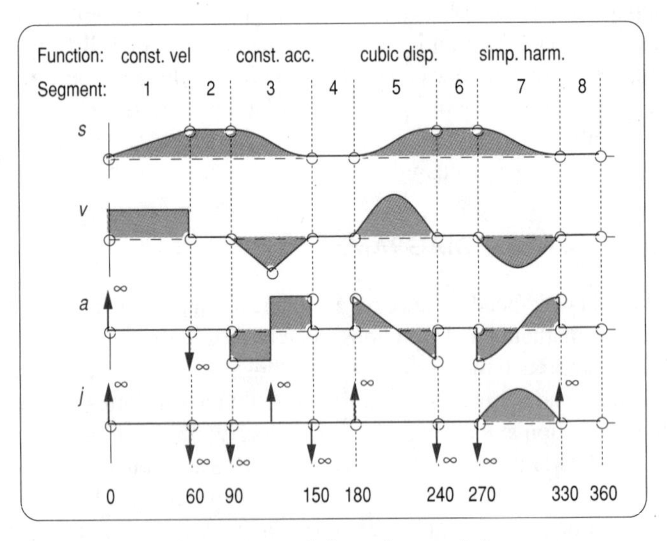

(b) Plots of cam-follower's s v a j diagrams

FIGURE 2-1

Unacceptable motion functions for a multi-dwell cam

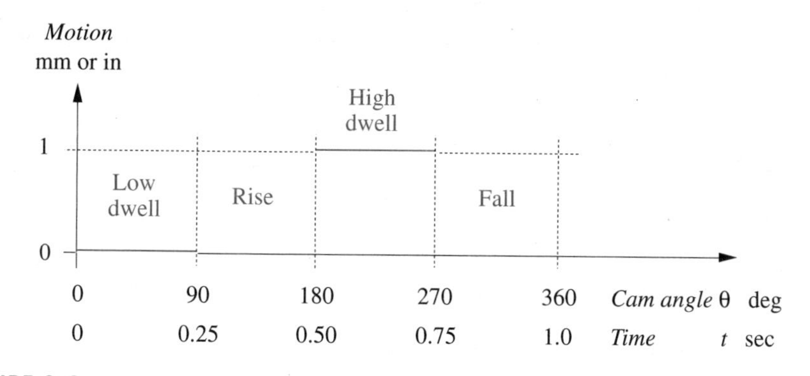

FIGURE 2-2

A cam timing diagram

tions and their *s v a j* diagrams. Functions for the non-dwell cam segments should be chosen based on their velocity, acceleration, and jerk characteristics and the relationships at the interfaces between adjacent segments including the dwells. The functions shown in Figure 2-1 are all **dynamically unacceptable** when used between dwells for reasons that will now be explored.

2.2 DOUBLE-DWELL CAM DESIGN—CHOOSING S V A J FUNCTIONS

Many cam design applications require multiple dwells and the double-dwell case is quite common. Perhaps a **double-dwell** cam is driving a part feeding station (also called a *pick and place* mechanism) on a production machine that makes a high-volume consumer product. This hypothetical cam's follower is fed a part during the low dwell and its articulated gripper grabs the part. The cam then moves the part into the assembly (during the rise), holding it absolutely still in a **critical extreme position** (CEP) during the high dwell while articulated tooling (separately driven) fastens the part into the assembly. The grippers open and the follower returns (falls) to the low dwell position to pick up another part and repeat the process.

Cam specifications such as this are often depicted on a timing diagram as shown in Figure 2-2 which is a graphical representation of the specified events in the machine cycle. A **machine's cycle** is defined as *one revolution of its master driveshaft*. In a complicated machine there will be a **timing diagram** for each subassembly in the machine (Figure 1-13, p. 12). The time relationships among all subassemblies are defined by their timing diagrams which are all drawn on a common time (or angle) axis. Obviously all these operations must be kept precisely synchronous and in time phase for the machine to work.

The simple example in Figure 2-2 is a critical extreme position (CEP) case, because nothing is specified about the functions to be used to get from the low dwell position (one extreme) to the high dwell position (other extreme). The designer is free to choose any function that will do the job. Note that these specifications contain only information about the displacement function. The higher derivatives are not specifically constrained in this example. We will now use this problem to investigate several different ways to meet the specifications.

EXAMPLE 2-1

Naive Cam Design—A Bad Cam.

Problem: Consider the following cam design CEP specification:

dwell	at zero displacement for 90° (low dwell)
rise	1 in (25.4 mm) in 90°
dwell	at 1 in (25.4 mm) for 90° (high dwell)
fall	1 in (25.4 mm) in 90°
cam ω	2π rad/sec = 1 rev/sec

Solution:

1 The naive or inexperienced cam designer might proceed with a design as shown in Figure 2-3a. Taking the given specifications literally, it is tempting to merely "connect the dots" on the timing diagram to create the displacement (s) diagram. After all, when we wrap

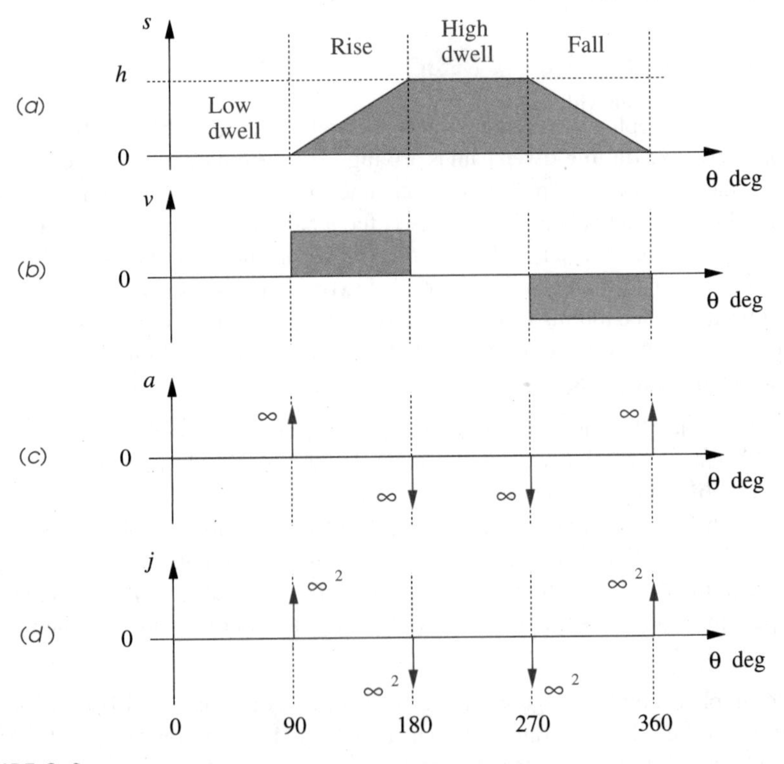

FIGURE 2-3

The *s v a j* diagrams of a "bad" cam design—pure constant velocity this s diagram around a circle to create the actual cam, it will look quite smooth despite the sharp corners on the s diagram. The mistake our beginning designer is making here is to ignore the effect on the higher derivatives of the displacement function that results from this simplistic approach.

2

2 Figures 2-3b, c, and d show the problem. Note that we have to treat each segment of the cam (rise, fall, dwell) as a separate entity in developing mathematical functions for the cam. Taking the rise segment first, the displacement function in Figure 2-3a during this portion is a straight line, or first-degree polynomial. The general equation for a straight line is:

$$y = mx + b \tag{2.2}$$

where m is the slope of the line and b is the y intercept. Substituting variables appropriate to this example in equation 2.2, angle θ replaces the independent variable x, and the displacement s replaces the dependent variable y. By definition, the constant slope m of the displacement is the velocity constant K_v.

3 For the rise segment, the y intercept b is zero because the low dwell position typically is taken as zero displacement by convention. Equation 2.2 then becomes:

$$s = K_v \theta \tag{2.3}$$

4 Differentiating with respect to θ gives a function for velocity during the rise.

$$v = K_v = \text{constant} \tag{2.4}$$

5 Differentiating again with respect to θ gives a function for acceleration during the rise.

$$a = 0 \tag{2.5}$$

This seems too good to be true—and it is. From Newton's second law, zero acceleration means zero dynamic force. This cam appears to have no dynamic forces or stresses in it!

Figure 2-3 shows what is really happening here. If we return to the displacement function and graphically differentiate it twice, we will observe that, from the definition of the derivative as the instantaneous slope of the function, the acceleration is in fact zero **during the interval**. *But, at the boundaries of the interval,* where rise meets low dwell on one side and high dwell on the other, note that *the velocity function is multivalued. There are discontinuities at these boundaries*. The effect of these discontinuities is to create a portion of the velocity curve which has **infinite slope** and zero duration. This results in the *infinite spikes of acceleration* shown at those points.

These spikes are more properly called **Dirac delta functions**. Infinite acceleration cannot really be obtained, as it requires infinite force. Clearly the dynamic forces will be very large at these boundaries and will create high stresses and rapid wear. In fact, if this cam were built and run at any significant speed, the sharp corners on the displacement diagram that are creating these theoretical infinite accelerations would be quickly worn to a different contour by the unsustainable stresses generated in the materials. *This is an unacceptable design.*

The unacceptability of this design is reinforced by the **jerk** diagram (Figure 2-3d), which shows theoretical values of **infinity squared*** at the discontinuities. The problem has been engendered by an inappropriate choice of displacement function. In fact, the cam designer generally should not be as concerned with the displacement function as with its higher derivatives.

2.3 THE FUNDAMENTAL LAW OF CAM DESIGN

Any cam designed for operation at other than very low speeds must be designed with the following constraints:

The cam-follower function must be continuous through the first and second derivatives of displacement (i.e., velocity and acceleration) across the entire interval.

Corollary:

The jerk function must be finite across the entire interval.

In any but the simplest of cams, the cam motion program cannot be defined by a single mathematical expression, but rather must be defined by several separate functions, each of which defines the follower behavior over one segment, or piece, of the cam. These expressions are sometimes called *piecewise functions*. These functions must have **third-order continuity** (the function plus two derivatives) at all boundaries. **The displacement, velocity and acceleration functions must have no discontinuities in them.**†

If any discontinuities exist in the acceleration function, then there will be infinite spikes, or Dirac delta functions, appearing in jerk, the derivative of acceleration. Thus the corollary merely restates the fundamental law of cam design. Our naive designer failed to recognize that by starting with a low-degree (linear) polynomial as the displacement function, discontinuities would appear in the upper derivatives.

Polynomial functions are among the better choices for some cams as we shall shortly see, but they do have some limitations that can lead to trouble in this application. One problem is that, each time they are differentiated, their order reduces by one degree. Eventually, after enough differentiations, a polynomial degenerates to zero degree (a constant value) as the velocity function in Figure 2-3b (p. 20) shows. Thus, by starting with a first-degree polynomial as a displacement function, it is inevitable that discontinuities will soon appear in its derivatives.

In order to obey the fundamental law of cam design, one must start with at least a fifth-degree polynomial (quintic) as the displacement function. This will degenerate to a cubic function in the acceleration. The parabolic jerk function will have discontinuities at the junctions with the dwells, and the derivative of jerk, which we will call *ping*. will have infinite spikes in it. This is nevertheless acceptable, as the jerk is still finite.

2.4 SIMPLE HARMONIC MOTION (SHM)

Our naive cam designer recognized his mistake in choosing a straight-line function for the displacement. He also remembered a family of functions he had met in a calculus

* Since infinity is defined as a number larger than any number you can think of, infinity squared cannot really be larger than infinity, but the exponent is mathematically legitimate from the differentiation. Any way you state it, it is a very big number.

† This rule is stated by Neklutin [2] but is disputed by some other authors.[3],[4] Despite the minor controversy over this issue, this author believes that it is a good (and simple) rule to follow in order to get acceptable dynamic results with high-speed cams.

course that have the property of remaining continuous throughout any number of differentiations. These are the harmonic functions. On repeated differentiation, sine becomes cosine, which becomes negative sine, which becomes negative cosine, etc., ad infinitum. One never runs out of derivatives with the harmonic family of curves. In fact, differentiation of a harmonic function really only amounts to a 90° phase shift of the function. It is though, as you differentiated it, you cut out with a scissors a different portion of the same continuous sine wave function, which is defined from minus infinity to plus infinity. The equations of simple harmonic motion (SHM) for a rise motion are:

$$s = \frac{h}{2}\left[1 - \cos\left(\pi\frac{\theta}{\beta}\right)\right] \tag{2.6a}$$

$$v = \frac{\pi}{\beta}\frac{h}{2}\sin\left(\pi\frac{\theta}{\beta}\right) \tag{2.6b}$$

$$a = \frac{\pi^2}{\beta^2}\frac{h}{2}\cos\left(\pi\frac{\theta}{\beta}\right) \tag{2.6c}$$

$$j = -\frac{\pi^3}{\beta^3}\frac{h}{2}\sin\left(\pi\frac{\theta}{\beta}\right) \tag{2.6d}$$

where h is the total rise or lift in length units, θ is the camshaft angle, and β is the total angle of the rise interval, the last two both in radians.

We have here introduced a notation to simplify the expressions. The independent variable in our cam functions is θ, the camshaft angle. The period of any one segment is defined as the angle β. Its value can, of course, be different for each segment. We normalize the independent variable θ by dividing it by the period of the segment β. Both θ and β are measured in radians (or both in degrees). The value of θ/β will then vary from 0 to 1 over any segment. It is a dimensionless ratio. Equations 2.6 define simple harmonic motion and its derivatives for this rise segment in terms of θ/β.

This family of harmonic functions appears, at first glance, to be well suited to our cam design problem above. If we define the displacement function to be one of the harmonic functions, we should not run out of derivatives before reaching the acceleration.

EXAMPLE 2-2

Sophomoric* Cam Design—Simple Harmonic Motion—Still a Bad Cam.

Problem: Consider the same cam design CEP specification as in Example 2-1:

dwell	at zero displacement for 90° (low dwell)
rise	1 in (25.4 mm) in 90°
dwell	at 1 in (25.4 mm) for 90° (high dwell)
fall	1 in (25.4 mm) in 90°
cam ω	2π rad/sec = 1 rev/sec

Solution:

* **Sophomoric**, from sophomore, *def. Wise fool*, from the Greek, *sophos* = *wisdom*, *moros* = *fool.*

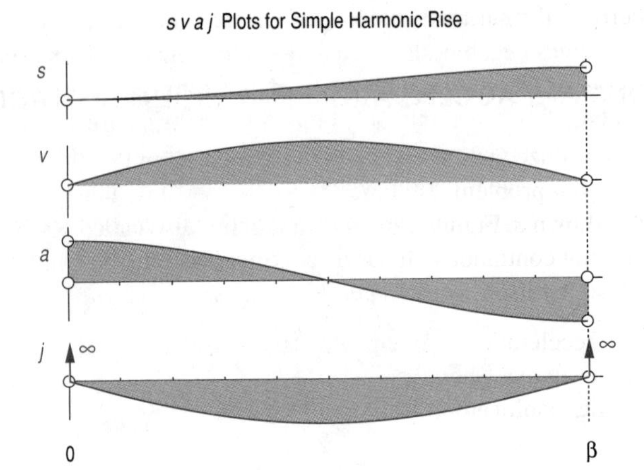

FIGURE 2-4

Simple harmonic motion with dwells has discontinuous acceleration—infinite jerk

1 Figure 2-4 shows a full-rise simple harmonic function* applied to the rise segment of our cam design problem.

2 Note that the velocity function is continuous, as it matches the zero velocity of the dwells at each end. The peak value is 6.28 in/sec (160 mm/sec) at the midpoint of the rise.

3 The acceleration function, however, is **not** continuous. It is a half-period cosine curve and has nonzero values at start and finish which are ± 78.8 in/sec^2 (2.0 m/sec^2).

4 Unfortunately, the dwell functions, which adjoin this rise on each side, have zero acceleration as can be seen in Figure 2-1 (p. 18). Thus there are **discontinuities in the acceleration at each end of the interval** which uses this simple harmonic displacement function.

5 This violates the fundamental law of cam design and creates **infinite spikes of jerk** at the ends of this fall interval. *This is also an unacceptable design.*

What went wrong? While it is true that harmonic functions are differentiable ad infinitum, we are not dealing here with single harmonic functions. Our cam function over the entire interval is a **piecewise function**, (Figure 2-1, p. 18) made up of several segments, some of which may be dwell portions or other functions. A dwell will always have zero velocity and zero acceleration. Thus we must match the dwells' zero values at the ends of those derivatives of any non-dwell segments that adjoin them. The simple harmonic displacement function, when used with dwells, does **not** satisfy the fundamental law of cam design. Its second derivative, acceleration, is nonzero at its ends and thus does not match the dwells required in this example.

The only case in which the simple harmonic displacement function will satisfy the fundamental law is the non-quick-return RF case, i.e., rise in 180° and fall in 180° with

* Though this is actually a half-period cosine wave, we will call it a *full-rise* (or *full-fall*) simple harmonic function to differentiate it from the *half-rise* (and *half-fall*) simple harmonic function which is actually a quarter-period cosine.

no dwells. Then the cam becomes an eccentric as shown in Figure 2-5. As a single continuous (not piecewise) function, its derivatives are continuous also.

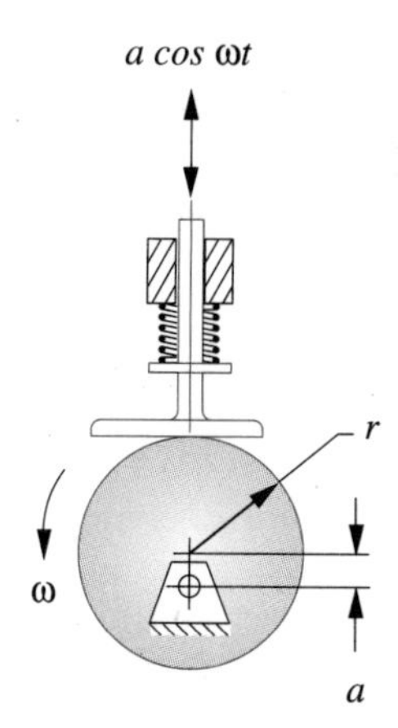

FIGURE 2-5

An eccentric cam has simple harmonic motion with no dwells and is the only dynamically acceptable application of that cam function

2.5 CONSTANT ACCELERATION (PARABOLIC DISPLACEMENT)

If we wish to minimize the theoretical peak value of the magnitude of the acceleration function for a given problem, the function that would best satisfy this constraint is the square wave as shown in Figure 2-6. This function is also called **constant acceleration**. This function is not continuous. It has discontinuities at the beginning, middle, and end of the interval, so, by itself, *is unacceptable as a cam acceleration function.*

If constant acceleration is integrated twice to obtain the displacement function it yields a parabolic curve as shown in Figure 2-1 (p. 18). Thus, this cam function is known both as **parabolic displacement** and as constant acceleration.

2.6 CUBIC DISPLACEMENT

A cubic function used as a displacement curve between two dwells is also an unacceptable choice. Differentiation gives a parabolic velocity function and a linear acceleration function with discontinuities at beginning and end that give spikes of infinite jerk at beginning and end of the interval. See Figure 2-1, p. 18.

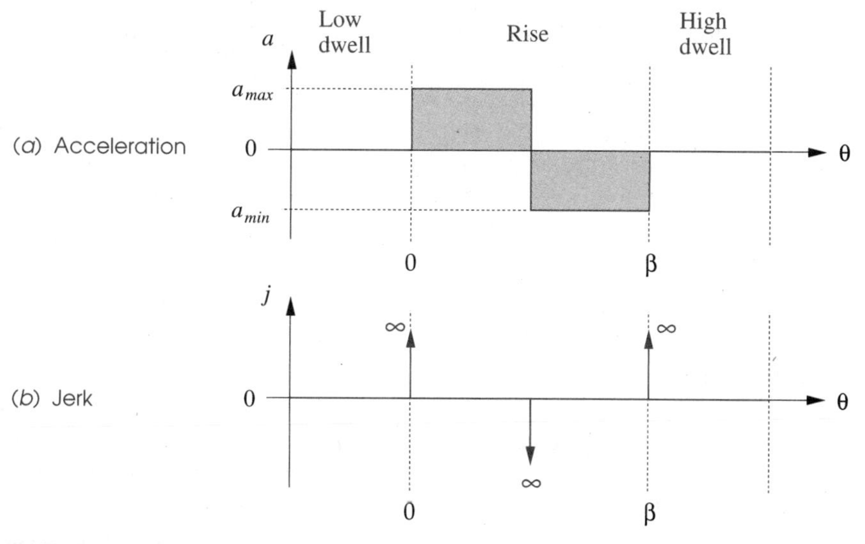

FIGURE 2-6

Constant acceleration gives infinite jerk

2.7 SUMMARY

This chapter has presented a fundamental law of cam design that requires continuity over the entire interval of all functions (including dwells) used for displacement, velocity, and acceleration. This leads to a condition known also as finite jerk. The law points out the folly of looking only at the displacement function when designing a cam. The cam designer needs to be most concerned with the derivatives of displacement, i.e., velocity, acceleration, and jerk.* The displacement function serves only to define the cam contour for manufacturing purposes. It has little influence on the follower's dynamic behavior. From Newton's second law, $F = ma$, the acceleration function has a large effect on the dynamic force, especially if the follower mass m is large. The follower velocity, v, affects the kinetic energy stored in the follower since $E = 0.5\ mv^2$. The jerk function has an effect on vibrations in the follower system as will be discussed in later chapters.

Several cam functions that have been used (historically) for multi-dwell applications have been shown to violate this fundamental law, and so should not, in general, be used in such applications at moderate to high speeds. These double-dwell cam functions to be avoided are *pure constant velocity*, *constant acceleration (parabolic displacement)*, *cubic displacement*, and *simple harmonic motion*. The *s v a j* diagrams of these four unacceptable multi-dwell cam functions are shown in Figure 2-1 (p. 18).

We will see in succeeding chapters that any of these functions can be made to work in a more sophisticated cam program by using them only for portions of a rise or fall and combining them with other functions to make acceptable composite cam functions. However, when used alone between dwells, none of these functions will give acceptable dynamic behavior. An exception to this statement is any situation in which the speed of the cam is so low as to render it a static rather than a dynamic case. One such example would be cams used in timer mechanisms (e.g., washing machine or dishwasher timers) to close switches over long time intervals. Cams in such applications turn at about two revolutions per hour. At such speeds, dynamic forces are not an issue and the cam contour is uncritical. Another example is that of cams used on weight training equipment where the manually applied rotation speed is so slow that it is a static system. Then any cam contour that provides the desired change in mechanical advantage is acceptable. In this book we will be concerned only with cams that run at moderate to high speed.

2.8 REFERENCES

1 **Neklutin, C. N.** (1952). "Designing Cams for Controlled Inertia and Vibration". *Machine Design*, June, 1952, pp. 143-160.

2 **Neklutin, C. N.** (1954). "Vibration Analysis of Cams." *Proc. of Second Conference on Mechanisms*, Purdue University, pp. 6-14.

3 **Wiederrich, J. L., and B. Roth**, eds. (1978). "Design of Low Vibration Cam Profiles." *Cams and Cam Mechanisms*, Jones, J. R., ed., Institution of Mechanical Engineers: London, pp. 3-8.

4 **Chew, M., and C. H. Chuang**. (1995). "Minimizing Residual Vibrations in High Speed Cam-Follower Systems over a Range of Speeds." *Journal of Mechanical Design*, **117**(1), p. 166.

* Neklutin[1] was the first to point out the need to focus on the acceleration function rather than on the displacement function as had previously been the standard approach.

Chapter 3

DOUBLE-DWELL CAM CURVES

3.0 INTRODUCTION

The previous chapter discussed a number of cam functions that proved to be dynamically unacceptable. This chapter will describe a set of functions that behave in an acceptable manner at moderate to high speeds for double-dwell applications, i.e., any rise or fall that lies between dwells. The variables used in this chapter are the same as those defined in Table 2-1 (p. 18). Note that in the form presented, with θ (in radians) as the independent variable, and constant cam angular velocity, the units of the expressions in equations for *s, v, a, j* are length, length/rad, length/rad^2, and length/rad^3 respectively. To convert these equations to a time base, multiply velocity v by the camshaft angular velocity ω (in rad/sec), acceleration a by ω^2, and jerk j by ω^3.

3.1 CYCLOIDAL DISPLACEMENT FOR DOUBLE DWELLS

The bad examples of cam design described in the previous chapter should lead the cam designer to the conclusion that consideration of only the displacement function when designing a cam is erroneous. The better approach is to start with consideration of the higher derivatives, especially acceleration. The acceleration function, and to a lesser extent the jerk function, should be the principal concern of the designer. In some cases, when there is a specification on velocity, the velocity function must be carefully designed as well.

With this in mind, we will redesign the cam for the same example specifications as used in the previous chapter. But this time, we will start with the acceleration function. The harmonic family of functions still have advantages which make them attractive for these applications. Figure 3-1 shows a full-period sinusoid applied as the acceleration function. It meets the constraint of zero magnitude at each end to match the dwell segments which adjoin it. The equation for a sine wave acceleration is:

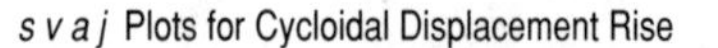

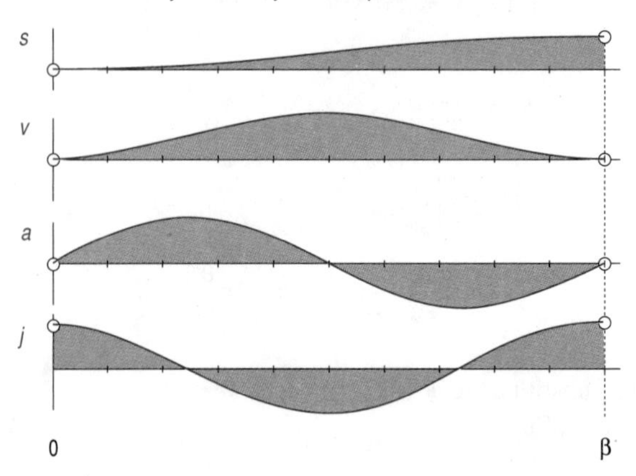

FIGURE 3-1

Sinusoidal acceleration gives cycloidal displacement

$$a = C\sin\left(2\pi\frac{\theta}{\beta}\right) \tag{3.1}$$

We have again normalized the independent variable θ by dividing it by the period of the segment β with both θ and β measured in radians. The value of θ/β ranges from 0 to 1 over any segment and is a dimensionless ratio. Since we want a full-cycle sine wave, we must multiply the argument by 2π. The argument of the sine function will then vary between 0 and 2π regardless of the value of β. The constant *C* defines the amplitude of the sine wave.

Integrate to obtain velocity, assuming a constant cam angular velocity ω,

$$a = \frac{dv}{d\theta} = C\sin\left(2\pi\frac{\theta}{\beta}\right)$$

$$\int dv = \int C\sin\left(2\pi\frac{\theta}{\beta}\right)d\theta \tag{3.2}$$

$$v = -C\frac{\beta}{2\pi}\cos\left(2\pi\frac{\theta}{\beta}\right) + k_1$$

where k_1 is the constant of integration. To evaluate k_1, substitute the boundary condition $v = 0$ at $\theta = 0$, since we must match the zero velocity of the dwell at that point. The constant of integration is then:

$$k_1 = C\frac{\beta}{2\pi}$$

and:

$$v = C\frac{\beta}{2\pi}\left[1 - \cos\left(2\pi\frac{\theta}{\beta}\right)\right] \tag{3.3}$$

Portions of this chapter were adapted from R. L. Norton, *Design of Machinery* 2ed. McGraw-Hill, 2001, with permission.

Note that substituting the boundary values at the other end of the interval, $v = 0$, $\theta = \beta$, will give the same result for k_1. Integrate again to obtain displacement:

$$v = \frac{ds}{d\theta} = C\frac{\beta}{2\pi}\left[1 - \cos\left(2\pi\frac{\theta}{\beta}\right)\right]$$

$$\int ds = \int\left\{C\frac{\beta}{2\pi}\left[1 - \cos\left(2\pi\frac{\theta}{\beta}\right)\right]\right\}d\theta \tag{3.4}$$

$$s = C\frac{\beta}{2\pi}\theta - C\frac{\beta^2}{4\pi^2}\sin\left(2\pi\frac{\theta}{\beta}\right) + k_2$$

To evaluate k_2, substitute the boundary condition $s = 0$ at $\theta = 0$, since we must match the zero displacement of the dwell at that point. To evaluate the amplitude constant C, substitute the boundary condition $s = h$ at $\theta = \beta$, where h is the maximum follower rise (or lift) required over the interval and is a constant for any one cam specification.

$$k_2 = 0$$

$$C = 2\pi\frac{h}{\beta^2} \tag{3.5}$$

Substituting the value of the constant C in equation 3.1 for acceleration gives:

$$a = 2\pi\frac{h}{\beta^2}\sin\left(2\pi\frac{\theta}{\beta}\right) \tag{3.6a}$$

Differentiating with respect to θ gives the expression for jerk.

$$j = 4\pi^2\frac{h}{\beta^3}\cos\left(2\pi\frac{\theta}{\beta}\right) \tag{3.6b}$$

Substituting the values of the constants C and k_1 in equation 3.3 for velocity gives:

$$v = \frac{h}{\beta}\left[1 - \cos\left(2\pi\frac{\theta}{\beta}\right)\right] \tag{3.6c}$$

This velocity function is the sum of a negative cosine term and a constant term. The coefficient of the cosine term is equal to the constant term. This results in a velocity curve which starts and ends at zero and reaches a maximum magnitude at $\beta/2$ as can be seen in Figure 3-1. Substituting the values of the constants C, k_1, and k_2 in equation 3.4 for displacement gives:

$$s = h\left[\frac{\theta}{\beta} - \frac{1}{2\pi}\sin\left(2\pi\frac{\theta}{\beta}\right)\right] \tag{3.6d}$$

Note that this displacement expression is the sum of a straight line of slope h and a negative sine wave. The sine wave is, in effect, "wrapped around" the straight line as can be seen in Figure 3-1. Equation 3.6d is the expression for a cycloid. This cam function is referred to either as **cycloidal displacement** or **sinusoidal acceleration**.

In the form presented, with θ (in radians) as the independent variable, the units of equation 3.6d are length, of equation 3.6c length/rad, of equation 3.6a length/rad^2, and of equation 3.6b length/rad^3. To convert these equations to a time base, multiply velocity v by the camshaft angular velocity ω (in rad/sec), multiply acceleration a by ω^2, and jerk j by ω^3.

EXAMPLE 3-1

Cycloidal Displacement—An Acceptable Cam.

Problem: Consider the same cam design CEP specification as in Examples 2-1 and 2-2:

rise	1 in (25.4 mm) in 90° with cycloidal motion
dwell	at 1 in (25.4 mm) for 90° (high dwell)
fall	1 in (25.4 mm) in 90° with cycloidal motion
dwell	at zero displacement for 90° (low dwell)
cam ω	2π rad/sec = 1 rev/sec

Solution:

1. The cycloidal displacement function is an acceptable one for this double-dwell cam specification. Its derivatives are continuous through the acceleration function as seen in Figure 3-1. The peak acceleration is 100.5 in/sec^2 (2.55 m/sec^2).

2. The jerk curve in Figure 3-1 is discontinuous at its boundaries but is of finite magnitude, and this is acceptable. Its peak value is 2523 in/sec^3 (64 m/sec^3).

3. The velocity is smooth and matches the zeros of the dwell at each end. Its peak value is 8 in/sec (0.2 m/sec).

4. The only drawback to this function is that it has relatively large magnitudes of peak acceleration and peak velocity compared to some other possible functions for the double-dwell case.

The reader may open the file Ex_03-01.cam in program DYNACAM to investigate this example in more detail.

3.2 COMBINED FUNCTIONS FOR DOUBLE DWELLS

Dynamic force is proportional to acceleration. We generally would like to minimize dynamic forces, and thus should be looking to minimize the magnitude of the acceleration function as well as to keep it continuous. Kinetic energy is proportional to velocity squared. With large mass follower trains, we also would like to minimize stored kinetic energy to minimize angular velocity changes, and so may be concerned with the magnitude of the velocity function as well.

CONSTANT ACCELERATION If we wish to minimize the peak value of the magnitude of the acceleration function for a given problem, the function that would best satisfy this constraint is the square wave as shown in Figure 3-2. This function is also called **constant acceleration**. The square wave has the property of minimum peak value for a

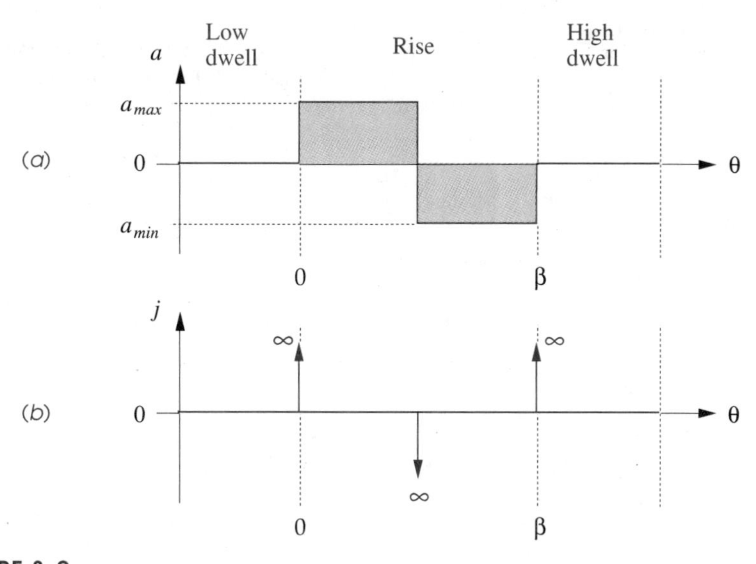

FIGURE 3-2

Constant acceleration gives infinite jerk

given area in a given interval. However, as described in the previous chapter, this function is not continuous. It has discontinuities at the beginning, middle, and end of the interval, so, by itself between dwells, *this is unacceptable as a cam acceleration function.*

TRAPEZOIDAL ACCELERATION The square wave's discontinuities can be removed by simply "knocking the corners off" the square wave function and creating the **trapezoidal acceleration** function shown in Figure 3-3a. The area lost from the "knocked off corners" must be replaced by increasing the peak magnitude above that of the original square wave in order to maintain the required specifications on lift and duration. But this increase in peak magnitude is small, and the theoretical maximum acceleration can be significantly less than the theoretical peak value of the sinusoidal acceleration (cycloidal displacement) function. One disadvantage of this trapezoidal function is its very discontinuous jerk function, as shown in Figure 3-3b. Ragged jerk functions such as this tend to excite vibratory behavior in the follower train due to their high harmonic content. The cycloidal's sinusoidal acceleration has a relatively smoother cosine jerk function with only two discontinuities in the interval and is preferable to the trapezoid's square waves of jerk. But the cycloidal's theoretical peak acceleration will be larger, which is not desirable. So, trade-offs must be made in selecting the cam functions.

MODIFIED TRAPEZOIDAL ACCELERATION An improvement can be made to the trapezoidal acceleration function by substituting pieces of sine waves for the sloped sides of the trapezoids as shown in Figure 3-4. This function is called the **modified trapezoidal acceleration** curve.* This function is a marriage of the sine acceleration and constant acceleration curves. Conceptually, a full period sine wave is cut into fourths and "pasted into" the square wave to provide a smooth transition from the zeros at the endpoints to the maximum and minimum peak values, and to make the transition from max-

* Developed by C. N. Neklutin of the Universal Match Corp. See Neklutin.[2]

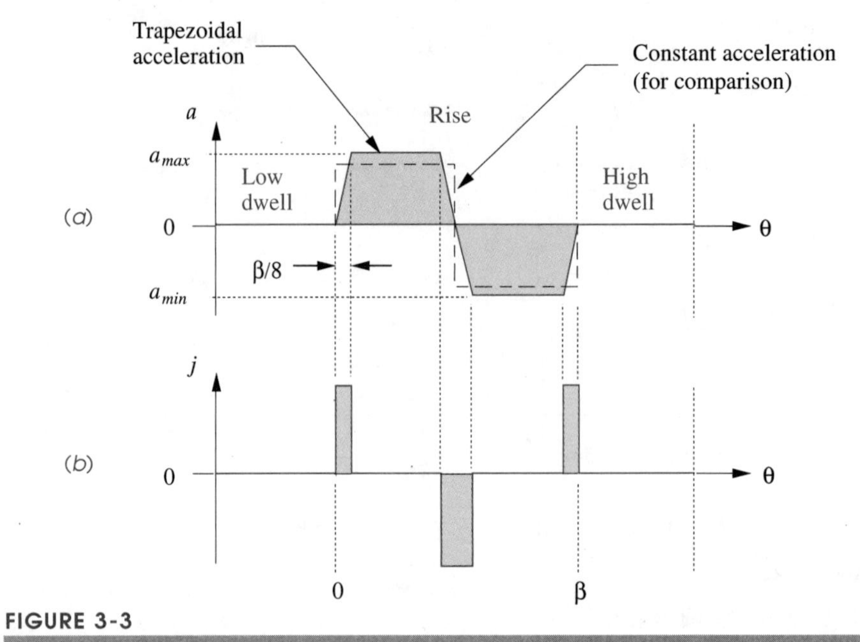

FIGURE 3-3

Trapezoidal acceleration gives finite jerk

imum to minimum in the center of the interval. The portions of the total segment period (β) used for the sinusoidal parts of the function can be varied. The most common arrangement* is to cut the square wave at β/8, 3β/8, 5β/8, and 7β/8 to insert the pieces of sine wave as shown in Figure 3-4. The *s v a j* formulas for that arrangement of a modified trapezoidal rise are:

for $0 \le \theta < \frac{1}{8}\beta$

$$s = h\left[0.38898448\frac{\theta}{\beta} - 0.0309544\sin\left(4\pi\frac{\theta}{\beta}\right)\right]$$

$$v = 0.38898448\frac{h}{\beta}\left[1 - \cos\left(4\pi\frac{\theta}{\beta}\right)\right] \tag{3.7a}$$

$$a = 4.888124\frac{h}{\beta^2}\sin\left(4\pi\frac{\theta}{\beta}\right)$$

$$j = 61.425769\frac{h}{\beta^3}\cos\left(4\pi\frac{\theta}{\beta}\right)$$

* As recommended by Neklutin.[2]

for $\frac{1}{8}\beta \le \theta < \frac{3}{8}\beta$

$$
\begin{aligned}
s &= h\left[2.44406184\left(\frac{\theta}{\beta}\right)^2 - 0.22203097\left(\frac{\theta}{\beta}\right) + 0.00723407\right] \\
v &= \frac{h}{\beta}\left[4.888124\left(\frac{\theta}{\beta}\right) - 0.22203097\right] \qquad (3.7b) \\
a &= 4.888124\frac{h}{\beta^2} \\
j &= 0
\end{aligned}
$$

for $\frac{3}{8}\beta \le \theta < \frac{5}{8}\beta$

$$
\begin{aligned}
s &= h\left[1.6110154\frac{\theta}{\beta} - 0.0309544\sin\left(4\pi\frac{\theta}{\beta} - \pi\right) - 0.3055077\right] \\
v &= \frac{h}{\beta}\left[1.6110154 - 0.38898448\cos\left(4\pi\frac{\theta}{\beta} - \pi\right)\right] \qquad (3.7c) \\
a &= 4.888124\frac{h}{\beta^2}\sin\left(4\pi\frac{\theta}{\beta} - \pi\right) \\
j &= 61.425769\frac{h}{\beta^3}\cos\left(4\pi\frac{\theta}{\beta} - \pi\right)
\end{aligned}
$$

for $\frac{5}{8}\beta \le \theta < \frac{7}{8}\beta$

$$
\begin{aligned}
s &= h\left[-2.44406184\left(\frac{\theta}{\beta}\right)^2 + 4.6660917\left(\frac{\theta}{\beta}\right) - 1.2292648\right] \\
v &= \frac{h}{\beta}\left[-4.888124\left(\frac{\theta}{\beta}\right) + 4.6660917\right] \qquad (3.7d) \\
a &= -4.888124\frac{h}{\beta^2} \\
j &= 0
\end{aligned}
$$

for $\frac{7}{8}\beta \le \theta \le \beta$

$$
\begin{aligned}
s &= h\left[0.6110154 + 0.38898448\frac{\theta}{\beta} + 0.0309544\sin\left(4\pi\frac{\theta}{\beta} - 3\pi\right)\right] \\
v &= 0.38898448\frac{h}{\beta}\left[1 + \cos\left(4\pi\frac{\theta}{\beta} - 3\pi\right)\right] \qquad (3.7e) \\
a &- -4.888124\frac{h}{\beta^2}\sin\left(4\pi\frac{\theta}{\beta} - 3\pi\right) \\
j &= -61.425769\frac{h}{\beta^3}\cos\left(4\pi\frac{\theta}{\beta} - 3\pi\right)
\end{aligned}
$$

(*a*) Take a sine wave

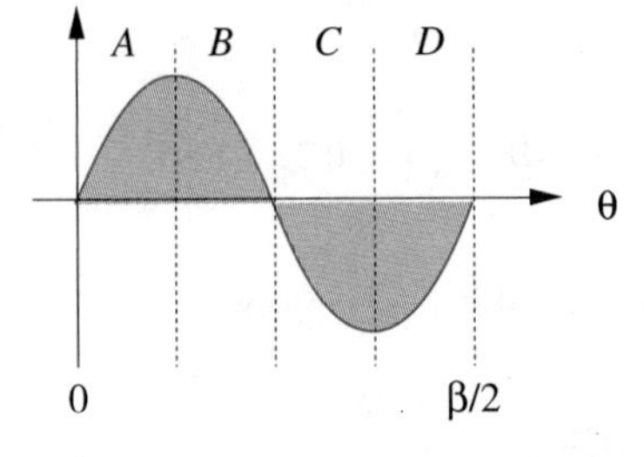

(*b*) Split the sine wave apart

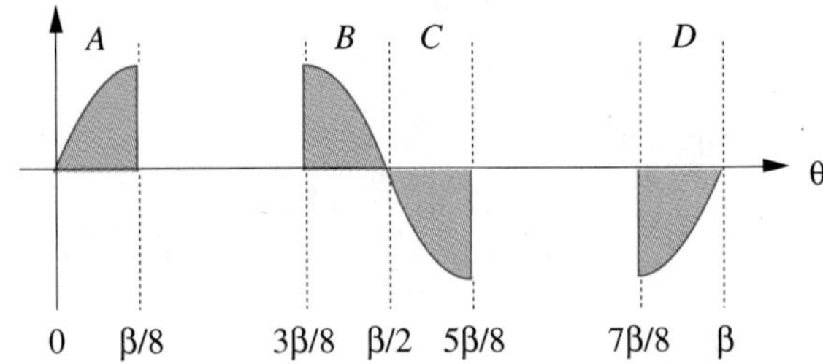

(*c*) Take a constant acceleration square wave

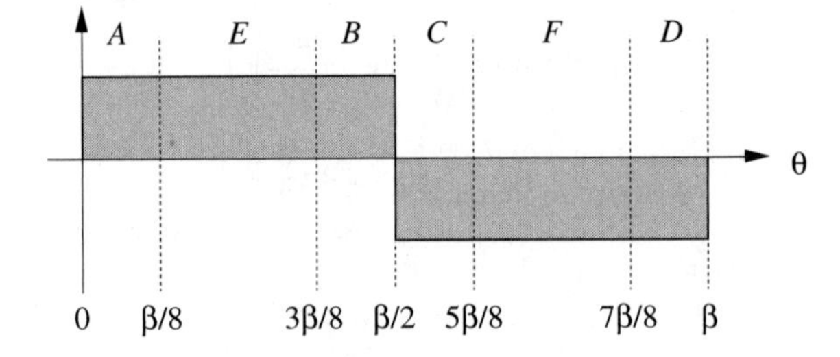

(*d*) Combine the two

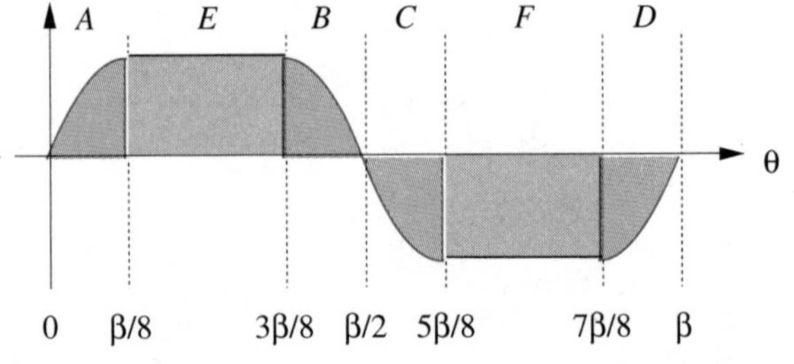

(*e*) Modified trapezoidal acceleration

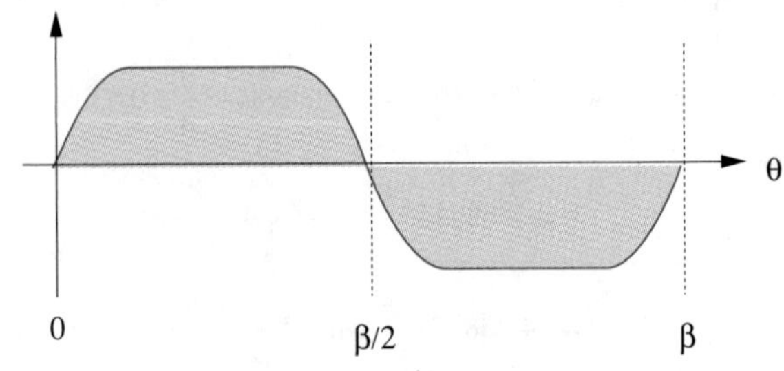

FIGURE 3-4

Creating the modified trapezoidal acceleration function

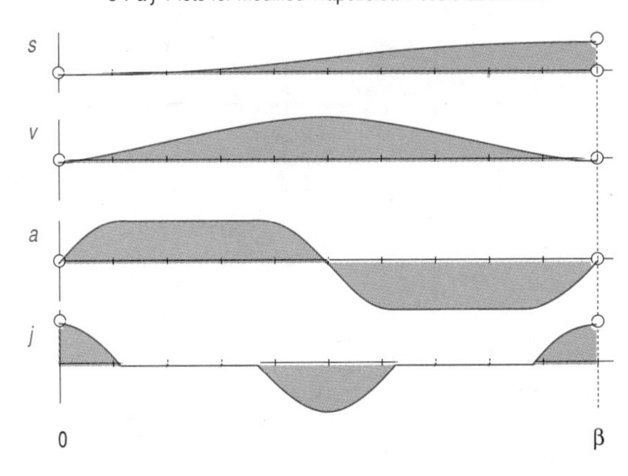

FIGURE 3-5
Modified trapezoidal acceleration

The modified trapezoidal function defined above in equations 3.7 is one of many combined functions created for cams by piecing together various standard functions, while being careful to match the values of the *s*, *v*, and *a* curves at all the interfaces between the joined functions. Its *s v a j* curves are shown in Figure 3-5.

It has the advantage of relatively low theoretical peak acceleration, and reasonably rapid, relatively smooth transitions at the beginning and end of the interval. The modified trapezoidal cam function is a popular and often used program for double-dwell cams, though it is not the best choice for that application, as will be shown in later chapters. It can have higher vibration than other possible double-dwell functions to be described later.

MODIFIED SINUSOIDAL ACCELERATION* The sine acceleration curve (cycloidal displacement) has the advantage of smoothness (less ragged jerk curve) compared to the modified trapezoid but has higher theoretical peak acceleration. By combining two harmonic (sinusoid) curves of different frequencies, we can retain some of the smoothness characteristics of the cycloid and also reduce the peak acceleration. As an added bonus, we will find that the peak velocity is also lower than in either the cycloidal or modified trapezoid.

Figure 3-6 shows how the modified sine acceleration curve is made up of pieces of two sinusoid functions, one of higher frequency than the other. The first and last quarter of the high-frequency (short period, $\beta/2$) sine curve is used for the first and last eighths of the combined function. The center half of the low-frequency (long period, $3\beta/2$) sine wave is used to fill in the center three-fourths of the combined curve. Obviously, the magnitudes of the two curves and their derivatives must be matched at their interfaces in order to avoid discontinuities. The equations for the **modified sine** curve for a rise of height h over a period β, with the functions joined at the $\beta/8$ and $7\beta/8$ points are as follows:

* Developed by E. H. Schmidt of DuPont.

(*a*) Sine wave #1 of period β/2

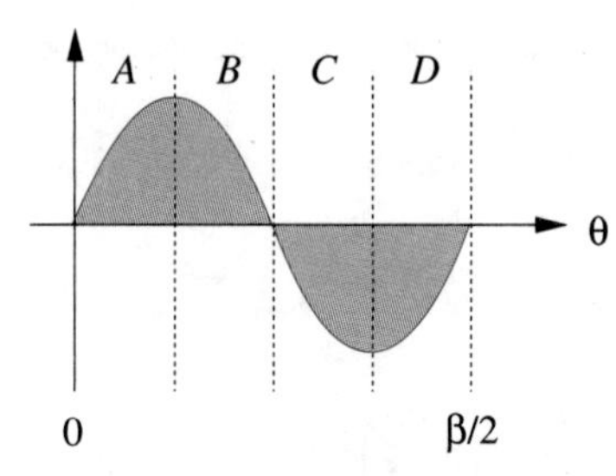

(*b*) Sine wave #2 of period 3β/2

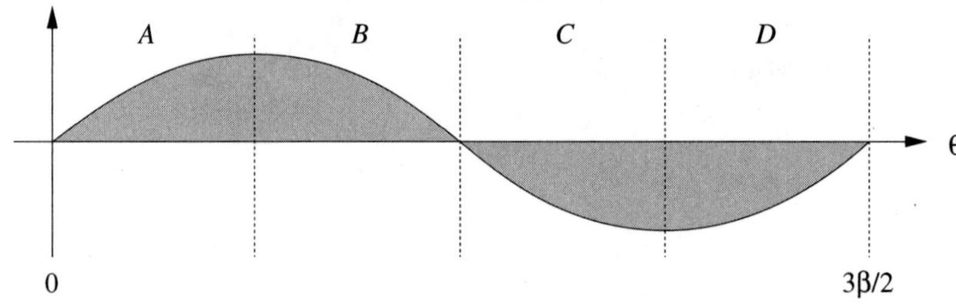

(*c*) Take 1st and 4th quarters of #1

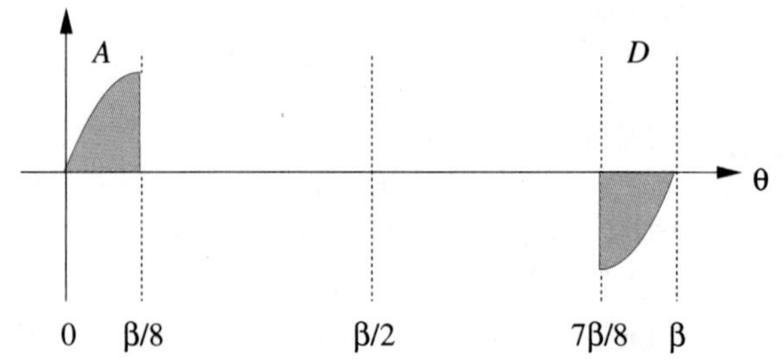

(*d*) Take 2nd and 3rd quarters of #2

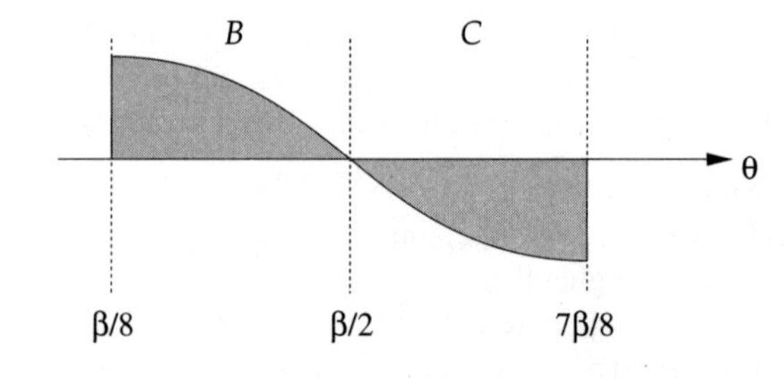

(*e*) Combine to get modified sine

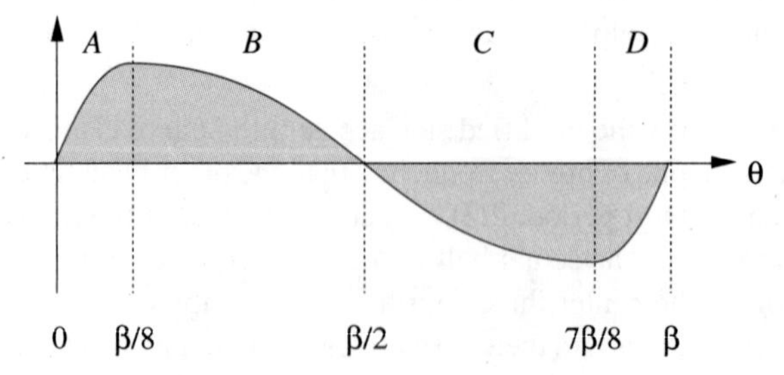

FIGURE 3-6

Creating the modified sine acceleration function

for $0 \le \theta < \frac{1}{8}\beta$

$$s = h\left[0.43990085\frac{\theta}{\beta} - 0.0350062\sin\left(4\pi\frac{\theta}{\beta}\right)\right]$$

$$v = 0.43990085\frac{h}{\beta}\left[1 - \cos\left(4\pi\frac{\theta}{\beta}\right)\right] \tag{3.8a}$$

$$a = 5.5279571\frac{h}{\beta^2}\sin\left(4\pi\frac{\theta}{\beta}\right)$$

$$j = 69.4663577\frac{h}{\beta^3}\cos\left(4\pi\frac{\theta}{\beta}\right)$$

for $\frac{1}{8}\beta \le \theta < \frac{7}{8}\beta$

$$s = h\left[0.28004957 + 0.43990085\frac{\theta}{\beta} - 0.31505577\cos\left(\frac{4\pi}{3}\frac{\theta}{\beta} - \frac{\pi}{6}\right)\right]$$

$$v = 0.43990085\frac{h}{\beta}\left[1 + 3\sin\left(\frac{4\pi}{3}\frac{\theta}{\beta} - \frac{\pi}{6}\right)\right] \tag{3.8b}$$

$$a = 5.5279571\frac{h}{\beta^2}\cos\left(\frac{4\pi}{3}\frac{\theta}{\beta} - \frac{\pi}{6}\right)$$

$$j = -23.1553\frac{h}{\beta^3}\sin\left(\frac{4\pi}{3}\frac{\theta}{\beta} - \frac{\pi}{6}\right)$$

for $\frac{7}{8}\beta \le \theta \le \beta$

$$s = h\left\{0.56009915 + 0.43990085\frac{\theta}{\beta} - 0.0350062\sin\left[2\pi\left(2\frac{\theta}{\beta} - 1\right)\right]\right\}$$

$$v = 0.43990085\frac{h}{\beta}\left\{1 - \cos\left[2\pi\left(2\frac{\theta}{\beta} - 1\right)\right]\right\} \tag{3.8c}$$

$$a = 5.5279571\frac{h}{\beta^2}\sin\left[2\pi\left(2\frac{\theta}{\beta} - 1\right)\right]$$

$$j = 69.4663577\frac{h}{\beta^3}\cos\left[2\pi\left(2\frac{\theta}{\beta} - 1\right)\right]$$

Let us again try to improve the double-dwell cam example with these combined functions of modified trapezoid and modified sine acceleration.

EXAMPLE 3-2

Combined Functions—Acceptable Double-Dwell Cams.

Problem: Consider the same cam design CEP specification as in Example 3-1:

rise	1 in (25.4 mm) in 90° with modified trapezoidal acceleration
dwell	at 1 in (25.4 mm) for 90° (high dwell)
fall	1 in (25.4 mm) in 90° with modified sine acceleration
dwell	at zero displacement for 90° (low dwell)
cam ω	2π rad/sec = 1 rev/sec

Solution:

1 The modified trapezoidal function is an acceptable one for this double-dwell cam specification. Its derivatives are continuous through the acceleration function as shown in Figures 3-4 and 3-5 (pp. 34-35). The peak acceleration is 78.2 in/sec^2 (1.98 m/sec^2).

2 The modified trapezoidal jerk curve in Figure 3-5 is discontinuous at its boundaries, but has finite magnitude of 3931 in/sec^3 (100 m/sec^3), and this is acceptable.

3 The modified trapezoidal velocity in Figure 3-5 is smooth and matches the zeros of the dwell at each end. Its peak magnitude is 8 in/sec (0.2 m/sec).

4 The advantage of this modified trapezoidal function is that it has smaller theoretical peak acceleration than the cycloidal, but its theoretical peak velocity is identical to that of the cycloidal.

5 The modified sinusoid function is also an acceptable one for this double-dwell cam specification. Its derivatives are also continuous through the acceleration function as shown in Figures 3-7 to 3-9. Its theoretical peak acceleration is 88.4 in/sec^2 (2.24 m/sec^2).

6 The modified sine jerk curve in Figure 3-8 is discontinuous at its boundaries, but is of finite magnitude. It is larger in magnitude at 4446 in/sec^3 (113 m/sec^3), but smoother than that of the modified trapezoid.

7 The modified sine velocity (Figure 3-9) is smooth, matches the zeros of the dwell at each end, and is lower in theoretical peak magnitude than either the cycloid or modified trapezoidal at 7 in/sec (0.178 m/sec). This is an advantage for high-mass follower systems as it reduces the kinetic energy. This, coupled with a peak acceleration lower than the cycloidal (but higher than the modified trapezoidal), is its chief advantage.

Figure 3-7 shows a comparison of the shapes and relative magnitudes of three cam acceleration programs: the cycloidal, modified trapezoid, and modified sine acceleration curves. The cycloidal curve has a theoretical peak acceleration which is approximately 1.3 times that of the modified trapezoid's peak value for the same cam specification. The peak value of acceleration for the modified sine is between the cycloidal and modified trapezoid. The figure also shows the peak values of acceleration, velocity, and jerk for these functions in terms of the total rise h and period β.

Figure 3-8 compares the jerk curves for the same functions. The modified sine jerk is somewhat less ragged than the modified trapezoid jerk but not as smooth as that of the

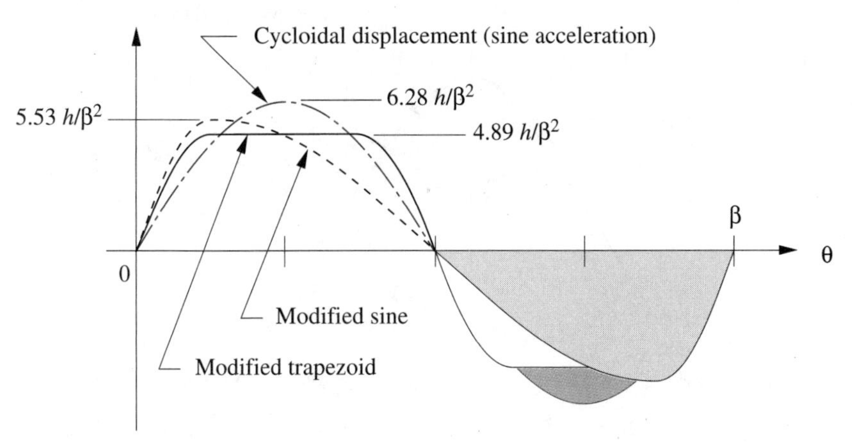

FIGURE 3-7

Comparison of three double-dwell cam acceleration functions

cycloid, which is a full-period cosine. Figure 3-9 compares their velocity curves. The peak velocities of the cycloidal and modified trapezoid functions are the same, so each will store the same peak kinetic energy in the follower train. The peak velocity of the modified sine function is the lowest of the three functions shown. This is an advantage of the modified sine acceleration curve and the reason that it is often chosen for applications in which the follower mass (or mass moment of inertia) is very large.

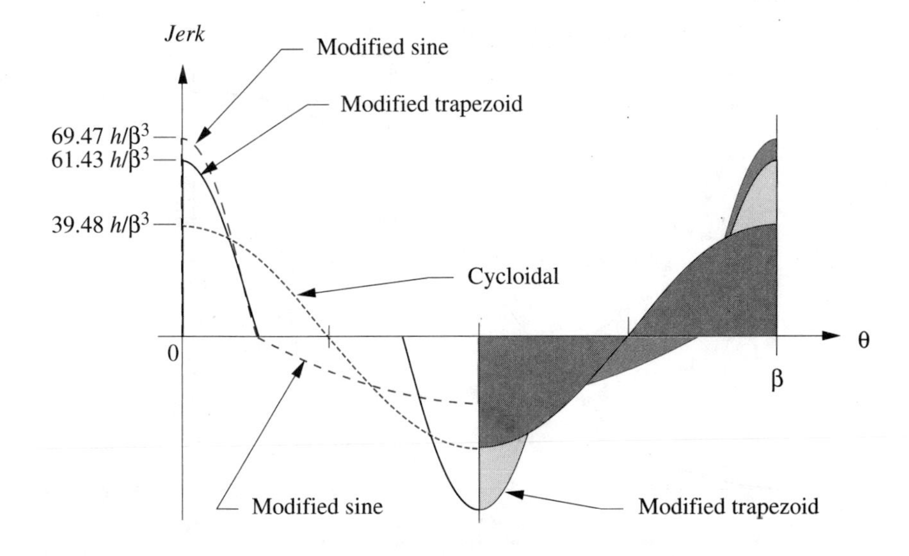

FIGURE 3-8

Comparison of three double-dwell cam jerk functions

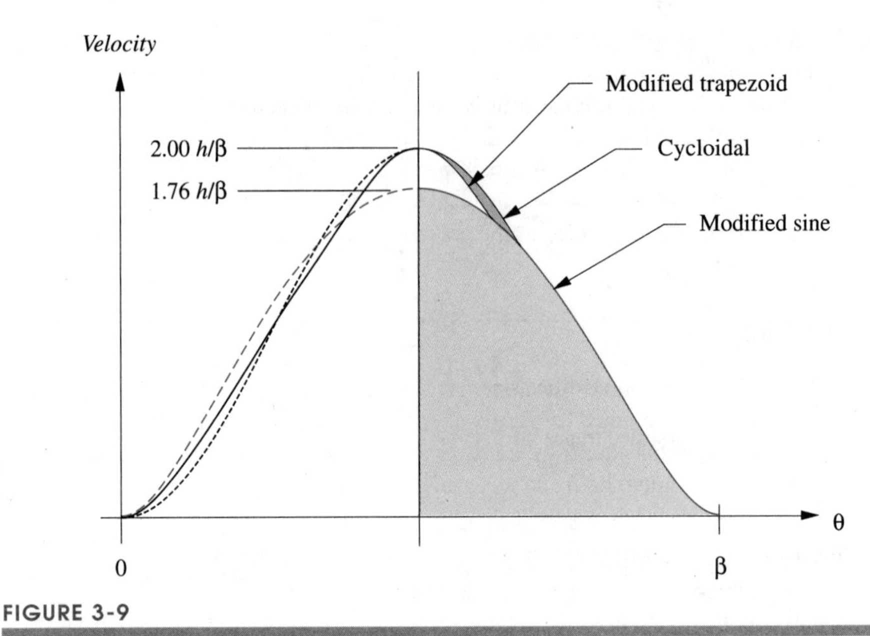

FIGURE 3-9

Comparison of three double-dwell cam velocity functions

Figure 3-10 shows the displacement curves for these three cam programs. Note how little difference there is between the displacement curves despite the large differences in their acceleration waveforms in Figure 3-7. This is evidence of the smoothing effect of

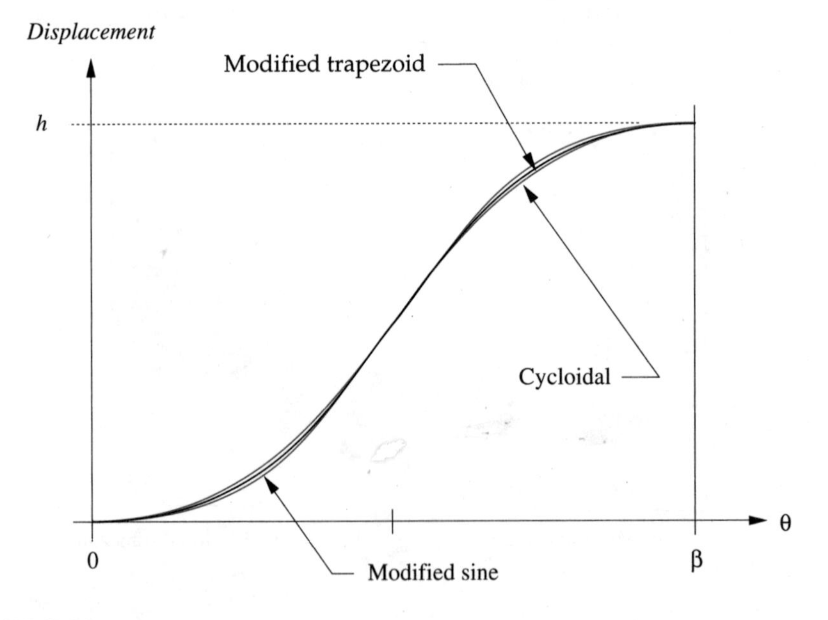

FIGURE 3-10

Comparison of three double-dwell cam displacement functions

the integration process. Differentiating any two functions will exaggerate their differences. Integration tends to mask their differences. It is nearly impossible to recognize these very differently behaving cam functions by looking only at their displacement curves. This is further evidence of the folly of our earlier naive approach to cam design which dealt exclusively with the displacement function. The cam designer must be concerned with the higher derivatives of displacement. The displacement function is primarily of value to the manufacturer of the cam who needs its coordinate information in order to cut the cam.

Fall Functions

We have used only the rise portion of the cam for these examples. The fall is handled similarly. The rise functions presented here are applicable to the fall with slight modification. To convert rise equations to fall equations, it is only necessary to subtract the rise displacement function s from the maximum lift h and to negate the higher derivatives, v, a, and j.

3.3 THE SCCA FAMILY OF DOUBLE-DWELL FUNCTIONS

SCCA stands for *Sine-Constant-Cosine-Acceleration* and refers to a family of acceleration functions that includes constant acceleration, simple harmonic, modified trapezoid, modified sine, and cycloidal curves.[1] These very different looking curves can all be defined by the same equation with only a change of numeric parameters. In like fashion, the equations for displacement, velocity, and jerk for all these SCCA functions differ only by their parametric values.

To reveal this similitude, it is first necessary to normalize the variables in the equations. We have already normalized the independent variable, cam angle θ, dividing it by the interval period β. We will further simplify the notation by defining

$$x = \frac{\theta}{\beta} \tag{3.9a}$$

The normalized variable x then runs from 0 to 1 over any interval. The normalized follower displacement is

$$y = \frac{s}{h} \tag{3.9b}$$

where s is the actual displacement and h is the total lift. The normalized variable y then runs from 0 to 1 over any follower displacement.

The general shapes of the $s\ v\ a\ j$ functions of the SCCA family are shown in Figure 3-11. The interval β is divided into five zones. Zones 0 and 6 represent the dwells on either side of the rise (or fall). The widths of zones 1 - 5 are defined in terms of β and one of three parameters, a, b, c. The values of these three parameters define the shape of the curve and define its identity within the family of functions. The normalized velocity, acceleration, and jerk are denoted, respectively, as:

$$y' = \frac{dy}{dx} \qquad y'' = \frac{d^2y}{dx^2} \qquad y''' = \frac{d^3y}{dx^3} \tag{3.10}$$

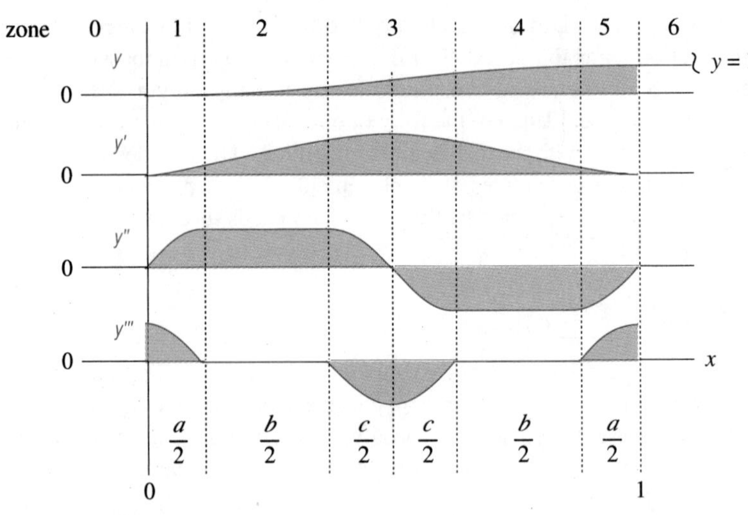

FIGURE 3-11

Parameters for the normalized SCCA family of curves

In zone 0, all functions are zero. The expressions for the functions within each other zone of Figure 3-11 are as follows:

Zone 1: $0 \le x \le \frac{a}{2}$: $\quad a \ne 0$

$$y = C_a\left[\frac{a}{\pi}x - \left(\frac{a}{\pi}\right)^2 \sin\left(\frac{\pi}{a}x\right)\right] \tag{3.11a}$$

$$y' = C_a\left[\frac{a}{\pi} - \frac{a}{\pi}\cos\left(\frac{\pi}{a}x\right)\right] \tag{3.11b}$$

$$y'' = C_a \sin\left(\frac{\pi}{a}x\right) \tag{3.11c}$$

$$y''' = C_a \frac{\pi}{a}\cos\left(\frac{\pi}{a}x\right) \tag{3.11d}$$

Zone 2: $\frac{a}{2} \le x \le \frac{1-c}{2}$

$$y = C_a\left[\frac{x^2}{2} + a\left(\frac{1}{\pi} - \frac{1}{2}\right)x + a^2\left(\frac{1}{8} - \frac{1}{\pi^2}\right)\right] \tag{3.12a}$$

$$y' = C_a\left[x + a\left(\frac{1}{\pi} - \frac{1}{2}\right)\right] \tag{3.12b}$$

$$y'' = C_a \tag{3.12c}$$

$$y''' = 0 \tag{3.12d}$$

Zone 3: $\frac{1-c}{2} \le x \le \frac{1+c}{2}: \quad c \ne 0$

$$y = C_a\left\{\left(\frac{a}{\pi}+\frac{b}{2}\right)x+\left(\frac{c}{\pi}\right)^2+a^2\left(\frac{1}{8}-\frac{1}{\pi^2}\right)-\frac{(1-c)^2}{8}-\left(\frac{c}{\pi}\right)^2\cos\left[\frac{\pi}{c}\left(x-\frac{1-c}{2}\right)\right]\right\} \quad (3.13a)$$

$$y' = C_a\left\{\frac{a}{\pi}+\frac{b}{2}+\frac{c}{\pi}\sin\left[\frac{\pi}{c}\left(x-\frac{1-c}{2}\right)\right]\right\} \quad (3.13b)$$

$$y'' = C_a \cos\left[\frac{\pi}{c}\left(x-\frac{1-c}{2}\right)\right] \quad (3.13c)$$

$$y''' = -C_a\frac{\pi}{c}\sin\left[\frac{\pi}{c}\left(x-\frac{1-c}{2}\right)\right] \quad (3.13d)$$

Zone 4: $\frac{1+c}{2} \le x \le 1-\frac{a}{2}$

$$y = C_a\left[-\frac{x^2}{2}+\left(\frac{a}{\pi}+1-\frac{a}{2}\right)x+\left(2c^2-a^2\right)\left(\frac{1}{\pi^2}-\frac{1}{8}\right)-\frac{1}{4}\right] \quad (3.14a)$$

$$y' = C_a\left(-x+\frac{a}{\pi}+1-\frac{a}{2}\right) \quad (3.14b)$$

$$y'' = -C_a \quad (3.14c)$$

$$y''' = 0 \quad (3.14d)$$

Zone 5: $1-\frac{a}{2} \le x \le 1: \quad a \ne 0$

$$y = C_a\left\{\frac{a}{\pi}x+\frac{2\left(c^2-a^2\right)}{\pi^2}+\frac{(1-a)^2-c^2}{4}-\left(\frac{a}{\pi}\right)^2\sin\left[\frac{\pi}{a}(x-1)\right]\right\} \quad (3.15a)$$

$$y' = C_a\left\{\frac{a}{\pi}-\frac{a}{\pi}\cos\left[\frac{\pi}{a}(x-1)\right]\right\} \quad (3.15b)$$

$$y'' = C_a \sin\left[\frac{\pi}{a}(x-1)\right] \quad (3.15c)$$

$$y''' = C_a\frac{\pi}{a}\cos\left[\frac{\pi}{a}(x-1)\right] \quad (3.15d)$$

Zone 6: $x > 1$

$$y = 1, \qquad y' = y'' = y''' = 0 \quad (3.16)$$

The coefficient C_a is a dimensionless peak acceleration factor. It can be evaluated from the fact that, at the end of the rise in zone 5 when $x = 1$, the expression for displacement (equation 3.15a) must have $y = 1$ to match the dwell in zone 6. Setting the right side of equation 3.15a equal to 1 gives:

$$C_a = \frac{4\pi^2}{\left(\pi^2 - 8\right)\left(a^2 - c^2\right) - 2\pi(\pi - 2)a + \pi^2} \tag{3.17a}$$

We can also define dimensionless peak factors (coefficients) for velocity (C_v), and jerk (C_j) in terms of C_a. The velocity is a maximum at $x = 0.5$. Thus C_v will equal the right side of equation 3.13b when $x = 0.5$.

$$C_v = C_a\left(\frac{a+c}{\pi} + \frac{b}{2}\right) \tag{3.17b}$$

The jerk is a maximum at $x = 0$. Setting the right side of equation 3.11d to zero gives:

$$C_j = C_a \frac{\pi}{a} \qquad a \neq 0 \tag{3.17c}$$

Table 3-1 shows the values of *a, b, c* and the resulting factors C_v, C_a, and C_j for the five standard members of the SCCA family. There is an infinity of related functions with values of these parameters between those shown. Figure 3-12 shows these 5 members of the " acceleration family" superposed with their design parameters noted. Note that all the functions shown in Figure 3-12 were generated with the same set of equations (3.11 through 3.17) with only changes to the values of the parameters *a, b*, and *c*. A *TKSolver* file (SCCA.tk) that is provided on the CD-ROM calculates and plots any of the SCCA family of normalized functions, along with their coefficients C_v, C_a, and C_j, in response to the input of values for *a, b*, and *c*. Note that there is an infinity of family members as *a, b*, and *c* can take on any set of values that add to 1.

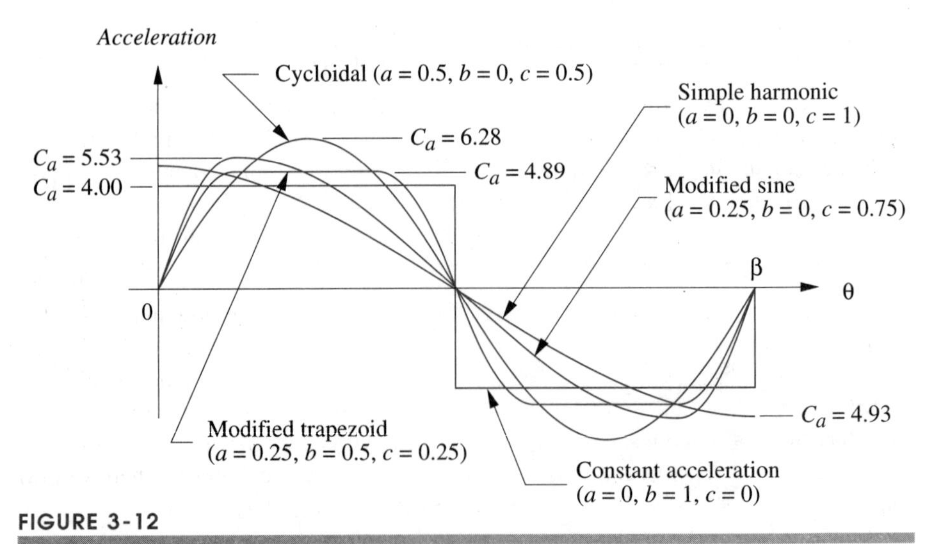

FIGURE 3-12

Comparison of five acceleration functions in the SCCA family

TABLE 3-1 Parameters and Coefficients for the SCCA Family of Functions

Function	a	b	c	C_v	C_a	C_j
constant acceleration	0.00	1.00	0.00	2.0000	4.0000	infinite
modified trapezoid	0.25	0.50	0.25	2.0000	4.8881	61.426
simple harmonic	0.00	0.00	1.00	1.5708	4.9348	infinite
modified sine	0.25	0.00	0.75	1.7596	5.5280	69.466
cycloidal displacement	0.50	0.00	0.50	2.0000	6.2832	39.478

To apply the SCCA functions to an actual cam design problem only requires that they be multiplied by factors appropriate to the particular problem, namely the actual rise h, the actual duration β (rad), and cam velocity ω (rad/sec).

$$\begin{array}{llll} s = hy & \text{length} & S = s & \text{length} \\ v = \dfrac{h}{B} y' & \text{length/rad} & V = v\omega & \text{length/sec} \\ a = \dfrac{h}{B^2} y'' & \text{length/rad}^2 & A = a\omega^2 & \text{length/sec}^2 \\ j = \dfrac{h}{B^3} y''' & \text{length/rad}^3 & J = j\omega^3 & \text{length/sec}^3 \end{array} \tag{3.18}$$

3.4 POLYNOMIAL FUNCTIONS

Another family of functions that is useful for cam programs is polynomials. These functions are among the most versatile for cam design. They are not limited to either single- or double-dwell applications and can be tailored to many design specifications. The general form of a polynomial function is:

$$s = C_0 + C_1 x + C_2 x^2 + C_3 x^3 + C_4 x^4 + C_5 x^5 + C_6 x^6 + \cdots + C_n x^n \tag{3.19}$$

where s is the follower displacement and x is the independent variable, which in our case will be replaced by either θ/β or time t. The constant coefficients C_n are the unknowns to be determined in our development of the particular polynomial equation to suit a design specification. The degree of a polynomial is defined as the highest power present in any term. Note that a polynomial of degree n will have $n + 1$ terms because there is an x^0 or constant term with coefficient C_0, as well as coefficients through and including C_n.

We structure a polynomial cam design problem by deciding how many boundary conditions (BCs) we want to specify on the $s\ v\ a\ j$ diagrams. The number of BCs then determines the degree of the resulting polynomial. We can write an independent equation for each BC by substituting it into equation 3.19 or one of its derivatives. We will then have a system of linear equations which can be solved for the unknown coefficients $C_0, \ldots, C_n$. If k represents the number of chosen boundary conditions, there will be k equations in k unknowns $C_0, \ldots, C_n$ and the **degree** of the polynomial will be $n = k - 1$. The **order** of the n-degree polynomial is equal to the number of terms, $k = n + 1$.

The 3-4-5 Polynomial

Let us return to the double-dwell problem of Example 3-2 (p. 38) and solve it with polynomial functions. Many different polynomial solutions are possible. We will start with the simplest one possible for the double-dwell case.

EXAMPLE 3-3

The 3-4-5 Polynomial for the Double-Dwell Case.

Problem: Consider the same cam design CEP specification as in Examples 3-1 and 3-2:

rise	1 in (25.4 mm) in 90° with 3-4-5 polynomial motion
dwell	at 1 in (25.4 mm) for 90° (high dwell)
fall	1 in (25.4 mm) in 90° with 3-4-5 polynomial motion
dwell	at zero displacement for 90° (low dwell)
cam ω	2π rad/sec = 1 rev/sec

Solution:

1 To satisfy the fundamental law of cam design, the values of the rise (and fall) functions at their boundaries with the dwells must match with no discontinuities in, at a minimum, *s*, *v*, and *a*.

2 Figure 3-13 shows the axes for the *s v a j* diagrams on which the known data have been drawn. The dwells are the only fully-defined segments at this stage. The requirement for continuity through the acceleration defines a minimum of **six boundary conditions** for the rise segment and six more for the fall in this problem. They are shown as filled circles on the plots. For generality, we will let the specified total rise be represented by the variable *h*. The minimum set of required BCs for this example is then:

for the rise:

$$\begin{aligned}&\text{when} \quad \theta = 0; \quad \text{then} \quad s = 0, \quad v = 0, \quad a = 0\\&\text{when} \quad \theta = \beta_1; \quad \text{then} \quad s = h, \quad v = 0, \quad a = 0\end{aligned} \tag{a}$$

for the fall:

$$\begin{aligned}&\text{when} \quad \theta = 0; \quad \text{then} \quad s = h, \quad v = 0, \quad a = 0\\&\text{when} \quad \theta = \beta_2; \quad \text{then} \quad s = 0, \quad v = 0, \quad a = 0\end{aligned} \tag{b}$$

3 We will use the rise for an example solution. (The fall is a similar derivation.) We have six BCs on the rise. This requires six terms in the equation. The highest term will be fifth degree. We will use the normalized angle θ/β as our independent variable, as before. Because our boundary conditions involve velocity and acceleration as well as displacement, we need to differentiate equation 3.19 versus θ to obtain expressions into which we can substitute those BCs. Rewriting equation 3.19 to fit these constraints and differentiating twice, we get:

$$s = C_0 + C_1\left(\frac{\theta}{\beta}\right) + C_2\left(\frac{\theta}{\beta}\right)^2 + C_3\left(\frac{\theta}{\beta}\right)^3 + C_4\left(\frac{\theta}{\beta}\right)^4 + C_5\left(\frac{\theta}{\beta}\right)^5 \tag{c}$$

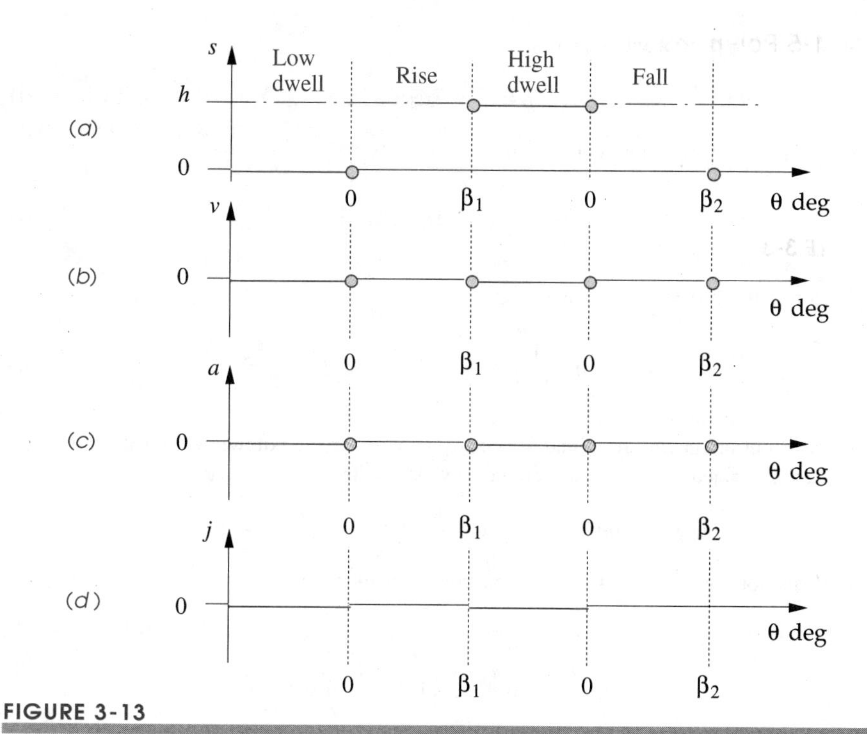

FIGURE 3-13

Minimum boundary conditions for the double-dwell case

$$v=\frac{1}{\beta}\left[C_1+2C_2\left(\frac{\theta}{\beta}\right)+3C_3\left(\frac{\theta}{\beta}\right)^2+4C_4\left(\frac{\theta}{\beta}\right)^3+5C_5\left(\frac{\theta}{\beta}\right)^4\right] \quad \text{(d)}$$

$$a=\frac{1}{\beta^2}\left[2C_2+6C_3\left(\frac{\theta}{\beta}\right)+12C_4\left(\frac{\theta}{\beta}\right)^2+20C_5\left(\frac{\theta}{\beta}\right)^3\right] \quad \text{(e)}$$

4 Substitute the boundary conditions $\theta=0$, $s=0$ into equation c:*

$$0=C_0+0+0+\cdots$$
$$C_0=0 \quad \text{(f)}$$

5 Substitute $\theta=0$, $v=0$ into equation d:*

$$0=\frac{1}{\beta}[C_1+0+0+\cdots]$$
$$C_1=0 \quad \text{(g)}$$

6 Substitute $\theta=0$, $a=0$ into equation e:*

$$0=\frac{1}{\beta^2}[C_2+0+0+\cdots]$$
$$C_2=0 \quad \text{(h)}$$

* Note the pattern here. Let $x=\theta/\beta$. If $x=0$ and $d^n x/dx^n=0$, then $C_n=0$.

7 Substitute $\theta = \beta$ $s = h$ into equation c:

$$h = C_3 + C_4 + C_5 \tag{i}$$

8 Substitute $\theta = \beta$ $v = 0$ into equation d:

$$0 = \frac{1}{\beta}\left[3C_3 + 4C_4 + 5C_5\right] \tag{j}$$

9 Substitute $\theta = \beta$ $a = 0$ into equation e:

$$0 = \frac{1}{\beta^2}\left[6C_3 + 12C_4 + 20C_5\right] \tag{k}$$

10 Three of our unknowns are found to be zero, leaving three unknowns to be solved for C_3, C_4, and C_5. Equations i, j, and k can be solved simultaneously to get:

$$C_3 = 10h; \qquad C_4 = -15h; \qquad C_5 = 6h \tag{l}$$

11 The equation for this cam design's displacement is then:

$$s = h\left[10\left(\frac{\theta}{\beta}\right)^3 - 15\left(\frac{\theta}{\beta}\right)^4 + 6\left(\frac{\theta}{\beta}\right)^5\right] \tag{3.20}$$

12 The expressions for velocity and acceleration can be obtained by substituting the values of C_3, C_4, and C_5 into equations d and e. This function is referred to as the **3-4-5 polynomial**, after its exponents. Open the file Ex_03-03.cam in program DYNACAM to investigate this example in more detail.

Figure 3-14 shows the resulting *s v a j* diagrams for a **3-4-5 polynomial rise** function. Note that the acceleration is continuous but the jerk is not, because we did not place any constraints on the boundary values of the jerk function. It is also interesting to note

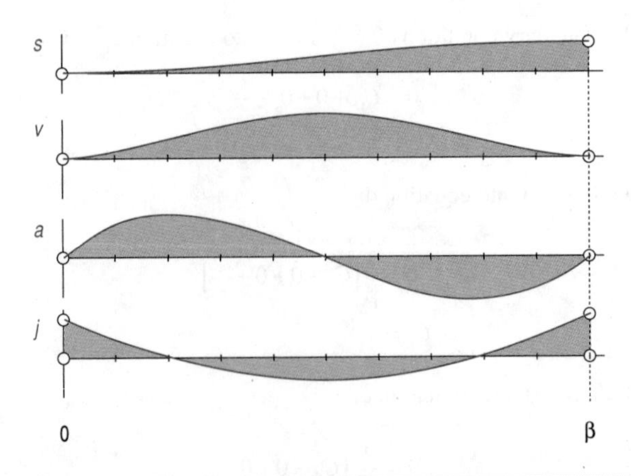

FIGURE 3-14

3-4-5 polynomial rise, its acceleration is very similar to the sinusoid of cycloidal motion

that the acceleration waveform looks very similar to the sinusoidal acceleration of the cycloidal function in Figure 3-1 (p. 28).

The 4-5-6-7 Polynomial

We left the jerk unconstrained in the previous example. We will now redesign the cam for the same specifications but will also constrain the jerk function to be zero at both ends of the rise. It will then match the adjacent dwells in the jerk function with no discontinuities. This gives eight boundary conditions and yields a seventh-degree polynomial. The solution procedure to find the eight unknown coefficients is identical to that used in the previous example. Write the polynomial with the appropriate number of terms. Differentiate it to get expressions for all orders of boundary conditions. Substitute the boundary conditions and solve the resulting set of simultaneous equations. This problem reduces to four equations in four unknowns, as the coefficients C_0, C_1, C_2, and C_3 turn out to be zero. For this set of boundary conditions the displacement equation for the rise is:

$$s = h\left[35\left(\frac{\theta}{\beta}\right)^4 - 84\left(\frac{\theta}{\beta}\right)^5 + 70\left(\frac{\theta}{\beta}\right)^6 - 20\left(\frac{\theta}{\beta}\right)^7\right] \tag{3.21}$$

This is known as the **4-5-6-7 polynomial**, after its exponents. Figure 3-15 shows the *s v a j* diagrams for this function. Compare them to the 3-4-5 polynomial functions shown in Figure 3-14. Note that the acceleration of the 4-5-6-7 starts off slowly, with zero slope (as we demanded with our zero jerk BC), and as a result goes to a larger peak value of acceleration in order to replace the missing area in the leading edge.

When compared to the **3-4-5 polynomial**, the **cycloidal**, and all other functions so far discussed, the **4-5-6-7 polynomial** function has the advantage of smoother jerk for better vibration control, but pays a price in the form of significantly higher theoretical acceleration than all those functions. Figure 3-16 shows the relative peak accelerations of the 3-4-5 and 4-5-6-7 polynomials compared to three other double-dwell functions

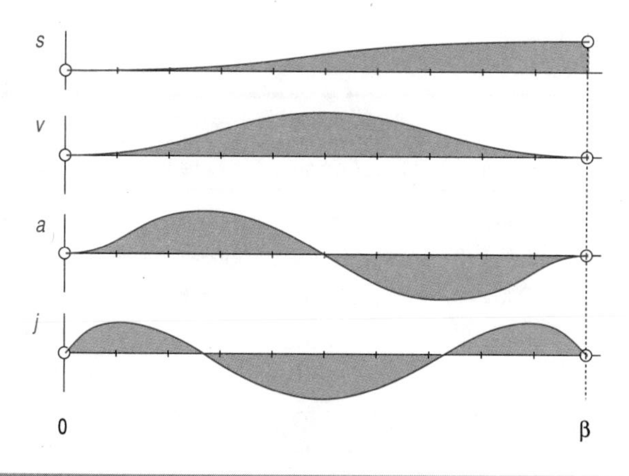

FIGURE 3-15

s v a j plots for a 4-5-6-7 polynomial rise, showing its jerk piecewise continuous with the dwells

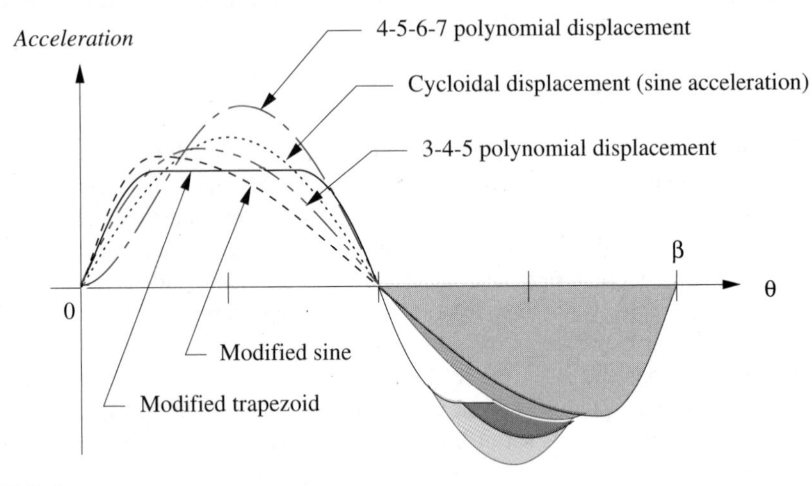

FIGURE 3-16

Comparison of five double-dwell cam acceleration functions

with the same h and β. Figure 3-17 compares their jerk functions and Figure 3-18 compares their velocity functions. Table 3-2 lists the C_v, C_a, and C_j factors for, respectively, the maximum velocity, acceleration and jerk of these functions normalized to the lift h and duration β.

Figure 3-10 (p. 40) shows the displacement functions for three of the five functions shown in Figures 3-16 through 3-18. Note their similarity. It is not useful to present another figure that adds the polynomial displacement curves to the trio of Figure 3-10 because they are all essentially indistinguishable. Except for the pure constant velocity function (Figure 2-1, p. 18), all the standard double-dwell function's displacement curves are difficult to tell apart when superposed. This is the result of the smoothing that results from mathematical integration. Two integrations are sufficient to mask the distinct

TABLE 3-2 Factors for Peak Velocity and Acceleration of Some Cam Functions

Function	Max. Veloc.	Max. Accel.	Max. Jerk	Comments
Constant accel.	2.000 h/β	4.000 h/β^2	Infinite	∞ jerk—not acceptable
Simp. Harm. disp.	1.571 h/β	4.935 h/β^2	Infinite	∞ jerk—not acceptable
Trapezoid accel.	2.000 h/β	5.300 h/β^2	44 h/β^3	Not as good as mod. trap.
Mod. trap. accel.	2.000 h/β	4.888 h/β^2	61 h/β^3	Low accel. but rough jerk
Mod. sine accel.	1.760 h/β	5.528 h/β^2	69 h/β^3	Low veloc., good accel
3-4-5 poly. disp.	1.875 h/β	5.777 h/β^2	60 h/β^3	Good compromise
Cycloidal disp.	2.000 h/β	6.283 h/β^2	40 h/β^3	Smooth accel. and jerk.
4-5-6-7 poly. disp.	2.188 h/β	7.526 h/β^2	52 h/β^3	Smooth jerk, high accel.

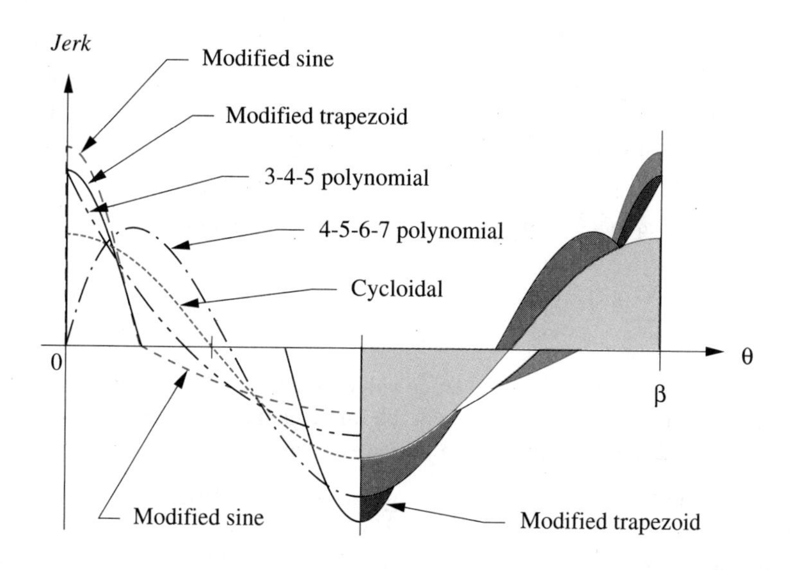

FIGURE 3-17

Comparison of five double-dwell cam jerk functions

differences among the acceleration functions of Figure 3-16. This is further evidence of the folly of designing cam functions based solely on their displacement curves.

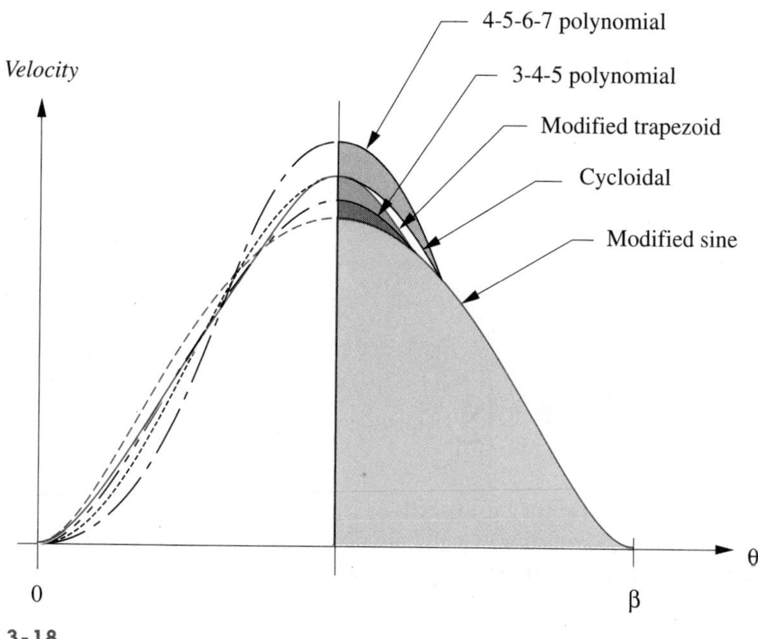

FIGURE 3-18

Comparison of five double-dwell cam velocity functions

3.5 FOURIER SERIES FUNCTIONS

Baron Jean Baptiste Fourier (1768-1830) published the description of the mathematical series that bears his name in *The Analytic Theory of Heat* in 1822. The Fourier series is widely used in harmonic analysis of all types of physical systems. Its general form is

$$y = \frac{a_0}{2} + (a_1 \sin x + b_1 \cos x) + (a_2 \sin 2x + b_2 \cos 2x) + \cdots + (a_n \sin nx + b_n \cos nx)$$

$$= \frac{a_0}{2} + \sum_{k=1}^{n} a_k \sin kx + b_k \cos kx \tag{3.22a}$$

The first term (a constant) is the average value, or DC component, of the function. The first nonconstant pair of terms represent the fundamental frequency or first harmonic of the function. Each succeeding pair represents the *j*th harmonic of the function.

Fourier discovered that, by choosing appropriate coefficients, this infinite series could be made to exactly represent any periodic function . Since, practically, it must be truncated to the summation of a finite number of terms, it then offers only an approximation of the original function. However with a relatively small number of terms (say 10) it can give quite a close approximation of most regular functions.

The Fourier series has been applied to the design of cam-follower programs, in particular for the purpose of controlling the higher harmonic content of the function. This is easily accomplished by omitting all terms above the *n*th harmonic, your choice of *n*. Expressed in more familiar terms for cam-follower functions, the series looks like:

$$s = h\left\{\frac{\theta}{\beta} - \frac{1}{2\pi}\left[\sum_{k=1}^{n} a_k \sin\left(2\pi k\frac{\theta}{\beta}\right) + b_k \cos\left(2\pi k\frac{\theta}{\beta}\right)\right]\right\} \tag{3.22b}$$

$$v = \frac{h}{\beta}\left\{1 - \left[\sum_{k=1}^{n} k a_k \cos\left(2\pi k\frac{\theta}{\beta}\right) - k b_k \sin\left(2\pi k\frac{\theta}{\beta}\right)\right]\right\} \tag{3.22c}$$

$$a = 2\pi\frac{h}{\beta^2}\left[\sum_{k=1}^{n} k^2 a_k \sin\left(2\pi j\frac{\theta}{\beta}\right) + k^2 b_k \cos\left(2\pi k\frac{\theta}{\beta}\right)\right] \tag{3.22d}$$

$$j = 4\pi^2\frac{h}{\beta^3}\left\{\sum_{k=1}^{n} k^3 a_k \cos\left(2\pi k\frac{\theta}{\beta}\right) - k^3 b_k \sin\left(2\pi k\frac{\theta}{\beta}\right)\right\} \tag{3.22e}$$

For symmetrical rise (or fall) functions (i.e., those for which the half rise point is reached at $\beta/2$), the Fourier series will contain only the a_k terms. If the half-period acceleration waveshape is symmetrical about the $\beta/4$ point, then its series will contain only odd harmonics.

Note that literally any set of coefficients can be used for a general function as long as the following relationships hold.

$$\sum_{k=1}^{n} k a_k = 1 \qquad \sum_{k=1}^{n} k b_k = 1 \tag{3.23}$$

Weber[3] derived the Fourier a_j coefficients for a number of standard symmetrical double-dwell functions to ten terms. Their coefficients are shown in Table 3-3 for the case of a normalized displacement function with a rise of 1 over a period of 1. In some cases, this number of terms is sufficient to give a very accurate approximation to the function. This can be seen in Figure 3-19, which shows the acceleration and jerk of a modified trapezoid function approximated with, respectively 3, 5, and 9 harmonic terms. (Differences in the displacement and velocity are more difficult to see.) The 9-harmonic approximation is essentially indistinguishable from the original function of Figure 3-7 (p. 39) or 3-16 (p. 50).

Figure 3-20 shows Fourier series approximations of both the constant acceleration (9-harmonics) and simple harmonic (10-harmonics) functions. These are not as close to the originals as is the modified trapezoid 9-harmonic approximation. Their discontinuities cause them to contain significant high harmonics. They thus require a much larger number of harmonic terms to represent them accurately. On the other hand, by approximating them to only a few harmonics, we essentially improve them by eliminating the high frequencies that can excite structural natural frequency vibrations in the follower system. For example, Figure 3-21 shows a 3-harmonic (2-term) Fourier series approximation of the constant acceleration function. Compare this to the 3-harmonic approximation of the modified trapezoid in Figure 3-19a. They are very similar. In fact, if you do a 1-term Fourier series approximation to any of the SCCA functions of Figure 3-12 (p. 44), you will get the cycloidal function, which has only a fundamental term.

Several researchers have suggested using Fourier series functions for the double-dwell case, truncating them to only a few terms (often 2) to control vibrations by limiting the harmonic content. Freudenstein[4] was the first to suggest this approach and defined several functions of up to ninth harmonic. Gutman[5] and Weber[3] each suggested

TABLE 3-3 Fourier Series Coefficients a_j for Several Standard Double-Dwell Functions

Harmonic	Cycloidal (exact)	Const Accel (approx)	Simp Harm (approx)	Mod trap (approx)	Mod sine (approx)	345 poly (approx)
1	1	0.84469100	0.70000000	0.93410900	0.84507100	0.92418300
2	0	0	0.07000000	0	0.05499850	0.02888050
3	0	0.03128500	0.02000000	0.02075800	0.01097500	0.00380333
4	0	0	0.00833333	0	0.00266775	0.00090250
5	0	0.00675760	0.00424242	0.00106760	0.00054620	0.00029580
6	0	0	0.00244750	0	0.00000000	0.00011883
7	0	0.00246271	0.00153843	-0.00018157	-0.00009214	0.00005500
8	0	0	0.00102938	0	-0.00006525	0.00002825
9	0	0.00115856	0.00072244	-0.00004989	-0.00002522	0.00001567
10	0	0	0.00052630	0	0.00000000	0.00000920

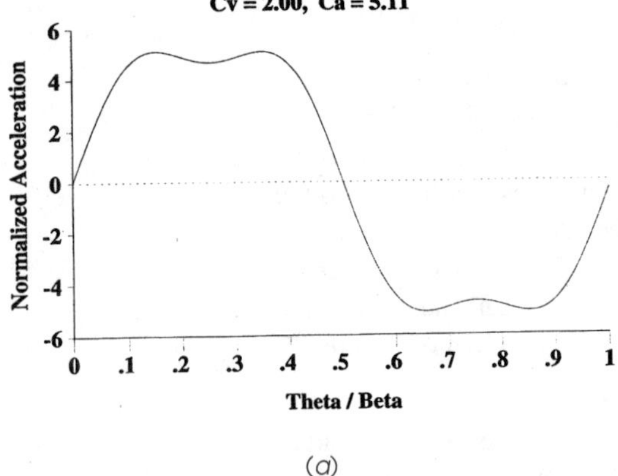

(a)

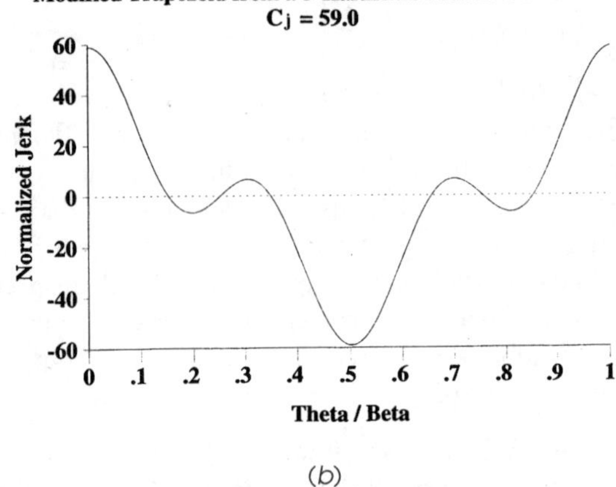

(b)

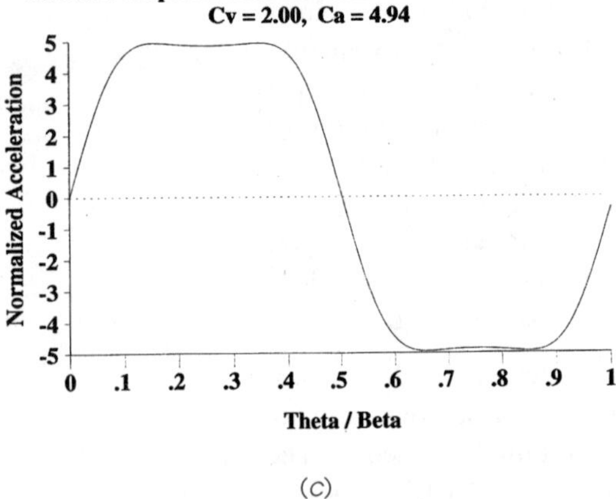

(c)

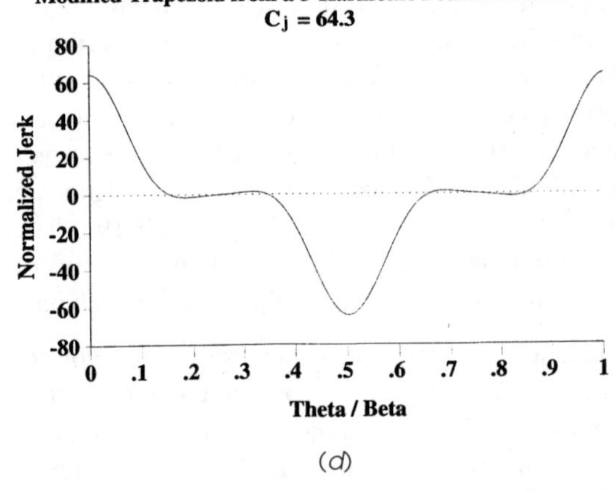

(d)

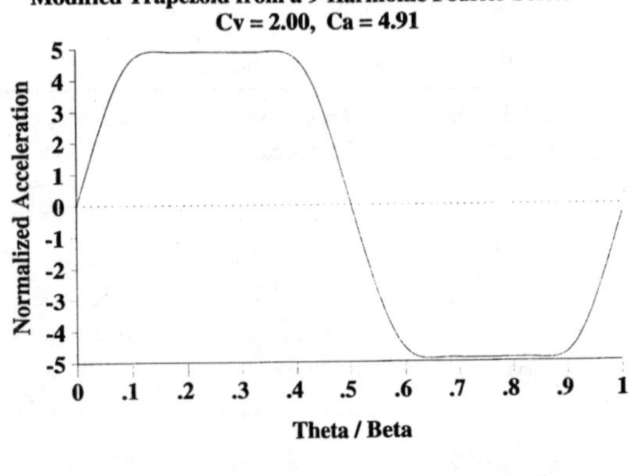

(e)

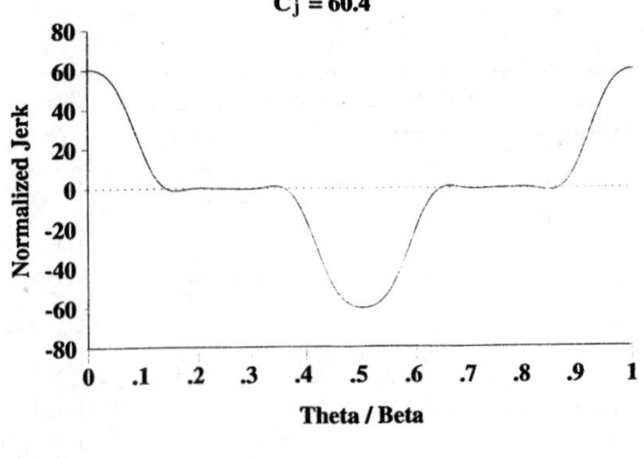

(f)

FIGURE 3-19

Fourier series approximations of the modified trapezoid acceleration function

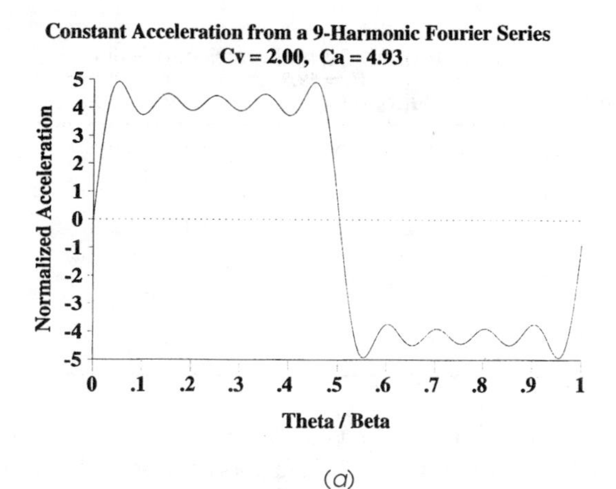

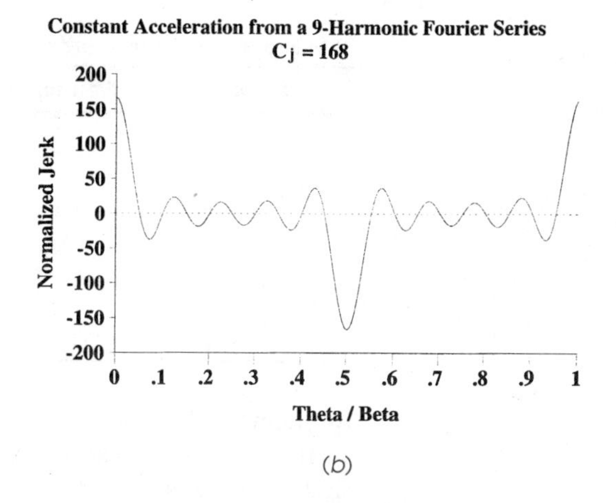

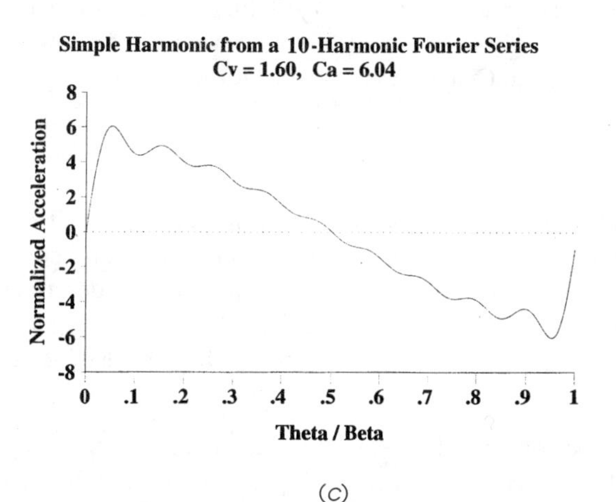

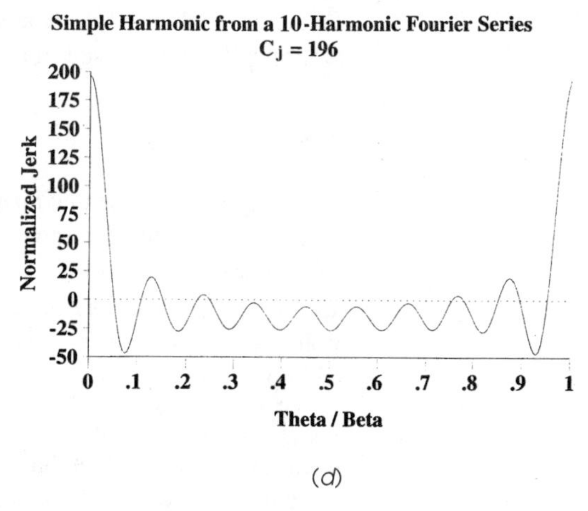

FIGURE 3-20

Fourier series approximations of the constant acceleration and simple harmonic functions

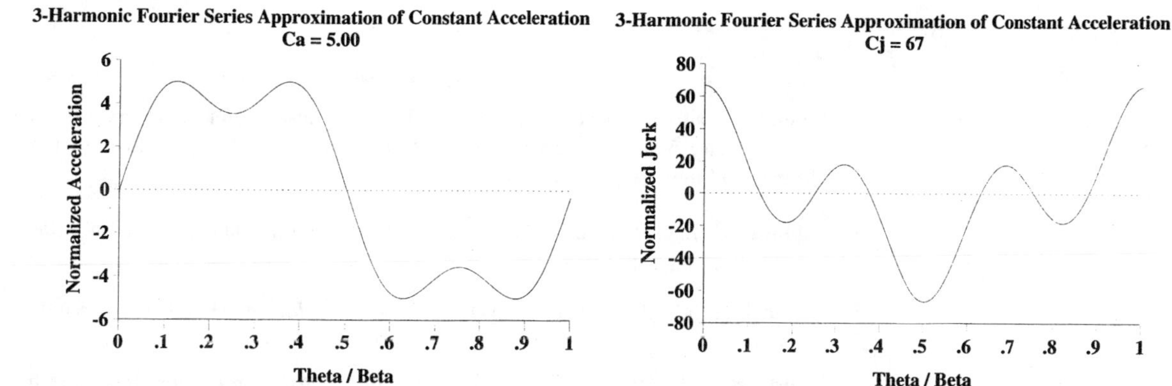

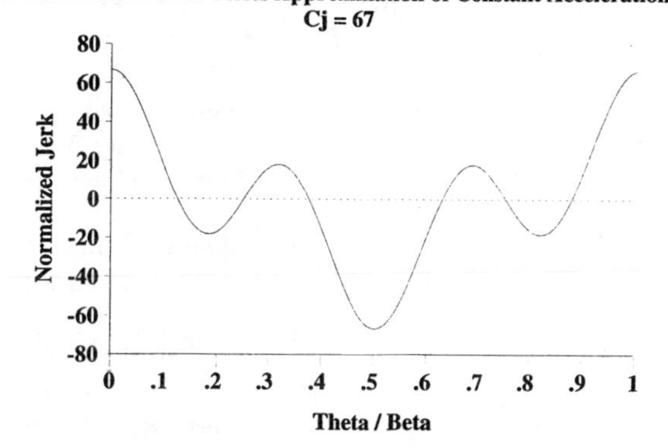

FIGURE 3-21

3-harmonic Fourier series approximation of the constant acceleration function

TABLE 3-4 Coefficients for Fourier Series Double-Dwell Functions

Harmonic	Freudenstein 1-3	Gutman F3	Weber 1-3
1	0.96428571	0.93750000	0.93545400
2	0	0	0
3	0.01190476	0.02083333	0.02151533
C_v	2.0	2.0	2.0
C_a	5.39	5.15	5.13

two-term (1-3 harmonic) functions that are similar to Freudenstein's 1-3 function. Their coefficients are shown in Table 3-4. Compare their C_v and C_a factors to those of the classical functions in Table 3-2 (p. 50). Their acceleration and jerk functions are similar to Figure 3-19a and 3-19b, respectively. These and other functions designed for low vibration will be discussed in more detail in Chapter 11.

3.6 SUMMARY

This chapter has presented an approach to the selection of appropriate double-dwell cam functions, using the common rise-dwell-fall-dwell cam as the example. It has also pointed out some of the pitfalls awaiting the cam designer. The particular functions described are only a few of the ones that have been developed for this double-dwell case over many years by many designers, and include those most commonly used. Most are also included in program DYNACAM.

There are many trade-offs to be considered in selecting a cam program for any application, some of which have already been mentioned, such as function continuity (the fundamental law), theoretical peak values of velocity and acceleration, and smoothness of jerk. There are other trade-offs still to be discussed in later chapters, involving cam sizing and the manufacturability of the cam as well as issues of dynamic forces, stresses, and vibrations.

3.7 REFERENCES

1 **Jones, J. R., and J. E. Reeve**, eds. (1978). "Dynamic Response of Cam Curves Based on Sinusoidal Segments." *Cams and Cam Mechanisms*, Jones, J. R., ed., Institution of Mechanical Engineers: London, pp. 14-24.

2 **Neklutin, C. J.**. (1952). "Designing Cams for Controlled Inertia and Vibration". *Machine Design*, June, 1952, pp. 143-160.

3 **Weber, T.** (1960). "Cam Dynamics via Filter Theory". *Machine Design*, v. 32, October 13, p. 160.

4 **Freudenstein, F.**, (1960) *On the Dynamics of High-Speed Cam Profiles*. Int. J. Mech. Sci., **1**: pp. 342 - 349.

5 **Gutman, A. S.**, (1961) "To Avoid Vibration, Try This New Cam Profile". *Product Engineering*, December 25.

Chapter 4

SINGLE-DWELL CAM CURVES

4.0 INTRODUCTION

The previous chapter discussed a number of acceptable cam functions for the double-dwell case, i.e., any rise or fall that lies between dwells. This chapter will describe a set of functions that behave in an acceptable manner at moderate to high speeds for single-dwell applications. The variables used in this chapter are the same as those defined in Table 2-1 (p. 18).

If your need is for a **rise-fall** (RF), critical extreme position (CEP) motion, with no dwell, then you should really be considering a crank-rocker linkage, rather than a cam-follower, to obtain all the linkage's relative advantages of reliability, ease of construction, and lower cost. If your needs for compactness outweigh those considerations, then the choice of a cam-follower in the RF case may be justified. Also, if you have a critical path motion (CPM) design specification, and the motion or its derivatives are defined over the interval, then a cam-follower system is the logical choice even in the RF case.

4.1 SINGLE-DWELL CAM DESIGN—CHOOSING *S V A J* FUNCTIONS

Many applications in machinery require a **single-dwell** cam program, **rise-fall-dwell** (RFD). Perhaps a single-dwell cam is needed to lift and lower a roller which carries a moving paper web on a production machine that makes envelopes. This cam's follower lifts the paper up to one critical extreme position at the right time to contact a roller which applies a layer of glue to the envelope flap. Without dwelling in the up position, it immediately retracts the web back to the starting (zero) position and holds it in this other critical extreme position (low dwell) while the rest of the envelope passes by. It repeats the cycle for the next envelope as it comes by. Another common example of a single-dwell application is the cam that opens the valves in your automobile engine. It lifts the valve open on the rise, immediately closes it on the fall, and then keeps the valve closed in a dwell while the compression and combustion take place.

If we attempt to use for a single-dwell application the same type of cam programs as were defined for the double-dwell case, we will achieve a solution which may work but will not be optimal. We will nevertheless do so here as an example in order to point out the problems that result. Then we will redesign the cam to eliminate those problems.

4

Portions of this chapter were adapted from R. L. Norton, *Design of Machinery* 2ed. McGraw-Hill, 2001, with permission.

EXAMPLE 4-1

Using Cycloidal Motion for the Single-Dwell Case.

Problem: Consider the following single-dwell cam specification:

rise	1 in (25.4 mm) in 90° with cycloidal motion
fall	1 in (25.4 mm) in 90° with cycloidal motion
dwell	at zero displacement for 180° (low dwell)
cam ω	15 rad/sec (143.24 rpm)

Solution:

1 Figure 4-1 shows a cycloidal displacement rise and separate cycloidal displacement fall applied to this single-dwell example. Note that the displacement (*S*) diagram looks acceptable in that it moves the follower from the low to the high position and back in the required intervals.

2 The velocity (*V*) also looks acceptable in shape in that it takes the follower from zero velocity at the low dwell to a peak value of 19.1 in/sec (0.49 m/sec) to zero again at the maximum displacement, where the glue is applied.

3 Figure 4-1 shows the acceleration function for this solution. Its maximum absolute value is about 573 in/sec^2.

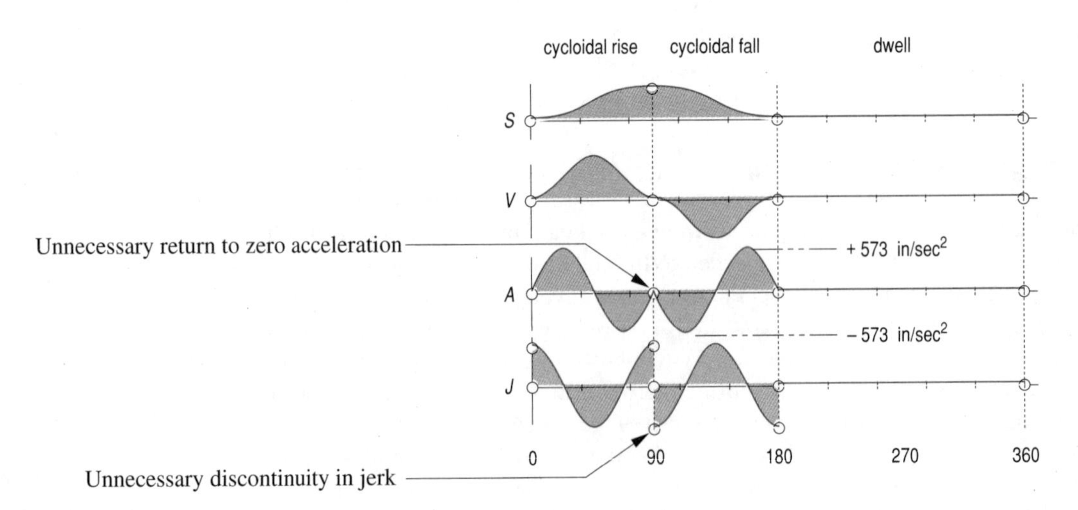

FIGURE 4-1

Cycloidal motion (or any double-dwell program) is a poor choice for the single-dwell case

4 The problem is that this acceleration curve has an **unnecessary return to zero** at the end of the rise. It is unnecessary because the acceleration during the first part of the fall is also negative. It would be better to keep it in the negative region at the end of the rise.

5 This unnecessary oscillation to zero in the acceleration causes the jerk to have more abrupt changes and discontinuities. The only real justification for taking the acceleration to zero is the need to change its sign (as is the case halfway through the rise or fall) or to match an adjacent segment which has zero acceleration.

The reader may input the file Ex_04-01.cam to program DYNACAM to investigate this example in more detail.

For the single-dwell case we would like a function for the rise which does not return its acceleration to zero at the end of the interval. This rules out all of the double-dwell functions described in Chapter 3. The function for the fall should begin with the same nonzero acceleration value as ended the rise, and then be zero at its terminus to match the dwell. One function which meets those criteria is the **double harmonic** which gets its name from its two cosine terms, one of which is a half-period harmonic and the other a full-period wave. The equations for the double harmonic functions are:

for the rise:

$$
\begin{aligned}
s &= \frac{h}{2}\left\{\left[1-\cos\left(\pi\frac{\theta}{\beta}\right)\right]-\frac{1}{4}\left[1-\cos\left(2\pi\frac{\theta}{\beta}\right)\right]\right\} \\
v &= \frac{\pi}{\beta}\frac{h}{2}\left[\sin\left(\pi\frac{\theta}{\beta}\right)-\frac{1}{2}\sin\left(2\pi\frac{\theta}{\beta}\right)\right] \\
a &= \frac{\pi^2}{\beta^2}\frac{h}{2}\left[\cos\left(\pi\frac{\theta}{\beta}\right)-\cos\left(2\pi\frac{\theta}{\beta}\right)\right] \\
j &= -\frac{\pi^3}{\beta^3}\frac{h}{2}\left[\sin\left(\pi\frac{\theta}{\beta}\right)-2\sin\left(2\pi\frac{\theta}{\beta}\right)\right]
\end{aligned}
\tag{4.1a}
$$

for the fall:

$$
\begin{aligned}
s &= \frac{h}{2}\left\{\left[1+\cos\left(\pi\frac{\theta}{\beta}\right)\right]-\frac{1}{4}\left[1-\cos\left(2\pi\frac{\theta}{\beta}\right)\right]\right\} \\
v &= -\frac{\pi}{\beta}\frac{h}{2}\left[\sin\left(\pi\frac{\theta}{\beta}\right)+\frac{1}{2}\sin\left(2\pi\frac{\theta}{\beta}\right)\right] \\
a &= -\frac{\pi^2}{\beta^2}\frac{h}{2}\left[\cos\left(\pi\frac{\theta}{\beta}\right)+\cos\left(2\pi\frac{\theta}{\beta}\right)\right] \\
j &= \frac{\pi^3}{\beta^3}\frac{h}{2}\left[\sin\left(\pi\frac{\theta}{\beta}\right)+2\sin\left(2\pi\frac{\theta}{\beta}\right)\right]
\end{aligned}
\tag{4.1b}
$$

Note that these double harmonic functions should **never** be used for the double-dwell case because their acceleration is nonzero at one end of the interval.

EXAMPLE 4-2

Double Harmonic Motion for the Symmetric Rise-Fall Single-Dwell Case.

Problem: Consider the same single-dwell cam specification as in Example 4-1:

rise	1 in (25.4 mm) in 90° with double-harmonic motion
fall	1 in (25.4 mm) in 90° with double-harmonic motion
dwell	at zero displacement for 180° (low dwell)
cam ω	15 rad/sec (143.24 rpm)

Solution:

1 Figure 4-2 shows a double harmonic rise and a double harmonic fall. The peak velocity is 19.5 in/sec (0.50 m/sec) which is similar to that of the cycloidal solution of Example 4-1.

2 Note that the acceleration of this double harmonic function does not return to zero at the end of the rise. This makes it more suitable for a symmetrical single-dwell case in that respect, but if the rise and fall durations are unequal, then their accelerations will not match.

3 The double harmonic jerk function peaks at 36 931 in/sec^3 (938 m/sec^3) and is quite smooth compared to the cycloidal solution.

4 Unfortunately, the peak negative acceleration is 900 in/sec^2, nearly twice that of the cycloidal solution. This is a smoother function but will develop higher dynamic forces. Open the diskfile Ex_04-02.cam in program DYNACAM to see this example in more detail.

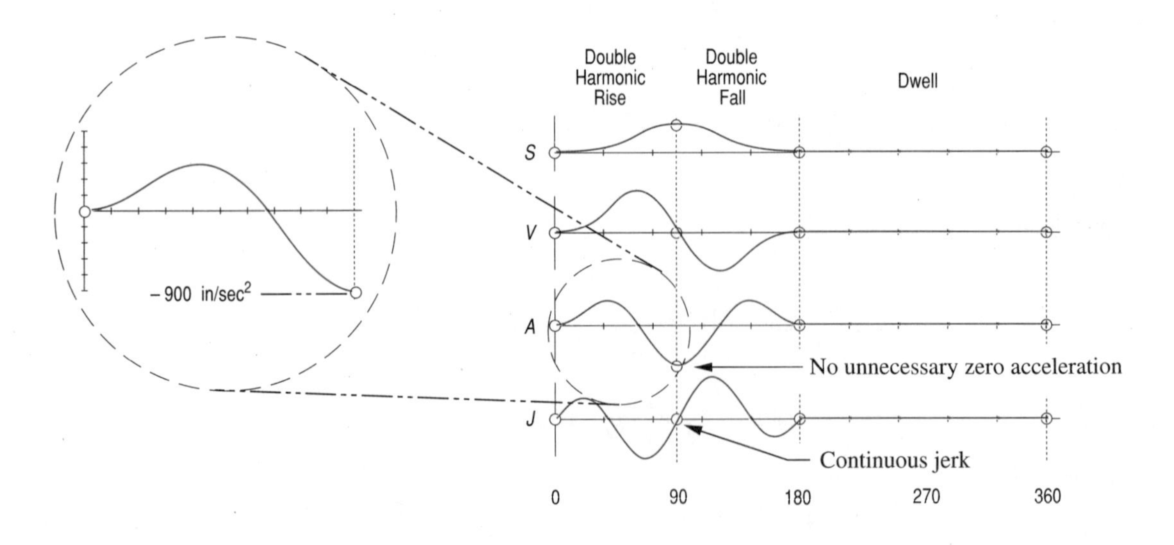

FIGURE 4-2

Double harmonic motion can be used for the single-dwell case if rise and fall durations are equal

Neither of the solutions in Examples 4-1 and 4-2 is optimal. We will redesign it using polynomial functions to improve its smoothness and reduce its peak acceleration.

4.2 SINGLE-DWELL APPLICATIONS OF POLYNOMIALS

Let us attempt to solve the symmetrical single-dwell problem of the previous two examples with a polynomial cam function. Restating the original problem for reference:

rise	1 in (25.4 mm) in 90°
fall	1 in (25.4 mm) in 90°
dwell	at zero displacement for 180° (low dwell)
cam ω	15 rad/sec

To solve this problem with a polynomial, we must decide on a suitable set of boundary conditions. We must also first decide how many segments to divide the cam cycle into. The problem statement seems to imply three segments—a rise, a fall, and a dwell. We could use those three segments to create the functions as we did in the two previous examples, but a better approach could be to use only **two segments**, one for the rise-fall combined and one for the dwell. *As a general rule we would like to minimize the number of segments in our polynomial cam functions*. Any dwell requires its own segment. So, the minimum number possible in this case is two segments.

Another rule of thumb in polynomial cam design is that *we would like to minimize the number of boundary conditions specified*, because the degree of the polynomial is tied to the number of BCs. As the degree of the function increases, so will the number of its **inflection points** and its number of **minima and maxima**. The polynomial derivation process will guarantee that the function will pass through all specified BCs but says nothing about the function's behavior between the BCs. *A high-degree function may have undesirable oscillations between its BCs.*

With these assumptions we can select a set of boundary conditions for a trial solution. First we will restate the problem to reflect our two-segment configuration.

EXAMPLE 4-3

Designing a Polynomial for the Symmetrical Rise-fall Single-Dwell Case.

Problem: Redefine the CEP specification from Examples 4-1 and 4-2.

rise-fall	rise 1 in (25.4 mm) in 90° and fall 1 in (25.4 mm) in 90° for 180°
dwell	at zero displacement for 180° (low dwell)
cam ω	15 rad/sec (143.24 rpm)

Solution:

1 Figure 4-3 shows the minimum set of seven BCs for this problem that will give a sixth-degree polynomial. The dwell on either side of the combined rise-fall segment has zero values of S, V, A, and J. The fundamental law of cam design requires that we match these zero values, through the acceleration function, at each end of the rise-fall segment.

Segment Number	Function Used	Start Angle	End Angle	Delta Angle
1	Poly 6	0	180	180

Boundary Conditions Imposed				Equation Resulting	
Function	Theta	θ / β	Boundary Cond.	Exponent	Coefficient
Displ	0	0	0	0	0
Veloc	0	0	0	1	0
Accel	0	0	0	2	0
Displ	180	1	0	3	64
Veloc	180	1	0	4	– 192
Accel	180	1	0	5	192
Displ	90	0.5	1	6	– 64

FIGURE 4-3

Boundary conditions and coefficients for a symmetrical rise-fall single-dwell polynomial application

2 The endpoints account for six BCs; $S = V = A = 0$ at each end of the rise-fall segment.

3 We also must specify a value of displacement at the 1-in peak of the rise that occurs at $\theta = 90°$. This is the seventh BC.

4 Figure 4-3 also shows the coefficients of the displacement polynomial which result from the simultaneous solution of the equations for the chosen BCs. For generality we have substituted the variable h for the specified 1-in rise. The function turns out to be a 3-4-5-6 polynomial whose equation is:

$$s = h\left[64\left(\frac{\theta}{\beta}\right)^3 - 192\left(\frac{\theta}{\beta}\right)^4 + 192\left(\frac{\theta}{\beta}\right)^5 - 64\left(\frac{\theta}{\beta}\right)^6\right] \tag{4.2}$$

5 Open the diskfile Ex_04-03.cam in program DYNACAM to see this example in more detail.

Figure 4-4 shows the *S V A J* diagrams for this solution with its maximum values noted. Compare these acceleration and *S V A J* curves to the double harmonic and cycloidal solutions to the same problem in Examples 4-1 and 4-2 (pp. 58-60). Note that this sixth-degree polynomial function is as smooth as the double harmonic functions (Figure 4-2 on p. 60) and does not unnecessarily return the acceleration to zero at the top of the rise as does the cycloidal solution (Figure 4-1, on p. 58). The polynomial has a peak acceleration of 547 in/sec^2, which is less than that of either the cycloidal or double harmonic solution. This 3-4-5-6 polynomial is a superior solution to either of the others presented for the same problem and is an example of how polynomial functions can be easily tailored to any particular design specifications. The reader may open the file Ex_04-03.cam in program DYNACAM to investigate this example in more detail.

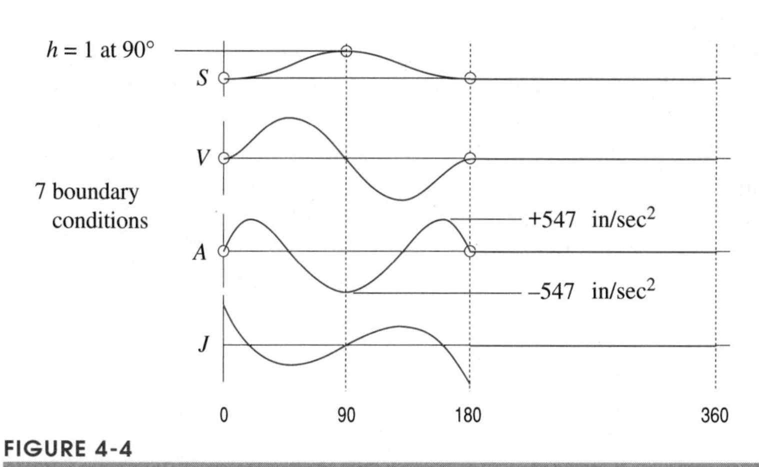

FIGURE 4-4

3-4-5-6 polynomial function for a two-segment symmetrical rise-fall, single-dwell cam

Effect of Asymmetry on the Rise-Fall Polynomial Solution

The examples so far presented in this chapter all had equal time for rise and fall, referred to as a symmetrical rise-fall curve. What will happen if we need an asymmetric program and attempt to use a single polynomial as was done in the previous example?

EXAMPLE 4-4

Designing a Polynomial for an Asymmetrical Rise-Fall Single-Dwell Case

Problem: Redefine the CEP specification from Example 4-3 as:

rise-fall	rise 1 in (25.4 mm) in 90° and fall 1 in (25.4 mm) in 90° for 180°
dwell	at zero displacement for 180° (low dwell).
cam ω	15 rad/sec (143.24 rpm)

Solution:

1 Figure 4-5 shows the minimum set of seven BCs for this problem that will give a sixth-degree polynomial. The dwell on either side of the combined rise-fall segment has zero values for S, V, A, and J. The fundamental law of cam design requires that we match these zero values, through the acceleration function, at each end of the rise-fall segment.

2 The endpoints account for six BCs; $S = V = A = 0$ at each end of the rise-fall segment.

3 We also must specify a value of displacement at the 1-in peak of the rise that occurs at $\theta = 45°$. This is the seventh BC.

4 Simultaneous solution of this equation set gives a 3-4-5-6 polynomial whose equation is:

$$s = h\left[151.704\left(\frac{\theta}{\beta}\right)^3 - 455.111\left(\frac{\theta}{\beta}\right)^4 + 455.111\left(\frac{\theta}{\beta}\right)^5 - 151.704\left(\frac{\theta}{\beta}\right)^6\right] \tag{4.3}$$

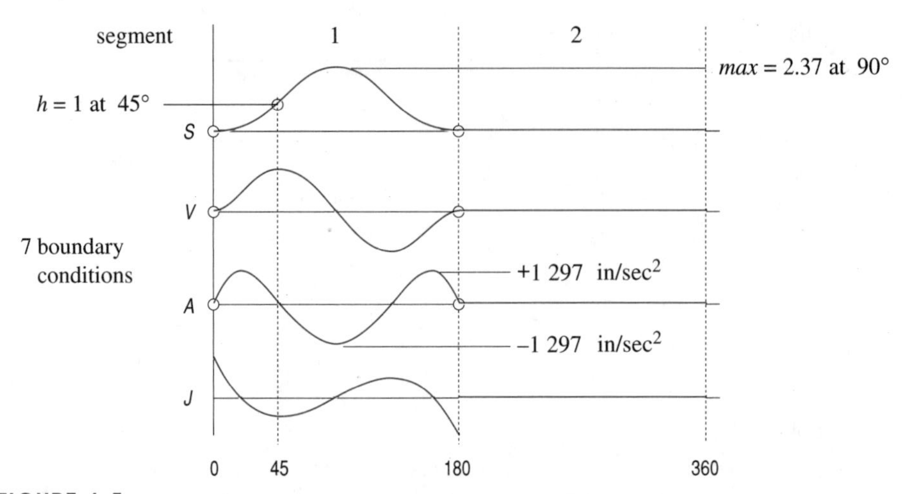

FIGURE 4-5

Unacceptable polynomial for a two-segment asymmetrical rise-fall, single-dwell cam

For generality we have substituted the variable h for the specified 1-in rise.

5 Figure 4-5 shows the $S\ V\ A\ J$ diagrams for this solution with its maximum values noted. Observe that the derived 6th degree polynomial has obeyed the stated boundary conditions and does in fact pass through a displacement of 1 unit at 45°. But note also that it overshoots that point and reaches a height of 2.37 units at its peak. The acceleration peak is also 2.37 times that of the symmetrical case of Example 4-4. Without any additional boundary conditions applied, the function seeks symmetry. Note that the zero velocity point is still at 90° when we would like it to be at 45°. We can try to force the velocity to zero with an additional boundary condition of $V = 0$ when $\theta = 45°$.

6 Figure 4-6 shows the $s\ v\ a\ j$ diagrams for a 7th degree polynomial having 8 BCs, $S = V = A = 0$ at $\theta = 0°$, $S = V = A = 0$ at $\theta = 180°$, $S = 1$, $V = 0$ at $\theta = 45°$. Note that the resulting function has obeyed our commands and goes through those points, but "does its own thing" elsewhere. It now plunges to a negative displacement of –3.93, and the peak acceleration is much larger. This points out an inherent problem in polynomial functions, namely that their behavior between boundary conditions is not controllable and may create undesirable deviations in the follower motion. This problem is exacerbated as the degree of the function increases since it then has more roots and inflection points, thus allowing more oscillations between the boundary conditions.

7 Open the diskfiles Ex_04-04a and b in program DYNACAM to see this example in more detail.

In this case, the rule of thumb to minimize the number of segments is in conflict with the rule of thumb to minimize the degree of the polynomial. One alternative solution to this asymmetrical problem is to use three segments, one for the rise, one for the fall, and one for the dwell. Adding segments will reduce the order of the functions and bring them under control.

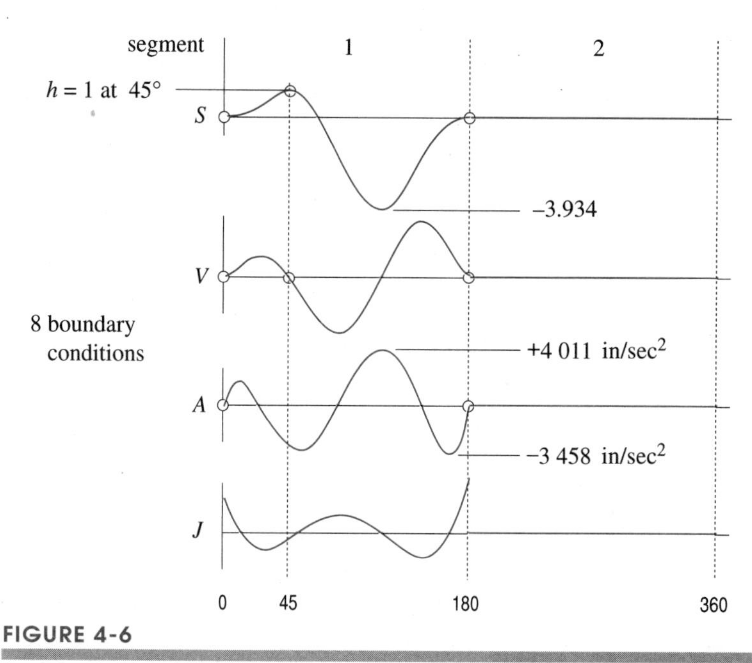

FIGURE 4-6

Unacceptable polynomial for a two-segment asymmetrical rise-fall, single-dwell cam

EXAMPLE 4-5

Designing a Three-Segment Polynomial for an Asymmetrical Rise-Fall Single-Dwell Case Using Minimum Boundary Conditions.

Problem: Redefine the CEP specification from Example 4-4 as:

rise	1 in (25.4 mm) in 45°
fall	1 in (25.4 mm) in 135°
dwell	at zero displacement for 180° (low dwell)
cam ω	15 rad/sec (143.24 rpm)

Solution:

1 The first attempt at this solution specifies 5 BCs; $S = V = A = 0$ at the start of the rise (to match the dwell), $S = 1$ and $V = 0$ at the end of the rise. Note that the rise segment BCs leave the acceleration at its end unspecified, but the fall's BCs must include the value of the acceleration at the end of the rise that results from the calculation of its acceleration. Thus, the fall requires one more BC than the rise.

2 This results in the following 4th degree equation for the rise segment:

$$s = h\left[4\left(\frac{\theta}{\beta}\right)^3 - 3\left(\frac{\theta}{\beta}\right)^4\right] \tag{4.4}$$

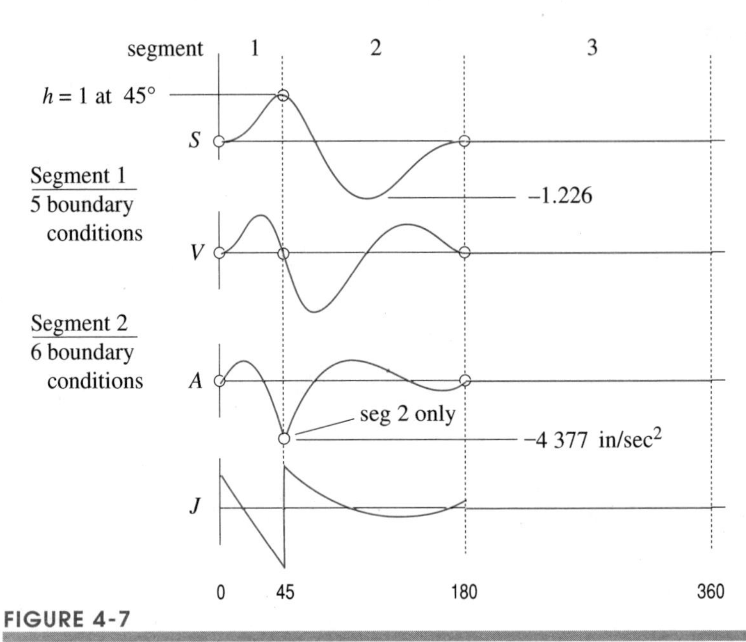

FIGURE 4-7

Uacceptable polynomials for a three-segment asymmetrical rise-fall, single-dwell cam

3 Evaluating the acceleration at the end of rise gives –4377.11 in/s^2. This value becomes a BC for the fall segment. The set of 6 BCs for the fall are then: $S = 1$, $V = 0$, $A = -4377.11$ at the start of the fall (to match the rise) and $S = V = A = 0$ at the end of the fall to match the dwell. The 5th degree equation for the fall is then:

$$s = h\left[1 - 54\left(\frac{\theta}{\beta}\right)^2 + 152\left(\frac{\theta}{\beta}\right)^3 - 147\left(\frac{\theta}{\beta}\right)^4 + 48\left(\frac{\theta}{\beta}\right)^5\right] \tag{4.5}$$

4 Figure 4-7 shows the $S\ V\ A\ J$ diagrams for this solution with its extreme values noted. Observe that this polynomial on the fall also has a problem—the displacement still goes negative.

5 The trick in this case is to first calculate the second segment as it will have a smaller negative acceleration due to its larger duration β. Then this smaller acceleration value will be used as a boundary condition on the first segment. The 5 BCs for segment 2 are then $S = 1$ and $V = 0$ at the start of the fall and $S = V = A = 0$ at the end of the fall (to match the dwell). These give the following 4th degree polynomial for the fall.

$$s = h\left[1 - 6\left(\frac{\theta}{\beta}\right)^2 + 8\left(\frac{\theta}{\beta}\right)^3 - 3\left(\frac{\theta}{\beta}\right)^4\right] \tag{4.6}$$

6 Evaluating the acceleration at the start of the fall gives –486.4 in/s^2. This value becomes a BC for the rise segment. The set of 6 BCs for the rise are then: $S = V = A = 0$ at the start

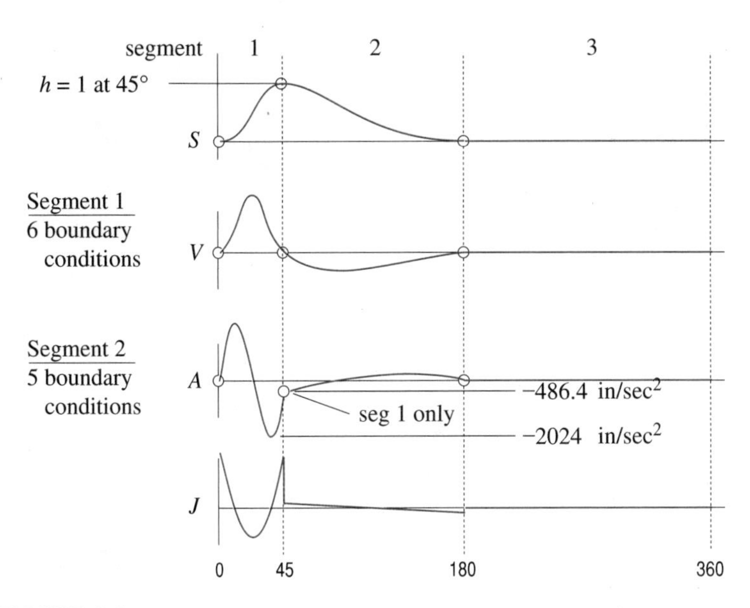

FIGURE 4-8

Acceptable polynomials for a three-segment asymmetrical rise-fall, single-dwell cam

of the rise to match the dwells, and $S = 1$, $V = 0$, $A = -486.4$ at the end of the rise (to match the fall). The 5th degree equation for the rise is then:

$$s = h\left[9.333\left(\frac{\theta}{\beta}\right)^3 - 13.667\left(\frac{\theta}{\beta}\right)^4 + 5.333\left(\frac{\theta}{\beta}\right)^5\right] \tag{4.7}$$

7 The resulting cam design is shown in Figure 4-8. The displacement is now under control and the peak acceleration is much less than the previous design at about 2 024 in/s^2.

8 A possible improvement to this design could be to specify the jerk to be zero at the beginning and end to match the dwells. This will smooth the jerk function to reduce potential vibration, but may come at the expense of an increase in peak acceleration. Using the same strategy as in step 5, the fall function is first derived with its beginning acceleration left unspecified, then that value of acceleration is used to define the first segment. The 6 BCs for segment 2 are then $S = 1$ and $V = 0$ at the start of the fall and $S = V = A = J = 0$ at end of the fall (to match the dwell). These give the following 5th degree polynomial for the fall.

$$s = h\left[1 - 10\left(\frac{\theta}{\beta}\right)^2 + 20\left(\frac{\theta}{\beta}\right)^3 - 15\left(\frac{\theta}{\beta}\right)^4 + 4\left(\frac{\theta}{\beta}\right)^5\right] \tag{4.8}$$

9 Evaluating the acceleration at the start of the fall gives –810.58 in/s^2. This value becomes a BC for the rise segment. The set of 7 BCs for the rise are then: $S = V = A = J = 0$ at start of rise to match the dwells, and $S = 1$, $V = 0$, $A = -810.58$ at the end of rise (to match the fall). The 6th degree equation for the segment 1 rise is then:

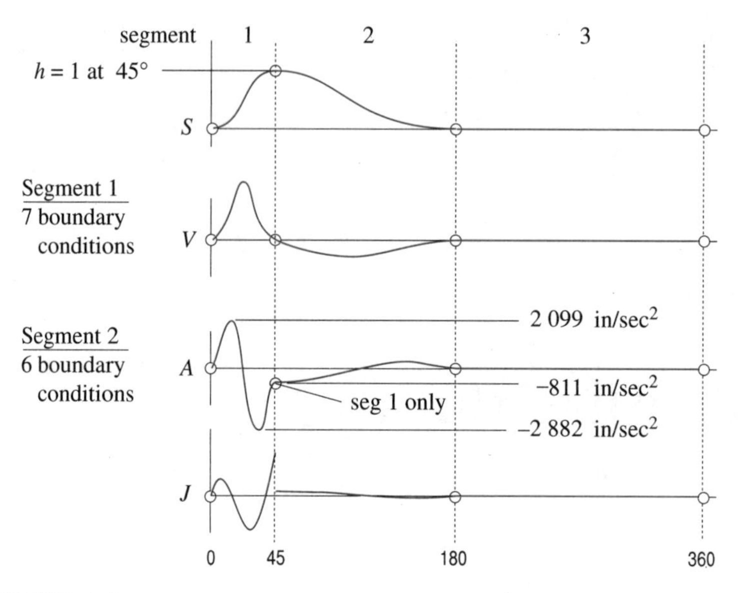

FIGURE 4-9

Acceptable polynomials for a three-segment asymmetrical rise-fall, single-dwell cam

$$s = h\left[13.889\left(\frac{\theta}{\beta}\right)^4 - 21.778\left(\frac{\theta}{\beta}\right)^5 + 8.889\left(\frac{\theta}{\beta}\right)^6\right] \tag{4.9}$$

10 The resulting cam design is shown in Figure 4-9. The displacement is still under control and the peak negative acceleration is greater than the previous design at –2 882 in/s^2.

11 Either of the designs of Figures 4.8 and 4.9 are acceptable for this example. Open the diskfiles Ex_04-05a , b, and c in program DYNACAM to see this example in more detail.

4.3 SUMMARY

This chapter has presented polynomial functions as a versatile approach to many cam design problems. It is only since the development and general availability of computers that these functions have become practical to use, as the computation to solve the simultaneous equations is often beyond hand calculation abilities. With the availability of a design aid to solve the equations such as program DYNACAM, polynomials have become a practical and preferable way to solve many cam design problems. **Spline functions**, of which polynomials are a subset, offer even more flexibility in meeting boundary constraints and other cam performance criteria. The next chapter introduces spline functions and shows how they can be applied to the cam designs shown in this and previous chapters to give superior results in many cases.

Chapter 5

SPLINE FUNCTIONS

Ronald G. Mosier

5.0 INTRODUCTION

In the days before computers and computer-aided design, draftsmen—wanting to draw a "fair" curve through prescribed points—would use a spline, a thin rod made of some flexible material such as bamboo. Weights, called ducks, would anchor the spline so that it passed over the points, as shown in Figure 5-1. The spline would assume an overall shape that would minimize its strain energy subject to the constraints caused by the ducks. The draftsman could then trace the shape of the rod to get a smooth curve through the points.

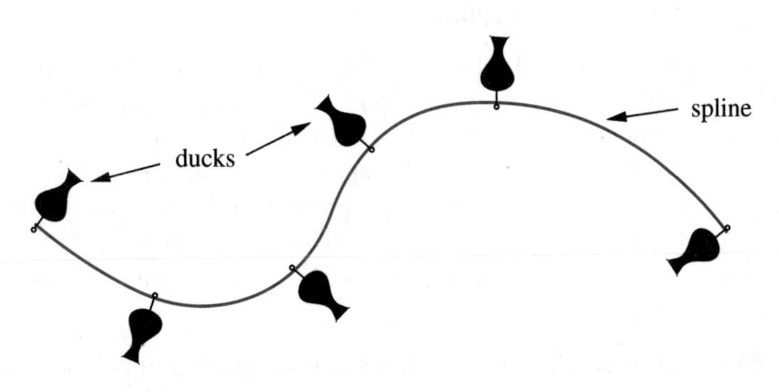

FIGURE 5-1

A draftsman's spline and ducks

Mathematical splines are an attempt to imitate this tool. As frequently happens, especially when the computer is involved, the imitators far exceed the original in versatility and in usefulness. This chapter explores how the engineer can use mathematical splines to design cams. Section 5.1 describes the classical splines, the closest imitators of the draftsman's spline. One of these, the cubic spline, is still the most commonly used of the mathematical splines. Section 5.2 discusses the general polynomial spline and Section 5.3 enlarges the engineer's spline toolbox by introducing B-splines and Section 5.4 discusses Bezier curves. A discussion of knot placement follows in Section 5.5. Knots are the mathematical version of the ducks. Section 5.6 introduces periodic splines and Section 5.7 trigonometric and rational splines. Section 5.8 summarizes everything and provides a list of advantages and disadvantages of various splines. From now on the term spline will be used to mean a mathematical spline and we will speak of spline curves as motions. The spline values will be *displacements*, the first derivative will be *velocity*, the second derivative will be the *acceleration*, the third will be referred to as *jerk*, the fourth derivative will be called *ping*, and the fifth derivative will be called *puff*.

5.1 CLASSICAL SPLINES

Knots

A classical spline of order m is a curve made up of polynomial pieces, each of degree $m - 1$, that are blended together at their ends in such a way that the curve is continuous and all of its derivatives are continuous, up to and including the derivative of order $m - 2$. This means, for example, that if a classical spline of order 5 is used to represent cam follower displacement, then the displacement will be made up of polynomial pieces of degree 4 and will be continuous. The velocity will be continuous, the acceleration will be continuous, and the jerk will be continuous. But the fourth derivative, ping, will not be continuous. The end points of the polynomial pieces are called **knots**. The knots are where the polynomial pieces are "tied together," which explains the name. The classical spline always interpolates prescribed values at the knots.

The classical spline of order 6 is one of the most useful for cam work and, for that reason, we will use it for purposes of illustration. The reader should be assured that everything we say will be true for splines of other orders.

We construct our classical spline on an interval from a to b. For purposes of discussion, we choose only three knots for the spline. There will be one knot at each end, (in fact there *must* be knots at both ends), and a third knot somewhere between a and b. We chose only three knots just to keep things simple and the equations short. We could have chosen any number from two on up, but we prefer that the reader pay attention to our arguments and not be fighting off an army of indices and a flotilla of equations. If we had used only the two end knots, the spline would have been a single polynomial, a curve that we hope is already familiar to the reader.

A little reflection and the reader will see that the number of polynomial pieces that make up the spline is just one less than the number of knots. Since we have three knots, we have two polynomial pieces. Each piece is a sixth order polynomial, which is to say that each has six coefficients. Let the polynomial pieces be

$$s(\theta) = \begin{cases} c_1 + c_2 \dfrac{\theta - a}{t - a} + c_3 \left(\dfrac{\theta - a}{t - a}\right)^2 + c_4 \left(\dfrac{\theta - a}{t - a}\right)^3 + c_5 \left(\dfrac{\theta - a}{t - a}\right)^4 + c_6 \left(\dfrac{\theta - a}{t - a}\right)^5, a \le \theta \le t \\ d_1 + d_2 \dfrac{\theta - t}{b - t} + d_3 \left(\dfrac{\theta - t}{b - t}\right)^2 + d_4 \left(\dfrac{\theta - t}{b - t}\right)^3 + d_5 \left(\dfrac{\theta - t}{b - t}\right)^4 + d_6 \left(\dfrac{\theta - t}{b - t}\right)^5, t \le \theta \le b \end{cases} \tag{5.1}$$

In equation 5.1, we designated the intermediate knot between a and b by t. We have written the equations so that, at the knots, the terms with θ evaluate to either zero or one. This is commonly done in spline work.

Smoothness Equations

The knot at t is called an interior knot, for obvious reasons. It is at the interior knots of a spline that the adjacent polynomial pieces must be matched to each other. According to the above definition, at the interior knot t we must match the values of the polynomials and the derivatives up to and including the fourth. We are indifferent to the fifth, as will be shown below. In the general case of a spline of order m, we have to match, at each interior knot, the values of the polynomials and their derivatives up to and including the $m - 2$ order derivative. In this example, we blend together the polynomial pieces by choosing values for the twelve coefficients so that

$$\begin{aligned} c_1 + c_2 + c_3 + c_4 + c_5 + c_6 &= d_1 \\ \frac{c_2}{t-a} + 2\frac{c_3}{t-a} + 3\frac{c_4}{t-a} + 4\frac{c_5}{t-a} + 5\frac{c_6}{t-a} &= \frac{d_2}{b-t} \\ \frac{c_3}{(t-a)^2} + 3\frac{c_4}{(t-a)^2} + 6\frac{c_5}{(t-a)^2} + 10\frac{c_6}{(t-a)^2} &= \frac{d_3}{(b-t)^2} \\ \frac{c_4}{(t-a)^3} + 4\frac{c_5}{(t-a)^3} + 10\frac{c_6}{(t-a)^3} &= \frac{d_4}{(b-t)^3} \\ \frac{c_5}{(t-a)^4} + 5\frac{c_6}{(t-a)^4} &= \frac{d_5}{(b-t)^4} \end{aligned} \tag{5.2}$$

These equations specify, respectively, continuity across the knot at t, continuity of the velocity at t, continuity of the acceleration at t, continuity of the jerk at t, and continuity of the ping at t. They are called the **smoothness equations** for the spline. The more perspicacious reader may have noticed that if we add one more equation, expressing continuity across the knot of the fifth derivative, the **puff**, then the polynomial on the right side of the knot at t would be the same as the polynomial on the left side of the knot at t. The knot itself would cease to be a knot. Classical splines are often called **complete splines** because it is true for them that adding one more continuity condition to any knot causes it to no longer be a knot. The condition that erases the knot is called the "*not a knot*" condition and below we will show when it can be useful.

For the general classical spline of order m, there are $m - 1$ smoothness equations at each interior knot. If there are k knots, then there are $k - 2$ interior knots for a total of $(k - 2)(m - 1)$ smoothness equations. The number of unknown coefficients will be m for every polynomial piece, there are $k - 1$ polynomial pieces, and so there must be $m(k - 1)$ unknowns. This works out in our example. We have $m = 6$ and $k = 3$, so we have five smoothness equations and twelve unknowns. For a spline of order eight and with ten knots, we would have 56 smoothness equations and 72 unknowns; you can understand and be glad that we are explaining classical splines using a sixth order spline with only three knots.

Interpolation Equations

Since we have twelve unknown coefficients and only five smoothness equations, it is clear that we need seven more equations. We can get one of them by interpolating, i.e., forcing the spline to have a given value, at an interior knot. In general this would give us as many equations as there are interior knots, $k - 2$. These equations are called, with an embarrassing lack of imagination, the **interpolation equations**. Here we have only one interpolation equation. Denoting the finished spline curve by $s(\theta)$ and denoting the desired interpolation value at the knot t by $s(t)$, we have the equation:

$$d_1 = s(t) \tag{5.3}$$

Boundary Conditions

We now need six more equations for our sixth order spline. For the general case, we would need m more equations for a spline of order m. These final equations are the **boundary conditions**. These are the same equations that the engineer would have to formulate for a polynomial or for any other type of cam follower displacement function, not just a spline.

In the case of the sixth order spline, and, indeed, every classical spline, two of the boundary condition equations are the equations to guarantee the desired end values. Denoting the desired values as $s(a)$ and $s(b)$, they are:

$$\begin{aligned} c_1 &= s(a) \\ d_1 + d_2 + d_3 + d_4 + d_5 + d_6 &= s(b) \end{aligned} \tag{5.4a}$$

Additional boundary conditions specify the derivatives at each end,

$$\begin{aligned} \frac{c_2}{t-a} &= s'(a) \\ \frac{d_2}{b-t} + \frac{2d_3}{b-t} + \frac{3d_4}{b-t} + \frac{4d_5}{b-t} + \frac{5d_6}{b-t} &= s'(b) \end{aligned} \tag{5.4b}$$

and specify the second derivatives at each end.

$$\begin{aligned} \frac{2c_3}{(t-a)^2} &= s''(a) \\ \frac{2d_3}{(b-t)^2} + \frac{6d_4}{(b-t)^2} + \frac{12d_5}{(b-t)^2} + \frac{20d_6}{(b-t)^2} &= s''(b) \end{aligned} \tag{5.4c}$$

Now is a good time to point out the fault with the popular fourth order or cubic spline as far as cam motion specification is concerned. It would, according to what we said above, satisfy only four boundary conditions: two end values and two derivatives. The derivatives could be both velocity or both acceleration or even velocity on one end and acceleration on the other. No matter what we do, however, it will not be enough to guarantee that we obey the fundamental law of cam design, which requires at least six boundary conditions, two end displacements, two end velocities, and two end accelerations. For this reason, fourth order classical splines are not used to design cam-follower displacement curves. Other order splines can and should be used. Table 5-1 shows the most commonly used classical splines and also shows the number of boundary conditions that each can satisfy. The reader has probably already noticed that the number of boundary conditions that the spline will match has nothing to do with how many knots it has. A classical spline of order m can only be made to satisfy at most m boundary conditions regardless of the number of knots. Curiously enough, the number of boundary values needed can be reduced by using the "not a knot" condition. This is equivalent to dropping a boundary condition equation and adding an extra equation to the smoothness set. Since the classical spline is a complete spline, this must result in removing a knot.

Notice also from Table 5-1 that the splines of odd order have an odd number of degrees of freedom with which to meet the derivative requirements at the two ends. This means that one of the ends will be shortchanged. Many cam designers refer to splines of odd order as problematic and avoid their use. This is a mistake as they can be very useful and have the advantage of being able to lower accelerations and velocities.

Another possible boundary condition set of equations is in imitation of the draftsman's spline. If one were to extend both ends of the draftsman's spline well past the end ducks, the added sections would be perfectly straight. We can impose this condition on the fourth order classical spline by setting the second derivatives at the ends equal to zero.

$$s''(a) = s''(b) = 0 \tag{5.5}$$

A fourth order spline with zero second derivatives at the ends is called a **natural spline** because it is a close imitation of the draftsman's spline.

The boundary conditions needed for natural splines of higher order are more elaborate. In the first place, only even order splines can be natural splines and in the second

TABLE 5-1 Properties of Classical Splines of Order 2 to 6

Spline Order	Degree of Each Polynomial Piece	Has Jump Discontinuities at	DOF at End Knots	Comments
2	1	Velocity	None	Not good for cam work
3	2	Acceleration	1	Not good for cam work
4	3	Jerk	2	Difficult to use for cams
5	4	Ping (1st Deriv. of Jerk)	3	Odd DOF for two ends
6	5	Puff (2nd Deriv. of Jerk)	4	Very useful, very smooth

place, many more end derivatives must be set to zero. Suppose the spline is of even order $m = 2p$. Then to be a natural spline, it must satisfy the following boundary conditions at each end:

$$\frac{d^{m-2}s}{d\theta^{m-2}} = \frac{d^{m-3}s}{d\theta^{m-3}} = \cdots = \frac{d^{p}s}{d\theta^{p}} = 0 \tag{5.6}$$

In the case of the fourth order spline, these conditions reduce to setting the second derivatives at the ends equal to zero just as we said before, since $p = 2$ and $m - 2 = 2$. Another natural spline would be the natural sixth order spline. For it, $p = 3$ and $m - 2 = 4$, so that the jerk and ping would be set to zero at each end. Since this would leave the all-important end accelerations and velocities unspecified, the **natural** sixth order spline is not used for cam work either.

5

We will now go over everything that was said above using a specific example.

EXAMPLE 5-1

Using a Classical Spline for a Single-Dwell Cam.

Problem: Consider the following single-dwell cam specification:

rise	1 in (25.4 mm) in 90°
fall	1 in (25.4 mm) in 90°
dwell	at zero displacement for 180° (low dwell)
cam ω	15 rad/sec

Solution:

1 The first step in constructing a classical spline is to pick knot locations. The first row of Table 5-2 shows our knot selections. We place knots at 0, 45, 90, 135, and 180°. The second row of the table shows the displacements we want at each of the knots. The knots at 45, 90 and 135° are interior knots. The knots at 0 and 180° are end knots. The knots at 0, 90, and 180° are demanded by the problem in that we are asked to design a cam motion that has a displacement of 0 at 0°, 1 inch at 90° and returns to 0 at 180°. The interior knots at 45 and 135° are not necessary. We have arbitrarily selected them and could have selected more knots or fewer knots or even no knots other than those demanded by the problem. The reason that we added the two extra knots was to give us control over the velocity, acceleration and jerk, as will soon be shown. We might also mention that we chose our knots to be equally spaced from each other. This will simplify our equations. Such a spline is called a **uniform spline**.

TABLE 5-2 Knot Locations and Boundary Conditions for 1st Try, Example 5-1

	Knot Locations (deg)				
Function	0	45	90	135	180
Displacement	0	0.5	1	0.5	0
Velocity	0	–	–	–	0
Acceleration	0	–	–	–	0

2 The next step in our classical spline construction is to choose the order of the spline. We choose a classical spline of order six to solve our problem. We therefore expect to have a very smooth curve with continuous derivatives up beyond jerk, except at the end knots. At the end knots we can match only the velocity and acceleration of the dwell. There will be a jump discontinuity at the end knots in the jerk, but this is acceptable.

3 Now the computer takes over. In truth, the computer will not be asked to solve the equations that we listed above, but rather an equivalent set that minimizes roundoff errors. Since this discussion is about cam design and not about numerical analysis, we will not derive the equations that are used by computer software packages such as IMSL or the NAG Library to construct classical splines. Nevertheless, for completeness, here is a sketch of what the computer does for a classical spline of order six. The values of the ping at the knots are chosen as the variables rather than the coefficients of the polynomial pieces. The ping function is then expressed as a second order spline and integrated four times to get the sixth order polynomial pieces that represent the displacement function. The constants of integration are determined by the interpolation equations. Finally the smoothness equations and boundary condition equations are solved for the new unknowns, the ping values at the knots. The main difference between the "raw" equations shown above and the ones used for computation is that much of their solution procedure, such as determining the constants of integration, can be done algebraically, without recourse to numerical computation. Also, the final set of equations solved is one-fourth the size of the original set. Thus the numerical work is minimized as is computer roundoff error. Of course, the final equations that are solved look so different from the initial problem that anyone seeing them for the first time without lengthy explanation might wonder about their purpose. Hence, for this example, we will stick to what we think are more obvious equations. Our example problem is small enough that we need not worry about numerical errors.

4 For those who want to see and possibly solve the intuitive form of the equations for our example we will now provide them. The polynomial pieces are:

$$
\begin{aligned}
p_1(\theta) &= a_1\left(\frac{\theta}{\pi/4}\right)^5 + b_1\left(\frac{\theta}{\pi/4}\right)^4 + c_1\left(\frac{\theta}{\pi/4}\right)^3 + d_1\left(\frac{\theta}{\pi/4}\right)^2 + e_1\left(\frac{\theta}{\pi/4}\right) + f_1 \\
p_2(\theta) &= a_2\left(\frac{\theta-\pi/4}{\pi/4}\right)^5 + b_2\left(\frac{\theta-\pi/4}{\pi/4}\right)^4 + c_2\left(\frac{\theta-\pi/4}{\pi/4}\right)^3 + d_2\left(\frac{\theta-\pi/4}{\pi/4}\right)^2 + e_2\left(\frac{\theta-\pi/4}{\pi/4}\right) + f_2 \\
p_3(\theta) &= a_3\left(\frac{\theta-\pi/2}{\pi/4}\right)^5 + b_3\left(\frac{\theta-\pi/2}{\pi/4}\right)^4 + c_3\left(\frac{\theta-\pi/2}{\pi/4}\right)^3 + d_3\left(\frac{\theta-\pi/2}{\pi/4}\right)^2 + e_3\left(\frac{\theta-\pi/2}{\pi/4}\right) + f_3 \\
p_4(\theta) &= a_4\left(\frac{\theta-3\pi/4}{\pi/4}\right)^5 + b_4\left(\frac{\theta-3\pi/4}{\pi/4}\right)^4 + c_4\left(\frac{\theta-3\pi/4}{\pi/4}\right)^3 + d_4\left(\frac{\theta-3\pi/4}{\pi/4}\right)^2 + e_4\left(\frac{\theta-3\pi/4}{\pi/4}\right) + f_4
\end{aligned}
\qquad \text{(a)}
$$

where the unknowns are the coefficients

$$a_1, b_1, c_1, d_1, e_1, f_1, a_2, b_2, c_2, d_2, e_2, f_2, a_3, b_3, c_3, d_3, e_3, f_3, a_4, b_4, c_4, d_4, e_4, f_4 \qquad \text{(b)}$$

a total of 24 in all.

The smoothness equations are

$$
\begin{aligned}
a_1 + b_1 + c_1 + d_1 + e_1 + f_1 &= f_2 \\
a_2 + b_2 + c_2 + d_2 + e_2 + f_2 &= f_3 \\
a_3 + b_3 + c_3 + d_3 + e_3 + f_3 &= f_4 \\
5a_1 + 4b_1 + 3c_1 + 2d_1 + e_1 &= e_2 \\
5a_2 + 4b_2 + 3c_2 + 2d_2 + e_2 &= e_3 \\
5a_3 + 4b_3 + 3c_3 + 2d_3 + e_3 &= e_4 \\
10a_1 + 6b_1 + 3c_1 + d_1 &= d_2 \\
10a_2 + 6b_2 + 3c_2 + d_2 &= d_3 \\
10a_3 + 6b_3 + 3c_3 + d_3 &= d_4 \\
10a_1 + 4b_1 + c_1 &= c_2 \\
10a_2 + 4b_2 + c_2 &= c_3 \\
10a_3 + 4b_3 + c_3 &= c_4 \\
5a_1 + b_1 &= b_2 \\
5a_2 + b_2 &= b_3 \\
5a_3 + b_3 &= b_4
\end{aligned}
\tag{c}
$$

These are a total of 15 equations. There are three interpolation equations:

$$
\begin{aligned}
a_1 + b_1 + c_1 + d_1 + e_1 + f_1 &= 0.5 \\
a_2 + b_2 + c_2 + d_2 + e_2 + f_2 &= 1 \\
a_3 + b_3 + c_3 + d_3 + e_3 + f_3 &= 0.5
\end{aligned}
\tag{d}
$$

and 6 boundary condition equations:

$$
\begin{aligned}
f_1 &= 0 \\
e_1 &= 0 \\
d_1 &= 0 \\
a_4 + b_4 + c_4 + d_4 + e_4 + f_4 &= 0 \\
5a_4 + 4b_4 + 3c_4 + 2d_4 + e_4 &= 0 \\
20a_4 + 12b_4 + 6c_4 + 2d_4 &= 0
\end{aligned}
\tag{e}
$$

The displacement that we obtain by solving these equations is

$$s(\theta)=\begin{cases} 0.2991\left(\frac{\theta}{\pi/4}\right)^5-1.2277\left(\frac{\theta}{\pi/4}\right)^4+1.4288\left(\frac{\theta}{\pi/4}\right)^3, \quad 0\le\theta\le\pi/4 \\ -0.0580\left(\frac{\theta-\pi/4}{\pi/4}\right)^5+0.2679\left(\frac{\theta-\pi/4}{\pi/4}\right)^4-0.4911\left(\frac{\theta-\pi/4}{\pi/4}\right)^3 \\ \qquad -0.0893\left(\frac{\theta-\pi/4}{\pi/4}\right)^2+0.8705\left(\frac{\theta-\pi/4}{\pi/4}\right)+0.5, \quad \pi/4\le\theta\le\pi/2 \\ 0.0580\left(\frac{\theta-\pi/4}{\pi/4}\right)^5-0.0223\left(\frac{\theta-\pi/4}{\pi/4}\right)^4-0.5357\left(\frac{\theta-\pi/4}{\pi/4}\right)^3+1, \quad \pi/2\le\theta\le3\pi/4 \\ -0.2991\left(\frac{\theta-3\pi/4}{\pi/4}\right)^5+0.2679\left(\frac{\theta-3\pi/4}{\pi/4}\right)^4+0.4911\left(\frac{\theta-3\pi/4}{\pi/4}\right)^3 \\ \qquad -0.0893\left(\frac{\theta-3\pi/4}{\pi/4}\right)^2-0.8705\left(\frac{\theta-3\pi/4}{\pi/4}\right)+0.5, \quad 3\pi/4\le\theta\le\pi \end{cases} \tag{f}$$

5 Figure 5-2 shows the displacement described by equation (f). The curves are very smooth, with very low values for velocity and acceleration. There is a jump discontinuity in jerk at the end knots, just as we knew there would be. This does not violate the fundamental law of cam design and is acceptable. The only complaint that might be suggested about this motion is that the jerk has a very high value at the jump, about 60 000 in/sec^3. Whether or not this is acceptable depends upon the application, of course, but assume it is not. This gives us a chance to show the real power of splines in cam design as well as what we mean by being able to control things with the extra knots. The jerk is high because we wanted to minimize the acceleration and the maximum velocity. We did this with those two interior knots at 45° and at 135°. The reasoning was that any displacement that starts at 0 at 0° and ends at 1 inch at 90° must have an average velocity of 1/90 = 0.01111 inches per degree. Any spline that we come up with will have this average. By placing a knot exactly half way between 0° and 90° with a displacement value equal to 0.5, we maintained that average both left and right of the knot. We did a similar thing at 135°.

6 We can slow things down for the initial and final parts of the motion by changing the values at those knots from 0.5 inches to something less, say 0.25 inches. Of course this means that the average velocity in the interval from 45° to 90° must increase as will the peak velocity from 90° to 135°. The accelerations will also increase. Table 5-3 lists these new conditions. Actually we only need to alter two equations from the set in equation (d).

$$\begin{aligned} a_1+b_1+c_1+d_1+e_1+f_1&=0.25 \\ a_2+b_2+c_2+d_2+e_2+f_2&=1 \\ a_3+b_3+c_3+d_3+e_3+f_3&=0.25 \end{aligned} \tag{g}$$

7 The results from the second trial spline can be seen in Figure 5-3. We have managed to lower the maximum jerk to about half of what it was and to make it continuous, just by tweaking a couple of knots. The acceleration curve now varies between –922 in/sec^2 and 500 in/sec^2, which may or may not be satisfactory. Again it depends on the application. It

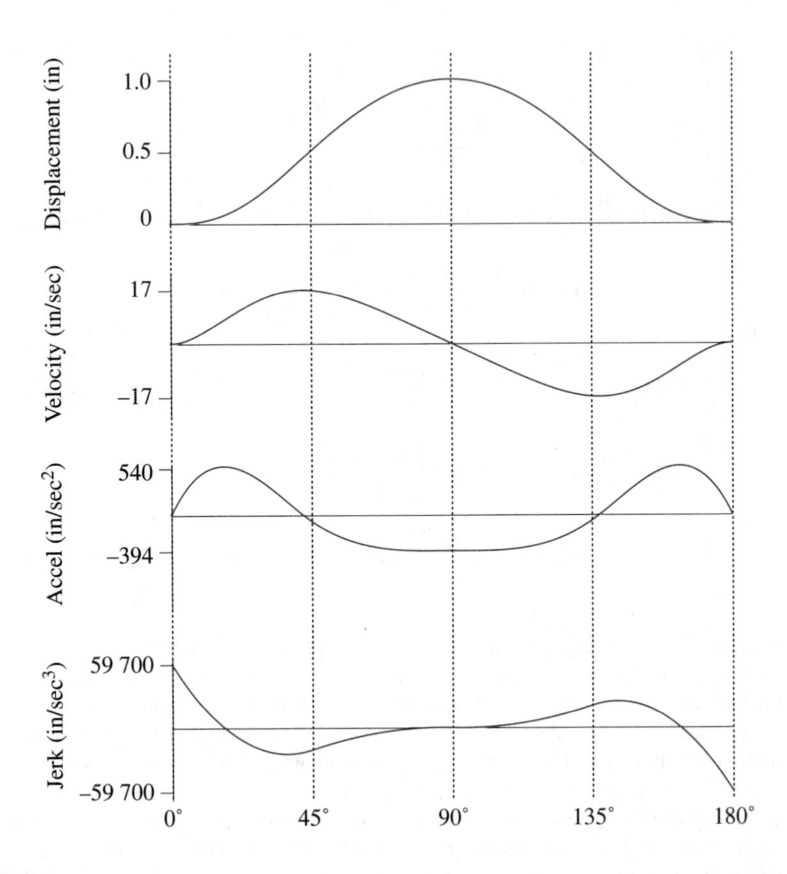

FIGURE 5-2

Spline functions for Example 5-1, first try

should be noted that the second trial spline has continuous jerk even at the ends. Lowering the value of the displacement at the knots caused the initial jerk to be zero, which matches the dwell. This solution is a very smooth one and should lead to a quiet cam. Now let us find one that has minimum acceleration, again by sliding the knots at 45° and 135° up or down.

TABLE 5-3 Knot Locations and Boundary Conditions for 2nd Try, Example 5-1

	Knot Locations (deg)				
Function	0	45	90	135	180
Displacement	0	0.25	1	0.25	0
Velocity	0	–	–	–	0
Acceleration	0	–	–	–	0

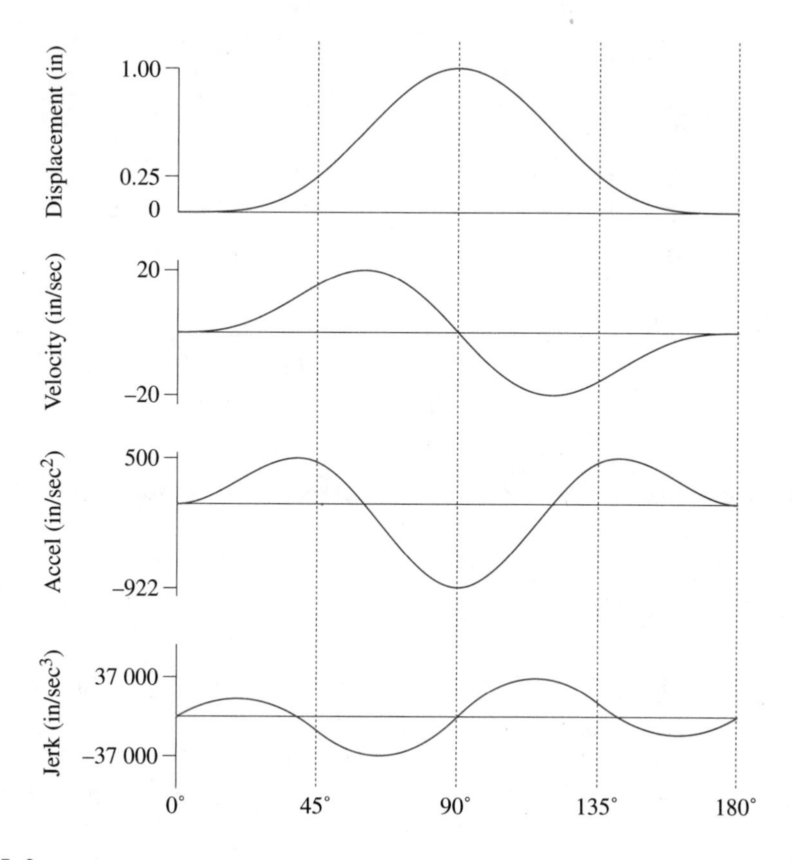

FIGURE 5-3

Spline functions for Example 5-1, second try

8 Figure 5-4 shows the absolute acceleration plotted against the values assigned to the displacement at the knots at 45° and 135°. We made this graph by having the computer solve for ten different displacement values at 45 and 135°. It is clear from the graph that we will obtain the minimum accelerations when the displacement takes on the value of 0.45 both at 45° and at 135°.

9 To see the entire motion, we change equation (d) again to be

$$\begin{aligned} a_1 + b_1 + c_1 + d_1 + e_1 + f_1 &= 0.45 \\ a_2 + b_2 + c_2 + d_2 + e_2 + f_2 &= 1 \\ a_3 + b_3 + c_3 + d_3 + e_3 + f_3 &= 0.45 \end{aligned} \tag{h}$$

and solve. Table 5-4 shows the knot specifications and Figure 5-5 shows our final result. The acceleration varies from –495 in/sec^2 up to 470 in/sec^2. It is just a guess, but the authors suspect that changing the knot value just a trifle more would cause the magnitudes of the negative and positive accelerations to be equal and to be somewhere between 470 in/sec^2 and 495 in/sec^2. This is as far as we will go however, since this is quite a satisfactory motion.

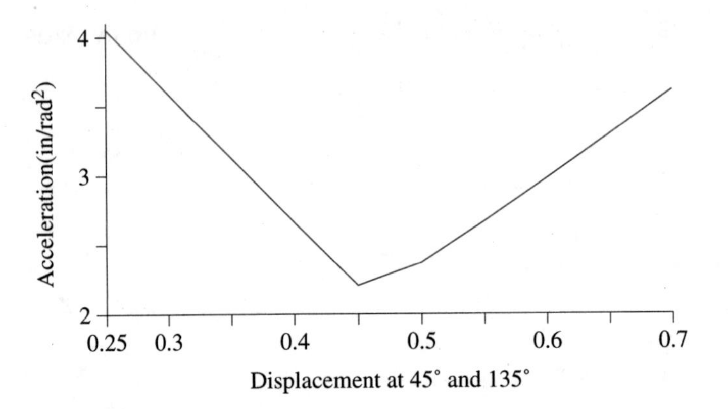

FIGURE 5-4

Absolute value of acceleration as a function of knot values for Example 5-1

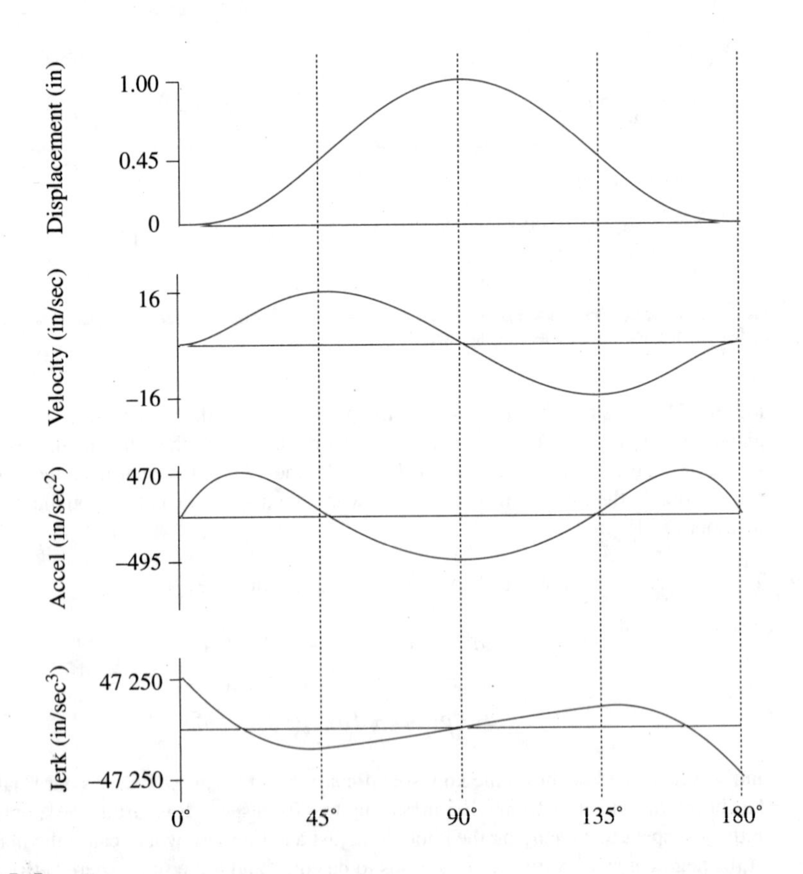

FIGURE 5-5

Spline functions for Example 5-1, third try

TABLE 5-4 Knot Locations and Boundary Conditions for 3rd Try, Example 5-1

	Knot Locations (deg)				
Function	0	45	90	135	180
Displacement	0	0.45	1	0.45	0
Velocity	0	–	–	–	0
Acceleration	0	–	–	–	0

10 Of the three motions, which one is the most desirable depends on the application. An automotive cam designer would prefer the motion seen in Figure 5-2. It has a "fat" displacement curve and a low, flat negative acceleration that would not challenge his valve spring and cause follower jump. The larger positive acceleration would mean nothing to him. Neither would the jerk, since there are a lot of noisy things in a car engine. Someone concerned with forces, say a cam designer who deals with heavy follower systems, would prefer the motion in Figure 5-5, where the accelerations are minimized. The motion in Figure 5-3 would appeal to an engineer concerned with smooth operation and low noise. As one can see, the spline is a valuable tool that lets one manipulate the motion curves to his advantage. The cost is the added discontinuities in the higher derivatives of the motion. We will discuss that later after we develop the general polynomial splines and the B-splines.

5.2 GENERAL POLYNOMIAL SPLINES

The spline is unlike every other cam displacement curve design tool in that it uses discontinuities in the higher derivatives of the follower motion for control. What were we doing in Example 5-1 to give us so many different, but good, solutions? At first glance, it would seem that we were just changing the displacement of the follower at a couple of angles of cam rotation, but careful examination will show something more. Figure 5-6 shows the end jerk, the jerk at 0°, plotted against the lift at the first knot, which was at 45°. The graph shows a straight line relationship. By altering the follower displacement at the knot, we were adjusting the amount of jerk used to pull the follower from the dwell. In essence here is how our order six spline of Example 5-1 worked. There was a jump in jerk at the first end knot because the boundary conditions set only the velocity and acceleration to zero. This got the follower moving. Further motion was controlled by "puffs," or discontinuities in the puff, at the knots. At the final end knot, another jerk brought it back to the dwell. This is how a spline works. Splines use discontinuities in the higher derivatives to control follower motion.

In Section 5.1, it was explained that there are three sets of equations to determine the spline. They are: the smoothness equations 5.2, the interpolation equations 5.3, and the boundary condition equations 5.4. Boundary conditions are used by every follower motion design tool. Polynomial fitting uses interpolation as well as boundary conditions. The spline is the only tool that uses the smoothness equations because of its reliance on using discontinuities for control.

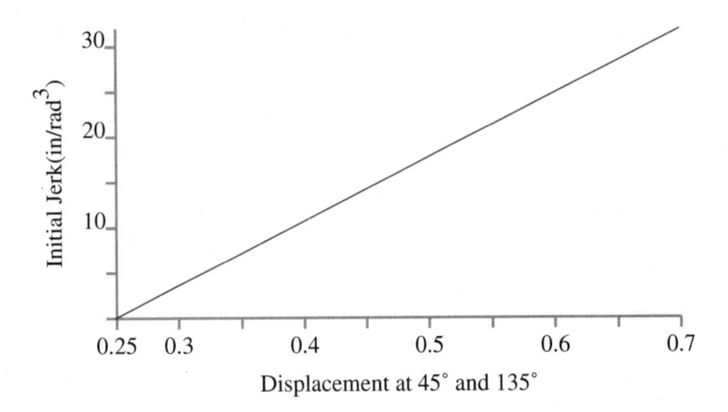

5

FIGURE 5-6

Relationship of jerk at zero degrees and displacement at the first knot

Another fact should be mentioned. The higher order derivatives have less immediate effect on follower motion than the lower ones because the "chain of command" is longer. The result is that very high order splines, like polynomials, have a mind of their own. The choice is up to the engineer, of course, but he will find that orders 5 and 6 are more user friendly than higher orders. Order 4 would give the engineer much more control of things, but the rough jerks would tell by creating increased vibration and noise due to their higher harmonic content.

Now to the task of making the spline more powerful. What if the engineer wants to match jerk at the boundaries? The classical spline by its definition has no way to do this. The engineer must make a new spline with higher order. But matching the jerk at the boundaries can be done otherwise, by dropping one of the interpolation equations or one of the smoothness equations and adding a boundary condition equation. We should mention that when a smoothness equation is dropped, it has to be one that insures smoothness at the knot for the highest derivative. It makes no sense to allow a discontinuity in jerk while attempting to get continuity in ping.

There is also no reason why the spline cannot interpolate at some other point that is not designated to be a knot. As a matter of fact, there are reasons why this is preferable. First, the cam designer is usually given specifications as to the boundary conditions and interpolation points. Designers are, however, free to choose any knots that they please. If the interpolation conditions are tied to the knots, they lose some or all of that freedom.

The engineer has three ways of placing interpolation points elsewhere than at the knots: adding a new interpolation equation that does not involve a knot and dropping one that does; dropping a boundary condition and adding a new interpolation equation; dropping a smoothness equation and adding an interpolation equation. If he drops one of the interpolation equations so that he no longer specifies a value at the knot, it changes the concept of "knot" as used by the classical spline. The smoothness equations will guarantee that the two adjacent polynomial pieces have a common value at the knot, but that value will not be specified by the engineer.

All of these things, although contrary to the definition of the classical spline, are useful to the engineer and so we define a **general polynomial spline** as a motion represented by polynomial pieces with knots, points where the polynomial pieces are blended together. We no longer insist that the spline function interpolate given values at the knots. We also no longer insist that the smoothness be the same at each interior knot. We no longer insist on satisfying any particular number of boundary conditions. We do require that the number of equations equals the number of unknowns, but we allow trade-offs among the smoothness equations, interpolation equations, and boundary conditions.

There are, as the reader already suspects, some costs. The first is what might be called intellectual manageability. It is very easy for an engineer to assemble a set of equations that has no solution. It is even easier to put together a set of equations that is very sensitive to the interpolation values, so that a small change in derivative or displacement results in a radical change in total follower motion. An example will show what we mean.

5

EXAMPLE 5-2

Using a General Polynomial Spline for Constant Velocity Critical Path Motion.

Problem: Consider the following statement of a critical path motion problem:

Accelerate the follower from zero to 10 in/sec
Maintain a constant velocity of 10 in/sec for 0.5 sec
Decelerate the follower to zero velocity without letting the follower rise more than 0.25 inches from the height at the top of the constant velocity ramp
Return the follower to start position
Cycle time exactly 1 sec

Solution:

1 If we make the bottom of the ramp a reference and assign it zero displacement, the problem specifications can be interpreted to say that the displacement cannot exceed 5.25 inches. The ramp velocity in inches per radian is $10 / 2\pi = 5 / \pi$. In addition to using the ramp bottom as the zero displacement reference, we will assign the ramp top to be the zero angle.

2 We will use a general polynomial spline to solve the problem. We will use a sixth order spline with knots at 0°, 90°, and 180°. At 0°, we have the boundary conditions that the displacement is 5 inches, the velocity is $5 / \pi$ and the acceleration is 0. At the knot at 180° we set the displacement to 0, the velocity to $5 / \pi$ and the acceleration 0. At the knot at 90° we will relax the continuity conditions so that there can be a jump in the ping. We will not interpolate at this knot. It remains to choose a cam angle for maximum displacement of 5.25 inches. If the velocity is not reduced at ramp end, the maximum allowable displacement will be obtained at 0.157 radians, which is about 9°. Hence our additional interpolation equations must be at some cam angle after 9°. If we wait too long after 9°, we will have less time to decelerate the follower. We must choose some angle slightly greater than 9°, then add to our set of equations two interpolation equations at that angle as replacements for the smoothness equation in the fifth derivative, the puff, and to replace the equation for displacement at the knot at 90°.

3 This problem is an example of the extreme sensitivity that can be encountered in deciding on the general spline equations. Figure 5-7 shows three different displacement curves, one resulting from forcing the maximum displacement to be at 13°, one for maximum displacement at about 14°,, and the last for maximum displacement at 14.5°. The graphs might be titled "What a Difference a Degree Makes." Apparently 13° is too soon for the maximum. We limit the upward motion at the cost of over 40 inches of motion below the start of the ramp. The mechanism requires 45.85 inches of travel to obtain a constant velocity ramp of 5 inches! The motion with maximum at 14.04° is a solution. All curves in Figure 5-7 are drawn to the same scale and the reasonable solution curve can not really be inspected there. The bottom curve in Figure 5-7 shows that 14.5° is too late to limit the upward motion of the follower. The displacement of 5.25 inches becomes a local minimum and the follower goes on up to about 10 inches eventually. If the restriction was placed on the motion to prevent collision, 5.25 inches at 14.5° will destroy the machine.

4 The solution is shown in Figure 5-8. The total travel is about 6.25 inches. The peak velocity is downward at – 40.2 in/sec. In order to prevent the follower from exceeding 5.25 in, we need a negative acceleration of –513 in/sec^2. In order to keep it from plunging on downward, as in case (*a*) of Figure 5-7, a positive acceleration of 374 in/sec^2 is required. The jerk is interesting as it shows the jump in ping at the knot at 90° that is necessary to keep the motion contained.

5 The equation set solved to get the final solution are the smoothness equations at the only interior knot, the knot at 90°,

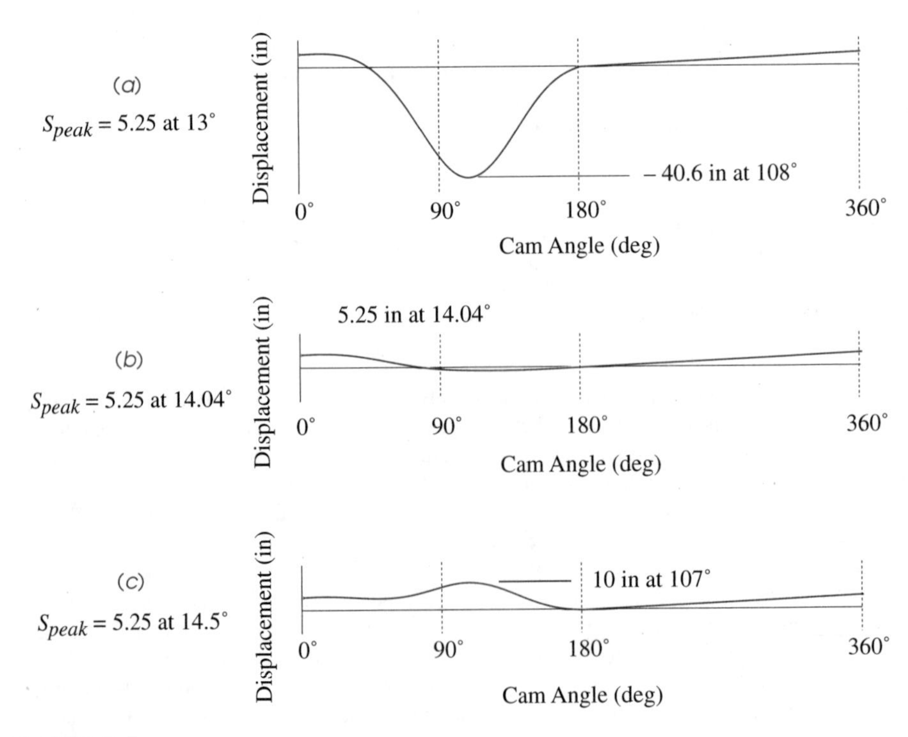

FIGURE 5-7

Sensitivity of displacement to cam angle location of peak S value (Example 5-2)

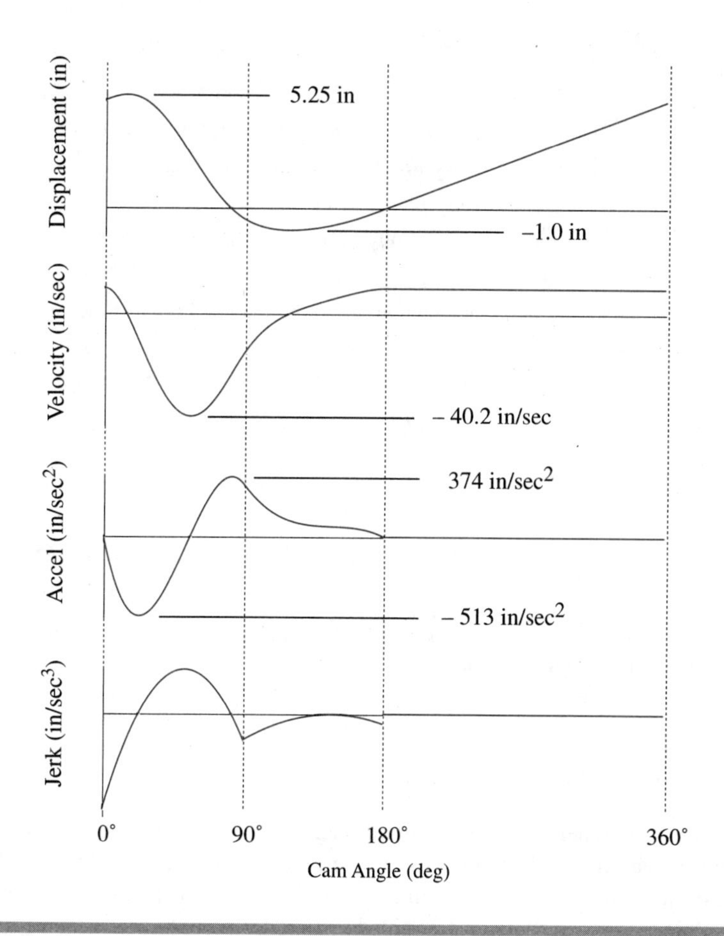

FIGURE 5-8

Spline functions for Example 5-2

$$\begin{aligned} a_1 + b_1 + c_1 + d_1 + e_1 + f_1 &= f_2 \\ 5a_1 + 4b_1 + 3c_1 + 2d_1 + e_1 &= e_2 \\ 10a_1 + 6b_1 + 3c_1 + d_1 &= d_2 \\ 10a_1 + 4b_1 + c_1 &= c_2 \end{aligned} \tag{a}$$

which represents the continuity of the motion, continuity of velocity, continuity of acceleration and continuity of jerk. The equation for continuity of ping has been dropped.

6 The boundary condition equations are

$$
\begin{aligned}
f_1 &= 5 \\
2e_1 &= 5 \\
d_1 &= 0 \\
a_2 + b_2 + c_2 + d_2 + e_2 + f_2 &= 0 \\
10a_2 + 8b_2 + 6c_2 + 4d_2 + 2e_2 &= 5 \\
10a_2 + 6b_2 + 3c_2 + d_2 &= 0
\end{aligned} \qquad \text{(b)}
$$

and the added interpolation equations are, noting that 14.05° = 0.245 radians,

$$
\begin{aligned}
a_1r^5 + b_1r^4 + c_1r^3 + d_1r^2 + e_1r + f_1 &= 5.25 \\
5a_1r^4 + 4b_1r^3 + 3c_1r^2 + 2d_1r + e_1 &= 0
\end{aligned} \qquad \text{(c)}
$$

where

$$r = \frac{0.245}{\pi/2} \qquad \text{(d)}$$

7 The equations in (c) are to make the lift 5.25 and make the velocity zero so that a maximum is obtained at that point.

5.3 B-SPLINES

The idea of a general spline is a good one, but formulating and solving the equations, trying to keep track of the indices and derivatives of various orders, while at the same time trying to design a decent cam is a daunting task. Despite its limitations, the classical spline at least presents an organized methodology. In this section we will construct a methodology for designing cam follower motion using general polynomial splines. Since classical splines fall under the heading of general polynomial splines, what we propose in this section will also apply to them.

We intend to imitate polynomials. When engineers use a single polynomial to design cam follower motion, they pick some general polynomial of suitable degree and determines the coefficients so that the boundary conditions are satisfied. We propose to do much the same, except that instead of a polynomial such as

$$a_5\theta^5 + a_4\theta^4 + a_3\theta^3 + a_2\theta^2 + a_1\theta + a_0 \qquad \text{(5.7a)}$$

we will use

$$a_5B_{m,5}(\theta) + a_4B_{m,4}(\theta) + a_3B_{m,3}(\theta) + a_2B_{m,2}(\theta) + a_1B_{m,1}(\theta) + a_0B_{m,0}(\theta) \qquad \text{(5.7b)}$$

where the B's are splines, called **B-splines**, of order m. These B-splines will take the place of the powers in the polynomial and the design of the displacement can proceed just as it did for the polynomial; the coefficients, the a's, can be determined by setting up equations for the boundary conditions. The B-splines themselves will have been constructed to guarantee the smoothness equations.

What can be used for B-splines? A first guess, not the final answer though, might be to try something like

$$g(m,t,\theta) = \begin{cases} 0, & \theta < t \\ (\theta - t)^{m-1}, & t \le \theta \end{cases} \tag{5.8}$$

which is a spline of order m with a single knot and a jump discontinuity in the $m - 1$ derivative. Figure 5-9 shows graphs of such splines with orders 1, 2, 3, and 4. These single-knot splines, called the *truncated power basis* (TPB), work theoretically. Any spline can be expressed as a sum of TPB functions. For a sequence of knots

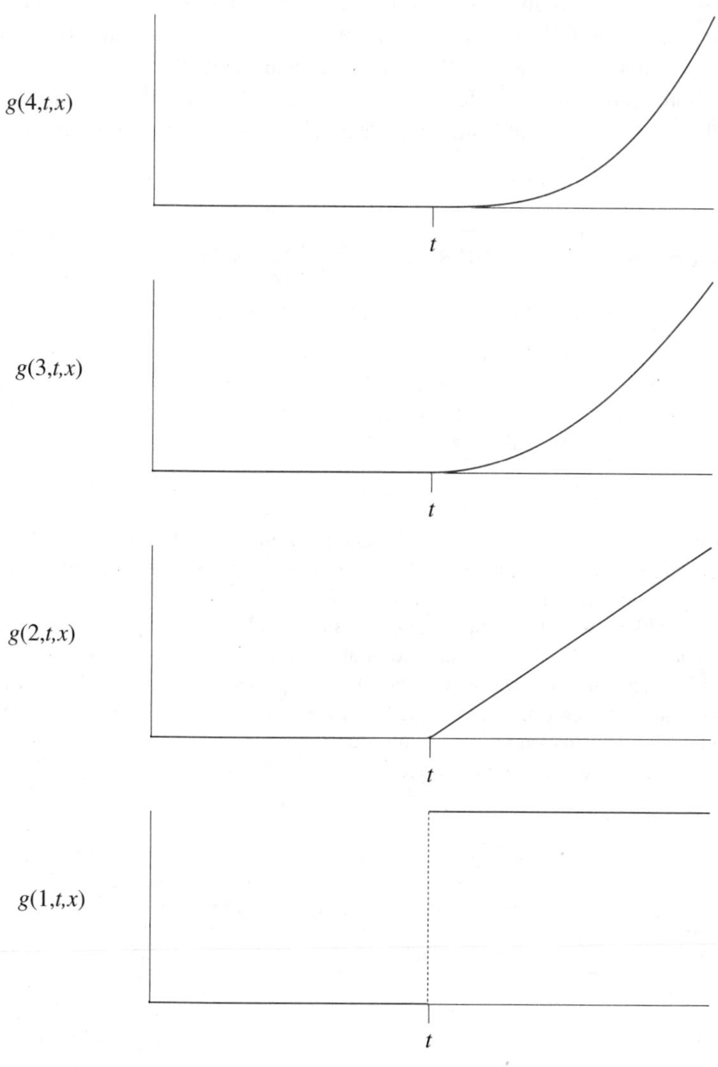

FIGURE 5-9

Truncated power basis (TPB) splines

$$t_1, t_2, \ldots, t_n \tag{5.9a}$$

a displacement expressed as a general polynomial spline of order m would be of the form

$$s(\theta) = a_1 g(m, t_1, \theta) + a_2 g(m, t_2, \theta) + \cdots + a_n g(m, t_n, \theta) \tag{5.9b}$$

There are valid reasons why the TPB is not used to build splines. A big problem is that for θ near t, $g(m, t, \theta)$ is very small but grows very quickly as θ increases. This causes numerical instability in solving the boundary value equations. Numerical errors will be especially bad when two knots are close to each other.

Nevertheless, engineers should not neglect the TPB as a tool in cam design. Using the TPB is a good way to gain insights into how the discontinuities at the knots of a spline can be used to control the follower motion and, if the number of knots are few and not relatively close to each other, there will be no numerical difficulties. An example will be instructive both as to how to build a spline with the TPB as well as what are their faults. We will reconstruct the cam from Example 5.1, but this time as a sum from the TPB.

EXAMPLE 5-3

Using the Truncated Power Basis (TPB) for a Single-Dwell Cam

Problem: Consider the following single-dwell specification:

rise	1 in (25.4 mm) in 90°
fall	1 in (25.4 mm) in 90°
dwell	at zero displacement for 180° (low dwell)
cam ω	15 rad/sec

Solution:

1 Table 5-5 shows the knot locations and boundary conditions. Just for variety, we have decided to ignore interpolating at 45 and 135° and instead have substituted two other boundary conditions, the jerk values at the ends of the spline. As in the previous consideration of this problem, we let symmetry insure that the displacement of 1 inch at 90° is a maximum and we do not impose the condition that the velocity is zero at 90°. The jerk values of 14 in/rad^3 and –14 in/rad^3 at 0° and 180° respectively are a more direct way to control maximum acceleration than we used for the classical spline. There, the reader may recall, we had to interpolate at the knots and were forced to specify values at 45 and 135°. Here we do not do so, using the more direct end knot jerks instead.

TABLE 5-5 Knot Locations and Boundary Conditions for Example 5-3

Knot Location	0°	45°	90°	135°	180°
Displacement BCs	0		1		0
Velocity BCs	0				0
Acceleration BCs	0				0
Jerk BCs	14				–14

2 We select order six for the spline; we want the spline to have a jump in the puff, the fifth derivative, at each interior knot. At the knots at 45, 90, and 135°, we use the functions $g(6, \pi/4, \theta)$, $g(6, \pi/2, \theta)$ and $g(6, 3\pi/4, \theta)$ respectively. For our computations, we naturally use radian measurements.

Now things get tricky. First notice that the function for the knot at 180°, $g(6, \pi, \theta)$, will always be 0 since the spline begins at 0 and ends at 180°. We do not need to use it at all. This leaves only one knot that is not represented yet, the one at 0. So we use $g(6, 0, \theta)$. However, we will have 9 equations for the boundary conditions. There are not enough unknowns. We need more knots.

There are two solutions possible. The first is to choose some phantom knots to the left of zero, say at –180, –135, –90, –45 and –30°, to bring the total number of knots to 9. We may then use the displacement

$$s(\theta) = a_1 g(6,-\pi,\theta) + a_2 g\left(6,-\frac{3\pi}{4},\theta\right) + a_3 g\left(6,-\frac{\pi}{2},\theta\right) + a_4 g\left(6,-\frac{\pi}{4},\theta\right)$$
$$+ a_5 g\left(6,-\frac{\pi}{6},\theta\right) + a_6 g(6,0,\theta) + a_7 g\left(6,\frac{\pi}{4},\theta\right) + a_8 g\left(6,\frac{\pi}{2},\theta\right) + a_9 g\left(6,\frac{3\pi}{4},\theta\right) \qquad \text{(a)}$$

set up the nine boundary condition equations

$$\begin{aligned} s(0) &= 0 \\ s'(0) &= 0 \\ s''(0) &= 0 \\ s'''(0) &= 14 \\ s(\pi/2) &= 1 \\ s(\pi) &= 0 \\ s'(\pi) &= 0 \\ s''(\pi) &= 0 \\ s'''(\pi) &= -14 \end{aligned} \qquad \text{(b)}$$

and then solve for the nine unknowns

$$a_1, a_2, a_3, a_4, a_5, a_6, a_7, a_8, a_9. \qquad \text{(c)}$$

3 This would work very well, but there is another way, the one that we shall use because it results in fewer equations and introduces the reader to the idea of multiple knots. First we note that every $g(m, t, \theta)$ with t larger than or equal to zero will have its value and first and second derivative equal to 0 at 0. So if we use only knots that are at zero or to the right of zero, we may drop the first three equations of the above set. This brings our number down to 6 equations. We propose the knot sequence 0, 0, 0, 45, 90, 135°, six in all as required by the number of equations. We treat the multiple knots at 0 with the following general convention:

For a general polynomial spline of order m, with knot t, one uses the function $g(m, t, \theta)$ for the first occurrence of the knot. For the second occurrence, one uses $g(m - 1, t, \theta)$. For the j-th occurrence the TPB function $g(m - j + 1, t, \theta)$ is used.

Notice that this rule will cause multiple knots to have jumps in lower derivatives. Multiple knots, for the general spline, are a mechanism with which the engineer can choose at what level to "jump" the motion, jerk, ping or puff.

4 According to the above convention, the displacement function becomes

$$s(\theta) = a_1 g(4,0,\theta) + a_2 g(5,0,\theta) + a_3 g(6,0,\theta) + a_4 g\left(6,\frac{\pi}{4},\theta\right) + a_5 g\left(6,\frac{\pi}{2},\theta\right) + a_6\left(6,\frac{3\pi}{4},\theta\right) \tag{d}$$

and the equations to solve, ignoring zero terms, are

$$\begin{aligned}
a_1 g'''(4,0,\theta) &= 14 \\
a_1 g\left(4,0,\frac{\pi}{2}\right) + a_2 g\left(5,0,\frac{\pi}{2}\right) + a_3 g\left(6,0,\frac{\pi}{2}\right) + a_4 g\left(6,\frac{\pi}{4},\frac{\pi}{2}\right) &= 1 \\
a_1 g(4,0,\pi) + a_2 g(5,0,\pi) + a_3 g(6,0,\pi) + a_4 g\left(6,\frac{\pi}{4},\pi\right) + a_5 g\left(6,\frac{\pi}{2},\pi\right) + a_6 g\left(6,\frac{3\pi}{4},\pi\right) &= 0 \\
a_1 g'(4,0,\pi) + a_2 g'(5,0,\pi) + a_3 g'(6,0,\pi) + a_4 g'\left(6,\frac{\pi}{4},\pi\right) + a_5 g'\left(6,\frac{\pi}{2},\pi\right) + a_6 g'\left(6,\frac{3\pi}{4},\pi\right) &= 0 \\
a_1 g''(4,0,\pi) + a_2 g''(5,0,\pi) + a_3 g''(6,0,\pi) + a_4 g''\left(6,\frac{\pi}{4},\pi\right) + a_5 g''\left(6,\frac{\pi}{2},\pi\right) + a_6 g''\left(6,\frac{3\pi}{4},\pi\right) &= 0 \\
a_1 g'''(4,0,\pi) + a_2 g'''(5,0,\pi) + a_3 g'''(6,0,\pi) + a_4 g'''\left(6,\frac{\pi}{4},\pi\right) + a_5 g'''\left(6,\frac{\pi}{2},\pi\right) + a_6 g'''\left(6,\frac{3\pi}{4},\pi\right) &= -14
\end{aligned} \tag{e}$$

The equations get longer as we move through the knots. The values also get larger, especially the derivatives. Evaluating the above equations numerically, we get

$$\begin{aligned}
6a_1 &= 14 \\
3.8758a_1 + 6.0881a_2 + 9.5631a_3 + 0.2988a_4 &= 1 \\
31.0063a_1 + 97.4091a_2 + 306.0197a_3 + 72.6199a_4 + 9.5631a_5 + 1.9025a_6 &= 0 \\
29.6088a_1 + 124.0251a_2 + 487.0455a_3 + 154.1042a_4 + 30.4403a_5 + 1.9025a_6 &= 0 \\
18.8496a_1 + 118.4353a_2 + 620.1255a_3 + 261.6155a_4 + 71.5157a_5 + 9.6895a_6 &= 0 \\
6a_1 + 75.3982a_2 + 592.1763a_3 + 333.0991a_4 + 148.0440a_5 + 37.0110a_6 &= -14
\end{aligned} \tag{f}$$

which illustrates the mix of large and small numbers. The largest number in the above equations is above 620 and the smallest is about 0.3. For more knots and for knots closer together, the situation is much worse. These equations are not unstable. They just hint about the possibility. We can solve equation (f) to get the final spline.

$$s(\theta) = 2.3333g(4,0,\theta) - 2.2960g(5,0,\theta) + 0.6392g(6,0,\theta) - 0.5965g\left(6,\frac{\pi}{4},\theta\right) - 0.0855g\left(6,\frac{\pi}{2},\theta\right) - 0.5965\left(6,\frac{3\pi}{4},\theta\right) \tag{g}$$

Figure 5-10 shows the displacement, velocity, acceleration and jerk for this TPB spline expression. The acceleration seems well balanced. A higher jerk lifting the follower from the dwell would have resulted in higher positive accelerations and a lower jerk would have caused the follower to take too much time to reach the necessary velocity. This would have meant that deceleration would be over a shorter time and therefore larger.

What can be done to make some spline functions that satisfy the smoothness equations, like the TPB do, but are more stable and result in sets of equations that are easier to solve? The source of the problems with the TPB functions is that to the right of the knot, these functions are never zero and continue to grow, thereby affecting all motion after the knot. What we want is a function that is zero just about everywhere except around a few knots. If each of our B-splines is nonzero for just one section of the follower displacement, then changing a coefficient will only change that section and leave the rest of the curve unchanged. The engineer will have local control. The interpolation equations and boundary value equations will be shorter and easier to solve.

If the order $m = 1$, then we would have no trouble finding what we want. Figure 5-11 shows a B-spline of order 1, that is, it is nonzero only between two adjacent knots. Its equation in terms of the TPB is

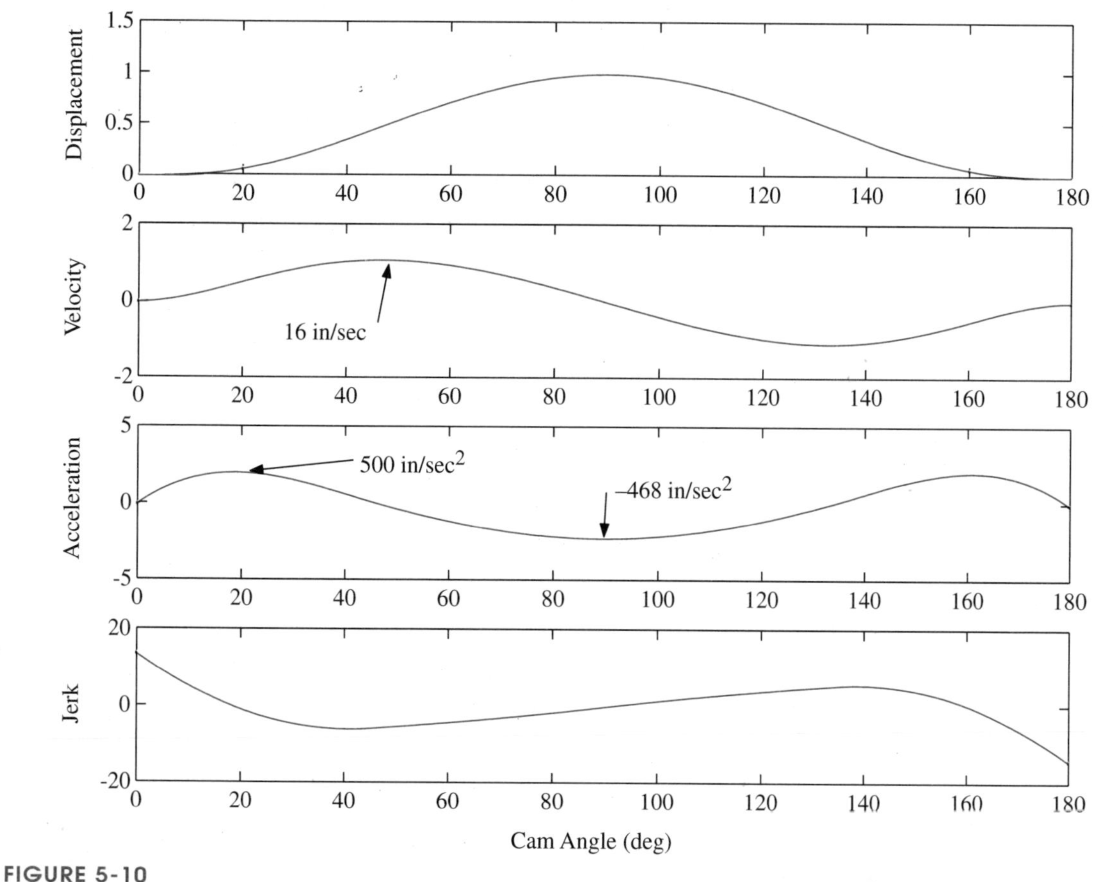

FIGURE 5-10

B-spline functions for Example 5-3

5

$$B_{1,j}(\theta) = g(1, t_j, \theta) - g(1, t_{j+1}, \theta). \tag{5.10}$$

For order two, we have to add one more knot. Figure 5-12 shows the "hat" functions, which have all the properties that we desire in a B-spline, namely zero most of the time except around one knot.

$$B_{2,j}(\theta) \begin{cases} = 0, & \theta \le t_j \\ \ne 0, & t_j < \theta < t_{j+2} \\ = 0, & \theta \ge t_{j+2} \end{cases} \tag{5.11}$$

The TPB form of this B-spline is

$$B_{2,j}(\theta) = \frac{g(2, t_j, \theta)}{(t_{j+1} - t_j)} - \frac{t_{j+2} - t_j}{(t_{j+1} - t_j)(t_{j+2} - t_{j+1})} g(2, t_{j+1}, \theta) + \frac{g(2, t_{j+2}, \theta)}{t_{j+2} - t_{j+1}} \tag{5.12}$$

A little bit of algebraic manipulation shows that this last (second order) spline, called the "hat" function can also be written as

$$B_{2,j}(\theta) = \frac{\theta - t_j}{t_{j+1} - t_j} B_{1,j}(\theta) + \frac{t_{j+2} - \theta}{t_{j+2} - t_{j+1}} B_{1,j+1}(\theta) \tag{5.13}$$

In fact, it turns out to be always true that the B-spline of order m, which begins at knot j can be computed by

$$B_{m,j}(\theta) = \frac{\theta - t_j}{t_{j+m-1} - t_j} B_{m-1,j}(\theta) + \frac{t_{j+m} - \theta}{t_{j+m} - t_{j+1}} B_{m-1,j+1}(\theta) \tag{5.14}$$

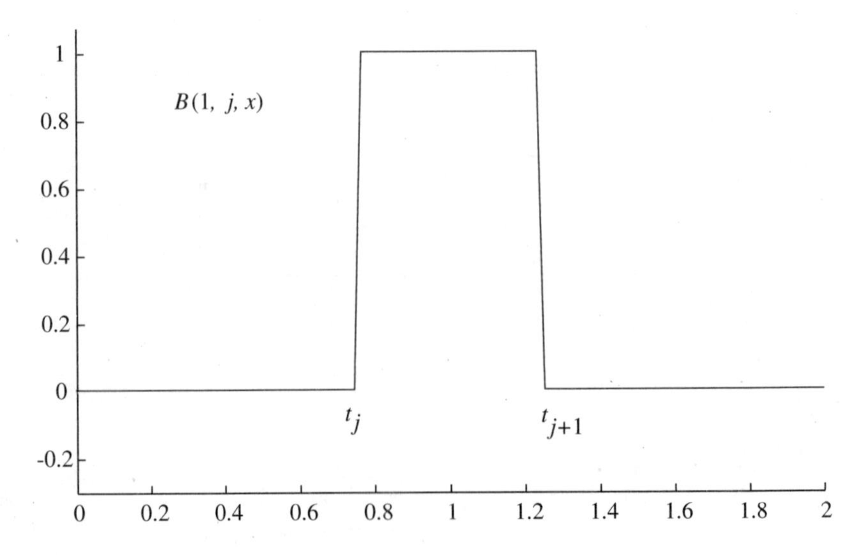

FIGURE 5-11

A B-spline of order 1

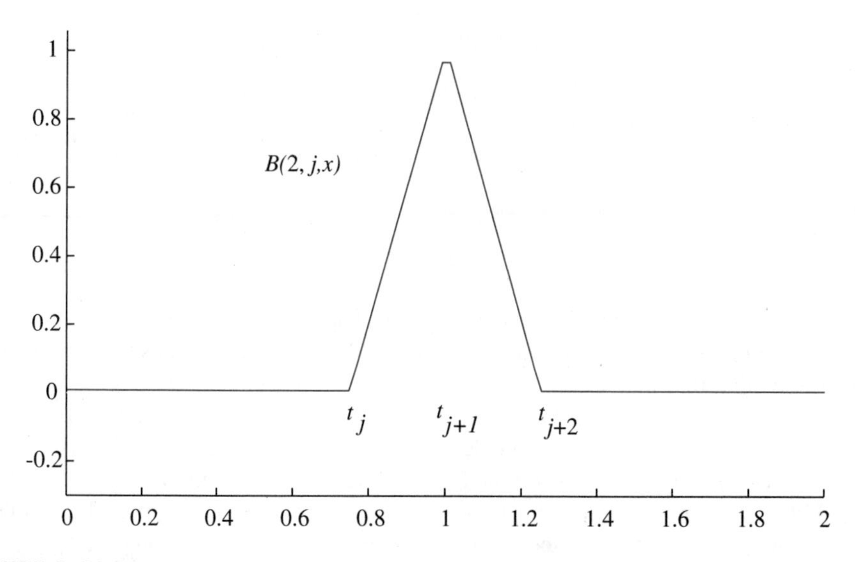

FIGURE 5-12

A B-spline of order 2, called the "hat" function

This makes the computation of B-splines very easy. To find the value of a B-spline of order m at θ, we just start with order 1 and keep multiplying and adding according to equation 5.14, until we reach order m. Furthermore most of the B-spline values will be zero. Here is what we mean:

Suppose it is desired to find a fifth order B-spline at t which is between the fourth and fifth knots. We want to find

$$B_{5,3}(t), \quad t_4 < t < t_5 \tag{5.15a}$$

According to equation 5.14, we must compute first

$$B_{4,3}(t), B_{4,4}(t) \tag{5.15b}$$

and for them we need all of the following

$$\begin{gathered} B_{3,3}(t), B_{3,4}(t), B_{3,5}(t) \\ B_{2,3}(t), B_{2,4}(t), B_{2,5}(t), B_{2,6}(t) \\ B_{1,3}(t), B_{1,4}(t), B_{1,5}(t), B_{1,6}(t), B_{1,7}(t) \end{gathered} \tag{5.15c}$$

But, an important point for the reader to notice, B-splines are local to the extent that

$$B_{m,j}(\theta) \begin{cases} = 0, & \theta \le t_j \\ \ne 0, & t_j < \theta < t_{j+m} \\ = 0, & \theta \ge t_{j+m} \end{cases} \tag{5.16}$$

and therefore, for the interval from knot j to knot $j+1$, the only B-splines that contribute anything are

$$B_{m,j-m+1}(\theta), B_{m,j-m+2}(\theta), \ldots, B_{m,j}(\theta). \tag{5.17}$$

so the only nonzero B-splines in the above stack will be

$$B_{4,3}(t), B_{3,3}(t), B_{2,3}(t) \tag{5.18a}$$

and

$$B_{1,3}(t) = 1 \tag{5.18b}$$

The reader is also probably wondering what happens when the knots are close to, or at, the end knots where $m + 1$ consecutive knots do not exist. The answer is that we do the same thing that was done in the case of the TPB. We either invent some knots outside the interval or else, as the authors prefer, use multiple knots at the end. In the case of multiple knots, at least one of the denominators in equation 5.14 will be zero. When this occurs we simply assume that the fraction itself is zero. Multiple knots are, as with the TPB, a way of controlling the jump discontinuities in the displacement. If a "ping" is not giving the engineer satisfactory motion, he can repeat the knot and try a "jerk".

The B-splines that we have defined are *normalized B-splines* because we have scaled them so that

$$\sum_j B_{m,j}(\theta) = 1 \tag{5.19}$$

Two other very convenient properties of B-splines are that the derivative of a B-spline of order m is a combination of B-splines of order $m - 1$ and the integral of a B-spline is a combination of B-splines of order $m + 1$. Therefore, boundary conditions pertaining to velocity, acceleration, or jerk can be easily set up.

Finally we repeat the most important property of the polynomial B-spline: every general polynomial spline can be expressed as a sum of B-splines. To convince the reader, we will redo Example 5-1 as a B-spline. Note how much easier things are here.

EXAMPLE 5-4

Using a B-Spline for a Single-Dwell Cam.

Problem: Consider the following single-dwell cam specification:

rise	1 in (25.4 mm) in 90°
fall	1 in (25.4 mm) in 90°
dwell	at zero displacement for 180° (low dwell)
cam ω	15 rad/sec

Solution:

1 We choose a fifth order spline and choose the knot sequence to be 0, 0, 0, 0, 0, 0, 45, 90, 135, 180, 180, 180, 180, 180, 180. The repeated knots are what we said we would have in order to have enough knots to define B-splines at the ends of the intervals. Again, we will

not choose, but will mention that we could have chosen phantom knots outside of the interval. Many engineers use this approach and many engineers use our approach.

2 Before we set up the interpolation equations and boundary value equations, we need to consider what B-splines we have. Table 5-6 shows which knots support which B-spline. There are nine in all since each one needs seven consecutive knots. The rule is that for order m and k knots, counting the multiples, there will be $k - m$ B-splines. In our case we have 15 knots and order 6 for a total of 9 B-splines as shown in Figure 5-13.

3 Before we do anything more however, we should inspect the first, second, third and the seventh, eighth, and ninth B-splines. Figure 5-14 shows the first, second, and third B-splines for our problem. Notice that the first B-spline is the only one that is nonzero at 0. The first and second B-splines are the only ones that have a nonzero derivative at 0. The first, second, and third B-splines are the only ones with a nonzero second derivative at 0. This will always be true and is a blessing to engineers designing cams with dwells. Our finished B-spline sum will satisfy the 0 degree boundary conditions simply by not using B-splines number 1, 2, and 3. The same can be said about the other end. We do not need to use B-splines 7, 8 or 9. More precisely, we assign them zero coefficients in our final expression. This means that the boundary value and interpolation set of equations is

$$\begin{aligned} a_4 B_{6,4}\left(\frac{\pi}{4}\right) + a_5 B_{6,5}\left(\frac{\pi}{4}\right) + a_6 B_{6,6}\left(\frac{\pi}{4}\right) &= 0.45 \\ a_4 B_{6,4}\left(\frac{\pi}{2}\right) + a_5 B_{6,5}\left(\frac{\pi}{2}\right) + a_6 B_{6,6}\left(\frac{\pi}{2}\right) &= 1 \\ a_4 B_{6,4}\left(\frac{3\pi}{4}\right) + a_5 B_{6,5}\left(\frac{3\pi}{4}\right) + a_6 B_{6,6}\left(\frac{3\pi}{4}\right) &= 0.45 \end{aligned} \qquad \text{(a)}$$

4 Evaluating the B-splines by equation 5.14, we get

$$\begin{aligned} 0.4109a_4 + 0.1111a_5 + 0.0104a_6 &= 0.45 \\ 0.2593a_4 + 0.4444a_5 + 0.2593a_6 &= 1 \\ 0.0104a_4 + 0.1111a_5 + 0.4109a_6 &= 0.45 \end{aligned} \qquad \text{(b)}$$

TABLE 5-6 B-spline Definition Knots

B-spline Number	Knots Used (deg)						
1	0	0	0	0	0	0	45
2	0	0	0	0	0	45	90
3	0	0	0	0	45	90	135
4	0	0	0	45	90	135	180
5	0	0	45	90	135	180	180
6	0	45	90	135	180	180	180
7	45	90	135	180	180	180	180
8	90	135	180	180	180	180	180
9	135	180	180	180	180	180	180

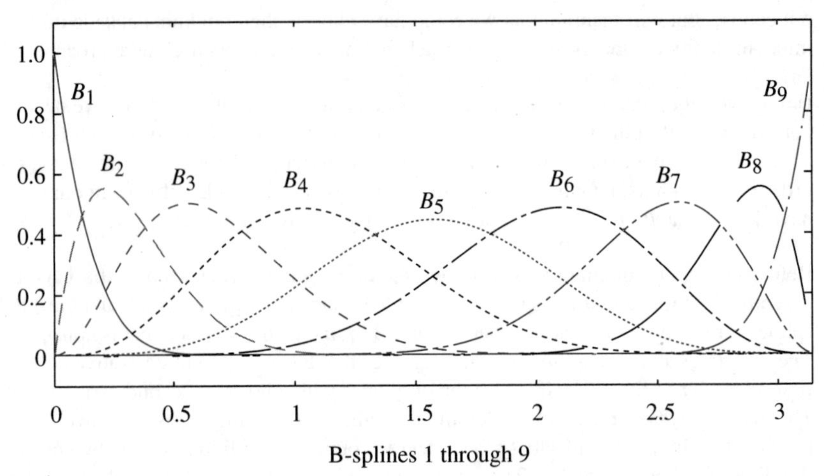

FIGURE 5-13

The nine B-splines for Example 5-4, these sum to the displacement function

that can be solved for our final answer

$$s(\theta) = 0.6857\, B_{6,4}(\theta) + 1.45\, B_{6,5}(\theta) + 0.6857\, B_{6,6}(\theta) \tag{c}$$

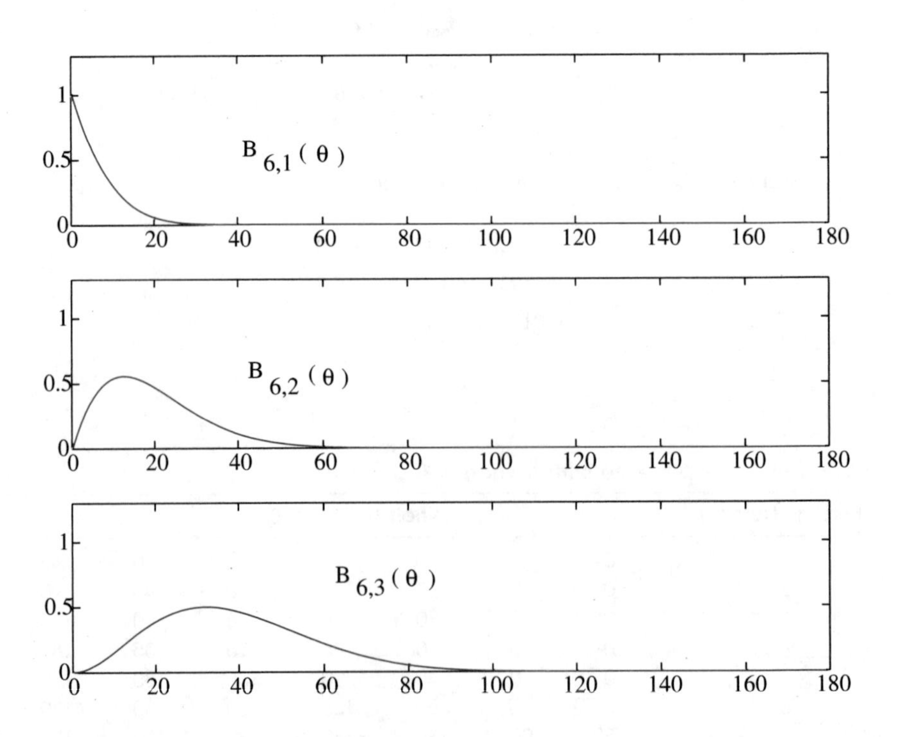

FIGURE 5-14

First, second, and third B-splines for Example 5-4

The graph of this equation and its derivatives through ping are shown in Figure 5-15. Compare these to the functions in Figure 5-5 (p. 80). They are exactly the same as the original solution that we obtained using the classical spline approach.

5.4 BEZIER CURVES

We said in an earlier section that every polynomial could be considered to be a spline without interior knots. Then we said in a later section that every spline could be written as a B-spline. Now seems to be a good time to say just how polynomials can be represented as B-splines. These special B-splines are called *Bezier curves*, named after one of their discoverers, Pierre Bezier, a mechanical engineer and son and grandson of engineers. Despite the different name, Bezier curves are just B-spline representations of polynomials. Since polynomials have no discontinuities internal to the interval on which they are defined, the knots for a Bezier curve are only at the endpoints. Remember, however, that by our convention the end points are repeated according to the order n of the polynomial. For example, a fifth-order polynomial defined on the interval from 0 to 90° written as a Bezier curve would have knots at 0, 0, 0, 0, 0, 90, 90, 90, 90, and 90°.

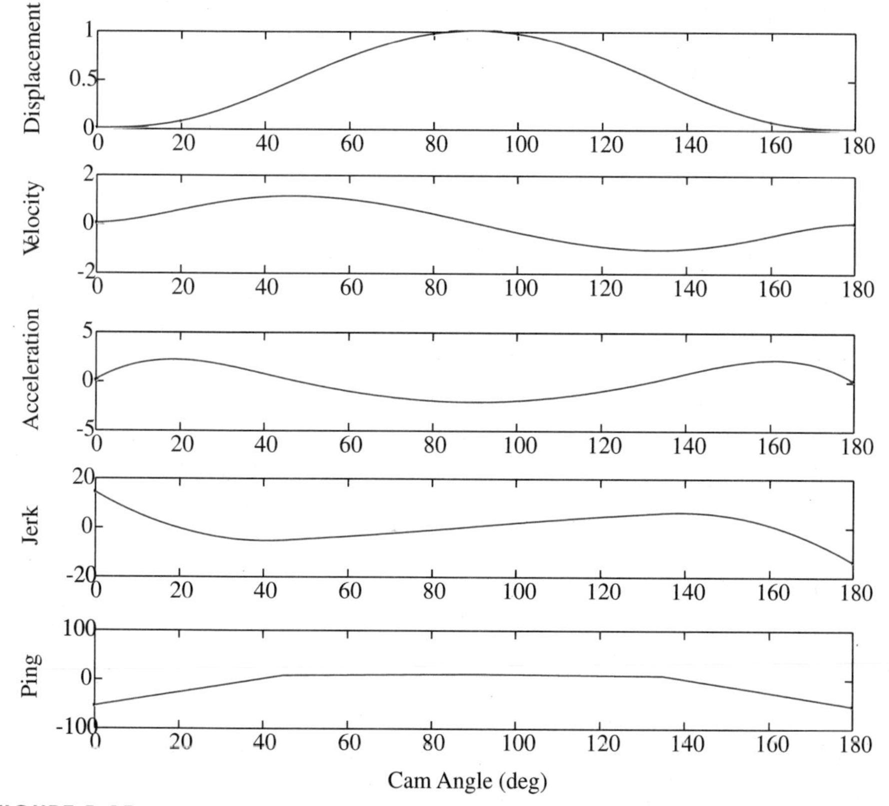

FIGURE 5-15

B-spline functions for Example 5-4

One builds Bezier curves just as B-spline curves are built, by using equation 5.14 (p. 92) to get the basis splines and then determining coefficients so as to satisfy the boundary conditions. To put it more mathematically, suppose we want to design cam follower motion over an interval from a to b and we want to use a polynomial of order n, or degree $n-1$. Further suppose that we want to represent the polynomial using Bezier curves rather than the polynomial methods that were discussed earlier in this book. There are no interior knots, just knots at the ends of the interval, so our knot structure is

$$a_1 = a_2 = a_3 = \ldots = a_n = a, \quad b_1 = b_2 = b_3 = \ldots = b_n = b \tag{5.20}$$

and using equation 5.14, the basis splines are the polynomials

$$Bz_{n,j}(\theta) = \frac{(n-1)(n-2)\cdots(n-j+1)}{(j-1)(j-2)\cdots(1)}\left(\frac{\theta-a}{b-a}\right)^{j-1}\left(1-\frac{\theta-a}{b-a}\right)^{n-j}, j = 1,2,\ldots,n \tag{5.21}$$

These polynomials are called Bernstein polynomials and are famous in mathematics. They are used extensively in probability and statistics. Figure 5-16 shows the Bernstein polynomials for $n = 6$. They have all of the properties of B-splines because they are B-splines. For instance, at any point the sum of all the Bernstein polynomials is one and each Bernstein polynomial is always positive. Because there are no interior knots, however, Bernstein polynomials are not local, i.e., they are nonzero at every point inside their interval of definition.

The final Bezier curve equation is

$$P(\theta) = \sum_{j=1}^{n} c_j Bz_{n,j}(\theta) \tag{5.22}$$

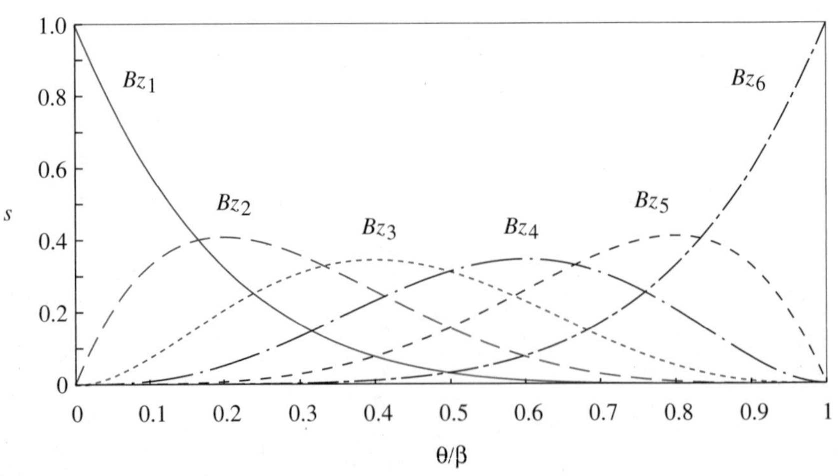

FIGURE 5-16

Bezier-Bernstein polynomials of order 6

and the method for determining the coefficients is the same as for other B-spline formulations; a linear equation is written for each boundary condition and the set of equations is solved for the coefficients.

To illustrate, we will redo Example 4-3 (p. 61), this time using a Bezier form.

EXAMPLE 5-5

Designing a Bezier Form Polynomial for the Symmetrical Rise-Fall Single-Dwell Case.

Problem: Consider the following single-dwell cam specification:

rise-fall	rise 1 in (25.4 mm) in 90° and fall 1 in (25.4 mm) in 90°
dwell	at zero displacement for 180° (low dwell)
cam ω	15 rad/sec

Solution:

1. The rise-fall part of the curve is from 0 to 180° and we have seven BCs, zero displacement, velocity, and acceleration at 0°, zero displacement, velocity, and acceleration at 180°, and 1-in displacement at 90°. We need a seventh-order Bezier curve. The knots will be 0, 0, 0, 0, 0, 0, 0, 180, 180, 180, 180, 180, 180, and 180°.

2. To determine the coefficients, we note, just as in all B-splines, that the first Bernstein polynomial is the only one that is nonzero at 0°, the second Bernstein and the first are the only ones with nonzero velocity at 0°, and the third, second, and first Bernstein polynomials are the only ones with nonzero acceleration at 0°. (See Figure 5-16.) Our first three coefficients, therefore, must be all be 0. After a similar observation about the Bernstein polynomials at 180°, we conclude that the last three coefficients are also all zero.

3. We have only one condition left, 1-in displacement at 90°. We need solve only one equation in one unknown.

$$\begin{aligned} c_4 Bz(90°) &= 0.3125c_4 = 1 \\ c_4 &= 3.2 \end{aligned} \tag{a}$$

4. The curves are the same as we obtained in Example 4-3 (p. 61). Only the displacement is drawn in Figure 5-17 and only to show that it is the same. To convince any remaining doubters, we will change the Bezier form to an ordinary polynomial form and show that it is exactly the same as equation 4.2 (p. 62).

$$\begin{aligned} s(\theta) &= 3.2Bz_{7,4}(\theta) \\ &= 3.2\left(\frac{6\cdot 5\cdot 4}{3\cdot 2}\right)\left(\frac{\theta}{180}\right)^3\left(1-\frac{\theta}{180}\right)^3 \\ &= 64\left(\frac{\theta}{180}\right)^3 - 192\left(\frac{\theta}{180}\right)^4 + 192\left(\frac{\theta}{180}\right)^5 - 64\left(\frac{\theta}{180}\right)^6 \end{aligned} \tag{b}$$

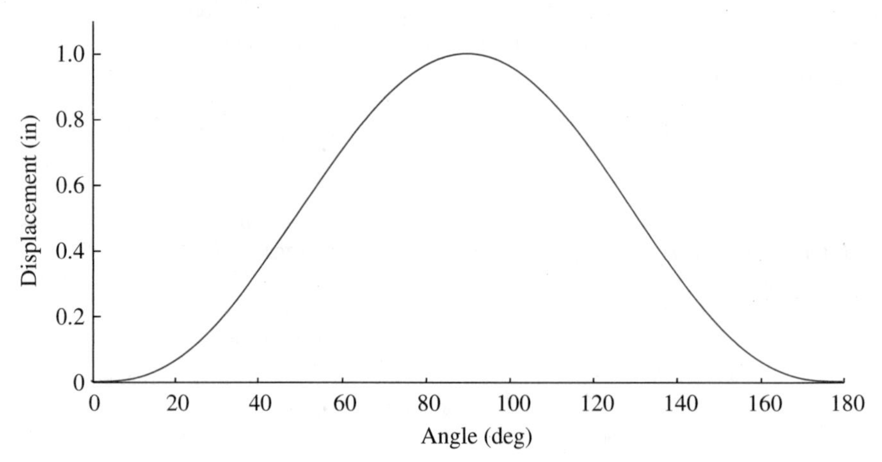

FIGURE 5-17

Example 5-5 Bezier curve solution to a symmetric rise-fall single-dwell case

The Bezier curve is more than just another way of writing a polynomial, however. Suppose that we have defined a Bezier curve of order m on some interval from a to b. Divide the interval into $m - 1$ pieces of equal length by locating the points

$$\theta_1 = a,\ \theta_2 = a + h,\ \theta_3 = a + 2h,\ \ldots\ ,\theta_m = b \tag{5.23}$$

where

$$h = \frac{b - a}{m - 1} \tag{5.24}$$

The points

$$(\theta_1, c_1), (\theta_2, c_2), \ldots, (\theta_m, c_m) \tag{5.25}$$

at which the c's are the coefficients of the Bezier curve are called *control points*. The polygonal curve formed by a set of straight line segments connecting the control points is called the *control polygon*. The control polygon has some very important relationships to the Bezier curve. For one, the control polygon is roughly the same shape as the Bezier curve. Four other important relationships are:

1 The end points of the control polygon, the first and last control points, are also the endpoints of the Bezier curve.

2 The straight line segments at the ends of the control polygon are tangent to the Bezier curve at its ends. This says that the slope of the line connecting the first two control points determines the beginning velocity. Similarly, the slope of the segment connecting the last two control points determines the ending velocity. Although not as apparent as are the slope conditions, the first three control points and the last three control points determine the beginning and ending accelerations, respectively. If the end three control points are on a horizontal line, then the end acceleration will be zero and the end velocity will be zero.

3 Any straight line that crosses the Bezier curve must cross the control polygon at least that many times. This is very important because it tells us among other things that our Bezier

defined displacement can never be negative unless the control polygon is negative. A Bezier curve always lives in the "shadow" of its control polygon.

4 The higher the degree of the polynomial, the more alike are the control polygon and the Bezier curve.

We can reshape a Bezier displacement just by moving the control points and reshaping the control polygon. This, in fact, is what is done by many of the commercial drawing programs including *Illustrator*, *Freehand*, *Canvas*, and others. Fourth-order Bezier curves are used and so there are four control points. The two end control points fix the curve ends and the other two define the curve end tangents. You have probably used a Bezier curve in a drawing program. Now you know how it worked.

Figures 5-18 through 5-21 show some examples of Bezier curve displacements and their control polygons so that you may observe what we have just said. Figure 5-18 shows the single-dwell displacement constructed in Example 5-5. At the beginning and end of the curve the lift, velocity, and acceleration are zero. Therefore, the first three control points and the last three control points are all on the horizontal axis. The fourth control point is raised high enough for the maximum displacement , directly beneath it, to be one inch. In Figure 5-19 we show the 3-4-5 polynomial that was used in Chapter 3 for a double-dwell cam. The sixth control point is at 1, which is the end dwell. The other control points are aligned so that the velocities and accelerations at the dwells are zero. Figure 5-20 is similar. It is a 4-5-6-7 polynomial, also discussed in Chapter 3. Two additional control points are used to ensure zero jerk at the ends. The reader can probably, by now, draw a control polygon for the double-dwell cam that will ensure zero velocity, acceleration, jerk, and ping at the dwells. Figure 5-21 shows Dudley's 2-10-12-14 polynomial which will be discussed in Chapter 10. We will not go into its details here. It is shown here to demonstrate that the higher the degree of the polynomial, the closer the control polygon is to the actual Bezier curve.

This discussion and these figures may have convinced readers that they can design cam motion using Bezier curves with no equations, just the control polygon. One can only design very simple motions that way. Most of the time things are not that easy. Cam follower motion must be precisely designed, too precisely designed to be done geometrically using control polygons. Nevertheless some knowledge of control polygons can greatly aid in cam motion design by exposing the pitfalls and suggesting improvements.

Here is how:

1 Draw a rough sketch of the desired displacement.

2 Divide the horizontal axis into a number of equal length segments equal to the degree of the polynomial. The ends of the segments will mark the horizontal coordinates of the control points.

3 Draw in, lightly, vertical lines at the ends of each segment. The control points must lie somewhere on these lines.

4 Mark in all the necessary and known control points. For instance, the end points of the curve must be control points; if the motion has zero velocity at one end, then at that end, two control points must align horizontally, and so forth.

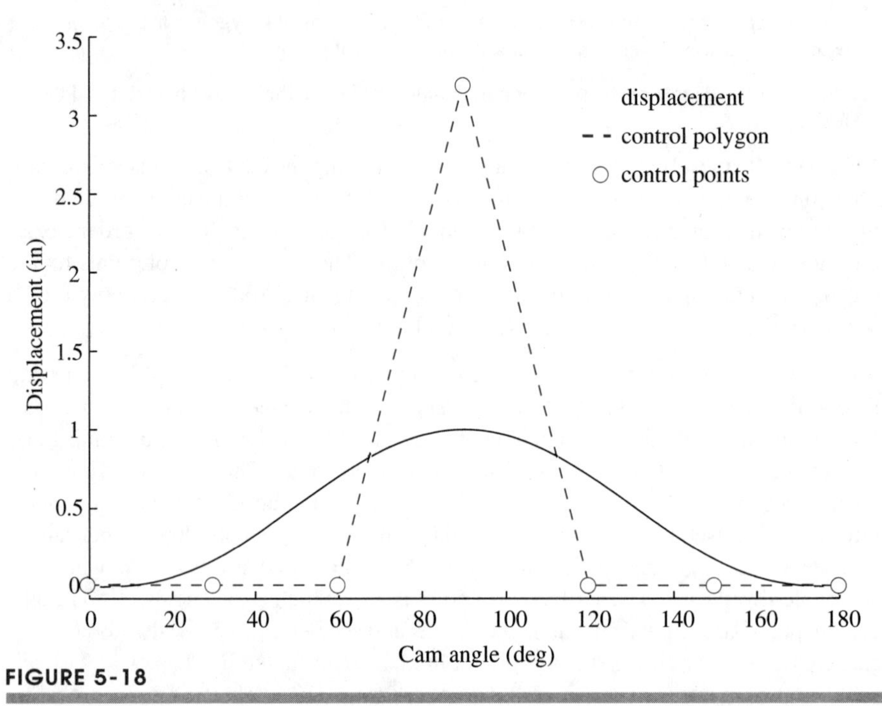

FIGURE 5-18

Bezier curve, control points, and control polygon for Example 5-5 displacement function

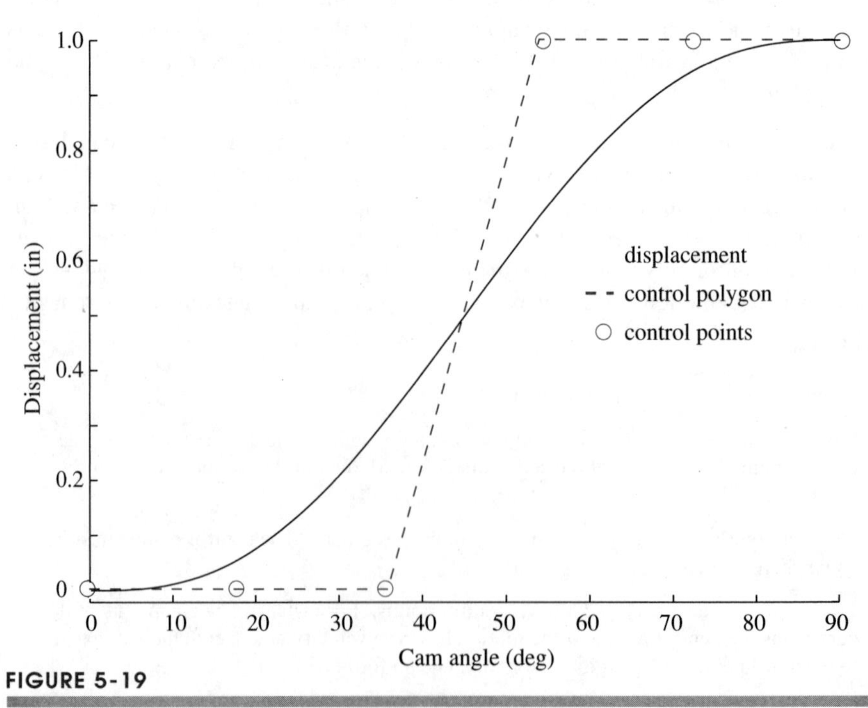

FIGURE 5-19

Bezier curve, control points, and control polygon for a 3-4-5 polynomial double-dwell rise

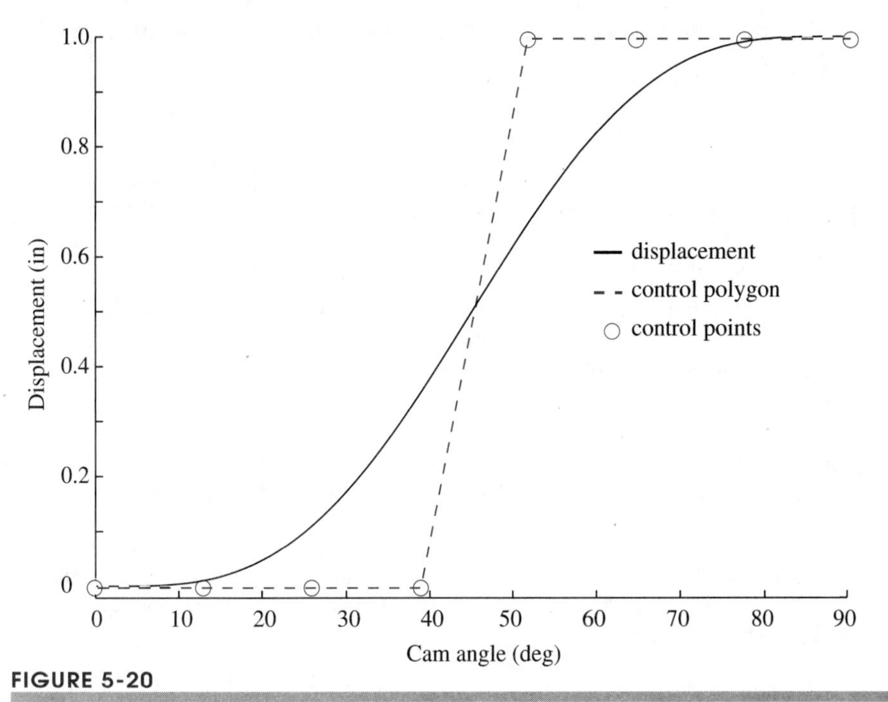

FIGURE 5-20

Bezier curve, control points, and control polygon for a 4-5-6-7 polynomial double-dwell rise

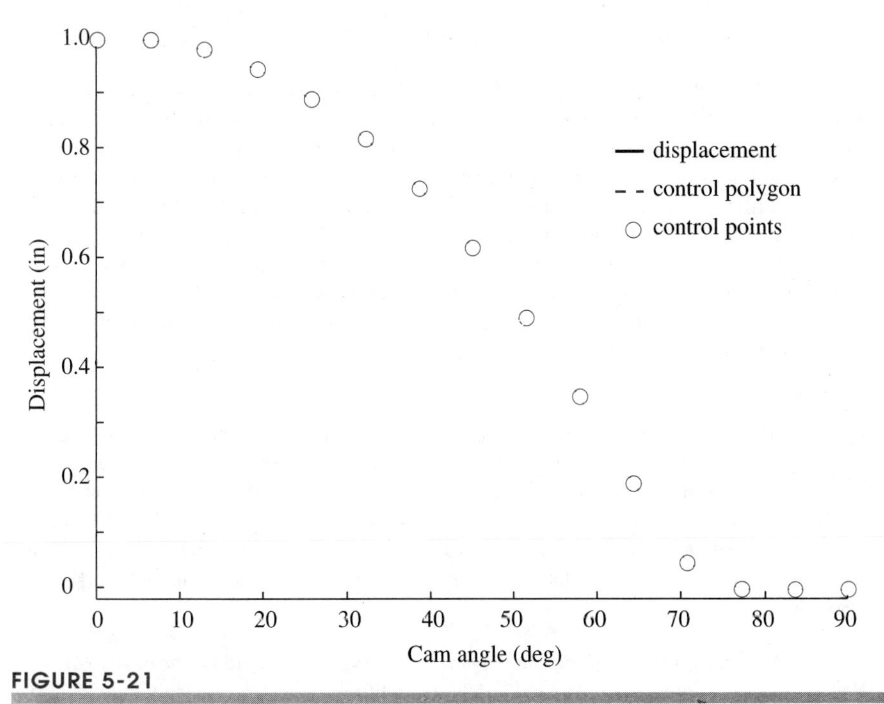

FIGURE 5-21

Bezier curve, control points, and control polygon for a Dudley 2-10-12-14 polynomial fall

5 Place the remaining control points in positions so that the control polygon has roughly the same shape as the curve. If you are unable to do so, then your Bezier will not work. You will have to change the degree or try some other curve.

EXAMPLE 5-6

Designing a Bezier Curve for an Asymmetrical Rise-Fall Single-Dwell Case.

Problem: Consider the following single-dwell cam specification:

rise-fall rise 1 in (25.4 mm) in 45° and fall 1 in (25.4 mm) in 135°
dwell at zero displacement for 180° (low dwell)
cam ω 15 rad/sec (143.24 rpm)

5

Solution:

1. There are eight boundary conditions. Setting $S = V = A = 0$ at each end of the rise-fall segment accounts for six conditions. The other two are maximum displacement and zero velocity at 45° to guarantee that it is a maximum there.

2. We must use a polynomial of order eight, so our knots are 0, 0, 0, 0, 0, 0, 0, 0, 180, 180, 180, 180, 180, 180, 180, 180°. We will have eight Bernstein polynomials and eight coefficients to determine.

3. Before setting up the BC equations, we first try to sketch the control polygon that we would expect for a good solution as in Figure 5-22. Breaking the interval from 0 to 180° into seven pieces we get control points roughly at

$$\theta_1 = 0,\ \theta_2 = 26,\ \theta_3 = 51,\ \theta_4 = 77,\ \theta_5 = 103,\ \theta_6 = 129,\ \theta_7 = 154,\ \theta_8 = 180 \qquad \text{(a)}$$

4. The first three and the last three control points have to be on the horizontal axis in order to have $S = V = A = 0$ at the beginning and end of the motion. This says that the coefficients for the first, second, third, sixth, seventh, and eighth Bernstein polynomials will be zero. The corresponding control points are therefore fixed to the θ axis. This leaves only the fourth and fifth control points. There is nowhere that we can place these two control points that will make the control polygon look anything like what we want. The eighth order Bezier curve fails to solve this problem. A glance at Figure 4-5 (p. 64), where we followed this problem to its tragic end, shows the eighth order attempt to be what our control polygon is predicting. The displacement maximum is not at 45° but is between the fourth and fifth control point.

5. What can we do? It is obvious that the critical point, the point where the curve is supposed to be a maximum, is caught between the second and third control points, which are fixed to the horizontal axis. For the end conditions and the maximum at 45° we need at least three, possibly four, control points prior to 45°. The minimum Bezier curve that might do would have control points at 0, 15, 30, 45, 60, 75, 90, 105, 120, 135, 150, 165, and 180°. This would mean that we need a polynomial of at least order 13. We hope that the reader notices that not only has the control polygon told us what would not work, it has also told us what might work.

6. The polynomial of order 13 or possibly 14 would work, but it would not be easy to do. We would need another six or seven boundary conditions. Chances are we would end up with slight oscillations in the displacement. Most certainly we would end up with

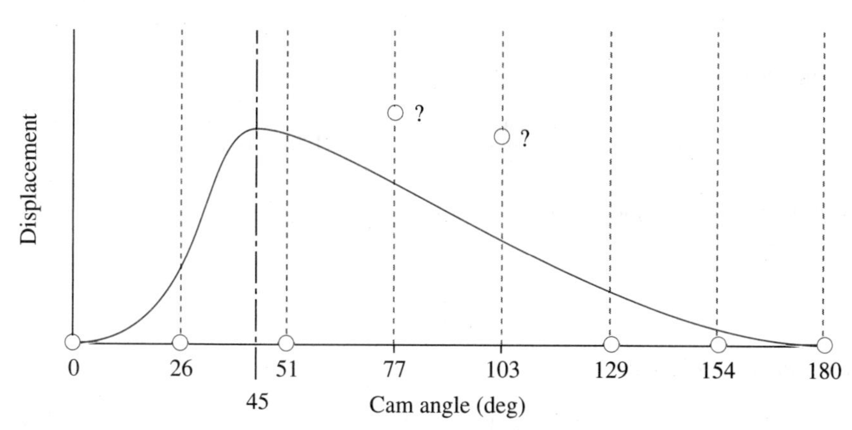

FIGURE 5-22

Desired curve and unknown control point locations for Example 5-6

oscillations in the velocity and acceleration. Another solution is to use two polynomial segments blended together. By now it is our fervent hope that the reader is saying to himself, "That is a proper B-spline!" It is indeed, and we will solve this problem in the next section where we discuss control polygons for B-splines and knot placement.

5.5 KNOT PLACEMENT

Control points can also be defined for general polynomial B-spline displacements. For the cam designer, their main use is in determining good knot locations. In designing a motion with B-splines, one of the most important and difficult tasks is to choose the locations of the interior knots. If there are q boundary conditions and we are using a B-spline of order m, then we must have $m + q$ knots. Of these, the number of end knots is equal to $2m$, twice the order. Thus the engineer must choose $q - m$ interior knots. Furthermore, the location of these knots is crucial to the characteristics of the motion that results. Control points and the control polygon are tools to help in picking those knots.

To compute the positions of the control points an engineer must first find the *knot averages*. Knot averages are just what the name suggests. To find the knot averages for a B-spline displacement of order m, ignore the first knot and take the average of the knots from the second to the m-th, next find the average of the third to the $m + 1$ knot, etc., ending when the next-to-the-last knot has been used. Essentially, what one is doing is taking the average of the knots for which each B-spline is nonzero. For illustration, Table 5-7 shows an example for the B-splines of order 6 with knot sequence 0, 0, 0, 0, 0, 0, 45, 90, 135, 180, 180, 180, 180, 180, 180.

There are the same number of knot averages as there are B-splines. To get the control points, associate the corresponding coefficient of the B-spline in our B-spline displacement with each knot average. We can state all of this mathematically by letting the knots of a B-spline displacement of order m be

$$t_1 \le t_2 \le \cdots \le t_n \tag{5.26}$$

Then the knot averages are

$$d_j = \frac{t_{j+1} + t_{j+2} + \cdots + t_{j+m-1}}{m-1}, \quad j = 1,2,\ldots,n-m \tag{5.27}$$

Now if the B-spline displacement is

$$s(\theta) = \sum_{j=1}^{n-m} c_j B_j(\theta) \tag{5.28}$$

then the control points are

$$\left(d_j, c_j\right), \quad j = 1,2,\ldots,n-m \tag{5.29}$$

The control polygon is the polygonal curve made by connecting adjacent pairs of control points with straight line segments. Figure 5-23 shows the B-spline displacement that was constructed in Example 5-6 along with its control points and control polygon.

If you compute control points for the Bezier curve, using the method just described, you will find that the knot averages come out to be spaced at equal intervals; you will get the same control points as have already been defined for the Bezier curve. Furthermore, everything said about the Bezier control points is also true for the general B-spline control points.

1 Both the B-spline displacement and the control polygon start and end at the same point.

2 The B-spline displacement is tangent to the control polygon at the curve ends.

3 Any straight line that crosses the B-spline curve must cross the control polygon at least as many times.

4 The more knots there are, the closer the control polygon mimics the B-spline displacement.

TABLE 5-7 B-spline Knot Averages for the Knot Sequence:
0 , 0 , 0, 0, 0, 0, 45, 90, 135, 180, 180, 180, 180, 180, 180°

Knot Index	Set 1	Set 2	Set 3	Set 4	Set 5	Knot Average
1	0	0	0	0	0	0
2	0	0	0	0	45	9
3	0	0	0	45	90	27
4	0	0	45	90	135	54
5	0	45	90	135	180	90
6	45	90	135	180	180	126
7	90	135	180	180	180	153
8	135	180	180	180	180	171
9	180	180	180	180	180	180

There are other important and interesting facts about the control points and control polygon. The derivative of a B-spline displacement is another B-spline curve to represent the velocity. It too must have control points and a control polygon. It turns out that the slopes of the line segments of the control polygon are the coefficients used in the B-spline velocity curve. Similarly, the coefficients used to specify acceleration as a B-spline curve are the slopes of the different line segments making up the velocity control polygon. Symbolically, if a B-spline displacement of order m is

$$s(\theta) = \sum_{j=1}^{n-m} c_j B_{m,j}(\theta) \tag{5.30}$$

with control points

$$(d_1, c_1), (d_2, c_2), \ldots, (d_{n-m}, c_{n-m}) \tag{5.31}$$

then the velocity is

$$s'(\theta) = \sum_{j=1}^{n-m-1} \frac{c_{j+1} - c_j}{d_{j+1} - d_j} B_{m-1j}(\theta) \tag{5.32}$$

This is how most computer programs calculate velocity, acceleration, jerk, ping, etc., for a B-spline displacement.

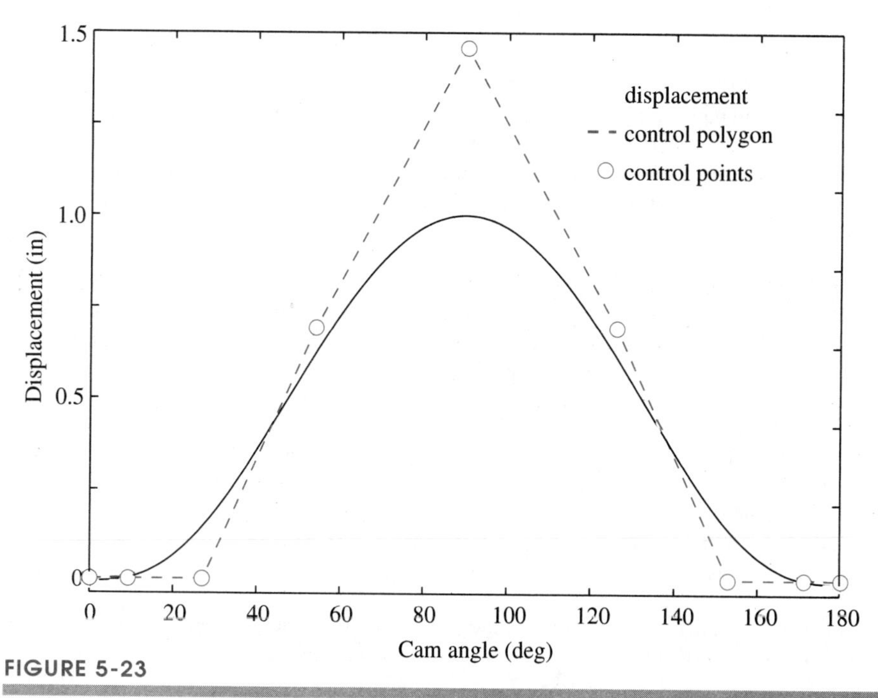

FIGURE 5-23

B-spline curve, control points, and control polygon for Example 5-6 displacement function

All of these properties of the control polygon suggest a procedure that can be used to help design follower motion. The procedure is to:

1 Make a rough sketch of the desired curve.

2 Using the tentative knot sequence, locate the knot averages.

3 Try to draw a control polygon using the knot averages that mimics the curve. If you cannot, then your knot sequence is unsatisfactory and you should change it. If you can, then you will likely find a suitable B-spline displacement for the problem. There is no absolute guarantee though, especially if interior derivatives are involved in the boundary conditions.

Fortunately with software such as program DYNACAM, one can interactively move the knots and the computer will draw not merely the control polygon, but the actual B-spline displacement that results from the knot sequence and the boundary conditions. Some examples will illustrate this. We will start with the same problem that defeated our Bezier curve in Example 5-6.

EXAMPLE 5-7

Designing an 8-Boundary-Condition B-spline Function for an Asymmetrical Rise-Fall Single-Dwell Case.

Problem: Using the CEP specification from Example 4-4:

rise-fall	rise 1 in (25.4 mm) in 45° and fall 1 in (25.4 mm) in 135° in 180°
dwell	at zero displacement for 180° (low dwell)
cam ω	15 rad/sec (143.24 rpm)

Solution:

1 The minimum number of BCs for this problem is eight. The dwell on either side of the combined rise-fall segment has zero values of *S, V, A,* and *J*. The fundamental law of cam design requires that we match these zero values, through the acceleration function, at each end of the rise-fall segment. These then account for six BCs; $S = V = A = 0$ at each end of the rise-fall segment.

2 We also must specify a value of displacement at the 1-in peak of the rise which occurs at $\theta = 45°$ and specify the velocity to be zero at that point. These are the seventh and eighth BCs. See Figure 5-24 for the set of boundary conditions for this problem. Note that these BCs are identical to those used in step 6 of Example 4-4, which failed in its attempt to put an acceptable single polynomial function through these points.

3 If we use a fifth-order spline, then we will need 5 + 8 = 13 knots. Ten of them will be end knots, leaving us to choose 3 interior knots. Figure 5-25 shows a rough sketch of what we anticipate and a control polygon that should more or less obtain it. The "knot symbols" are the knots and the circles mark possible control points using knot averages computed from the knot values. Note that at each end there are five repeated knots as shown.

4 Notice that we are forced to put six of the control points on the axis because the motion must begin and end at a dwell. Also our drawing makes clear that the knot averages are

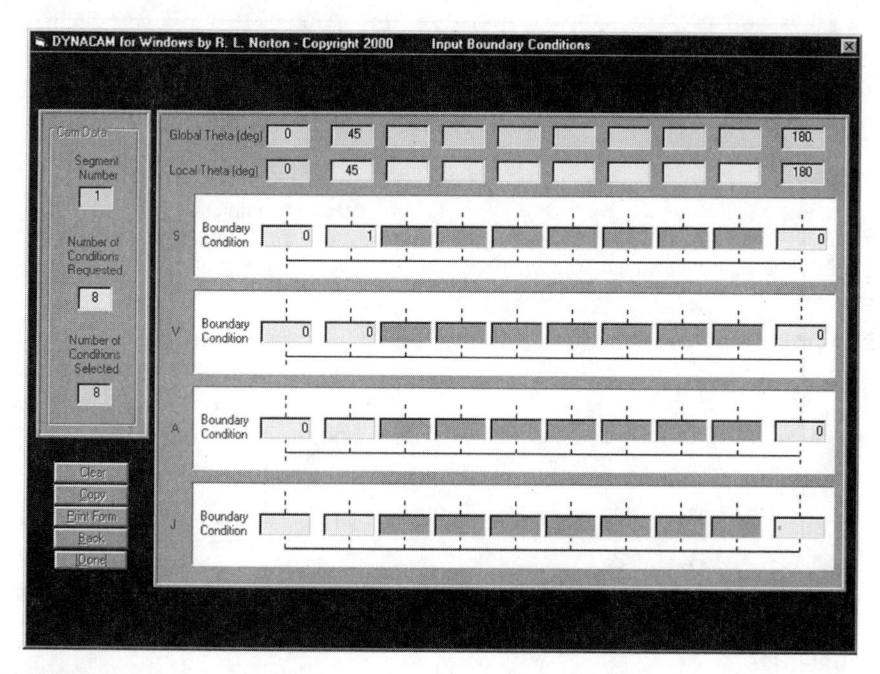

FIGURE 5-24

Boundary conditions for the B-splines of Example 5-7

closer at the beginning of the motion and farther away at the ends. Hence, feasible interior knots should also be shifted towards the beginning of the motion. Program DYNACAM lets us interactively place the knots. We know from the rough sketch that they should be placed in the first half of the motion, before 90°, except, of course, for the necessary end knots at 0° and 180°.

5 Using DYNACAM and sliding each initially equispaced knot to the left we get a very good solution with the knots at 0, 0, 0, 0, 0, 23, 42, 90, 180, 180, 180, 180, 180°. Figure 5-26 shows the results. The displacement B-spline functions are quartic (4th degree, 5th order), the velocity functions are cubics, the accelerations are quadratic (parabolic), and the jerks are linear and continuous within the motion event but discontinuous at the dwell boundaries (but still finite), which is acceptable.

6 Compare this solution to that of Example 4-5 in Figure 4-8 (p. 67), which was the first successful solution to this problem statement using polynomials that was shown. That solution required that the rise and fall be dealt with as separate segments and that the longer of the two segments be calculated first in order to obtain an acceleration value for use as a matching parameter for the other segment. The peak negative acceleration of the polynomial solution in Figure 4-8 is –2024 in/sec^2. The corresponding peak acceleration for this B-spline solution is –1921 in/sec^2, a 6% reduction. The peak-to-peak acceleration is 3841 in/sec^2. Note also the improved smoothness of the acceleration function in Figure 5-26 as compared to that of Figure 4-8. The jerk in Example 4-5 is discontinuous within

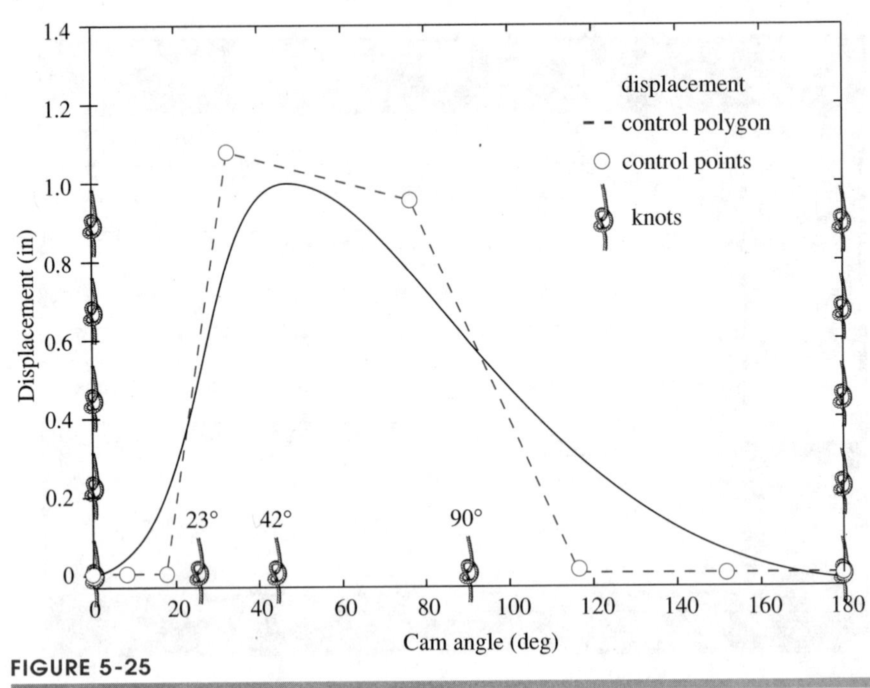

FIGURE 5-25

Sketched curve, control points, control polygon, and knots for Example 5-7 displacement

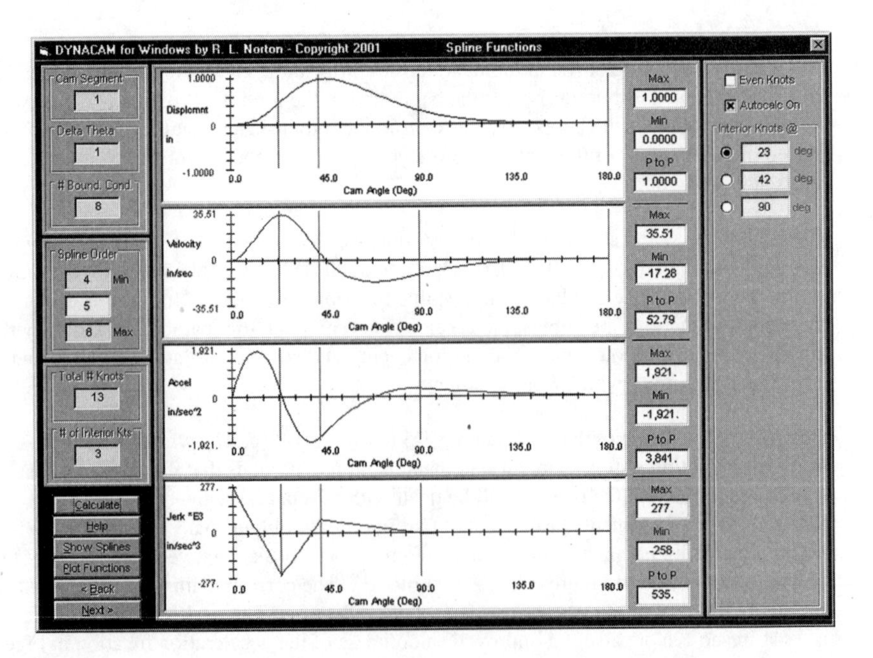

FIGURE 5-26

Spline order, knot locations and spline functions for the 5th-order B-spline of Example 5-7

the motion event, but here is continuous over that interval. The difference is due to the B-spline approach.

7 We also can solve this problem using a B-spline of order six as shown in Figure 5-27. The increased order will raise the degree of each function. The displacement will now be quintic (6th order), the velocity quartic, the acceleration cubic and the jerk will change from a second order spline to a third order, or parabolic spline. Since we now have sixth order and eight boundary conditions we need 6 + 8 = 14 knots. It is always the case that the number of boundary conditions plus the order of the spline is equal to the number of knots required. Note that increasing the order of the spline without changing the number of BCs reduces the number of available interior knots on a one-for-one basis. Since the number of end knots must be twice the order of the B-splines, we can now only choose two interior knots. Using DYNACAM we find a very smooth curve when we place the interior knots at 9.5° and 39°. The cost of the increased smoothness is increased peak acceleration, as we should have expected since increases in acceleration nearly always accompany improved smoothness. The positive peak acceleration is now 3292 in/sec^2 and the negative peak is -1312 in/sec^2, a peak-to-peak variation of 4604 in/sec^2. Compare this to the 5th order spline result above of 3841 in/sec^2 peak to peak. A positive trade-off for the increased acceleration is a very smooth return to the dwell where the jerk is now zero.

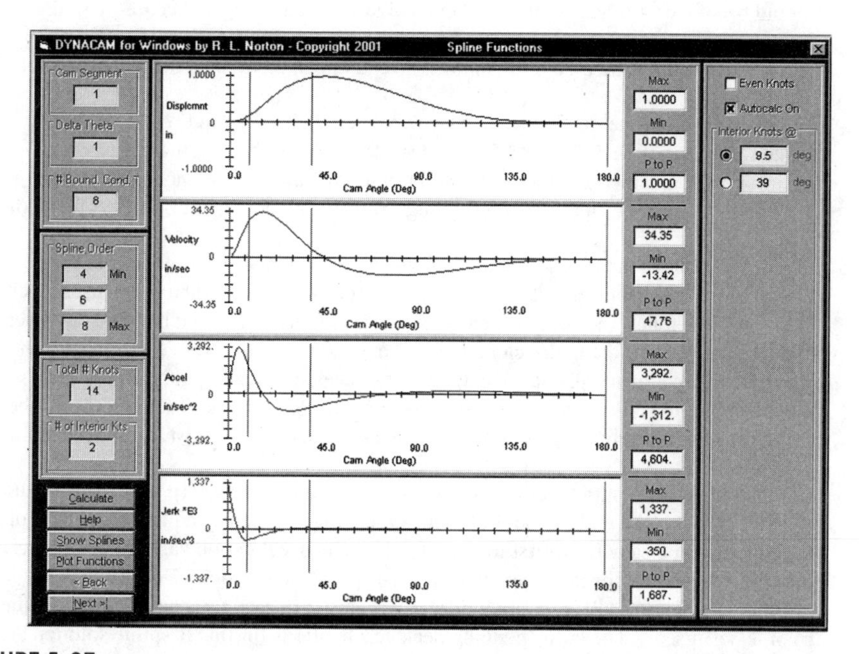

FIGURE 5-27

Spline order, knot locations and spline functions for the 6th-order B-spline of Example 5-7

EXAMPLE 5-8

Designing a 10-Boundary-Condition B-Spline Function for an Asymmetrical Rise-Fall Single-Dwell Case.

Problem: Using the CEP specification from Example 5-7:

rise-fall	rise 1 in (25.4 mm) in 45° and fall 1 in (25.4 mm) in 135° in 180°
dwell	at zero displacement for 180° (low dwell)
cam ω	15 rad/sec (143.24 rpm)

Solution:

5

1 A possible improvement to the designs of Example 5-7 could be to specify the jerk to be zero at the beginning and end of motion to match the dwells. This will smooth the jerk function to reduce potential vibration, but may come at the expense of an increase in peak acceleration. The dwell on either side of the combined rise-fall segment has zero values of *S, V, A,* and *J*. The fundamental law of cam design requires that we match these zero values, through the acceleration function, at each end of the rise-fall segment. With the additional constraint to match the jerk, these then account for 8 BCs; $S = V = A = J = 0$ at each end of the rise-fall segment.

2 We also must specify a value of displacement at the 1-in peak of the rise which occurs at $\theta = 45°$ and specify the velocity to be zero at that point in order to define it as the maximum displacement point. These are the 9th and 10th BCs. All are shown in Figure 5-28.

3 The order of the spline cannot exceed the number of boundary conditions specified and should be at least 4 to satisfy the fundamental law of cam design. According to the formulas given above, if we choose a fifth order B-spline displacement, we have 10 + 5 = 15 knots required. Since 10 of them must be end knots, there must be 5 interior knots.

4 Figure 5-29 shows the displacement curve obtained with program DYNACAM for the knot sequence 0, 0, 0, 0, 0, 5, 20, 45, 85, 105, 180, 180, 180, 180, 180. It also shows the control polygon, which verifies that with DYNACAM's interactive knot selection tool we made a good choice of interior knots, though we have no choice for number of end knots.

5 Figure 5-30 shows the displacement, velocity, acceleration, and jerk functions from DYNACAM. The displacement functions are quartic (4th degree, 5th order), the velocity functions are cubics, the accelerations are quadratic (parabolic), and the jerks are linear and continuous over the entire cam, which is quite acceptable. Notice that despite the smooth ends that match displacement, velocity, acceleration and jerk, the jerk curve, though continuous, is ragged. The ping has jump discontinuities. This will always be true for a fifth order B-spline. This lack of smoothness trades-off against lower acceleration.

6 Compare this solution to that of Example 4-5, step 9 in Figure 4-9 (p. 68). That solution required that the rise and fall be dealt with as separate segments and that the longer of the two segments be calculated first in order to obtain an acceleration value to be used as a matching parameter for the other segment. The peak acceleration values of the polynomial solution in Figure 4-9 are + 2099 in/sec^2 and –2882 in/sec^2 for a peak-to-peak value (p-p) of 4981 in/sec^2. The corresponding peak accelerations for this B-spline solution are +2264 in/sec^2 and –1974 in/sec^2 for a peak-to-peak of 4258 in/sec, a 15% reduction peak-to-peak. Note also the improved smoothness of the acceleration function in Figure 5-30 as compared to that of Figure 4-9 (p. 68).

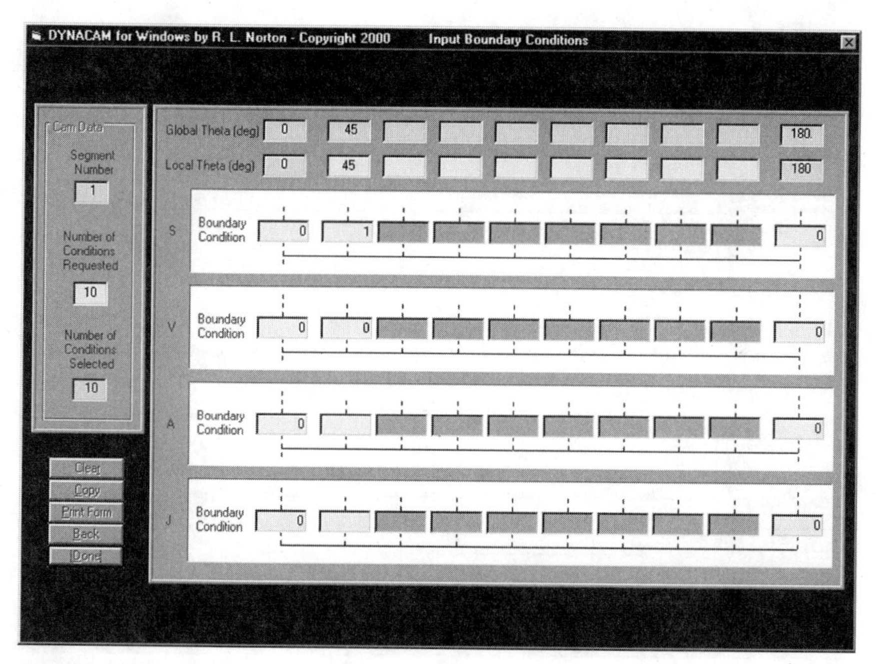

FIGURE 5-28

Boundary conditions for the B-spline of Example 5-8

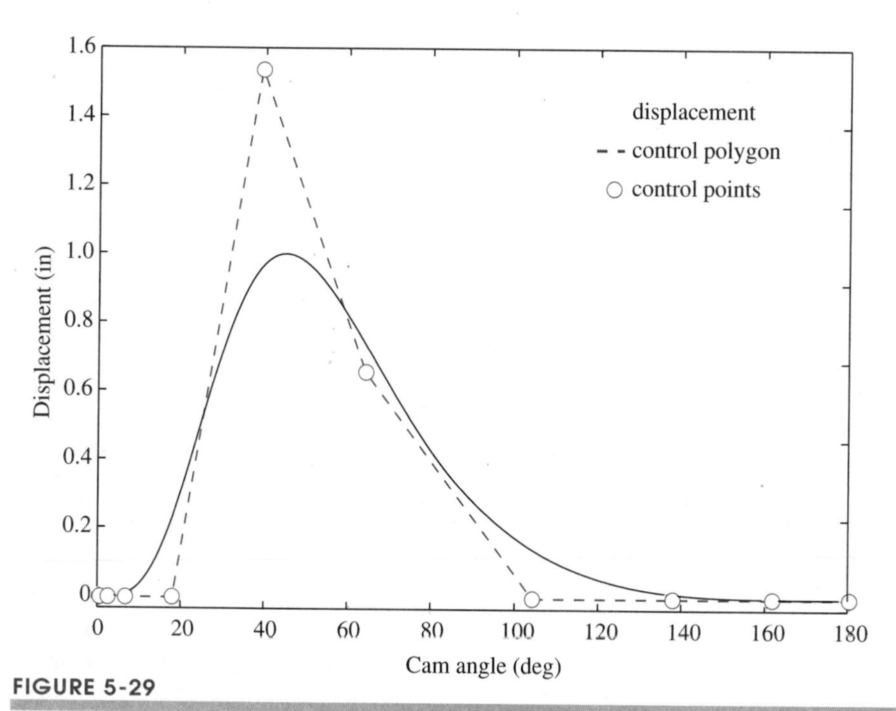

FIGURE 5-29

Fifth-order B-spline curve, control points, and control polygon for Example 5-8

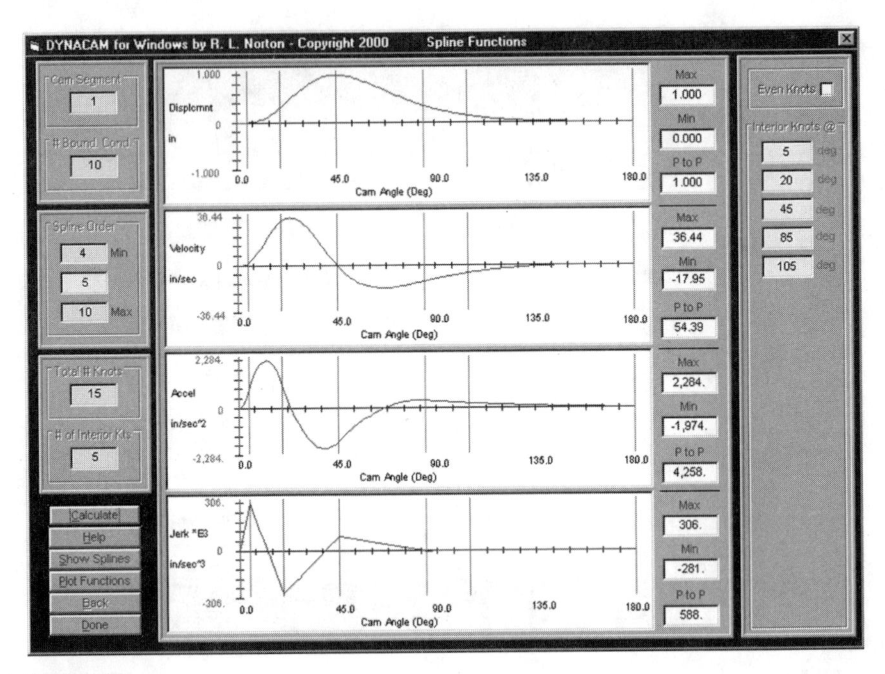

FIGURE 5-30

Spline order, knot locations and spline functions for the 5th-order B-spline of Example 5-8

7 Again, as done in Example 5-7, we can smooth the jerk from a linear spline to a quadratic spline by increasing the displacement to sixth order. Using our counting rule we will have 6 + 10 = 16 knots total. Of these, 12 will be end knots. Increasing the order of the spline without changing the number of BCs reduces the number of available interior knots on a one-for-one basis. This design has four interior knots, placed at 5, 37, 45, and 100°. The displacement is now quintic (6th order), the velocity is quartic, the acceleration is cubic, and the jerk is parabolic.

8 Figure 5-31 shows this 6th-order B-spline solution. This solution appears to be optimal. It is certainly the best of those so far investigated. In addition to its good smoothness we have low accelerations. The peak values of acceleration are now ±2046 in/sec^2 or 4093 in/sec^2 peak-to-peak. These are lower than either the 5th-order spline or the three-segment polynomial and are nearly as low as the solution of Figure 4-8 that has discontinuous jerk.

To summarize, finding the knot locations is one of the crucial tasks that the cam designer faces when using B-splines. B-splines are enormously powerful design tools and there is usually more than one B-spline solution to any motion design problem. The goal is to find the best set of knot locations, i.e., the best places to "jerk," "ping," or "puff" the motion in order to exercise dynamic control. A good software program using B-splines will have an interactive knot placement feature. Failing that, the engineer should use the control polygon as an aid to knot placement.

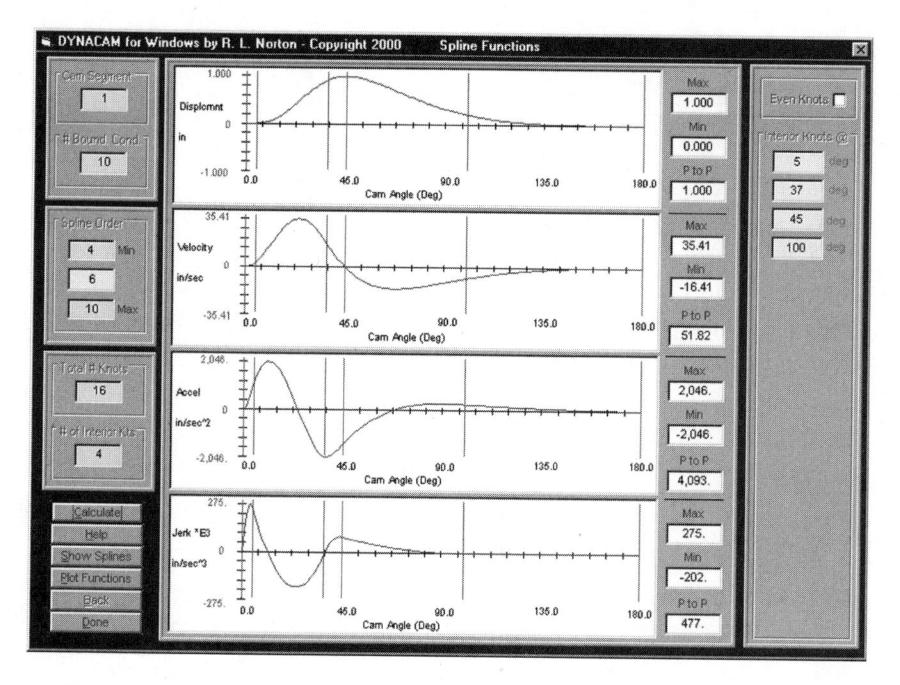

FIGURE 5-31

Spline order, knot locations and spline functions for the 6th-order B-spline of Example 5-8

5.6 PERIODIC SPLINES

So far, all of the splines that we have talked about were for only portions or segments of the follower motion. There are times, though, when the entire motion from 0 to 360° needs to be constructed as a single spline. The splines that do this are called *periodic splines* and have some special characteristics. Periodic splines have no end knots, only interior knots. The point, and it is only one point not two, at 0° and 360° may or may not be a knot. The periodic B-spline basis functions wrap around the circle; the B-spline that begins, for example, at 270 degrees may end at 45 degrees. If the number of knots is smaller than the order of the spline, the B-spline basis functions, as we have previously defined them, may actually overlap themselves, like the ancient drawings of snakes eating their own tails.

An entire mathematical theory for periodic splines exists and is interesting to study, but the practical engineer need not be concerned with it. For designing a follower motion as a single spline over the entire cam, the engineer can proceed exactly as for the nonperiodic spline, but with additional smoothness equations. Assume that the order of the spline is m and that the point at 0 and 360° is not a knot. Then the following smoothness equations must be included with the set of BC equations:

$$
\begin{aligned}
s(0) &= s(360) \\
s'(0) &= s'(360) \\
s''(0) &= s''(360) \\
s'''(0) &= s'''(360) \\
&\dots \\
s^{(m-1)}(0) &= s^{(m-1)}(360)
\end{aligned} \tag{5.33a}
$$

If there is a knot at 0°, the engineer need not include the last equation. The additional equations needed are therefore

$$
\begin{aligned}
s(0) &= s(360) \\
s'(0) &= s'(360) \\
s''(0) &= s''(360) \\
s'''(0) &= s'''(360) \\
&\dots \\
s^{(m-2)}(0) &= s^{(m-2)}(360)
\end{aligned} \tag{5.33b}
$$

Recall that for a spline of order m with k interior knots, there can be $k + m$ boundary condition equations. The additional smoothness equations means that the number of boundary condition equations must be reduced. In the case where 0° is not a knot, there are m additional smoothness equations, so there can be only k boundary condition equations, the same number as there are interior knots. In the case that 0° is a knot, it would seem at first that there can be one more boundary equation than there are interior knots, but that is not true because 0° is an interior knot. There are no end knots; the spline has no ends. So again, the number of boundary condition equations that can be satisfied with a periodic spline is equal to the number of interior knots or, better, equal to the number of knots, since all the knots of a periodic spline are interior knots.

EXAMPLE 5-9

Designing a Periodic Spline to Simulate a Quick Return Mechanism.

Problem:

fall	1 inch in 90°
rise	1 inch in 270°
ω	1 deg/sec

Solution:

1 Select the lowest position of the follower as a zero reference. The spline, then, must have a displacement of 1 inch at 0° which, of course, is also 360°, and a displacement of 0 at 90°. At both of these angles the velocity must be zero to guarantee the maximum at 0° and minimum at 90°. We choose a sixth order spline (degree 5) with discontinuities in the puff, the fifth derivative. The motion will be quite smooth.

2 There are four necessary boundary conditions and, according to the above discussion, we need four knots, one of which is at 0°. However, as in previous examples, we choose extra knots to give us some local control of the curve. The final knot selections are at 0, 90, 100, 110, 300, 320, 350, 359°. Some explanation is called for here. The knots are clustered close to the critical parts of the curve, i.e., the maximum and minimum. The "puffs" at 90, 100, 110° control the motion immediately after the minimum at 90° and the "puffs" at 300, 320, 350, 359 and 0° control the maximum at 0°. The authors do not pretend that this knot selection came to us intuitively. A great deal of "tuning" was necessary.

3 The boundary conditions are listed in Table 5-8. Again, they are the result of tuning the spline.

4 Although it possible to complete the problem using DYNACAM, we will write out the equations that must be solved in order that you may gain some knowledge about what goes on in the dark recesses of the computer when it computes a periodic spline. The knot set is the same as that used for a nonperiodic spline. There are two end knots with full multiplicity. The knots are at 0, 0, 0, 0, 0, 0, 90, 100, 110, 300, 320, 350, 359, 360, 360, 360, 360, 360, 360°. The B-spline basis is constructed in the usual way and the resulting spline will be

$$s(\theta) = a_1 B_1(\theta) + a_2 B_2(\theta) + a_3 B_3(\theta) + \cdots + a_{13} B_{13}(\theta) \tag{a}$$

where the coefficients, the a's, need to be determined. This is done with the following set of equations. For the boundary conditions, we have

$$\begin{aligned} s(0) &= 1 \\ s'(0) &= 0 \\ s'''(0) &= 0 \\ s(90) &= 0 \\ s'(90) &= 0 \\ s(150) &= 0.1 \\ s(232) &= 0.38 \\ s(285) &= 0.6 \end{aligned} \tag{b}$$

In addition, the following smoothness equations must be added

$$\begin{aligned} s(0) &= s(360) \\ s'(0) &= s'(360) \\ s''(0) &= s''(360) \\ s'''(0) &= s'''(360) \\ s^{(4)}(0) &= s^{(4)}(360) \end{aligned} \tag{c}$$

This gives us 13 equations in 13 unknowns and the resulting spline will be periodic even though built with nonperiodic B-spline basis functions.

5 Plots of the displacement, velocity, acceleration, and jerk can be seen in Figure 5-32. The curious shape of the jerk function shows the puffs at work. Notice the long smooth

TABLE 5-8 Boundary Conditions for Example 5-9

BC Location	0°	90°	150°	232°	285°
Displacement BCs	1	0	0.1	0.38	0.6
Velocity BCs	0				
Acceleration BCs					
Jerk BCs	0				

expanse between the knots at 110 and 350° and the smooth jerk from 0 to 110°. Apparently at 110° the spline is subjected to a large puff which virtually eliminates jerk. From 350 to 360° the spline is subjected to violent puffs in order to cause the maximum at 0°. Note, however, that the jerk is continuous. The ping, shown in Figure 5-32, is continuous and virtually zero everywhere except around 90° and approaching 360°. This is consistent with what we have pointed out in the jerk function.

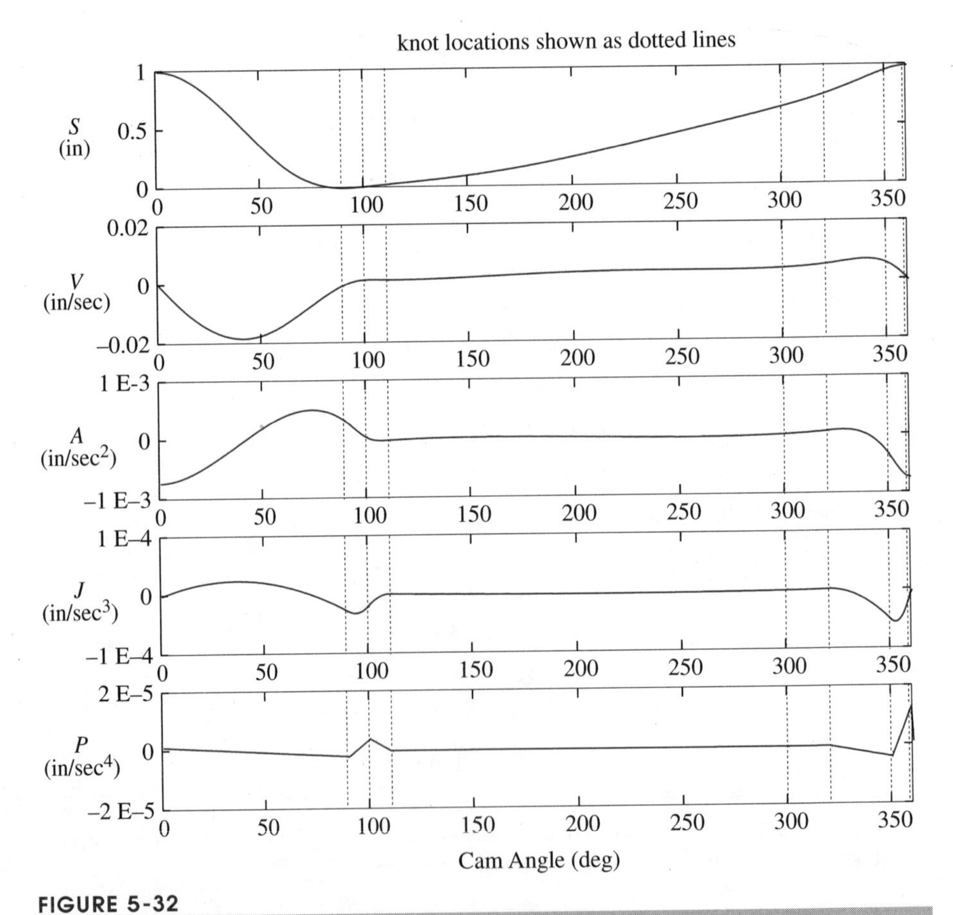

FIGURE 5-32
Displacement, velocity, acceleration, jerk, and ping of a quick-return periodic spline

5.7 SPLINES OTHER THAN POLYNOMIAL SPLINES

There are no particular reasons why splines should be piecewise polynomials and not piecewise other things such as exponentials, harmonics, or even hyperbolic functions. In fact, all of these types of spline have been developed and can be looked up in the references. Because of their versatility polynomial splines are the most frequently used. They are more than adequate for cam-follower motion design. Polynomial splines are simple to use and easy to understand. That is why we have spent so much time and effort on them and are just now mentioning other kinds of splines. There is not enough room here to fully explain all of the other kinds of splines that have been developed mathematically, but we will mention two that the reader may have heard about. Others are described in publications listed in the bibliography of this chapter.

Trigonometric Splines

A *trigonometric polynomial* is an expression of the form

$$a_0 + a_1 \cos\theta + a_2 \cos 2\theta + \cdots + a_r \cos r\theta + b_1 \sin\theta + b_2 \sin 2\theta + \cdots + b_r \sin r\theta \qquad (5.34a)$$

or of the form

$$a_1 \cos\frac{\theta}{2} + a_2 \cos\frac{3\theta}{2} + \cdots + a_r \cos\frac{2r-1}{2} + b_1 \sin\frac{\theta}{2} + b_2 \sin\frac{3\theta}{2} + \cdots + b_r \sin\frac{2r-1}{2} \qquad (5.34b)$$

Using our usual meaning of order, the number of coefficients to which we are free to assign values, the first expression has order $m = 2r + 1$, an odd number and the second has order $m = 2r$, an even number. *Trigonometric splines* are splines as we have defined them, except trigonometric polynomials are used in place of the conventional polynomials.

Modified Trapezoidal Acceleration as a Spline Trigonometric splines can be of great use to the cam engineer who has, in fact, been using them already. For example, the modified trapezoidal acceleration is a trigonometric spline with interior knots of order 3. It was defined in Chapter 3 by equations 3.7a through 3.7e (pp. 32-33). Below we again sketch the derivation of the modified trapezoidal acceleration, but this time as a trigonometric spline.

Trigonometric polynomials are most useful when the arguments are between 0 and 2π. If the θ in equation 5.34a or 5.34b goes beyond 2π the trigonometric polynomial will repeat itself. Thus, trigonometric splines must have arguments that are always in the range of 0 to 2π. This, of course, is no problem for cam designers. When we consider the curve we have in mind, the modified trapezoidal acceleration, we choose the full range. It resembles a sine function. Since, in general, we want the actual range to be from 0 to β, we scale our variable by defining

$$x = 2\pi\frac{\theta}{\beta} \qquad (5.35)$$

and let the trigonometric polynomials be of the form

$$a_0 + a_1 \cos x + a_2 \cos 2x + b_1 \sin x + b_2 \sin 2x \qquad (5.36)$$

We shall fit these so that the acceleration and the jerk are continuous at the knots which will be at

$$0, \frac{\pi}{4}, \frac{3\pi}{4}, \frac{5\pi}{4}, \frac{7\pi}{4}, 2\pi \tag{5.37a}$$

These values of x correspond to the values of θ given in Chapter 3 of

$$0, \frac{\beta}{8}, \frac{3\beta}{8}, \frac{5\beta}{8}, \frac{7\beta}{8}, \beta \tag{5.37b}$$

5

At the ends we will allow a jump in the jerk, and at the interior knots we will allow a jump in the ping. The interior knots will thus be third order. The trigonometric polynomials that we chose are fifth order, so this is not a complete spline. It is a general spline. If we were constructing it using B-splines (and trigonometric B-splines do exist), we would use multiple knots. Actually constructing this spline is quite easy because every other piece is constant. Consider just the first piece. It can be found using the left end boundary conditions and the smoothness requirements at the first interior knot. This is because the first interior knot is a constant function and we already know it. We have the equations:

$$\begin{aligned} a_0 + a_1 + a_2 &= 0 \\ a_0 + a_1 \cos\frac{\pi}{4} + b_1 \sin\frac{\pi}{4} + b_2 &= 0 \\ -a_1 \sin\frac{\pi}{4} + b_1 \cos\frac{\pi}{4} &= 0 \end{aligned} \tag{5.38}$$

The first equation says that the trigonometric polynomial is 0 at the left end. The second equation says that the trigonometric polynomial reaches maximum acceleration at the first interior knot. The third is a smoothness equation across the first interior knot. These are only three equations in all of seven unknowns and we can impose more conditions should we want, but a simple and immediate solution is

$$a_0 = a_1 = a_2 = b_1 = 0 \quad \text{and} \quad b_2 = A \tag{5.39}$$

where A is the maximum acceleration. It will have to be determined from the high dwell value h after we have an expression for the acceleration. If we formulate all of the boundary condition equations and the smoothness equations and solve, we get the following spline:

$$a(x) = \begin{cases} A\sin 2x & 0 \le x < \frac{\pi}{4} \\ A & \frac{\pi}{4} \le x < \frac{3\pi}{4} \\ -A\sin 2x & \frac{3\pi}{4} \le x < \frac{5\pi}{4} \\ -A & \frac{5\pi}{4} \le x < \frac{7\pi}{4} \\ A\sin 2x & \frac{7\pi}{4} \le x \le 2\pi \end{cases} \tag{5.40a}$$

where

$$x = 2\pi\frac{\theta}{\beta} \quad \text{and} \quad A = 4.888124\frac{h}{\beta^2} \tag{5.40b}$$

This acceleration is exactly the same as the one described in Chapter 3. If the formulas do not look the same, it is merely because these are in the trigonometric polynomial form of equation 5.34a. The maximum acceleration was determined just as in Chapter 3 by integrating twice, setting the resulting expression equal to the desired high dwell displacement, h, and solving for the maximum acceleration. Readers who glance at Chapter 3, equations 3.7, will notice that unlike the polynomial spline, the integral of a trigonometric spline is not always another trigonometric spline, but rather a spline made up of a mixture of polynomials and trigonometric functions. Accordingly, the modified trapezoidal displacement is not a trigonometric spline.

MODIFIED SINE ACCELERATION AS A SPLINE The modified sine function is also a trigonometric spline. A glance at Figure 3-6b (p. 36) shows that it is advantageous to take

$$x = 2\pi\frac{2\theta}{3\beta} = \frac{4\pi\theta}{3\beta} \tag{5.41}$$

The knots are

$$0, \frac{\pi}{6}, \frac{7\pi}{6}, \frac{4\pi}{3} \tag{5.42a}$$

corresponding to the θ values in equations 3.8a through 3.8c of

$$0, \frac{\beta}{8}, \frac{7\beta}{8}, \beta \tag{5.42b}$$

The trigonometric spline for the modified sine acceleration is then

$$a(x) = \begin{cases} A\sin 3x & 0 \le x < \dfrac{\pi}{6} \\ A\dfrac{\sqrt{3}}{2}\cos x + \dfrac{A}{2}\sin x & \dfrac{\pi}{6} \le x < \dfrac{7\pi}{6} \\ A\sin 3x & \dfrac{7\pi}{6} \le x < \dfrac{4\pi}{3} \end{cases} \tag{5.43a}$$

where x is as defined in equation 5.41 and

$$A = 5.5279571\frac{h}{\beta^2} \tag{5.43b}$$

Again, despite the fact that they look simpler, these expressions are exactly the same as those derived for the modified sine acceleration in Chapter 3.

CYCLOIDAL DISPLACEMENT AS A SPLINE As a last example, the cycloidal displacement's sine acceleration function is also a trigonometric spline even though it has no interior knots and is a single trigonometric polynomial. Just as polynomials are also splines, trigonometric polynomials are also trigonometric splines.

TRIGONOMETRIC B-SPLINES We said that there were trigonometric B-splines just as there are polynomial B-splines.

Suppose that

$$0 \le \theta_1 \le \theta_2 \le \cdots \le \theta_m \le 2\pi \tag{5.44a}$$

is a knot sequence. Then the trigonometric B-splines of order 1 are

$$T_j^1(\theta) = \begin{cases} \dfrac{1}{\sin\left[\dfrac{\theta_{j+1}-\theta_j}{2}\right]}, & \theta_j \le \theta < \theta_{j+1} \\ 0 & \text{otherwise} \end{cases} \tag{5.44b}$$

and the trigonometric B-splines of order k are

$$T_j^k(\theta) = \frac{\sin\left(\dfrac{\theta-\theta_j}{2}\right) T_j^{k-1}(\theta) + \sin\left(\dfrac{\theta_{j+k}-\theta}{2}\right) T_{j+1}^{k-1}(\theta)}{\sin\left(\dfrac{\theta_{j+k}-\theta_j}{2}\right)} \tag{5.44c}$$

The same convention on multiple knots used for the polynomial B-splines should be used in this formula. If the argument of the sine function in the denominator is zero, the whole expression is taken to be zero.

It would more than double the length of this chapter to give all the details of trigonometric B-splines. Once the reader has mastered the polynomial ones, he may wish to consult the literature, particularly papers and books by L. L. Schumaker and T. Lyche.

Rational Splines

The rational spline is made up of piecewise rational functions, i.e. polynomials divided by polynomials (ratios). Rational splines are much more powerful than polynomial splines, but with that power comes additional complexity and unwelcome features for the cam designer. If the denominator of the rational spline ever goes to zero, the curve will take off to infinity, hardly desirable for a follower displacement. Their derivatives also can be very troublesome. The rational spline that the reader probably has heard of is the NURB, or nonuniform rational B-spline. The NURB is a plane curve or a space curve with each of its coordinates being a rational spline in some parameter. The nonuniform term in the name means that the knots, which are defined for the parameter, are not uniformly spaced.* The profile (contour) of a cam, as opposed to the follower motion function, might be well described with a NURB.

Let the parameter be the cam angle θ and the (x, y) points be

* Note that most useful B-spline functions designed for cam follower motions also have nonuniform knot spacing as seen in the examples of this chapter. Perhaps we should call them NUBS as they lack only the ratios of NURBS.

$$x(\theta) = \frac{\sum_{j=1}^{n} a_j B_{mj}(\theta)}{\sum_{j=1}^{n} b_j B_{mj}(\theta)} \tag{5.45a}$$

$$y(\theta) = \frac{\sum_{j=1}^{n} c_j B_{mj}(\theta)}{\sum_{j=1}^{n} b_j B_{mj}(\theta)} \tag{5.45b}$$

the $B(\theta)$ refer to polynomial B-splines on the knot sequence

$$\theta_1 \leq \theta_2 \leq \theta_3 \leq \cdots \theta_n \tag{5.45c}$$

If the knots are not uniformly spaced, then the profile is said to be described by a NURB. NURBs have become almost standard in CAD work for designing curves and shapes. Of course, planar curves and space curves are more complicated than the functions that we have been discussing for describing follower motion. Consider the continuity condition at the knots. The derivative of a planar curve or a space curve is a vector and has two aspects, direction and magnitude. If the direction of the derivative on one side of the knot is the same as on the other side of the knot, which means that the curve has a continuous tangent, then the NURB is said to have G1 continuity. If it has G1 continuity and the magnitudes are also the same, the NURB is said to have C1 continuity. For cam profiles, C1 continuity is a must. A similar situation exists for the second derivative and the radius of curvature. Geometric continuity of the second derivative is called G2 and both geometric and magnitude continuity is called C2. A discontinuity in the radius of curvature anywhere on a cam profile can cause disastrous follower motion. Experienced cam designers can see such discontinuities in radial cams by training themselves to observe the light reflections from the cam surface.

5.8 SUMMARY

We have barely touched on all the things that there are to know about splines. Perhaps this chapter should have been titled "What Every Engineer Should Know About Splines." The reader has probably noticed that this chapter is different than most of the other chapters in this book. No "recipes" were given; nowhere in the chapter does it say what type of spline to use for what type of problem. No formulae are given so that simple substitution of parameters will always result in desirable follower motions. Instead we have given the reader the theory of splines, that spline motion is controlled by selected discontinuities in the jerk, ping, puff, and possibly higher derivatives. The engineer who masters this theory will have a higher level of competence and a deeper insight into the science of cam design. The fact is that splines are not really new; every traditional follower motion is spline motion in the sense that the motion is controlled by jumps in the higher derivatives. It can be mathematically proven that only the simplest cams produce follower motions that are not splines. Every follower motion with a dwell must be a spline motion, unless the motion is only a dwell, in which case the cam is circular in shape. What is new and what the chapter explains is spline theory. An analogy might be made with the theory of gravity. Before Newton, no one recognized any theory of gravity, but engineers most certainly used the force of gravity in the design of machinery such as pile drivers, water wheels, etc. After Newton, however, a whole new world of possibilities was realized. Spline theory is certainly not as far reaching as Newton's theory of gravitation, nevertheless, recognizing the theory of splines opens the way to many more design possibilities and to better solutions of design problems.

This chapter has used a historical approach to explaining splines. The first splines that were discussed were the classical splines. The original idea for a set of basis splines with which to build all the polynomial splines was the truncated power basis, which the chapter discussed after classical splines. The truncated power basis as a means to design follower motion was quickly displaced in practice by the B-splines which today dominate the field. We have tried to give the reader an understanding of all of these ideas and how they lead from one to the other. Except for a brief mention of trigonometric splines and NURBs, the chapter concentrated on B-splines. The use of B-splines simplifies the use of polynomial splines and expands a cam design engineer's capabilities. Even the old-fashioned polynomial function is more easily handled and applied when considered to be a B-spline or Bezier curve.

We also hope that the reader understands that splines are a product of the computer age and that good software is a necessity for up-to-date cam design. Today, the demands made on the cam designer are severe because the power of splines and of computers is recognized and appreciated. An engineer who neglects either will suffer from inferior cam designs.

5.9 BIBLIOGRAPHY

Ault, H. K., and **R. L. Norton**. (1993). "Spline Based Cam Functions for Minimum Kinetic Energy Follower Motion." *Proc. of 3rd Conference on Applied Mechanisms and Robotics*, Cincinnati OH, pp. 2.1-2.6.

Butterfield, K. R. (1976). "The Computation of All the Derivatives of a B-Spline Basis." *J. Inst. Maths Applics*, 17, pp. 15-25.

Chenard, J. W., and **T. J. Chlupsa**. (1990). "Computer Implementation of Spline-Based Cam Design." Major Qualifying Project, Worcester Polytechnic Institute.

Cox, M. G. (1972). "The Numerical Evaluation of B-Splines." *J. Inst. Maths Applics*, 10, pp. 134-149.

Cox, M. G., ed. (1981). "Practical Spline Approximation." *Lecture Notes in Mathematics* 965, Dold, A. and B. Eckmann, eds., Springer-Verlag: Berlin, pp. 79-112.

deBoor, C. (1972). "On Calculating with B-Splines." *J. Approximation Theory*, 6, pp. 50-62.

deBoor, C. (1977). "Package for Calculating with B-Splines." *SIAM J. of Numerical Analysis*, 14(3), pp. 441-472.

deBoor, C. (1978). *A Practical Guide to Splines*. Springer-Verlag: Berlin.

deBoor, C., ed. (1981). "Topics in Multivariate Approximation Theory." *Lecture Notes in Mathematics* 965, Dold, A. and B. Eckmann, eds., Springer-Verlag: Berlin, pp. 39-78.

Ge, Q. J., and **B. Ravani.** (1994). "Geometric Construction of Bézier Motions." *Journal of Mechanical Design*, 116, pp. 749-755.

Ge, Q. J., and **L. Srinivasan.** (1996). "C2 Piecewise Bezier Harmonics for Motion Specifications of High Speed Cam Mechanisms" 96-DETC/MECH-1173, ASME Design Engineering Technical Conference: Irvine, CA.

Ge, Q. J., et al. (1997). "Low Harmonic Rational Bezier Curves for Trajectory Generation of High Speed Machinery." *Computer Aided Geometric Design*, 14, pp. 251-271.

Greville, T. N. E., ed. (1969). "Introduction to Spline Functions." *Theory and Application of Spline Functions*, Greville, T. N. E., ed., Academic Press: New York, pp. 1-35.

Hoskins, W. D. (1970). "Interpolating Quintic Splines on Equidistant Knots. Algorithm 62." *The Computer J.*, 13(4), pp. 437-438.

Li, Y. M., and **V. Y. Hsu.** (1998). "Curve Offsetting Based on Legendre Series." *Computer Aided Geometric Design*, 15, pp. 711-720.

Lyche, T., and **R. Winther** (1979) "A Stable Recurrence Relation for Trigonometric B-splines," *J. Approximation Theory*, 25, 266-279.

MacCarthy, B. L. (1988). "Quintic Splines for Kinematic Design." *Computer-Aided Design*, 20(7), pp. 406-415.

MacCarthy, B. L., and **N. D. Burns.** (1985). "An Evaluation of Spline Functions for Use in Cam Design." *IMechE*, 199(C3), pp. 239-248.

McAllister, D. F., and **J. A. Roulier.** (1981). "An Algorithm for Computing a Shape-Preserving Osculatory Quadratic Spline." *ACM Trans. Mathematical Software*, 7(3), pp. 331-347.

Mosier, R. G. (2000). "Modern Cam Design." *Int. J. Vehicle Design*, 23(1/2), pp. 38-55.

Neamtu, M., et al. (1998). "Designing Nurbs Cam Profiles Using Trigonometric Splines." *Journal of Mechanical Design*, 120(2), pp. 175-180.

Nutbourne, A. W., et al. (1972). "A Cubic Spline Package: Part 1 - The User's Guide, Part 2 the Mathematics." *Computer Aided Design*, 4, 5, pp. Part 1: .228-238, Part 2: 7-13.

Pham, B. (1992). "Offset Curves and Surfaces: A Brief Survey." *Computer Aided Design*, 24, pp. 223-229.

Sanchez, M. N., and **J. G. deJalon.** (1980). "Application of B-Spline Functions to the Motion Specifications of Cams." *Proc. of ASME Design Engineering Technical Conference*, Beverly Hills, CA.

Sandgren, E., and **R. L. West.** (1990). "Shape Optimization of Cam Profiles Using B-Spline Representation." *ASME Journal of Mechanisms, Transmissions, and Automation in Design,* 111, pp. 195-201.

Schumaker, L. (1983). "On Shape Preserving Quadratic Spline Interpolation." *SIAM J. Numer. Anal.,* 20(4), pp. 854-864.

Schumaker, L. L., ed. (1969). "Some Algorithms for the Computation and Approximating Spline Functions." *Theory and Application of Spline Functions*, Greville, T. N. E., ed., Academic Press: New York, pp. 87-102.

Schumaker, L. L. (1981). *Spline Functions: Basic Theory*. John Wiley & Sons: New York.

Spath, H. (1974). *Spline Algorithms for Curves and Surfaces,* Hoskins, W. D. and H. W. Sager, translator. Utilitas Mathematica: Winnipeg, p. 198.

Srinivasan, L. N., and **Q. J. Ge.** (1996). "Parametric Continuous and Smooth Motion Interpolation." *Journal of Mechanical Design*, 116(4), pp. 494-498.

Srinivasan, L. N., and **Q. J. Ge.** (1998). "Designing Dynamically Compensated and Robust Cam Profiles with Bernstein-Bezier Harmonic Curves." *Journal of Mechanical Design*, 120(1), pp. 40-45.

Srinivasan, L. N., and **Q. J. Ge.** (1998). "Fine Tuning of Rational B-Spline Motions." *Journal of Mechanical Design*, 120(1), pp. 46-51.

Ting, K. L., et al. (1990). "Bezier Motion Programs in Cam Design." *Proc. of ASME 21st Biennial Mechanisms Conference,* Chicago, pp. 141-148.

Ting, K. L., et al. (1994). "Synthesis of Polynomial and Other Curves with the Bezier Technique." *Mechanism and Machine Theory*, 29(6), pp. 887-903.

Tsay, D. M., and **C. O. Huey.** (1986). "Cam Profile Synthesis Using Spline Functions to Approximately Satisfy Displacement, Velocity, and Acceleration Constraints" 86-DET-109, ASME: Columbus OH, Oct 5, 1986.

Tsay, D. M., and **C. O. Huey**. (1987). "Cam Motion Using Spline Functions: Part I - Basic Theory and Elementary Applications." ASME Advances in Design Automation, 2, pp. 143-150.

Tsay, D. M., and **C. O. Huey.** (1987). "Cam Motion Using Spline Functions: Part II - Applications." ASME Advances in Design Automation, 2, pp. 151-157.

Tsay, D. M., and **C. O. Huey.** (1988). "Cam Motion Synthesis Using Spline Functions." *Journal of Mechanisms, Transmissions, and Automation in Design*, 110, pp. 161-165.

Tsay, D. M., and **C. O. Huey.** (1990). "Spline Functions Applied to the Synthesis and Analysis of Non-Rigid Cam-Follower Systems." *Journal of Mechanisms, Transmissions, and Automation in Design,* 111, pp. 561-569.

Tsay, D. M., and **C. O. Huey**. (1993). "Application of Rational B-Splines to the Synthesis of Cam-Follower Motion Programs." *Journal of Mechanical Design*, 115(3), pp. 621-626.

Wang, F. C., and **D. C. H. Yang**. (1993). "Nearly Arc-Length Parameterized Quintic Spline Interpolation for Precision Machining." *Computer Aided Design*, 25(5), pp. 281-288.

Yang, D. C. H., and **F. C. Wang**. (1994). "A Quintic Spline Interpolator for Motion Command Generation of Computer-Controlled Machines." *Journal of Mechanical Design,* 116(1), pp. 226-231.

Yoon, K., and **S. S. Rao.** (1993). "Cam Motion Synthesis Using Cubic Splines." *Journal of Mechanical Design,* 115(3), pp. 441-446.

Chapter 6

CRITICAL PATH MOTION CAM CURVES

6.0 INTRODUCTION

The previous chapters discussed a number of acceptable cam functions for the critical extreme position (CEP) case, for double- and single-dwell cases, respectively. This chapter will discuss the design of custom cam functions to provide control of displacement, velocity, or acceleration over some range of motion as opposed to just the endpoints of motion. This case is referred to as critical path motion or CPM. The variables used in this chapter are the same as those defined in Table 2-1 (p. 18).

6.1 CONSTANT VELOCITY MOTION

Probably the most common application of **critical path motion** (CPM) specifications in production machinery design is the need for **constant velocity motion**. There are two general types of automated production machinery in common use, **intermittent motion** assembly machines and **continuous motion** assembly machines.

Intermittent motion assembly machines carry the manufactured goods from work station to work station, stopping the workpiece or subassembly at each station while another operation is performed upon it. The throughput speed of this type of automated production machine is typically limited by the dynamic forces that are due to accelerations and decelerations of the mass of the moving parts of the machine and its workpieces. The workpiece motion may be either in a straight line (as on a conveyor) or in a circle (as on a rotary table).

Continuous motion assembly machines never allow the workpiece to stop and thus are capable of higher throughput speeds. All operations are performed on a moving target. Any tools which operate on the product have to "chase" the moving assembly line to do their job. Since the assembly line (often a conveyor belt or chain, or a rotary table) is moving at some constant velocity, there is a need for mechanisms to provide constant velocity motion, matched exactly to the conveyor, in order to carry the tools alongside for a long enough time to do their job. These cam driven "chaser" mechanisms must then return the tool quickly to its start position in time to meet the next part or subassembly on the conveyor (quick-return). There is a motivation in manufacturing to convert from intermittent motion machines to continuous motion in order to increase production rates. Thus there is considerable demand for this type of constant velocity mechanism. The cam-follower system is well suited to this problem, and the polynomial cam function is particularly adaptable to the task.

Polynomials Used for Critical Path Motion

EXAMPLE 6-1

Designing a Polynomial for Constant Velocity Critical Path Motion.

Problem: Consider the following statement of a critical path motion (CPM) problem:

Accelerate the follower from zero to 10 in/sec
Maintain a constant velocity of 10 in/sec for 0.5 sec
Decelerate the follower to zero velocity
Return the follower to start position
Cycle time exactly 1 sec

Solution:

1 This unstructured problem statement is typical of real design problems. No information is given as to the means to be used to accelerate or decelerate the follower or even as to the portions of the available time to be used for those tasks. A little reflection will cause the engineer to recognize that the specification on total cycle time in effect defines the camshaft velocity to be its reciprocal or one revolution per second. Converted to appropriate units, this is an angular velocity of 2π rad/sec.

2 The constant velocity portion uses half of the total period of 1 sec in this example. The designer must next decide how much of the remaining 0.5 sec to devote to each other phase of the required motion.

3 The problem statement seems to imply that four segments are needed. Note that the designer has to somewhat arbitrarily select the lengths of the individual segments (except the constant velocity one). Some iteration may be required to optimize the result. Program DYNACAM makes the iteration process quick and easy, however.

4 Assuming four segments, the timing diagram in Figure 6-1 shows an acceleration phase, a constant velocity phase, a deceleration phase, and a return phase, labeled as segments 1 through 4.

5 The segment angles (β's) are assumed, for a first approximation, to be 30° for segment 1, 180° for segment 2, 30° for segment 3, and 120° for segment 4 as shown in Figure 6-1. These angles may need to be adjusted in later iterations, except for segment 2, which is rigidly constrained in the specifications.

Portions of this chapter were adapted from R. L. Norton, *Design of Machinery* 2ed, McGraw-Hill, 2001, with permission.

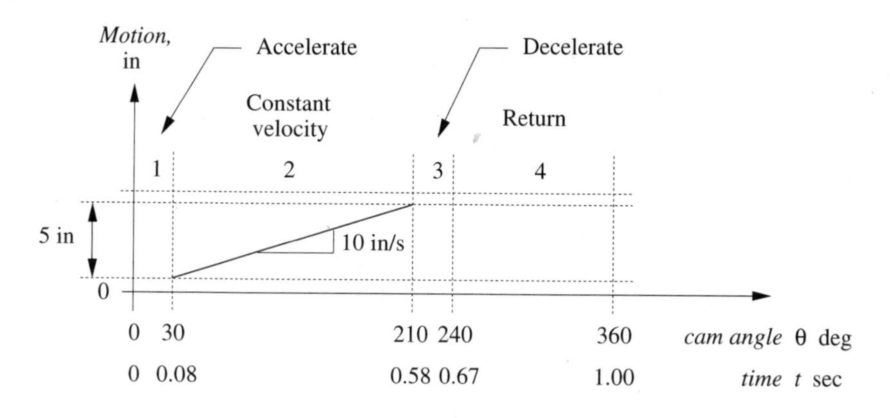

FIGURE 6-1

One possible constant velocity cam timing diagram

6 Figure 6-2 shows a tentative *S V A J* diagram. The solid circles indicate a set of boundary conditions that will constrain the continuous function to these specifications. These are for segment 1:

when $\theta = 0°$	$S = 0$	$V = 0$	*none*
when $\theta = 30°$	*none*	$V = 10$	$A = 0$

7 Note that the displacement at $\theta = 30°$ is left unspecified. The resulting polynomial function will provide us with the values of displacement at that point, which can then be used as a boundary condition for the next segment, in order to make the overall functions continuous as required. The acceleration at $\theta = 30°$ must be zero in order to match that of the constant velocity segment 2. The acceleration at $\theta = 0$ is left unspecified. The resulting value will be used later to match the end of the last segment's acceleration.

8 Putting these four BCs for segment 1 into program DYNACAM yields a cubic function whose *S V A J* plots are shown in Figure 6-3. Its equation is:

$$S = 0.83376\left(\frac{\theta}{\beta}\right)^2 - 0.27792\left(\frac{\theta}{\beta}\right)^3 \tag{6.1a}$$

The maximum displacement of 0.556 in occurs at $\theta = 30°$. This will be used as one BC for segment 2. The entire set for segment 2 is:

when $\theta = 30°$	$S = 0.556$	$V = 10$
when $\theta = 210°$	*none*	*none*

9 Note that in the derivations and in the DYNACAM program each segment's local angles run from zero to the β for that segment. Thus, segment 2's local angles are 0° to 180°, which correspond to 30° to 210° globally in this example. We have left the displacement, velocity, and acceleration at the end of segment 2 unspecified. They will be determined by the computation.

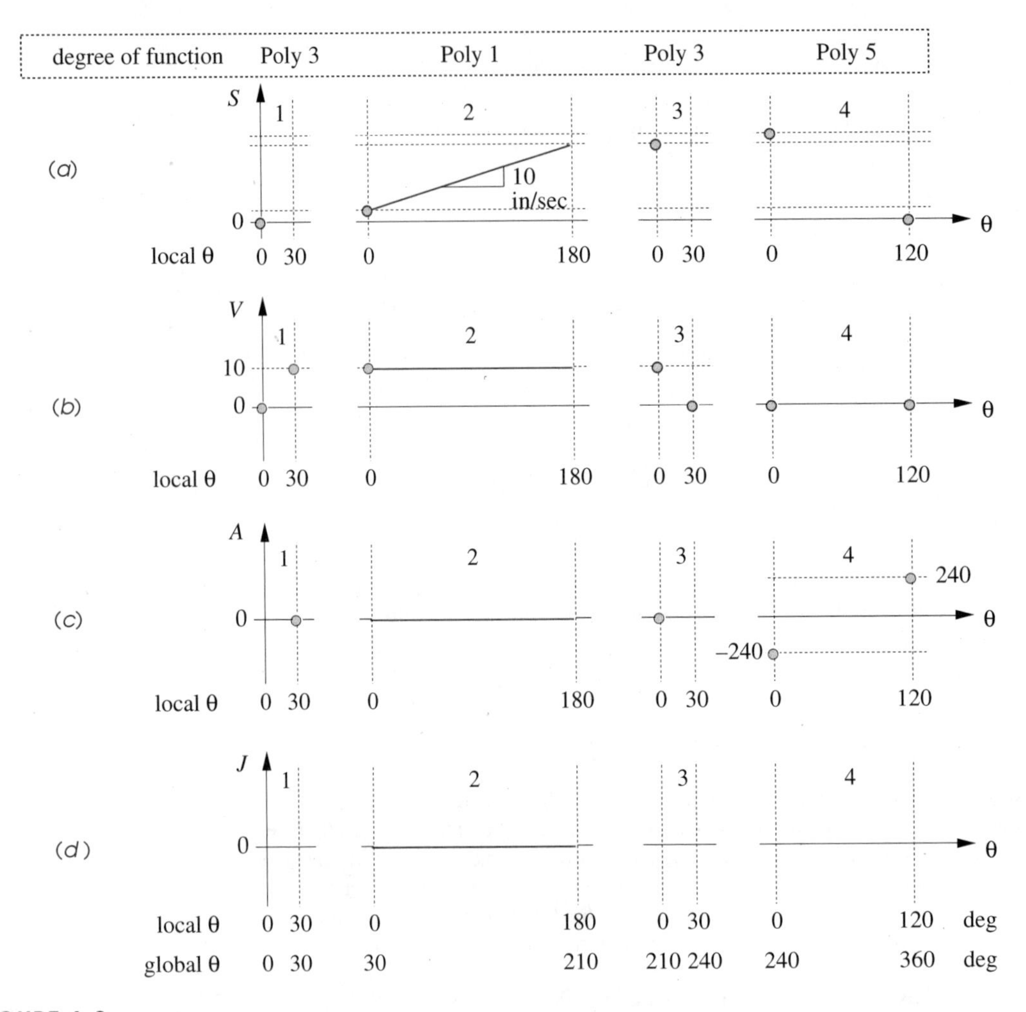

FIGURE 6-2

A possible set of boundary conditions for the 4-segment constant velocity solution

10 Since this is a constant velocity segment, its integral, the displacement function, must be a polynomial of degree one, i.e., a straight line. If we specify more than two BCs we will get a function of higher degree than one that will pass through the specified endpoints but which may also oscillate between them and deviate from the desired constant velocity. Thus we can *only* provide two BCs, a slope and an intercept, as defined in equation 2.2 (p. 21). But, we must provide at least one displacement boundary condition in order to compute the coefficient C_0 from equation 2.2 (p. 21). Specifying the two BCs at only one end of the interval is perfectly acceptable. The equation for segment 2 is:

$$S = 5\left(\frac{\theta}{\beta}\right) + 0.556 \tag{6.1b}$$

11 Figure 6-4 shows the displacement and velocity plots of segment 2. The acceleration and jerk are both zero. The resulting displacement at $\theta = 210°$ is 5.556.

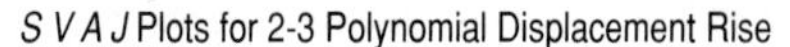

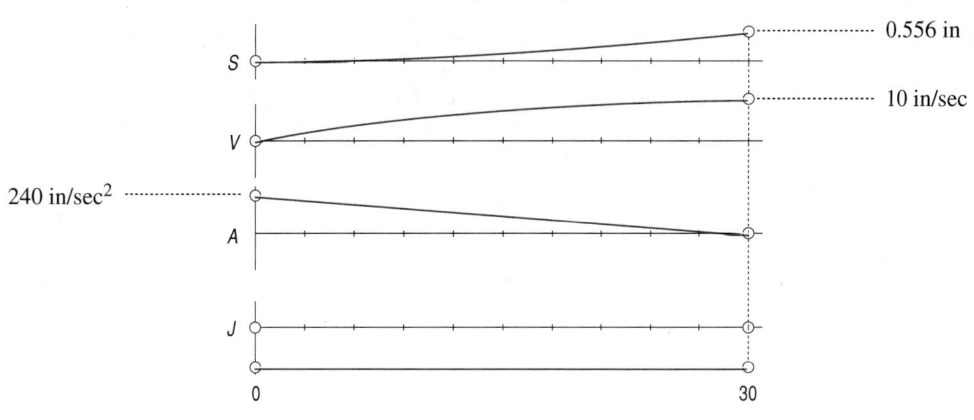

FIGURE 6-3

Segment 1 for the 4-segment solution to the constant velocity problem

12 The displacement at the end of segment 2 is now known from its equation. The four boundary conditions for segment 3 are then:

when $\theta = 210°$	$S = 5.556$	$V = 10$	$A = 0$
when $\theta = 240°$	*none*	$V = 0$	*none*

13 This generates a cubic displacement function as shown in Figure 6-5. Its equation is:

$$S = -0.27792\left(\frac{\theta}{\beta}\right)^3 + 0.83376\left(\frac{\theta}{\beta}\right) + 5.556 \tag{6.1c}$$

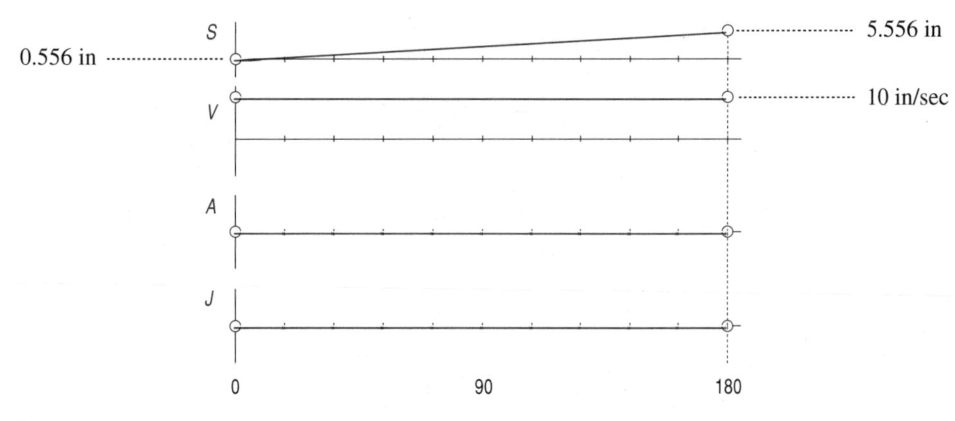

FIGURE 6-4

Segment 2 for the 4-segment solution to the constant velocity problem

14 The boundary conditions for the last segment 4 are now defined, as they must match those of the end of segment 3 and the beginning of segment 1. The displacement at the end of segment 3 is found from the computation in DYNACAM to be $S = 6.112$ at $\theta = 240°$ and the acceleration at that point is –240 in/sec^2. We left the acceleration at the beginning of segment 1 unspecified. From the second derivative of the equation for displacement in that segment we find that the acceleration is 240 in/sec^2 at $\theta = 0°$. The BCs for segment 4 are then:

$$\text{when} \quad \theta = 240° \qquad s = 6.112 \qquad v = 0 \qquad a = -240$$

$$\text{when} \quad \theta = 360° \qquad s = 0 \qquad v = 0 \qquad a = 240$$

15 The equation for segment 4 is then:

$$s = -9.9894\left(\frac{\theta}{\beta}\right)^5 + 24.9735\left(\frac{\theta}{\beta}\right)^4 - 7.7548\left(\frac{\theta}{\beta}\right)^3 - 13.3413\left(\frac{\theta}{\beta}\right)^2 + 6.112 \tag{6.1d}$$

16 Figure 6-5 shows the *S V A J* plots for the complete cam. It obeys the fundamental law of cam design because the piecewise functions are continuous through the acceleration. The maximum value of acceleration is 257 in/sec^2. The maximum negative velocity is –29.4 in/sec. We now have four piecewise and continuous functions, equations 6.1, that will meet the performance specifications for this problem.

The reader may open the file Ex_06-01.cam in program DYNACAM to investigate this example in more detail.

While this design is acceptable, it can be improved. One useful strategy in designing polynomial cams is to minimize the number of segments, provided that this does not

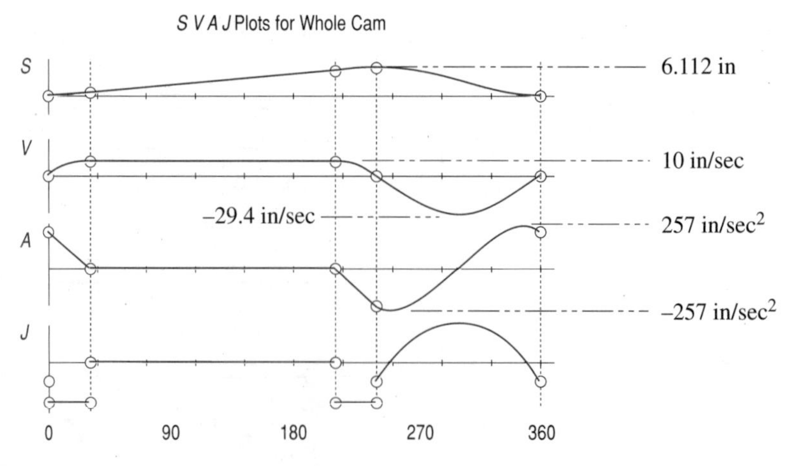

FIGURE 6-5

Four-segment solution to the constant velocity problem showing maximum values result in functions of such high degree that they misbehave between boundary conditions. Another strategy is to always start with the segment for which you have the most information. In this example, the constant velocity portion is the most constrained and must be a separate segment, just as a dwell must be a separate segment. The rest of the cam motion exists only to return the follower to the constant velocity segment for the next cycle. If we start by designing the constant velocity segment, it may be possible to complete the cam with only one additional segment. We will now redesign this cam, to the same specifications but with only two segments, as shown in Figure 6-6.

EXAMPLE 6-2

Designing an Optimum Polynomial for Constant Velocity Critical Path Motion.

Problem: Redefine the problem statement of Example 6-1 to have only two segments.

Maintain a constant velocity of 10 in/sec for 0.5 sec
Decelerate to zero velocity and **accelerate** to constant velocity
Cycle time exactly 1 sec

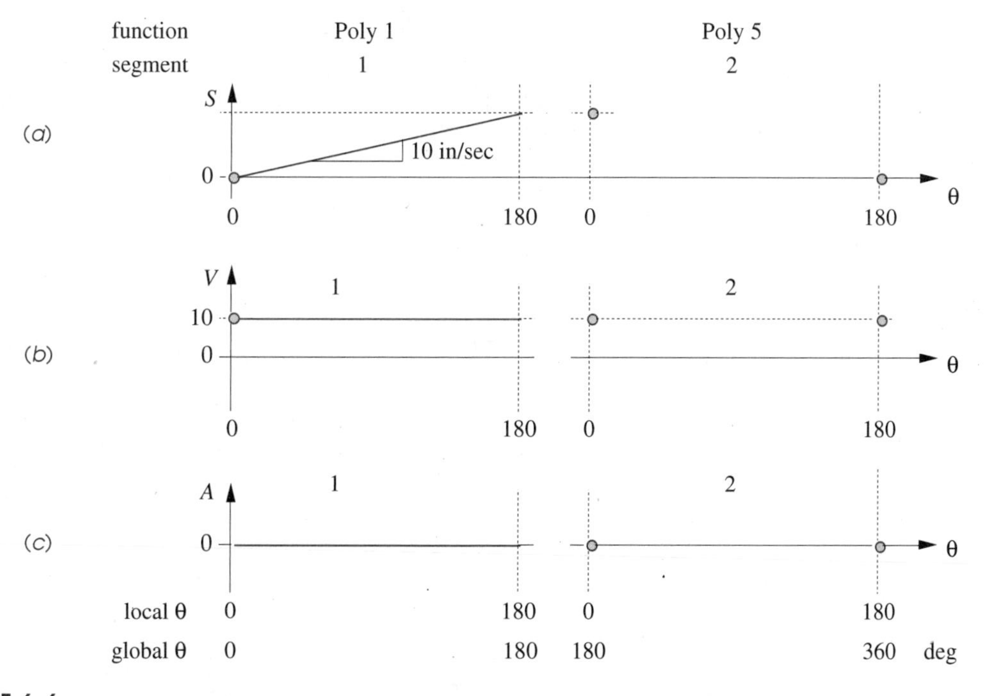

FIGURE 6-6

Boundary conditions for the 2-segment constant velocity solution

Solution:

1 The BCs for the first segment, constant velocity, will be similar to our previous solution except for the global values of its angles and the fact that we will start at zero displacement. They are:

when $\theta = 0°$	$S = 0$	$V = 10$
when $\theta = 180°$	*none*	*none*

2 The displacement and velocity plots for this segment are identical to those in Figure 6-4 (p. 131) except that the displacement starts at zero. The equation for segment 1 is:

$$s = 5\left(\frac{\theta}{\beta}\right) \tag{6.2a}$$

3 The program calculates the displacement at the end of segment 1 to be 5.00 in. This defines that BC for segment 2. The set of BCs for segment 2 is then:

when $\theta = 180°$	$S = 5.00$	$V = 10$	$A = 0$
when $\theta = 360°$	$S = 0$	$V = 10$	$A = 0$

The equation for segment 2 is:

$$s = -60\left(\frac{\theta}{\beta}\right)^5 + 150\left(\frac{\theta}{\beta}\right)^4 - 100\left(\frac{\theta}{\beta}\right)^3 + 5\left(\frac{\theta}{\beta}\right)^1 + 5 \tag{6.2b}$$

4 The *S V A J* diagrams for this design are shown in Figure 6-7. Note that they are much smoother than the 4-segment design. The maximum acceleration in this example is now 230 in/sec^2, and the maximum negative velocity is –27.5 in/sec. These are both less than in the previous design of Example 6-1.

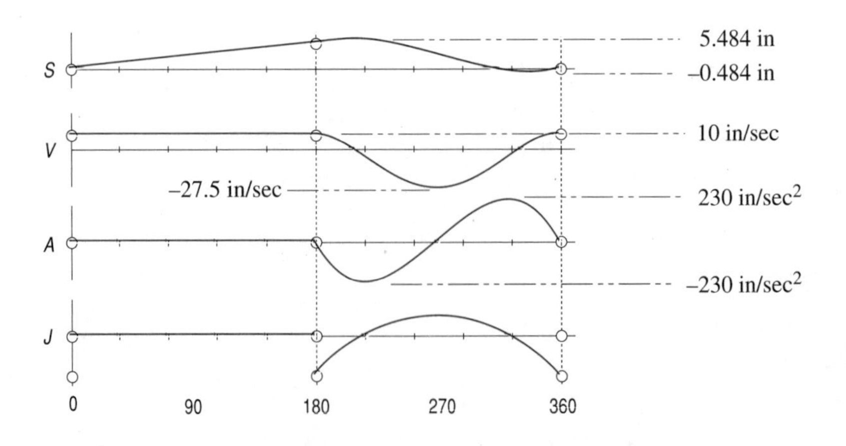

FIGURE 6-7

Two-segment solution to the constant velocity problem showing maximum values

5 The fact that our displacement in this design contains negative values as shown in the *S* diagram of Figure 6-7 is of no real concern. This is due to our starting with the beginning of the constant velocity portion as zero displacement. The follower has to go to a negative position in order to have distance to accelerate up to speed again. We will simply shift the displacement coordinates by that negative amount to make the cam. To do this, simply calculate the displacement coordinates for the cam. Note the value of the largest negative displacement. Add this value to the displacement boundary conditions for all segments and recalculate the cam functions with DYNACAM. (Do not change the BCs for the higher derivatives.) The finished cam's displacement profile will be shifted up such that its minimum value will now be zero.

Not only do we now have a smoother cam, but the dynamic forces and stored kinetic energy are both lower. Note that we did not have to make any assumptions about the portions of the available nonconstant velocity time to be devoted to speeding up or slowing down. This all happened automatically from our choice of only two segments and the specification of the minimum set of necessary boundary conditions. This is clearly a superior design to the previous attempt and is, in fact, an optimal polynomial solution to the given specifications. The reader is encouraged to read the file Ex_06-02.cam into program DYNACAM to investigate this example in more detail.

Half-Period Harmonic Family Functions

The full-rise cycloidal and the full-rise modified sine functions are generally suited only to the double-dwell cases as they have zero acceleration at each end. However, pieces of these functions can be used to match other functions, such as constant velocity segments, in a similar fashion to that used to piece together the modified sine from two harmonics of different frequency. The full-rise functions mentioned above (except simple harmonic) contain one complete period in their velocity and acceleration. The simple harmonic function (Figure 2-1, p. 18) does not have zero acceleration at its ends, but the modified sine's and cycloid's accelerations (Figure 3-7, p. 39) do start and end at zero. In order to match a nonzero velocity, as in the example above, we could use half of any of these harmonic family functions and design them to match the desired constant velocity of the adjacent segment.

As an example of this approach, Figure 6-8 shows the *S V A J* functions for a **half-cycloid** rise function #1 that has zero velocity at the beginning of the interval and nonzero velocity at the end. Note that its displacement starts at zero and ends at some positive value, but its acceleration is zero at both extremes. This makes it possible to mate this function to a constant velocity segment and match both the desired velocity and its zero acceleration at the boundary. The total displacement required of the half-cycloid will "come out in the wash" when the required boundary conditions of velocity and duration are applied to the particular case. The equations for this half-cycloid #1 are:

$$s = L\left[\frac{\theta}{\beta} - \frac{1}{\pi}\sin\left(\pi\frac{\theta}{\beta}\right)\right] \tag{6.3a}$$

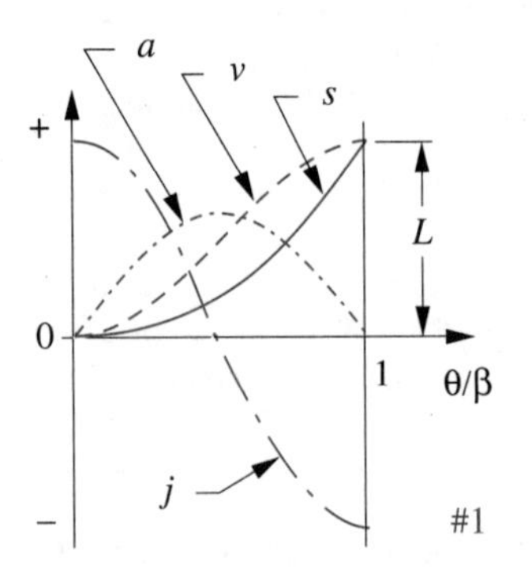

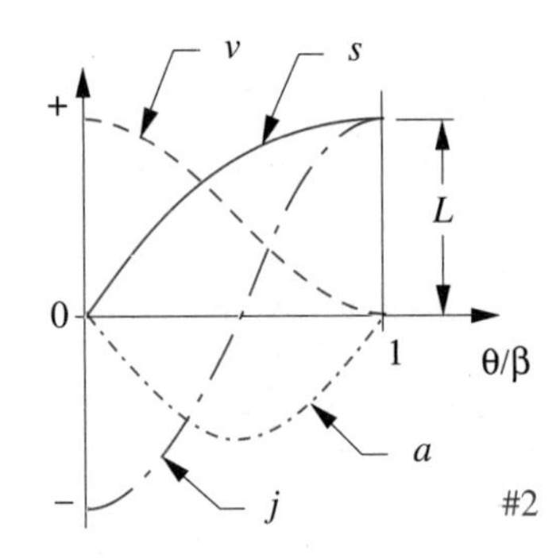

6

FIGURE 6-8

Half-cycloidal functions for use on a rise segment

$$v = \frac{L}{\beta}\left[1 - \cos\left(\pi\frac{\theta}{\beta}\right)\right] \tag{6.3b}$$

$$a = \frac{\pi L}{\beta^2}\sin\left(\pi\frac{\theta}{\beta}\right) \tag{6.3c}$$

$$j = \frac{\pi^2 L}{\beta^3}\cos\left(\pi\frac{\theta}{\beta}\right) \tag{6.3d}$$

Figure 6-8 also shows the *S V A J* functions for the **half-cycloid** rise function #2 with nonzero velocity at the beginning of the interval and zero velocity at the end. The equations for this half-cycloid #2 are:

$$s = L\left[\frac{\theta}{\beta} + \frac{1}{\pi}\sin\left(\pi\frac{\theta}{\beta}\right)\right] \tag{6.4a}$$

$$v = \frac{L}{\beta}\left[1 + \cos\left(\pi\frac{\theta}{\beta}\right)\right] \tag{6.4b}$$

$$a = -\frac{\pi L}{\beta^2}\sin\left(\pi\frac{\theta}{\beta}\right) \tag{6.4c}$$

$$j = -\frac{\pi^2 L}{\beta^3}\cos\left(\pi\frac{\theta}{\beta}\right) \tag{6.4d}$$

For a fall instead of a rise, subtract the rise displacement expressions from the total rise L and negate all the higher derivatives.

To fit these functions to a particular constant velocity situation, solve either equation 6.3b or 6.4b (depending on which function is desired) for the value of L that results from the specification of the known constant velocity V to be matched at $\theta = \beta$ or $\theta = 0$ You will have to choose a value of β for the interval of this half-cycloid that is appropriate to the problem. In this example, the value of $\beta = 30°$ used for the first segment of the four-piece polynomial could be tried as a first iteration or another value could be chosen. Once L and β are known, all the functions are defined.

The same approach can be taken with the **modified sine** and the **simple harmonic** functions. Either half of their full-rise functions can be sized to match with a constant velocity segment. The half-modified sine function mated with a constant velocity segment has the advantage of low peak velocity, useful with large inertia loads. When matched to a constant velocity, the half simple harmonic has the same disadvantage of infinite jerk as its full-rise counterpart does when matched to a dwell, so it is not recommended.

We will now solve the previous constant velocity problem of Example 6-1 using half-cycloid, constant velocity, and full-fall modified sine functions.

EXAMPLE 6-3

Using Half-Cycloids to Match Constant Velocity Critical Path Motion.

Problem: Consider the same problem statement as Example 6-1.

Accelerate	the follower from zero to 10 in/sec
Maintain	a constant velocity of 10 in/sec for 0.5 sec
Decelerate	the follower to zero velocity
Return	the follower to start position
Cycle time	exactly 1 sec

Solution:

1 We must express the specified constant velocity in units of length per rad. The angular velocity is 2π rad/sec.

$$v = 10\frac{\text{in}}{\text{sec}}\left(\frac{1 \text{ sec}}{2\pi \text{ rad}}\right) = \frac{5}{\pi}\frac{\text{in}}{\text{rad}} \tag{a}$$

2 The constant velocity portion uses half of the total period of 1 sec, or π rad, in this example. The designer must decide how much of the remaining 0.5 sec to devote to each other phase of the required motion. The segment angles (β's) are assumed for a first approximation to be 25° for segment 1, 180° for segment 2, 25° for segment 3, and 130° for segment 4. These angles may need to be adjusted in later iterations to balance and minimize the accelerations (except for segment 2 which is rigidly constrained in the specifications).

3 The segments will consist of:

Segment	β (deg)	β (rad)	Function	Motion
1	25	0.43633	Half-cycloid #1	Rise
2	180	3.14159	1° polynomial	Rise
3	25	0.43633	Half-cycloid #2	Rise
4	130	2.26893	Modified sine	Fall

4 To determine the total rise L of half-cycloid #1 needed to match the specified constant velocity, solve equation 6.3b for L at $\theta = \beta$ where it must match the constant velocity V.

$$L = \frac{\beta v}{1 - \cos\left(\pi \frac{\theta}{\beta}\right)} = \frac{0.43633\left(\frac{5}{\pi}\right)}{1 - \cos\left(\pi \frac{\beta}{\beta}\right)} = 0.34722 \tag{b}$$

5 Substitute this value of L in equation 6.3a to get the displacement for the first segment:

$$s = 0.3472\left[\frac{\theta}{\beta} - \frac{1}{\pi}\sin\left(\pi \frac{\theta}{\beta}\right)\right] \tag{c}$$

6 The constant velocity segment is found in the same way as was done in Example 6-1 (p. 128). The initial displacement for segment 2 in this case is the value of L, and the equation for segment 2 is:

$$s = 5\left(\frac{\theta}{\beta}\right) + 0.3472 \tag{d}$$

The total lift within this segment is 5 in as before.

7 Segment 3 is half-cycloid #2. Its coefficient L is identical to that of segment 1 because we used the same β for both. But, we must offset it by the sum of the displacement of segments 1 and 2 or $L + 5$. The lift for this segment is 0.3472 and the offset is 5.3472. The equation for segment 3 (from equation 6.4a) is then:

$$s = 0.3472\left[\frac{\theta}{\beta} + \frac{1}{\pi}\sin\left(\pi \frac{\theta}{\beta}\right)\right] + 5.3472 \tag{e}$$

8 Segment 4 is a full-period modified sine to return the follower from its maximum displacement of $h = 0.3472 + 5.0 + 0.3472 = 5.6944$. See equations 3.8 (p. 37).

9 The complete set of data needed to compute these functions in (or out of) DYNACAM is:

Segment	β (deg)	β (rad)	Function	Start (in)	Motion	Move (in)
1	25	0.43633	Half-cycloid #1	0	Rise	0.3472
2	180	3.14159	1° polynomial	0.3472	Rise	5.0000
3	25	0.43633	Half-cycloid #2	5.3472	Rise	0.3472
4	130	2.26893	Modified sine	5.6944	Fall	5.6944

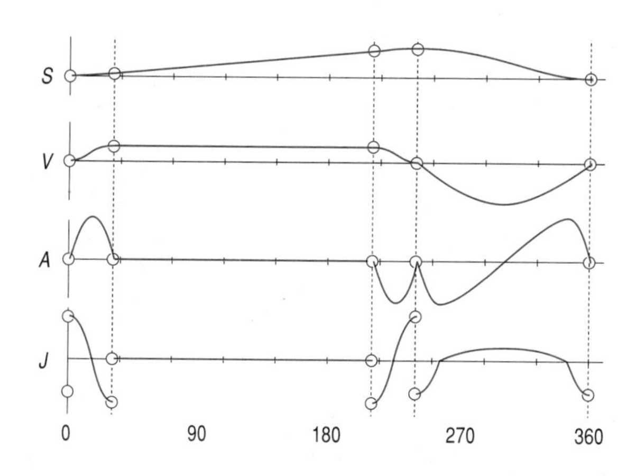

FIGURE 6-9

Half-cycloidal functions used as transitions to constant velocity

10 The resulting *S V A J* diagrams are shown in Figure 6-9. The peak acceleration is 241 in/sec^2 and peak velocity is –28 in/sec.

These results are nearly as low as the values from the two-segment polynomial solution in Example 6-2 (p. 133). The factor that makes this an inferior cam design to that of Example 6-2 is the unnecessary returns to zero in the acceleration waveform. This creates a more "ragged" or rough jerk function which will increase vibration problems. The polynomial approach of Example 6-2 is superior to the other solutions so far presented in this case. The reader may open the file Ex_06-03.cam in program DYNACAM to investigate this example in more detail.

6.2 COMBINED DISPLACEMENT AND VELOCITY CONSTRAINTS

Many machine design situations demand some particular combination of displacement, velocity, and acceleration constraints over one or more segments of motion. These often are the most difficult cam design problems encountered in practice. This subject is best addressed with some examples taken from actual machine designs of the author's experience.

The previous three examples addressed the case of a constant velocity segment in combination with a CEP return motion. The next example involves a combination of constant velocity and specified displacement constraints.

EXAMPLE 6-4

Constant Velocity Critical Path Motion With Specified Displacements.

Problem: The tooling for one station of a production machine is required to have the following timing diagram specifications as shown in Figure 6-10.

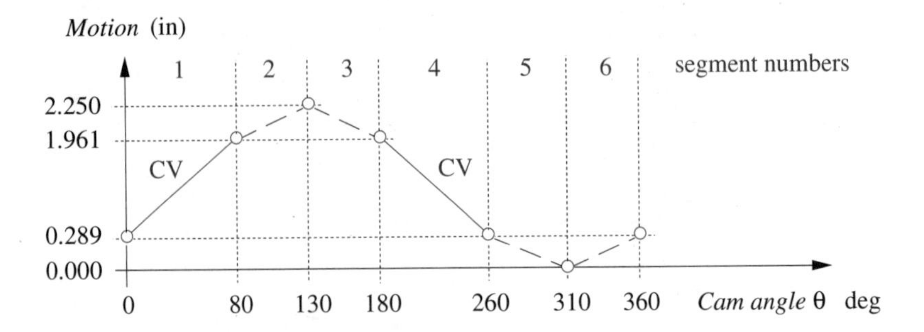

FIGURE 6-10

Timing diagram for Example 6-4

Maintain	a positive constant velocity over 0 – 80° from a displacement of 0.289 in to 1.961 in
Achieve	a maximum displacement of 2.25 in at 130°
Maintain	a negative constant velocity over 180 – 260° from a displacement of 1.961 to 0.289 in
Achieve	a minimum displacement of 0 in at 310°
Return	the follower to start position
Cam speed	250 rpm

Solution:

1 Note that the specification does not start at a zero displacement. As was seen in Example 6-2, it makes sense to start with one of the constant velocity segments, which are fully defined.

2 This problem naturally divides into four segments: the two constant velocity ones and the two others that connect them. We will start by attempting to design four polynomials.

3 Segment 1 is constant velocity, so requires only two BCs; $S = 0.289$ at 0° and $S = 1.961$ at 80°. These yield the linear polynomial:

$$s = 0.289 + 1.672\frac{\theta}{\beta} \tag{a}$$

At 250 rpm, the constant velocity of this segment is 31.4 in/sec.

4 Segment 2 needs seven BCs; $S = 1.961$, $V = 31.4$, $A = 0$ at 80° (local $\theta/\beta = 0$), $S = 2.250$ at 130° (local $\theta/\beta = 0.5$), and $S = 1.961$, $V = -31.4$, $A = 0$ at 180° (local $\theta/\beta = 1$). It is not necessary to set $V = 0$ at 130° due to symmetry. These yield the 6th-degree polynomial:

$$s = 1.961 + 2.093\frac{\theta}{\beta} - 27.557\left(\frac{\theta}{\beta}\right)^3 + 72.205\left(\frac{\theta}{\beta}\right)^4 - 70.112\left(\frac{\theta}{\beta}\right)^5 + 23.371\left(\frac{\theta}{\beta}\right)^6 \tag{b}$$

5 Segment 3 is the second constant velocity piece and requires two BCs; $S = 1.961$ at180° and $S = 0.289$ at 260°. These yield the linear polynomial:

$$s = 1.961 - 1.672\frac{\theta}{\beta} \tag{c}$$

At 250 rpm, the constant velocity of this segment is –31.4 in/sec.

6 Segment 4 requires 7 BCs, S =0.289, V = –31.4, A = 0 at 260° (local θ/β = 0), S = 0 at 310° (local θ/β = 0.5) , and S =0.289, V = 31.4, A = 0 at 360° (local θ/β = 1). It is not necessary to set V = 0 at 130° due to symmetry. These yield the 6th-degree polynomial:

$$s = 0.289 - 2.093\frac{\theta}{\beta} + 27.557\left(\frac{\theta}{\beta}\right)^3 - 72.205\left(\frac{\theta}{\beta}\right)^4 + 70.112\left(\frac{\theta}{\beta}\right)^5 - 23.371\left(\frac{\theta}{\beta}\right)^6 \tag{d}$$

7 Putting these together gives the *S V A J* diagrams of Figure 6-11. It is not very apparent at the small scale of this displacement plot, but the function both overshoots and undershoots the desired limits as can be seen in Figure 6-12. This can be confirmed from the values printed in the *Max* and *Min* boxes to the right of the displacement plot. It reaches 2.28 in and drops to –0.03 in. The sign reversals in the velocity and acceleration plots also are

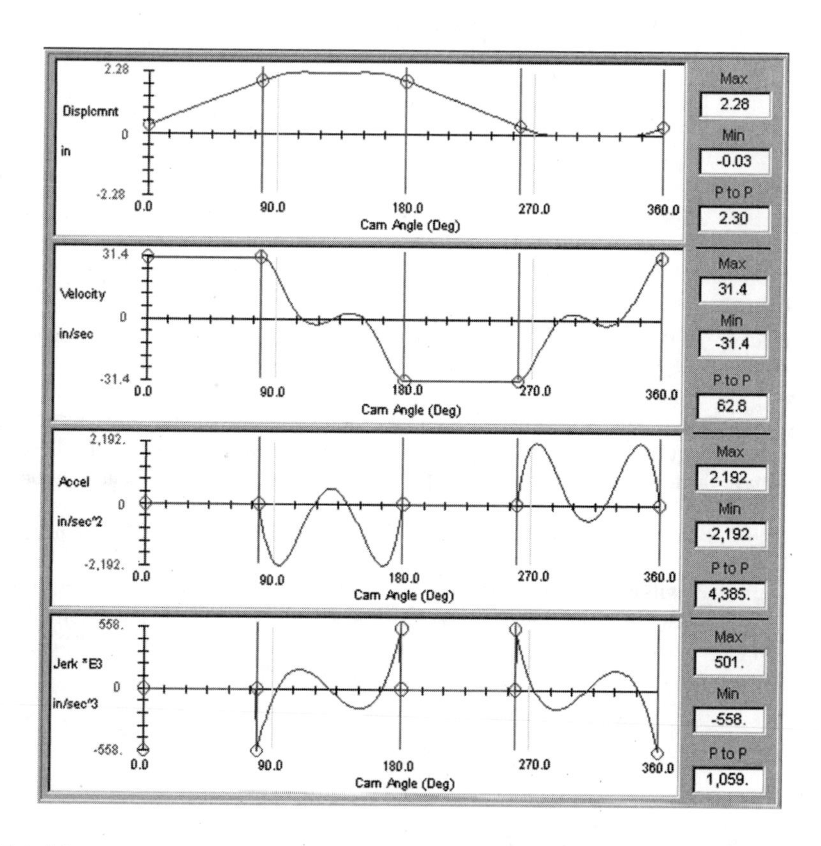

FIGURE 6-11

S V A J diagrams for a 4-segment polynomial (unsuccessful) solution to Example 6-4

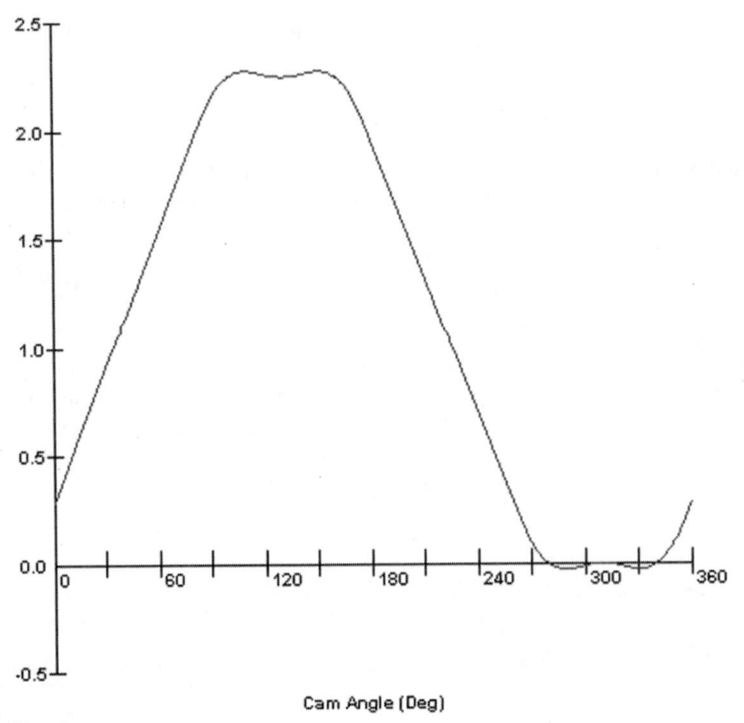

FIGURE 6-12

Displacement diagram of a 4-segment polynomial for Example 6-4 shows overshoot and undershoot

evidence of the overshoot and undershoot. This is an unacceptable design for this problem.

8 As we did in Example 4-5, we can attempt to add segments, splitting the nonconstant velocity portions in two in the hope that the lower order polynomials that result may behave themselves and not overshoot the target. So, let's try a 6-segment solution.

9 Segment 1 will be identical to that of the previous attempt—see step 3 and equation (a).

10 Segment 2 will now run only from 80° to 130° and will need only 5 BCs: S =1.961, V = 31.4, A = 0 at 80°, S = 2.250, V = 0 at 130°. Acceleration will be left unspecified at 130° and its resulting value used as a BC on the next segment. These yield the 4th-degree polynomial:

$$s = 1.961 + 1.047\frac{\theta}{\beta} - 1.984\left(\frac{\theta}{\beta}\right)^3 + 1.226\left(\frac{\theta}{\beta}\right)^4 \tag{e}$$

The acceleration at the end of this segment is 2531 in/sec^2.

11 The new segment 3 runs from 130° to 180° and has 6 BCs: S =2.250, $V = 0$, $A = 2531$ at 130° to match the previous segment, $S = 1.961$, $V = -31.4$, $A = 0$ at 180°. These yield the 5th-degree polynomial:

$$s = 2.250 + 1.406\left(\frac{\theta}{\beta}\right)^2 - 2.922\left(\frac{\theta}{\beta}\right)^3 + 1.227\left(\frac{\theta}{\beta}\right)^4 + 0\left(\frac{\theta}{\beta}\right)^5 \qquad \text{(f)}$$

Note that the 5th-degree term has a zero coefficient, making it actually a quartic function.

12 Segment 4 is the negative constant velocity portion and is identical to segment 3 of the previous design—see step 5 and equation (*c*).

13 Segment 5 runs from 260° to 310° and will need only 5 BCs: S = 0.289, $V = -31.4$, $A = 0$ at 260°, S = 0, $V = 0$ at 310°. Acceleration will be left unspecified at 310° and its resulting value used as a BC on the next segment. These yield the 4th-degree polynomial:

$$s = 0.289 - 1.047\frac{\theta}{\beta} + 1.984\left(\frac{\theta}{\beta}\right)^3 - 1.226\left(\frac{\theta}{\beta}\right)^4 \qquad \text{(g)}$$

The acceleration at the end of this segment is –2531 in/sec^2.

14 Segment 6 runs from 310° to 360° and has 6 BCs: S = 0, $V = 0$, $A = -2531$ at 310° to match the previous segment, S = 0.289, $V = 31.4$, $A = 0$ at 360°. These yield the 5th-degree polynomial:

$$s = -1.406\left(\frac{\theta}{\beta}\right)^2 + 2.922\left(\frac{\theta}{\beta}\right)^3 - 1.227\left(\frac{\theta}{\beta}\right)^4 + 0\left(\frac{\theta}{\beta}\right)^5 \qquad \text{(h)}$$

Note that the 5th-degree term has a zero coefficient, making it actually a quartic function.

15 Figure 6-13 shows the displacement function for this design. It also has overshoot and undershoot. In fact, the error is greater than that of the previous 4-segment design, rising to 2.32 in and falling to –0.07 in. This is also an unacceptable solution to this problem.

16 Polynomials cannot solve this problem. It is an example that requires the use of spline functions. We will now design it using B-splines for the nonconstant velocity segments.

17 We return to the 4-segment approach. The constant velocity segments (1 and 3) are identical to both previous designs as defined in steps 3 and 5.

18 Segment 2 will use a B-spline having eight boundary conditions, the same seven as the polynomial of step 4 plus $V = 0$ at the midpoint. We now have the additional degrees of freedom associated with knot locations.

19 Figure 6-14 shows the *S V A J* splines that result from the default condition of evenly spaced interior knots for a 6th-order B-spline. Since we specified eight BCs, a 6th-order B-spline will have two interior knots available for manipulation. This number comes about from the fact that the total number of knots is the sum of the number of BCs and the order of the spline chosen—here 8 + 6 = 14. A B-spline requires as many knots at each end as its order, so a 6th-order B-spline uses 12 knots at its segment ends. With 14 total knots in

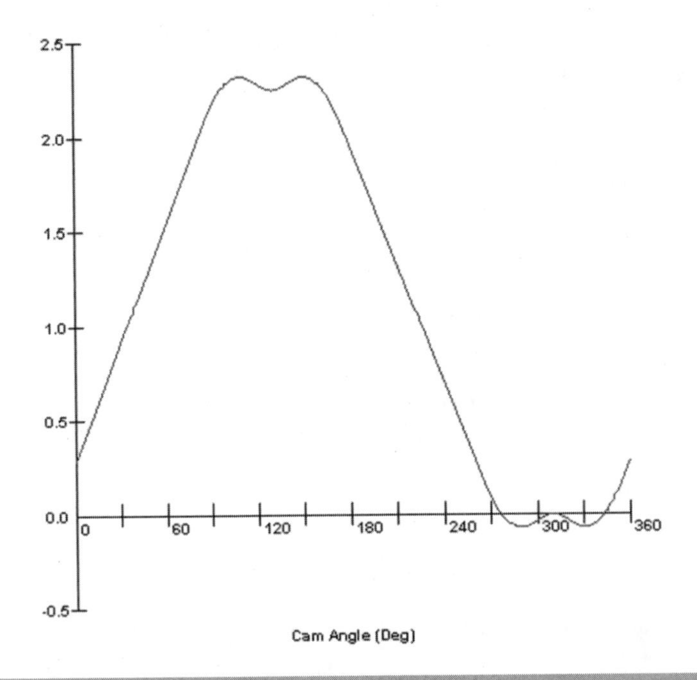

6

FIGURE 6-13

Displacement diagram of a 6-segment polynomial for Example 6-4 shows overshoot and undershoot

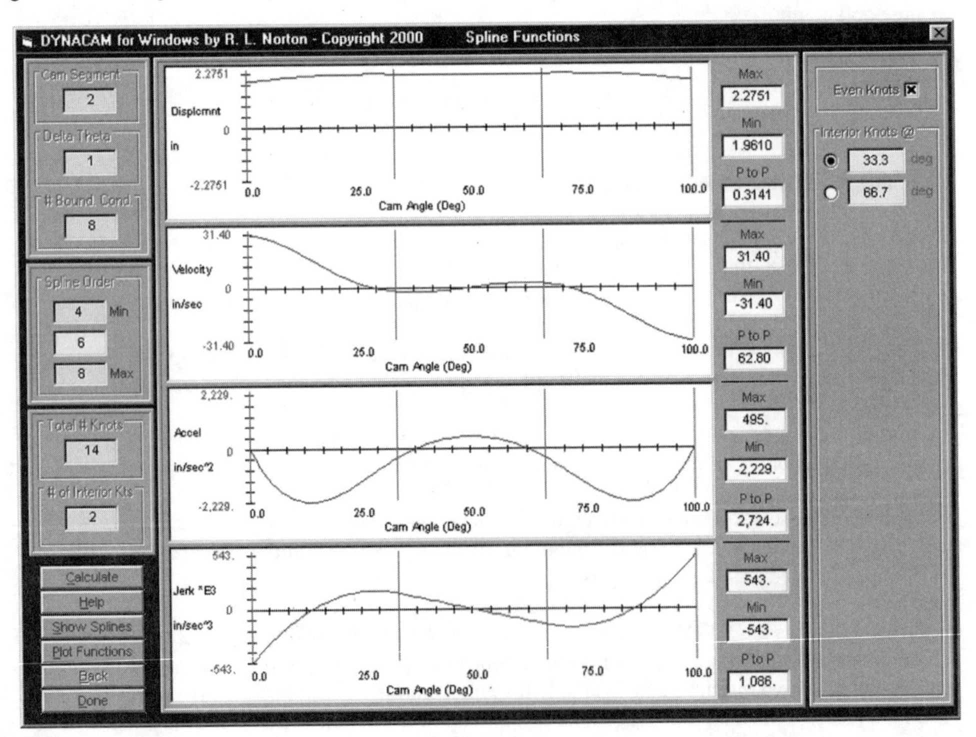

FIGURE 6-14

S V A J diagrams of a B-spline function for segment 2 of Example 6-4, even knots give overshoot in this case

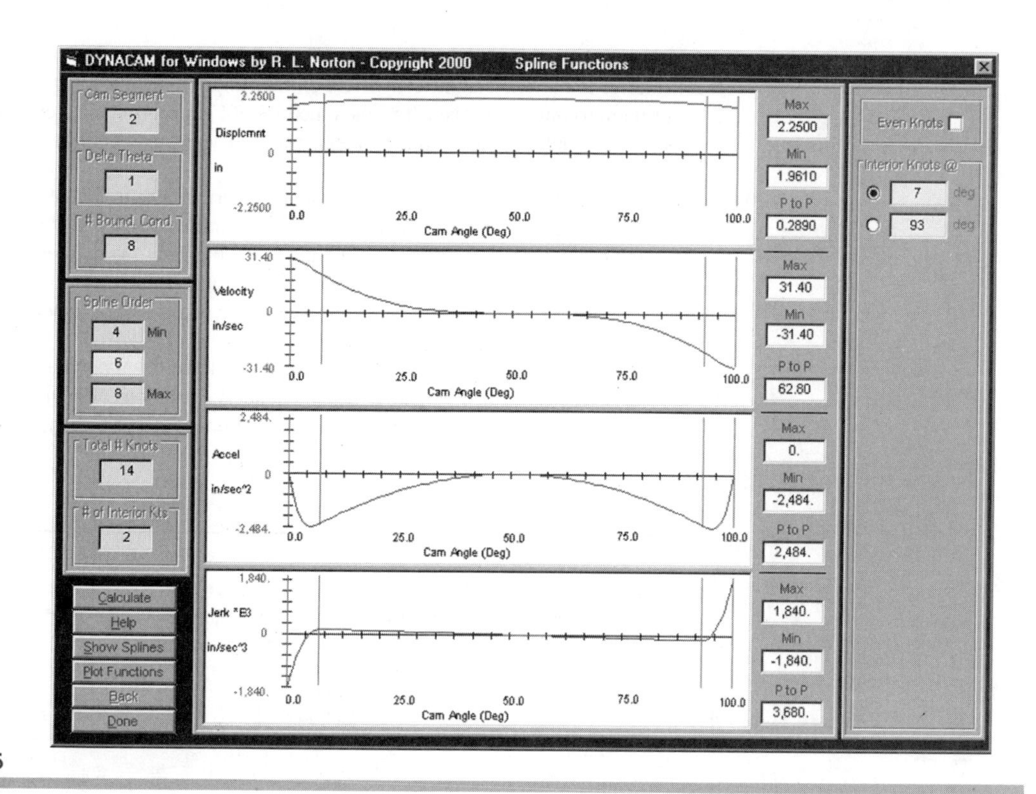

FIGURE 6-15

S V A J diagrams of a B-spline function for segment 2 of Example 6-4, knots placed at 7 and 93 degrees eliminate the overshoot—an acceptable solution

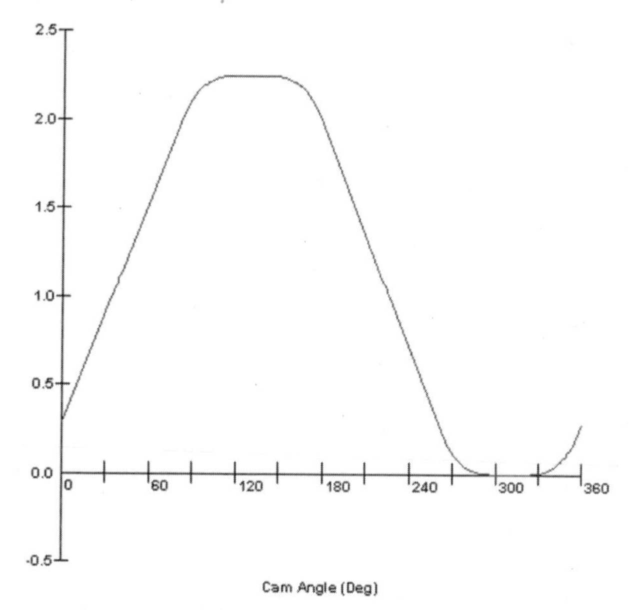

FIGURE 6-16

Displacement diagram of a successful B-spline/polynomial solution to Example 6-4—no overshoot

this case, that leaves 2 knots to place anywhere within the segment. The placement of these interior knots has a significant effect on the shape of the curves and allows us to contour them to our needs. Note that the functions of Figure 6-14 (using evenly spaced knots) are not an acceptable solution to this problem. The displacement overshoots to 2.275 in.

20 Figure 6-15 (p. 145) shows the result of moving the interior knots to 7 and 93 (local) degrees. This has "stretched" the function flat in the center and eliminated the overshoot. Compare the acceleration functions between Figures 6-14 and 6-15. Figure 6-14's acceleration rises above zero at the center while Figure 6-15's is essentially zero at the center, indicating that the displacement has zero curvature at that point—necessary to prevent overshoot.

21 The complete design uses B-splines for segments 2 and 4 and linear polynomial functions for the constant velocity segments 1 and 3. Seven boundary conditions of segments 2 and 4 are identical to those of the unsuccessful polynomial solutions of step 4 with the addition of an

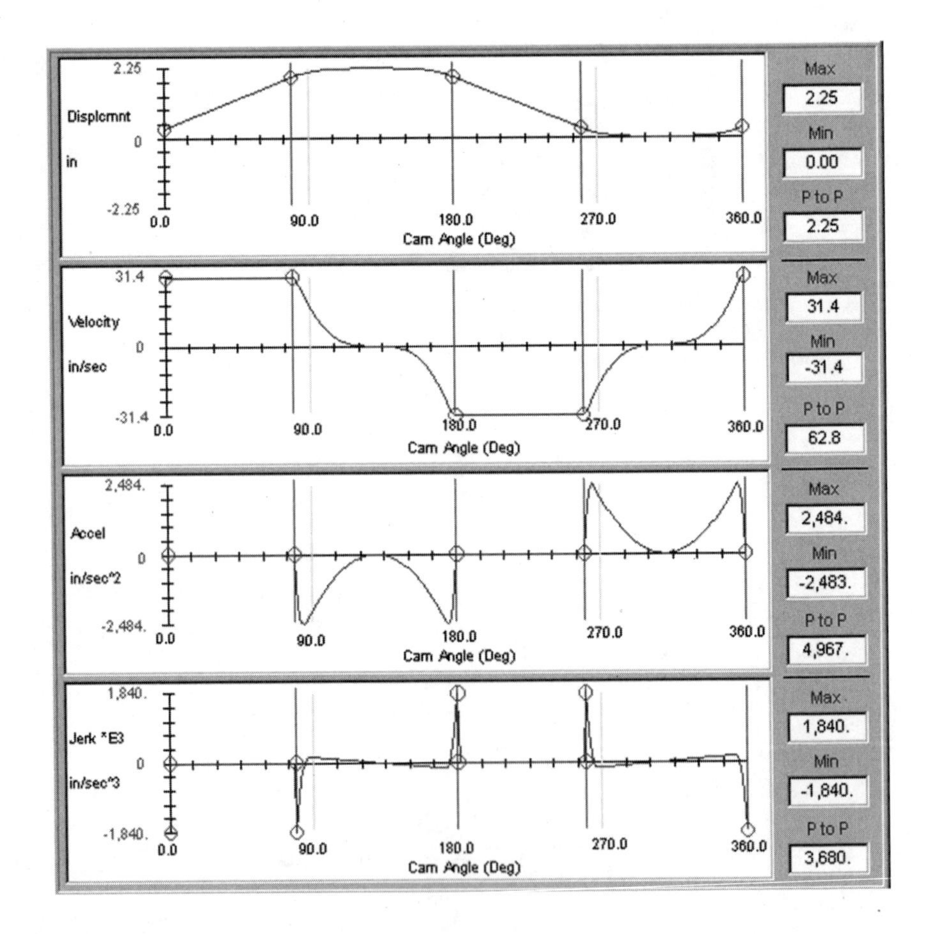

FIGURE 6-17

S V A J diagrams of a successful B-spline/polynomial solution to Example 6-4 with no overshoot

eighth BC to increase the number of available interior knots. By correctly positioning the interior knots, the B-spline functions have been made to conform to the desired constraints. Figure 6-16 (p. 145) shows an enlarged view of the resulting displacement function showing that it has no overshoot or undershoot.

22 Figure 6-17 shows the *S V A J* diagrams for this solution. Note the correct maximum and minimum values for displacement.

Open the files Ex_06-04a.cam, Ex_06-04ba.cam, and Ex_06-04c.cam in program DYNACAM to see more detail of this example.

Another case of a combination of displacement and velocity constraints, this time without any constant velocity portions, is developed in the next example. This is also taken from an actual design problem that involved a tooling cam for a high-production assembly machine. While it is possible to solve this problem with polynomials, a superior solution is found using B-splines.

EXAMPLE 6-5

Combined Velocity and Displacement Specifications.

Problem: The tooling for one station of a production machine is required to have the following timing diagram specifications as shown in Figure 6-18.

1 Dwell	at 12.1 mm from 0 to 30°
2 Fall	to 7.9 mm at 50°
3 Dwell	at 7.9 mm from 50 to 60°
4 Fall	to 0.5 mm at 110°
5 Dwell	at 0.5 mm from 110 to 125°
6 Fall	to 0.0 mm at 130°
7 Dwell	at 0.0 mm from 130 to 161°
8 Move	to 0.5 mm and come to an instantaneous zero velocity at 166°, then move to 4.0 mm at 178.5° and come to zero velocity, then move back to 1.1 mm at 191° and come to zero velocity, then move to 14.7 mm at 220°
9 Dwell	at 14.7 mm from 220 to 327°
10 Fall	to 12.1 mm at 337°
11 Dwell	at 12.1 mm from 337 to 360°
Cam Speed	30 rpm*

Solution:

1 With the exception of segment 8, all the other motions (i.e., segments 2, 4, 6, and 10) are simple double-dwell cases. Any suitable double-dwell function such as a 3-4-5 polynomial or a modified sine can be used for these. These will not be addressed here. See Chapter 3.

2 Segment 8 poses an interesting problem. If we attempt to deal with this collection of motions as one segment, then the only possible functions will be either polynomials or splines. We will first attempt to solve it with a polynomial.

3 The set of twelve BCs for a single polynomial over segment 8 will be: $S = 0$, $V = 0$, $A = 0$ at 161° to match the previous segment, $S = 0.5$, $V = 0$ at 166°, $S = 4.0$, $V = 0$, at 178.5°,

* While 30 rpm may not sound like high speed, this machine has multiple sets of tools running against each cam. Its throughput exceeds 10 parts per second.

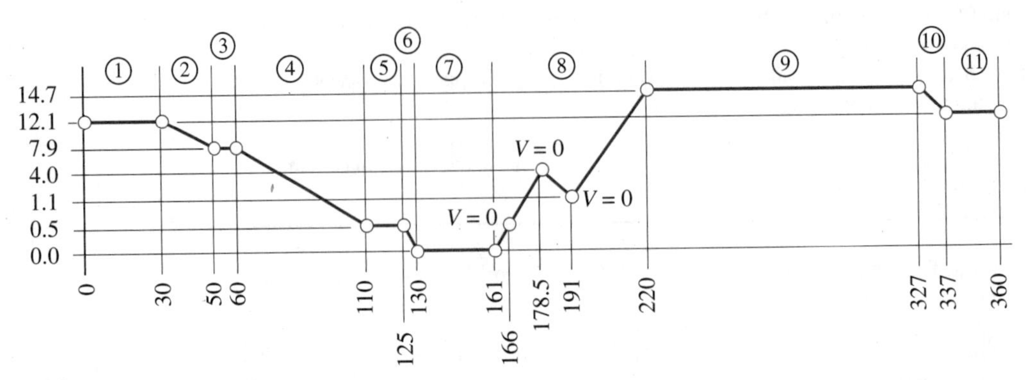

FIGURE 6-18

Timing diagram for Example 6-5 (not to scale)

$S = 1.1$, $V = 0$ at 191°, and $S = 14.7$, $V = 0$, $A = 0$ at 220° to match the following segment. This yields the 11th-degree polynomial:

$$s = 8784.9\left(\frac{\theta}{\beta}\right)^3 - 196\,562\left(\frac{\theta}{\beta}\right)^4 - 1\,779\,829\left(\frac{\theta}{\beta}\right)^5 - 8\,408\,374\left(\frac{\theta}{\beta}\right)^6 + 22\,851\,230\left(\frac{\theta}{\beta}\right)^7$$
$$-36\,938\,040\left(\frac{\theta}{\beta}\right)^8 + 34\,975\,710\left(\frac{\theta}{\beta}\right)^9 - 17\,873\,490\left(\frac{\theta}{\beta}\right)^{10} + 3\,800\,931\left(\frac{\theta}{\beta}\right)^{11} \qquad \text{(a)}$$

Even before plotting this function one can see from the size of its coefficients that it is going to be out of control. This is confirmed when its *S V A J* functions are plotted as shown in Figure 6-19. It plunges to –224 mm, and its velocities and accelerations are outrageous. The extremely large coefficients result in a displacement discontinuity of 0.4 mm at the segment end as well. This is another example of the problems that can occur with high-order polynomials.

4 It is possible to solve this problem with a series of piecewise continuous, low-order polynomials running between the various points within segment 8, and this was done successfully. But a superior solution can be obtained using a B-spline for this segment.

5 The set of boundary conditions for the B-spline solution is identical to that of step 3. However, we now have the flexibility of moving interior knots around to shape the *S V A J* curves to suit our needs. Figure 6-20 (p. 150) shows a B-spline solution that works. The knot locations were determined in a matter of a few minutes in program DYNACAM by trial and error. The program allows any knot to be moved with a mouse click and the resulting *S V A J* curves are immediately redrawn on screen. The six interior knots for this 6th-order B-spline are placed at 4, 5, 13, 21, 25, and 40 degrees, respectively. The displacement function has no overshoot or undershoot, the velocity is zero where required and is reasonable elsewhere, and the acceleration is smooth, low, and well balanced, The jerk is also reasonably smooth.

Open the files Ex_06-05a.cam and Ex_06-05b.cam in program DYNACAM to see more detail of this example.

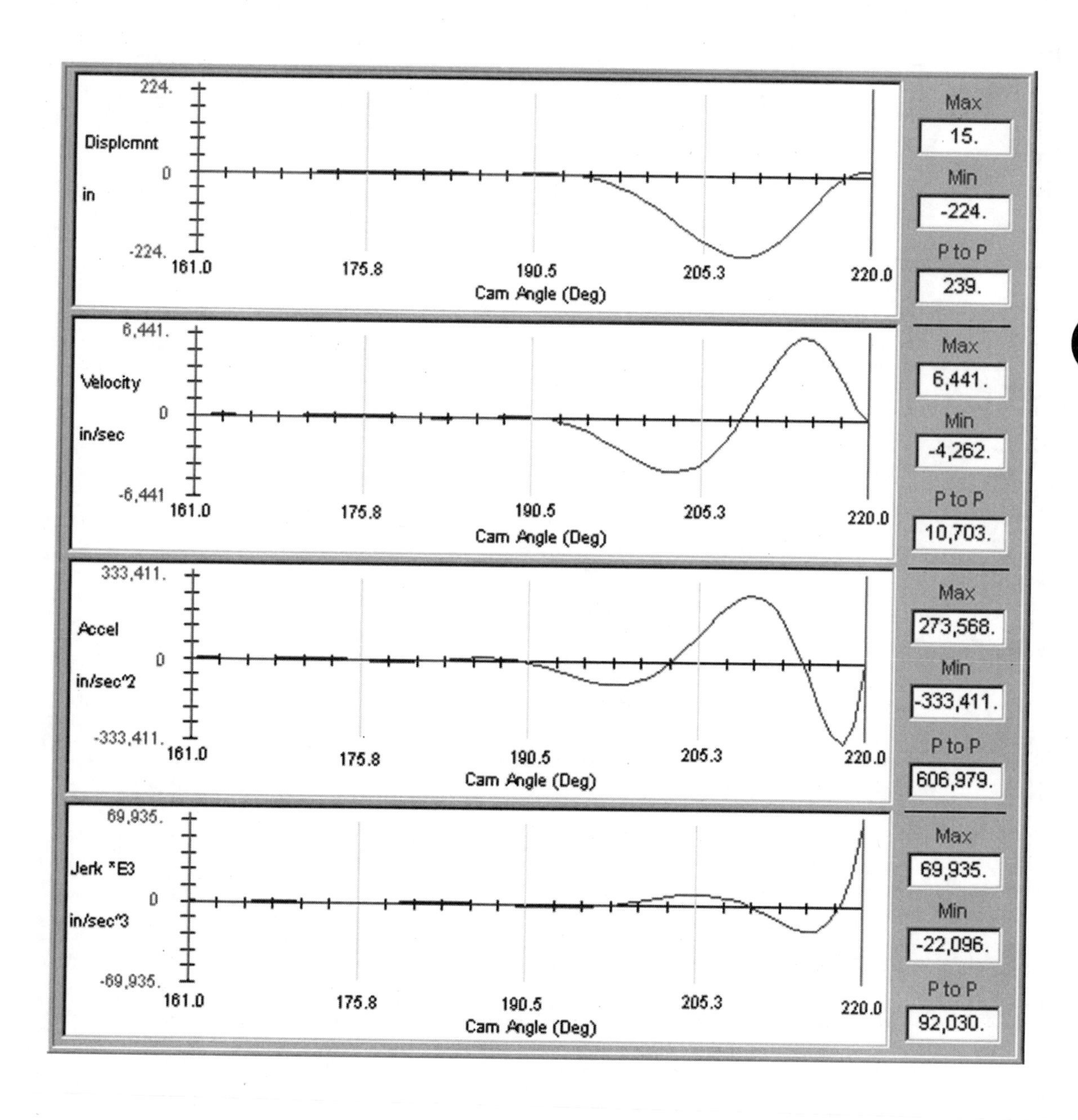

FIGURE 6-19

S V A J diagrams of an unsuccessful polynomial solution to segment 8 of Example 6-5

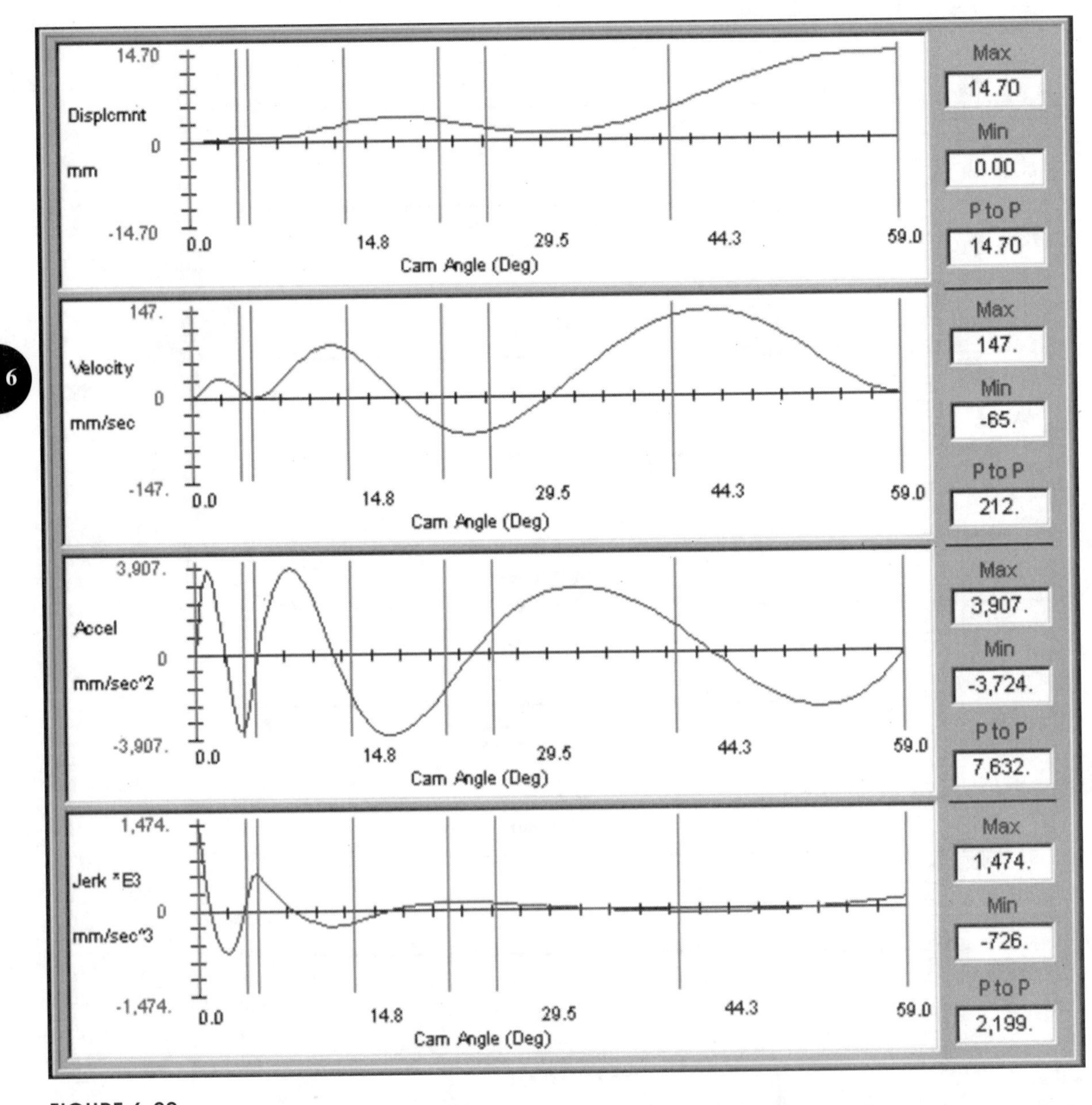

FIGURE 6-20

S V A J diagrams of a successful 6th-order B-spline for segment 8 of Example 6-5, knots at 4, 5, 13, 21, 25, and 40°

6.3 SUMMARY

The examples in this chapter have shown the power of spline functions to solve difficult cam design tasks. When the motions and their derivatives are tightly constrained, splines are often the only way that the desired results can be achieved.

Chapter 7

CAM SIZE DETERMINATION

7.0 INTRODUCTION

Once the *s v a j* functions have been defined, the next step is to size the cam. There are two major factors which affect cam size, the **pressure angle** and the **radius of curvature**. Both of these involve either the **base circle radius** on the cam (R_b) when using flat-faced followers, or the **prime circle radius** on the cam (R_p) when using roller or curved followers.

The base circle's and prime circle's centers are at the center of rotation of the cam. The base circle of a radial cam is defined as *the smallest circle which can be drawn tangent to the physical cam surface* as shown in Figure 7-1. All radial cams will have a base circle, regardless of the follower type used.

The prime circle is only applicable to cams with roller followers or radiused (mushroom) followers and is measured to the center of the roller follower. The **prime circle** is defined as *the smallest circle which can be drawn tangent to the locus of the centerline of the follower* as shown in Figure 7-1 for a radial cam. *The locus of the centerline of the follower* is called the **pitch curve**. Cams with roller followers are in fact defined for manufacture with respect to the pitch curve rather than with respect to the cam's physical surface. Radial cams with flat-faced followers must be defined for manufacture with respect to their physical surface, as there is no pitch curve.

The process of creating a radial cam from the *s* diagram can be visualized conceptually by imagining the *s* diagram to be cut out of a flexible material such as rubber. The *x* axis of the *s* diagram represents the circumference of a circle, which could be either the **base circle**, or the **prime circle**, around which we will "wrap" our "rubber" *s* diagram. We are free to choose the initial length of our *s* diagram's *x* axis, though the height of the displacement curve is fixed by the cam displacement function we have chosen. In effect we will choose the base or prime circle radius as a design parameter and stretch the length of the *s* diagram's axis to fit the circumference of the chosen circle.

TABLE 7-1 Notation Used in This Chapter

t = time, seconds

θ = camshaft angle, degrees (deg) or radians (rad)

ω = camshaft angular velocity, rad/sec

β = total angle of any segment, rise, fall, or dwell, deg or rad

h = total lift (rise or fall) of any one segment, length units, or angle units (deg or rad)

s or S = follower displacement, length units, or angle units (deg or rad)

$v = ds/d\theta$ = follower velocity, length/rad (translating follower) or rad/rad (oscillating follower)

$V = dS/dt$ = follower velocity, length/sec (translating follower) or rad/sec (oscillating follower)

$a = dv/d\theta$ = follower acceleration, length/rad^2 (translating follower) or rad/rad^2 (oscillating follower)

$A = dV/dt$ = follower acceleration, length/sec^2 (translating follower) or rad/sec^2 (oscillating follower)

$j = da/d\theta$ =follower jerk, length/rad^3 (translating follower) or rad/rad^3 (oscillating follower)

$J = dA/dt$ = follower jerk, length/sec^3 (translating follower) or rad/sec^3 (oscillating follower)

$s\ v\ a\ j$ refer to the group of diagrams, length or angle units versus radians

$S\ V\ A\ J$ refer to the group of diagrams, length or angle units versus time

Rb = base circle radius, length units

Rp = prime circle radius, length units

Rf = roller follower radius, length units

ε = eccentricity of cam-follower, length units

ϕ = pressure angle, deg or radians, and a signed quantity

ρ = radius of curvature of cam pitch curve, length units

ρ_c = radius of curvature of cam surface, length units

ρ_{pitch} = radius of curvature of pitch curve, length units

ρ_{min} = minimum radius of curvature of pitch curve or cam surface, length units

F, N = applied and normal force, respectively, force units

Portions of this chapter were adapted from R. L. Norton, *Design of Machinery* 2ed. McGraw-Hill, 2001, with permission.

7.1 PRESSURE ANGLE—RADIAL CAM WITH TRANSLATING ROLLER FOLLOWER

The **pressure angle** is defined as shown in Figure 7-2 that depicts a radial or disk cam with a translating roller follower. Force can only be transmitted from cam to follower or vice versa along the *common normal* or **axis of transmission** which is perpendicular to the **axis of slip**, or *common tangent* as shown in Figure 7-2. The **pressure angle** ϕ is *the angle between the direction of motion (velocity) of the follower and the direction of the axis of transmission.** When $\phi = 0$, all the transmitted force goes into motion of the follower and none into slip velocity. When ϕ becomes 90° there will be no motion of the follower. As a rule of thumb, we would like the pressure angle to be between zero and about ±30° for translating followers to avoid excessive side load on the sliding follower. If the follower is oscillating on a pivoted arm, a pressure angle up to about ±35° is acceptable. Values of ϕ greater than these can increase the follower sliding or pivot friction to undesirable levels and may tend to jam a translating follower in its guides.

* Dresner and Buffington point out [5] that this definition is only valid for single-input systems. For multi-input systems, a more complicated definition and calculation of pressure angle is needed. See the reference for more information.

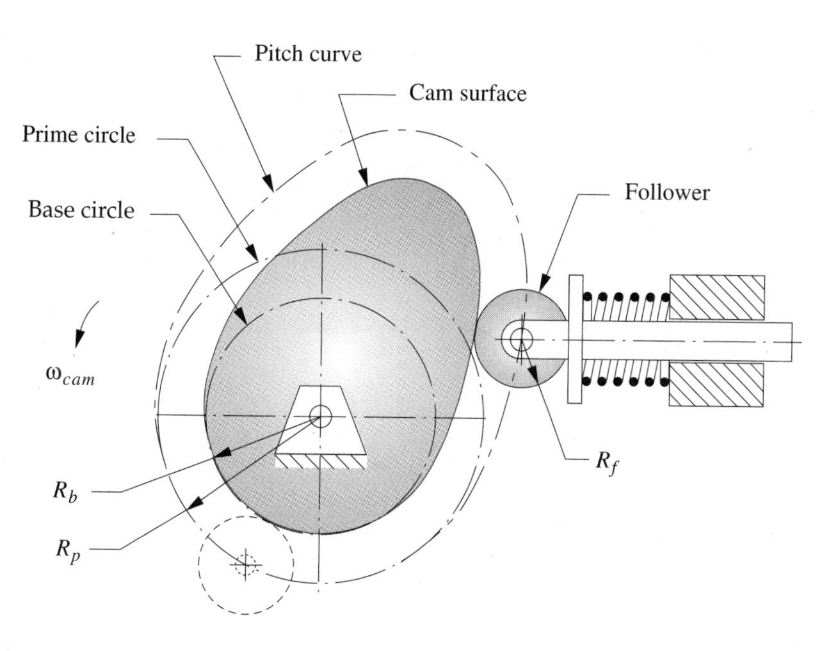

FIGURE 7-1

Base circle R_b, prime circle R_p, and pitch curve of a radial cam with roller follower

7

FORCES The forces acting on this system are shown in Figure 7-2. Summing forces and moments for equilibrium conditions:

$$\begin{aligned}
\Sigma F_x &= 0 = R_2 - R_1 - N\sin\phi \\
\Sigma F_y &= 0 = N\cos\phi - \mu(R_1 + R_2) - F \\
\Sigma M_{R_1} &= 0 = R_2(b + \mu d) + \frac{d}{2}(F - N\cos\phi) - (a+b)N\sin\phi
\end{aligned} \tag{7.1a}$$

We can define a force magnification factor that relates the normal force N and the follower force F.

$$\frac{N}{F} = \frac{b}{b\cos\phi - \left(2\mu a + \mu b - \mu^2 d\right)\sin\phi} \tag{7.1b}$$

If there were no friction in the follower guide ($\mu = 0$) then this ratio would become infinite at $\phi = 90°$. A nonzero μ makes it infinite at an angle somewhat less than 90°. When ϕ is 0, the ratio is 1. The pressure angle needs to be kept to a low value to prevent the normal force from becoming so large as to fail the surface of either cam or roller.

Eccentricity

Figure 7-3 shows the geometry of a radial cam and translating roller follower in an arbitrary position. This shows the general case in that the axis of motion of the follower does

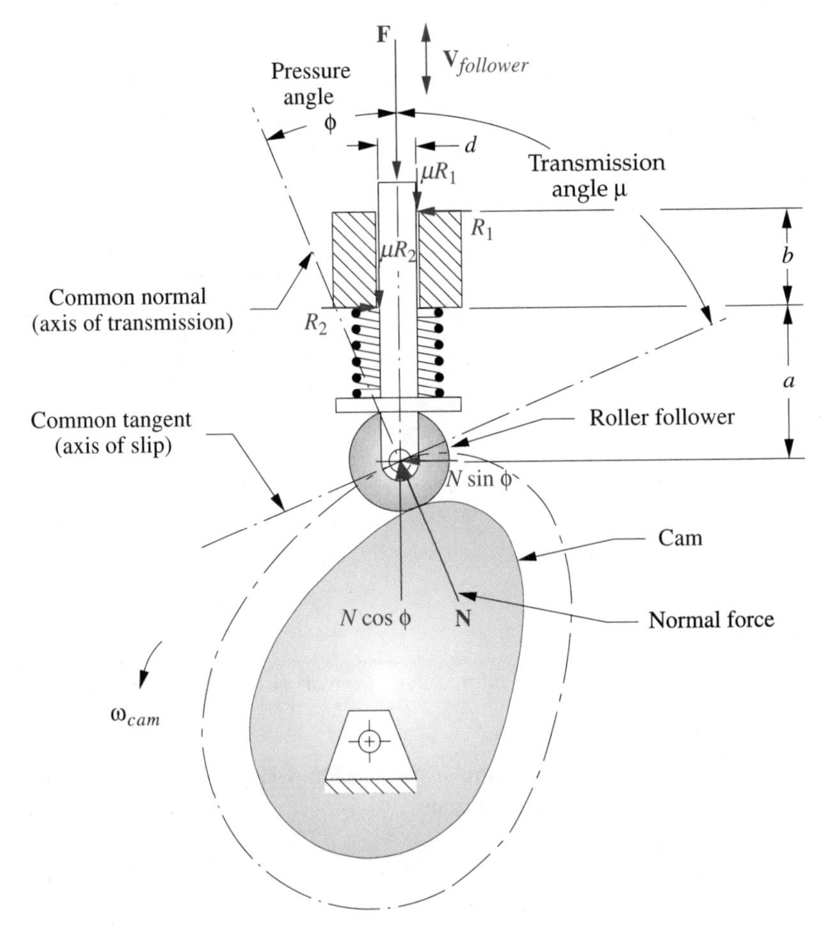

FIGURE 7-2

Cam pressure angle and forces in a radial cam with offset translating roller follower

7

* An instant center of velocity is defined as a point, common to two rigid bodies (links) in plane motion that has the same instantaneous velocity in each body. In other words, there is no relative velocity between these two points at that instant. Thus, one body can be considered to be in pure rotation with respect to the other about their common instant center. The notation is *Ixy* where *x* and *y* denote the reference numbers of the two links "connected" by the instant center.

† This link numbering omits link 3, that number being reserved for the (physically nonexistent, but kinematically valid) effective coupler of the equivalent fourbar linkage. This cam-follower system has three physical links, the ground (link 1), the cam (link 2), and the follower slider link 4). The roller is a kinematically redundant link that serves only to reduce friction and increase the area of contact between cam and follower.

not intersect the center of the cam. There is an *offset* or **eccentricity** ε defined as *the perpendicular distance between the follower's axis of motion and the center of the cam.* Often this eccentricity ε will be zero, making it an **aligned follower** (also called a radial follower), which is the special case.

In the figure, the axis of transmission is extended to intersect effective link 1, which is the ground link. (See Section 1.1 and Figure 1-4, p. 3 for a discussion of effective links in cam systems.) This intersection is instant center $I_{2,4}$ (labeled *B*) which, by definition*, has the same velocity in link 2 (the cam) and in link 4 (the follower).† Because link 4 is in pure translation, all points on it have identical velocities $V_{follower}$, which are equal to the velocity of $I_{2,4}$ in link 2. We can write an expression for the velocity of $I_{2,4}$ in terms of an assumed constant cam angular velocity ω and the radius b from cam center to $I_{2,4}$,

$$V_{I_{2,4}} = b\omega = \dot{s} \tag{7.2a}$$

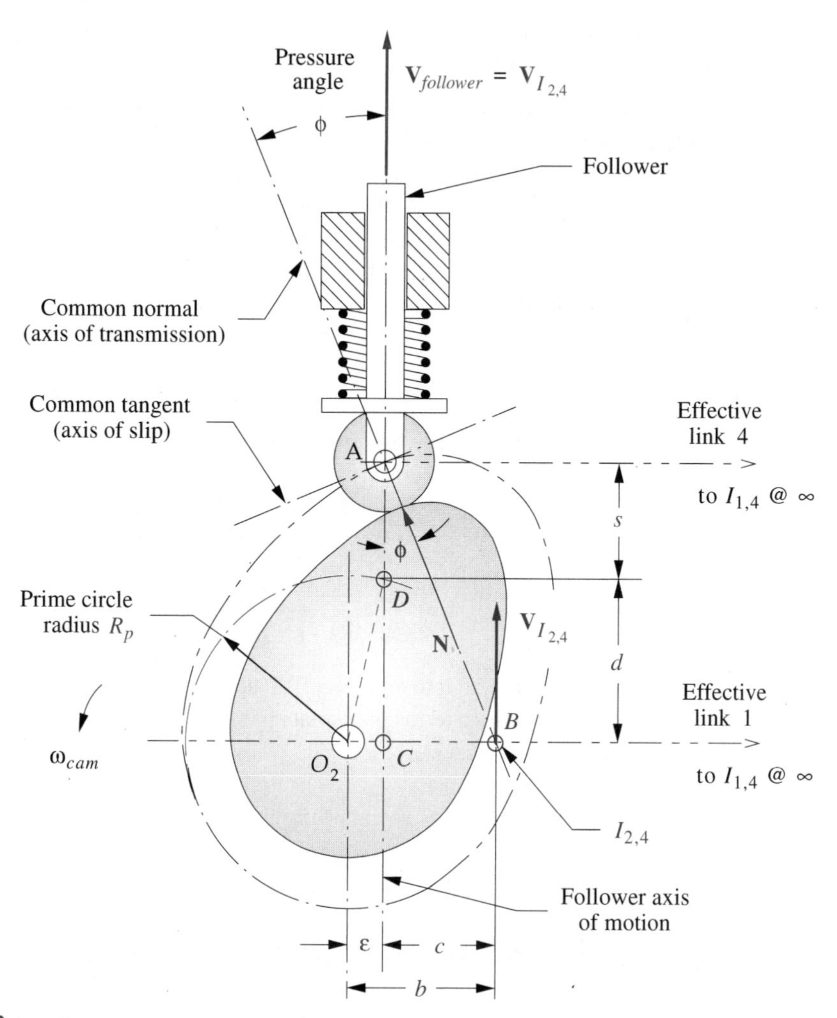

FIGURE 7-3

Geometry for the derivation of the equation for pressure angle in a radial cam with offset translating roller follower

where s is the instantaneous displacement of the follower from the s diagram and $\dot{s}$ is its time derivative in units of length/sec. Now applying the chain rule with ω constant:

given
$$b\omega = \dot{s} = \frac{ds}{dt}$$

but
$$\frac{ds}{dt}\frac{d\theta}{d\theta} = \frac{ds}{d\theta}\frac{d\theta}{dt} = \frac{ds}{d\theta}\omega = v\omega$$

so
$$b\omega = v\omega$$

then
$$b = v \tag{7.2b}$$

This is an interesting relationship that says that the **distance** b **to the instant center** $\mathbf{I_{2,4}}$ **is numerically equal to the velocity of the follower** v in units of length per radian as derived in previous sections. We have reduced this expression to pure geometry, independent of the angular velocity ω of the cam.

Note that we can express the distance b in terms of the prime circle radius R_p and the eccentricity ε, by the construction shown in Figure 7-3. Swing the arc of radius R_p until it intersects the axis of motion of the follower at point D. This defines the length of line d from effective link 1 to this intersection. This is constant for any chosen prime circle radius R_p. Points A, C, and $I_{2,4}$ form a right triangle whose upper angle is the pressure angle ϕ and whose vertical leg is $(s + d)$, where s is the instantaneous displacement of the follower. From this triangle:

$$c = (s + d)\tan\phi$$

and (7.3a)

$$b = (s + d)\tan\phi + \varepsilon$$

Then from equation 7.2,

$$v = (s + d)\tan\phi + \varepsilon \tag{7.3b}$$

and from triangle CDO_2,

$$d = \sqrt{R_P^2 - \varepsilon^2} \tag{7.3c}$$

Substituting equation 7.3c into equation 7.3b and solving for ϕ gives an expression for the signed pressure angle in terms of displacement s, velocity v, offset or eccentricity ε, and the prime circle radius R_p.

$$\phi = \arctan\frac{v - \varepsilon}{s + \sqrt{R_P^2 - \varepsilon^2}} \tag{7.3d}$$

The velocity v in this expression is in units of length/rad, and all other quantities are in compatible length units. We have typically defined s and v by this stage of the cam design process and wish to manipulate R_p and ε to get an acceptable maximum pressure angle ϕ. As R_p is increased, ϕ will be reduced. The only constraints against large values of R_p are the practical ones of package size and cost. Often there will be some upper limit on the size of the cam-follower package dictated by its surroundings. There will always be a cost constraint and bigger = heavier = more expensive.

Choosing a Prime Circle Radius

Both R_p and ε are within a transcendental expression in equation 7.3d, so they cannot be conveniently solved for directly. The simplest approach is to assume a trial value for R_p and an initial eccentricity of zero. Next use program DYNACAM, your own program, or an equation solver such as *Matlab*, *TKSolver** or *Mathcad* to quickly calculate the values of ϕ for the entire cam. Then adjust R_p and repeat the calculation until an acceptable arrangement is found. Figure 7-4 shows the calculated pressure angles for a four-dwell radial cam with translating roller follower. Note their similarity in shape to velocity functions as that term is dominant in equation 7.3d.

* *TKSolver* is particularly adept at solving this kind of problem as it will automatically iterate to a value of *Rp* for any chosen values of ϕ and ε in equation 7.3d.

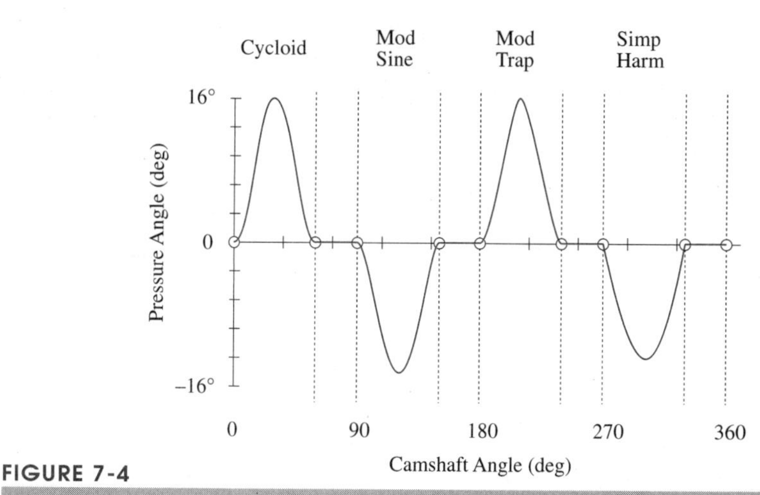

FIGURE 7-4

Pressure angle functions are similar in shape to velocity functions

USING ECCENTRICITY If a suitably small cam cannot be obtained with acceptable pressure angle, then eccentricity can optionally be introduced to change the pressure angle. Using eccentricity to control the pressure angle has its limitations. For a positive ω, a positive value of eccentricity will *decrease the pressure angle on the rise* but will *increase it on the fall.* Negative eccentricity does the reverse.

This is of little value with a form-closed (groove or track) cam, as it is driving the follower in both directions. For a force-closed cam with spring return, you can sometimes afford to have a larger pressure angle on the fall than on the rise because the stored energy in the spring is attempting to speed up the camshaft on the fall, whereas the cam is storing that energy in the spring on the rise. The limit of this technique can be the degree of overspeed attained with a larger pressure angle on the fall. The resulting variations in cam angular velocity may be unacceptable.*

The most value gained from adding eccentricity to a follower comes in situations where the cam program is asymmetrical and significant differences exist (with no eccentricity) between maximum pressure angles on rise and fall. Introducing eccentricity can balance the pressure angles in this situation and create a smoother running cam.

If adjustments to R_p or ε do not yield acceptable pressure angles, the only recourse is to return to an earlier stage in the design process and redefine the problem. Less lift or more time to rise or fall will reduce the causes of the large pressure angle. Design is, after all, an iterative process.

7.2 PRESSURE ANGLE—BARREL CAM WITH TRANSLATING ROLLER FOLLOWER

It is not possible to use a flat-faced follower on an axial or barrel cam as there will always be regions of negative radius of curvature regardless of the barrel cylinder diameter. Thus a roller follower is required for a barrel cam, especially if it is form closed,

* Note that this situation will only obtain if the camshaft drive motor is not speed-controlled. If a servomotor is used to drive the camshaft, then it will compensate for position and velocity errors due to torque variation.

which is common for barrel cams. The calculation of pressure angle for a barrel cam with translating roller follower is much simpler than for the radial cam. Unlike the radial cam, there is no distortion of the *s* diagram when it is wrapped around a cylindrical "barrel." The barrel diameter is analogous to the prime circle of a radial cam and we will call it the *prime cylinder* radius *Rp*, typically taken to the centerline of the follower as shown in Figure 7-5. The pressure angle $\phi(\theta)$ of a barrel cam with translating follower is directly proportional to the slope of the *s* diagram, i.e., the velocity of the follower expressed in units of length/radian. The actual slope of the displacement function at any

7

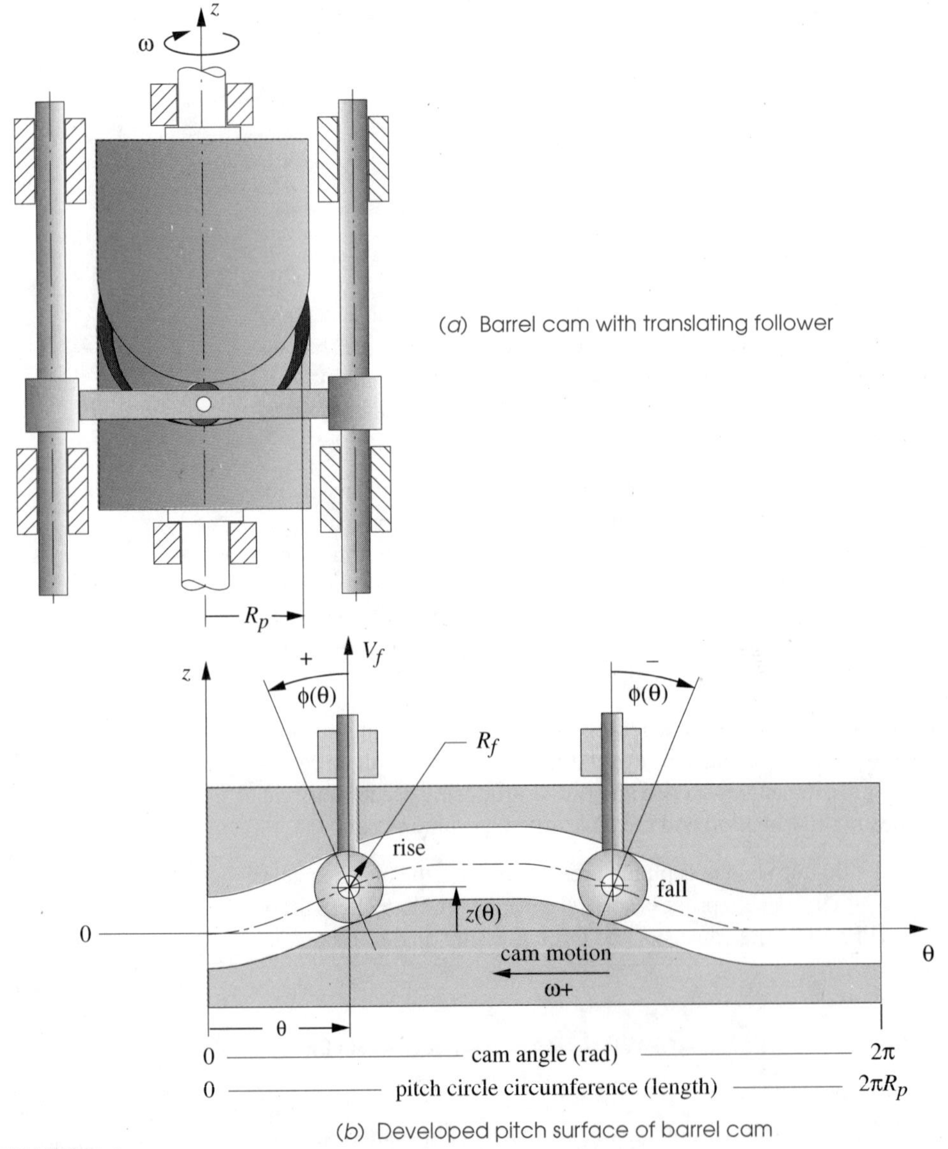

FIGURE 7-5

Pressure angle of a barrel cam with translating follower

point depends on the circumference (thus radius) of the prime cylinder. The expression for pressure angle for follower motion parallel to the cam axis is:

$$\phi(\theta) = \tan^{-1}\left[\frac{V(\theta)}{\omega R_p}\right] \tag{7.4}$$

where $V(\theta)$ is the follower velocity in units of length/sec, ω is the cam angular velocity in rad/sec, and Rp is the prime cylinder radius in length units. Note that this equation assumes no follower offset; that is, the follower is positioned with its roller axis intersecting the cam axis. While it is possible to offset the roller, there is no advantage to doing so and it complicates machining of the cam surface. Offsetting an axial translating follower is generally unnecessary and is not recommended.

7.3 PRESSURE ANGLE—BARREL CAM WITH OSCILLATING ROLLER FOLLOWER

Figure 7-6a shows a barrel cam with oscillating roller follower. The follower arm has radius R_a and pivots about point P that lies in and defines the datum plane, which is orthogonal to the cam rotation axis z and through the arm pivot. The initial position of the roller with respect to the datum plane is defined by the x and z offsets, x_0 and $z(0)$, which can each be either positive or negative. Note that x_0 is a constant while z is a function of θ. Depending on CW or CCW initial arm rotation ω_{arm} on the rise, in combination with x_0, $z(0)$, the roller's arc motion may or may not cross the datum plane or cam axis. For best balance of forces and pressure angle, good design suggests that the arm's motion should be symmetrical about the datum plane and cam axis as shown in the figure.

For any oscillating arm follower, the displacement $s(\theta)$ is defined in relative angular measure as a positively increasing angle starting from zero at the minimum displacement position (typically the low dwell). The initial angle of the arm $\psi(0)$ at $s(0)$ with respect to the positive x-axis of the follower datum plane is defined as

$$\psi(0) = \sin^{-1}\left(\frac{z(0)}{R_a}\right) \qquad x_0 \le 0$$

$$\psi(0) = \pi - \sin^{-1}\left(\frac{z(0)}{R_a}\right) \qquad x_0 > 0 \tag{7.5a}$$

Figure 7-6b shows the developed pitch surface of the cam and several inverted positions of the follower. The length of the developed surface is the circumference of the prime cylinder $2\pi Rp$. Note that this is dimensioned from the initial to final positions of the arm pivot P. Point $P(0)$ is offset a distance along the surface equal to the arc subtended by the horizontal component of the initial position of the follower at radius R_p, expressed in terms of cam angle as:

$$\gamma(0) = \frac{R_a}{R_p}\cos[\psi(0)] \quad \text{radians} \tag{7.5b}$$

At an arbitrary cam angle θ, the inverted arm's pivot location is at $P(\theta)$ and the angle of the arm with respect to the positive x-axis of the datum plane is:

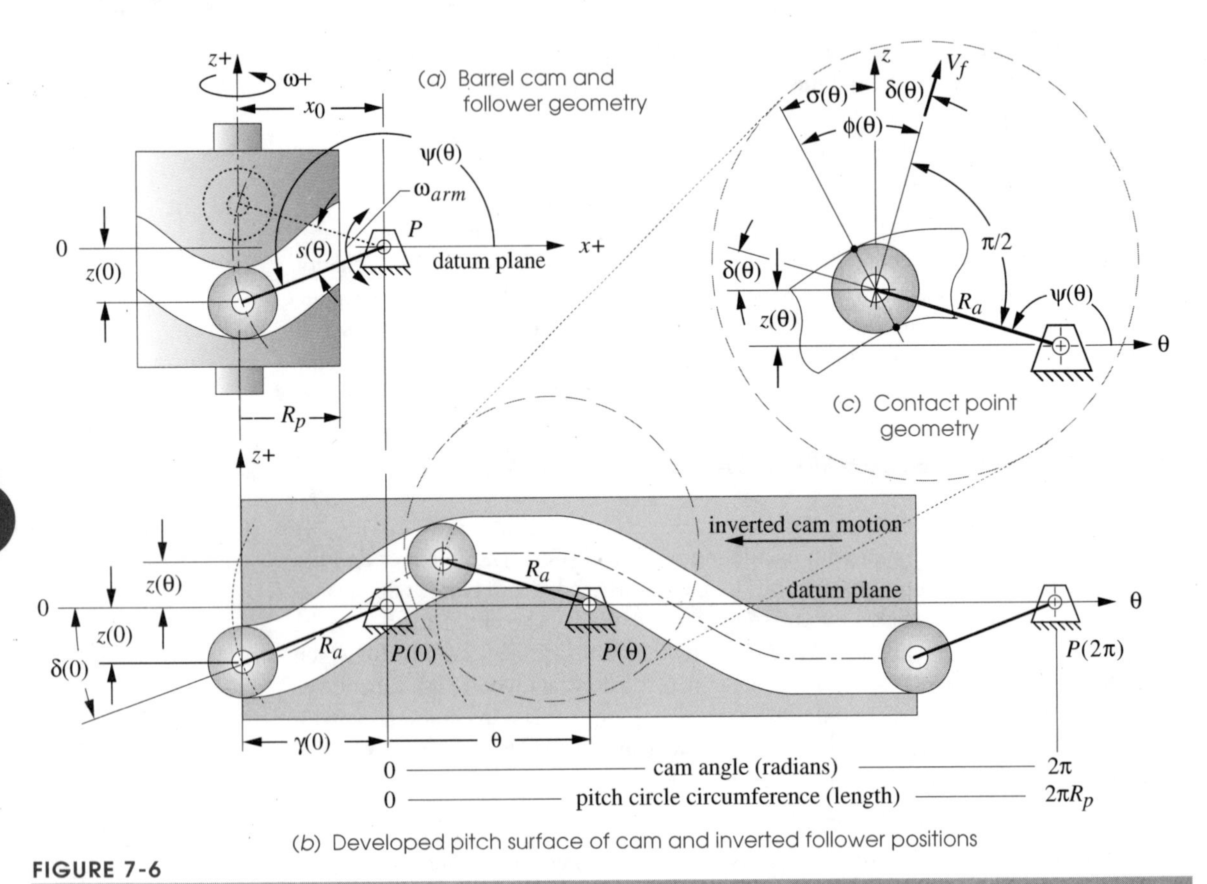

FIGURE 7-6

Pressure angle of a barrel cam with oscillating roller follower and its developed surface at the pitch radius

$$\psi(\theta) = \psi_0 + \operatorname{sgn}(\omega_{arm})s(\theta) \tag{7.5c}$$

Figure 7-6c shows a detail of the follower at an arbitrary cam angle θ. In the case of an oscillating arm follower, the direction of follower velocity V is not always parallel to the cam axis as it was for the translating follower. Rather, it is now always orthogonal to the arm radius *Ra*.

The pressure angle for an oscillating arm follower on a barrel cam is then:

$$\phi(\theta) = -\operatorname{sgn}(\omega)\tan^{-1}\left\{\frac{\operatorname{sgn}(\omega)\operatorname{sgn}(x_0)R_a\dfrac{V}{\omega} - R_p\sin[\psi(\theta)]}{R_p\cos[\psi(\theta)]}\right\} \tag{7.5d}$$

The various *sgn* functions in equations 7.5 account for cam and arm rotation directions, and the location of the arm pivot to the left or right of the cam axis, and ω is camshaft rotation in rad/sec. *Rp* is the nominal cam prime cylinder radius. The arm angular velocity V must be in rad/sec.

The above derivation does not account for error associated with the arc motion of the oscillating roller that results from its contact point moving out of the plane of the axis of the cam during its sweep and effectively increasing the cam radius at which the center of the roller contacts the slot. If the arm radius is reasonably long in respect to the lift of the cam and has a small angular excursion, and/or the pitch cylinder is large, then this error may be small. If these conditions do not obtain, then the nominal pitch cylinder radius R_p needs to be corrected for each position of the roller follower as:

$$R'_p = \sqrt{R_p^2 + \left\{R_a \cos\left[\psi(\theta)\right] + x_0\right\}^2} \tag{7.5e}$$

Substitute Rp' for Rp in equations 7.5b and d.

7.4 OVERTURNING MOMENT—RADIAL CAM WITH TRANSLATING FLAT-FACED FOLLOWER

Figure 7-7 shows a translating, flat-faced follower running against a radial cam. The pressure angle can be seen to be zero for all positions of cam and follower. This seems

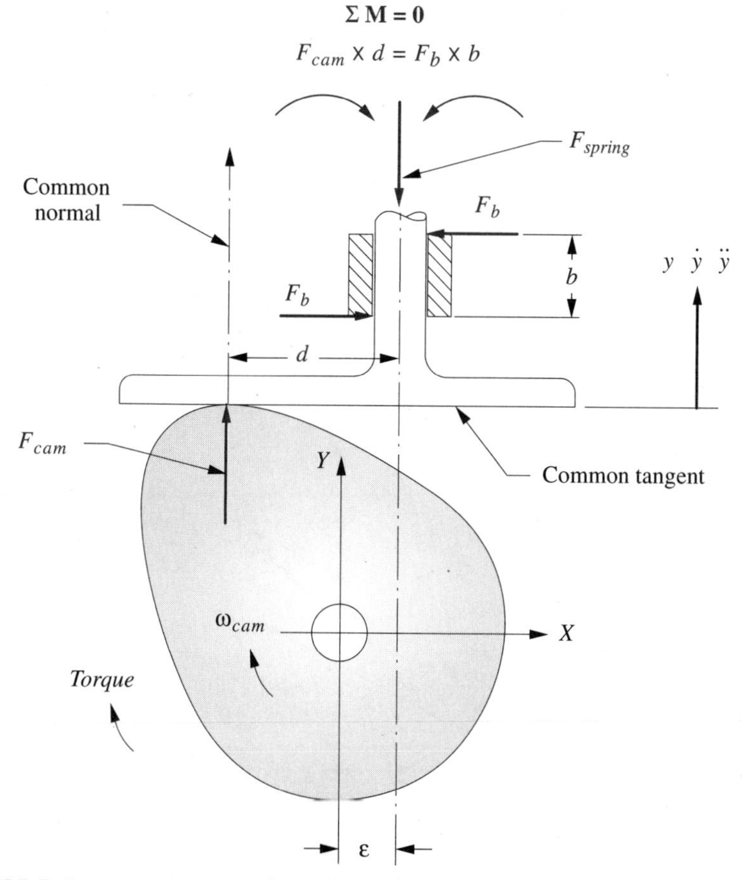

FIGURE 7-7

Overturning moment on a flat-faced follower

to be giving us something for nothing, which can't be true. As the contact point moves left and right, the point of application of the force between cam and follower moves with it. There is an overturning moment on the follower associated with this off-center force that tends to jam the follower in its guides, just as did too large a pressure angle in the roller follower case. In this case, we would like to keep the cam as small as possible in order to minimize the moment arm of the force. Eccentricity will affect the average value of the moment, but the peak-to-peak variation of the moment about that average is unaffected by eccentricity. Considerations of a too-large pressure angle do not limit the size of this cam, but other factors do. The minimum radius of curvature of the cam surface must be kept large enough to avoid undercutting (see Figure 7-10 and its discussion on p. 166). This is true regardless of the type of follower used.

7.5 PRESSURE ANGLE—RADIAL CAM WITH OSCILLATING ROLLER FOLLOWER

The geometry of a radial cam with an oscillating arm roller follower is shown in Figure 7-8. Derivations for the pressure angle in this case have been developed by Baxter,[1]

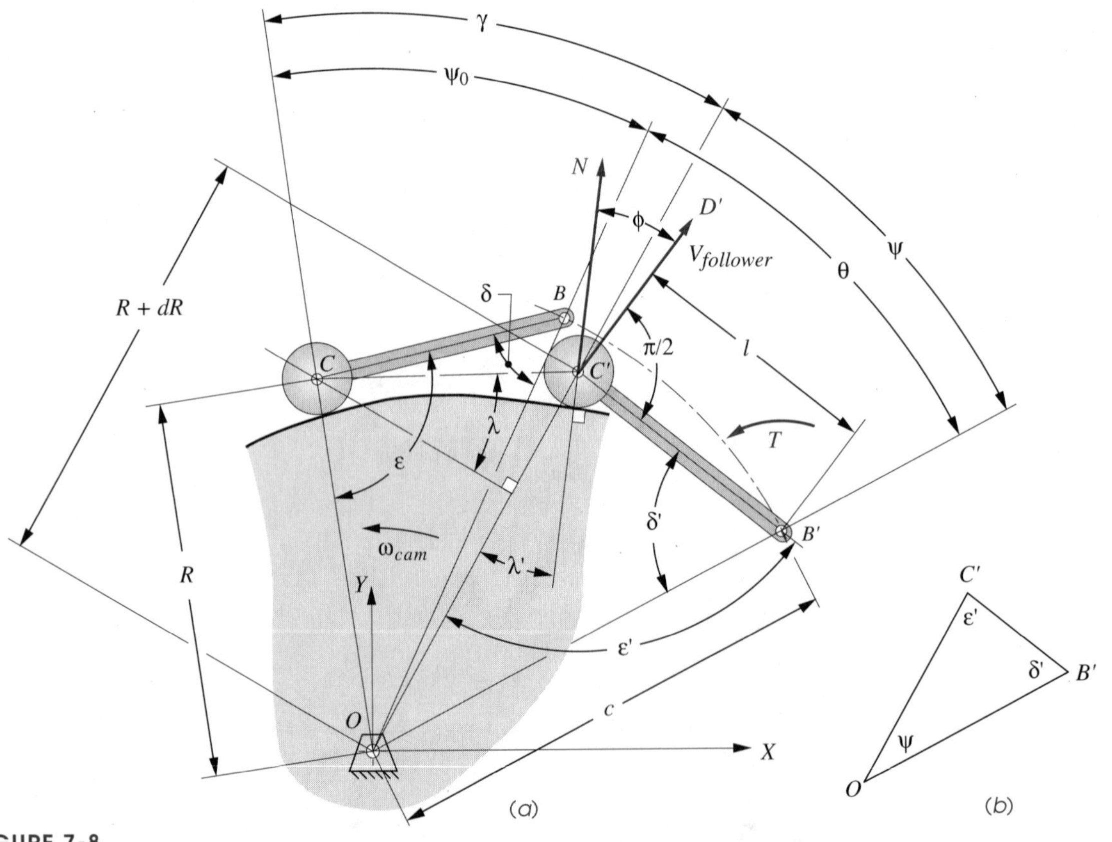

FIGURE 7-8

Geometry for derivation of pressure angle in a radial cam with oscillating roller follower

Kloomock and Muffley,[2] and Raven.[3] We will follow Kloomock and Muffley's presentation here. Figure 7-8 shows two positions of the follower arm BC being rotated around a "stationary" cam in the typical inversion of the motion for analysis purposes. (Typically, the follower arm pivot B remains stationary and the cam rotates.) The initial position BC becomes $B'C'$ at a later time after the cam has rotated through the angle γ. Though these positions are shown widely separated for clarity, the analysis considers them to be an infinitesimal angle $d\gamma$ apart.

The pressure angle ϕ is defined as the angle between the normal force N applied at the cam-roller interface, shown as vector $C'N$, and the direction of the velocity of the roller center, shown as $C'D'$. Neglecting friction and taking moments about the arm pivot B' gives

$$\frac{Nl}{T} = \frac{1}{\cos\phi} \tag{7.6}$$

where l is the length of the arm and T is the applied load torque on the follower arm. The torque ratio Nl/T is similar to the force magnification factor N/F of equation 7.1b for a radial cam with translating roller follower.

From the geometry of Figure 7-8, note that as $d\gamma$ approaches zero, γ' approaches γ, δ' approaches δ, and ε' approaches ε. An expression for pressure angle ϕ can be written as:

$$\phi = \frac{\pi}{2} - (\varepsilon - \lambda) \tag{7.7a}$$

$$\lambda = \tan^{-1}\frac{1}{R}\frac{dR}{d\gamma} \tag{7.7b}$$

The triangle $OB'C'$ in Figure 7-8a (and shown separately in Figure 7-8b) can be solved for R, ε, and ψ.

$$R = \sqrt{l^2 + c^2 - 2lc\cos\delta} \tag{7.7c}$$

$$\varepsilon = \sin^{-1}\left(\frac{c}{R}\sin\delta\right) \tag{7.7d}$$

$$\psi = \cos^{-1}\left(\frac{c^2 + R^2 - l^2}{2Rc}\right) \tag{7.7e}$$

Also from Figure 7-8 it can be seen that

$$\gamma = \psi_0 - \psi + \theta \tag{7.7f}$$

Differentiating equation 7.7f with respect to R:

$$\frac{d\gamma}{dR} = \frac{d\theta}{dR} - \frac{d\psi}{dR} \tag{7.7g}$$

Differentiating equation 7.7c with respect to q and reciprocating:

$$\frac{d\theta}{dR} = \frac{R}{lc\sin\delta\frac{d\delta}{d\theta}} \tag{7.7h}$$

Differentiating equation 7.7e with respect to R:

$$\frac{d\psi}{dR} = \frac{c^2 - R^2 - l^2}{2R^2 c\sin\psi} \tag{7.7i}$$

Collecting terms from equations 7.7a, 7.7d, 7.7g, 7.7h, and 7.7i, and substituting in equation 7.7b gives the following expressions for pressure angle during a rise or fall.

$$\phi_1 = \frac{\pi}{2} - \sin^{-1}\left(\frac{c}{R}\sin\delta\right) + \tan^{-1}\left(\frac{1}{\dfrac{R^2}{lc\sin\delta\dfrac{d\delta}{d\theta}} - \dfrac{c^2 - R^2 - l^2}{2Rc\sin\psi}}\right) \tag{7.8a}$$

$$\phi_2 = -\frac{\pi}{2} + \sin^{-1}\left(\frac{c}{R}\sin\delta\right) + \tan^{-1}\left(\frac{1}{\dfrac{R^2}{lc\sin\delta\dfrac{d\delta}{d\theta}} + \dfrac{c^2 - R^2 - l^2}{2Rc\sin\psi}}\right) \tag{7.8b}$$

where δ is the lift angle of the arm for each cam position with respect to the line of centers CB between cam and follower arm pivot (in radians), and $d\delta/d\theta$ is the angular velocity v of the follower arm for each cam position (in rad/rad, i.e., dimensionless). The angle ψ varies with each cam position due to the arc motion of the follower tip. This geometry must be computed for each position of the cam for which a value of the pressure angle is desired.

The first expression for ϕ (equation 7.8a) is used for the case of the cam rotating away from the follower arm pivot as depicted in Figure 7-8. If the cam shown in that figure were rotating clockwise, it would be rotating toward the follower arm pivot and then equation 7.8b must be used. Note that these two equations differ only by three signs.

When the follower is on a dwell, the equations for pressure angle become:

$$\phi_1 = \frac{\pi}{2} - \sin^{-1}\left(\frac{c}{R}\sin\delta\right) \tag{7.8c}$$

$$\phi_2 = -\frac{\pi}{2} + \sin^{-1}\left(\frac{c}{R}\sin\delta\right) \tag{7.8d}$$

with equation 7.8c used for the case of the cam rotating away from, and 7.8d for the case of the cam rotating toward, the follower arm pivot.

7.6 RADIUS OF CURVATURE—RADIAL CAM WITH TRANSLATING ROLLER FOLLOWER

The **radius of curvature** is a *mathematical property of a function.* Its value and use is not limited to cams but has great significance in their design. The concept is simple. No matter how complicated a curve's shape may be, nor how high the degree of the function which describes it, the curve will have an instantaneous radius of curvature at every point on the curve. These radii of curvature will have instantaneous centers (which may be at infinity), and the radius of curvature of any function is itself a function which can be computed and plotted. For example, the radius of curvature of a straight line is infinity everywhere; that of a circle is a constant value. A parabola has a constantly changing radius of curvature which approaches infinity along the parabola's asymptotes. A cubic curve will have radii of curvature that are sometimes positive (convex) and sometimes negative (concave). The higher the degree of a function, in general, the more potential variety in its radius of curvature. The **radius of curvature** is the reciprocal of the **curvature** of the function, also a mathematical property of the function and its derivatives.

Cam contours are usually functions of high degree. When they are wrapped around their base or prime circles, they may have portions which are concave, convex, or flat. Infinitesimally short flats of infinite radius will occur at all inflection points on the cam surface where it changes from concave to convex or vice versa.

The radius of curvature of the finished cam is of concern regardless of the follower type, but the concerns are different for different followers. Figure 7-9 shows an obvious problem with a roller follower (greatly exaggerated) whose own (constant) radius of curvature R_f is too large to follow the locally smaller concave (negative) radius $-\rho$ on the cam.

A more subtle problem occurs when the roller follower radius R_f is larger than the smallest positive (convex) local radius $+\rho$ on the cam pitch curve. This problem is called

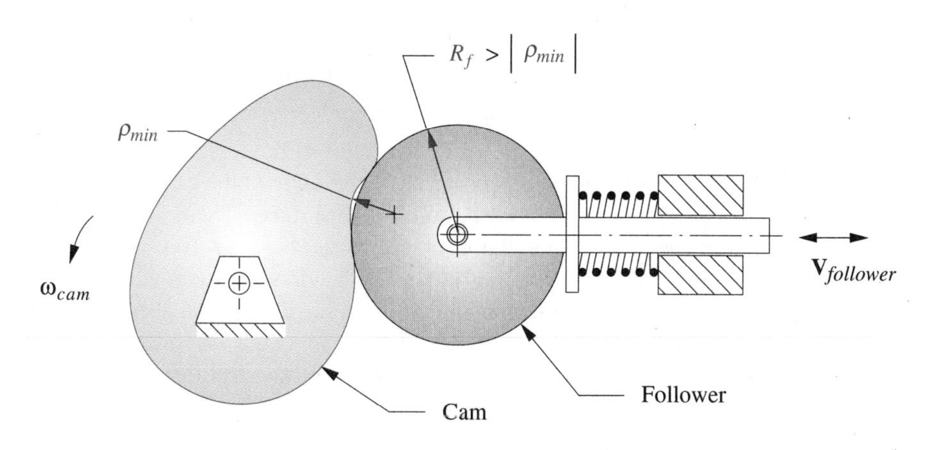

FIGURE 7-9

The result of using a roller follower larger than the one for which the cam was designed

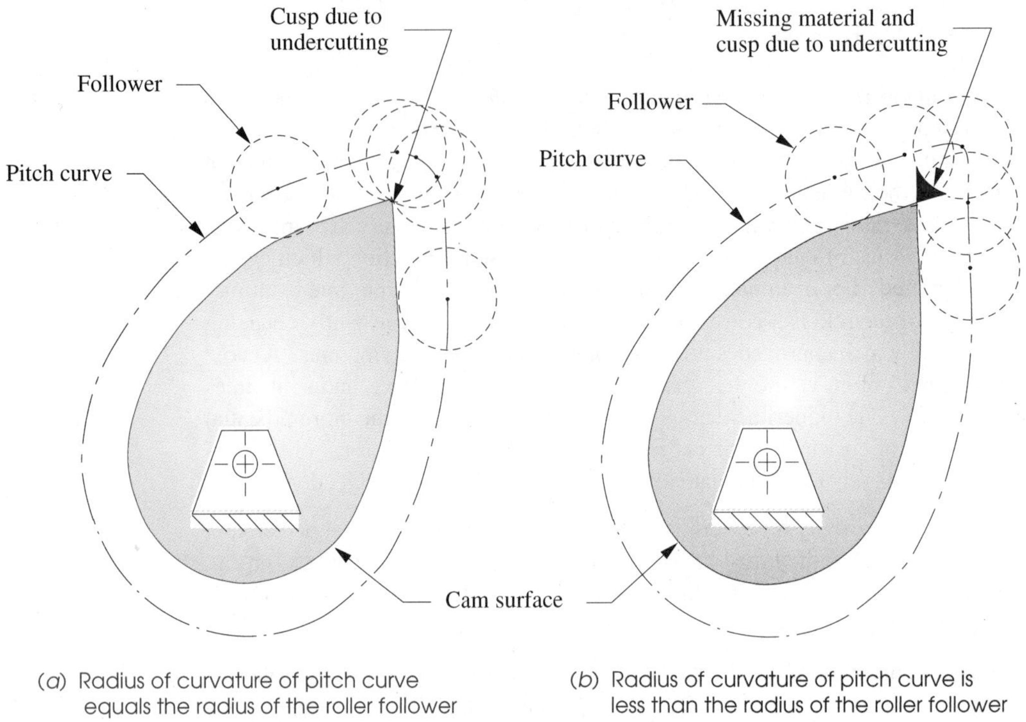

Cusp due to undercutting

Follower

Pitch curve

Missing material and cusp due to undercutting

Follower

Pitch curve

Cam surface

(a) Radius of curvature of pitch curve equals the radius of the roller follower

(b) Radius of curvature of pitch curve is less than the radius of the roller follower

FIGURE 7-10

Small positive radius of curvature can cause undercutting

undercutting and is depicted in Figure 7-10. Recall that for a roller follower cam, the cam contour is actually defined as the locus of the center of the roller follower, or the **pitch curve**. The machinist is given these *x, y* coordinate data (as a file on computer tape or disk) and also told the radius of the follower R_f. The machinist will then cut the cam with a cutter of the same effective radius as the follower, following the pitch curve coordinates with the center of the cutter.*

Figure 7-10a shows the situation in which the follower (cutter) radius R_f is at one point exactly equal to the minimum convex radius of curvature of the cam ($+\rho_{min}$). The cutter creates a perfect sharp point, or **cusp**, on the cam surface. There will be infinite contact stress at the cusp. This cam will not run very well. Figure 7-10b shows the situation in which the follower (cutter) radius is greater than the minimum convex radius of curvature of the cam. The cutter now undercuts or removes material needed for cam contours in different locations and also creates a sharp point or cusp on the cam surface. This cam no longer has the same displacement function you so carefully designed.

The rule of thumb is to keep the absolute value of the minimum radius of curvature ρ_{min} of the cam pitch curve preferably at least 1.5 to 3 times as large as the radius of the roller follower R_f. The radii of curvature of both cam and follower have a significant effect on the local surface stresses as will be discussed in Chapter 12.

* If the actual cutter diameter differs from that of the follower, a so-called cutter compensation algorithm will be executed to recalculate the path of the center of the cutter (or grinding wheel) in order to create the proper cam surface profile. This will be explored in detail in Chapter 14.

$$|\rho_{min}| >> R_f \tag{7.9}$$

A derivation for radius of curvature can be found in any calculus text. For our case of a roller follower, we can write the equation for the radius of curvature of the pitch curve of the cam as:

$$\rho_{pitch} = \frac{\left[(R_P + s)^2 + v^2\right]^{3/2}}{(R_P + s)^2 + 2v^2 - a(R_P + s)} \tag{7.10}$$

In this expression, s, v, and a are the displacement, velocity, and acceleration of the cam program as defined in a previous section. Their units are length, length/rad, and length/rad^2, respectively. R_p is the prime circle radius. **Do not confuse** this *prime circle radius* R_p with the *radius of curvature*, ρ_{pitch}. R_p is a **constant value** which you choose as a design parameter and ρ_{pitch} is the constantly changing radius of curvature which results from your design choices.

Also do not confuse R_p, the *prime circle radius* with R_f, the *radius of the roller follower*. See Figure 7-1 (p. 153) for definitions. You can choose the value of R_f to suit the problem, so you might think that it is simple to satisfy equation 7.9 by just selecting a roller follower with a small value of R_f. Unfortunately it is more complicated than that, as a small roller follower may not be strong enough to withstand the dynamic forces from the cam. The radius of the pin on which the roller follower pivots is substantially smaller than R_f because of the space needed for roller or ball bearings within the follower. Dynamic forces will be addressed in later chapters where we will revisit this problem.

We can solve equation 7.10 for ρ_{pitch} since we know s, v, and a for all values of θ and can choose a trial R_p. If the pressure angle has already been calculated, the R_p found for its acceptable values should be used to calculate ρ_{pitch} as well. If a suitable follower radius cannot be found which satisfies equation 7.4 for the minimum values of ρ_{pitch} calculated from equation 7.10, then further iteration will be needed, possibly including a redefinition of the cam specifications.

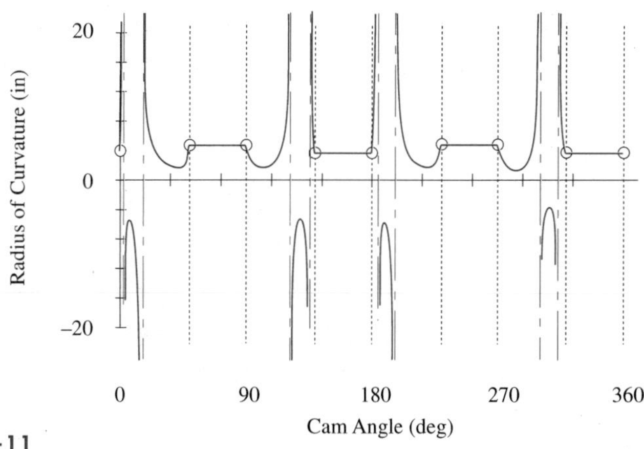

FIGURE 7-11

Radius of curvature of a four-dwell radial cam with translating roller follower

Program DYNACAM calculates ρ_{pitch} for all values of θ for a user supplied prime circle radius R_p. Figure 7-11 shows ρ_{pitch} for a four-dwell cam. Note that this cam has both positive and negative radii of curvature. The large values of radius of curvature are truncated at arbitrary levels on the plot as they are heading to infinity at the inflection points between convex and concave portions. Note that the radii of curvature go out to positive infinity and return from negative infinity or vice versa at these inflection points (perhaps after a round trip through the universe?). The radii of curvature are constant during the dwells.

Once an acceptable prime circle radius and roller follower radius are determined based on pressure angle and radius of curvature considerations and other design constraints, the cam can be drawn in finished form and subsequently manufactured. Figure 7-12 shows the profile of the four dwell cam from Figure 7-11. The cam surface contour is swept out by the envelope of follower positions just as the cutter will create the cam in metal. The sidebar shows the parameters for the design, which is an acceptable one. The ρ_{min} is 1.7 times R_f and the pressure angles are less than 30°. The contours on the cam surface appear smooth, with no sharp corners. Figure 7-13 shows the same cam with only one change. The radius of follower R_f has been made the same as the minimum radius of curvature, ρ_{min}. The sharp corners or cusps in several places indicate that undercutting has occurred. This has now become an **unacceptable cam**, *simply because of a roller follower that is too large.*

The coordinates for the cam contour, measured to the locus of the center of the roller follower, or the **pitch curve** as shown in Figure 7-11, are defined by the following expressions (equations 7.11), referenced to the center of rotation of the cam. See Figure 7-3 (p. 155) for nomenclature. The subtraction of the cam input angle θ from 2π is necessary because the relative motion of the follower versus the cam is opposite to that of the cam versus the follower. In other words, to define the x, y, coordinates of the contour

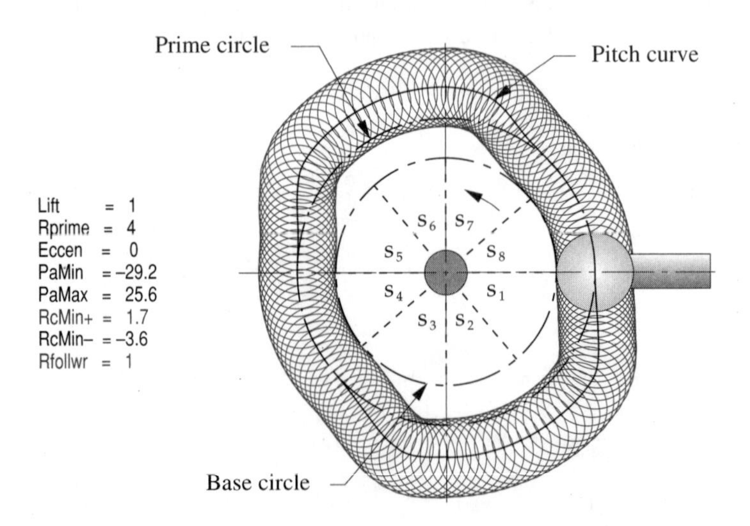

FIGURE 7-12

Radial plate cam profile is generated by the locus of the roller follower (or cutter)

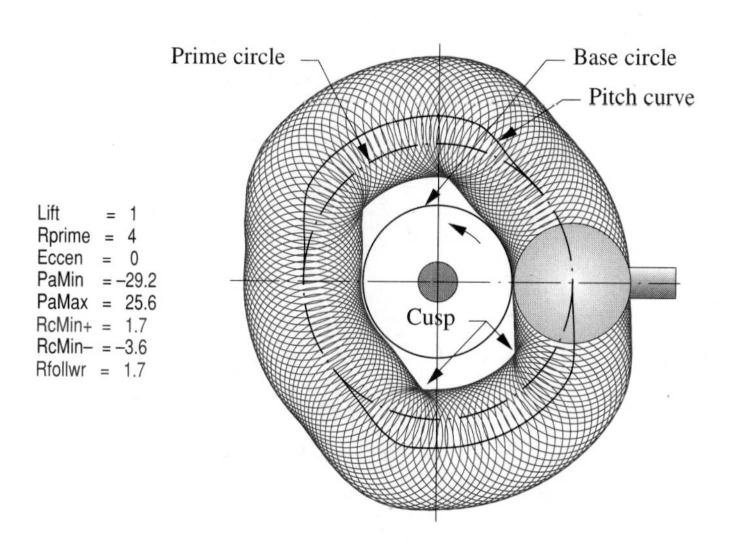

FIGURE 7-13

Cusps formed by undercutting due to radius of follower $R_f \geq$ cam radius of curvature ρ

of the centerline of the follower's path around a stationary cam, we must move the follower (and also the cutter to make the cam) in the opposite direction of cam rotation.

$$x = \cos\lambda\sqrt{(d+s)^2+\varepsilon^2}$$
$$y = \sin\lambda\sqrt{(d+s)^2+\varepsilon^2} \tag{7.11}$$

where:

$$\lambda = (2\pi - \theta) - \arctan\left(\frac{\varepsilon}{d+s}\right)$$

7.7 RADIUS OF CURVATURE—RADIAL CAM WITH TRANSLATING FLAT-FACED FOLLOWER

The situation with a flat-faced follower is different to that of a roller follower.* A negative radius of curvature on the cam cannot be accommodated with a flat-faced follower. The flat follower obviously cannot follow a concave cam. Undercutting will occur when the radius of curvature becomes negative if a cam with that condition is made.

Figure 7-14 shows a cam and flat-faced follower in an arbitrary position. The origin of the global *XY* coordinate system is placed at the cam's center of rotation, and the *X* axis is defined parallel to the common tangent, which is the surface of the flat follower. The vector **r** is attached to the cam, rotates with it, and serves as the reference line to which the cam angle θ is measured from the *X* axis. The point of contact *A* is defined by the position vector $\mathbf{R}_A$. The instantaneous center of curvature is at *C* and the radius of

* Derivations can be found in Baxter[1] and Raven.[3]

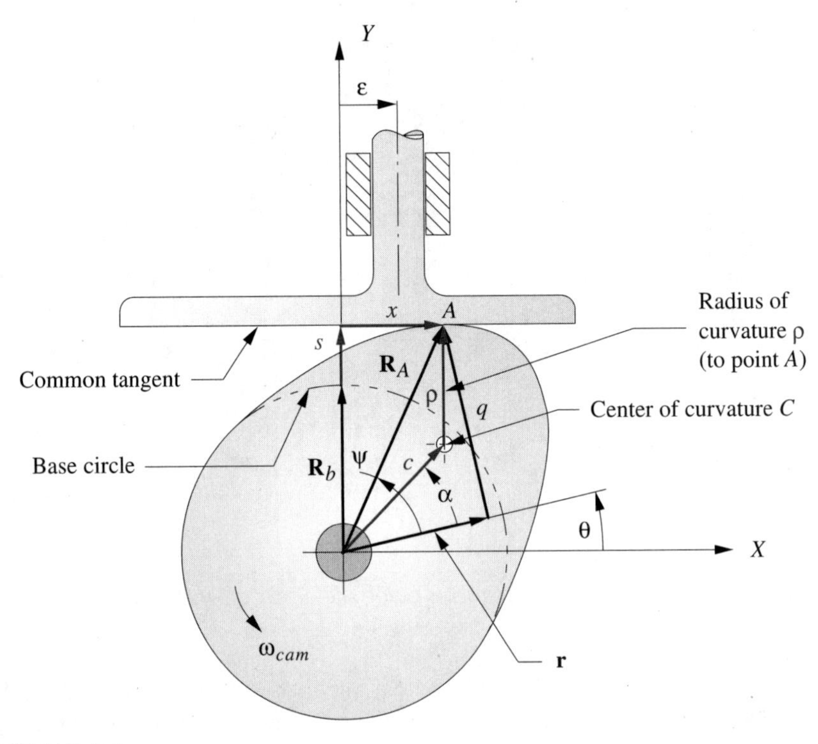

FIGURE 7-14

Geometry for derivation of radius of curvature and cam contour with flat-faced follower

curvature is ρ. R_b is the radius of the base circle and s is the displacement of the follower for angle θ. The eccentricity is ε.

We can define the location of contact point A from two vector loops (in complex number notation).

$$\mathbf{R}_A = x + j\left(R_b + s\right)$$

and

$$\mathbf{R}_A = ce^{j(\theta+\alpha)} + j\rho$$

so:

$$ce^{j(\theta+\alpha)} + j\rho = x + j\left(R_b + s\right) \tag{7.12a}$$

Substitute the Euler equation ($e^{j\tau} = \cos\tau + j\sin\tau$, where τ is a dummy variable) in equation 7.12a and separate the real and imaginary parts.

real:

$$c\cos(\theta+\alpha) = x \tag{7.12b}$$

imaginary:

$$c\sin(\theta+\alpha)+\rho = R_b + s \tag{7.12c}$$

The center of curvature C is **stationary** on the cam, meaning that the magnitudes of c and ρ, and angle α do not change for small changes in cam angle θ. (These values are not constant but are at stationary values. Their first derivatives with respect to θ are zero, but their higher derivatives are not zero.)

Differentiating equation 7.12a with respect to θ then gives:

$$jce^{j(\theta+\alpha)} = \frac{dx}{d\theta} + j\frac{ds}{d\theta} \tag{7.13}$$

Substitute the Euler equation in equation 7.13 and separate the real and imaginary parts.

real:

$$-c\sin(\theta+\alpha) = \frac{dx}{d\theta} \tag{7.14}$$

imaginary:

$$c\cos(\theta+\alpha) = \frac{ds}{d\theta} = v \tag{7.15}$$

Inspection of equations 7.12b and 7.15 shows that:

$$x = v \tag{7.16}$$

This is an interesting relationship that says the x position of the contact point between cam and follower is numerically equal to the velocity of the follower in length/rad. This means that the v diagram gives a direct measure of the necessary minimum face width of the flat follower.

$$facewidth > v_{max} - v_{min} \tag{7.17}$$

If the velocity function is asymmetric, then a minimum-width follower will have to be asymmetric also, in order not to fall off the cam.

Differentiating equation 7.16 with respect to θ gives:

$$\frac{dx}{d\theta} = \frac{dv}{d\theta} = a \tag{7.18}$$

Equations 7.12c and 7.14 can be solved simultaneously and equation 7.18 substituted in the result to yield:

$$\rho = R_b + s + a \tag{7.19}$$

BASE CIRCLE Note that equation 7.19 defines the radius of curvature in terms of the base circle radius and the displacement and acceleration functions from the $s\ v\ a\ j$ diagrams only. Because ρ cannot be allowed to become negative with a flat-faced follower, we can formulate a relationship from this equation that will predict the minimum base circle radius R_b needed to avoid undercutting. The only factor on the right side of equation 7.19 that can be negative is the acceleration, a. We have defined s to be always

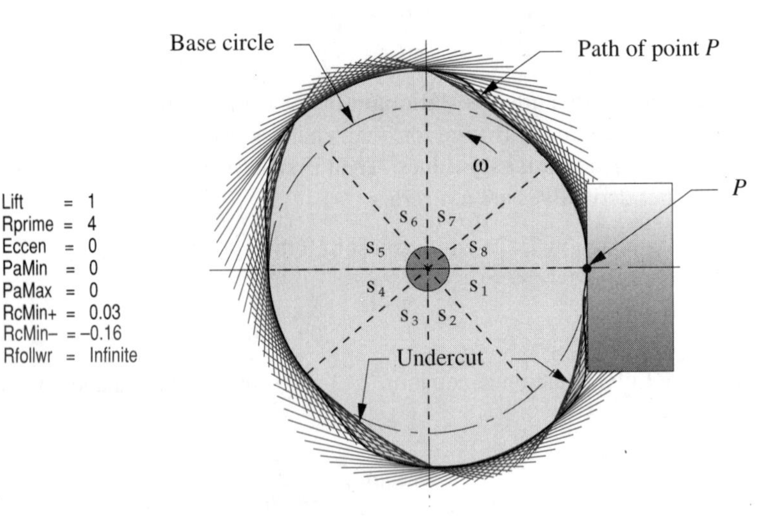

7

FIGURE 7-15

Undercutting due to negative radius of curvature used with flat-faced follower

positive, as is R_b. Therefore, the worst case for undercutting will occur when a is near its **largest negative value**, a_{min}, whose value we know from the a diagram. The minimum base circle radius can then be defined as:

$$R_{b_{\min}} > \rho_{\min} - s_{@a_{\min}} - a_{\min} \tag{7.20}$$

Note that the value of s in this equation is taken at the cam angle θ corresponding to that of a_{min}. Because the value of a_{min} is negative and it is also negated in equation 7.20, it dominates the expression. To use this relationship, we must choose some minimum radius of curvature ρ_{min} for the cam surface as a design parameter. Since the hertzian contact stresses at the contact point are a function of local radius of curvature, that criterion can be used to select ρ_{min}. That topic is discussed in a later chapter.

Note that none of the equations developed above for this case involves the **eccentricity**, ε. It is only a factor in cam size when a roller follower is used. It does not affect the geometry of a flat follower cam.

Figure 7-15 shows the result of trying to use a flat-faced follower on a cam whose theoretical path of follower point P has negative radius of curvature due to a base circle radius that is too small. If the follower tracked the path of P as is required to create the motion function defined in the s diagram, the cam surface would actually be as developed by the envelope of straight lines shown. But, these loci of the follower face are cutting into cam contours that are needed for other cam angles. The line running through the forest of follower loci is the path of point P needed for this design. The undercutting can be clearly seen as the crescent-shaped missing pieces at four places between the path of P and the follower face loci. Note that if the follower were zero width (at point P), it would work kinematically, but the stresses would be a bit high, in fact infinite.

7.8 RADIUS OF CURVATURE—BARREL CAM WITH TRANSLATING ROLLER FOLLOWER

The radius of curvature of a barrel cam with a translating follower is the same as the radius of curvature of the *s* diagram when scaled to the prime cylinder circumference since there is no distortion of that function in the direction of follower motion when "wrapped" around the curved-surface plane of a cylinder. It is found from:

$$\rho = -\frac{\left[1+\left(\frac{V}{\omega R_p}\right)^2\right]^{\frac{3}{2}}}{\frac{A}{\omega^2 R_p^2}} \tag{7.21}$$

where ω is camshaft angular velocity in rad/sec, *Rp* is the barrel prime cylinder radius in length units, *V* is the follower velocity in length/sec and *A* is the follower acceleration in length/sec^2. On a dwell, and anywhere else the acceleration is zero, equation 7.21 cannot be numerically evaluated, and $\rho = \infty$.

7.9 RADIUS OF CURVATURE—BARREL CAM WITH OSCILLATING ROLLER FOLLOWER

The radius of curvature of a barrel cam with an oscillating follower is affected by the radius and direction of initial motion of the follower arm in addition to the factors in equation 7.21, which is for the translating follower case. (See Figure 7-6 on p. 160).

$$\rho = \frac{\operatorname{sgn}(x_0)\operatorname{sgn}(\omega_{arm})\left\{R_p^2\omega^2 + 2\operatorname{sgn}(\omega_{arm})R_p R_a V\omega_{arm}\sin[\psi(\theta)] + R_a^2 V^2\right\}^{\frac{3}{2}}}{R_a\left\{-R_p\omega\sin[\psi(\theta)]V^2 + \operatorname{sgn}(\omega_{arm})R_a V^3 + \operatorname{sgn}(\omega_{arm})R_p\omega\cos[\psi(\theta)]A\right\}} \tag{7.22}$$

where ω is camshaft angular velocity in rad/sec, *Rp* is the barrel prime cylinder radius in length units, *Ra* is the follower arm radius in length units, *V* is the follower arm angular velocity in rad/sec, *A* is the follower arm angular acceleration in rad/sec^2 and $\psi(\theta)$ is the angle of the follower arm as defined in equation 7.5c. The *sgn* function defines the arm motion as CCW or CW on the rise. On a dwell, or at any other point where the acceleration and velocity are simultaneously zero, equation 7.22 cannot be numerically evaluated, and $\rho = \infty$. Note that elsewhere the radius of curvature will vary radially as well as circumferentially over the surface of the cam because it is a function of both cam radius and cam angle θ. Equation 7.22 calculates it at radius *Rp*, defined at the radial center of the cam track.

Note that the radius of curvature at the centerline of the roller follower will also vary due to the arc motion of the oscillating arm that moves out of the plane of the axis of the cam during its sweep, effectively changing the cam pitch radius at which the center of the roller contacts the slot. If the arm radius is reasonably long in respect to the lift of the cam and has a small angular excursion, then this error will be small. If not then the nominal pitch cylinder radius needs to be corrected by substituting *Rp'* for *Rp* from equation 7.5e in equation 7.22.

7.10 RADIUS OF CURVATURE—RADIAL CAM WITH OSCILLATING ROLLER FOLLOWER

Derivations for the radius of curvature in the case of a radial cam with an oscillating arm roller follower have been developed by Baxter,[1] Kloomock and Muffley,[4] and Raven.[3] We will use Baxter's method here. See that reference for the complete derivation.

The geometry of a radial cam with an oscillating arm roller follower is shown in Figure 7-8 (p. 162). The radius of curvature of the pitch curve ρ when the follower velocity is nonzero is found from

$$\rho = \frac{c^2 \sin^2 \delta}{\cos\phi\left[c \sin\delta - l\cos\phi\left(a\cos\phi + v(v-1)\sin\phi\right)\right]} \tag{7.23a}$$

where δ, c, and l, are defined in Figure 7-8, ϕ is the signed pressure angle, v is the follower velocity in units of rad/rad, and a is the follower acceleration in units of rad/rad^2. When $v = 0$, the radius of curvature of the pitch curve is found from the law of cosines

$$\rho = \sqrt{c^2 + l^2 - 2cl\cos\delta} \tag{7.23b}$$

The radius of curvature of the cam surface ρ_c is

$$\rho_c = \rho - R_f \tag{7.24}$$

where Rf is the radius of the roller follower.

7.11 RADIUS OF CURVATURE—RADIAL CAM WITH OSCILLATING FLAT-FACED FOLLOWER

Derivations for the radius of curvature in the case of a radial cam with an oscillating flat-faced follower have been developed by Baxter[1] and Raven.[3] We will use Baxter's method here. See that reference for the complete derivation.

The geometry of a radial cam with an oscillating flat-faced follower is shown in Figure 7-16. The radius of curvature of the cam surface is found from

$$\rho = \frac{c}{(1+v)^2}\left[(1-2v)\sin\delta + \frac{a}{(1-v)}\cos\delta\right] + f \tag{7.25a}$$

where c and δ are as defined in Figure 7-16, v is the follower velocity in units of rad/rad, and a is the follower acceleration in units of rad/rad^2. When $v = 0$, the radius of curvature of the cam surface becomes

$$\rho = c(\sin\delta + a\cos\delta) + f \tag{7.25b}$$

Note that any offset f of the follower face from its centerline is positive if directed away from the cam center and negative if toward the cam center.

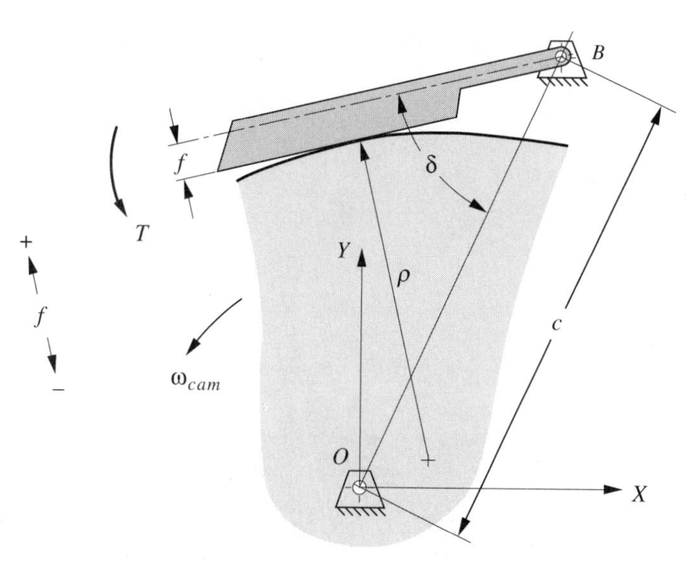

FIGURE 7-16

Geometry for radius of curvature in a radial cam with oscillating flat-faced follower

Undercutting of Radial Cams with Oscillating Flat-Faced Followers

A radial cam with translating flat-faced follower that has positive radius of curvature of the theoretical follower locus, at all points will not show undercutting. This is not necessarily true of the radial cam with an oscillating flat-faced follower. It is necessary but not sufficient to have positive radius of curvature to avoid undercutting in this case.

Figure 7-17 shows a cam that has positive radius of curvature ρ everywhere on its theoretical follower locus but nevertheless shows undercutting in two regions. This is due to the geometry of the arm pivot location in combination with the chosen base circle radius. It is sometimes necessary to increase the base circle radius beyond the point that eliminates negative radius of curvature on the follower path in order to eliminate undercutting. In effect, the straight line from the desired cam-follower contact point for a given cam angle back to the arm pivot sometimes cuts through a portion of cam contour needed to correctly position the follower at an earlier or later angular position of the cam.

On balance, there is little if anything to recommend the use of an oscillating arm flat-faced follower over an oscillating arm roller follower. The roller follower will in general have less friction and can usually be designed for very long life. It is also very easily replaced when its surface or bearing fails and is a purchased item, whereas flat followers are typically custom made and so are more expensive in small quantities. Flat faced followers are very seldom used in industrial machinery for these reasons. They are sometimes used in IC engine valve trains where volume production makes them inexpensive, but this author has yet to see one used in an industrial machine in his experience.

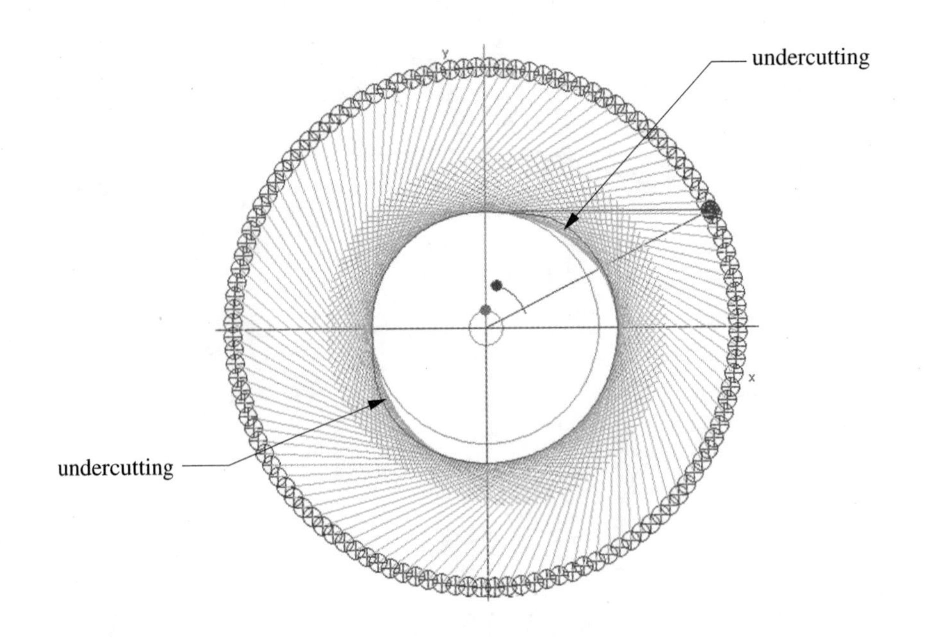

FIGURE 7-17

Undercutting on a radial cam with oscillating flat-faced follower and positive ρ

7.12 REFERENCES

1 **Baxter, M. L.** (1948). "Curvature-Acceleration Relations for Plane Cams." *Trans ASME*, pp. 483-489.

2 **Kloomok, M., and R. V. Muffley**. (1955). "Plate Cam Design-Pressure Angle Analysis". *Product Engineering*, v. 26, May, 1955, pp. 155-171.

3 **Raven, F. H.** (1959). "Analytical Design of Disk Cams and Three-Dimensional Cams by Independent Position Equations." *Trans ASME*, **26 Series E**(1), pp. 18-24.

4 **Kloomok, M., and R. V. Muffley**. (1955). "Plate Cam Design-Radius of Curvature". *Product Engineering*, v. 26, September, 1955, pp. 186-201.

5 **Dresner, T. L.**, and **K. W. Buffington**. (1991). "Definition of Pressure and Transmission Angles Applicable to Multi-Input Mechanisms." *Journal of Mechanical Design*, 113(4), p. 495.

Chapter 8

DYNAMICS OF CAM SYSTEMS—MODELING FUNDAMENTALS

8.0 INTRODUCTION

This chapter presents a review of the fundamentals of dynamic modeling in order to establish a base of information on which to develop the tools for the dynamic analysis of cam-follower systems in succeeding chapters.

8.1 NEWTON'S LAWS OF MOTION

Dynamic force analysis involves the application of **Newton's** three **laws of motion** which are:

1 A body at rest tends to remain at rest and a body in motion will tend to maintain its velocity unless acted upon by an external force.

2 The time rate of change of momentum of a body is equal to the magnitude of the applied force and acts in the direction of the force.

3 For every action force, there is an equal and opposite reaction force.

The second law is expressed in terms of rate of change of *momentum*, $\mathbf{P} = m\mathbf{v}$, where m is mass and $\mathbf{v}$ is velocity. Mass m is assumed to be constant in this analysis. The time rate of change of $m\mathbf{v}$ is $m\mathbf{a}$, where $\mathbf{a}$ is the acceleration of the mass center.

$$\mathbf{F} = m\mathbf{a} \tag{8.1}$$

$\mathbf{F}$ is the resultant of all forces on the system acting at the mass center.

We can differentiate between two subclasses of dynamics problems depending upon which quantities are known and which are to be found. The "**forward dynamics problem**" is the one in which we know everything about the external loads (forces and/or torques) being exerted on the system, and we wish to determine the accelerations, velocities, and displacements that result from the application of those forces and torques. This subclass is typical of problems such as determining the acceleration of a block sliding down a plane, acted upon by gravity. Given $\mathbf{F}$ and m, solve for $\mathbf{a}$.

The second subclass of dynamics problem, called the "**inverse dynamics problem**," is one in which we know the (desired) accelerations, velocities, and displacements to be imposed upon our system and wish to solve for the magnitudes and directions of the forces and torques that are necessary to provide the desired motions and which result from them. This inverse dynamics case is sometimes also called **kinetostatics**. Given **a** and *m*, solve for **F**.

Whichever subclass of problem is addressed, it is important to realize that they are both dynamics problems. Each merely solves **F** = *m***a** for a different variable. To do so we should first review some fundamental geometric principles and mass properties that are needed for the calculations.

8.2 DYNAMIC MODELS

It is often convenient in dynamic analysis to create a simplified model of a complicated part. These models are sometimes considered to be a collection of point masses connected by massless rods, referred to as a lumped-parameter model, or just lumped model. For a lumped model of a rigid body to be dynamically equivalent to the original body, three things must be true:

1 The mass of the model must equal that of the original body.

2 The center of gravity must be in the same location as that of the original body.

3 The mass moment of inertia must equal that of the original body.

8.3 MASS

Mass is not weight. Mass is an invariant property of a rigid body. The weight of the same body varies depending on the gravitational system in which it sits. We will assume the mass of our parts to be constant over time in our calculations.

When designing cam-follower systems (or any machinery), we must first do a complete kinematic analysis of our design in order to obtain information about the rigid body accelerations of the moving parts. We can then use Newton's second law to calculate the dynamic forces. But to do so, we need to know the masses of all the moving parts that have these known accelerations. If we have a design of the follower train done in a CAD program that will calculate masses and mass moments of inertia, then we are in good shape, as the data needed for dynamic calculations are available. Lacking that luxury, we will have to calculate estimates of the mass properties of the follower train to do the dynamics calculations.

Absent a solids-modeler representation of your design, a first estimate of your parts' masses can be obtained by assuming some reasonable shapes and sizes for all the parts and choosing appropriate materials. Then calculate the volume of each part and multiply its volume by the material's **mass density** (not weight density) to obtain a first approximation of its mass. These mass values can then be used in Newton's equation to estimate the dynamic forces.

How will we know whether our chosen sizes and shapes of links are even acceptable, let alone optimal? Unfortunately, we will not know until we have carried the com-

Portions of this chapter were adapted from R. L. Norton, *Design of Machinery* 2ed, McGraw-Hill, 2001, with permission.

putations all the way through a complete stress and deflection analysis of the parts. It is often the case, especially with long, thin elements such as shafts or slender links, that the deflections of the parts under their dynamic loads will limit the design even at low stress levels. In other cases the stresses at design loads will be excessive.

If we discover that the parts fail or deflect excessively under the dynamic forces, then we will have to go back to our original assumptions about the shapes, sizes, and materials of these parts, redesign them, and repeat the force, stress, and deflection analyses. Design is, unavoidably, an **iterative process**.

We need the dynamic forces to do the stress analyses on our parts. (Stress analysis is addressed in a later chapter.) It is also worth noting that, unlike a static force situation in which a failed design might be fixed by adding more mass to the part to strengthen it, to do so in a dynamic-force situation can have a deleterious effect. More mass with the same acceleration will generate even higher forces and thus higher stresses. The machine designer often needs to remove mass (in the right places) from parts in order to reduce the stresses and deflections due to $\mathbf{F} = m\mathbf{a}$. The designer needs to have a good understanding of both material properties and stress and deflection analysis to properly shape and size parts for minimum mass while maximizing strength and stiffness to withstand dynamic forces.

8

8.4 MASS MOMENT AND CENTER OF GRAVITY

When the mass of an object is distributed over some dimensions, it will possess a moment with respect to any axis of choice. Figure 8-1 shows a mass of general shape in an *xyz* axis system. A differential element of mass is also shown. The **mass moment (first moment of mass)** of the differential element is equal to the **product of its mass and its distance** along the axis of interest. With respect to the *x*, *y*, and *z* axes these are:

$$dM_x = r_x^1 dm \tag{8.2a}$$

$$dM_y = r_y^1 dm \tag{8.2b}$$

$$dM_z = r_z^1 dm \tag{8.2c}$$

The radius from the axis of interest to the differential element is shown with an exponent of 1 to emphasize the reason for this property being called the first moment of mass. To obtain the mass moment of the entire body we integrate each of these expressions.

$$M_x = \int r_x \, dm \tag{8.3a}$$

$$M_y = \int r_y \, dm \tag{8.3b}$$

$$M_z = \int r_z \, dm \tag{8.3c}$$

If the mass moment with respect to a particular axis is numerically zero, then that axis passes through the **center of mass (*CM*)** of the object, which for earthbound systems is coincident with its **center of gravity (*CG*)**. By definition, the summation of first moments about all axes through the center of gravity is zero. We will need to locate the

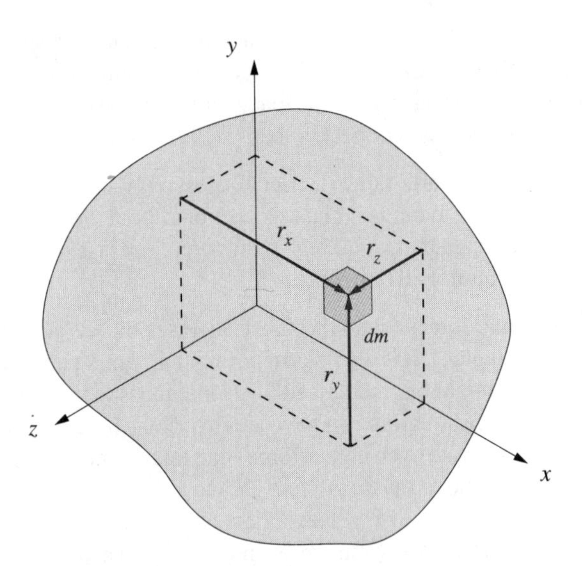

FIGURE 8-1

A generalized mass element in a 3-D coordinate system

8

CG of all moving bodies in our designs because the linear acceleration component of each body is calculated as acting at that point.

It is often convenient to model a complicated shape as several interconnected simple shapes whose individual geometries allow easy computation of their masses and the locations of their local *CGs*. The global *CG* can then be found from the summation of the first moments of these simple shapes set equal to zero. Appendix C contains formulas for the volumes and locations of centers of gravity of some common shapes. Of course, if the system is designed in a solids modeling CAD package, then the mass and other properties can be automatically calculated.

Figure 8-2 shows a simple model of a mallet broken into two cylindrical parts, the handle and the head, which have masses m_h and m_d, respectively. The individual centers of gravity of the two parts are at l_d and $l_h/2$, respectively, with respect to the axis *ZZ*. We want to find the location of the composite center of gravity of the mallet with respect to *ZZ*. Summing the first moments of the individual components about *ZZ* and setting them equal to the moment of the entire mass about *ZZ* gives.

$$\sum M_{ZZ} = m_h \frac{l_h}{2} + m_d l_d = (m_h + m_d) d \tag{8.3d}$$

This equation can be solved for the distance d along the x axis, which, in this symmetrical example, is the only dimension of the composite *CG* not discernible by inspection. The y and z components of the composite *CG* are both zero.

$$d = \frac{m_h \frac{l_h}{2} + m_d l_d}{(m_h + m_d)} \tag{8.3e}$$

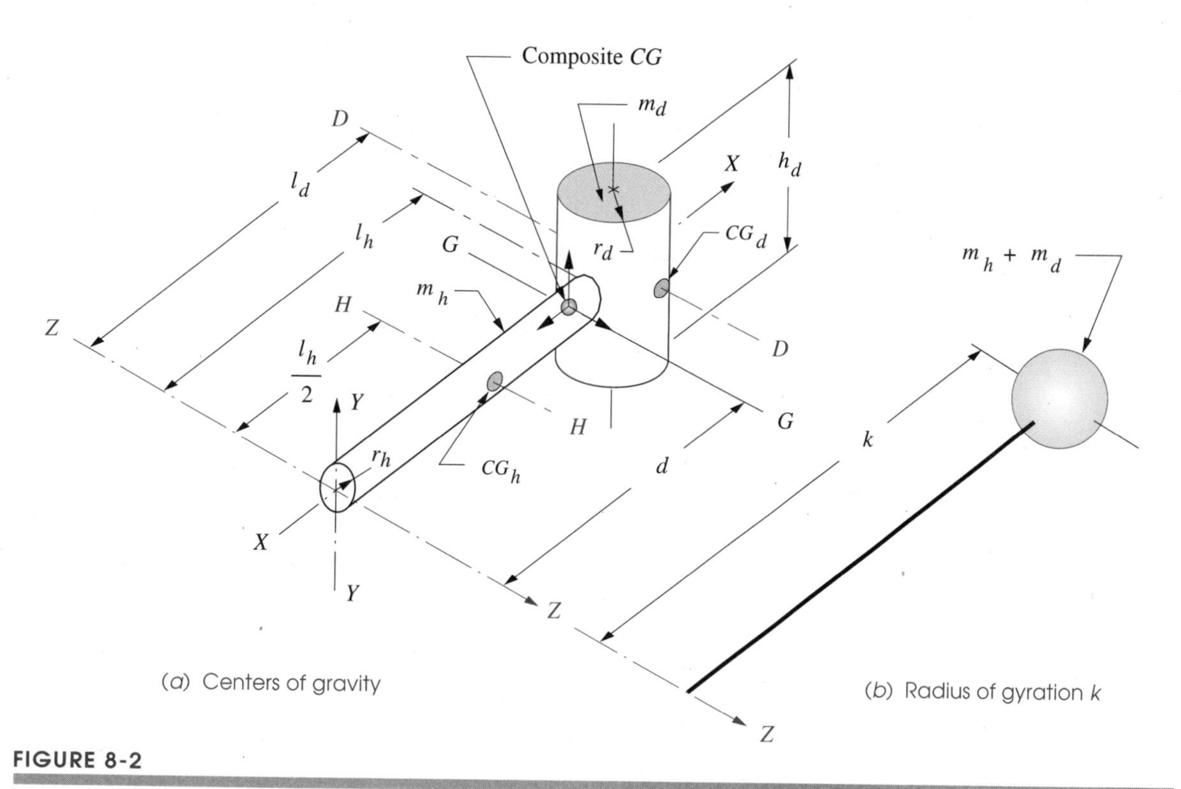

FIGURE 8-2

Dynamic models, composite center of gravity, and radius of gyration of a mallet

8.5 MASS MOMENT OF INERTIA (SECOND MOMENT OF MASS)

Newton's law applies to systems in rotation as well as to those in translation. The rotational form of Newton's second law is:

$$\mathbf{T} = I\alpha \tag{8.4}$$

where **T** is resultant torque about the mass center, α is angular acceleration, and I is mass moment of inertia about an axis through the mass center.

The mass moment of inertia is referred to some axis of rotation, usually one through the *CG*. Look again at Figure 8-1, which shows a mass of general shape and an *xyz* axis system. A differential element of mass is also shown. The **mass moment of inertia** of the differential element is equal to the **product of its mass and the square of its distance** from the axis of interest. With respect to the *x, y,* and *z* axes they are:

$$dI_x = r_x^2\,dm = (y^2 + z^2)\,dm \tag{8.5a}$$

$$dI_y = r_y^2\,dm = (x^2 + z^2)\,dm \tag{8.5b}$$

$$dI_z = r_z^2\,dm = (x^2 + y^2)\,dm \tag{8.5c}$$

The exponent of 2 on the radius term gives this property its other name of **second moment of mass**. To obtain the mass moments of inertia of the entire body we integrate each of these expressions.

$$I_x = \int (y^2 + z^2)dm \tag{8.6a}$$

$$I_y = \int (x^2 + z^2)dm \tag{8.6b}$$

$$I_z = \int (x^2 + y^2)dm \tag{8.6c}$$

While it is fairly intuitive to appreciate the physical significance of the first moment of mass, it is more difficult to do the same for the second moment, or moment of inertia.

Consider equation 8.4. It says that torque is proportional to angular acceleration, and the constant of proportionality is this moment of inertia, *I*. Picture a common hammer or mallet as depicted in Figure 8-2 (p. 181). The head, made of steel, has large mass compared to the light wooden handle. When gripped properly, near the end of the handle, the radius to the mass of the head is large. Its contribution to the total *I* of the mallet is proportional to the square of the radius from the axis of rotation (your wrist at axis *ZZ*) to the head. Thus it takes considerably more torque to swing (and thus angularly accelerate) the mallet when it is held properly than if held near the head. In a translating system, kinetic energy (KE) is:

8

$$KE = \frac{1}{2}mv^2 \tag{8.7a}$$

and in a rotating system kinetic energy is:

$$KE = \frac{1}{2}I\omega^2 \tag{8.7b}$$

Thus the kinetic energy stored in the mallet is also proportional to its moment of inertia *I* and to ω^2. You can see that holding the mallet close to its head reduces the *I* and lowers the energy available for driving the nail.

Moment of inertia, then, is one indicator of the ability of the rotating body to store kinetic energy. It is also an indicator of the amount of torque that will be needed to rotationally accelerate the body. Unless you are designing a device intended for the storage and transfer of large amounts of energy (punch press, drop hammer, rock crusher, etc.), you will probably be trying to minimize the moments of inertia of the rotating parts. Just as mass is a measure of resistance to linear acceleration, moment of inertia is a measure of resistance to angular acceleration. A large *I* will require a large driving torque and thus a larger and more powerful motor to obtain the same acceleration. Later we will see how to make moment of inertia work for us in rotating machinery by using flywheels with large *I*. The units of moment of inertia can be determined by doing a unit balance on either equation 8.4 or equations 8.7. In the **ips** system, they are lb-in-sec^2 or bl-in. In the **SI** system, they are N-m-sec^2 or kg-m^2.

8.6 PARALLEL AXIS THEOREM (TRANSFER THEOREM)

The moment of inertia of a body with respect to any axis *(ZZ)* can be expressed as the sum of its moment of inertia about an axis *(GG)* parallel to *ZZ* through its *CG*, and the square of the perpendicular distance between those parallel axes times the body mass:

$$I_{ZZ} = I_{GG} + md^2 \tag{8.8}$$

where *ZZ* and *GG* are parallel axes, *GG* goes through the *CG* of the body or assembly, m is the mass of the body or assembly, and d is the perpendicular distance between the parallel axes. This property is most useful when computing the moment of inertia of a complex shape which has been broken into a collection of simple shapes as shown in Figure 8-2a, which represents a simplistic model of a mallet. The mallet was broken into two cylindrical parts, the handle and the head, which have masses m_h and m_d, and radii r_h and r_d, respectively. The expressions for the mass moments of inertia of a cylinder with respect to axes through its *CG* can be found in Appendix C. For the handle about its *CG* axis *HH* the expression is:

$$I_{HH} = \frac{m_h\left(3r_h^2 + l_h^2\right)}{12} \tag{8.9a}$$

and for the head about its *CG* axis *DD*:

$$I_{DD} = \frac{m_d\left(3r_d^2 + h_d^2\right)}{12} \tag{8.9b}$$

Using the parallel axis theorem to transfer the moment of inertia to the axis *ZZ* at the end of the handle:

$$I_{ZZ} = \left[I_{HH} + m_h\left(\frac{l_h}{2}\right)^2\right] + \left[I_{DD} + m_d l_d^{\,2}\right] \tag{8.9c}$$

8.7 RADIUS OF GYRATION

The **radius of gyration** of a body is defined as the radius at which the entire mass of the body could be concentrated such that the resulting model will have the same moment of inertia as the original body. The mass of this model must be the same as that of the original body. Let I_{ZZ} represent the mass moment of inertia about *ZZ* from equation 8.9c and m the mass of the original body. From the parallel axis theorem, a concentrated mass m at a radius k will have a moment of inertia:

$$I_{ZZ} = mk^2 \tag{8.10a}$$

Since we want I_{ZZ} to be equal to the original moment of inertia, the required **radius of gyration** at which we will concentrate the mass m is then:

$$k = \sqrt{\frac{I_{ZZ}}{m}} \tag{8.10b}$$

Note that this property of radius of gyration allows the construction of an even simpler dynamic model of the system in which all the system mass is concentrated in a "point mass" at the end of a massless rod of length k. Figure 8-2b shows such a model of the mallet in Figure 8-2a.

By comparing equation 8.10a with equation 8.8, it can be seen that the radius of gyration k will always be larger than the radius to the composite CG of the original body.

$$I_{CG} + md^2 = I_{ZZ} = mk^2 \qquad \therefore k > d \tag{8.10c}$$

8.8 MODELING ROTATING LINKS

Many cam-follower trains contain relatively long rotating links such as the ones shown in Figure 8-10 (p. 198). As a first approximation, it is possible to model these links as lumped masses in translation. The error in so doing will be acceptably small if the angular rotation of the link is small. Then the difference between the length of the arc over a small angle and its chord is small.

The goal is to model the distributed mass of the rotating link as a lumped, point mass placed at the point of attachment to its adjacent link, connected to the pivot by a massless rod. Figure 8-3 shows a link, rotating about an axis *ZZ*, and its lumped dynamic model. The mass of the lump placed at the link radius r must have the same moment of inertia about the pivot as the original link. The mass moment of inertia I_{ZZ} of the original link must be known or calculated. This is most easily done in a solids modeling CAD system. Lacking that facility, I_{ZZ} can be found by breaking the link into regular shapes for which the I_{GG} is easily calculated (see Appendix C). These portions of the link are then combined using the transfer theorem (equation 8.8) to obtain an estimate of the overall I_{ZZ}. The mass moment of inertia of a point mass at a radius is found from the transfer theorem. Since a point mass, by definition, has no dimension, its moment of inertia I_{GG} about its center of mass is zero and equation 8.8 reduces to

$$I_{ZZ} = mr^2 \tag{8.11a}$$

The effective mass m_{eff} to be placed at the radius r is then

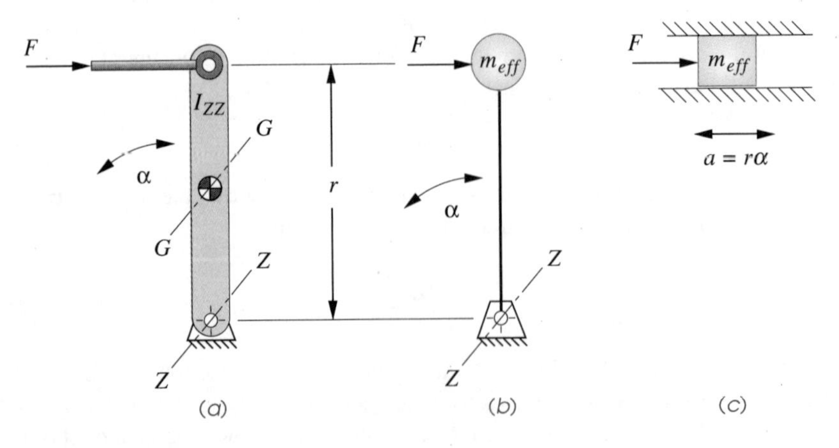

FIGURE 8-3

Modeling a rotating link as a translating mass

$$m_{eff} = \frac{I_{ZZ}}{r^2} \tag{8.11b}$$

For small angles of rotation, the rotating link can then be modeled as a mass m_{eff} in pure rectilinear translation for inclusion in a model such as that shown in Figure 8-6 (p. 189).

8.9 LUMPED PARAMETER DYNAMIC MODELS

Figure 8-4a shows a simple plate or disk cam driving a spring-loaded roller follower. This is a force-closed system which depends on the spring force to keep the cam and follower in contact at all times. Figure 8-4b shows a lumped parameter model of this system in which all the **mass** that moves with the follower train is lumped together as m, all the springiness in the system is lumped within the **spring rate** k, and all the **damping** or resistance to movement is lumped together as a damper with coefficient c. The sources of mass which contribute to m are fairly obvious. The mass of the follower stem, the roller, its pivot pin, and any other hardware attached to the moving assembly all add together to create m. Figure 8-4c shows the free-body diagram of the system acted upon by the cam force F_c, the spring force F_s, and the damping force F_d. There will, of course, also be the effects of mass times acceleration on the system.

Spring Rate

We have been assuming all links and parts to be rigid bodies in order to do the kinematic analyses, but to do a more accurate force analysis we need to recognize that these bodies are not truly rigid. The springiness in the system is assumed to be linear, thus describable by a spring rate k. A spring rate is defined as the force per unit deflection.

$$k = \frac{F_s}{x} \tag{8.12}$$

The total spring rate k of the system is a combination of the spring rates of the actual coil spring, plus the spring rates of all other parts that are deflected by the forces. The roller, its pin, and the follower stem are all springs themselves as they are made of elastic materials. The spring rate for any part can be obtained from the equation for its deflection under the applied loading. Any deflection equation relates force to displacement and can be algebraically rearranged to express a spring rate. An individual part may have more than one k if it is loaded in several modes as, for example, a camshaft with a spring rate in bending and also one in torsion. We will discuss the procedures for combining these various spring rates in the system together into a combined, effective spring rate k in the next section. For now, let us just assume that we can so combine them for our analysis and create an overall k for our lumped parameter model.

Damping

The friction, more generally called **damping**, is the most difficult parameter of the three to model. It needs to be a combination of all the damping effects in the system. These may be of many forms. **Coulomb friction** results from two dry or lubricated surfaces rubbing together. The contact surfaces between cam and follower and between the follower and its sliding joint can experience coulomb friction. It is generally considered to be independent of velocity magnitude, but has a different, larger value when velocity is zero (static friction force F_{st} or *stiction*) than when there is relative motion between the parts (dynamic friction F_d). Figure 8-5a shows a plot of coulomb friction force versus relative velocity v at the contact surfaces. Note that friction always opposes motion, so the friction force abruptly changes sign at $v = 0$. The stiction F_{st} shows up as a larger spike at zero v than the dynamic friction value F_d. Thus, this is a **nonlinear** friction function. It is multivalued at zero. In fact, at zero velocity, the friction force can be any value between $-F_{st}$ and $+F_{st}$. It will be whatever force is needed to balance the system forces and create equilibrium. When the applied force exceeds F_{st}, motion begins and the friction force suddenly drops to F_d. This nonlinear damping creates difficulties in our simple model since we want to describe our system with linear differential equations having known solutions.

Other sources of damping may be present besides coulomb friction. **Viscous damping** results from the shearing of a fluid (lubricant) in the gap between the moving parts and is considered to be a linear function of relative velocity as shown in Figure 8-5b. **Quadratic damping** results from the movement of an object through a fluid medium as with an automobile pushing through the air or a boat through the water. This factor is a fairly negligible contributor to a cam-follower's overall damping unless the speeds are very high or the fluid medium very dense. Quadratic damping is a function of the square of the relative velocity as shown in Figure 8-5c. The relationship of the dynamic damping force F_d as a function of relative velocity for all these cases can be expressed as:

$$F_d = cv|v|^{r-1} \tag{8.13a}$$

where c is the constant damping coefficient, v is the relative velocity, and r is a constant which defines the type of damping.

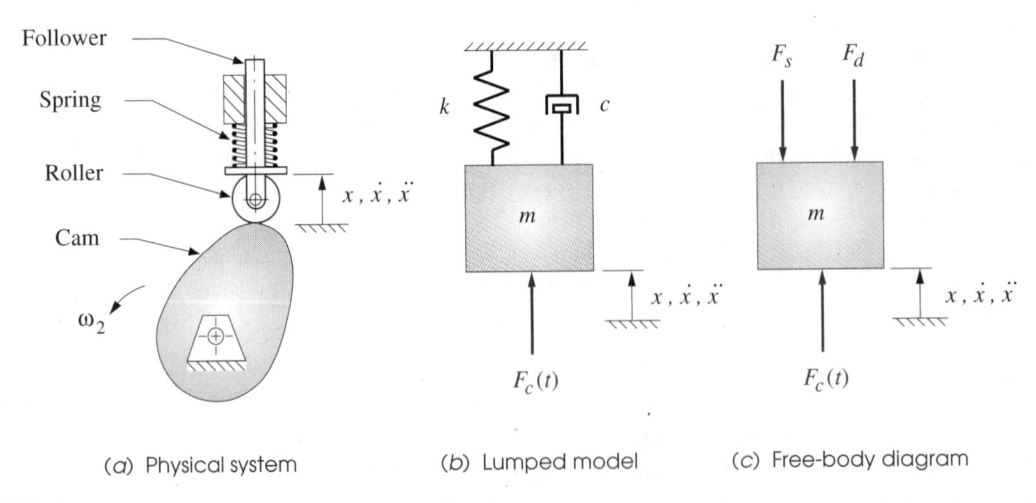

FIGURE 8-4

One-*DOF* lumped parameter model of a cam-follower system

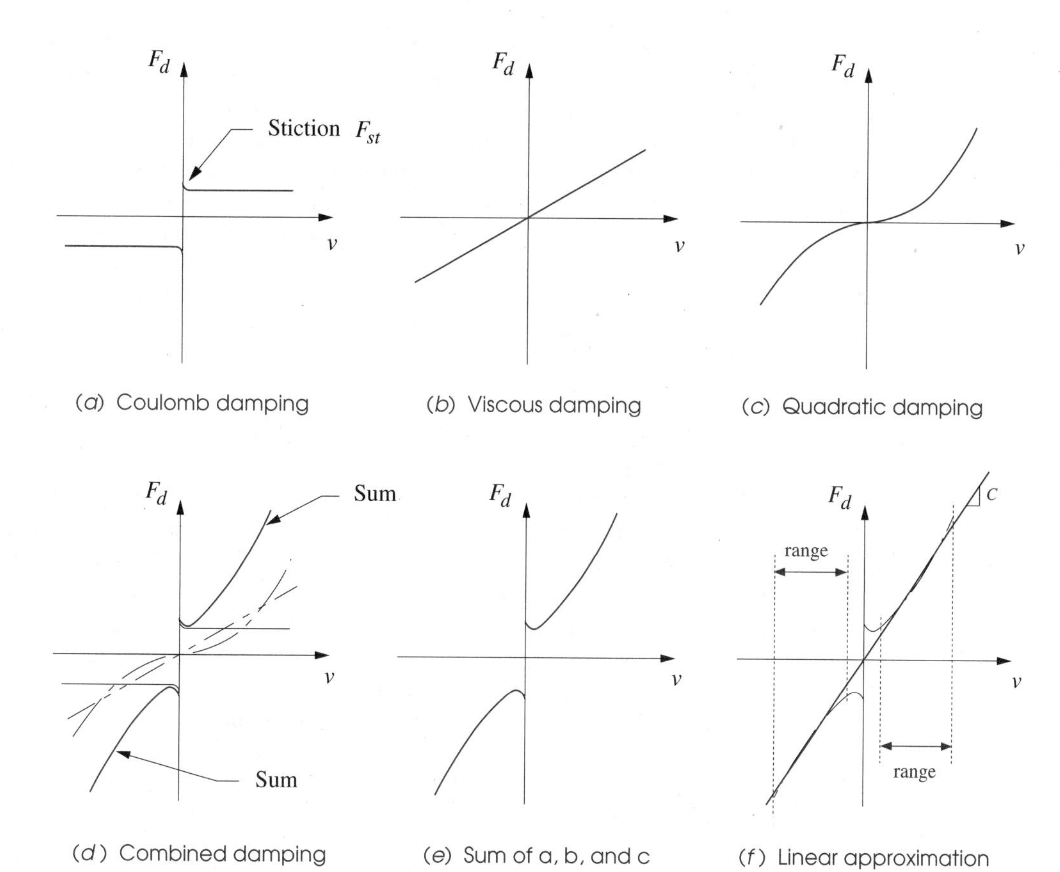

FIGURE 8-5
Modeling damping

For coulomb damping, $r = 0$ and:

$$F_d = \pm c \tag{8.13b}$$

For viscous damping, $r = 1$ and:

$$F_d = cv \tag{8.13c}$$

For quadratic damping, $r = 2$ and:

$$F_d = \pm cv^2 \tag{8.13d}$$

If we combine these three forms of damping, their sum will look like Figure 8-5d and 8.5e. This is obviously a nonlinear function. But we can approximate it over a reasonably small range of velocity as a linear function with a slope c which is then a *pseudo-viscous damping coefficient*. This is shown in Figure 8-5f. While not an exact method to account for the true damping, this approach has been found to be acceptably accurate for a first approximation during the design process.[2] The damping in these kinds of mechanical systems can vary quite widely from one design to the next due to different geometries, pressure or transmission angles, types of bearings, lubricants or their absence, etc. It is very difficult to accurately predict the level of damping (i.e., the value of c) in advance of the construction and testing of a prototype, which is the best way to determine the damping coefficient. If similar devices have been built and tested, their history can provide a good prediction. In a later chapter, a simple method for measuring the damping in an existing cam-follower system will be presented. For the purpose of our dynamic modeling, we will assume *pseudo-viscous damping* and some value for c.

8.10 EQUIVALENT SYSTEMS

More complex systems than that shown in Figure 8-4 (p. 186) will have multiple masses, springs, and sources of damping connected together as shown in Figure 8-6. These models can be analyzed by writing dynamic equations for each subsystem and then solving the set of differential equations simultaneously. This allows a multi-degree-of-freedom analysis, with one-*DOF* for each subsystem included in the analysis. Koster[1] found in his extensive study of vibrations in cam mechanisms that a five-*DOF* model—including the effects of both torsional and bending deflection of the camshaft, backlash in the driving gears, squeeze effects of the lubricant, nonlinear coulomb damping, and motor speed variation—gave a very good prediction of the actual, measured follower response. But he also found that a single-*DOF* model as shown in Figure 8-4 gave a reasonable simulation of the same system. We can then take the simpler approach and lump all the subsystems of Figure 8-6 together into a single-*DOF* **equivalent system** as shown in Figure 8-4 (p. 186). The combining of the various springs, dampers, and masses must be done carefully to properly approximate their dynamic interactions with each other.

There are only two types of variables active in any dynamic system. These are given the general names of *through variable* and *across variable*. These names are descriptive of their actions within the system. A **through variable** *passes through the system*. An **across variable** *exists across the system*. The power in the system is the product of the through and across variables. Table 8-1 lists the through and across variables for various types of dynamic systems.

We commonly speak of the voltage across a circuit and the current flowing through it. We also can speak of the velocity across a mechanical "circuit" or system and the force that flows through it. Just as we can connect electrical elements such as resistors, capacitors, and inductors together in series or parallel or a combination of both to make an electrical circuit, we can connect their mechanical analogs, dampers, springs, and masses together in series, parallel, or a combination thereof to make a mechanical system.

Table 8-2 shows the analogs among three types of physical systems. The fundamental relationships between through and across variables in electrical, mechanical, and fluid systems are shown in Table 8-3.

Recognizing series or parallel connections among elements in an electrical circuit is fairly straightforward, as their interconnections are easily seen. Determining how mechanical elements in a system are interconnected is more difficult as their interconnections are sometimes hard to see. The test for series or parallel connection is best done by ex-

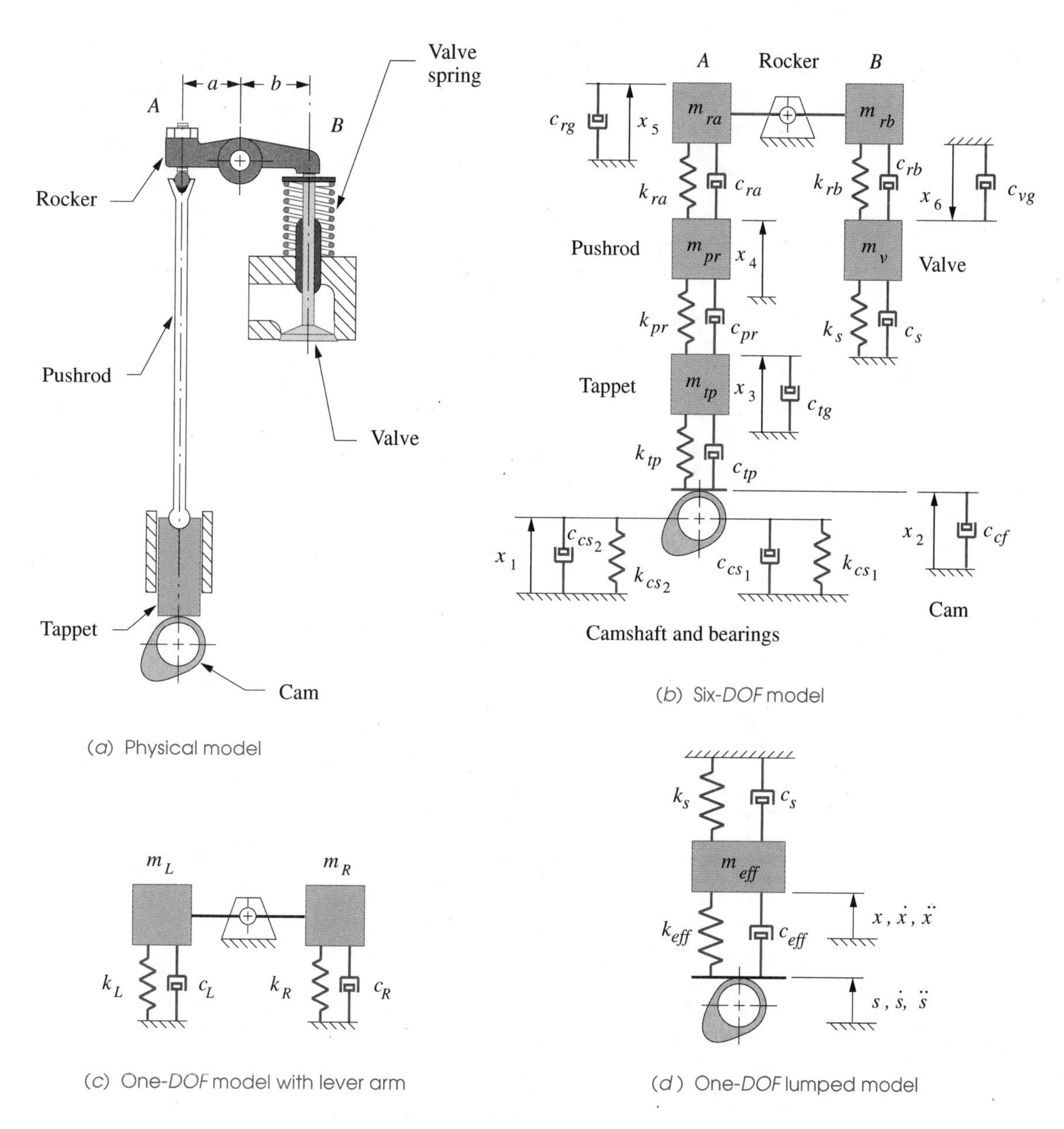

FIGURE 8-6

Lumped parameter models of an overhead valve engine cam-follower system

amining the forces and velocities (or the integral of velocity, displacement) that exist in the particular elements. If two elements have the same force passing through them, they are in series. If two elements have the same velocity or displacement, they are in parallel.

TABLE 8-1 Through and Across Variables in Dynamic Systems

System Type	Through Variable	Across Variable	Power Units
Electrical	Current (i)	Voltage (e)	ei = watts
Mechanical	Force (F)	Velocity (v)	Fv = (in-lb)/sec
Fluid	Flow (Q)	Pressure (P)	PQ = (in-lb)/sec

TABLE 8-2 Physical Analogs in Dynamic Systems

System Type	Energy Dissipator	Energy Storage	Energy Storage
Electrical	Resistor (R)	Capacitor (C)	Inductor (L)
Mechanical	Damper (c)	Mass (m)	Spring (k)
Fluid	Fluid resistor (R_f)	Accumulator (C_f)	Fluid inductor (L_f)

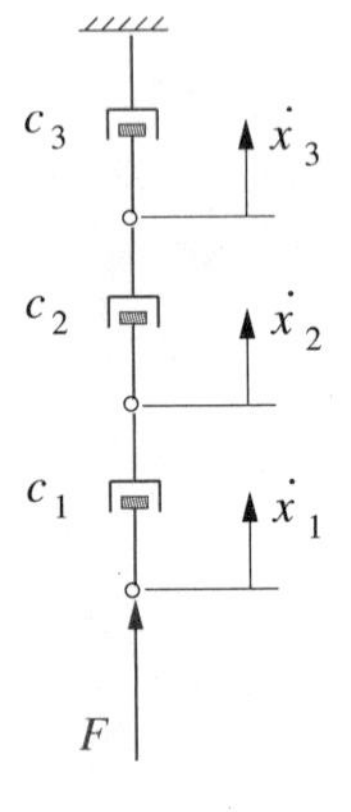

(a) Series

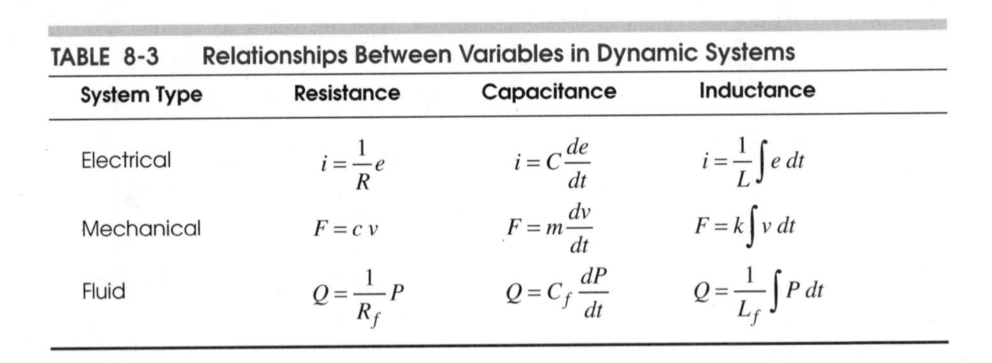

TABLE 8-3 Relationships Between Variables in Dynamic Systems

System Type	Resistance	Capacitance	Inductance
Electrical	$i = \dfrac{1}{R} e$	$i = C \dfrac{de}{dt}$	$i = \dfrac{1}{L} \int e \, dt$
Mechanical	$F = c\,v$	$F = m \dfrac{dv}{dt}$	$F = k \int v \, dt$
Fluid	$Q = \dfrac{1}{R_f} P$	$Q = C_f \dfrac{dP}{dt}$	$Q = \dfrac{1}{L_f} \int P \, dt$

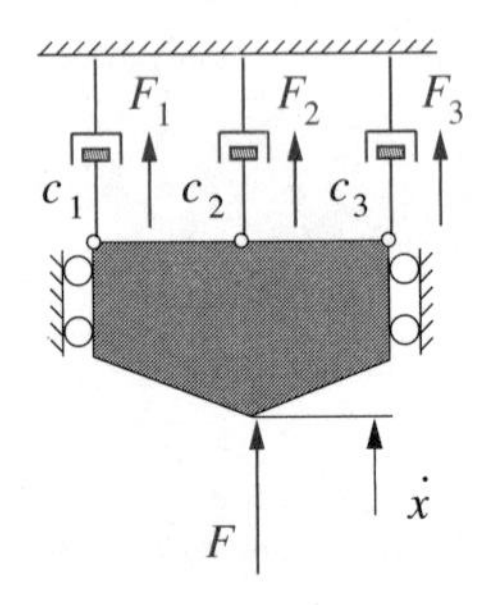

(b) Parallel

Combining Dampers

DAMPERS IN SERIES Figure 8-7a shows three dampers in series. The force passing through each damper is the same, and their individual displacements and velocities are different.

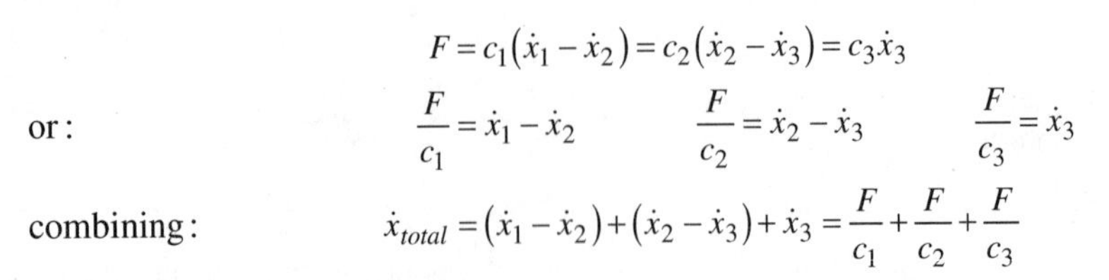

$$F = c_1(\dot{x}_1 - \dot{x}_2) = c_2(\dot{x}_2 - \dot{x}_3) = c_3\dot{x}_3$$

or: $$\frac{F}{c_1} = \dot{x}_1 - \dot{x}_2 \qquad \frac{F}{c_2} = \dot{x}_2 - \dot{x}_3 \qquad \frac{F}{c_3} = \dot{x}_3$$

combining: $$\dot{x}_{total} = (\dot{x}_1 - \dot{x}_2) + (\dot{x}_2 - \dot{x}_3) + \dot{x}_3 = \frac{F}{c_1} + \frac{F}{c_2} + \frac{F}{c_3}$$

FIGURE 8-7
Dampers in series and in parallel

then:
$$\dot{x}_{total} = F\frac{1}{c_{eff}} = F\left(\frac{1}{c_1}+\frac{1}{c_2}+\frac{1}{c_3}\right)$$
$$\frac{1}{c_{eff}} = \frac{1}{c_1}+\frac{1}{c_2}+\frac{1}{c_3}$$
$$c_{eff} = \frac{1}{\dfrac{1}{c_1}+\dfrac{1}{c_2}+\dfrac{1}{c_3}} \tag{8.14a}$$

The reciprocal of the effective damping of the dampers in series is the sum of the reciprocals of their individual damping coefficients.*

* Note in Table 8-3 that electrical resistance R is defined as the reciprocal of mechanical damping C, thus accounting for the reversal in the equations for parallel and series circuits between the two systems.

Dampers in Parallel Figure 8-7b shows three dampers in parallel. The force passing through each damper is different, and their displacements and velocities are all the same.

$$F = F_1 + F_2 + F_3$$
$$F = c_1\dot{x} + c_2\dot{x} + c_3\dot{x}$$
$$F = (c_1 + c_2 + c_3)\dot{x}$$
$$F = c_{eff}\dot{x}$$
$$c_{eff} = c_1 + c_2 + c_3 \tag{8.14b}$$

The effective damping of the three is the sum of their individual damping coefficients.

Combining Springs

Springs are the mechanical analog of electrical inductors. Figure 8-8a shows three springs in series. The force passing through each spring is the same, and their individual displacements are different. A force F applied to the system will create a total deflection that is the sum of the individual deflections. The spring force is defined from the relationship in equation 8.12 (p. 185):

$$F = k_{eff}x_{total}$$
where:
$$x_{total} = (x_1 - x_2) + (x_2 - x_3) + x_3 \tag{8.15a}$$

$$(x_1 - x_2) = \frac{F}{k_1} \qquad (x_2 - x_3) = \frac{F}{k_2} \qquad x_3 = \frac{F}{k_3} \tag{8.15b}$$

Substituting, we find that the reciprocal of the effective k of **springs in series** is the sum of the reciprocals of their individual spring rates.

$$\frac{F}{k_{eff}} = \frac{F}{k_1}+\frac{F}{k_2}+\frac{F}{k_3}$$
$$k_{eff} = \frac{1}{\dfrac{1}{k_1}+\dfrac{1}{k_2}+\dfrac{1}{k_3}} \tag{8.15c}$$

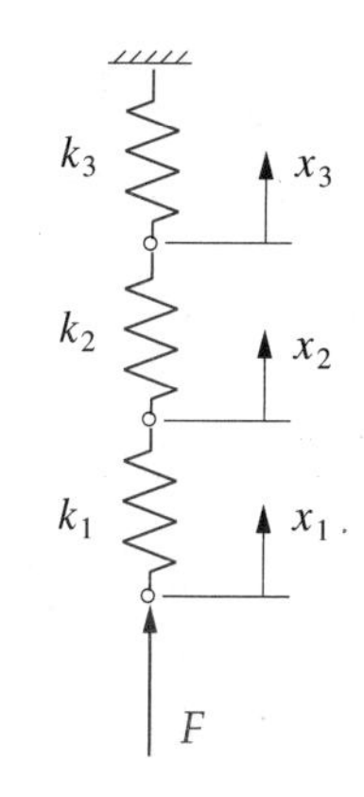

(a) Series

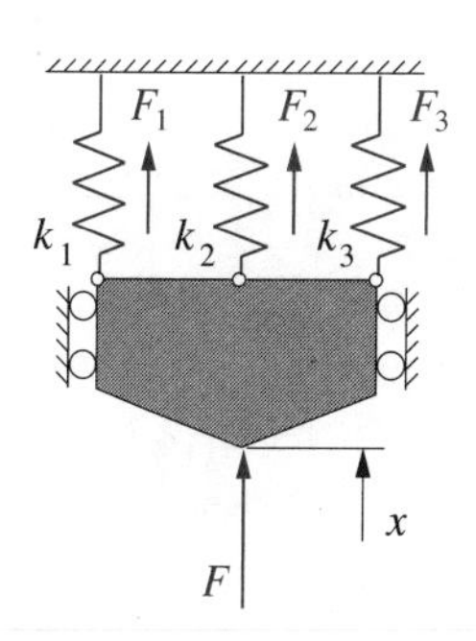

(b) Parallel

FIGURE 8-8

Springs in series and in parallel

Figure 8-8b shows three springs in parallel. The force passing through each spring is different, and their displacements are all the same. The total force is the sum of the individual forces.

$$F_{total} = F_1 + F_2 + F_3 \tag{8.16a}$$

Substituting equation 8.15a we find that the effective k of **springs in parallel** is the sum of the individual spring rates.

$$\begin{aligned} k_{eff}x &= k_1x + k_2x + k_3x \\ k_{eff} &= k_1 + k_2 + k_3 \end{aligned} \tag{8.16b}$$

Combining Masses

Masses are the mechanical analog of electrical capacitors. The inertial forces associated with all moving masses are referenced to the ground plane of the system because the acceleration in $\mathbf{F} = m\mathbf{a}$ is absolute. Thus, all masses are connected in parallel and combine in the same way as do capacitors in parallel with one terminal connected to a common ground.

8

$$m_{eff} = m_1 + m_2 + m_3 \tag{8.17}$$

Lever and Gear Ratios

Whenever an element is separated from the point of application of a force or from another element by a **lever ratio** or **gear ratio**, its effective value will be modified by that ratio. Figure 8-9a shows a spring at one end (A) and a mass at the other end (B) of a lever. We wish to model this system as a single-*DOF* lumped parameter system assuming small angular rotation of the link. There are two possibilities in this case. We can either transfer an equivalent mass m_{eff} to point A and attach it to the existing spring k_A, as shown in Figure 8-9b, or we can transfer an equivalent spring k_{eff} to point B and attach it to the existing mass m_B as shown in Figure 8-9c. In either case, for the lumped model to be equivalent to the original system, it must have the same energy in it.

Let's first find the effective mass that must be placed at point A to eliminate the lever. Equating the kinetic energies in the masses at points A and B:

$$\frac{1}{2}m_B v_B^2 = \frac{1}{2}m_{eff} v_A^2 \tag{8.18a}$$

The velocities at each end of the lever can be related by the lever ratio:

$$v_A = \left(\frac{a}{b}\right) v_B$$

substituting:

$$m_B v_B^2 = m_{eff}\left(\frac{a}{b}\right)^2 v_B^2$$

$$m_{eff} = \left(\frac{b}{a}\right)^2 m_B \tag{8.18b}$$

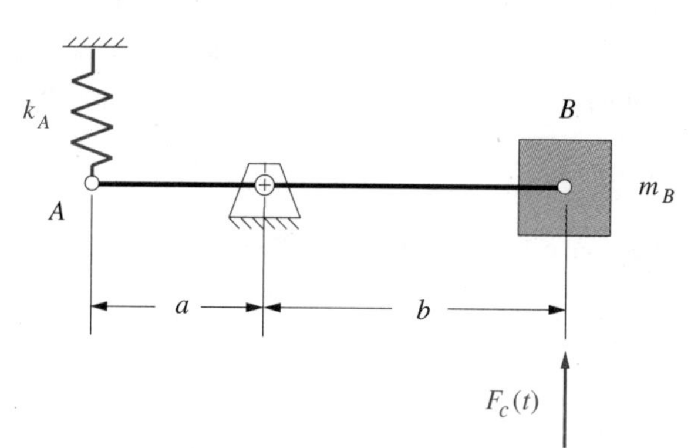

(*a*) Physical system

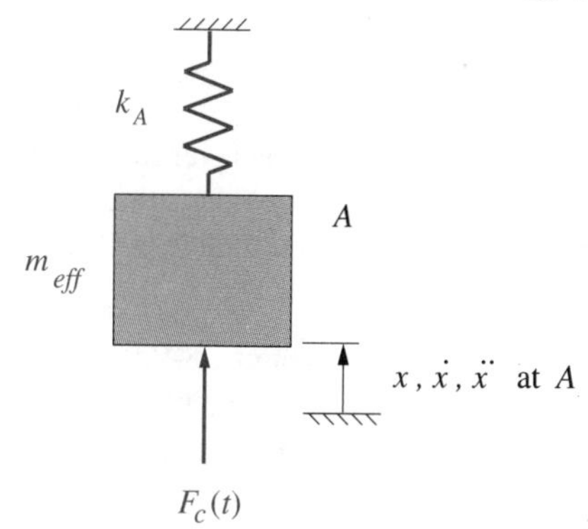

(*b*) Equivalent mass at point *A*

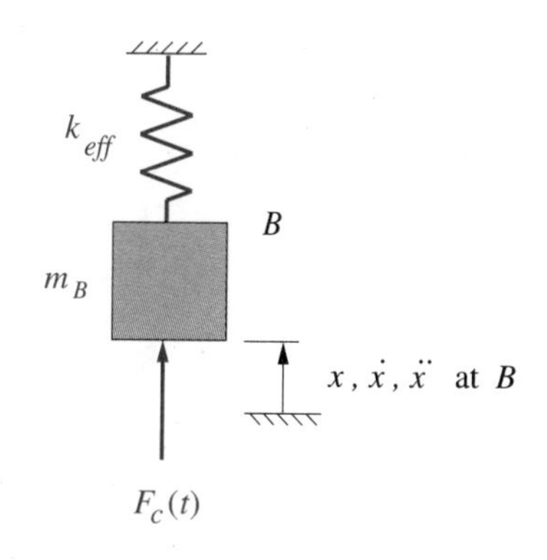

(*c*) Equivalent spring at point *B*

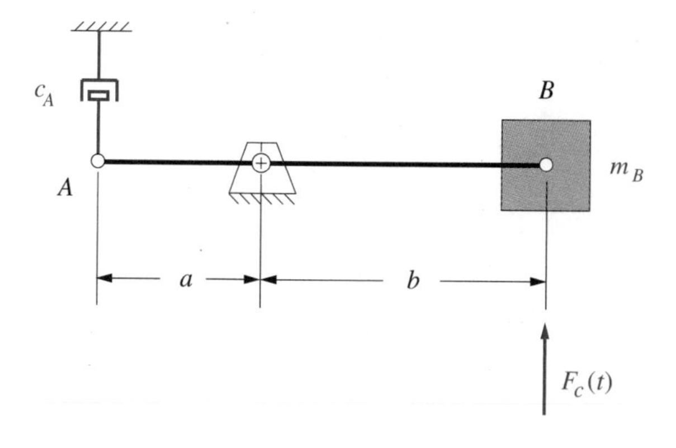

(*d*) Physical system

(*e*) Equivalent damper at point *B*

FIGURE 8-9

Lever or gear ratios affect the equivalent system

Note that the effective mass varies from the original mass by the square of the lever ratio.

Now find the effective spring that would have to be placed at B to eliminate the lever. Equating the potential energies in the springs at points A and B:

$$\frac{1}{2}k_A x_A^2 = \frac{1}{2}k_{eff} x_B^2 \tag{8.19a}$$

The deflection at B is related to the deflection at A by the lever ratio:

$$x_B = \left(\frac{b}{a}\right)x_A$$

substituting:

$$k_A x_A^2 = k_{eff}\left(\frac{b}{a}\right)^2 x_A^2$$

$$k_{eff} = \left(\frac{a}{b}\right)^2 k_A \tag{8.19b}$$

The effective k varies from the original k by the square of the lever ratio. If the lever were instead a pair of gears of radii a and b, the result would be the same. So, gear or lever ratios can have a large effect on the lumped parameters' values in the simplified model.

Damping coefficients are also affected by the lever ratio. Figure 8-9d shows a damper and a mass at opposite ends of a lever. If the damper at A is to be replaced by a damper at B, then the two dampers must produce the same moment about the pivot, thus:

$$F_{d_A} a = F_{d_B} b \tag{8.19c}$$

Substitute the product of the damping coefficient and velocity for force:

$$\left(c_A \dot{x}_A\right)a = \left(c_{B_{eff}} \dot{x}_B\right)b \tag{8.19d}$$

The velocities at points A and B in Figure 8-9d can be related from kinematics:

$$\omega = \frac{\dot{x}_A}{a} = \frac{\dot{x}_B}{b}$$

$$\dot{x}_A = \dot{x}_B \frac{a}{b} \tag{8.19e}$$

Substituting in equation 8.19d, we get an expression for the effective damping coefficient at B resulting from a damper at A.

$$\left(c_A \dot{x}_B \frac{a}{b}\right)a = \left(c_{B_{eff}} \dot{x}_B\right)b$$

$$c_{B_{eff}} = c_A\left(\frac{a}{b}\right)^2 \tag{8.19f}$$

Again, the square of the lever ratio determines the effective damping. The equivalent system is shown in Figure 8-9e.

EXAMPLE 8-1

Creating a Single-*DOF* Equivalent System Model of an Automotive Valve Train.

Given: An automotive valve cam with translating flat follower, long pushrod, rocker arm, valve, and valve spring as shown in Figure 8-6a (p. 189).

Problem: Create a suitable, approximate, single-*DOF*, lumped parameter model of the system. Define its effective mass, spring rate, and damping in terms of the individual elements' parameters.

Assumptions: Angular motions of rotating links are small enough to model in translation.

Solution:

1 Break the system into individual elements as shown in Figure 8-6b. Each significant moving part is assigned a lumped mass element which has a connection to ground through a damper. There is also elasticity and damping within the individual elements, shown as connecting springs and dampers. The rocker arm is modeled as two lumped masses at its ends, connected with a rigid, massless rod for the crank and connecting rod of the slider-crank linkage. The breakdown shown represents a six-*DOF* model as there are six independent displacement coordinates, x_1 through x_6.

2 Define the individual spring rates of each element which represents the elasticity of a lumped mass from the elastic deflection formula for the particular part. For example, the pushrod is loaded in compression, so its relevant deflection formula and its k are,

$$x = \frac{Fl}{AE} \qquad \text{and} \qquad k_{pr} = \frac{F}{x} = \frac{AE}{l} \tag{a}$$

where A is the cross-sectional area of the pushrod, l is its length, and E is Young's modulus for the material. The k of the tappet element will have the same expression. The expression for the k of a helical coil compression spring, as used for the valve spring, can be found in any spring design manual or machine design text and is:

$$k_{sp} = \frac{d^4 G}{8D^3 N} \tag{b}$$

where d is the wire diameter, D is the mean coil diameter, N is the number of coils, and G is the modulus of rupture of the material.

The rocker arm also acts as a spring, as it is a beam in bending. It can be modeled as a double cantilever beam with its deflection on each side of the pivot considered separately. These spring effects are shown in the model as if they were compression springs, but that is just schematic. They really represent the bending deflection of the rocker arms. From the deflection formula for a cantilever beam with concentrated load:

$$x = \frac{Fl^3}{3EI} \qquad \text{and} \qquad k_{ra} = \frac{3EI}{l^3} \tag{c}$$

where I is the cross-sectional second moment of area of the beam, l is its length, and E is Young's modulus for the material. The spring rates of any other elements in a system can be obtained in similar fashion from their deflection formulas.

8

3 The dampers shown connected to ground represent the friction or viscous damping at the interfaces between the elements and the ground plane. The dampers between the masses represent the internal damping in the parts, which typically is quite small as will be shown in the next chapter. These values will either have to be estimated from experience or measured in prototype assemblies.

4 The rocker arm provides a lever ratio which must be taken into account. The strategy will be to combine all elements on each side of the lever separately into two lumped parameter models as shown in Figure 8-6c, and then transfer one of those across the lever pivot to create one, single-*DOF* model as shown in Figure 8-6d.

5 The next step is to determine the types of connections, either series or parallel, between the elements. The masses are all in parallel as they each communicate their inertial force directly to ground and have independent displacements. On the left and right sides, respectively, the effective masses are:

$$m_L = m_{tp} + m_{pr} + m_{ra} \qquad m_R = m_{rb} + m_v \tag{d}$$

Note that m_v includes about one-third of the spring's mass to account for that portion of the spring that is moving. The two springs shown representing the bending deflection of the camshaft split the force between them, so they are in parallel and thus add directly.

$$k_{cs} = k_{cs_1} + k_{cs_2} \tag{e}$$

Note that, for completeness, the torsional deflection of the camshaft should also be included but is omitted in this example to reduce complexity. The combined camshaft spring rate and all the other springs shown on the left side are in series as they each have independent deflections and the same force passes through them all. On the right side, the spring of the rocker arm is in series with that of the left side, but the physical valve spring is in parallel with the effective spring of the follower train elements as it has a separate path from the effective mass at the valve to ground. (The follower-train elements all communicate back to ground through the cam pivots.) The effective spring rates of the follower-train elements for each side of the rocker arm are then:

$$k_L = \cfrac{1}{\cfrac{1}{k_{cs}} + \cfrac{1}{k_{tp}} + \cfrac{1}{k_{pr}} + \cfrac{1}{k_{ra}}} \qquad k_R = k_{rb} \tag{f}$$

The dampers are in a combination of series and parallel. The pair of dampers c_{cs1} and c_{cs2} shown supporting the camshaft represent the friction in the two camshaft bearings and are in parallel.

$$c_{cs} = c_{cs_1} + c_{cs_2} \tag{g}$$

The ones representing internal damping are in series with one another and with the combined shaft damping.*

$$c_{in_L} = \cfrac{1}{\cfrac{1}{c_{tp}} + \cfrac{1}{c_{pr}} + \cfrac{1}{c_{ra}} + \cfrac{1}{c_{cs}}} \qquad c_{in_R} = c_{rb} \tag{h}$$

where c_{in_L} is all internal damping on the left side and c_{in_R} is all internal damping on the right side of the rocker arm pivot. The combined internal damping c_{in_L} goes to ground through c_{rg} and the combined internal damping c_{in_R} goes to ground through the valve

* This analysis assumes that the internal damping values (c's) of the elements are very small and vary approximately proportionally to the stiffnesses (k's) of the respective elements to which they apply. Because damping is typically small in these systems, its effect on the equivalent spring rate is small, but the reverse is not true since high stiffness will affect damping levels. A very stiff element will deflect less under a given load than a less stiff one. If damping is proportional to velocity across the element, then a small deflection will have small velocity. Even if the damping coefficient of that element is large, it will have little effect on the system due to the element's relatively high stiffness. A more accurate way to estimate damping must take the interaction between the k's and c's into account. For n springs $k_1, k_2, \ldots, k_n$ in series, placed in parallel with n dampers $c_1, c_2, \ldots, c_n$ in series, the effective damping can be shown to be:

$$c_{eff} = k_{eff} \sum_{i=1}^{n} \frac{c_i}{k_i^2}$$

As a practical matter, however, it is usually quite difficult to determine the values of the individual damping elements that are needed to do a calculation such as shown above and in equation (h). Section 9.3 (p. 227) presents a relatively simple method to experimentally determine an approximate overall damping value for an entire system.

8

spring c_s. These two series combinations are then in parallel with all the other dampers that go to ground. The combined damping for each side of the system are then:

$$c_L = c_{tg} + c_{rg} + c_{in_L} \qquad\qquad c_R = c_{vg} + c_{in_R} \tag{i}$$

6 The system can now be reduced to a single-*DOF* model with masses and internal springs lumped on either end of the rocker arm as shown in Figure 8-6c. We will bring the elements at point *B* across to point *A*. Note that we have reversed the sign convention across the pivot so that positive motion on one side results also in positive motion on the other. The damper, mass, and spring rate are affected by the square of the lever ratio as shown in equations 8.18 (p. 192) and 8.19 (p. 194).

$$\begin{aligned} m_{eff} &= m_L + \left(\frac{b}{a}\right)^2 m_R \\ k_{eff} &= k_L + \left(\frac{b}{a}\right)^2 k_R \\ c_{eff} &= c_L + \left(\frac{b}{a}\right)^2 c_R \end{aligned} \tag{j}$$

These are shown in Figure 8-6d on the final, one-*DOF* lumped model of the system. It shows all the elasticity of the follower train elements lumped into the effective spring k_{eff} and the damping as c_{eff}. The cam's displacement input $s(t)$ acts against a massless but rigid shoe. The valve spring and the valve's damping against ground provide forces that keep the joint between cam and follower closed. If the cam and follower separate, the system changes dynamically from that shown. The dynamic analysis of this system will be explored in Chapter 10.

Note that these one-*DOF* models provide only an approximation of a complex system's behavior. Even though the model may be an oversimplification, it is nevertheless still very useful as a first approximation and serves in this context as an example of the general method involved in modeling dynamic systems. A more complex model with multiple degrees of freedom can provide a better approximation of the dynamic system's behavior at the expense of greater modeling and computational effort, as will be discussed in later chapters.

8.11 MODELING NONLINEAR SPRINGS

It is quite common in industrial machines to use air cylinders as springs to close a cam-follower joint. In fact, many automated machines use them exclusively rather than use mechanical springs. An air cylinder has several advantages over a mechanical spring in these applications. The air pressure can be quickly adjusted to control both the preload and the effective spring rate. Most importantly, the follower train can be "locked out" of service while the camshaft continues to turn by retracting the (double-acting) cylinder, which pulls the follower entirely off the cam. This can be used as an emergency stop for one or all cam-driven mechanisms in the case of a jam or accident. Lockout of one, several, or all follower trains can be invoked during debugging of an automated machine, or during servicing.

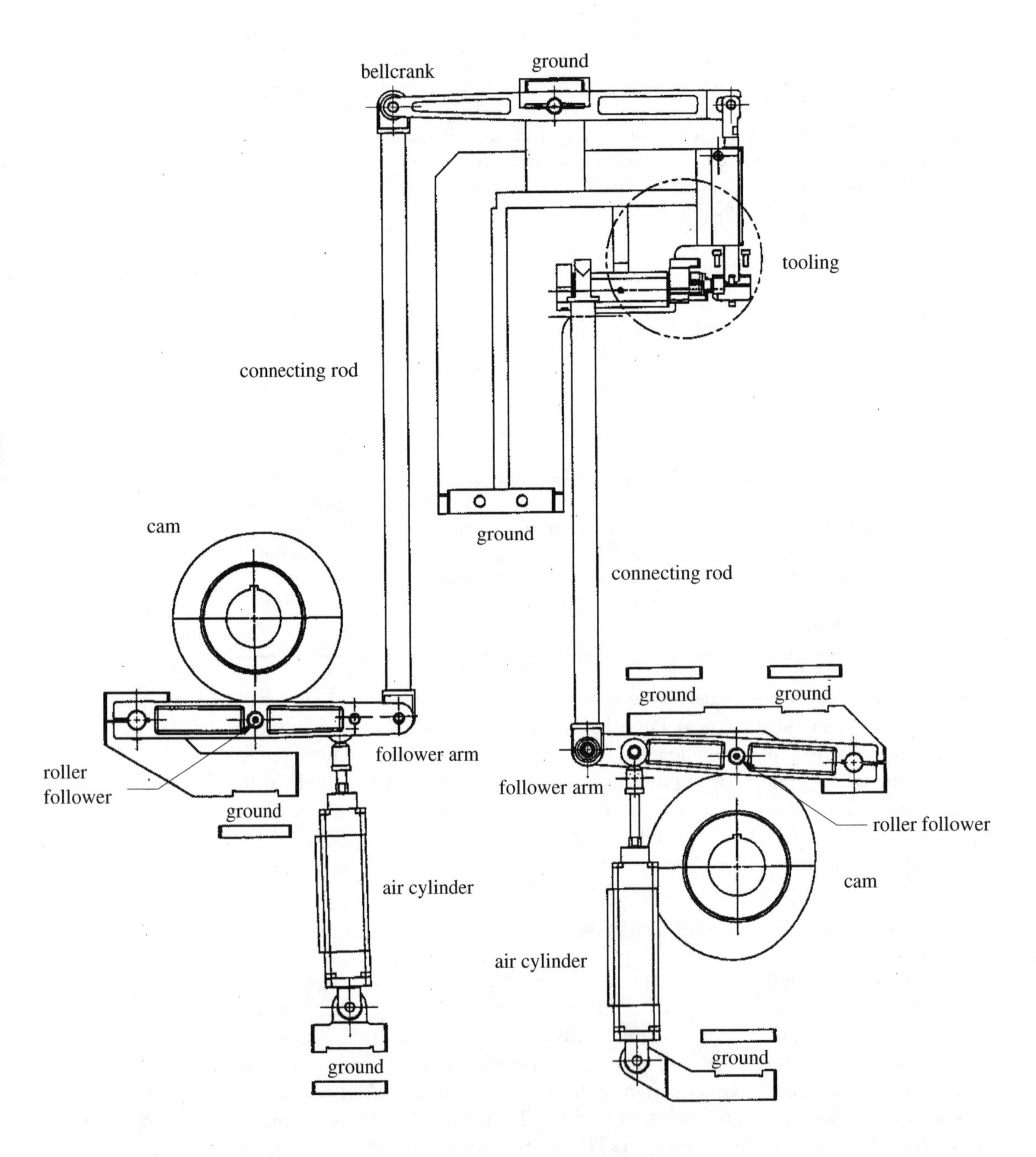

FIGURE 8-10

Air-cylinder closed cam-follower mechanisms for one station of an automated assembly machine (Courtesy of the Gillette Co., Boston, MA)

A typical arrangement is shown in Figure 8-10 with the air cylinder placed between ground and the follower arm to force the roller against the cam. Double-acting cylinders are controlled by 4-way solenoid valves that switch the line air pressure from one side of the piston to the other while venting its opposite side to atmosphere, as shown in Figure 8-11. Thus the piston is air driven in both directions and no internal spring is needed to return it.

Line air in U. S. factories typically is limited to a maximum of 80-90 psig for safety reasons. A pressure regulator can be used to adjust the working pressure to any value up to that maximum. An accumulator, as shown in Figure 8-11, can also be fitted (preferably near the cylinder, and necessarily upstream of the solenoid valve) to provide higher transient flows than available from long air feed lines of limited diameter. This will improve the transient response of the system to an emergency lockout and to high accelerations demanded by the cam profile. While there is no accumulator shown in Figure 8-10, that machine has a 1700 in^3, 80-psi accumulator built into hollow sections of its welded steel frame.

During the motion of the cam-driven follower arm, the solenoid valve is directing continuous air pressure to the side of the piston necessary to close the joint. Absent an accumulator close to (and with a low impedance connection to) the cylinder, a rapid motion of the cam follower arm will compress the trapped air behind the piston, reducing its volume and increasing its pressure and temperature, and attempt to flow air back through the open solenoid valve, the pressure regulator, and into the accumulator or feed lines. If the impedance of this escape path is high, as it will tend to be without an accu-

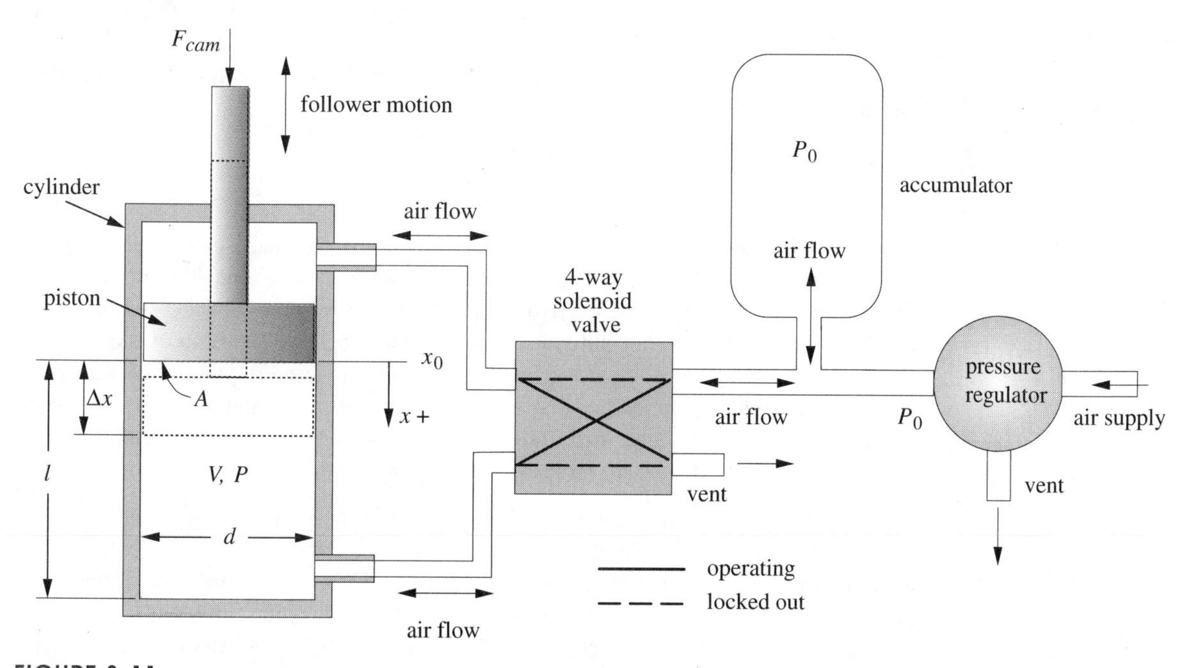

FIGURE 8-11

Air cylinder used as a spring to close a cam-follower joint

mulator, there may not be time for much air to exit the cylinder before the cam reverses motion and expands the cylinder volume back to its rest condition, especially at high speeds.

The advantage of adding an accumulator, which is merely a single-ended pressure vessel charged with a significant volume of air at pressure, i.e., a fluid capacitor, is that the air displaced rapidly from the cylinder during the stroke has a place to go through a relatively low impedance path (if properly designed). This, in the extreme, can make the pressure in the cylinder during the stroke approach the constant value set in the accumulator by the pressure regulator. This is the analog of a very low-rate mechanical spring. To achieve this condition requires using a solenoid valve with large diameter internal ports and using large-diameter, short-length tubes for all connections to minimize the impedance to flow. Air cylinders with integral solenoid valves attached to the cylinder side are available and are sometimes used in this application to minimize impedance.

This kind of accumulator arrangement is common on cam-driven industrial production machinery. However, some industrial machines that use air cylinders as follower springs are fitted with small-port solenoid valves located away from the cylinder and connected to the regulated air source by relatively small diameter tubing of significant length. A high impedance flow path can make the air cylinder a potentially stiff spring that is inherently nonlinear.

If the impedance of the lines, solenoid valve, and pressure regulator feeding the cylinder is high, then little of the air volume displaced by the piston may make it through the high impedance path in the short cycle time of the cam motion. The air cylinder effectively becomes a trapped volume acted upon by the piston motion. This volume includes the air in the lines back to valve and regulator . Adding an accumulator will increase the trapped volume significantly. Such a trapped volume of gas obeys the perfect gas law relationship:

$$\frac{PV^{\alpha}}{T} = C \tag{8.20}$$

where P is the absolute pressure in the cylinder, V is the volume, T is absolute temperature, and C is a constant. The exponent α takes on different values depending on the thermodynamic assumptions. For isothermal (constant temperature) conditions in air, $\alpha = 1.0$. For adiabatic (constant heat) conditions in air, $\alpha = 1.4$. Both of these conditions describe ideal states that are seldom exactly encountered in practice. **Isothermal** assumes that all heat generated is lost to the surroundings, resulting in the system's temperature remaining constant. **Adiabatic** assumes that all heat generated remains within the system with a concomitant increase in system temperature.

In this example, a case can be made that the system, at steady state, will operate at some elevated temperature that will remain essentially constant over time, thus closely approaching the isothermal condition. Measurements made of the pressure, volume displacement, and temperature of an air cylinder on such a machine show this to be true.[3] If the cylinder's temperature were to increase without limit (the adiabatic case), it would soon fail. Moreover, as the air heats and expands, thus tending to increase pressure, the regulator will allow the expanded air to equilibrate back to its set pressure over time.

Thus the temperature T of the system can be considered to be essentially constant, and the initial pressure at the maximum volume of the stroke, assumed to be regulated to the set pressure at the operating temperature. This allows equation 8.20 to be reduced to

$$PV^{\alpha} = C \tag{8.21}$$

with α assumed to be close to 1.0.

Figure 8-12 shows three pressure-volume (p-v) curves as measured on a test apparatus similar to that shown in Figure 8-10, driven by a multi-dwell cam.[3] Three modalities were tested at various initial pressures and camshaft speeds. Figure 8-12a shows the results of a test in which the air cylinder was charged from a regulated supply and then isolated from the regulator and air supply by a manual valve during the test (i.e., deadheaded). The system had reached a steady-state operating temperature at the time that the data were taken. A power function was fitted to the averaged data yielding an exponent for equation 8.21 of $\alpha = 1.023$. Thus, this relatively low-speed system is essentially isothermal even when deadheaded (an unrealistic but extreme condition done for test purposes only). Tests of a more realistic situation, with the cylinder connected to the air supply through a pressure regulator, gave similar results. Note the clipping of the pressure at upper left in Figure 8-12b due to the regulator bleed-off. The vertical slope portions at left and right sides occur during the dwells. Figure 8-12c shows the p-v curve for the same system with a substantial accumulator charged to 60 psig connected to the air cylinder through a 15-in-long, 0.375-in-dia hose. The pressure variation during the cycle is now minimal and the air cylinder acts like a "spring" with $k_{eff} \cong 0$, providing a nearly constant force on the cam. The exponent α in equation 8.21 is now essentially zero. Increasing the impedance of the connection between cylinder and accumulator will make the shape of the p-v curve more like Figure 8-12b.

8

Determining the Effective Spring Rate of an Air Cylinder

Figure 8-11 (p. 199) shows a double-acting air cylinder, fed by a regulated supply through a 4-way solenoid valve. The piston is shown in two positions. The upper position corresponds to the low dwell of the cam and the lower position corresponds to the high dwell. This is consistent with the geometry of the cam-follower linkage on the left side of Figure 8-10. At the piston-up position, the pressure in the cylinder will be at the regulated value. When the cam drives the follower downward through a rise, the piston will move rapidly to the lower position, compressing the air and increasing the pressure according to equation 8.21.

The initial volume V_0 of the cylinder is

$$V_0 = Al \tag{8.22a}$$

where A is piston area and l is the distance from piston face to the bottom of cylinder.

The constant C for the system is found from equation 8.21 using V_0 and the initial cylinder pressure set equal to the regulated value P_0, plus atmospheric pressure P_{atm}:

$$C = \left(P_0 + P_{atm}\right)V_0^{\alpha} \tag{8.22b}$$

(*a*) Charged to 55 psia and deadheaded

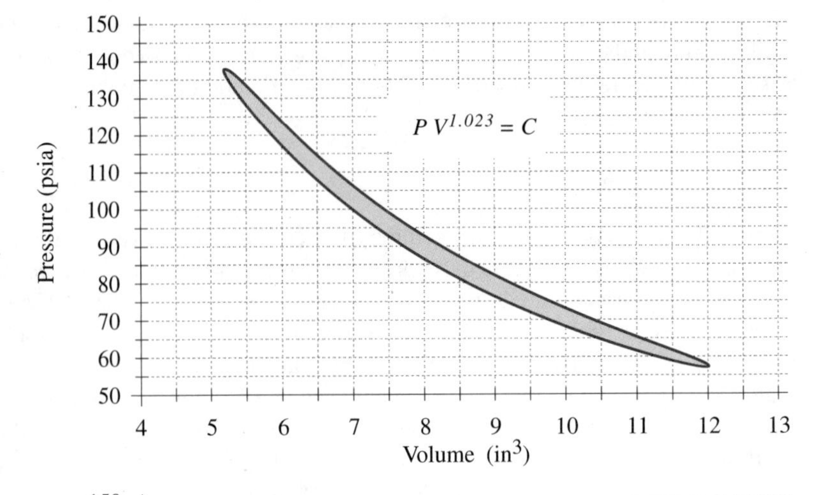

(*b*) Connected to 75 psia regulated air supply

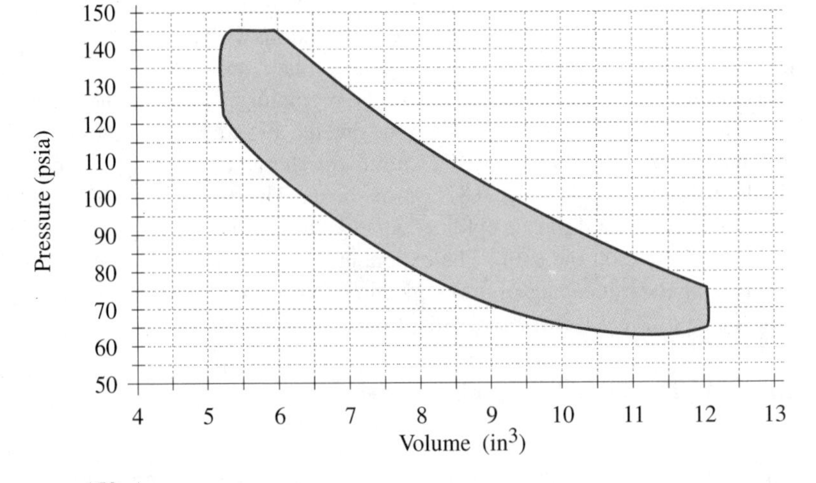

(*c*) Connected to 75 psia accumulator

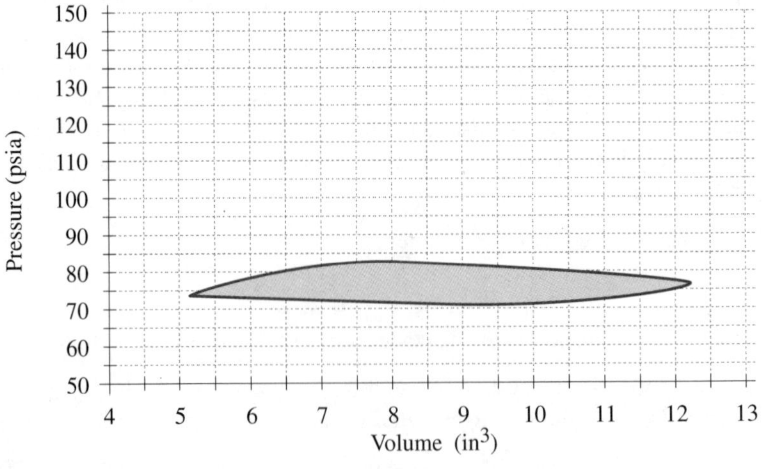

FIGURE 8-12

Experimentally measured pressure-volume characteristics of an air cylinder spring operated in 3 different modes [3]

Substitute equation 8.22b in equation 8.21 to get an expression for the pressure P at any point in the cycle.

$$P = \frac{(P_0 + P_{atm})V_0^\alpha}{V^\alpha} - P_{atm} \tag{8.22c}$$

Figure 8-13a shows a schematic plot of this function. For $\alpha = 1.0$ it is an hyperbola.

What we really want is a relationship between force on the piston F and its displacement x. The slope of that function is the effective spring rate of the cylinder. At any piston position x, the volume V is:

$$V = V_0 - Ax \tag{8.23a}$$

Substitute equation 8.23a in equation 8.22c to get a relationship between instantaneous pressure P and displacement x.

$$P = \frac{(P_0 + P_{atm})V_0^\alpha}{(V_0 - Ax)^\alpha} - P_{atm} \tag{8.23b}$$

Force is related to pressure from

$$P = \frac{F}{A} \tag{8.23c}$$

Substitute 8.23c in 8.23b to get force F as a function of displacement x.

$$F = \frac{A(P_0 + P_{atm})V_0^\alpha}{(V_0 - Ax)^\alpha} - P_{atm}A \tag{8.23d}$$

A schematic of this function is plotted in Figure 8-13b.

Differentiating equation 8.23d with respect to x gives an expression for the spring rate k for the general case.

$$k = \frac{dF}{dx} = A^2 \alpha V_0^\alpha (P_0 + P_{atm})(V_0 - Ax)^{(-\alpha-1)} \tag{8.24}$$

If we assume $\alpha = 1.0$, equation 8.23d becomes

$$F = \frac{A(P_0 + P_{atm})V_0}{(V_0 - Ax)} - P_{atm}A \tag{8.25}$$

Differentiating equation 8.25 with respect to x gives an expression for the spring rate k for the case of $\alpha = 1.0$.

$$k = \frac{dF}{dx} = \frac{A^2 V_0 (P_0 + P_{atm})}{(Ax - V_0)^2} \tag{8.26}$$

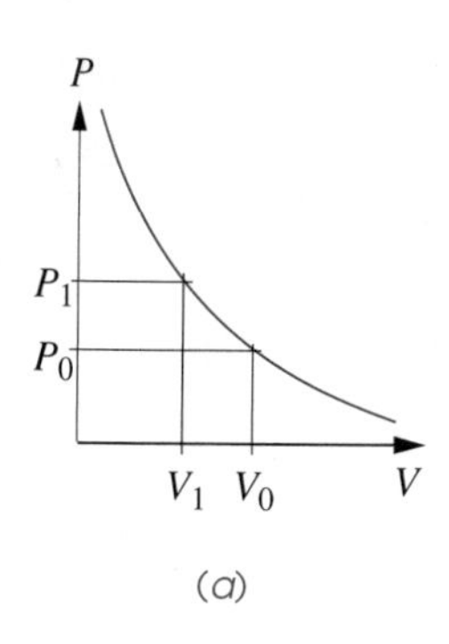

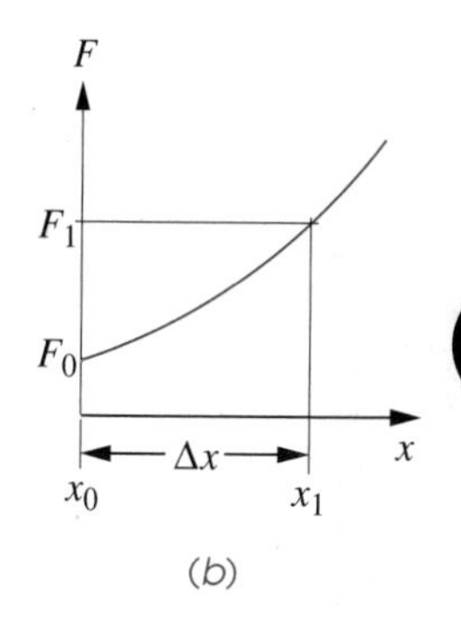

FIGURE 8-13

Pressure-volume and force-displacement relationships in an air cylinder

EXAMPLE 8-2

Determining the Effective Spring Rate of an Air Cylinder.

Given: The cam-follower system shown in the left side of Figure 8-10 (p. 198) has the following parameters as defined in Figure 8-11 (p. 199):

Maximum piston displacement, $\Delta x = 0.646$ in
Air cylinder diameter, $d = 1.25$ in, area $A = 1.227\ \text{in}^2$
Maximum length of air chamber, $l = 2$ in at low dwell
Maximum regulated air pressure, $P_0 = 80$ psig
Machine is at sea level, $P_{atm} = 14.7$ psig

Problem: Find the effective spring rate as a function of displacement. Also explore the sensitivity of this function to variation in the parameters of cylinder diameter, cylinder initial length, and the addition of an accumulator.

Assumptions: The cylinder operates at essentially isothermal conditions, $\alpha = 1.0$.

Solution: See Figures 8-10 and 8-11.

1 The initial volume of the cylinder from equation 8.22a is

$$V_0 = Al = 1.227(2) = 2.454\ \text{in}^3 \tag{a}$$

2 The gas constant from equation 8.22b is

$$C = \left(P_0 + P_{atm}\right)V_0^{\alpha} = (80 + 14.7)(2.454)^{1.0} = 232.4\ \text{lb - in} \tag{b}$$

3 The cylinder's force-displacement characteristic can be calculated from equation 8.25

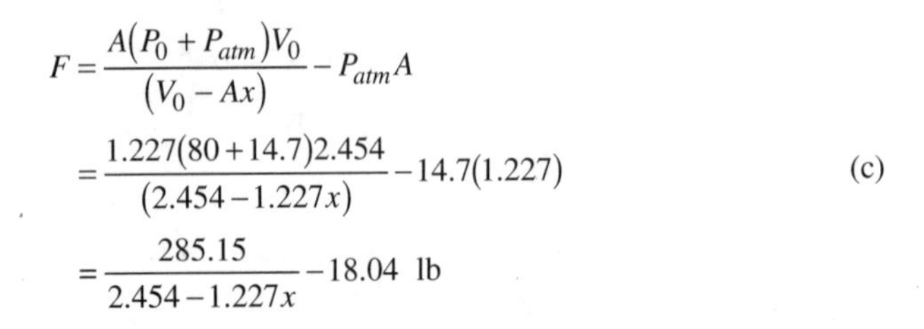

$$\begin{aligned} F &= \frac{A\left(P_0 + P_{atm}\right)V_0}{\left(V_0 - Ax\right)} - P_{atm}A \\ &= \frac{1.227(80 + 14.7)2.454}{(2.454 - 1.227x)} - 14.7(1.227) \\ &= \frac{285.15}{2.454 - 1.227x} - 18.04\ \text{lb} \end{aligned} \tag{c}$$

which is plotted over the working range of x for this problem in Figure 8-14a.

4 The cylinder's effective spring rate can be calculated from equation 8.26

$$\begin{aligned} k &= \frac{A^2V_0\left(P_0 + P_{atm}\right)}{\left(Ax - V_0\right)^2} \\ &= \frac{1.227^2(2.454)(80 + 14.7)}{(1.227x - 2.454)^2} \\ &= \frac{98.39}{(1.227x - 2.454)^2}\ \text{lb/in} \end{aligned} \tag{d}$$

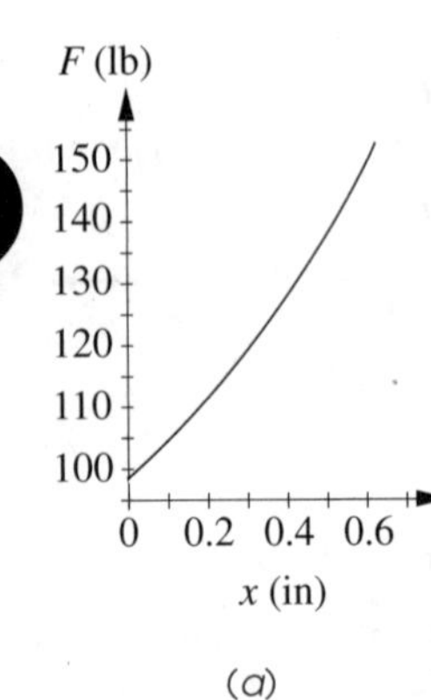

(a)

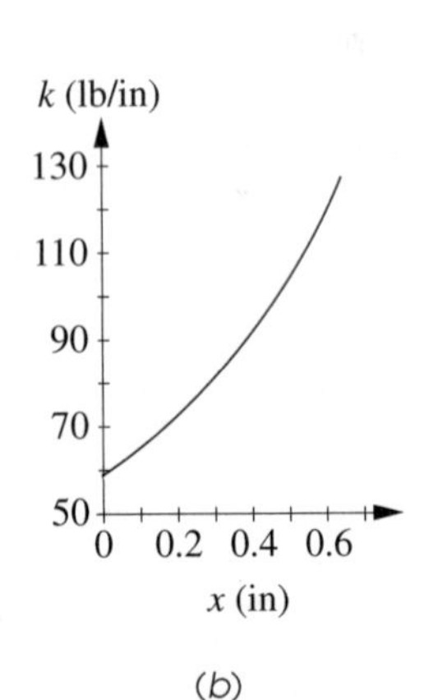

(b)

FIGURE 8-14

Force-displacement and spring rate-displacement plots for Example 8-2

which is plotted over the working range of x for this problem in Figure 8-14b. Note the large change in spring rate over the working range. It runs from under 60 lb/in at low dwell to over 120 lb/in at high dwell, a variation of ± 33% around the mean.

5 To investigate the effect of cylinder diameter on spring rate in this example, the cylinder length was held constant at its design value and the displacement was held constant at the midpoint ($x = 0.32$ in), while the cylinder diameter was varied from about 0.5 in to 5 in. The result is shown in Figure 8-15a. The design point is a 1.25 in-dia cylinder that gives $k \cong 82$ lb/in at that displacement. Increasing cylinder diameter has a strongly nonlinear and positive effect on spring rate. Larger diameters displace more air for the same stroke, making it a stiffer system.

6 To investigate the effect of cylinder length on spring rate in this example, the cylinder diameter was held constant at its design value and the displacement was held constant at the midpoint ($x = 0.32$ in), while the cylinder length was varied from about 0.4 in to 2.4 in.

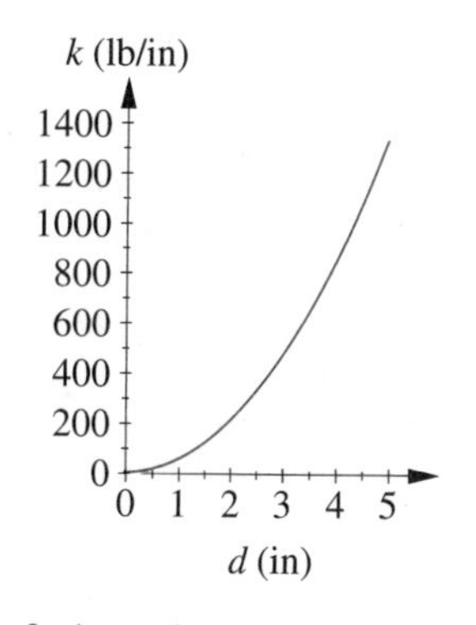

(a) Spring rate vs cylinder diameter

(b) Spring rate vs cylinder length

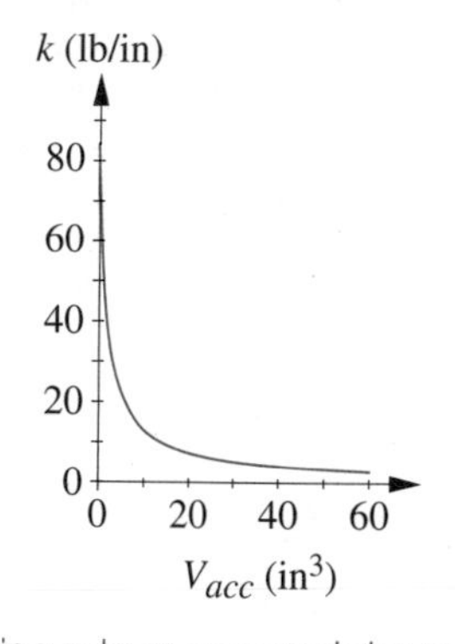

(c) Spring rate vs accumulator volume

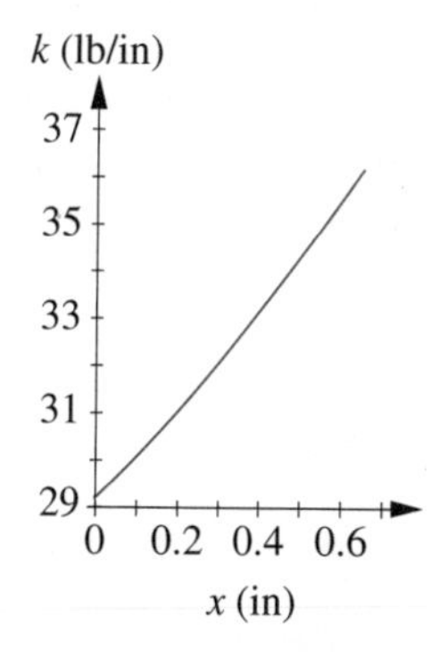

(d) Spring rate vs piston displacement with 21 in^3 accumulator and 2.5-in-dia air cylinder regulated to 22 psig

FIGURE 8-15

Variation of spring rate with cylinder parameters and accumulator volume (Example 8-2)

The result is shown in Figure 8-15b. The design point is a 2 in-long cylinder that gives k = 82 lb/in at that displacement. Decreasing cylinder length has a strongly nonlinear and positive effect on spring rate. Shorter cylinders have smaller initial volumes, making them a stiffer system and vice versa.

7 To investigate the effect of the addition of an accumulator on spring rate in this example, the cylinder length and diameter were held constant at their design values and the displacement was held constant at the midpoint (x = 0.32 in), while the initial volume was varied by adding accumulator volume ranging from zero (no accumulator) to 60 in^3. The result is shown in Figure 8-15c. The design point has no accumulator and gives k = 82 lb/in at that displacement. Adding accumulator volume has a strongly nonlinear and negative effect on spring rate. Larger accumulators make it a softer system, just as does lengthening the cylinder. If the force associated with the original design parameters were needed to prevent cam follower jump, then just adding an accumulator will not be effective as it will reduce the force. However, a combination of accumulator and cylinder diameter increase can result in a more nearly constant spring rate and still provide sufficient force.

8 Figure 8-15d shows the result of adding a 21 in^3 accumulator and increasing the cylinder diameter to 2.5 in. The force at the midpoint of follower motion is approximately the same as before at around 120 lb, but the variation in spring rate over the operating range is now only about 29 lb/in to 36 lb/in, or ± 11% around the mean. The larger cylinder area does not require as much pressure to achieve the needed force on the cam follower and the regulator pressure can now be reduced to about 22 psig. Over one revolution of the cam, the cam follower force ranges from 110 to 131 lb for this redesign as compared to 98 to 153 lb for the original design. There is still no follower jump and the peak stresses on the cam and follower are 15% lower than before. The pneumatic connections between cylinder and accumulator, including the solenoid valve, need to be of low impedance to achieve these results. Note that the addition of an accumulator will significantly reduce the natural frequency of the uncoupled system (after separation) as that depends in part on the air cylinder's effective spring rate.

8.12 MODELING AN INDUSTRIAL CAM-FOLLOWER SYSTEM

We will now put together the concepts of this chapter in an example that is representative of a common cam-follower system as used in industrial machinery. We will use a follower train similar to the one shown on the left side of Figure 8-10 (p. 198).

EXAMPLE 8-3

Creating a One-*DOF* Equivalent System Model of an Industrial Cam-Follower Train.

Given: A cam-follower system as is shown in Figure 8-16

Problem: Create a suitable, approximate, one-*DOF*, lumped-parameter model of the system as shown in Figure 8-4 (p. 186). Define its effective mass and effective spring rate at the roller follower in terms of the individual elements' parameters.

Assumptions: The air cylinder is connected to a large accumulator that gives it an effective spring rate that is close to zero. The preload on the cam follower due to air pressure is sufficient to prevent follower jump. All parts are steel.

Solution:

1 For this example, only the linkage shown on the left side of Figure 8-10 will be modeled, as can be seen in Figure 8-16. The approach would be the same for the companion linkage on the right side of Figure 8-10, or for any other similar linkage.

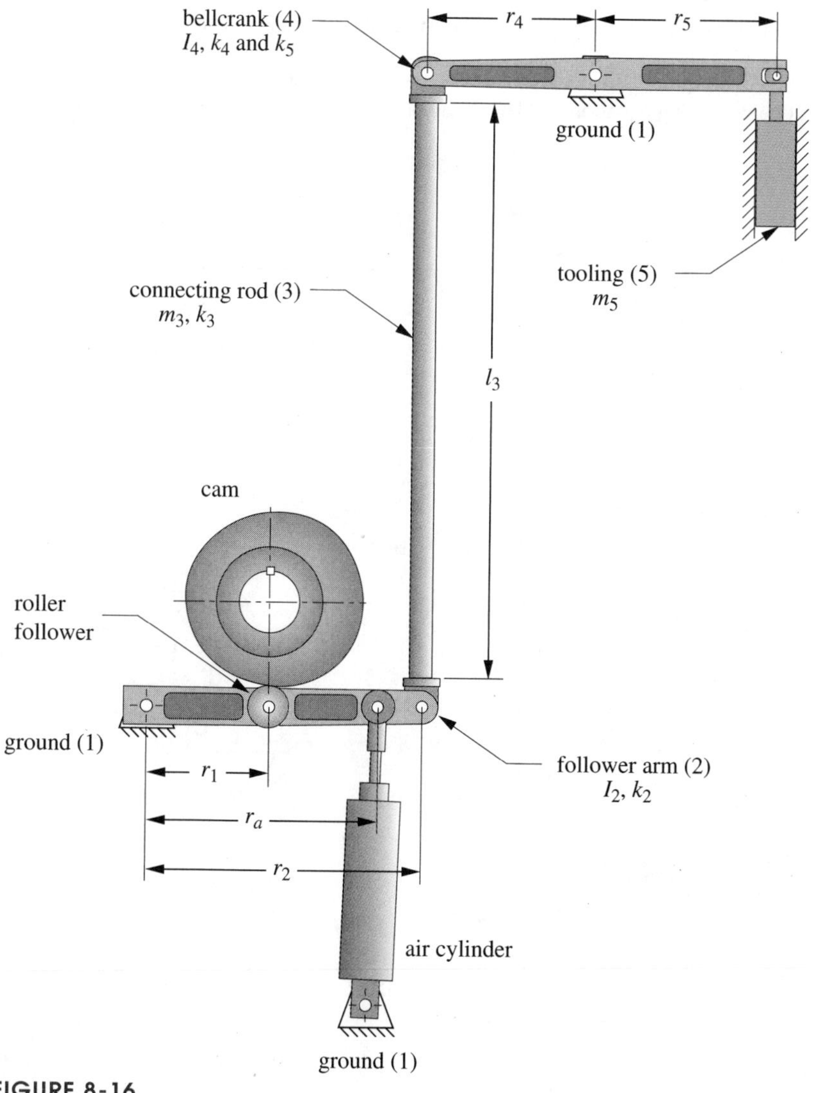

FIGURE 8-16

Cam-follower mechanism for Example 8-3

2 The first step is to lump the masses of the links as shown in Figure 8-17a. The tooling mass m_5 is lumped at the right end of link 4, at point *D*. It has a value of 0.9 kg in this linkage.

3 The bellcrank, link 4, is in rotation about O_4 and so must be converted to an equivalent mass at point *C* using equations 8.11 with radius $r_4 = 0.173$ m. The mass moment of inertia of link 4 about pivot O_4 was determined from a CAD solid model of the link to be 0.0087 kg-m^2. The effective mass of link 4 placed at point *C* is then

$$m_{4_{eff}} = \frac{I_{ZZ}}{r^2} = \frac{0.0087}{0.173^2} = 0.2908 \text{ kg} \tag{a}$$

4 The mass of the connecting rod, link 3, is found from the CAD solid model to be 0.8338 kg.

5 The follower arm, link 2, is in rotation about O_2 and so must be converted to an equivalent mass at point *B* using equations 8.11 with radius $r_2 = 0.283$ m. The mass moment of

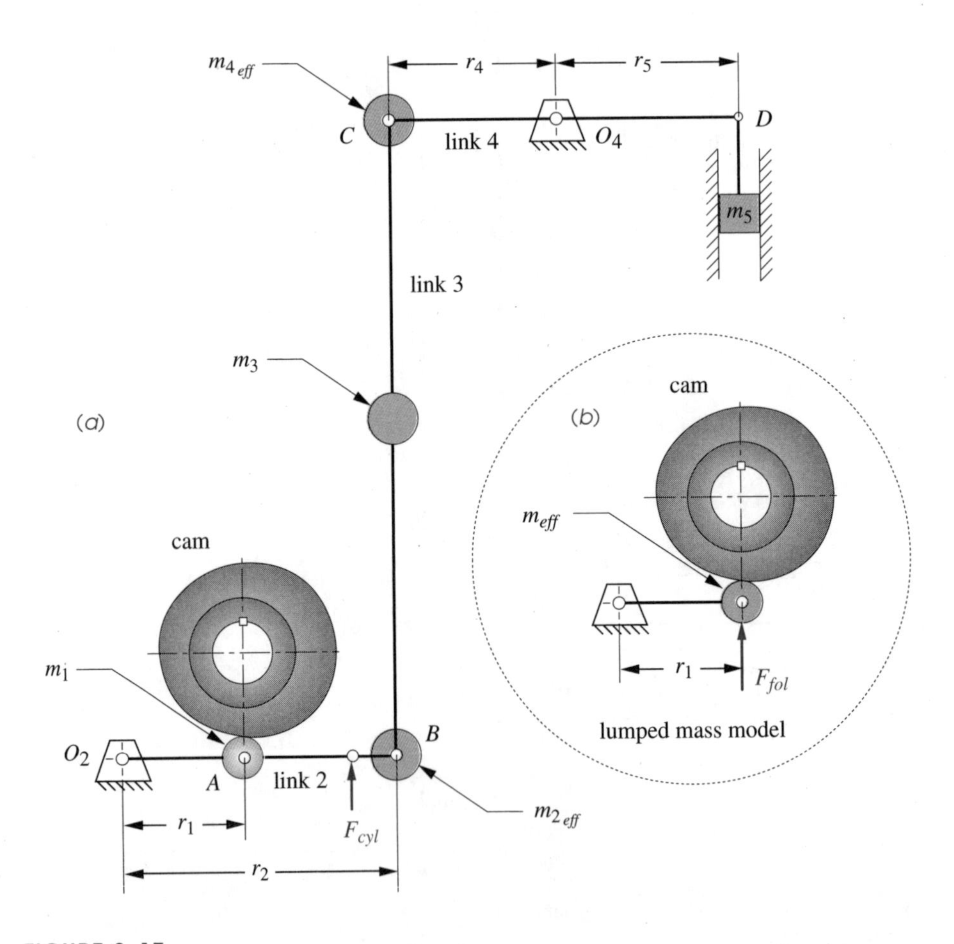

FIGURE 8-17

Combining lumped masses for Example 8-3

inertia of link 2 about pivot O_2 was determined from a CAD solid model of the link to be 0.0325 kg-m^2. The effective mass of link 2 placed at point B is then

$$m_{2_{eff}} = \frac{I_{ZZ}}{r^2} = \frac{0.0325}{0.283^2} = 0.4058 \text{ kg} \tag{b}$$

6 The mass m_1 of the roller follower is 0.196 kg obtained from the manufacturer's catalog information.

7 The next step is to bring all these lumped masses back to the roller follower location at point A where we wish to place the effective mass of the system. This will require the application of any link ratios that may be present.

8 First, bring mass m_5 across the bellcrank from point D to C using the radii of link 4 in equation 8.18b.

$$m_{5@C} = m_5\left(\frac{r_5}{r_4}\right)^2 = 0.9\left(\frac{0.185}{0.173}\right)^2 = 0.1029 \text{ kg} \tag{c}$$

9 Add the effective masses of links 4 and 5 at point C.

$$m_C = m_{4_{eff}} + m_{5@C} = 0.2908 + 0.1029 = 0.3937 \text{ kg} \tag{d}$$

10 Bring the mass m_C and the mass of the connecting rod m_3 down to point B and add them to the effective mass of link 2 at that point.

$$m_B = m_C + m_3 + m_{2_{eff}} = 0.3937 + 0.8338 + 0.4058 = 1.6357 \text{ kg} \tag{e}$$

11 Bring the total mass lumped at point B back to the follower at point A with equation 8.18b and add it to the mass of the roller that is there.

$$m_{eff} = m_A = m_B\left(\frac{r_2}{r_1}\right)^2 + m_1 = 1.6357\left(\frac{0.283}{0.127}\right)^2 + 0.196 = 8.318 \text{ kg} \tag{f}$$

This is the effective mass to be used in the 1-DOF model of the system.

12 Next, the compliances of each element must be combined to find the overall effective spring constant of the system as felt at the cam follower, point A. Figure 8-18 shows a schematic of the various compliant elements that comprise this system. Note that the air cylinder does not contribute to this compliance. It essentially provides a near constant force as shown in Figure 8-12c, due to the accumulator. As will be seen in Chapter 10, as long as the spring or air cylinder used to close the cam follower joint in this type of "industrial" cam-follower system has sufficient force to prevent follower jump, then its spring constant will not affect the overall compliance of the system.

13 The bellcrank, link 4, can best be modeled as a double-cantilever beam. Standard beam equations available from references such as [4] can be used in combination with the beam's cross-section geometry to determine the deflection and thus the spring rate of each half of the double-cantilever beam. However, in this case, the cross section of the beam is

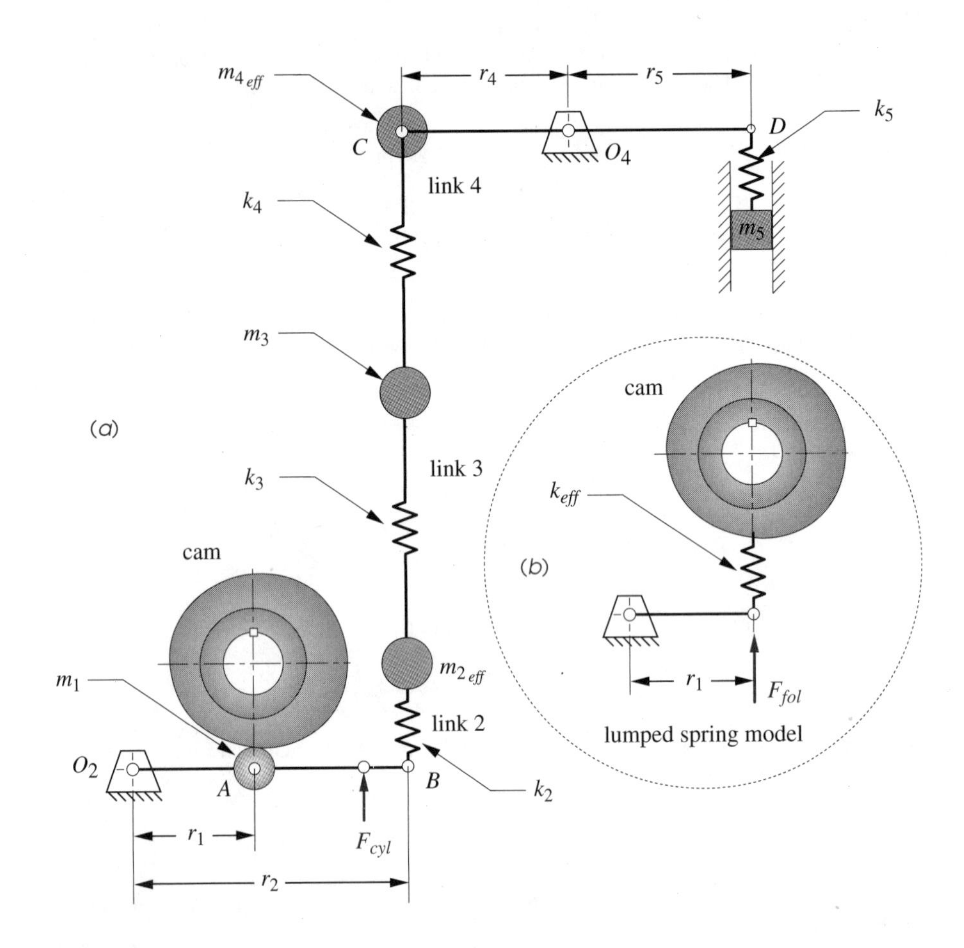

FIGURE 8-18

Combining lumped springs for Example 8-3

not uniform along its length, causing variation in its area moment of inertia that complicates a classic computation of its deflection. The best approach is to use a CAD solids modeler and Finite Element Analysis (FEA) software to calculate the deflection at the locations where the loads are applied. Figure 8-19a shows the result of such an FEA analysis of link 4. Two calculations were done, one for each half of the beam. In each case the elements at the section containing the pivot axis were fixed and a 1000 N load applied at one pin joint. The calculated deflections were then divided into the applied load to get a spring rate. This analysis gives k_4 = 2.474E6 N/m and k_5 = 2.242E6 N/m.

14 The follower arm, link 2, is a simply-supported, overhung beam. Its deflection was analyzed by fixing the pivot locations and applying a 1000 N load to its pin center at point *B* of Figure 8-18. The resulting deflection shown in Figure 8-19b indicates a spring rate of k_2 = 7.813E6 N/m.

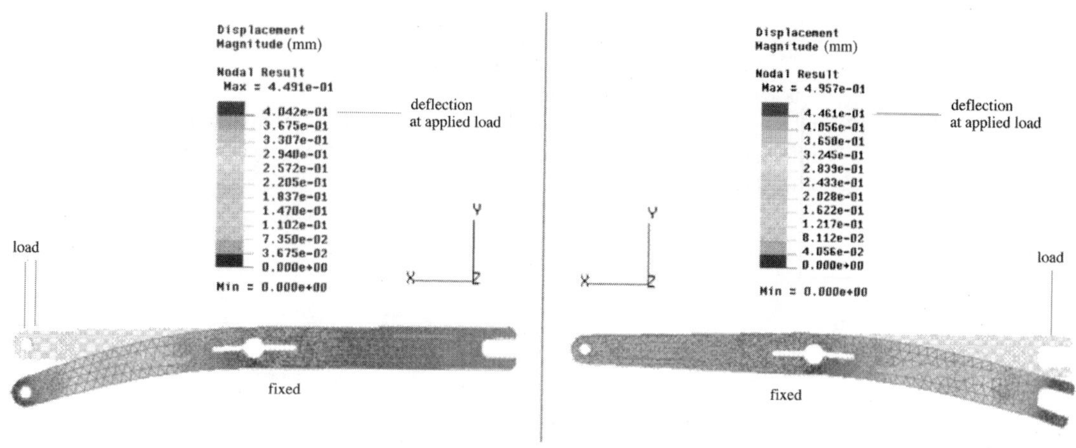

(a) Link 4 analyzed as a double-cantilever beam

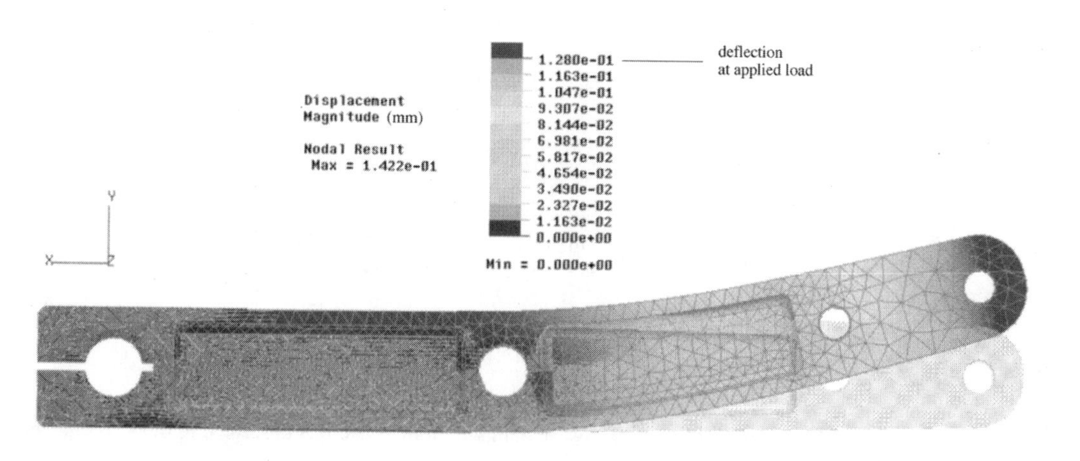

(b) Link 2 analyzed as a simply-supported ovehung beam

FIGURE 8-19

Finite element analysis of deflections of beam-levers for Example 8-3

15 The connecting rod is in axial compression, and assuming no buckling, its spring rate can be found from the equation for axial deflection of a uniform rod. In this case, the rod is a steel tube of 25.5 mm OD and 22.1 mm ID, giving a cross-sectional area of 1.231E-4 m^2. The length l_3 is 0.576 m. Its spring rate is then:

$$k_3 = \frac{AE}{l} = \frac{1.231E-4 \text{ m}^2 (207E9) \text{ Pa}}{0.576 \text{ m}} = 4.424E7 \text{ N/m} \tag{g}$$

16 Now the compliances can be combined to form an effective spring rate at the roller follower. First bring the right hand side of link 4, k_5, across to point C.

$$k_{5@C} = k_5\left(\frac{r_5}{r_4}\right)^2 = 2.242E6\left(\frac{0.185}{0.173}\right)^2 = 2.564E6 \text{ N/m} \tag{h}$$

17 All the compliances are in series because they pass a common force and each has a different deflection. Combine springs k_2, k_3, k_4, and $k_{5@C}$ according to equation 8.15c and bring them to point B.

$$k_B = \frac{1}{\frac{1}{k_2}+\frac{1}{k_3}+\frac{1}{k_4}+\frac{1}{k_{5@C}}}$$

$$= \frac{1}{\frac{1}{7.813E6}+\frac{1}{4.424E7}+\frac{1}{2.474E6}+\frac{1}{2.564E6}} = 1.058E6 \text{ N/m} \tag{i}$$

18 Now bring the combined compliance at B back to point A with equation 8.19b.

$$k_{eff} = k_A = k_b\left(\frac{r_2}{r_1}\right)^2 = 1.058E6\left(\frac{0.283}{0.127}\right)^2 = 5.255E6 \text{ N/m} \tag{j}$$

This is the effective spring rate to be used in the 1-DOF model of the system.

19 The air cylinder force in this example is applied at radius r_a on the follower arm as shown in Figure 8-16 (p. 207). This force must be brought back to the roller follower by the ratio of the radii of follower and cylinder to the first power.

$$F_{fol} = F_{cyl}\left(\frac{r_a}{r_1}\right) = 498 \text{ N}\left(\frac{0.235 \text{ m}}{0.127 \text{ m}}\right) = 921.5 \text{ N} \tag{k}$$

20 An approach for estimating the damping in the system is described in the next chapter.

8.13 REFERENCES

1 **Koster, M. P.** (1974). *Vibrations of Cam Mechanisms*. Phillips Technical Library Series, Macmillan Press Ltd.: London.

2 **Hundal, M. S.** (1963). "Aid of Digital Computer in the Analysis of Rigid Spring-Loaded Valve Mechanisms." *SAE Progress in Technology*, **5**, pp. 4-9, 57.

3 **O'Brien, C** and **P. Duperry**. (2001). "Modeling a Cam-Driven Linkage with an Air-Cylinder Spring", Major Qualifying Project, Worcester Polytechnic Institute.

4 **Norton, R. L.** (2000). *Machine Design: An Integrated Approach*, 2ed. Prentice-Hall, Upper Saddle River, NJ.

Chapter 9

DYNAMICS OF CAM SYSTEMS—FORCE, TORQUE, VIBRATION

9.0 INTRODUCTION

There are two approaches to dynamic analysis, commonly called the **forward** and the **inverse** dynamics problems. The forward problem assumes that the forces acting on the system are known and solves for the resulting displacements, velocities, and accelerations. This requires solving one or more differential equations. The inverse problem is, as its name says, the inverse of the other. In this approach, the displacements, velocities, and accelerations are known or assumed, and we solve for the forces and torques that result. This requires only solving an algebraic equation. In this chapter we will set up both methods for cam-follower analysis. Sections 9.1 to 9.3 present fundamental dynamic principles needed to understand both analyses. The rest of the chapter presents the inverse or **kinetostatic** solution whose main value is to predict cam-follower separation. The forward or dynamic solution will be fully developed in Chapter 10. Both approaches are useful in this application of a force-closed (spring-loaded) cam-follower system, but the forward solution is capable of giving more information about the system's vibratory behavior both before and after separation than is the kinetostatic solution.

9.1 DYNAMIC FORCE ANALYSIS OF THE FORCE-CLOSED CAM-FOLLOWER

Figure 9-1a shows a simple plate or disk cam driving a spring-loaded, roller follower. This is a force-closed system which depends on the spring force to keep the cam and follower in contact at all times. It is assumed that the moving follower parts are significantly stiffer than the follower joint-closure spring. Figure 9-1b shows a lumped parameter model of this system in which all the **mass** that moves with the follower train is lumped together as m, all the springiness in the system is lumped within the **spring constant** k, and all the **damping** is lumped together as a damper with coefficient c. The

TABLE 9-1 Notation Used in This Chapter

c = damping coefficient
c_c = critical damping coefficient
k = spring constant
F_c = force of cam on follower
F_s = force of spring on follower
F_d = force of damper on follower
m = mass of moving elements
t = time in seconds
T_c = torque on camshaft
θ = camshaft angle, in degrees or radians
ω = camshaft angular velocity, rad/sec
ω_d = damped circular natural frequency, rad/sec
ω_f = forcing frequency, rad/sec
ω_n = undamped circular natural frequency, rad/sec
x = follower displacement, length units
$\dot{x} = v$ = follower velocity, length/sec
$\ddot{x} = a$ = follower acceleration, length/sec^2
ζ = damping ratio

Portions of this chapter were adapted from R. L. Norton, *Design of Machinery* 2ed. McGraw-Hill, 2001, with permission.

sources of mass which contribute to m are fairly obvious. The mass of the follower stem, the roller, its pivot pin, and any other hardware attached to the moving assembly all add together to create m. Figure 9-1c shows the free-body diagram of the system acted upon by the cam force F_c, the spring force F_s, and the damping force F_d. There will of course also be the effects of mass times acceleration on the system.

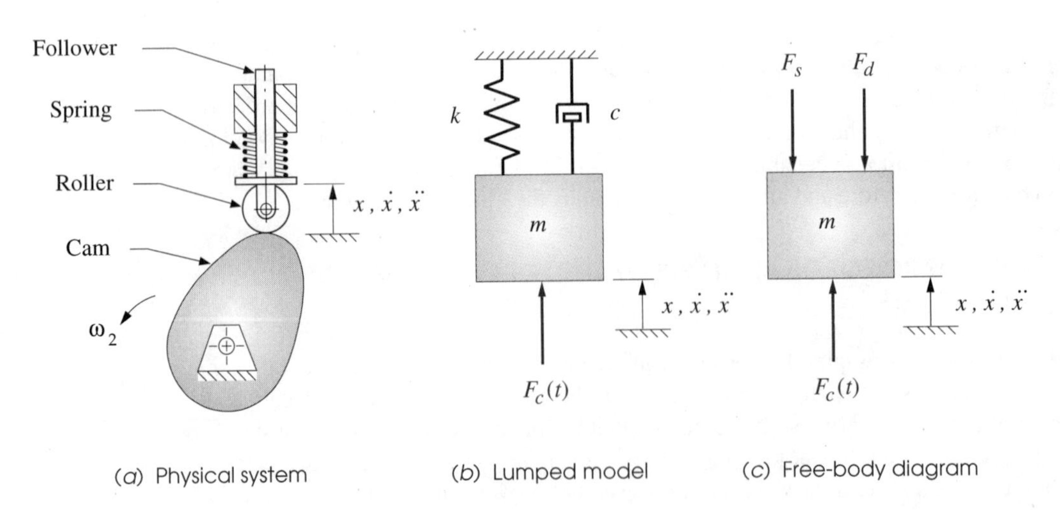

FIGURE 9-1

One-*DOF* lumped parameter model of a cam-follower system including damping

Undamped Response

Figure 9-2 shows an even simpler lumped parameter model of the same system as in Figure 9-1, but one that omits the damping altogether. This is referred to as a *conservative model* since it conserves energy with no losses. This is not an accurate assumption in this case, but will serve a purpose in the path to a better model that will include the damping. In fact, these systems tend to have very low damping, so the error introduced by its elimination is small. The free-body diagram for this mass-spring model is shown in Figure 9-2c. We can write Newton's equation for this one-*DOF* system:

$$\sum F = ma = m\ddot{x}$$

$$F_c(t) - F_s = m\ddot{x}$$

From equation 8.12 (p. 185):

$$F_s = kx$$

then:

$$m\ddot{x} + kx = F_c(t) \qquad (9.1a)$$

This is a second-order ordinary differential equation (ODE) with constant coefficients. The complete solution will consist of the sum of two parts, the transient (homogeneous) and the steady state (particular). The homogeneous ODE is,

$$m\ddot{x} + kx = 0$$

$$\ddot{x} = -\frac{k}{m}x \qquad (9.1b)$$

that has the well-known solution,

$$x = A\cos\omega t + B\sin\omega t \qquad (9.1c)$$

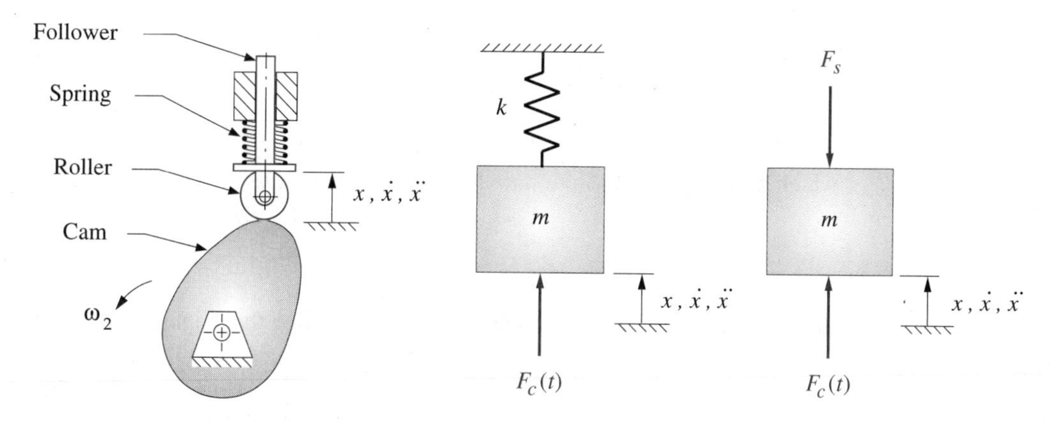

(*a*) Physical system (*b*) Lumped model (*c*) Free-body diagram

FIGURE 9-2

One-*DOF* lumped parameter model of a cam-follower system without damping

where A and B are the constants of integration to be determined by the initial conditions. To check the solution, differentiate it twice, assuming constant ω, and substitute in the homogeneous ODE.

$$-\omega^2(A\cos\omega t + B\sin\omega t) = -\frac{k}{m}(A\cos\omega t + B\sin\omega t)$$

This is a solution provided that:

$$\omega^2 = \frac{k}{m} \qquad \omega_n = \sqrt{\frac{k}{m}} \tag{9.1d}$$

The quantity ω_n (rad/sec) is called the *circular natural frequency* of the system and is the frequency at which the system wants to vibrate if left to its own devices. This represents the *undamped natural frequency* since we ignored damping. The *damped natural frequency* will be slightly lower than this value. Note that ω_n is a function only of the physical parameters of the system m and k; thus it is completely determined and unchanging with time, once the system is built. By creating a one-*DOF* model of the system, we have limited ourselves to one natural frequency that is an "average" natural frequency usually close to the lowest, or fundamental, frequency.

Any real physical system will also have higher natural frequencies that in general will not be integer multiples of the fundamental. In order to find them, we need to create a multi-degree-of-freedom model of the system. The fundamental tone at which a bell rings when struck is its natural frequency defined by this expression. The bell also has overtones which are its other, higher, natural frequencies. The fundamental frequency tends to dominate the transient response of the system.[1]

The circular natural frequency ω_n (rad/sec) can be converted to cycles per second (hertz) by noting that there are 2π radians per revolution and one revolution per cycle:

$$f_n = \frac{1}{2\pi}\omega_n \quad \text{hertz} \tag{9.1e}$$

The constants of integration, A and B in equation 9.1c, depend on the initial conditions. A general case can be stated as,

When $t = 0$, let $x = x_0$ and $v = v_0$, where x_0 and v_0 are constants

which gives a general solution to the homogeneous ODE 9.1b of:

$$x = x_0 \cos\omega_n t + \frac{v_0}{\omega_n}\sin\omega_n t \tag{9.1f}$$

Equation 9.1f can be put into polar form by computing the magnitude and phase angle:

$$X_0 = \sqrt{x_0^2 + \left(\frac{v_0}{\omega_n}\right)^2} \qquad \phi = \arctan\left(\frac{v_0}{x_0\omega_n}\right)$$

then:

$$x = X_0\cos(\omega_n t - \phi) \tag{9.1g}$$

Note that this is a pure harmonic function whose amplitude X_0 and phase angle ϕ are a function of the initial conditions and the natural frequency of the system. It will oscillate forever in response to a single, transitory input if there is truly no damping present.

Damped Response

If we now reintroduce the damping of the model in Figure 9-1b and draw the free-body diagram as shown in Figure 9-1c, the summation of forces becomes:

$$F_c(t) - F_d - F_s = m\ddot{x} \tag{9.2a}$$

Substituting equations 8.12 (p. 185) and 8.13c (p. 187):

$$m\ddot{x} + c\dot{x} + kx = F_c(t) \tag{9.2b}$$

Homogeneous Solution We again separate this differential equation into its homogeneous and particular components. The homogeneous part is:

$$\ddot{x} + \frac{c}{m}\dot{x} + \frac{k}{m}x = 0 \tag{9.2c}$$

The solution to this ODE is of the form:

$$x = Re^{st} \tag{9.2d}$$

where R and s are constants. Differentiating versus time:

$$\dot{x} = Rse^{st}$$
$$\ddot{x} = Rs^2e^{st}$$

and substituting in equation 9.2c:

$$Rs^2e^{st} + \frac{c}{m}Rse^{st} + \frac{k}{m}Re^{st} = 0$$

$$\left(s^2 + \frac{c}{m}s + \frac{k}{m}\right)Re^{st} = 0 \tag{9.2e}$$

For this solution to be valid, either R or the expression in parentheses must be zero as e^{st} is never zero. If R were zero, then the assumed solution, in equation 9.2d, would also be zero and thus not be a solution. Therefore, the quadratic equation in parentheses must be zero.

$$\left(s^2 + \frac{c}{m}s + \frac{k}{m}\right) = 0 \tag{9.2f}$$

This is called the characteristic equation of the ODE and its solution is:

$$s = \frac{-\frac{c}{m} \pm \sqrt{\left(\frac{c}{m}\right)^2 - 4\frac{k}{m}}}{2}$$

which has the two roots:

$$s_1 = -\frac{c}{2m} + \sqrt{\left(\frac{c}{2m}\right)^2 - \frac{k}{m}}$$

$$s_2 = -\frac{c}{2m} - \sqrt{\left(\frac{c}{2m}\right)^2 - \frac{k}{m}} \tag{9.2g}$$

These two roots of the characteristic equation provide two independent terms of the homogeneous solution:

$$x = R_1 e^{s_1 t} + R_2 e^{s_2 t} \qquad \text{for } s_1 \neq s_2 \tag{9.2h}$$

If $s_1 = s_2$, then another form of solution is needed. The quantity s_1 will equal s_2 when:

$$\sqrt{\left(\frac{c}{2m}\right)^2 - \frac{k}{m}} = 0 \qquad \text{or:} \qquad \frac{c}{2m} = \sqrt{\frac{k}{m}}$$

and:

$$c = 2m\sqrt{\frac{k}{m}} = 2m\omega_n = c_c \tag{9.2i}$$

This particular value of c is called the **critical damping** and is labeled c_c. The system will behave in a unique way when critically damped, and the solution must be of the form:

$$x = R_1 e^{s_1 t} + R_2 t e^{s_2 t} \qquad \text{for } s_1 = s_2 = -\frac{c}{2m} \tag{9.2j}$$

It will be useful to define a dimensionless ratio called the **damping ratio** ζ which is the actual damping divided by the critical damping.

$$\zeta = \frac{c}{c_c}$$

$$\zeta = \frac{c}{2m\omega_n} \tag{9.3a}$$

and then:

$$\zeta\omega_n = \frac{c}{2m} \tag{9.3b}$$

The damped natural frequency ω_d is slightly less than the undamped natural frequency ω_n and is:

$$\omega_d = \sqrt{\frac{k}{m} - \left(\frac{c}{2m}\right)^2} \tag{9.3c}$$

We can substitute equations 9.1d and 9.3b into equations 9.2g to get an expression for the characteristic equation in terms of dimensionless ratios:

$$s_{1,2} = -\omega_n \zeta \pm \sqrt{(\omega_n \zeta)^2 - \omega_n^{\,2}}$$

$$s_{1,2} = \omega_n \left(-\zeta \pm \sqrt{\zeta^2 - 1}\right) \tag{9.4a}$$

This shows that the system response is determined by the damping ratio ζ that dictates the value of the discriminant. There are three possible cases:

CASE 1:	$\zeta > 1$	Roots real and unequal	
CASE 2:	$\zeta = 1$	Roots real and equal	(9.4b)
CASE 3:	$\zeta < 1$	Roots complex conjugate	

Let's consider the response of each of these cases separately.

CASE 1: $\zeta > 1$ *overdamped*

The solution is of the form in equation 9.2h and is:

$$x = R_1 e^{\left(-\zeta + \sqrt{\zeta^2 - 1}\right)\omega_n t} + R_2 e^{\left(-\zeta - \sqrt{\zeta^2 - 1}\right)\omega_n t} \tag{9.5a}$$

Note that since $\zeta > 1$, both exponents will be negative making x the sum of two decaying exponentials as shown in Figure 9-3. This is the transient response of the system to a disturbance and dies out after a time. There is no oscillation in the output motion. An example of an overdamped system which you may have encountered is the tone arm on a good-quality record turntable with a "cueing" feature.* The tone arm can be lifted up, then released, and it will slowly "float" down to the record. This is achieved by putting a large amount of damping in the system, at the arm pivot. The arm's motion follows an exponential decay curve such as in Figure 9-3.

CASE 2: $\zeta = 1$ *critically damped*

The solution is of the form in equation 9.2j and is:

$$x = R_1 e^{-\omega_n t} + R_2 t e^{-\omega_n t} = \left(R_1 + R_2 t\right) e^{-\omega_n t} \tag{9.5b}$$

This is the product of a linear function of time and a decaying exponential function and can take several forms, depending on the values of the constants of integration, R_1 and R_2, which in turn depend on initial conditions. A typical transient response might

* Readers below "a certain age" may be unfamiliar with records and turntables, being more used to CD players and MP3. This quaint technology (originally invented by Thomas Edison in the 19th century) used vinyl discs with grooves in which a sensitive needle, carried on the tone arm, ran as the disk turned and reproduced the sound wave information that had been captured as microscopic surfaceundulations in the sides of the groove. It worked amazingly well.

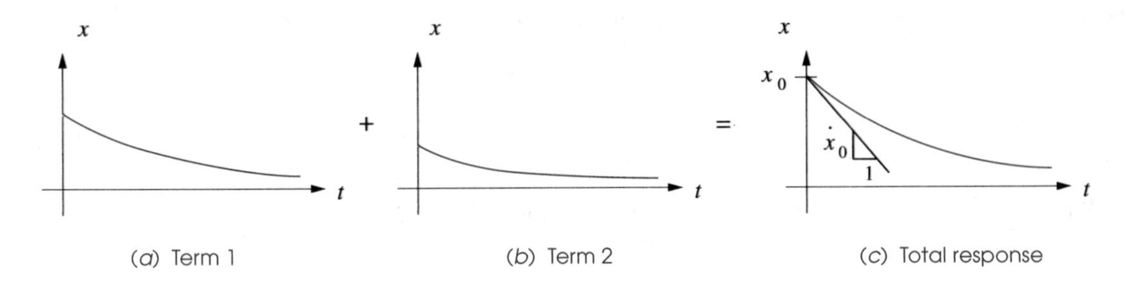

FIGURE 9-3

Transient response of an overdamped system

look like Figure 9-4. This is the transient response of the system to a disturbance, which response dies out after a time. There is fast response but no oscillation in the output motion. An example of a critically damped system is the suspension system of a new sports car in which the damping is usually made close to critical in order to provide crisp handling response without either oscillating or being slow to respond. A critically damped system will, when disturbed, return to its original position within one bounce. It may overshoot but will not oscillate and will not be sluggish.

CASE 3: $\zeta < 1$ *underdamped*

The solution is of the form in equation 9.2h and s_1, s_2 are complex conjugate. Equation 9.4a can be rewritten in a more convenient form as:

$$s_{1,2} = \omega_n\left(-\zeta \pm j\sqrt{1-\zeta^2}\right); \qquad j = \sqrt{-1} \tag{9.5c}$$

Substituting in equation 9.2h:

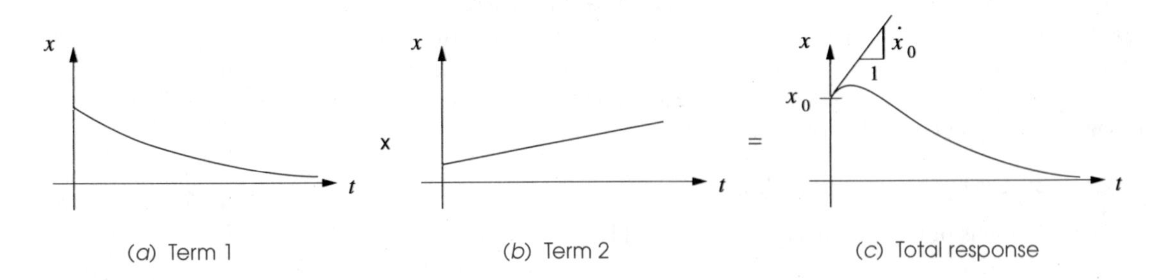

FIGURE 9-4

Transient response of a critically damped system

$$x = R_1 e^{\left(-\zeta + j\sqrt{1-\zeta^2}\right)\omega_n t} + R_2 e^{\left(-\zeta - j\sqrt{1-\zeta^2}\right)\omega_n t}$$

and noting that:

$$y^{a+b} = y^a y^b$$

$$x = R_1\left[e^{-\zeta\omega_n t} e^{\left(j\sqrt{1-\zeta^2}\right)\omega_n t}\right] + R_2\left[e^{-\zeta\omega_n t} e^{\left(-j\sqrt{1-\zeta^2}\right)\omega_n t}\right]$$

factor:

$$x = e^{-\zeta\omega_n t}\left[R_1 e^{\left(j\sqrt{1-\zeta^2}\right)\omega_n t} + R_2 e^{\left(-j\sqrt{1-\zeta^2}\right)\omega_n t}\right] \tag{9.5d}$$

Substitute the Euler equation:

$$x = e^{-\zeta\omega_n t}\left\{\begin{aligned} &R_1\left[\cos\left(\sqrt{1-\zeta^2}\,\omega_n t\right) + j\sin\left(\sqrt{1-\zeta^2}\,\omega_n t\right)\right] \\ &+ R_2\left[\cos\left(\sqrt{1-\zeta^2}\,\omega_n t\right) - j\sin\left(\sqrt{1-\zeta^2}\,\omega_n t\right)\right]\end{aligned}\right\}$$

and simplify:

$$x = e^{-\zeta\omega_n t}\left\{(R_1 + R_2)\left[\cos\left(\sqrt{1-\zeta^2}\,\omega_n t\right) + (R_1 - R_2)\,j\sin\left(\sqrt{1-\zeta^2}\,\omega_n t\right)\right]\right\} \tag{9.5e}$$

Note that R_1 and R_2 are just constants yet to be determined from the initial conditions, so their sum and difference can be denoted as some other constants:

$$x = e^{-\zeta\omega_n t}\left\{A\cos\left(\sqrt{1-\zeta^2}\,\omega_n t\right) + B\sin\left(\sqrt{1-\zeta^2}\,\omega_n t\right)\right\} \tag{9.5f}$$

We can put this in polar form by defining the magnitude and phase angle as:

$$X_0 = \sqrt{A^2 + B^2} \qquad \phi = \arctan\frac{B}{A}$$

then:

$$x = X_0 e^{-\zeta\omega_n t}\cos\left[\left(\sqrt{1-\zeta^2}\,\omega_n t\right) - \phi\right] \tag{9.5g}$$

This is the product of a harmonic function of time and a decaying exponential function where X_0 and ϕ are the constants of integration determined by the initial conditions.

Figure 9-5 shows the transient response for this **underdamped case**. The response overshoots and oscillates before finally settling down to its final position. Note that if the damping ratio ζ is zero, equation 9.5g reduces to equation 9.1g, which is a pure harmonic.

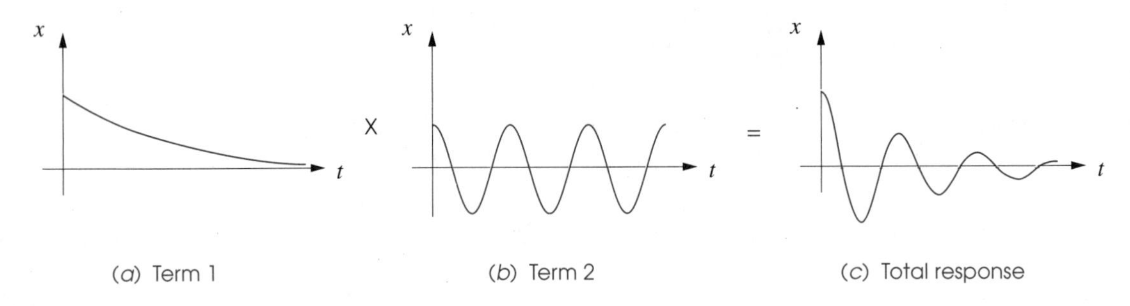

FIGURE 9-5

Transient response of an underdamped system

An example of an **underdamped system** is a diving board that continues to oscillate after the diver has jumped off, finally settling back to zero position. *Many real systems in machinery are underdamped, including the typical cam-follower system.* This often leads to **vibration problems**. It is not usually a good solution simply to add damping to the system as this causes heating and is very energy inefficient. It is better to design the system to avoid the vibration problems.

9

PARTICULAR SOLUTION Unlike the homogeneous solution which is always the same regardless of the input, the particular solution to equation 9.2b (p. 217) will depend on the forcing function $F_c(t)$ that is applied to the cam-follower from the cam. In general, the output displacement x of the follower will be a function of similar shape to the input function, but will lag the input function by some phase angle. It is quite reasonable to use a sinusoidal function as an example since any periodic function can be represented as a Fourier series of sine and cosine terms of different frequencies (see equation 16.1 on p. 486).

Assume the forcing function to be:

$$F_c(t) = F_0 \sin \omega_f t \tag{9.6a}$$

where F_0 is the amplitude of the force and ω_f is its circular frequency. Note that ω_f is unrelated to ω_n or ω_d and may be any value. The system equation then becomes:

$$m\ddot{x} + c\dot{x} + kx = F_0 \sin \omega_f t \tag{9.6b}$$

The solution must be of harmonic form to match this forcing function and we can try the same form of solution as used for the homogeneous solution:

$$x_f(t) = X_f \sin\left(\omega_f t - \psi\right) \tag{9.6c}$$

where:

X_f = amplitude

ψ = phase angle between applied force and displacement

ω_f = angular velocity of forcing function

The factors X_f and ψ are not constants of integration here. They are constants determined by the physical characteristics of the system and the forcing function's frequency and magnitude. They have nothing to do with the initial conditions. To find their values, differentiate the assumed solution twice, substitute in the ODE, and get:

$$X_f = \frac{F_0}{\sqrt{\left(k - m\omega_f^2\right)^2 + \left(c\omega_f\right)^2}}$$

$$\psi = \arctan\left[\frac{c\omega_f}{\left(k - m\omega_f^2\right)^2}\right] \tag{9.6d}$$

Substitute equations 9.1d, 9.2i, and 9.3a and put in dimensionless form:

$$\frac{X_f}{\left(\dfrac{F_0}{k}\right)} = \frac{1}{\sqrt{\left[1 - \left(\dfrac{\omega_f}{\omega_n}\right)^2\right]^2 + \left(2\zeta\dfrac{\omega_f}{\omega_n}\right)^2}}$$

$$\psi = \arctan\left[\frac{2\zeta\dfrac{\omega_f}{\omega_n}}{1 - \left(\dfrac{\omega_f}{\omega_n}\right)^2}\right] \tag{9.6e}$$

The ratio ω_f / ω_n is called the **frequency ratio**. Dividing X_f by the static deflection F_0 / k creates the **amplitude ratio** which defines the relative dynamic displacement compared to the static.

Complete Response The complete solution to our system differential equation for a sinusoidal forcing function is the sum of the homogeneous and particular solutions:

$$x = X_0 e^{-\zeta\omega_n t} \cos\left[\left(\sqrt{1-\zeta^2}\,\omega_n t\right) - \phi\right] + X_f \sin\left(\omega_f t - \psi\right) \tag{9.7}$$

The homogeneous term represents the **transient response** of the system that will die out in time, but is reintroduced any time the system is again disturbed. The **particular term** represents the **forced response** or **steady-state response** to a sinusoidal forcing function that will continue as long as the forcing function is present.

Note that the solution to this equation, shown in equations 9.5g and 9.6e, depends only on two ratios, the damping ratio ζ which relates the actual damping relative to the critical damping, and the *frequency ratio* ω_f / ω_n which relates the forcing frequency to the natural frequency of the system. Koster[2] found by experiment that a typical value for the damping ratio in cam-follower systems is $\zeta = 0.06$, so they are underdamped and can **resonate** (see Section 9.2) if operated at frequency ratios close to 1.

The initial conditions for the specific problem are applied to equation 9.7 to determine the values of X_0 and ϕ. Note that these constants of integration are contained within the homogeneous part of the solution.

9.2 RESONANCE

The natural frequency (and its overtones) are of great interest to the designer as they define the frequencies at which the system will **resonate**. The single-*DOF* lumped parameter systems shown in Figures 9-1 and 9-2 (pp. 214-215) are the simplest possible to describe a dynamic system, yet they contain all the basic dynamic elements. Masses and springs are energy storage elements. A mass stores kinetic energy, and a spring stores potential energy. The damper is a dissipative element. It uses energy and converts it to heat. Thus all the losses in the model of Figure 9-1 occur through the damper.

These are "pure" idealized elements which possess only their own special characteristics. That is, the spring has no damping and the damper no springiness, etc. Any system that contains more than one energy storage device, such as a mass and a spring, will possess at least one natural frequency. If we excite the system at its natural frequency, we will set up the condition called resonance in which the energy stored in the system's elements will oscillate from one element to the other at that frequency. The result can be violent oscillations in the displacements of the movable elements in the system as the energy moves from potential to kinetic form and vice versa.

Figure 9-6a shows a plot of the amplitude and phase angle of the displacement response Y of the system to a sinusoidal input forcing function at various frequencies ω_f. In our case, the forcing frequency is the angular velocity at which the cam is rotating. The plot normalizes the forcing frequency as the frequency ratio ω_f / ω_n. The forced response amplitude Y is normalized by dividing the dynamic deflection y by the static deflection F_0 / k that the same force amplitude would create on the system. Thus at a frequency of zero, the output is one, equal to the static deflection of the spring at the amplitude of the input force. As the forcing frequency increases toward the natural frequency ω_n, the amplitude of the output motion, for zero damping, increases rapidly and becomes theoretically infinite when $\omega_f = \omega_n$. Beyond this point the amplitude decreases rapidly and asymptotically toward zero at high frequency ratios. Figure 9-6c shows that the phase angle between input and output of a forced system switches abruptly at resonance.

Figure 9-6b shows the amplitude response of a "self-excited" system for which there is no externally applied force. An example might be a shaft coupling a motor and a generator. The loading is theoretically pure torsion. However, if there is any unbalance in the rotors on the shaft, the centrifugal force will provide a forcing function proportional to angular velocity. Thus, when stopped there is no dynamic deflection, so the amplitude is zero. As the system passes through resonance, the same large response as the forced case is seen. At large frequency ratios, well above critical, the deflection becomes static at an amplitude ratio of 1. Cam-follower systems are subject to both of these types of vibratory behavior. An unbalanced camshaft will self-excite, and the follower force creates a forced response.

The effects of damping ratio ζ can best be seen in Figure 9-6d, which shows a 3-D plot of forced vibration amplitude as a function of both frequency ratio and damping ratio. The addition of damping reduces the amplitude of vibration at the natural frequency, but very large damping ratios are needed to keep the output amplitude less than or equal to the input amplitude. This is much more damping than is found in cam-follower systems and most machinery. About 50 to 60% of critical damping will eliminate the resonance peak. Unfortunately,

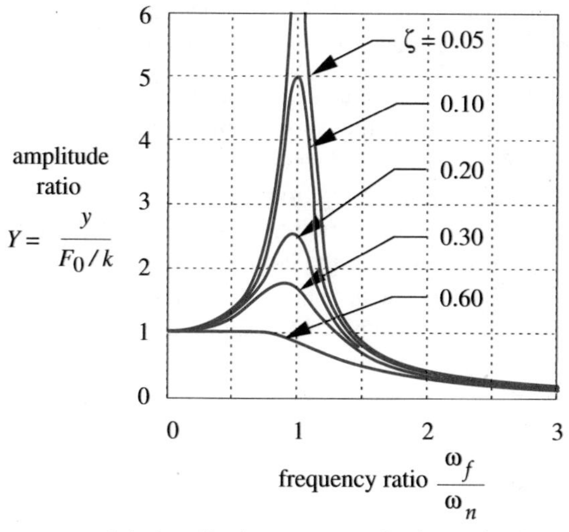

(a) Amplitude response of a forced system

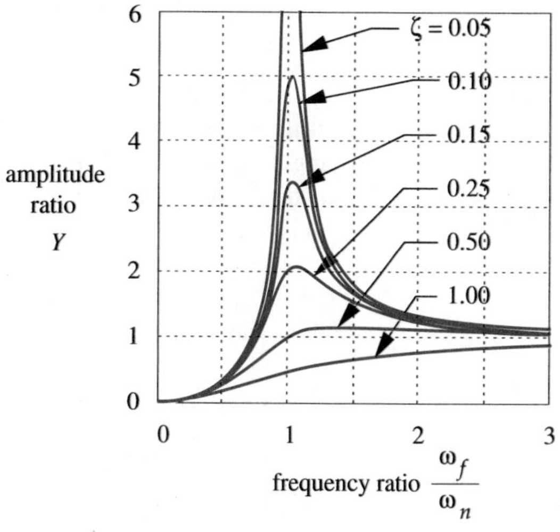

(b) Amplitude response of a self-excited system

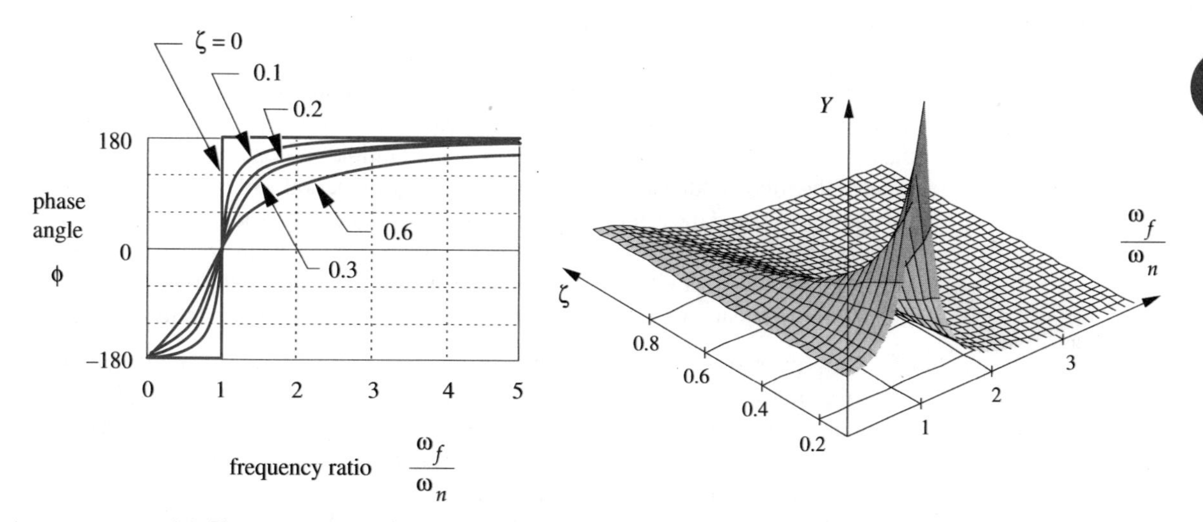

(c) Phase response of a system

(d) 3-D plot of forced system amplitude response

FIGURE 9-6

Amplitude and phase response of a one-degree-of-freedom system to forcing or self-excitation frequencies

most cam follower systems have damping ratios of less than about 10% of critical. At those damping levels, the response at resonance is about five times the static response. This will create unsustainable stresses in most systems if allowed to occur.

It is obvious that we must avoid driving an underdamped system at or near its natural frequency. One result of operation of an underdamped cam-follower system near ω_n can be **follower jump**. The system of follower mass and spring can oscillate violently at its natural frequency and leave contact with the cam. When it does reestablish contact, it may do so with severe impact loads that can quickly fail the materials.

The designer has a degree of control over resonance in that the system's mass m and stiffness k can be tailored to move its natural frequency away from any required operating frequencies. A common rule of thumb is to design the system to have a fundamental natural frequency ω_n at least ten times the highest forcing frequency expected in service, thus keeping all operation well below the resonance point. This is often difficult to achieve in mechanical systems. One tries to achieve the largest ratio ω_n / ω_f possible nevertheless. As will be shown in later chapters, the high harmonic content of cam functions can require that even higher ratios than 10 are needed in some cases.

Follower Rise Time

For cams with finite jerk having a dwell after the rise of the same order of duration as the rise, Koster[2] defines a dimensionless ratio,

$$\tau = \frac{T_n}{T_r} \qquad T_n = \frac{1}{\omega_n} \tag{9.8}$$

where T_n is the period of the follower system's natural frequency and T_r is the time to complete the rise portion of the follower motion. To minimize the effects of the transient vibrations on the system, this ratio should be as small as possible and always less than 1. A small value of τ is equivalent to a small ratio of ω_f / ω_n. The error in acceleration of the follower due to vibrations during the dwell period will be proportional to the first power of τ, and the error in follower position will be proportional to the third power of τ, for values of $\tau < 0.5$. If the dwell's duration is about as long or longer than that of the rise, these transient vibrations will tend to die out by the end of the dwell.[2] The next fall or rise will again provide an input to the system and cause a new transient response.

Thus, the response of a cam-follower system of this type will be dominated by the recurring transient responses rather than by the forced, or steady-state, response. It is important to adhere to the fundamental law of cam design and use cam programs with finite jerk in order to minimize these residual vibrations in the follower system. Koster[2] reports that the cycloidal and 3-4-5 polynomial programs both gave low residual vibrations in the double-dwell cam. The modified sine acceleration program will also give good results. All these have finite jerk. Other functions that offer even lower residual vibration levels will be introduced in later chapters.

Some thought and observation of equation 9.1d (p. 216) will show that we would like our system members to be both light (low m) and stiff (high k) to get high values for ω_n and thus low values for τ. Unfortunately, the lightest materials are seldom also the stiffest. Aluminum is one-third the weight of steel but is also about a third as stiff. Titanium is about half the weight of steel but also about half as stiff. Some of the exotic composite materials such as carbon fiber/epoxy offer better stiffness-to-weight ratios, but their cost is high and processing is difficult.

Note in Figure 9-6 (p. 225) that the amplitude of vibration at large frequency ratios approaches zero for forced and one for self-excited systems. So, if the system can be brought up to speed through the resonance point without damage and then kept operating at a large frequency ratio, the vibration will be minimal, especially if it is a forced system. An example of systems designed to be run this way are large self-excited devices that must run at higher speed such as electrical power generators. Their large mass creates a lower natural frequency than their required operating speeds. They are "run up" as quickly as possible through the resonance region to avoid damage from their vibrations and "run down" quickly through resonance when stopping them. They also have the advantage of long duty cycles of constant speed operation in the safe frequency region between infrequent starts and stops.

9.3 ESTIMATING DAMPING

While it is relatively easy to model the masses and springs in a cam-follower system, it is difficult to accurately estimate its damping at the design stage by any calculation method. As noted above, experimenters such as Koster[2] have measured the damping ratio ζ in actual cam-follower systems and found it to be typically less than 0.1. This author has also measured the damping ratio in many cam-follower systems and found it to be consistent with Koster's data. Luckily, it is a fairly simple matter to measure the damping ratio in an existing physical system. To the degree that a newly designed system is similar to one in which the damping has been experimentally measured, those data can be extrapolated to the new design. Once a prototype of a new design is built, the same simple test can be done to confirm the earlier assumption regarding its damping.

Logarithmic Decrement

Figure 9-7 shows the response of an underdamped system to an impulse (hammer blow). As shown in Figure 9-5 and defined in equation 9.5g, this response is the product of a decaying exponential function and a cosine. Figure 9-7a shows the decaying exponential as a dotted line enveloping the response. Figure 9-7b plots this exponential term by itself as was done in Figure 9-5a. These are linear plots. Figure 9-7c plots the same exponential function on a natural log (ln) axis. It plots as a straight line. The slope of this line δ is called the **logarithmic decrement** and is proportional to the damping in the system.

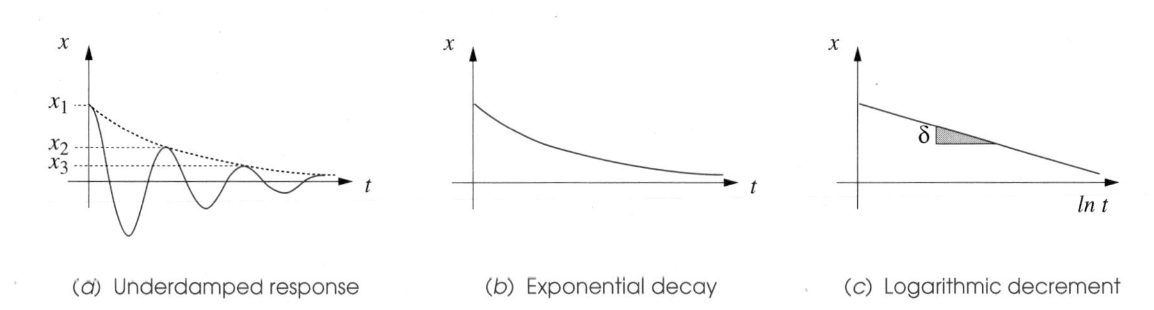

(*a*) Underdamped response (*b*) Exponential decay (*c*) Logarithmic decrement

FIGURE 9-7

Logarithmic decrement of an underdamped system

The value of the logarithmic decrement can be found from a simple test of the system. A transducer, such as an accelerometer, is placed on the output end of the system and the system is struck lightly with a hammer or other hard object. The transducer measures the transient response of the system, which will look something like the solid line in Figure 9-7a when viewed on an oscilloscope. The values of the successive peaks of the response, x_1, x_2, x_3, etc., are then read from the plot. Any two successive values are sufficient to compute the logarithmic decrement from:

$$\delta = \ln \frac{x_1}{x_2} \tag{9.9a}$$

A potentially more accurate estimate of δ can be found by measuring a pair of peaks that are further apart in time to obtain an average value from

$$\delta = \frac{1}{m-1} \ln\left(\frac{x_1}{x_m}\right) \tag{9.9b}$$

where x_m is the mth peak measured.

The damping ratio ζ is related to the logarithmic decrement as

$$\zeta = \frac{\delta}{\sqrt{(2\pi)^2 + \delta^2}} \tag{9.9c}$$

For small values of damping ratio ($\zeta < 0.3$), an approximate value can be found from

$$\zeta \cong \frac{\delta}{2\pi} \tag{9.9d}$$

This technique provides a simple means to measure the damping in a system.

EXAMPLE 9-1

Measuring the Damping of a Cam-Follower System.

Given: A cam-follower system as shown on the left side of Figure 9-8

Problem: Measure the response of the system to an excitation and calculate an estimate of the system damping ratio ζ.

Solution:

1 A small, sensitive, piezoelectric accelerometer was attached with a magnet to the output arm of the follower train at point *B* and the follower arm was struck with a small ball-peen hammer at point *A*, near where the cam contacts the roller follower. The machine power was off, but the air cylinder that closes the cam joint was pressurized.

2 Figure 9-9a shows the impulse response of the entire follower train at point *B* to an impact at point *A*. Note its general similarity to the theoretical response of a one-DOF system shown in Figure 9-5c. However, this actual response is not as "clean" as the theoretical. Moreover, it does not decay smoothly. The reason is that this is a multi-DOF system and

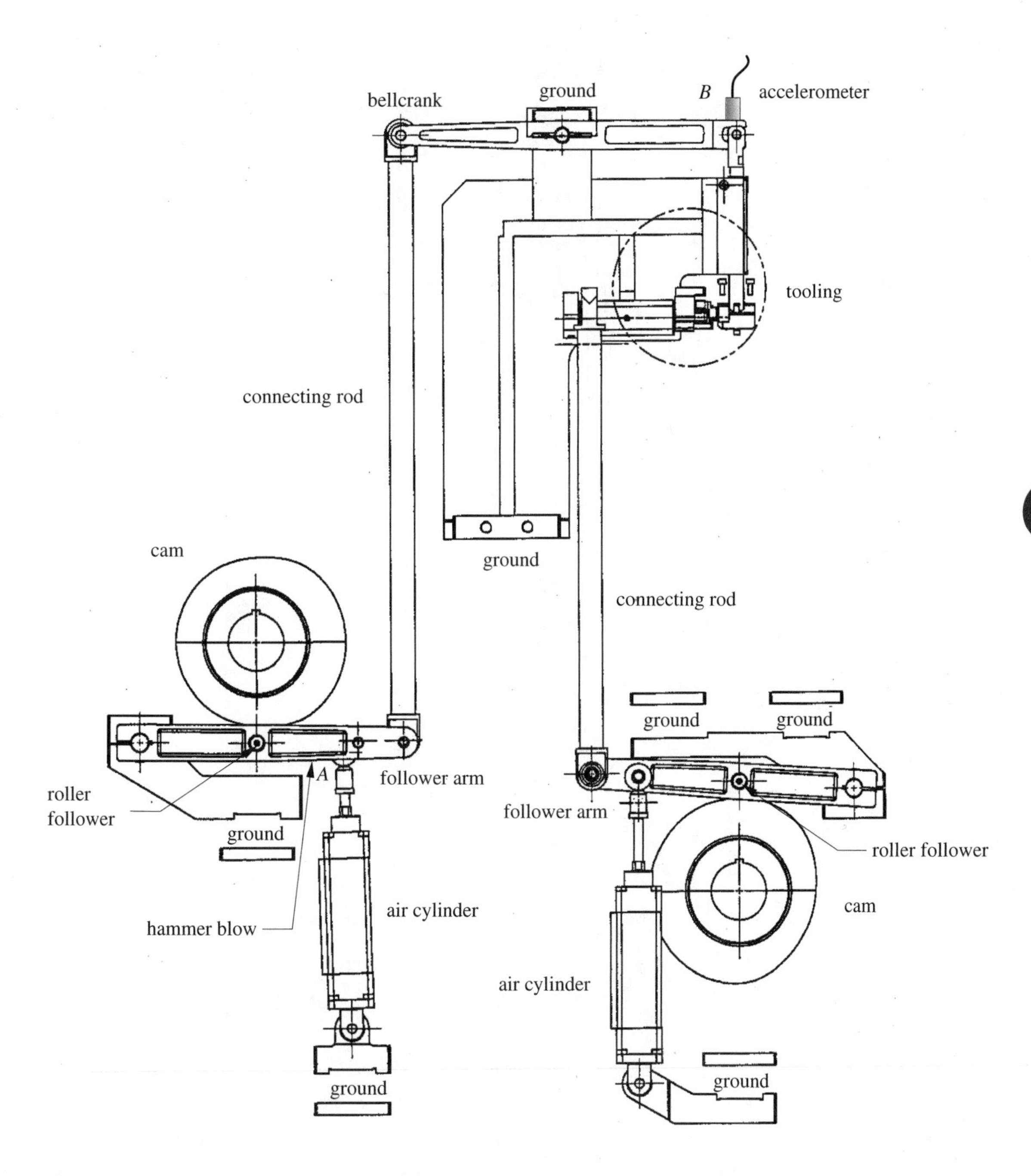

FIGURE 9-8

Testing the damping of a cam-follower system for Example 9-1 *(Courtesy of The Gillette Company, Boston, MA)*

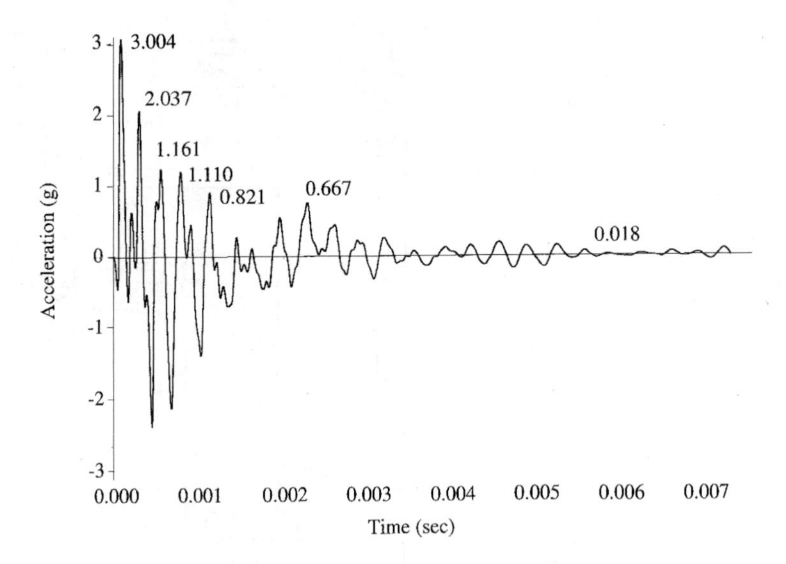

(*a*) Impulse response of system at point B

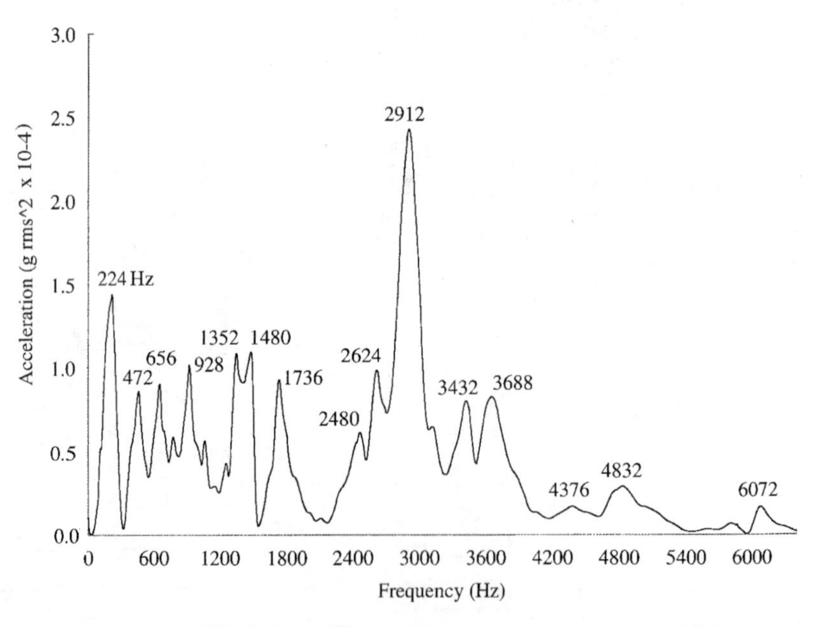

(*b*) Autospectrum of frequency response at point B

FIGURE 9-9

Impulse response and frequency spectrum of the left follower train in Figure 9-8

so has many natural frequencies. Each of these resonances is decaying with a different periodicity. These functions combine causing the irregularities in the response.

3 This can be seen more clearly in Figure 9-9b, which shows the Fourier transform* of the impulse response out to 6400 Hz. There are about 20 peaks (natural frequencies) seen here. Many of them have been tagged with their corresponding frequency in hertz. The lowest is 224 Hz, which has a period of 0.0045 sec. The spacing of the first two peaks in the impulse response of Figure 9-9a is about 0.2 ms which corresponds to a frequency of about 4800 Hz. A peak appears in the spectrum near that frequency.

4 We would like to estimate the damping in this system from the logarithmic decrement of the envelope of the impulse response in Figure 9-9a. The difficulty comes in deciding which peaks are members of the same decaying resonance. The first three peaks appear to be spaced evenly, which implies that they are of the same family. Their peak values are shown on the plot and are $x_1 = 3.004$, $x_2 = 2.037$, $x_3 = 1.161$. Using the first two of these in equations 9.9 a and d gives:

$$\delta = \ln\frac{x_1}{x_2} = \ln\frac{3.004}{2.037} = 0.3885 \tag{a}$$

$$\zeta \cong \frac{\delta}{2\pi} = \frac{0.3885}{2\pi} = 0.06 \tag{b}$$

5 Using the first and third values in equations 9.9b and d gives:

$$\delta = \frac{1}{m-1}\ln\left(\frac{x_1}{x_m}\right) = \frac{1}{2}\ln\left(\frac{x_1}{x_3}\right) = \frac{1}{2}\ln\left(\frac{3.004}{1.161}\right) = 0.4753 \tag{c}$$

$$\zeta \cong \frac{\delta}{2\pi} = \frac{0.4753}{2\pi} = 0.08 \tag{d}$$

6 We can continue this process, using any number of oscillations in the calculation. It sometimes becomes difficult to decide what the number of an oscillation is when the waveform is as irregular as this one. If we take the first peak ($x_1 = 3.004$) and what appears to be the 18th peak ($x_{18} = 0.018$) in equations 9.9b and d we get

$$\delta = \frac{1}{m-1}\ln\left(\frac{x_1}{x_m}\right) = \frac{1}{17}\ln\left(\frac{x_1}{x_{18}}\right) = \frac{1}{17}\ln\left(\frac{3.004}{0.018}\right) = 0.3010 \tag{e}$$

$$\zeta \cong \frac{\delta}{2\pi} = \frac{0.3010}{2\pi} = 0.05 \tag{f}$$

7 Another way to get a measure of system damping is to analyze acceleration measurements made while a machine is running. Figure 9-10 shows the system of Figure 9-8 running at 100 rpm. The accelerometer is again at point B. This cam is a double-dwell with modified trapezoid acceleration on both rise and fall. It can be seen that the measured accelerations look very little like the theoretical curves shown in Figure 3-5 (p. 35). The presence of vibration in the follower train has severely distorted the waveforms. In fact the impulse response of the system is present in the acceleration and that is why we can get a measure

* See "Spectrum Analysis" in Section 16.3 (p. 486) for an explanation of the Fourier transform.

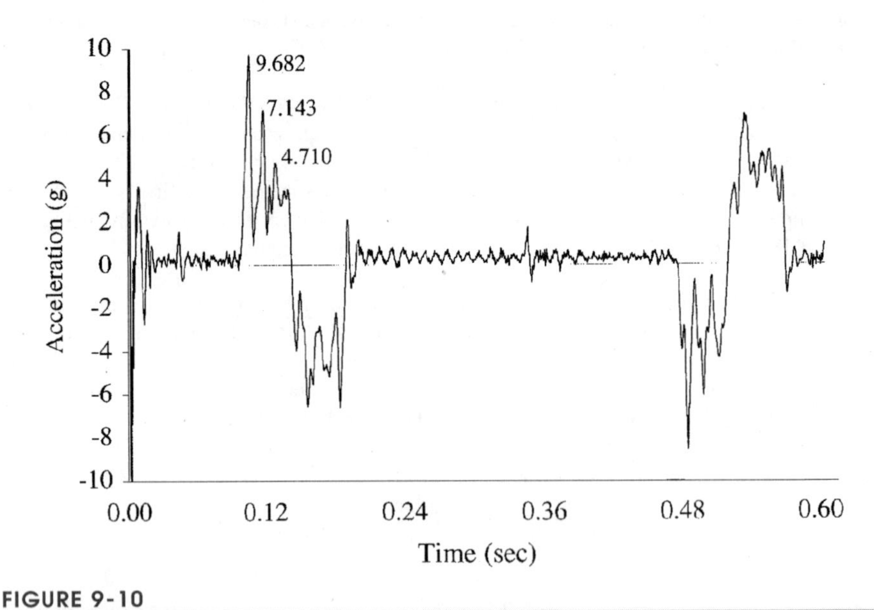

FIGURE 9-10

Acceleration of follower arm in Figure 9-8 at point B—machine running at 100 rpm

of damping from these data as well. The values of the first three peaks during the positive acceleration pulse of the rise acceleration are noted in the figure. They are $x_1 = 9.682$, $x_2 = 7.143$, $x_3 = 4.710$. Using the first two of these in equations 9.9 a and d gives:

$$\delta = \ln\frac{x_1}{x_2} = \ln\frac{9.682}{7.143} = 0.3041 \tag{g}$$

$$\zeta \cong \frac{\delta}{2\pi} = \frac{0.3041}{2\pi} = 0.05 \tag{h}$$

8 Using the first and third values in equations 9.9b and d gives:

$$\delta = \frac{1}{m-1}\ln\left(\frac{x_1}{x_m}\right) = \frac{1}{2}\ln\left(\frac{x_1}{x_3}\right) = \frac{1}{2}\ln\left(\frac{9.862}{4.710}\right) = 0.3695 \tag{i}$$

$$\zeta \cong \frac{\delta}{2\pi} = \frac{0.3695}{2\pi} = 0.06 \tag{j}$$

9 If we average these five estimates, we conclude that this system has a damping ratio of about $\zeta = 0.06$—a very underdamped system—consistent with other data in the literature such as Koster.[2]

10 The natural frequencies of the system obtained from this measurement also will prove very useful in the analysis of vibrations in the cam-follower system, as will be discussed in a later chapter.

9.4 KINETOSTATIC FORCE ANALYSIS OF THE FORCE-CLOSED CAM-FOLLOWER

The previous sections introduced forward dynamic analysis and the solution to the system differential equation of motion (9.2b). The applied force $F_c(t)$ is presumed to be known, and the system equation is solved for the resulting displacement x from which its derivatives can also be determined. The **inverse dynamics**, or **kinetostatics**, approach provides a quick way to determine how much spring force is needed to keep the follower in contact with the cam at a chosen design speed. The displacement and its derivatives are defined from the kinematic design of the cam based on an assumed constant angular velocity ω of the cam. Equation 9.2b can be solved algebraically for the force $F_c(t)$ provided that values for mass m, spring constant k, preload F_{pl}, and damping factor c are known in addition to the kinematic displacement, velocity, and acceleration functions.

Figure 9-1a (p. 214) shows a simple plate or disk cam driving a spring-loaded, roller follower. This is a force-closed system which depends on the spring force to keep the cam and follower in contact at all times. Figure 9-1b shows a lumped parameter model of this system in which all the **mass** that moves with the follower train is lumped together as m, all the springiness in the system is lumped within the **spring constant** k, and all the **damping** or resistance to movement is lumped together as a damper with coefficient c.

9

The designer has a large degree of control over the system spring constant k_{eff} as it tends to be dominated by the k_s of the physical return spring in this model. The elasticities of the follower parts also contribute to the overall system k_{eff} but are usually much stiffer than the physical spring. If the follower stiffness is in series with the return spring, as it often is, equation 8.15c (p. 191) shows that the softest spring in series will dominate the effective spring constant. Thus, the return spring will virtually determine the overall k unless some parts of the follower train have similarly low stiffness.

The designer will choose or design the return spring and thus can specify both its k and the amount of preload to be introduced at assembly. Preload of a spring occurs when it is compressed (or extended if an extension spring) from its *free length* to its initial assembled length. This is a necessary and desirable situation as we want some residual force on the follower even when the cam is at its lowest displacement. This will help maintain good contact between the cam and follower at all times. This spring preload F_{pl} adds a constant term to equation 9.2b which becomes:

$$F_c(t) = m\ddot{x} + c\dot{x} + kx + F_{pl} \tag{9.10a}$$

or :

$$F_c(t) = m\ddot{x} + c\dot{x} + k(x + x_0) \tag{9.10b}$$

The value of m is determined from the effective mass of the system as lumped in the single-*DOF* model of Figure 9-1. The value of c for most cam-follower systems can be estimated for a first approximation to be about 0.06 of the critical damping c_c as defined in equation 9.2i (p. 218).

Calculating the damping c based on an assumed value of ζ requires specifying a value for the overall system k and for its effective mass. The choice of k will affect both the natural frequency of the system for a given mass and the available force to keep the joint closed. Some iteration will probably be needed to find a good compromise. A selection of data for commercially available helical coil springs is provided in Appendix D. Note in equations 9.10 that the terms involving acceleration and velocity can be either positive or negative. The terms involving the spring parameters k and F_{pl} are the only ones that are always positive. To keep the overall function always positive requires that the spring force terms be large enough to counteract any negative values in the other terms. Typically, the acceleration is larger numerically than the velocity, so the negative acceleration usually is the principal cause of a negative force F_c.

The principal concern in this analysis is to keep the cam force always positive in sign as its direction is defined in Figure 9-1. The cam force is shown as positive in that figure. In a force-closed system the cam can only push on the follower. It cannot pull. The follower spring is responsible for providing the force needed to keep the joint closed during the negative acceleration portions of the follower motion. The damping force has an effect, but the spring must supply the bulk of the force to maintain contact between the cam and follower. If the force F_c goes negative at any time in the cycle, the follower and cam will part company, a condition called **follower jump**. When they meet again, it will be with large and potentially damaging impact forces. The follower jump, if any, will usually occur near the point of maximum negative acceleration. Thus, we must select the spring constant and preload to guarantee a positive force at all points in the cycle. In automotive engine valve cam applications, follower jump is also called *valve float*, because the valve (follower) "floats" above the cam, also periodically impacting the cam surface (sometimes called *valve crash*). This will occur if the cam rpm is increased to the point that the larger negative acceleration makes the follower force negative. The "redline" maximum engine rpm often indicated on its tachometer is to warn of impending valve float above that speed that will damage the cam and follower.

Program DYNACAM allows the iteration of equation 9.10 to be done quickly for any cam whose kinematics have been defined in that program. The program's *Dynamics* button will solve equation 9.10 for all values of camshaft angle, using the displacement, velocity, and acceleration functions previously calculated for that cam design in the program. The program requires values for the effective system mass m, effective spring constant k, preload F_{pl}, and the assumed value of the damping ratio ζ. These values need to be determined for the model by the designer using the methods described in Sections 8.9 and 8.10 (pp. 185-197). The calculated force at the cam-follower interface can then be plotted or its values printed in tabular form. The system's first natural frequency, based on the model of Figure 9-1, is also reported when the tabular force data are printed.

EXAMPLE 9-2

Kinetostatic Force Analysis of a Force-Closed (Spring-Loaded) Cam-Follower System.

Given: A translating roller follower as shown in Figure 9-1 is driven by a force-closed radial plate cam which has the following program:

Segment 1: Rise 1 inch in 50° with modified sine acceleration
Segment 2: Dwell for 40°

Segment 3: Fall 1 inch in 50° with cycloidal displacement
Segment 4: Dwell for 40°
Segment 5: Rise 1 inch in 50° with 3-4-5 polynomial displacement
Segment 6: Dwell for 40°
Segment 7: Fall 1 inch in 50° with 4-5-6-7 polynomial displacement
Segment 8: Dwell for 40°
Camshaft angular velocity is 18.85 rad/sec
Follower effective mass is 0.0738 lb-sec^2/in (blobs)
Damping is 10% of critical ($\zeta = 0.10$)

Problem: Compute the necessary spring constant and spring preload to maintain contact between cam and follower, and calculate the dynamic force function for the cam. Calculate the system natural frequency with the selected spring. Keep the pressure angle under 30°.

Solution:

1 Calculate the kinematic data (follower displacement, velocity, acceleration, and jerk) for the specified cam functions. The acceleration for this cam is shown in Figure 9-11 and has a maximum value of 3504 in/sec^2.

2 Calculate the pressure angle and radius of curvature for trial values of prime circle radius, and size the cam to control these values. Figure 9-12 shows the pressure angle function and Figure 9-13 the radii of curvature for this cam with a prime circle radius of 4 in and zero eccentricity. The maximum pressure angle is 29.2° and the minimum radius of curvature is 1.7 in.

3 With the kinematics of the cam defined, we can address its dynamics. To solve equation 9.10 for cam force, we must assume values for the spring constant k and the preload F_{pl}.

Segment Number	Function Used	Start Angle	End Angle	Delta Angle
1	ModSine rise	0	50	50
2	Dwell	50	90	40
3	Cycloid fall	90	140	50
4	Dwell	140	180	40
5	345 poly rise	180	230	50
6	Dwell	230	270	40
7	4567 poly fall	270	320	50
8	Dwell	320	360	40

(a) Cam program specifications

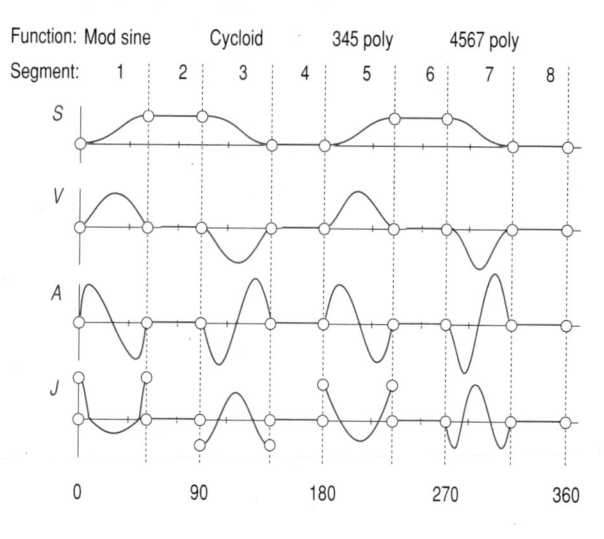

(b) Plots of cam-follower's *S V A J* diagrams

FIGURE 9-11

S V A J diagrams for Examples 9-2 and 9-3

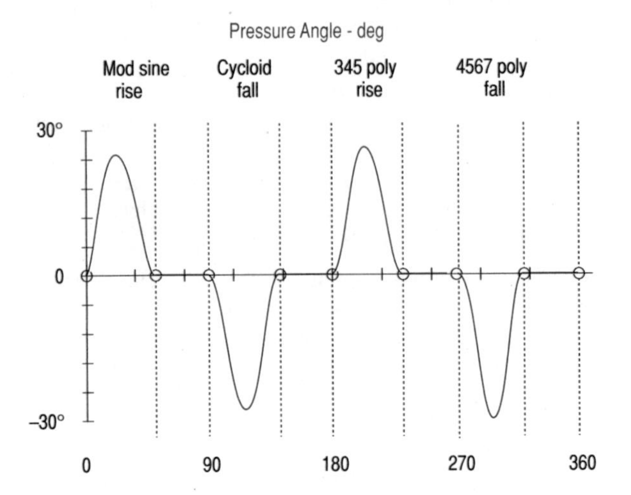

FIGURE 9-12

Pressure angle plot for Examples 9-2 and 9-3

The value of c can be calculated from equation 9.3a using the given mass m, the damping factor ζ, and assumed k. The kinematic parameters are known.

4 Program DYNACAM does this computation for you. The dynamic force that results from an assumed k of 150 lb/in and a preload of 75 lb is shown in Figure 9-14a. The damping coefficient $c = 0.998$. Note that the force dips below the zero axis in two places during negative acceleration. These are locations of follower jump. The follower has left the

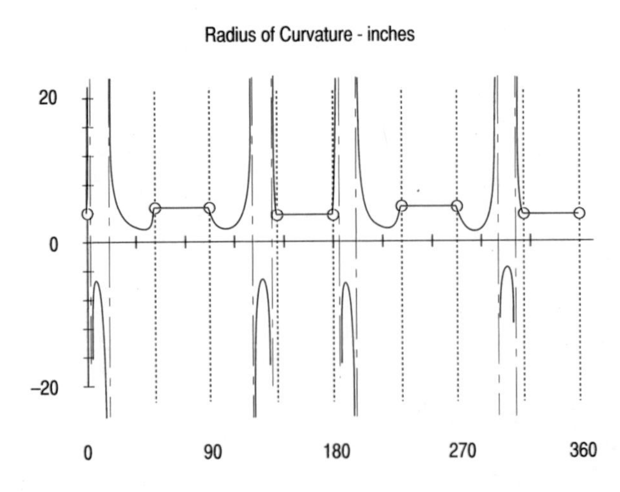

FIGURE 9-13

Radius of curvature of a four-dwell cam for Examples 9-2 and 9-3 cam during the fall because the spring does not have enough available force to keep the follower in contact with the rapidly falling cam. Open the file EX_09-01.cam in DYNACAM and provide the specified k and F_{pl} to see this example. Another iteration is needed to improve the design.

5 Figure 9-14b shows the dynamic force for the same cam with a spring constant of $k = 200$ lb/in and a preload of 150 lb. The damping coefficient $c = 1.153$. This additional force has lifted the function up sufficiently to keep it positive everywhere. There is no follower jump in this case. The maximum force during the cycle is 400.4 lb. A margin of safety has been provided by keeping the minimum force comfortably above the zero line at 36.9 lb. Run example #5 in the DYNACAM program, providing the specified spring constant and preload values to see this example.

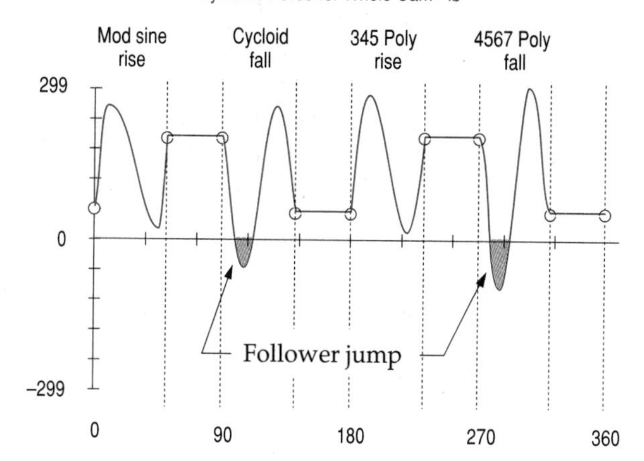

(a) Insufficient spring force allows follower jump

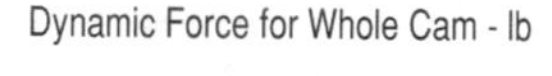

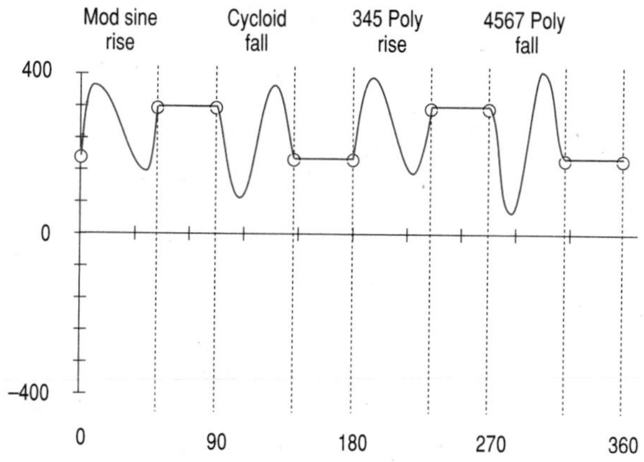

(b) Sufficient spring force keeps the dynamic force positive

FIGURE 9-14

Dynamic forces in a force-closed cam-follower system

6 The fundamental natural frequencies after jump, both undamped and damped, can be calculated for the system from equations 9.1d (p. 216) and 9.3c (p. 219) and are:

$\omega_n = 52.06$ rad/sec or $f_n = 8.28$ Hz $\qquad$ $\omega_d = 51.98$ rad/sec or $f_n = 8.27$ Hz

Open the files Ex_09-02a and Ex_09-02b in program DYNACAM for more detail.

9.5 KINETOSTATIC FORCE ANALYSIS OF THE FORM-CLOSED CAM-FOLLOWER

Section 8.1 described two types of joint closure used in cam-follower systems, **force closure** and **form closure**. Force closure uses an open joint and requires a spring or other force source to maintain contact between the elements. Form closure provides a geometric constraint at the joint such as the cam groove shown in Figure 9-15a or the conjugate cams of Figure 9-15b. No spring is needed to keep the follower in contact with these cams. The follower will run against one side or the other of the groove or conjugate pair as necessary to provide both positive and negative forces. Since there is no spring in this system, its dynamic force equation 9.10 (p. 233) simplifies to:

$$F_c(t) = m\ddot{x} + c\dot{x} \tag{9.11}$$

9

Note that there is now only one energy storage element in the system (the mass), so, theoretically, resonance is not possible. There is no natural frequency for it to resonate at. This is the chief advantage of a form-closed system over a force-closed one. Follower jump will not occur, short of complete failure of the parts, no matter how fast the system is run. This arrangement is sometimes used in high-performance or racing engine valve trains to allow higher redline engine speeds without valve float. In engine valve trains, a form-closed cam-follower valve train is called a *desmodromic* system.

As with any design, there are trade-offs. While the form-closed system typically allows higher operating speeds than a comparable force-closed system, it is not free of all vibration problems. Even though there is no physical return spring in the system, the follower train, the camshaft, and all other parts still have their own spring constants that combine to provide an overall effective spring constant for the system as shown in Section 8.9 (p. 185). The positive side is that this spring constant will typically be quite large (stiff) since properly designed follower parts are designed to be stiff. The effective natural frequency will then be high (see equation 9.1d, p. 216) and possibly well above the forcing frequency as desired.

Another problem with form-closed cams, especially the grooved or track type shown in Figure 9-15a, is the phenomenon of **crossover shock**. Every time the acceleration of the follower changes sign, the inertial force also does so. This causes the follower to abruptly shift from one side of the cam groove to the other. There cannot be zero clearance between the roller follower and the groove and still have it operate. Even if the clearance is very small, there will still be an opportunity for the follower to develop some velocity in its short trip across the groove, and it will impact the other side. Track cams of the type shown in Figure 9-15a typically fail at the points where the acceleration reverses sign, due to many cycles of crossover shock. Note also that the single roller fol-

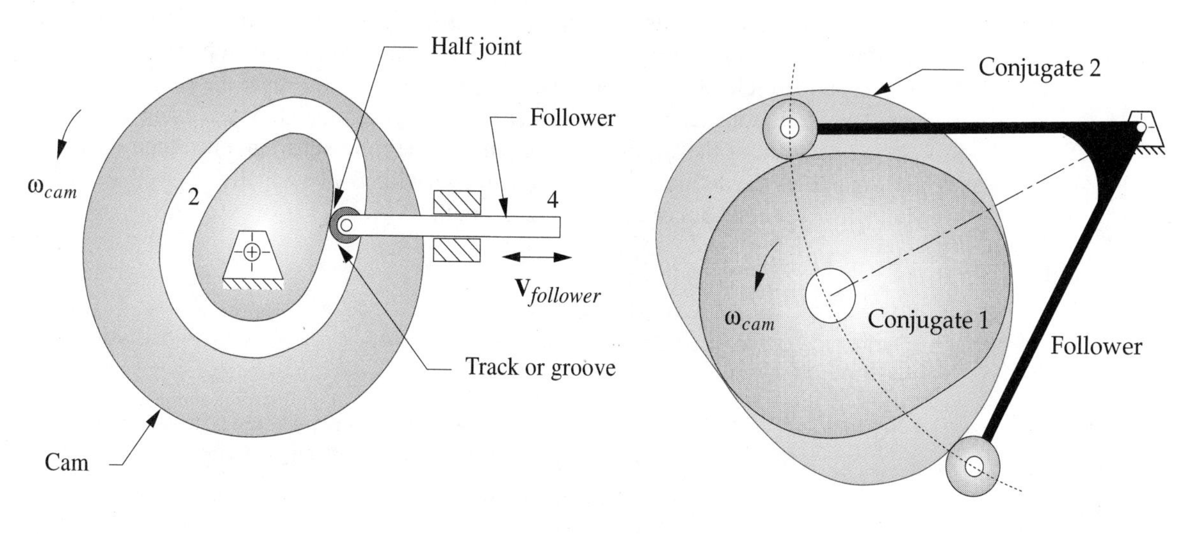

(*a*) Form-closed cam with translating follower

(*b*) Conjugate cams on common shaft

FIGURE 9-15

Form-closed cam-follower systems

lower has to reverse its angular velocity direction every time it crosses over to the other side of the groove. This causes significant follower slip and high wear on the follower compared to an open, force-closed cam where the follower typically will have less than 1% slip.

To avoid the roller reversal in a grooved cam such as that in Figure 9-15a, two rollers can be fitted to the end of the follower arm, each riding on a different side of the groove. Then each rotates in one direction. This solution adds complexity and takes more space as the groove must be wider and the follower arm larger to accommodate the twin rollers, or the groove sides must be offset axially.

Crossover shock can be reduced or eliminated by spring-loading one roller with respect to the arm such as to force it against the track through any tolerance variations present. This must be a stiff spring to avoid introducing resonance problems. The pair of conjugate follower arms of Figure 9-15b can also be sprung together to take up tolerance and wear of the cams. The same potential resonance issues appear in this case also.

Because there are two cam surfaces to machine and because the cam track, or groove, must be cut and ground to high precision to control the clearance, form-closed cams tend to be more expensive to manufacture than force-closed cams. Track cams usually must be ground after heat treatment to correct the distortion of the groove resulting from the high temperatures. Grinding significantly increases cost. Many force-closed cams are not ground after heat treatment and are used as milled. Though the conjugate cam approach avoids the groove tolerance and heat treat distortion problems, there are still two matched cam surfaces to be made per cam. Thus, the desmodromic cam's dynamic advantages come at a significant cost premium.

We will now repeat the cam design of Example 9-2, modified for desmodromic operation. This is simple to do with program DYNACAM. We will merely specify the spring constant and preload values to be zero, which then assumes that the follower train is a rigid body. A more accurate result can be obtained by calculating and using the effective spring constant of the combination of parts in the follower train, once their geometries and materials are defined. The dynamic forces will now be negative as well as positive, but a form-closed cam can both push and pull.

EXAMPLE 9-3

Dynamic Force Analysis of a Form-Closed (Desmodromic) Cam-Follower System.

Given: A translating roller follower as shown in Figure 9-15a is driven by a form-closed radial plate cam which has the following program:

Segment 1: Rise 1 inch in 50° with modified sine acceleration
Segment 2: Dwell for 40°
Segment 3: Fall 1 inch in 50° with cycloidal displacement
Segment 4: Dwell for 40°
Segment 5: Rise 1 inch in 50° with 3-4-5 polynomial displacement
Segment 6: Dwell for 40°
Segment 7: Fall 1 inch in 50° with 4-5-6-7 polynomial displacement
Segment 8: Dwell for 40°
Camshaft angular velocity is 18.85 rad/sec
Follower effective mass is 0.0738 in-lb-sec^2 (blobs)
Damping is 10% of critical ($\zeta = 0.10$)

Problem: Compute the dynamic force function for the cam. Keep the pressure angle under 30°.

Solution:

1 Calculate the kinematic data (follower displacement, velocity, acceleration, and jerk) for the specified cam functions. The acceleration for this cam is shown in Figure 9-11 (p. 235) and has a maximum value of 3504 in/sec^2.

2 Calculate radius of curvature and pressure angle for trial values of prime circle radius, and size the cam to control these values. Figure 9-12 shows the pressure angle function and Figure 9-13 the radii of curvature for this cam with a prime circle radius of 4 in and zero eccentricity. The maximum pressure angle is 29.2° and the minimum radius of curvature is 1.7 in.

3 With the kinematics of the cam defined, we can address its dynamics. To solve equation 9.10 (p. 233) for the cam force, we assume zero values for the spring constant k and the preload F_{pl}. The value of c is assumed to be 1.153, the same as in the previous example. The kinematic parameters are known.

4 Program DYNACAM does this computation for you. The dynamic force that results is shown in Figure 9-16. Note that the force is now more nearly symmetric about the axis and its peak absolute value is 289 lb. Crossover shock will occur each time the follower force changes sign. Open the file EX_09-03.cam in DYNACAM to see this example.

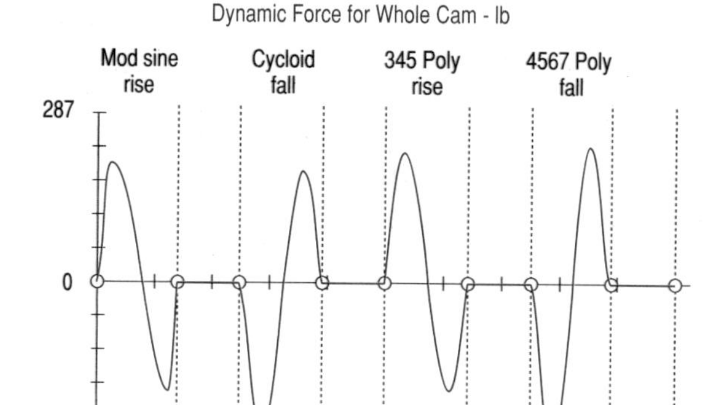

FIGURE 9-16
Dynamic force in a form-closed cam-follower system

Compare the dynamic force plots for the force-closed system (Figure 9-14b, p. 237) and the form-closed system (Figure 9-16). The absolute peak force magnitude on either side of the track in the form-closed cam is less than that on the spring-loaded one. This shows the penalty that the spring imposes on the system in order to keep the joint closed. Thus, either side of the cam groove will experience lower stresses than will the open cam, except for the areas of crossover shock mentioned on p. 238.

9.6 KINETOSTATIC CAMSHAFT TORQUE

A kinetostatic analysis assumes that the camshaft will operate at some constant speed ω. The input torque must vary over the cycle if the shaft velocity is to remain constant. The torque can be easily calculated from the power relationship, ignoring losses.

$$\begin{aligned} \text{Power in} &= \text{Power out} \\ T_c\omega &= F_c v \\ T_c &= \frac{F_c v}{\omega} \end{aligned} \tag{9.12}$$

Once the cam force has been calculated from either equation 9.10 or 9.11, the camshaft torque T_c is easily found since the follower velocity v and camshaft ω are both known. Figure 9-17a shows the camshaft input torque needed to drive the force-closed cam designed in Example 9-2 (p. 234). Figure 9-17b shows the camshaft input torque needed to drive the form-closed cam designed in Example 9-3. Note that the torque required to drive the force-closed (spring-loaded) system is significantly higher than that needed to drive the form-closed (track) cam. The spring force is also extracting a penalty here as energy must be stored in the spring during the rise portions that will tend to slow the camshaft. This stored energy is then returned to the camshaft during the fall

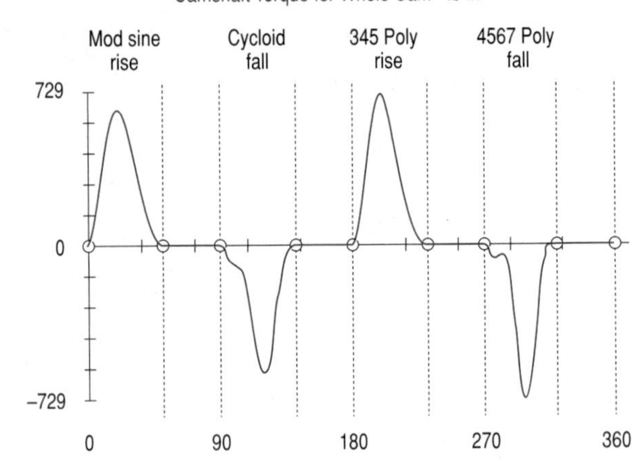

(*a*) Force-closed (spring-loaded) cam-follower

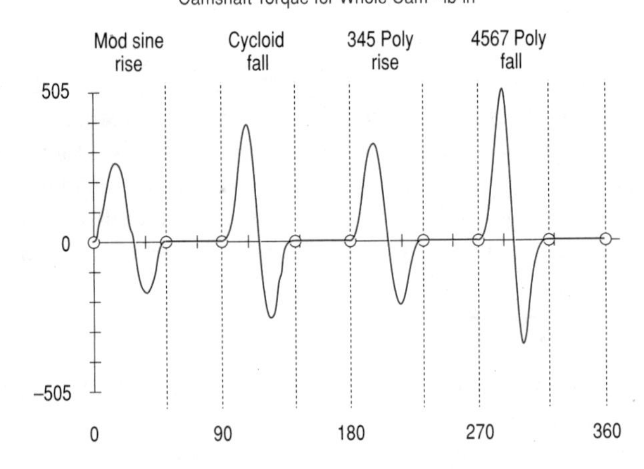

(*b*) Form-closed (desmodromic) cam-follower

FIGURE 9-17

Input torque in force- and form-closed cam-follower systems

portions, tending to speed it up. The spring loading causes larger oscillations in the torque. Section 9.8 (p. 251) discusses the use of flywheels to reduce torque oscillation.

One useful way to compare alternate cam designs is to look at the torque function as well as at the dynamic force. A smaller torque variation will require a smaller motor and/or flywheel and will run more smoothly. Three different designs for a symmetrical single-dwell cam were explored in Chapter 4. (See Examples 4-1, p. 58; 4-2, p. 60; and 4-3, p. 61.) All had the same lift and duration but used different cam functions. One was a double harmonic, one cycloidal, and one a sixth-degree polynomial. On the basis of their kinematic results, principally acceleration magnitude, we found that the polynomial design was superior. We will now revisit this cam as an example and compare its dynamic force and torque among the same three programs.

EXAMPLE 9-4

Comparison of Kinetostatic Torques and Forces Among Three Alternate Designs of the Same Cam.

Given: A translating roller follower as shown in Figure 9-1a (p. 214) is driven by a force-closed radial plate cam that has the following program:

Design 1

Segment 1: Rise 1 inch in 90° double harmonic displacement
Segment 2: Fall 1 inch in 90° double harmonic displacement
Segment 3: Dwell for 180°

Design 2

Segment 1: Rise 1 inch in 90° cycloidal displacement
Segment 2: Fall 1 inch in 90° cycloidal displacement
Segment 3: Dwell for 180°

Design 3

Segment 1: Rise 1 inch in 90° and fall 1 inch in 90° with polynomial displacement
Segment 2: Dwell for 180°

Camshaft velocity is 15 rad/sec (143.24 rpm); Follower effective mass is 0.0738 lb-sec^2/in (blobs); Damping is 10% of critical ($\zeta = 0.10$)

Problem: Find the dynamic force and torque functions for the cam. Compare their peak magnitudes for the same prime circle radius.

Solution:

1 Calculate the kinematic data (follower displacement, velocity, acceleration, and jerk) for each of the specified cam designs. See Chapter 8 to review this procedure.

2 Calculate the radius of curvature and pressure angle for trial values of prime circle radius, and size the cam to control these values. A prime circle radius of 3 in gives acceptable pressure angles and radii of curvature. See Chapter 7 to review these calculations.

3 With the kinematics of the cam defined, we can address its dynamics. To solve equation 9.1a (p. 215) for the cam force, we will assume a value of 50 lb/in for the spring constant k and adjust the preload F_{pl} for each design to obtain a minimum dynamic force of about 10 lb. For design 1, this requires a spring preload of 28 lb; for design 2, 13 lb; and for design 3, 10 lb.

4 The value of damping c is calculated from equation 9.2i (p. 218). The kinematic parameters x, v, and a are known from the prior analysis.

5 Program DYNACAM will do these computations for you. The dynamic forces that result from each design are shown in Figure 9-18 and the torques in Figure 9-19. Note that the force is largest for design 1 at 82 lb peak and least for design 3 at 53 lb peak. The same ranking holds for the torques, which range from 96 lb-in for design 1 to 52 lb-in for design 3. These represent reductions of 35% and 46% in the dynamic loading due to a change in the kinematic design. Not surprisingly, the sixth-degree polynomial design, which had the lowest acceleration, also has the lowest forces and torques and is the clear winner. Open the files E09-04a.cam, E09-04b.cam, and E09-04c.cam in program DYNACAM to see these results.

(*a*) Double harmonic rise - double harmonic fall

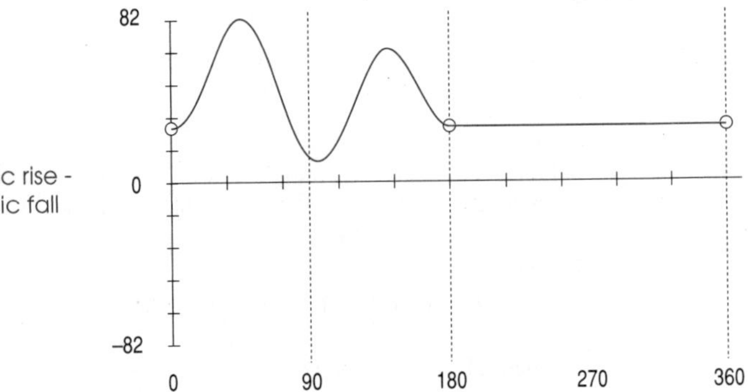

(*b*) Cycloidal rise - cycloidal fall

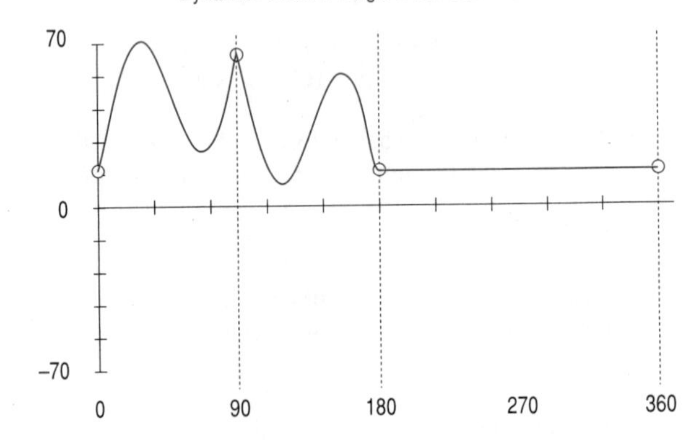

(*c*) Sixth-degree polynomial

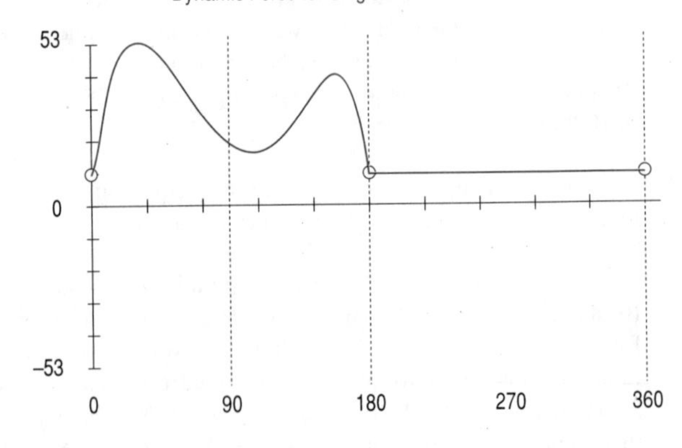

FIGURE 9-18

Dynamic forces in three different designs of a single-dwell cam

Camshaft Torque for Single-Dwell Cam - lb-in

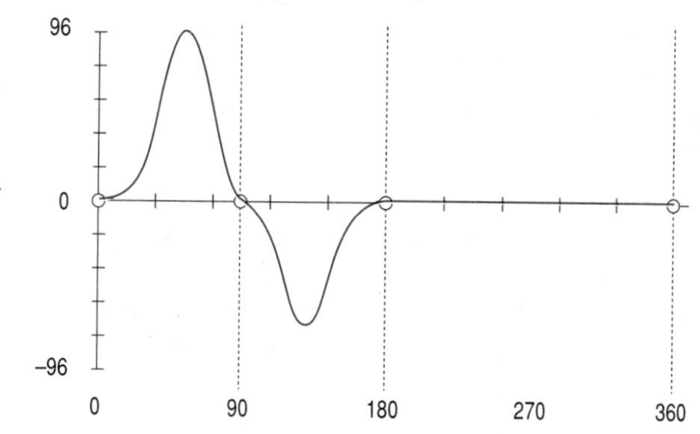

(*a*) Double harmonic rise - double harmonic fall

Camshaft Torque for Single-Dwell Cam - lb-in

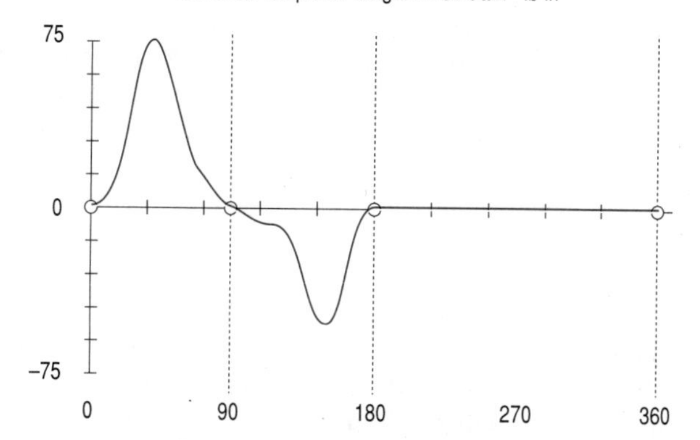

(*b*) Cycloidal rise - cycloidal fall

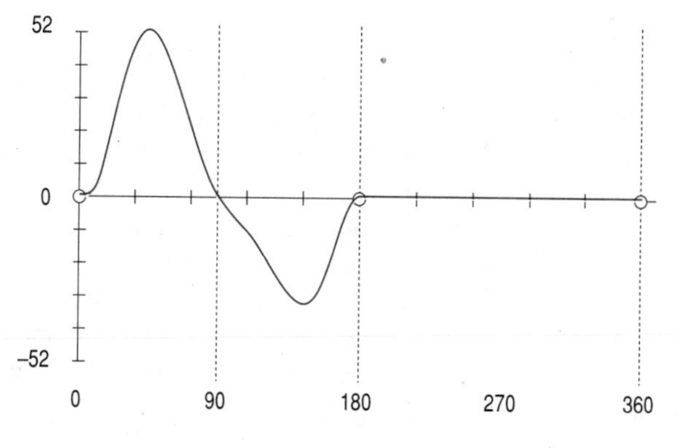

(*c*) Sixth-degree polynomial

FIGURE 9-19

Dynamic input torque in three different designs of a single-dwell cam

9.7 CONTROLLING CAM SPEED—MOTORS

Some type of driver device is needed to provide the input torque to the camshaft. A motor or engine* is the logical choice. Motors come in a wide variety of types. The most common energy source for a motor is electricity, but compressed air and pressurized hydraulic fluid are also used to power air and hydraulic motors. Gasoline or diesel engines are another possibility.

A conventionally driven industrial machine will use open-loop electric motors, often 3-phase AC, shunt-wound DC, or permanent magnet DC motors that allow some degree of speed control via circuitry that converts the readily available AC power from the line to DC and controls current to the motor for speed control. The torque-speed characteristic of these motors varies depending on the type of windings and field-armature connections.

Machine designers are increasingly using servomotors to power automated assembly equipment. Though the expense is high compared to other types of motors, the added cost is sometimes justified on the basis of the superior control and programming flexibility that results. The design of servo systems is well beyond the scope of this book and requires specialized knowledge not typically possessed by the machine designer. Servo engineers usually work as a team with machine design engineers to accomplish this task. We will limit our discussion to an overview of some of the differences, advantages, and disadvantages of using servomotors in lieu of conventional power sources. First, some definitions, terminology, and descriptions.

Electric Motors

Electric motors are classified both by their function or application and by their electrical configuration. Some functional classifications (described below) are **gearmotors**, **servomotors**, and **stepping motors**. Many different electrical configurations are also available, independent of their functional classifications. The main electrical configuration division is between **AC** and **DC** motors, though one type, the **universal motor** is designed to run on either AC or DC.

AC and **DC** refer to *alternating current* and *direct current* respectively. AC is typically supplied by the power companies and, in the U. S., alternates sinusoidally at 60 hertz (Hz), at about 120, 240, or 480 volts (V) rms. Many other countries supply AC at 50 Hz. Single-phase AC provides a single sinusoid varying with time, and 3-phase AC provides three sinusoids at 120° phase angles. DC current is constant with time, supplied from generators or battery sources and is most often used in vehicles, such as ships, automobiles, and aircraft. Lead-acid batteries are made in multiples of 2 V,† with 6, 12, and 24 V being the most common. Both AC and DC motors are designed to provide continuous rotary output. While they can be stalled momentarily against a load, they can not tolerate a full-current, zero-velocity stall for more than a few minutes without overheating.

DC MOTORS These motors are made in different electrical configurations, such as *permanent magnet (PM), shunt-wound, series-wound, and compound-wound.* The names refer to the manner in which the rotating armature coils are electrically connected

* The terms motor and engine are often used interchangeably, but they do not mean the same thing. Their difference is largely semantic, but the "purist" reserves the term *motor* for electrical, hydraulic, and pneumatic motors and the term *engine* for thermodynamic devices such as steam engines and internal combustion engines. Thus, your automobile is powered by an engine, but its windshield wipers and window lifts are run by motors.

† Other battery types have different cell voltages. Carbon-zinc batteries are 1.5 V/cell, alkaline batteries are 1.3 or 1.55V/cell, and nickel-cadmium batteries are 1.2 V/cell.

to the stationary field coils—in parallel (shunt), in series, or in combined series-parallel (compound). Permanent magnets replace the field coils in a PM motor. Each configuration provides different *torque-speed* characteristics. The *torque-speed* curve of a motor describes how it will respond to an applied load and is of great interest to the mechanical designer as it predicts how the mechanical-electrical system will behave when the load varies dynamically with time.

PERMANENT MAGNET DC MOTORS Figure 9.20a shows a torque-speed curve for a permanent magnet (PM) motor. Note that its torque varies greatly with speed, ranging from a maximum (stall) torque at zero speed to zero torque at maximum (no-load) speed. This relationship comes from the fact that *power* = *torque* x *angular velocity*. Since the power available from the motor is limited to some finite value, an increase in torque requires a decrease in angular velocity and vice versa. Its torque is maximum at stall (zero velocity), which is typical of many electric motors. This is an advantage when starting heavy loads: e.g., an electric-motor-powered vehicle needs no clutch, unlike one powered by an internal combustion engine that cannot start from stall under load. An engine's torque increases rather than decreases with increasing angular velocity.

Figure 9-20b shows a family of **load lines** superposed on the *torque-speed* curve of a PM motor. These load lines represent a time-varying load applied to the driven mechanism. The problem comes from the fact that *as the required load torque increases, the motor must reduce speed to supply it*. Thus, the input speed will vary in response to load variations in most motors, regardless of their design.* If constant speed is required, this may be unacceptable. Other types of DC motors have either more or less speed sensitivity to load than the PM motor. A motor is typically selected based on its torque-speed curve.

SHUNT-WOUND DC MOTORS These motors have a torque speed curve like that shown in Figure 9-21a. Note the flatter slope around the rated torque point (at 100%) compared to Figure 9-20. The shunt-wound motor is less speed-sensitive to load varia-

* Synchronous AC motors, servomotors, and speed-controlled DC motors are exceptions.

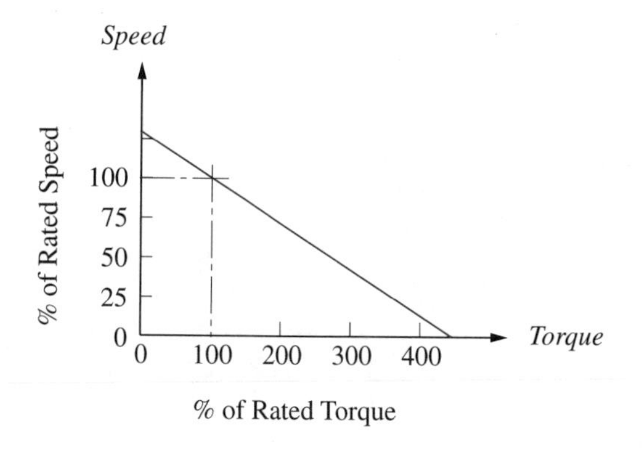

(a) Speed–torque characteristic of a PM electric motor

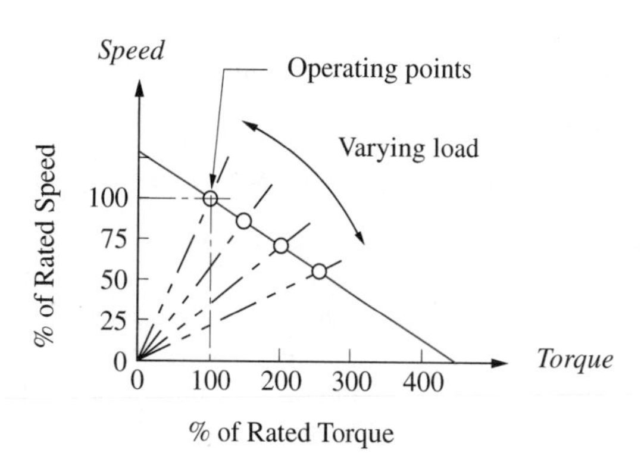

(b) Load lines superposed on speed–torque curve

FIGURE 9-20

DC permanent magnet (PM) electric motor's typical speed-torque characteristic

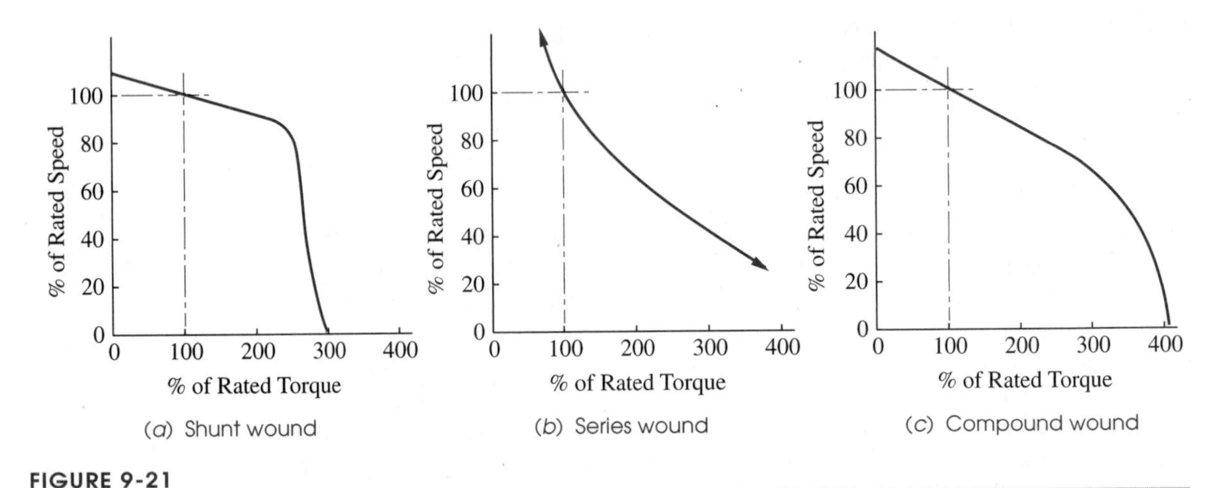

FIGURE 9-21

Torque-speed curves for three types of DC motor

tion in its operating range, but stalls very quickly when the load exceeds its maximum overload capacity of about 250% of rated torque.

Series-Wound DC Motors These have a torque-speed characteristic like that shown in Figure 9-21b. This type is more speed-sensitive than the shunt or PM configurations. However, its starting torque can be as high as 800% of full-load rated torque. It also does not have any theoretical maximum no-load speed that makes it tend to run away if the load is removed. Actually, friction and windage losses will limit its maximum speed, which can be as high as 20 000 to 30 000 revolutions per minute (rpm). Overspeed detectors are sometimes fitted to limit its unloaded speed. Series-wound motors are used in sewing machines and portable electric drills where their speed variability can be an advantage as it can be controlled, to a degree, with voltage variation. They are also used in heavy-duty applications such as vehicle traction drives where their high starting torque is an advantage. Also, their speed sensitivity (large slope) is advantageous in high-load applications as it gives a "soft-start" when moving high-inertia loads. The motor's tendency to slow down when the load is applied cushions the shock that would be felt if a large step in torque were suddenly applied to the mechanical elements. Series-wound motors are seldom if ever used for camshaft drives.

Compound-Wound DC Motors This group have their field and armature coils connected in a combination of series and parallel. As a result, their torque-speed characteristic has aspects of both the shunt-wound and series-wound motors as shown in Figure 9-21c. Their speed sensitivity is greater than a shunt-wound but less than a series-wound motor and it will not run away when unloaded. This feature, plus its high starting torque and soft-start capability, make it a good choice for cranes and hoists which experience high inertial loads and can suddenly lose the load due to cable failure, creating a potential runaway problem if the motor does not have a self-limited no-load speed.

AC Motors These are the least expensive way to get continuous rotary motion. They can be had with a variety of *torque-speed* curves to suit various load applications.

They are limited to a few standard speeds that are a function of the AC line frequency (60 Hz in North America, 50 Hz elsewhere). The synchronous motor speed n_s is 120 times the line frequency f divided by the number of magnetic poles p present in the rotor.

SYNCHRONOUS MOTORS These motors "lock on" to the AC line frequency and run exactly at synchronous speed. They are used for clocks and timers. Nonsynchronous AC motors have a small amount of slip that makes them lag the line frequency by about 3 to 10%.

Table 9-2 shows synchronous and nonsynchronous speeds for various AC motor-pole configurations. The most common AC motors have four poles, giving nonsynchronous *no-load speeds* of about 1725 rpm, which reflects slippage from the 60-Hz synchronous speed of 1800 rpm.

Figure 9-22 shows typical torque-speed curves for single-phase (1ϕ) and three-phase (3ϕ) AC motors of various designs. The single-phase shaded pole and permanent split capacitor designs have a starting torque lower than their full-load torque. To boost the start torque, the split-phase and capacitor-start designs employ a separate starting circuit that is cut off by a centrifugal switch as the motor approaches operating speed. The broken curves indicate that the motor has switched from its starting circuit to its running circuit. The NEMA* three-phase motor designs B, C, and D in Figure 9-22b differ mainly in their starting torque and in speed sensitivity (slope) near the full-load point.

SPEED-CONTROLLED MOTORS If speed control is needed, as is often the case in production machinery, another solution is to use a speed-controlled motor. Both DC and AC types are available. The DC type operates from an open-loop controller that increases and decreases the voltage to the motor in the face of changing load to try to maintain constant speed. These speed-controlled (typically PM) DC motors will run from an AC source since the controller also converts AC to DC. The AC types are variable frequency (VF) drives in which the controller varies the frequency and pulse width of a synthe-

* National Electrical Manufacturers Association

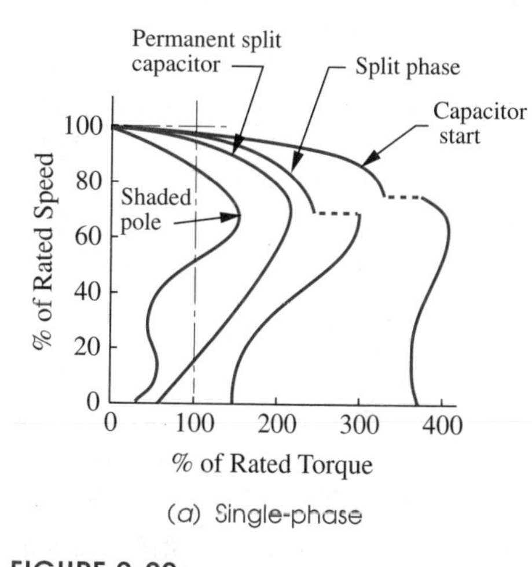

(a) Single-phase

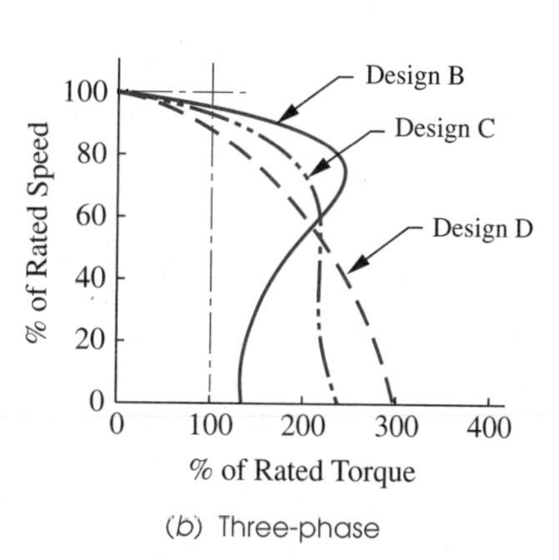

(b) Three-phase

FIGURE 9-22

Torque-speed curves for single- and three-phase AC motors

TABLE 9-2
AC Motor Speeds

Poles	Sync rpm	Async rpm
2	3600	3450
4	1800	1725
6	1200	1140
8	900	850
10	720	690
12	600	575

sized AC waveform to change motor speed. The cost of this solution is relatively high;* nevertheless, these are a common choice for camshaft drives.

GEARMOTORS If different single (as opposed to variable) output speeds than the standard ones of Table 9-1 are needed, a gearbox speed reducer (gearhead) can be attached to the motor's output shaft, or a gearmotor can be purchased that has an integral gearhead. Gearmotors are commercially available in a large variety of output speeds and power ratings, can be any variety of AC or DC motor, and are frequently used for camshaft drives.

STEPPER MOTORS These are brushless permanent magnet, variable reluctance, or hybrid-type motors designed to position an output device. Unlike servomotors, they run **open loop**, meaning they *receive no feedback as to whether the output device has responded as requested.* Thus, they can get out of phase with the desired program. They will, however, happily sit energized for an indefinite period, holding the output in one position (though they get hot—100-150°F). Their internal construction consists of a number of magnetic strips arranged around the circumference of both the rotor and stator. When energized, the rotor will move one step, to the next magnet, for each pulse received. Thus, these are **intermittent motion** devices and do not provide continuous rotary motion like other motors. The number of magnetic strips and controller type determines their resolution (typically 200 steps/rev, but a microstepper drive can increase this to 2000 or more steps/rev). They are relatively small compared to AC/DC motors and have low drive torque capacity but high holding torque. They are moderately expensive and require special controllers. Stepper motors are seldom used to drive camshafts, though they can sometimes be used instead of a cam to directly drive a follower train linkage in low-load applications.

SERVOMOTORS These are fast-response, closed-loop-controlled motors capable of providing a programmed function of acceleration or velocity, providing position control, and of holding a fixed position against a load. **Closed loop** means that *sensors (typically shaft encoders) on the output device being moved feed back information on its position and velocity.* Circuitry in the motor controller responds to the fed back information by reducing or increasing (or reversing) the current flow (and/or its frequency) to the motor. Precise positioning of the output device is then possible, as is control of the speed and shape of the motor's response to changes in load or input commands. These are relatively expensive devices* that are commonly used in applications such as moving the flight control surfaces in aircraft and guided missiles, in numerically-controlled machining centers, and in controlling robots, for example.

Servomotors are made in both AC and DC configurations, with the AC type currently becoming more popular. These achieve speed control by the controller generating a variable frequency current that the synchronous AC motor locks onto. The controller first rectifies the AC line current to DC and then "chops" it into the desired frequency, a common method being pulse-width modification. They have high torque capability and a flat torque-speed curve similar to Figure 9-21a. Also, they will typically provide as much as three times their continuous rated torque for short periods such as intermittent overloads.

Servomotors have several advantages as drives for assembly machines. With a conventional electric motor drive, one large motor typically powers, via gearboxes or toothed-belt drives, one or more line shafts that run the length of the machine. All the

* Costs of all electronic devices seem to continuously fall as technology advances and motor controllers are no exception.

cams in the machine are mounted on these shafts. The timing of the machine is then determined by the mechanical phasing of the cams on the shafts. Torsional deflections and vibrations within the shafts and their interconnecting gearboxes and timing belts can cause dynamic phase errors in the face of the severe time-varying torque that is typical of cam-follower systems (see Figure 9-17, p. 242). With a servomotor drive system, individual motors are fitted to each station and may drive only one or a few cams for that station. This is sometimes referred to as an "electronic line shaft" since the only interconnection between the various stations of the machine now comes from the electronic coupling off all "slave" servos to the one axis chosen as "master." The timing pulses from the shaft encoder on the master shaft are used to synchronize all the slaves dynamically. Digital shaft encoders that provide hundreds to millions of pulses per revolution are available and can be either relative or absolute.

Other advantages of servomotors include their ability to do programmed "soft starts," hold any speed to a close tolerance in the face of variation in the load torque, and make a rapid emergency stop using dynamic braking. It is common for machines of this type to be required to come to a stop from full speed within one product cycle, which may be a tenth of a second or less in high speed machines.

Perhaps the greatest advantage of servomotors is their inherent programmability, hence flexibility. Without making any mechanical changes to the machine, it is a simple task to adjust the phasing of any cam within the machine if it is driven by its own servomotor. It is even possible to change the dynamic motion of the follower by programming the servo to rotate the cam with a nonconstant pattern of angular velocity each revolution such that the output motion becomes the combination of the mechanical program within the cam shape and the velocity pattern imposed by the servomotor.*

All this flexibility and adjustability comes at a price as servomotors and their controllers are significantly more expensive than conventional electric drives. Nevertheless, many cam-driven industrial machines are being so equipped as these expensive solutions can sometimes be cost effective when their performance advantages are considered.

9.8 CONTROLLING CAM SPEED—FLYWHEELS

As shown in Figure 9-23, the typically large variation in accelerations within a cam-follower system can cause significant oscillations in the torque required to drive it at a constant or near constant speed. The peak torques needed may be so high as to require an overly large motor to deliver them. However, the average torque over the cycle, due mainly to losses and external work done, may often be much smaller than the peak torque.

Unless servomotors are used, we may need to provide some means to smooth out these oscillations in torque during the cycle. This will allow us to size the motor to deliver the average torque rather than the peak torque. One convenient and relatively inexpensive means to this end is the addition of a **flywheel** to the system. A flywheel can be sized, designed, and fitted to the camshaft to smooth variations in torque. Program DYNACAM integrates the camshaft torque function pulse by pulse and prints those areas to the screen. These energy data can be used to calculate the required flywheel size for any selected coefficient of fluctuation.

* Note, however, that imposing an angular acceleration on the camshaft will change the dynamic force and torque of the system. All the analysis presented here assumes a constant shaft angular velocity, or zero angular acceleration. If a servomotor is used to provide a pattern of angular acceleration, then this must be accounted for mathematically in its dynamic analysis.

Note that if a servomotor is used to drive the cam, then a flywheel should, in general, not be fitted to the camshaft as its added inertia will make it more difficult for the servo system to accelerate and decelerate the shaft to maintain near-constant velocity in the face of torque variations.

TORQUE VARIATION Figure 9-23 shows the variation in the input torque for a cam-follower system over one full revolution of the camshaft. It is running at a constant angular velocity of 50 rad/sec. The torque varies a great deal within one cycle of the mechanism, going from a positive peak of 341.7 lb-in to a negative peak of –166.4 lb-in. The **average value of this torque** over the cycle is only 70.2 lb-in, being due to the *external work done plus losses*. The large variations in torque are evidence of the kinetic energy that is stored in the follower system as it moves. We can think of the positive pulses of torque as representing energy delivered by the driver (motor) and stored temporarily in the moving follower train as kinetic energy, and the negative pulses of torque as kinetic energy attempting to return from the follower to the camshaft. Unfortunately, most motors are designed to deliver energy but not to take it back. Thus the "returned energy" has no place to go.

Figure 9-20 (p. 247) shows the speed torque characteristic of a non-speed-controlled permanent magnet (PM) DC electric motor. Other types of motors have differently shaped functions that relate motor speed to torque, but all drivers (sources) will have some such characteristic curve as shown in Figure 9-21 and 9-22 (p. 248-249). As the torque demands on the motor change, the motor's speed must also change according to its inherent characteristic unless a speed-controller compensates for the variation. This means that the torque curve being demanded in Figure 9-23 will be very difficult for a standard (non-servo) motor to deliver without drastic changes in its speed.

The computation of the torque curve in Figure 9-23 was made on the assumption that the camshaft (thus the motor) speed was a constant value. All the kinematic data used in the force and torque calculation was generated on that basis. With the torque

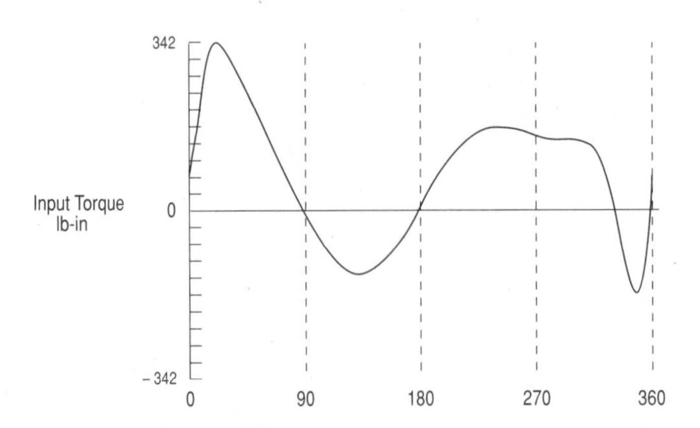

FIGURE 9-23

Input torque curve for a cam-follower mechanism

variation shown, we would have to use a large-horsepower motor (or a servomotor) to provide the power required to reach that peak torque at the design speed:

$$\begin{aligned} Power &= torque \times angular\ velocity \\ Peak\ power &= 341.7 \text{ lb-in} \times 50 \frac{\text{rad}}{\text{sec}} = 17{,}085 \frac{\text{in-lb}}{\text{sec}} = 2.59 \text{ hp} \end{aligned}$$

The power needed to supply the average torque is much smaller.

$$Average\ power = 70.2 \text{ lb-in} \times 50 \frac{\text{rad}}{\text{sec}} = 3{,}510 \frac{\text{in-lb}}{\text{sec}} = 0.53 \text{ hp}$$

It would be extremely inefficient to specify a motor based on the peak demand of the system, as most of the time it will be underutilized. We need something in the system which is capable of storing kinetic energy. One such kinetic energy storage device is called a **flywheel**.

FLYWHEEL ENERGY Figure 9-24 shows a **flywheel**, designed as a flat circular disk, attached to a motor shaft that might also be the camshaft. The motor supplies a torque magnitude T_M that we would like to be as constant as possible, i.e., to be equal to the average torque T_{avg}. The load (the follower system), on the other side of the flywheel, demands a torque T_L which is time varying, as shown in Figure 9-17 (p. 242). The kinetic energy in a rotating system is:

$$E = \frac{1}{2} I \omega^2 \tag{9.13}$$

where I is the moment of inertia of all rotating mass on the shaft. This includes the I of the motor rotor and of the cams and camshaft plus that of the flywheel. We want to determine how much I we need to add in the form of a flywheel to reduce the speed variation of the camshaft to an acceptable level. We begin by writing Newton's law for the free-body diagram in Figure 9-24.

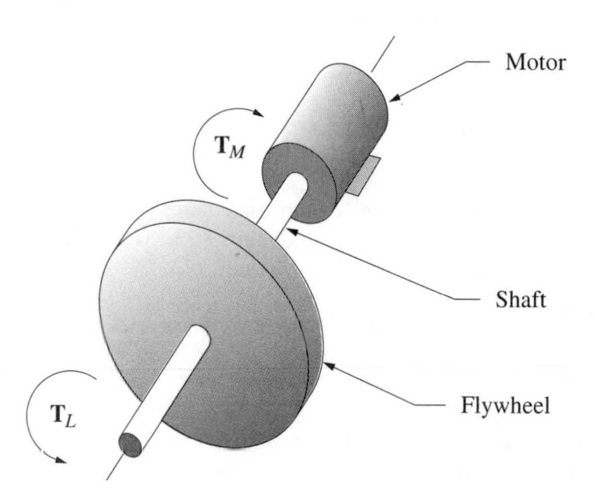

FIGURE 9-24
Flywheel on a camshaft

$$\sum T = I\alpha$$

$$T_L - T_M = I\alpha$$

but we want: $$T_M = T_{avg}$$

so: $$T_L - T_{avg} = I\alpha \qquad (9.14a)$$

substituting: $$\alpha = \frac{d\omega}{dt} = \frac{d\omega}{dt}\left(\frac{d\theta}{d\theta}\right) = \omega\frac{d\omega}{d\theta}$$

gives: $$T_L - T_{avg} = I\omega\frac{d\omega}{d\theta}$$

$$\left(T_L - T_{avg}\right)d\theta = I\omega\, d\omega \qquad (9.14b)$$

and integrating:

$$\int_{\theta @ \omega_{min}}^{\theta @ \omega_{max}} \left(T_L - T_{avg}\right)d\theta = \int_{\omega_{min}}^{\omega_{max}} I\omega\, d\omega$$

$$(9.14c)$$

$$\int_{\theta @ \omega_{min}}^{\theta @ \omega_{max}} \left(T_L - T_{avg}\right)d\theta = \frac{1}{2}I\left(\omega_{max}^2 - \omega_{min}^2\right)$$

The left side of equation 9.14c represents the change in energy E between the maximum and minimum shaft ω's and is equal to the *area under the torque-time diagram** between those extreme values of ω. The right side of equation 9.14c is the change in energy stored in the flywheel. The only way we can extract energy from the flywheel is to slow it down, as indicated by equation 9.13 (p. 253). Adding energy will speed it up. Thus it is impossible to obtain exactly constant camshaft velocity in the face of changing energy demands by the follower load. The best we can do is to minimize the speed variation ($\omega_{max} - \omega_{min}$) by providing a flywheel with sufficiently large I.

* There is often confusion between torque and energy because they appear to have the same units of *lb-in (in-lb)* or *N-m (m-N)*. This leads some students to think that they are the same quantity, but they are not: torque ≠ energy. The **integral** of torque with respect to angle, measured in radians, **is** equal to energy. This integral has the units of *in-lb-rad*. The radian term is usually omitted since it is in fact unity. Power in a rotating system is equal to torque x angular velocity (measured in *rad/sec*), and the power units are then *(in-lb-rad)/sec*. When power is integrated versus time to get energy, the resulting units are *in-lb-rad*, the same as the integral of torque versus angle. The radians are again usually dropped, contributing to the confusion.

EXAMPLE 9-5

Determining the Energy Variation in a Torque-Time Function.

Given: An input torque-time function that varies over its cycle; Figure 9-25 shows the input torque curve from Figure 9-23. The torque is varying during the 360° cycle about its average value.

Problem: Find the total energy variation over one cycle.

Solution:

1 Calculate the average value of the torque-time function over one cycle, which in this case is 70.2 lb-in.

2 Note that the *integration on the left side of equation 9.14c is done with respect to the average line of the torque function, not with respect to the θ axis.* (From the definition of

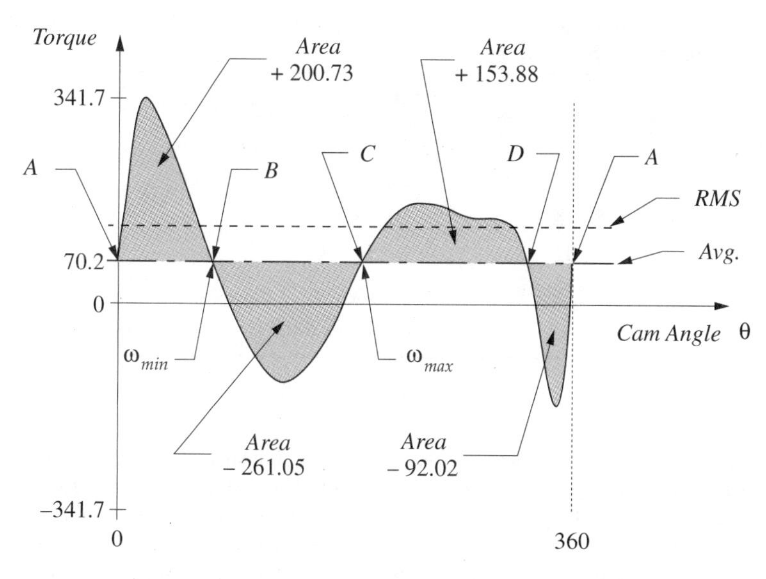

Areas of Torque Pulses in Order Over One Cycle

Order	Neg Area	Pos Area
1	– 261.05	200.73
2	– 92.02	153.88

Energy Units are lb–in–rad

FIGURE 9-25

Integrating the pulses above and below the average value in the input torque function

the average, the sum of positive area above an average line is equal to the sum of negative area below that line.) The integration limits in equation 9.14c are from the shaft angle θ at which the shaft ω is a minimum to the shaft angle θ at which ω is a maximum.

3 The minimum ω will occur after the maximum positive energy has been delivered from the motor to the load, i.e., at a point (θ) where the summation of positive energy (area) in the torque pulses is at its largest positive value.

4 The maximum ω will occur after the maximum negative energy has been returned to the load, i.e., at a point (θ) where the summation of energy (area) in the torque pulses is at its largest negative value.

5 To find these locations in θ corresponding to the maximum and minimum ω's and thus find the amount of energy needed to be stored in the flywheel, we need to numerically integrate each pulse of this function from crossover to crossover with the average line. The crossover points in Figure 9-25 have been labeled *A, B, C,* and *D*. (Program DYNACAM does this integration for you numerically, using a trapezoidal rule.)

6 The DYNACAM program prints the table of areas shown in Figure 9-25. The positive and negative pulses are separately integrated as described above. Reference to the plot of the torque function will indicate whether a positive or negative pulse is the first encountered in a particular case. The first pulse in this example is a positive one.

7 The remaining task is to accumulate these pulse areas beginning at an arbitrary crossover (in this case point *A*) and proceeding pulse by pulse across the cycle. Table 9-3 shows this process and the result. There is some numerical error in the integration. The last value should be zero.

8 Note in Table 9-3 that the minimum shaft speed occurs after the largest accumulated positive energy pulse (+200.73 in-lb) has been delivered from the driveshaft to the system.

TABLE 9-3 Integrating the Torque Function

From	Δ Area = ΔE	Accum. Sum = E	
A to B	+200.73	+200.73	ω_{min} @B
B to C	−261.05	−60.32	ω_{max} @C
C to D	+153.88	+93.56	
D to A	−92.02	+1.54	
	Total Δ Energy	$= E @ \omega_{max} - E @ \omega_{min}$	
		$= (-60.32) - (+200.73) = -261.05$ in - lb	

This delivery of energy slows the motor down. The maximum shaft speed occurs after the largest accumulated negative energy pulse (−60.32 in-lb) has been received back from the system by the driveshaft. This return of stored energy will speed up the motor. The total energy variation is the algebraic difference between these two extreme values, which in this example is −261.05 in-lb. This negative energy coming out of the system needs to be absorbed by the flywheel and then returned to the system *during each cycle* to smooth the variations in shaft speed.

9

SIZING THE FLYWHEEL We now must determine how large a flywheel is needed to absorb this energy with an acceptable change in speed. The change in shaft speed during a cycle is called its *fluctuation (Fl)* and is equal to:

$$Fl = \omega_{max} - \omega_{min} \tag{9.15a}$$

We can normalize this to a dimensionless ratio by dividing it by the average shaft speed. This ratio is called the *coefficient of fluctuation (k)*.

$$k = \frac{(\omega_{\max} - \omega_{\min})}{\omega_{avg}} \tag{9.15b}$$

This *coefficient of fluctuation* is a design parameter to be chosen by the designer. It typically is set to a value between 0.01 and 0.05, which correspond to a 1 to 5% fluctuation in shaft speed. The smaller this chosen value, the larger the flywheel will have to be. This presents a design trade-off. A larger flywheel will add more cost and weight to the system, which factors have to be weighed against the smoothness of operation desired.

We found the required change in energy E by integrating the torque curve

$$\int_{\theta @ \omega_{min}}^{\theta @ \omega_{max}} \left(T_L - T_{avg}\right) d\theta = E \tag{9.16a}$$

and can now set it equal to the right side of equation 9.14c (p. 254):

$$E = \frac{1}{2} I \left(\omega_{max}^2 - \omega_{min}^2\right) \tag{9.16b}$$

Factoring this expression:

$$E = \frac{1}{2} I \left(\omega_{max} + \omega_{min} \right) \left(\omega_{max} - \omega_{min} \right) \tag{9.16c}$$

If the torque-time function were a pure harmonic, then its average value could be expressed exactly as:

$$\omega_{avg} = \frac{\left(\omega_{max} + \omega_{min} \right)}{2} \tag{9.17}$$

Our torque functions will seldom be pure harmonics, but the error introduced by using this expression as an approximation of the average is acceptably small. We can now substitute equations 9.15b and 9.17 into equation 9.16c to get an expression for the mass moment of inertia I_s of the system flywheel needed.

$$E = \frac{1}{2} I \left(2\omega_{avg} \right) \left(k\,\omega_{avg} \right)$$

$$I_s = \frac{E}{k\,\omega_{avg}^2} \tag{9.18}$$

Equation 9.18 can be used to design the physical flywheel by choosing a desired coefficient of fluctuation k, and using the value of E from the numerical integration of the torque curve (see Table 9-3) and the average shaft ω to compute the needed system I_s. The physical flywheel's mass moment of inertia I_f is then set equal to the required system I_s. But if the moments of inertia of the other rotating elements on the same driveshaft (such as the motor) are known, the physical flywheel's required I_f can be reduced by those amounts.

The most efficient flywheel design in terms of maximizing I_f for minimum material used is one in which the mass is concentrated in its rim and its hub is supported on spokes, like a carriage wheel. This puts the majority of the mass at the largest radius possible and minimizes the weight for a given I_f. Even if a flat, solid circular disk flywheel design is chosen, either for simplicity of manufacture or to obtain a flat surface for other functions (such as an automobile clutch), the design should be done with an eye to reducing weight and thus cost. Since in general $I = mr^2$, a thin disk of large diameter will need fewer pounds of material to obtain a given I than will a thicker disk of smaller diameter. Dense materials such as cast iron and steel are the obvious choices for a flywheel. Aluminum is not used. Though many metals (lead, gold, silver, platinum) are more dense than iron and steel, their high cost prohibits use for a flywheel.

Figure 9-26 shows the change in the input torque $\mathbf{T}_{12}$ for the cam-follower system of Figure 9-25 after the addition of a flywheel sized to provide a coefficient of fluctuation of 0.05. The oscillation in torque about the unchanged average value is now 5%, much less than what it was without the flywheel. The angular velocity of the shaft will also vary 5% as the flywheel must change velocity to deliver and absorb energy as can be seen in equation 9.16b. A much lower power (non-servo) motor can now be used because the flywheel is available to absorb the energy returned from the follower during its cycle, though the added inertia of the flywheel will require sufficient motor torque to accelerate it up to speed in a reasonably short time.

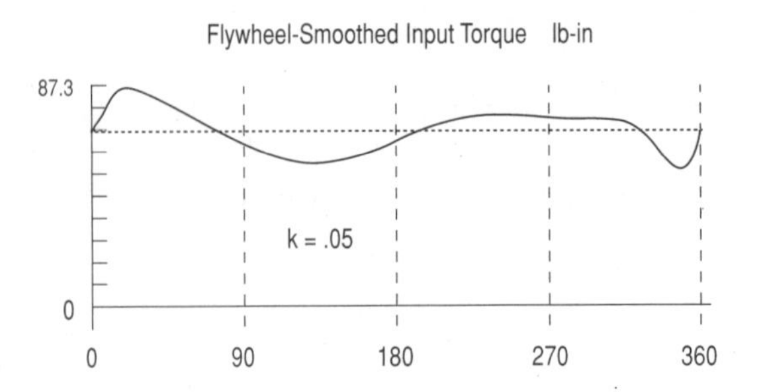

FIGURE 9-26

Input torque curve after smoothing with a flywheel

9.9 TORQUE COMPENSATION CAMS

Another way to balance torque on a camshaft is to add one or more cams driving dummy loads (essentially springs) that are designed to provide an approximately equal and opposite torque to the shaft, as is generated by the actual driving cams that do the machine's

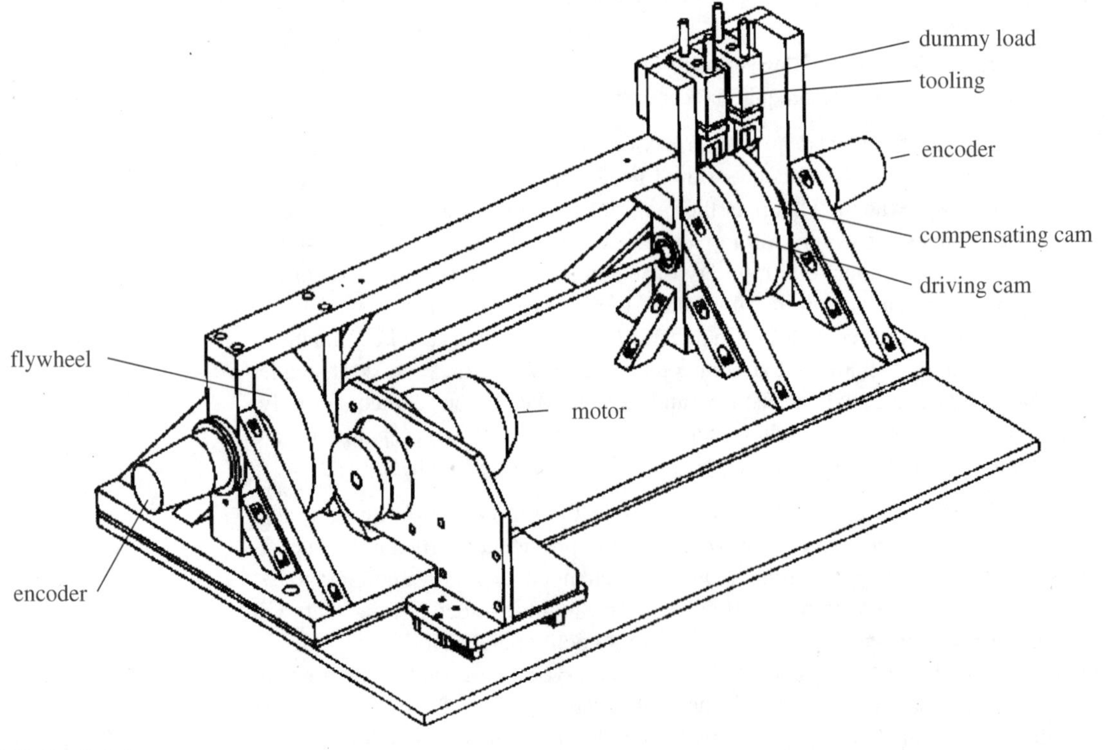

FIGURE 9-27

Compensating cam test fixture [3]

function. If the motivation is simply to hide the cyclic torque fluctuations from the driving motor (as with a flywheel), then only one torque compensation cam is needed to counteract the sum of the torques of all the driving cams on a given camshaft. If on the other hand, one wishes to reduce the torsional stresses and deflections in the camshaft, then it may be necessary to add a compensating cam adjacent to each driving cam, thus cancelling the torque locally before it has an effect on the relative angular positions between driving cams.

Figure 9-27 shows such an arrangement in the form of a test fixture designed and built to test the efficacy of this approach to torque compensation.[3] The two cams, one representing a "driving" cam and the other a "compensating" cam are seen at the right end of the camshaft each moving a translating, spring-loaded follower train. In a real situation, one of these would be driving tooling and the other driving a dummy load. At the left end of the shaft is a flywheel serving as an inertial reference, friction driven by a speed-controlled DC motor. Each free end of the shaft is fitted with a 5000 count per revolution shaft encoder. Their difference measures the instantaneous end-to-end torsional deflection of the shaft. The shaft is made in three sections. The two ends that, respectively, support the cams and flywheel are large in diameter for good torsional stiffness, and the long center section is small in diameter, thus torsionally compliant to exaggerate the cam-induced torsional deflections for ease of measurement.

9

Equation 9.12 (p. 241) gives the applied torque T_c in a camshaft. Equation 9.10b (p. 233) gives the follower force for a force-closed system. Combining them gives

$$T_c = \frac{\dot{x}}{\omega}\left[m\ddot{x} + c\dot{x} + k\left(x + x_0\right)\right] \tag{9.19}$$

The compensating cam needs to provide an equal and opposite torque T_b to cancel the applied torque

$$T_b = \frac{\dot{y}}{\omega}\left[m_b\ddot{y} + c_b\,\dot{y} + k_b\left(y + y_0\right)\right] \tag{9.20}$$

where the displacement of the compensating dummy follower is designated as y to distinguish it from the primary follower's displacement x.

Equating the two torques gives

$$T_c = -T_b$$
$$\frac{\dot{x}}{\omega}\left[m\ddot{x} + c\dot{x} + k\left(x + x_0\right)\right] = -\frac{\dot{y}}{\omega}\left[m_b\ddot{y} + c_b\,\dot{y} + k_b\left(y + y_0\right)\right] \tag{9.21}$$

The terms on the left-hand side (LHS) of equation 9.21 are all known. The factors m_b, c_b, k_b, and the initial displacement y_0 on the right-hand side (RHS) are all under the control of the designer. The goal is to solve for the displacement function y of the compensating cam. To do so requires iteration. Aviza[3] found that for realistic values of driving cam parameters, if the acceleration of the compensating cam follower train were included in equation 9.21, the numerical method failed to converge. The equation was successfully solved by assuming a zero mass for the compensating follower, thus eliminating the acceleration term on the RHS.

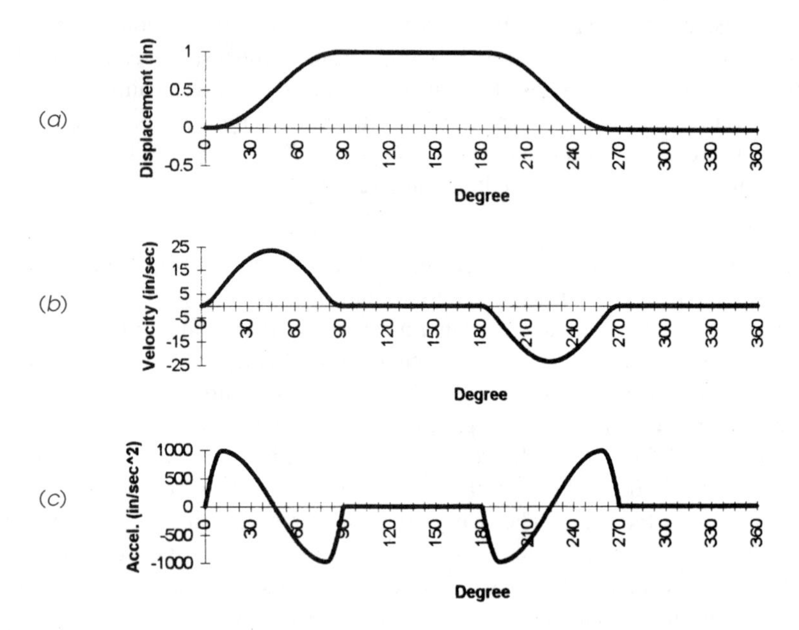

FIGURE 9-28

Displacement, velocity, and acceleration of a test cam for torque compensation[3]

Figure 9-28 shows the s, v, and a diagrams of a test cam designed and built for this investigation. It is a simple, double-dwell design with a 1-in modified sine rise over 90°, dwell for 90°, 1-in modified sine fall over 90°, and dwell for 90°. Figure 9-29 shows the camshaft torque needed to drive a follower mass of 0.004 bl against a follower spring of 30.23 lb/in with a preload of 17.4 lb and 5% of critical damping.

Figure 9-30 shows the resulting s, v, and a diagrams of the required compensating cam to cancel the torque assuming zero follower mass. Figure 9-31a shows the original driving torque superposed on the countertorque from the compensating cam. Figure 9-31b shows their difference—the residual torque in the shaft—which is essentially zero.

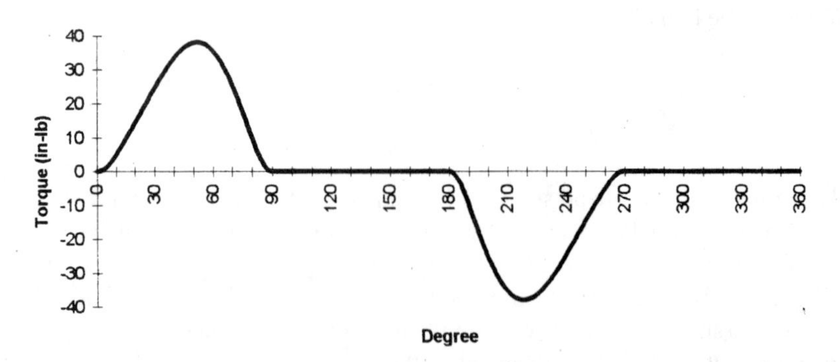

FIGURE 9-29

Driving torque required for test cam[3]

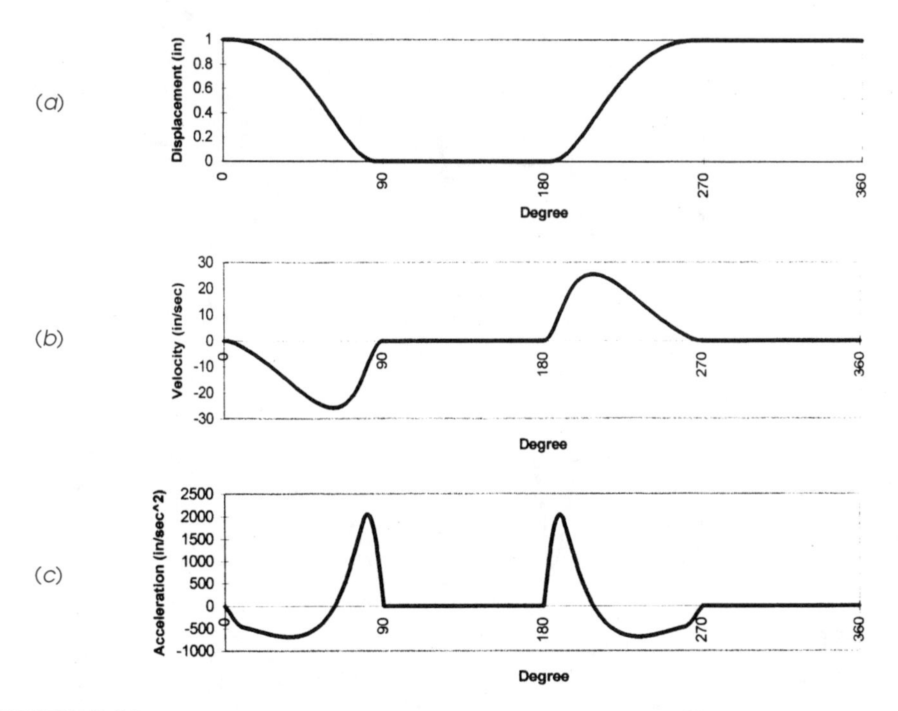

FIGURE 9-30

Displacement, velocity, and acceleration required for torque compensation cam[3]

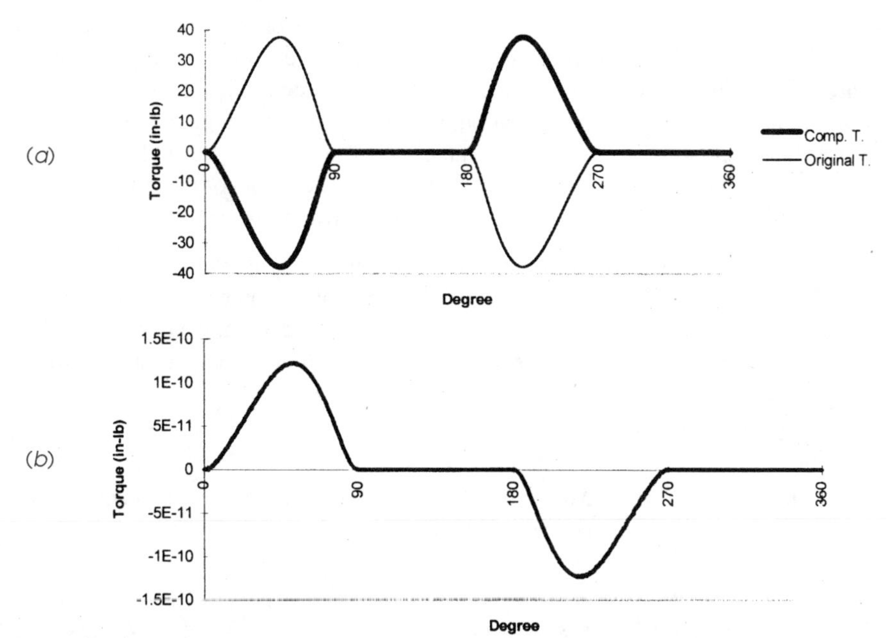

FIGURE 9-31

Original driving torque, compensating cam torque, and net torque on camshaft with zero mass dummy follower[3]

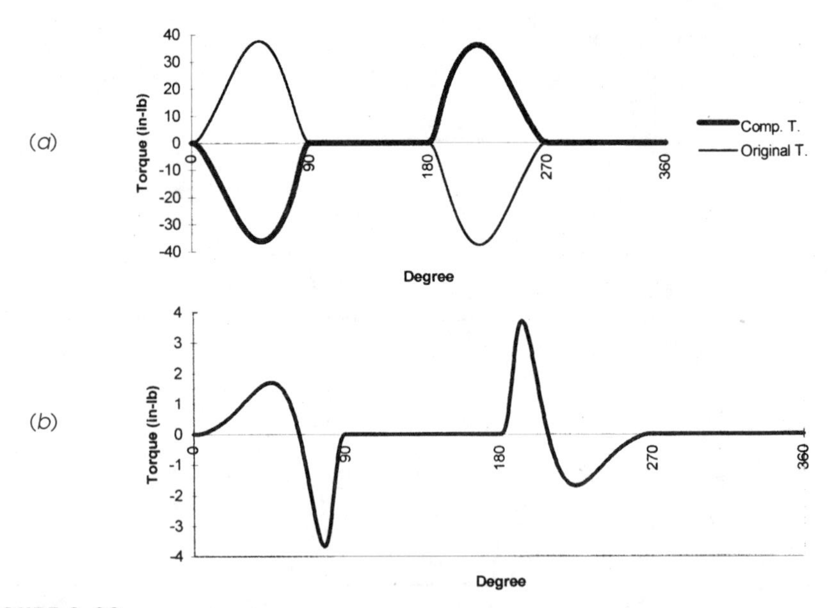

FIGURE 9-32

Original driving torque, compensating cam torque, and net torque on camshaft with nonzero mass dummy follower [3]

However, this computation results from the unrealistic assumption of a zero-mass compensating cam follower train. When a realistic follower train mass for this system of 0.003 bl is applied to the compensating cam train, the resulting torque is as shown in Figure 9-32a, superposed on the original driving cam torque. Figure 9-32b shows their difference or residual torque in the shaft, now about 10% of the original, uncompensated torque. This technique, though approximate, has resulted in a 90% reduction of peak torque in the camshaft. These results are from a dynamic model of the system and were demonstrated experimentally using the apparatus shown in Figure 9-27.[3]

Note that it is possible in some cases for the contour of the calculated compensating cam to have unacceptable pressure angles and/or radii of curvature and thus be impractical to implement. In this study, the driving cam's radii of curvature and pressure angles were of reasonable values and so resulted in a manufacturable compensating cam.

9.10 REFERENCES

1 **Barkan, P., and R. V. McGarrity**. (1965). "A Spring Actuated, Cam-Follower System: Design Theory and Experimental Results." *ASME J. Engineering for Industry (August)*, pp. 279-286.

2 **Koster, M. P.** (1974). *Vibrations of Cam Mechanisms*. Phillips Technical Library Series, Macmillan Press Ltd.: London.

3 **Aviza, G. D.** (1997). "An Experimental Investigation of Torque Balancing to Reduce the Torsional Vibration of Camshafts." Masters Thesis, Worcester Polytechnic Institute.

Chapter 10

MODELING CAM-FOLLOWER SYSTEMS

10.0 INTRODUCTION

In Chapter 9, a simple kinetostatic model of a cam-follower system was presented. That model is sufficient for determining (and thus avoiding by design) the condition of gross follower jump due to inadequate spring force and/or preload. Properly designed cam-follower systems for industrial machine applications usually do not have follower jump problems, in part because they are operated at controlled speeds. Internal combustion engine valve trains, on the other hand, can experience follower jump (or toss) if the "nut behind the wheel" revs the engine beyond its redline rpm. The "redline" on the tachometer is there to indicate the maximum engine speed allowed before the cam-followers will leave contact with the cams. Modern engines with electronically controlled ignition and fuel injection are usually equipped with a rev limiter that cuts the ignition or fuel supply if the driver tries to exceed the redline rpm.

All cam-follower systems have sufficient elasticity in their components to present the possibility of residual vibrations in operation. These are oscillations of relatively small magnitude within the various links and levers that comprise the follower train. Though these oscillations are small compared to the potential deviation in follower motion that accompanies a gross jump phenomenon from over-revving, they may nevertheless create dynamic problems. In automotive valve trains, vibration of the coils of the return spring, called spring surge, is a common problem. The harmonic content of the cam profile can interact with the natural frequencies of the spring coils at particular engine speeds, causing the coils to vibrate so violently that they impact one another, a condition known as coil clash. In industrial machinery, residual vibrations in the follower train can introduce significant positional error, especially during dwells when the continuing vibration induced by the rise or fall event compromises follower position accuracy. Vibration can also disrupt the phasing of follower motions creating the potential for interference between closely timed and spaced followers driven from different cams.

This chapter presents several dynamic models that are suitable for the determination of dynamic behavior and residual vibrations within cam-follower systems, as well as providing the means to predict follower jump. Some of these models are linear and so their differential equations of motion can be solved analytically. Others include nonlinear or nonconstant terms and so must be solved numerically, typically with a Runge-Kutta, Adams, Bulirsch-Stoer, or other algorithm. In either case, a computer solution is needed. Many commercially available computer programs will solve these equations, e.g., *TKSolver*,* *Mathcad*,† and MATLAB/SIMULINK.‡

10.1 DEGREES OF FREEDOM

A great deal of work has been done in developing dynamic models of cam systems and comparing their simulated data to experimental data over the past fifty years. There is general agreement[1],[2],[3],[4] that a lumped parameter single degree of freedom (SDOF) model is adequate to model most aspects of dynamic behavior of these systems. Koster[2] developed one-, two-, four-, and five-DOF models of a cam-follower system, using the additional DOF to include the effects of camshaft torsion and bending, backlash, squeeze of lubricant in bearings, camshaft angular velocity variation, and the drive motor characteristic. He concludes that the SDOF model (which correlated well with experiment)

> *proves to be an adequate tool for predicting the amplitude of the residual vibration of a cam follower driven by a relatively flexible shaft, despite the fact that considerable simplifications have been introduced . . ..*[5]

10

Nevertheless, others have developed multi-DOF lumped-parameter models for this application. Seidlitz[6] describes a 21-DOF model of a pushrod overhead valve train as shown in Figure 10-1 in which he used one DOF for the camshaft, two for the hydraulic tappet, two for the pushrod, four for the rocker arm, one for the spring retainer, two for the valve, and nine for the valve spring. Two of the dampers were nonlinear. From this extensive numerical simulation that correlated well with experimental measurements he concludes that:

> *In spite of the presence of some nonlinear damping and three-dimensional motions, a valve train can be viewed as a linear, one-dimensional mechanical system with two resonant frequencies. The first mode is predominately the first axial mode of the valve spring with the highest modal motion at the middle of the valve spring. The second mode is predominately the remainder of the valve train stretching and compressing along its axial length with the highest modal motion at the valve head.*[7]

This recommends a two-DOF model that can separate the spring and follower train.

Pisano[8] used a combined lumped and distributed parameter model of a valve train and confirmed its accuracy experimentally. Regarding the model, he concludes:

> *The incorporation of a distributed parameter element (helical valve spring) is crucial to the formulation of an accurate predictive model of a high-speed cam-follower system . . . The careful modeling of the valve spring as a distributed parameter (continuum) component is the key to success of the complete dynamic model . . .*[9]

* Universal Technical Services, 1220 Rock St., Rockford, IL 61101. www.uts.com

† Mathsoft Inc., 101 Main St. Cambridge, MA 02142. www.mathsoft.com

‡ The MathWorks Inc., 24 Prime Park Way, Natick, Massachusetts, 01760. www.mathworks.com

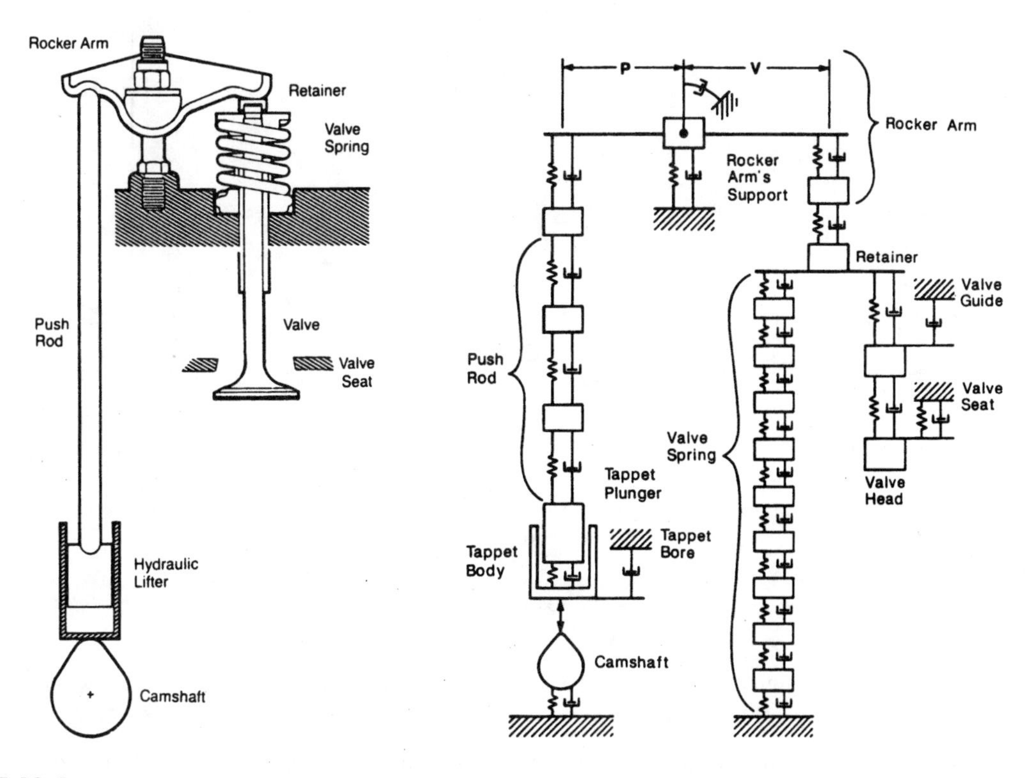

FIGURE 10-1

21-DOF model of an overhead valve train (6)

10.2 SINGLE-MASS SDOF LINEAR DYNAMIC MODELS

Figure 10-2a shows the simplest SDOF model that will account for residual vibration as well as follower jump in a force-closed cam-follower system of the type shown in Figure 10-1. Note that this system has the spring that closes the cam joint at the end effector (valve) with most of the follower train elasticity between it and the cam. All the mass in the follower train is lumped into a single effective mass m by the method shown in Chapter 8. The cam acts against a massless "cam shoe" that connects to m through a spring k_1 and damper c_1. These represent the idealized spring and viscous damping in the follower train lumped into effective values by the methods shown in Chapter 8. Spring k_2 and damper c_2 connect m to ground and represent, respectively, the physical valve spring used to close the joint between cam and follower, and the effective viscous damping of the valve versus ground. As long as the physical spring provides sufficient force to prevent gross follower jump, the system will be as shown in Figure 10-2a. If the cam and shoe separate, the system changes to that shown in Figure 10-2b. It then has two DOF.

Figure 10-2c shows the simplest SDOF model that will account for residual vibration in a form-closed cam-follower system. It is the same as the force-closed system model except for the elimination of spring k_2. There still could be external damping to ground represented by damper c_2. No gross follower jump is possible in a form-closed system,

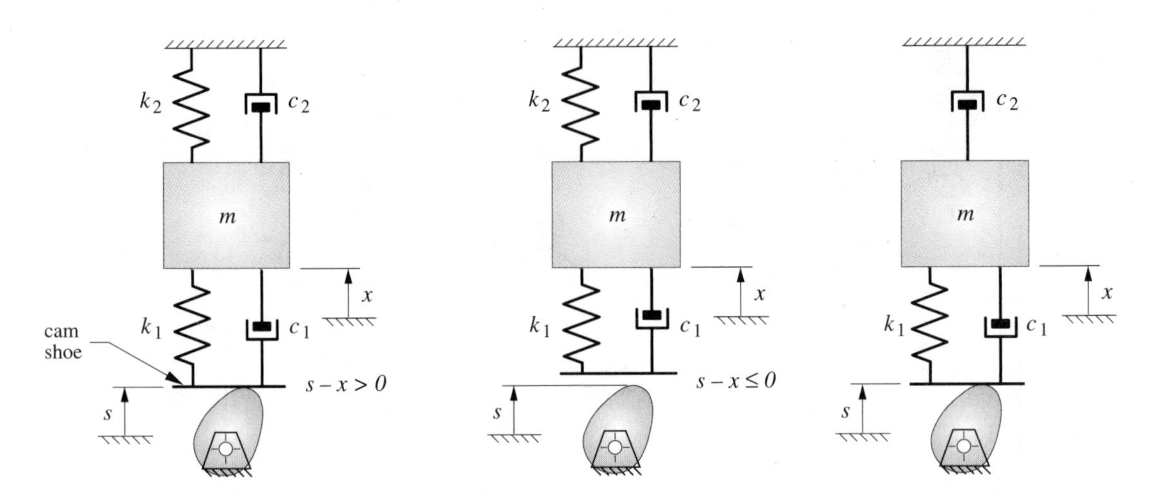

FIGURE 10-2

Simplest SDOF lumped parameter models of a cam-follower system with return spring (k_2) at end effector

though the presence of backlash allows impact between cam and follower whenever the sign of the cam-follower force changes sign. This is termed crossover shock.

In both cases the input to the system is the designed cam displacement $s(t)$ and the output is the motion $x(t)$ of the mass. Note that as long as the force between cam and shoe is > 0 in the force-closed system the cam and follower remain in contact. If not, the follower will jump off the cam.

Force-Closed Models

Figure 10-3 shows the model of Figure 10-2a and its free body diagram. The equation of motion for the system is found from Newton's second law:

$$\sum F = m\ddot{x}$$

$$k_1(s-x)+c_1(\dot{s}-\dot{x})-k_2x-c_2\dot{x}=m\ddot{x}$$

$$\ddot{x}+\frac{c_1+c_2}{m}\dot{x}+\frac{k_1+k_2}{m}x=\frac{k_1}{m}s+\frac{c_1}{m}\dot{s} \tag{10.1a}$$

where s = cam profile input displacement and x = cam follower output displacement.

With the following substitutions, equation 10.1a can be put in a form that involves natural frequencies and damping ratios.

$$\frac{c_1+c_2}{m}=2\zeta_2\omega_2 \qquad \frac{k_1+k_2}{m}=\omega_2^2 \qquad \frac{k_1}{m}=\omega_1^2 \qquad \frac{c_1}{m}=2\zeta_1\omega_1$$

$$\ddot{x}+2\zeta_2\omega_2\dot{x}+\omega_2^2x=\omega_1^2s+2\zeta_1\omega_1\dot{s} \tag{10.1b}$$

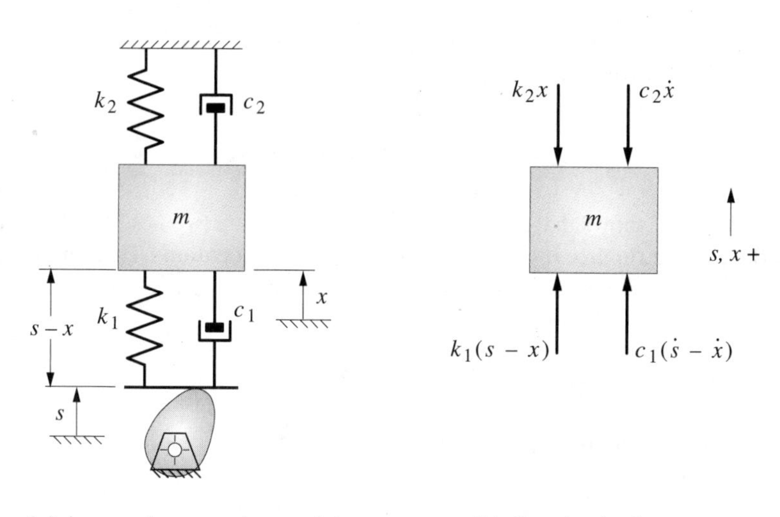

FIGURE 10-3

SDOF lumped parameter model and free-body diagram of a force-closed cam-follower system with return spring k_2 at end effector

10

Note that in this model, the springs k_1 and k_2 combine in parallel to compute ω_2, because each has an independent path to ground, k_2 directly and k_1 through the cam (as long as no follower jump occurs). Program DYNACAM solves the model in Figure 10-3a.

After jump occurs, the system m, k_2, c_2 is in free vibration according to:

$$\ddot{x} + \frac{c_2}{m}\dot{x} + \frac{k_2}{m}x = 0 \tag{10.1c}$$

the solution to which was derived in Chapter 9 (see equation 9.5).

Wiederrich[10] points out that the vibration of the system can be more clearly defined by creating a relative vibration parameter r that defines the deviation of the instantaneous position x versus its static deflection x_s from

$$r = x_s - x \tag{10.2a}$$

The static deflection can be found by setting the time derivatives in equation 10.1b to zero

$$x_s = \frac{\omega_1^2}{\omega_2^2}s \tag{10.2b}$$

Combine 10.2a and 10.2b to get an expression for relative vibration in terms of s and x

$$r = \frac{\omega_1^2}{\omega_2^2}s - x \tag{10.2c}$$

Substitute equation 10.2c and its derivatives in 10.1b to obtain the equation of motion in terms of the relative vibration r.

$$\ddot{r}+2\zeta_2\omega_2\dot{r}+\omega_2^2 r=\frac{\omega_1^2}{\omega_2^2}\ddot{s}+2(\omega_1\zeta_2-\omega_2\zeta_1)\frac{\omega_1}{\omega_2}\dot{s} \tag{10.2d}$$

Equation 10-2d shows that for systems with low damping (as these are) **the factor that has the most influence on vibration is the cam acceleration**. This further reinforces the points made in the early chapters that the higher derivatives of follower motion are the most important in respect to system dynamics.

Form-Closed Model

The equation of motion for the model shown in Figure 10-2c is:

$$\ddot{x}+\frac{c_1+c_2}{m}\dot{x}+\frac{k_1}{m}x=\frac{k_1}{m}s+\frac{c_1}{m}\dot{s} \tag{10.3a}$$

With the following substitutions, it can be put in a different form that involves natural frequencies and damping ratios.

$$\frac{k_1}{m}=\omega_n^2 \qquad \frac{c_1+c_2}{m}=2\zeta_2\omega_n \qquad \frac{c_1}{m}=2\zeta_1\omega_n$$

$$\ddot{x}+2\zeta_2\omega_n\dot{x}+\omega_n^2 x=\omega_n^2 s+2\zeta_1\omega_n\dot{s} \tag{10.3b}$$

10

The damping ratio ζ_2 in a form-closed (track) cam is typically so low (being mainly internal material damping) that it can be considered to be zero. Then:

$$\ddot{x}+\omega_n^2 x=\omega_n^2 s+2\zeta_1\omega_n\dot{s} \tag{10.3c}$$

BACKLASH In a form closed cam, the backlash will be taken up whenever the force between cam and follower reverses direction. This is called crossover shock and can create potentially severe impact forces. With negligible damping, an impact force F_i can be roughly estimated by conservation of energy as[11]

$$F_i \cong v_i\sqrt{mk} \tag{10.4a}$$

where v_i is the impact velocity, m is the striking mass, and k is the spring constant of the struck system.

Koster[2] reports on a method (attributed to Van der Hoek) to estimate the impact velocity in a track cam from:

$$v_i \cong \frac{1}{2}\left(j_{t_i}\right)^{1/3}\left(6y_i\right)^{2/3} \tag{10.4b}$$

where j_{t_i} is the value of the jerk of the follower motion at the time of impact t_i, and y_i is the amount of backlash present in the cam track at that position. The times of impact (force reversals) can be determined from the simulation using equation 10.3. They typically will be close to points of acceleration reversal. The amount of backlash can be estimated at the design stage from nominal fits in joints and expected tolerances of manufacture of cam and roller follower. For an existing system, backlash is easily measured with a dial indicator. It can be seen from equations 10.4 that for any amount of irreducible backlash, the impact force can be reduced by choosing a cam function with lower values of jerk at the points of acceleration reversal. Also, wear will increase backlash over time.

10.3 TWO-MASS, ONE- OR TWO-DOF, NONLINEAR DYNAMIC MODEL OF A VALVE TRAIN

The simple models described in the previous section are considered to be adequate for many cam-follower systems such as valve trains where the spring is at the end effector, particularly if the mass at the end effector is relatively large compared to the mass of the follower train components closer to the cam, if jump is unlikely, damping is not dominated by coulomb friction, and if the follower is never deliberately held off the cam (such as during a dwell). If these conditions are not true, then a more elaborate model is needed to give a better prediction of follower dynamics.

Barkan[1] developed a model for the automotive valve train that, with slight refinements by others,[3],[4] has proven to quite accurately predict the behavior of this more dynamically complicated cam-follower system.

Barkan showed that including a coulomb damper as well as a viscous damper significantly improved the model. The viscous damping ratio was assumed to be between 0.02 and 0.15, based on the literature. A coulomb friction coefficient of 0.2 was used. Barkan also includes the gas force acting on the valve head.

Akiba added a second lumped mass at the cam.[3] This better predicts cam contact force, and thus jump, than the single mass model.[4] The two-mass model also becomes a 2-DOF system after jump occurs, giving a more realistic and accurate simulation. The model also provides for contact between valve and valve seat at closure.

For proper engine operation, a valve train must ensure tight closure of the valve during dwell, both to eliminate blowby and to give the valves (especially the exhaust valve) good coupling for heat transfer to the valve seat. This is accomplished by either maintaining a deliberate clearance (valve lash) between the cam follower and cam during the dwell, or providing a hydraulic lash adjuster as a compliant link between the two. This also compensates for the thermal expansion of the follower train at operating temperature.

An analogous condition is often encountered in industrial machinery when a follower train is brought back to a "hard stop" on the ground plane just short of the low dwell. A hard stop is sometimes needed when the cam is interrupted and leaves the follower for part of the low dwell or it may be needed to guarantee follower train position during the low dwell to a greater mechanical accuracy than can be obtained from the cam, given the stack-up of assembly tolerances and dynamic deflections.

In both of these situations, there is an impact event when the follower train hits the hard stop (or valve seat) with some velocity, and again at the beginning of the rise as the cam encounters the stationary follower with some velocity. The cam design in these cases often contains constant velocity "ramps" at the beginning and end of the rise to control the magnitude of the impact velocities over a range of tolerance of the impact point.

The 1- or 2-DOF, two-mass model[4] is shown in Figure 10-4a. Its equation of motion for the condition of continuous contact between mass m_1 and the cam is:

$$m_2\ddot{x}+c\dot{x}+(k_1+k_2)x-k_1\frac{\dot{y}}{|\dot{y}|}\mu|x|=m_2\ddot{z}+k_2(z+z_i) \tag{10.5a}$$

where $x = z - y$ and z_i is the initial deflection of return spring k_2 at assembly (preload deflection).

When mass m_1 and the cam separate, equation 10.5a still applies, but z is unknown. A second equation is then needed, involving the force between cam and follower:

$$F_c = m_1\ddot{z}+c\dot{x}+k_1x \tag{10.5b}$$

F_c will be zero when mass m_1 is not in contact with the cam.

The simplified two-mass model of Figure 10-4b eliminates the nonlinear coulomb damper and the valve seat impact. Its equations for the condition of contact between cam and follower mass m_1 ($z = s$) are:

$$\ddot{x}=\ddot{z}+\frac{k_2}{m_2}(z-z_i)+\frac{c_2}{m_2}\dot{z}-\frac{c_1+c_2}{m_2}\dot{x}-\frac{k_1+k_2}{m_2}x \tag{10.5c}$$

$$F_c = m_1\ddot{z}+c_1\dot{x}+k_1x \tag{10.5d}$$

10

To solve either of the systems in Figure 10-4, equation 10.5a or 10.5c is first solved for displacement x, then equation 10.5b or 10.5d is solved for F_c. If F_c is positive, then the value of x is valid (Case 1). If F_c is negative, then separation has occurred and equations 10.5a and b (or 10.5c and d) are solved simultaneously with F_c set to zero at that time step (Case 2), maintaining the initial velocity and displacement conditions. As the

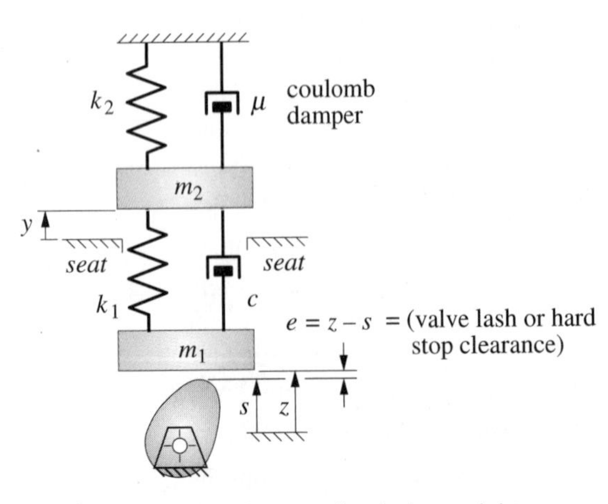

(a) Dresner & Barkan's two-mass valve-train model

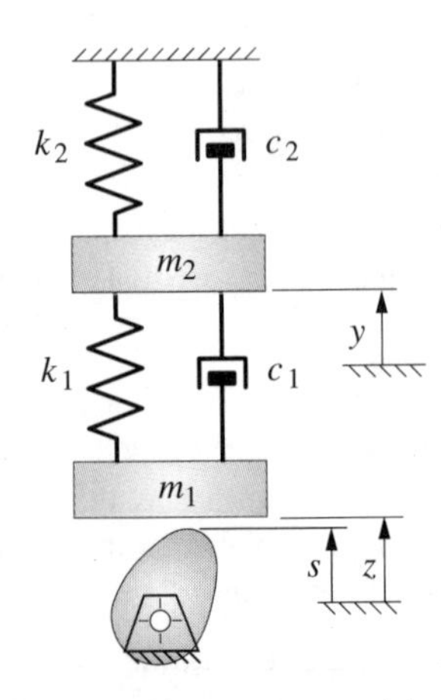

(b) Simplified two-mass model

FIGURE 10-4

One- or two-DOF two-mass model of a cam-follower system with return spring (k_2) at end effector (4)

solution continues for additional time steps, s and z are compared at each step. When z becomes $\leq s$, the cam and follower have regained contact and the solution reverts to Case 1, maintaining the initial condition for dx/dt but changing the initial condition for dz/dt to match the cam velocity. This assumes a coefficient of restitution of zero on impact, which does not accurately model the impact event but is valid after the follower stops bouncing on the cam. When a transition is made from one case to the other, the time at which F_c changes sign must be found by iteration in order to maximize accuracy.[4]

Figure 10-5 shows displacement and acceleration data from a valve train modeled as in Figure 10-4a. The simulated, theoretical, and experimentally measured curves are shown to be quite similar.[1] Program DYNACAM solves the model in Figure 10-4b, but does not attempt to track the follower motion after separation because that is an unacceptable condition requiring redesign of the joint closure system.

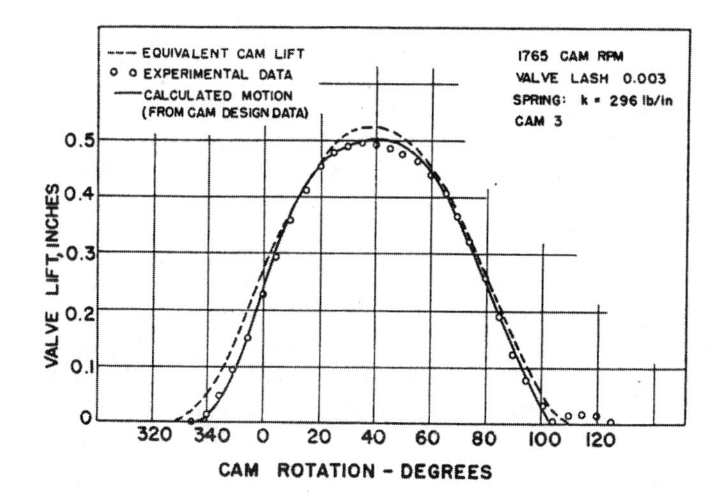

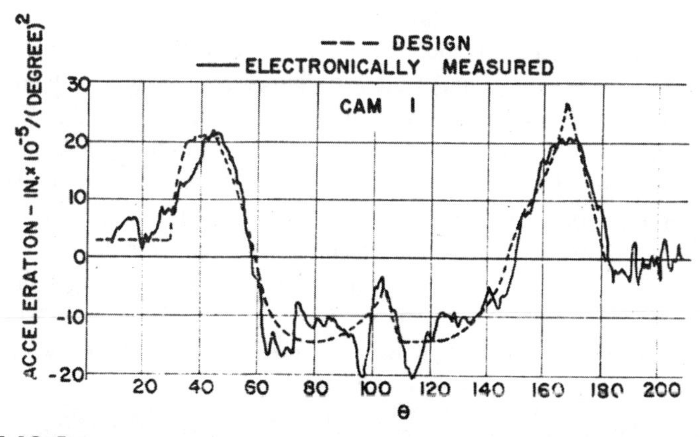

FIGURE 10-5

Simulated, theoretical, and experimental valve train displacement and acceleration [1]

10

10.4 MULTI-DOF DYNAMIC MODEL OF A VALVE TRAIN

Seidlitz[6] built the 21-DOF model of a valve train shown in Figure 10-1 (p. 265). A brief description of his modeling techniques and assumptions will be given here as a guide to the reader wishing to create a multi-DOF model of a different cam-follower system. More detail can be found in the original reference, which is quite complete.

Springs and/or dampers attach the model to ground at the camshaft, hydraulic tappet, rocker arm center, valve center, valve head, and the bottom of the valve spring. The camshaft was modeled as one mass, spring, and damper system, with its mass being that of one cam lobe plus half of the remaining mass between the cam bearings. Its spring rate was that of the camshaft in bending at the cam lobe.

The hydraulic tappet model had two masses, one spring, and two dampers. The spring rate of the tappet's oil column was "backed out" of the model (when run) by adjusting its value to get good correlation with experimental data. This method is commonly used in dynamic modeling to estimate values of elements (particularly dampers) that are difficult to either calculate or measure directly. A dead band was also built into the model calculation to simulate the lost motion associated with the tappets's internal check valve.

The pushrod was modeled as two masses, each half its total mass, three springs, and three dampers. The end springs each had the same stiffness but were twice the stiffness of the center spring. The combined spring rate of the three springs in series equaled that of the actual pushrod.

10

The rocker arm was modeled as three masses, one inertia, three springs, and four dampers. One mass was placed at each end of the arm and one at its center. The rotational inertia value applied at the center mass was the actual arm's inertia minus the effective inertia of the two end masses. Each end spring represented the effective spring rate from the center of the arm to that end. These values were determined by a static force-deflection measurement on the rocker arm. (Nowadays, if a solids model of a proposed design exists, it is a short step to make FEA models of its elements such as rocker arms and determine their spring rates before any metal is cut.)

The valve spring was modeled as nine masses, ten springs, and ten dampers. The total spring mass was divided by the number of coils, and one coil's worth assigned to each of the end masses. The remaining mass was divided equally among the remaining seven lumped masses. The two end springs were made ten times stiffer than the middle springs. This made the middle springs 8.2 times as stiff as that of the actual valve spring. Each value could be entered separately in the model allowing any combination of masses, springs, and dampers to mimic alternate spring designs. Coil-to-coil contact was simulated and when coils touched in the simulation, their spring and damper values were increased by a hundredfold. By varying clearances between coils in the model, a variable rate spring could be simulated. (A variable-rate valve spring typically is wound with a varying coil pitch. As the spring compresses, the close-spaced coils contact one another and thus increase the spring rate dynamically.)

Camshaft speed was assumed constant. The cam lift profile was entered into the model as an array of lift values at half-degree intervals and smoothed with a 5th-order polynomial for sampling at the model's calculation times. The system equation was written in matrix form:

$$[\mathbf{M}]\ddot{x}+[\mathbf{C}]\dot{x}+[\mathbf{K}]x=F(t) \tag{10.6}$$

where **M** is the mass matrix, **C** is the damping matrix, and **K** is the stiffness matrix for the system. This matrix equation was solved with a custom-written computer program. Damping coefficients that are difficult to determine independently were adjusted to obtain good correlation of the simulation with experimental measurements made of the actual valve train. Once a model such as this is "tuned" to match experimental data, it becomes a tool that can be used to investigate the dynamic effects of proposed design changes, or entirely new but similar designs, in advance of their implementation in hardware.

Figure 10-6 shows a sample of the simulated and experimental data, superposed. Good correlation between the simulation and experiment can be seen in the acceleration plots of valve head and pushrod. The pushrod force shows reasonable correlation, as does the acceleration at the middle of the spring. While a great deal of effort is required to create and verify a model of this complexity, when done, it can give a wealth of dynamic information about that system and about similar systems yet to be built.

10.5 ONE-MASS MODEL OF AN INDUSTRIAL CAM-FOLLOWER SYSTEM

The models described in the previous sections are all directed at the automotive valve train application, particularly the overhead-valve pushrod type. They have been shown to agree quite closely with experimental data from those mechanisms. However, these models may not provide a good representation of the typical industrial cam-follower train as used in automated machinery. These systems often look like Figure 10-7 where the follower spring (or air-cylinder "spring") connects the follower to ground close to the cam rather than at the end effector, as it is in the case of the overhead valve train of Figure 10-1.* Figure 10-8 shows SDOF models and free-body diagrams of this system for the form-closed and force-closed cam cases, their difference being the additional spring needed to maintain contact of the cam joint in the latter case, along with its associated damping.

THE FORM-CLOSED CASE Figure 10-8a shows an example of the classic, base-excited SDOF system found in any introductory text on vibration theory if we assume that the damping between the mass and ground is negligible. Its system equation is found from summing forces on the free-body diagram (FBD) of Figure 10-8a.

$$\ddot{x}+\frac{c}{m}\dot{x}+\frac{k}{m}x=\frac{k}{m}s+\frac{c}{m}\dot{s} \tag{10.7a}$$

where:

$$\frac{k}{m}=\omega_n^2, \qquad \frac{c}{m}=2\zeta\omega_n$$

then:

$$\ddot{x}+2\zeta\omega_n\dot{x}+\omega_n^2x=\omega_n^2s+2\zeta\omega_n\dot{s} \tag{10.7b}$$

Its response to an harmonic excitation of the base, $s(t) = h\sin(\omega t)$, is given by:

* Note that some industrial cam-follower trains have the joint closure spring placed at the end effector in order to load all the joint clearances in one direction and take out the backlash in the system. If this is done, then the valve train model gives a better representation of its dynamic behavior. Program DYNACAM provides both dynamic models for analysis.

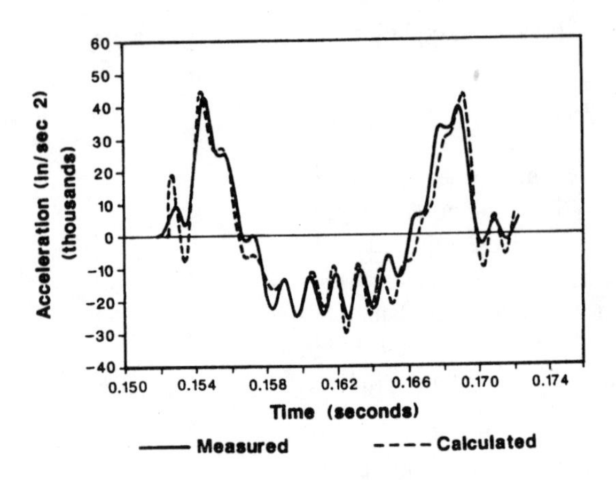

(a) Acceleration of valve head

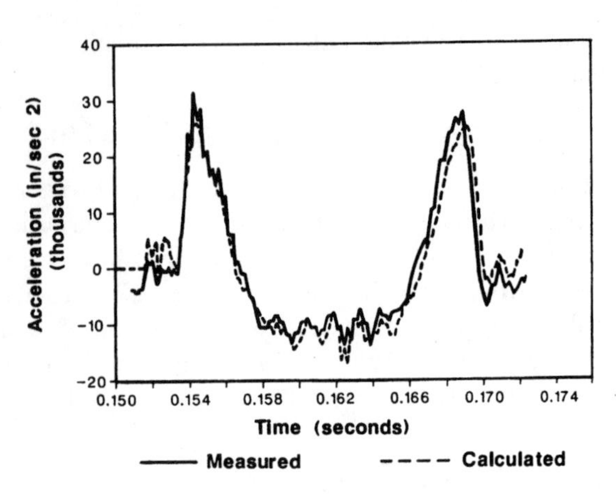

(b) Acceleration at bottom of pushrod

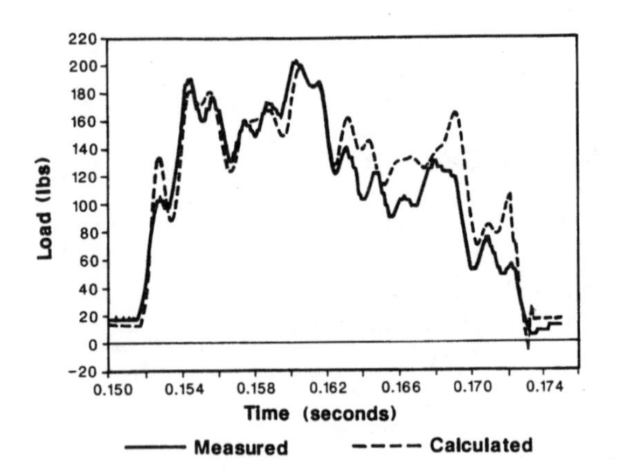

(c) Force in pushrod

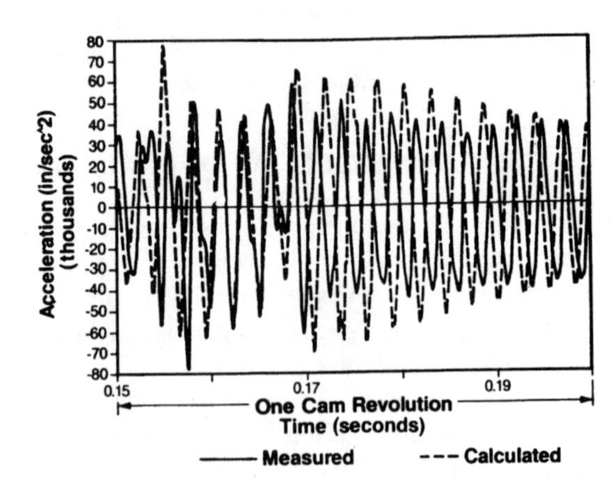

(d) Acceleration at middle of valve spring

FIGURE 10-6

Simulated and experimental dynamics of an overhead valve train at 2400 engine rpm [6]

$$x = A\,h\sin(\omega t - \phi)$$

where: $A = \text{amplitude ratio} = \dfrac{x}{h} = \sqrt{\dfrac{1+(2F_r\zeta)^2}{\left(1-r^2\right)^2+(2F_r\zeta)^2}}$ (10.7c)

$$\phi = \text{phase angle} = \arctan\left[\frac{2\zeta F_r^{\,3}}{1+\left(4\zeta^2-1\right)F_r^{\,2}}\right]$$

where: $F_r = \text{frequency ratio} = \dfrac{\omega}{\omega_n}$

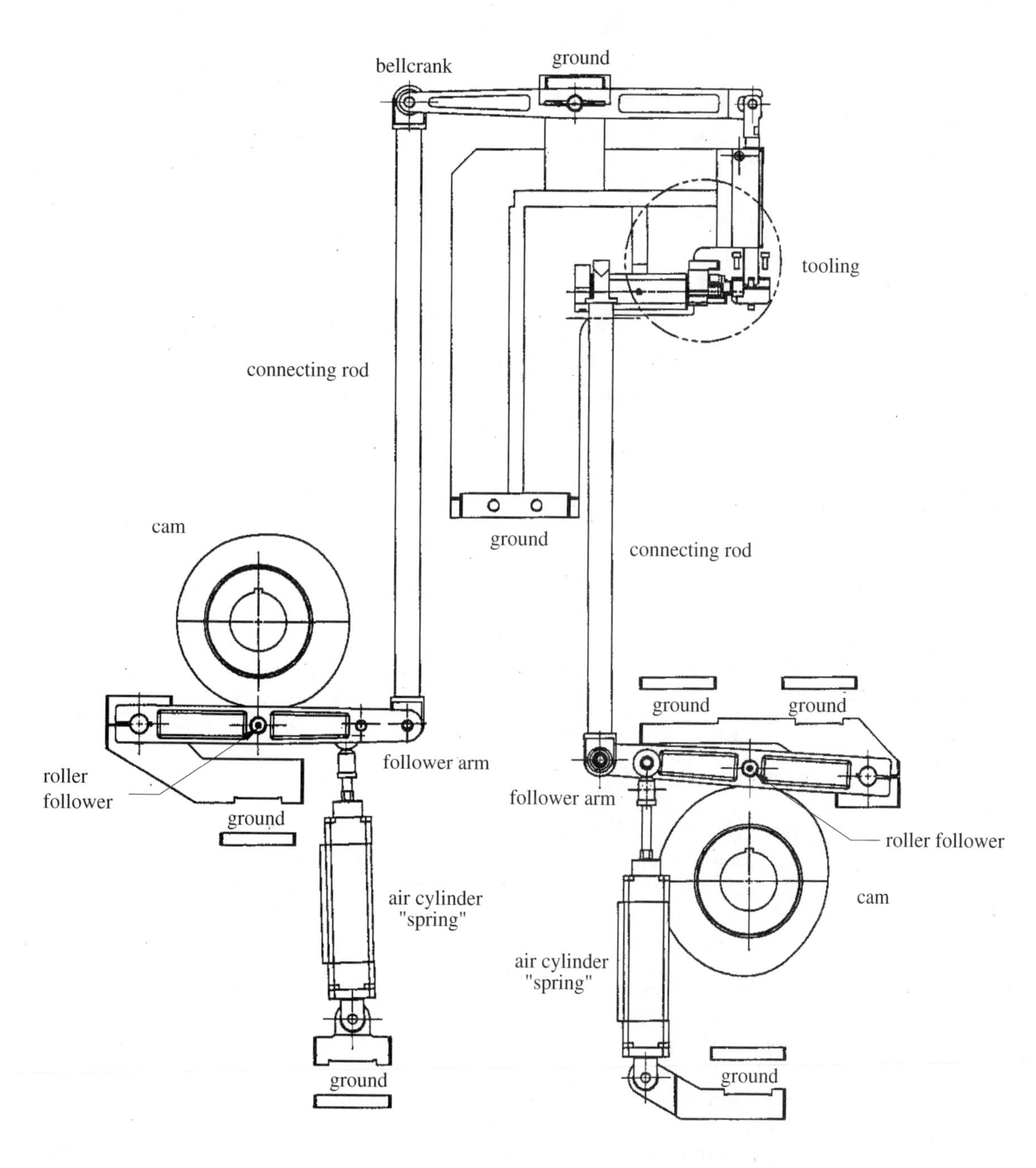

FIGURE 10-7

Typical industrial automated machinery cam-follower systems, two shown *(Courtesy of the Gillette Company, Boston, MA)*

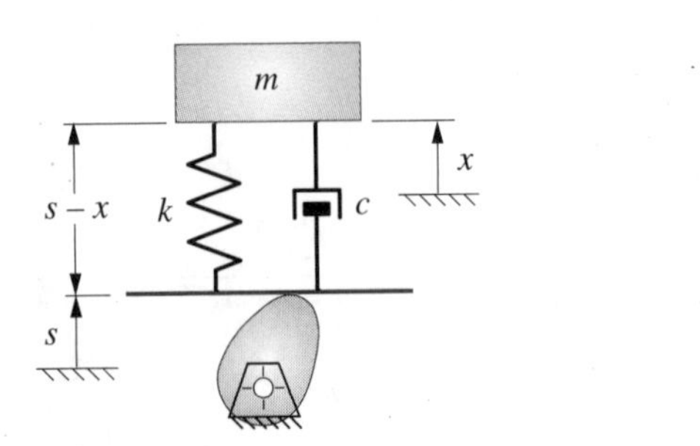

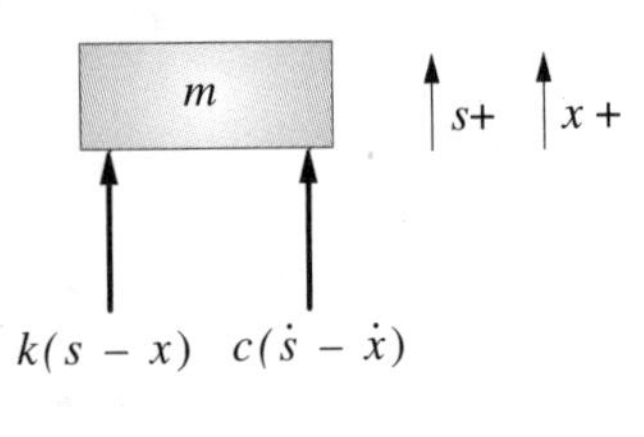

(a) Form-closed system and free-body diagram

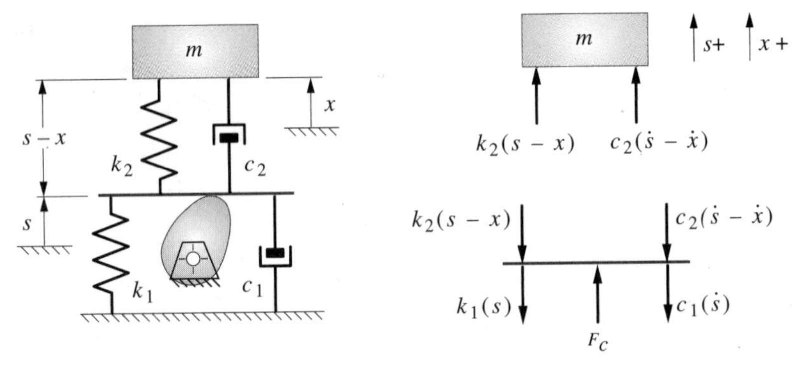

(b) Force-closed system and free-body diagrams

FIGURE 10-8

SDOF models and free-body diagrams (FBD) of a typical industrial cam-follower system

These functions are plotted in Figure 10-9 over a range of frequency ratios F_r with several different damping ratios ζ.

Note in Figure 10-9 that at the typical damping ratio for these systems of $\zeta = 0.05$, the amplitude ratio, or system gain at resonance is about 10. Note also that with essentially zero damping, the phase relationship between the forcing function s and the response x abruptly shifts from in-phase ($\phi = 0°$) to opposite phase ($\phi = 180°$) at resonance. With low damping ratios, there is a more gradual phase change as the system approaches and passes through resonance, shifting 160° at $\zeta = 0.05$ and 100° at $\zeta = 0.25$.

THE FORCE-CLOSED CASE Figure 10-8b shows a force-closed system that requires an additional spring k_1 to keep the joint from separating. We assume that the damping between the mass and ground is negligible. This system has two possible operational modes. As long as contact is maintained at the cam joint, it behaves essentially similar to the form-closed system of Figure 10-8a. In that condition, the joint-closure spring k_1 does not contribute to its vibratory behavior and the system's natural frequency is only a function of m and k_2. If the joint separates, then the system natural frequency abruptly switches to a lower value since k_1 is usually significantly smaller than k_2. Because the two springs are then in series, the lower k value dominates and k_1 is typically much less than k_2.

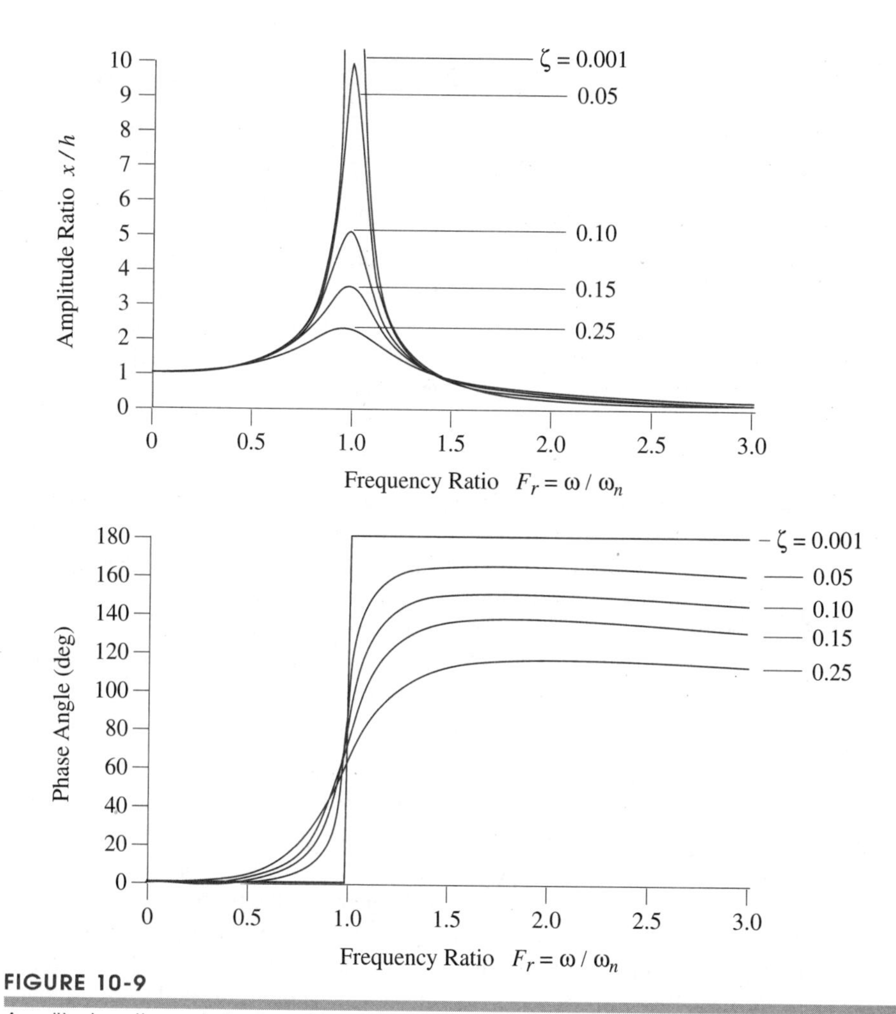

FIGURE 10-9

Amplitude ratio and phase angle of a harmonically base-excited SDOF system

If we assume that contact between cam and follower is somehow maintained at all times, then the system equation is the same as that of equation 10.7, repeated here using notation consistent with Figure 10-8b.

$$\ddot{x} + \frac{c_2}{m}\dot{x} + \frac{k_2}{m}x = \frac{k_2}{m}s + \frac{c_2}{m}\dot{s} \tag{10.8a}$$

where:

$$\frac{k_2}{m} = \omega_2^2, \qquad \frac{c_2}{m} = 2\zeta_2\omega_2$$

then:

$$\ddot{x} + 2\zeta_2\omega_2\dot{x} + \omega_2^2 x = \omega_2^2 s + 2\zeta_2\omega_2\dot{s} \tag{10.8b}$$

If the joint separates, then the system equation becomes:

$$\ddot{x} + \frac{c_{eff}}{m}\dot{x} + \frac{k_{eff}}{m}x = \frac{k_{eff}}{m}s + \frac{c_{eff}}{m}\dot{s} \tag{10.9a}$$

* This analysis assumes that the internal damping values (c's) of the elements are very small and vary approximately proportionally to the stiffnesses (k's) of the respective elements to which they apply. Because damping is typically small in these systems, its effect on the equivalent spring rate is small, but the reverse is not true since high stiffness will affect damping levels. A very stiff element will deflect less under a given load than a less stiff one. If damping is proportional to velocity across the element, then a small deflection will have small velocity. Even if the damping coefficient of that element is large, it will have little effect on the system due to the element's relatively high stiffness. A more accurate way to estimate damping must take the interaction between the k's and c's into account. For n springs $k_1, k_2, \ldots, k_n$ in series, placed in parallel with n dampers $c_1, c_2, \ldots, c_n$ in series, the effective damping can be shown to be:

$$c_{eff} = k_{eff} \sum_{i=1}^{n} \frac{c_i}{k_i^2}$$

As a practical matter, however, it is usually quite difficult to determine the values of the individual damping elements that are needed to do a calculation such as shown above and in equation (h) on p. 196. Section 9.3 (p. 227) presents a relatively simple method to expermentally determine an approximate overall damping value for an entire system.

where:*
$$k_{eff} = \frac{k_1 k_2}{k_1 + k_2} \qquad c_{eff} = \frac{c_1 c_2}{c_1 + c_2} \qquad \frac{k_{eff}}{m} = \omega_1^2 \qquad \frac{c_{eff}}{m} = 2\zeta_1\omega_1$$

then:
$$\ddot{x} + 2\zeta_1\omega_1\dot{x} + \omega_1^2 x = \omega_1^2 s + 2\zeta_1\omega_1\dot{s} \tag{10.9b}$$

Program DYNACAM solves both of the models in Figure 10-8 (p. 276).

This system, though of only one DOF, has two natural frequencies, one when in contact, and another when separated. The question then becomes: under what conditions will it separate? Looking at Figure 10-9 (p. 277), which depicts the in-contact case, it would appear that, with the damping typical in these systems ($\zeta = 0.05$ to 0.10), one could operate at frequency ratios lower than about $\omega_2 / 3$ and expect no separation, provided that the closure spring rate k_1 and preload had been properly selected by the kinetostatic technique shown in Chapter 9 in order to avoid gross kinematic jump.

But what will happen if the operating speed ω chosen on that basis happens to be close to, or above, the lower, separation natural frequency ω_1 of the system? This is entirely possible if the follower train is stiff compared to the closure spring ($k_2 >> k_1$). This situation is both common and desirable in cam-follower systems. We want the stiffest possible follower train in order to minimize the dynamic error $(s - x)$. If k_2 were infinite, then x would equal s, giving zero error. We also do not want the closure spring stiffness k_1 to be any greater than is needed to prevent kinematic jump, because its force, in part, determines the stresses in the cam and follower.

Figure 10-10 shows the two natural frequencies of a system like that in Figure 10-8b (p. 276). The in-contact natural frequency ω_2 is shown higher than the separation natural frequency ω_1 as described above. If the operating frequency ω is made about 1/3 of ω_2, then, as depicted in the figure, the system has to pass through ω_1 on start-up; but, ω_1 does not exist as long as contact is maintained. However, if the follower separates from the cam even for an instant as it passes through ω_1, the phase of the output response x will shift from 0 to 100° or greater (depending on ζ_1) with respect to the input excitation s as shown in Figure 10-9b, and it will begin to resonate at ω_1. This will cause follower bounce. Therefore, the design constraint for this system is that it should not be allowed to operate too close to the separation natural frequency ω_1. The smaller the frequency

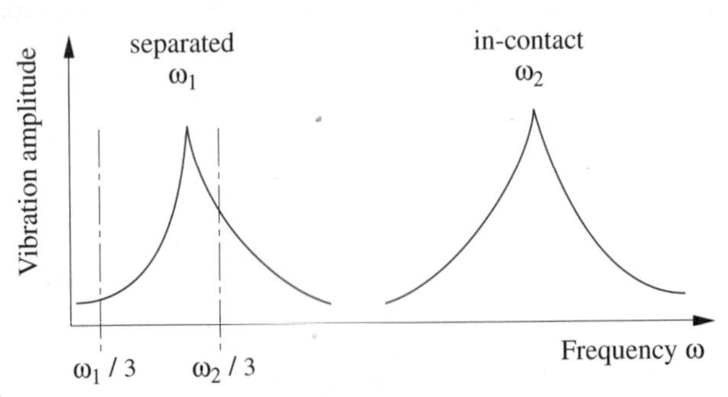

FIGURE 10-10
Separated and in-contact natural frequencies of the system of Figure 10-8b

ratio ω / ω_1 the better. Usually a ratio of about $\omega_1 / 3$ or less will be sufficient to avoid separation and follower jump. The follower spring stiffness k_1, rather than the follower train stiffness k_2, then limits the system operating speed ω.

10.6 TWO-MASS MODEL OF AN INDUSTRIAL CAM-FOLLOWER SYSTEM

A better model for the type of system shown in Figure 10-7 is the two-mass model shown in Figure 10-11. This model takes into account that the mass of the system is roughly divided between the cam-follower arm and the end effector, which are connected by relatively lighter connecting rods. The model has two masses, m_1 and m_2, one at the cam and one at the end effector. Note that m_1 is directly connected to ground by the follower spring k_1 and damper c_1, while mass m_2 is "piggybacked" on mass m_1 through the stiffness of the follower train k_2 and damper c_2.

It requires some engineering judgment to appropriately lump the distributed mass of all the links in the system into m_1 and m_2. Let common sense prevail. For example, in Figure 10-7 (p. 275), lump the mass of the follower arm, roller follower, and half the mass of the connecting rod into m_1 at the cam, properly taking into account all lever ratios as described in Chapter 8. Then lump the mass of the bellcrank, all end effector mass (tooling), and half the connecting rod into m_2. The spring in the system is easily divided. The actual return spring is k_1 and the combined effective stiffness of the follower train components is k_2. The damping is typically estimated by selecting appropriate damping ratios based on experience or on measurement as described in Chapter 9.

10

As with the previous one-mass model, if contact between cam and follower is maintained, this displacement-driven system has one DOF, expressed as x. However, if the cam and follower separate, the system then has two DOF. Moreover, the two DOFs are coupled. Note that when cam and follower separate, this system is similar to the classic vibration absorber wherein a mass-spring combination is added to a (presumed) SDOF

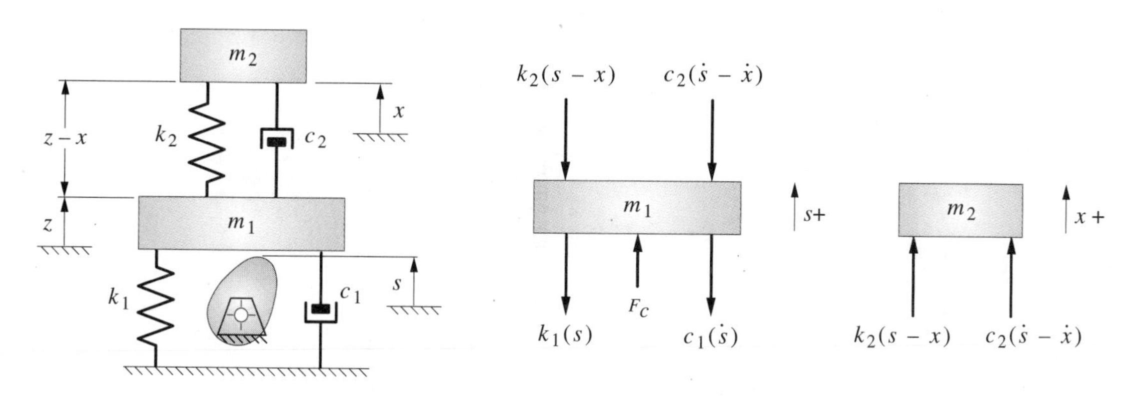

(a) Lumped parameter model (b) Free-body diagrams

FIGURE 10-11

Two-*DOF* lumped parameter model and free-body diagrams of the cam-follower system in Figure 10-7

system (see Figure 9-1 on p. 214) in order to split its original single natural frequency into two frequencies Ω_1 and Ω_2, one above and one below the original natural frequency as shown in Figure 10-12a. In that application, the parameters m_2 and k_2 are carefully selected to "tune" the system such that the vibratory response at the original SDOF natural frequency ω_0 becomes essentially zero, its energy having been "absorbed" into the two new frequencies Ω_1 and Ω_2 as shown in Figure 10-12b.

In our case, the parameters m_1, m_2, k_1, k_2 are essentially "what they are" within the range of our design freedom, and so we will not have a system "tuned" to any particular frequency. However, ω_1, the lower of the two natural frequencies of this two-DOF system, will always be smaller than the ω_0 of an SDOF model of the same system that assumed the follower train to be rigid (i.e., $k_2 = \infty$), and ω_2 will always be larger than ω_0.

$$\omega_0 = \sqrt{\frac{k_1}{m_1 + m_2}} \tag{10.10}$$

Thus, a similar situation exists with this two-mass model as with the one-mass model of the previous section. That is, when firmly in contact at the cam joint, it has one, relatively high natural frequency (due to the follower train stiffness k_2), but if perturbed to separate, it immediately shifts to a two-DOF mode in which both new frequencies are different, and one is lower, than the in-contact natural frequency,

Each stage of the model, if independent of the other, would have its own undamped natural frequency as defined by:

10

$$\omega_1 = \sqrt{\frac{k_1}{m_1}} \qquad \omega_2 = \sqrt{\frac{k_2}{m_2}} \tag{10.11a}$$

The two undamped natural frequencies of the coupled system after separation are then:

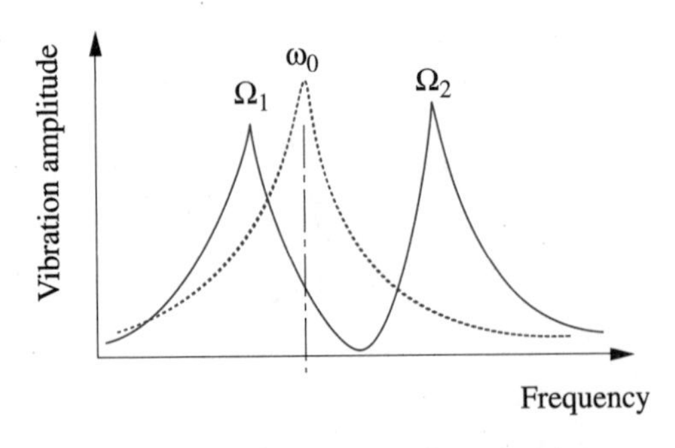

(a) An untuned vibration absorber

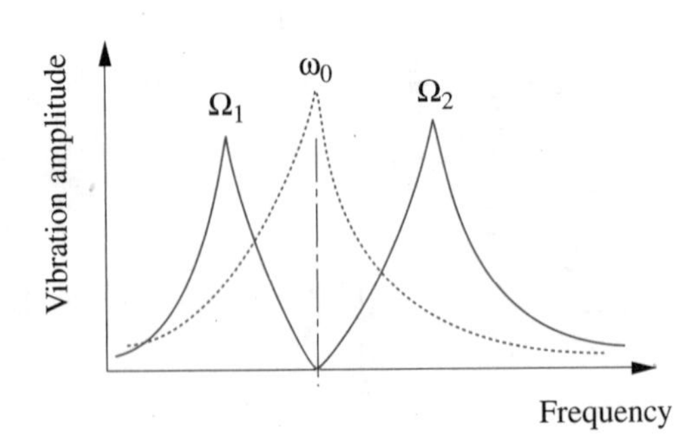

(b) A tuned vibration absorber

FIGURE 10-12

Frequency response of an SDOF system, with and without a vibration absorber, untuned (a) and tuned (b)

$$\Omega_{1,2} = \frac{\omega_1}{\sqrt{2}}\sqrt{1 + q^2(1+u) \pm \sqrt{q^4(1+u)^2 + 2(u-1)q^2 + 1}} \tag{10.11b}$$

where: $q = \dfrac{\omega_2}{\omega_1}$ $\qquad u = \dfrac{m_2}{m_1}$

Note that both Ω_1 and Ω_2 are functions of the mass ratio m_2 / m_1 and the frequency ratio ω_2 / ω_1, which includes the spring rates k_1 and k_2.*

Summing forces and applying Newton's 2nd law to the free-body diagrams of Figure 10-11 gives, for the in-contact case:

$$m_2\ddot{x} = c_2(\dot{z} - \dot{x}) + k_2(z - x)$$

$$\ddot{x} = \frac{c_2}{m_2}\dot{z} + \frac{k_2}{m_2}z - \frac{c_2}{m_2}\dot{x} - \frac{k_2}{m_2}x$$

$$\ddot{x} = 2\zeta_2\omega_2\dot{z} + \omega_2^2 z - 2\zeta_2\omega_2\dot{x} - \omega_2^2 x \tag{10.12a}$$

where, if contact is maintained, $z = s$. The contact force F_c is found from:

$$m_1\ddot{z} = F_c - c_1\dot{z} - k_1 z - c_2(\dot{z} - \dot{x}) - k_2(z - x)$$

$$F_c = m_1\ddot{z} + (c_2 + c_1)\dot{z} + (k_2 + k_1)z - c_2\dot{x} - k_2 x \tag{10.12b}$$

When separation occurs, the contact force F_c becomes zero.

To solve this system, first assume that cam and follower are in contact, ($z = s$), and solve equation 10.12a for x. Use the value for x at that time step to find F_c from equation 10.12b. If $F_c > 0$, then the cam and follower are in contact at that time and the computed value of x is valid. If $F_c < 0$, then separation has occurred, $z > s$ and the computed value of x is invalid. Then set F_c to zero and solve equations 10.12a and b simultaneously for x and z. Continue this second solution approach, testing z against s at each time step. When $z \le s$, contact has reoccurred, and equation 10.12a can again be used alone to solve for x until a negative value of F_c is again encountered. Thus the solution must switch between the two solution cases based on the sign of F_c. Note also that when the solution is switched from one stage to the other, the most recent values of x, z and their derivatives must be used as initial conditions for the next stage of the solution.

From a practical standpoint, since separation of cam and follower is generally unacceptable in these systems, the solution process could be aborted when the force goes negative and the designer then prompted to modify the system parameters to correct the problem. For example, increasing the stiffness and/or reducing the mass of the follower system may cure the problem as will increasing the stiffness or preload of the closure spring. Redesigning the cam function to reduce the negative peak acceleration will also improve the situation. Program DYNACAM solves the two-mass model of Figure 10-11 as described in this paragraph.

* As an aside, the deliberate addition of tuned vibration absorbers to a cam-follower system such as is shown in Figure 10-7 can be a very effective way to reduce vibration at any operating speeds that are close to the system's natural frequencies. In fact, one could introduce two such mass-spring absorbers to the system of Figure 10-7, as modeled in Figure 10-11, one mounted on mass m_1 and another mounted on mass m_2 to tune their respective responses to zero at the operating speed.

10.7 SOLVING SYSTEM DIFFERENTIAL EQUATIONS

The closed-form solution to the classic linear, second-order, constant-coefficient ordinary differential equation (ODE) was presented in Chapter 9. That solution is only valid for that particular type of ODE. Though equations 10.1 are in that category, other models of cam follower systems involve nonconstant and nonlinear coefficients and so cannot be solved by traditional closed-form analytical methods. For such equations, which occur frequently in the modeling of dynamic systems, a numerical method of solution is needed. Many such algorithms exist. The Runge-Kutta methods are perhaps the best known, but there also are other methods that may be superior in some instances.[21]

Block Diagram Solution—Simulink/MatLab

Many computer packages are now available that will solve even complicated differential equations numerically (and symbolically in some cases). One such is MATLAB which contains a simulation package called SIMULINK. This "front-end" to the MATLAB computational engine provides a set of graphical icons for the assembly of a block-diagram solution to any ODE, including ones with nonlinear or nonconstant coefficients. Execution of the model from SIMULINK causes the compilation of the necessary MATLAB commands to solve the system using one of several selectable numerical integration algorithms. In this case, the Adams algorithm appeared to give the best results.

A block diagram is a graphical depiction of the differential equation. To solve an ODE with a numerical algorithm, it is necessary to rearrange the equation so as to be explicit in the highest derivative. Equations 10.1a and b then become, respectively:

$$\ddot{x} = \frac{k_1}{m}s + \frac{c_1}{m}\dot{s} - \frac{c_1 + c_2}{m}\dot{x} - \frac{k_1 + k_2}{m}x \tag{10.13a}$$

$$\ddot{x} = \omega_1^2 s + 2\zeta_1\omega_1\dot{s} - 2\zeta_2\omega_2\dot{x} - \omega_2^2 x \tag{10.13b}$$

The first two terms on the right side of these equations are the input to the system, containing the designed cam displacement s and velocity *sdot*.

Figure 10-13 shows two such block diagrams, created in SIMULINK. Figure 10-13a is a block diagram for equation 10-13a; it expresses the system motion in terms of the basic parameters m, k_1, k_2, c_1, and c_2. Figure 10-13b shows a block diagram for equation 10-13b; it defines the same system in terms of the natural frequencies ω_1, ω_2, and the dimensionless damping ratios ζ_1 and ζ_2. If one wants to work in terms of the fundamental parameters of the system rather than normalized values, then Figure 10-13a would be preferred. Both models give identical results when solved numerically.

The block diagram for the normalized equation 10.13b in Figure 10-13b is somewhat "cleaner" and so is easier to follow. At point 1 (circled), a signal generator is used to provide an input function. This SIMULINK icon provides sine, sawtooth, square wave, and random noise functions. A different icon than shown allows the input of a tabulated function that can be the actual displacement and/or velocity data (including dwells) for the selected cam program, such as modified sine, cycloidal, or a custom spline program, for example.

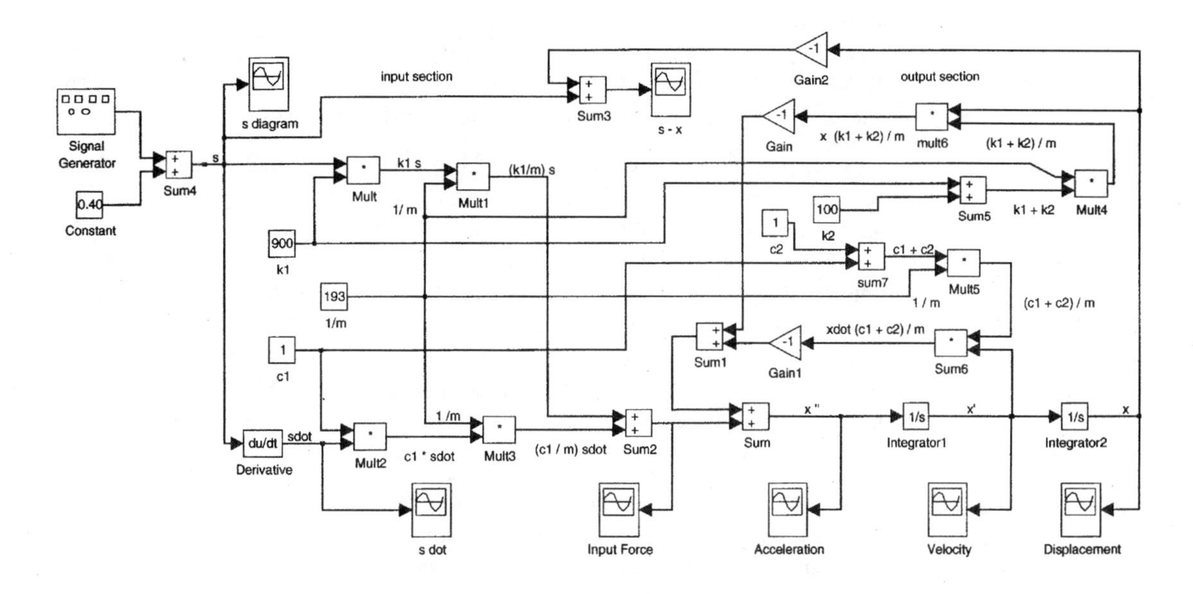

(*a*) Block diagram for the solution of equation 10.13a (Simulink/Matlab)

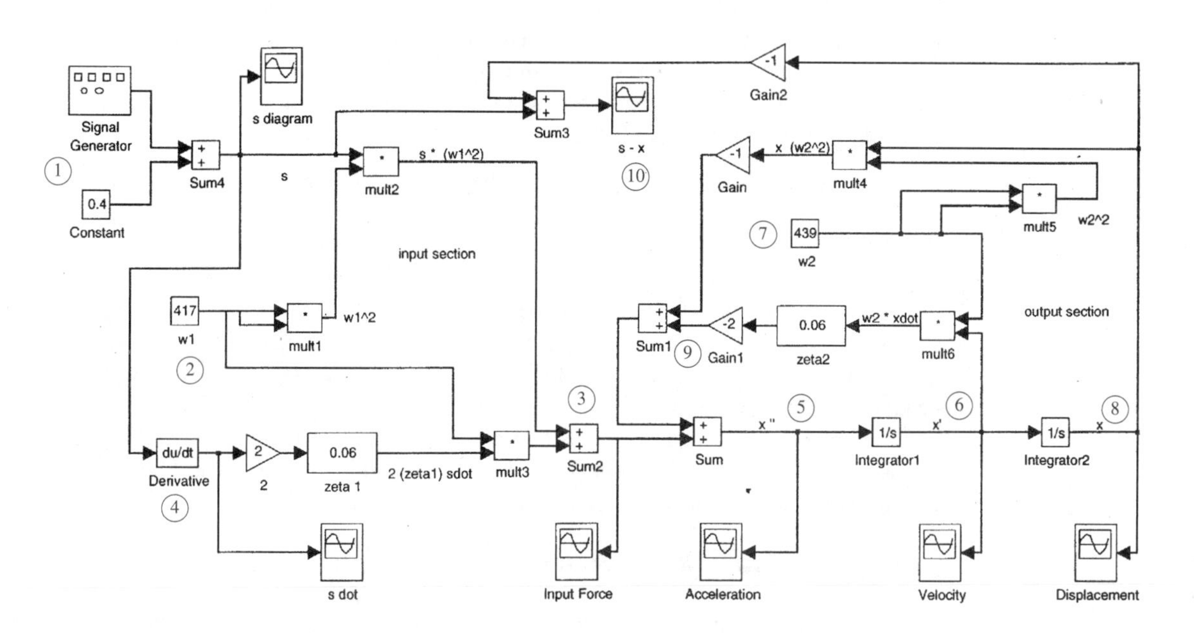

(*b*) Block diagram for the solution of equation 10.13b (Simulink/Matlab)

FIGURE 10-13

Block diagrams for the solution of differential equation models of a cam-follower system

A constant is added to the sine wave in the *Sum4* block to force it positive for all values of time. In this simulation, the amplitude of the sine function for s is 0. 25 in, and the offset is 0.4 in, making its displacement range from 0.15 to 0.65 in as shown in Figure 10-14a. The additional offset (0.4 – 0.25 = 0.15 in) acts to create the necessary spring preload on the system, defined as F_{pl} in equation 9.10 (p. 233).

This displacement s is fed to the *mult2* block where it is multiplied by ω_1^2, which was created by multiplying the user-input value of ω_1 (point 2) by itself in the *mult1* block. This product is fed to the *sum2* block at point 3. The displacement s is also fed to the derivative block that creates $\dot{s}$ (point 4), which is the second part of the input portion of equation 10.13b. *Velocity* $\dot{s}$ is multiplied by a fixed gain of 2, then by an adjustable value for ζ_1 (using a graphical slider not shown). This product is then fed to *mult3* where it is multiplied by ω_1 to create the second term in equation 10.13b (p. 282), which is then added to the first term of the equation in block *sum2* at point 3. The output of *sum2* is fed to the block *sum* that combines it with the fed-back output from the calculation to create the value of $\ddot{x}$ at point 5.

The value of $\ddot{x}$ is passed through *integrator1* to create $\dot{x}$ at point 6. This value is fed to block *mult6* to be multiplied by the value that was input for ω_2 at point 7. This result is multiplied by the selected value of ζ_2 and then by –2 to complete the third term of equation 10.13b at point 9.

The value of $\dot{x}$ is passed through *integrator2* to create x (point 8), whose value is fed to block *mult4* where it is combined with ω_2^2, and multiplied by –1 to create the fourth term in equation 10.13b (p. 282). This result is added to the third term of the equation in *sum1* at point 9. This result is fed into the block *sum* to be added to the input and close the loop to create $\ddot{x}$. An error function is created at point 10 by subtracting the calculated value of x from the input value of s.

Several "storage scopes" are placed to display the values of the variables as the simulation proceeds. The simulation was run for 2 sec using an Adams integration algorithm. The input was a 3 Hz sine wave. Figure 10-14 shows the results for the period from 1 to 2 sec, after the initial transients have died out. Figure 10-14a shows the input displacement s. Figure 10-14b shows the cam-follower force that results. Figure 10-14c shows the output displacement x of the mass m. Figure 10-14d shows the error function ($s - x$). Figure 10-14e shows the velocity of the follower mass $\dot{x}$, which appears well behaved. Figure 10-14f shows the acceleration of the follower mass $\ddot{x}$, which shows considerable vibration activity. This reinforces the point made in the previous section that acceleration is the bellwether of vibration in dynamic systems. The MATLAB files for these examples are included on the attached CD-ROM.

Ordinary Differential Equation Solution—Using Mathcad

Mathcad 2000® has a very simple-to-use ODE solver. It is only necessary to write the ODE (of any order) as part of a "solve block" as shown in Figure 10-15. The equation can be expressed in implicit form as shown. Any parameters needed for its computation must be defined in *Mathcad* above the solve block. One such example is shown in the figure as a calculation of the natural frequency η_1. The forcing functions $V_{pw}(t)$ and

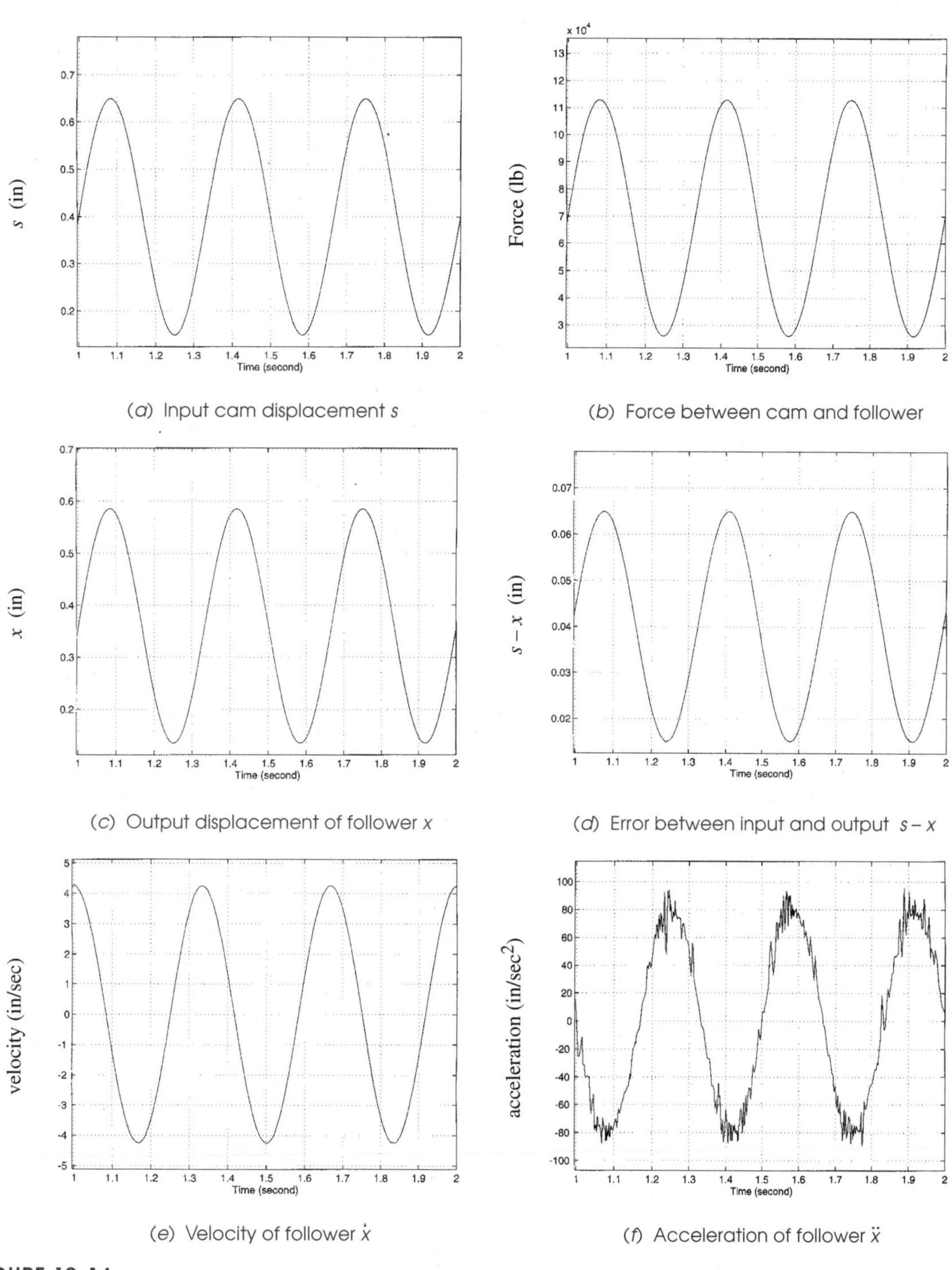

(*a*) Input cam displacement *s*

(*b*) Force between cam and follower

(*c*) Output displacement of follower *x*

(*d*) Error between input and output $s - x$

(*e*) Velocity of follower $\dot{x}$

(*f*) Acceleration of follower $\ddot{x}$

FIGURE 10-14

Steady-state solution to equation 10.13 for an arbitrary cam-follower system driven sinusoidally at 3 Hz

$$\eta_1 := \sqrt{\frac{k_{fol}}{m}}$$

The following solve block uses the Mathcad ODESOLVE function to solve for the follower response.

Given

$$\frac{d^2}{dt^2}x(t) + 2\cdot\zeta\cdot\eta_1\cdot\frac{d}{dt}x(t) + \eta_1{}^2\cdot x(t) = 2\cdot\zeta\cdot\eta_1\cdot V_{pw}(t)\cdot\frac{\omega}{1000} + \eta_1{}^2\cdot\frac{S_{pw}(t)}{1000}$$

$x(0) = 0 \qquad x'(0) = 0$

$R := \text{Odesolve}(t, t_{end}, inc_t)$

FIGURE 10-15

Mathcad solve block to numerically integrate a differential equation

$S_{pw}(t)$, the forcing frequency ω, and the damping ratio ζ were all defined in the *Mathcad* file upstream of the solve block, but are not shown in Figure 10-15.

The two initial conditions needed for this second order ODE are then defined, here x and x' at time $t = 0$ are both set to zero. Then the built-in numerical solver ODESOLVE is called and passed values for start time t, end time, t_{end}, and a starting time increment inc_t. This solver uses a 4th-order Runge-Kutta algorithm with adaptive step size control and places the result in the variable R. The solution took only a few seconds to execute on a Pentium PC.

Figure 10-16 shows the result of the *Mathcad* solution to a cam-follower system run at 180 rpm with the following specifications: Dwell at 15 mm for 150°, modsine fall to zero in 45°, dwell at zero for 120°, and modsine rise to 15 mm in 45°. The follower effective mass is 10 kg, the return spring has $k_2 = 15\,000$ N/m and the follower train has $k_1 = 1E6$ N/m.

State Space Solutions

Most numerical ODE solver algorithms require that an ODE of order higher than one be converted to a set of simultaneous first-order ODE's for solution. This is called the *state space* form of the system equations. The solver algorithm is structured to do a single numerical integration of each equation in the set and iterate until it converges to a solution at each time step. Most solvers also implement a so-called *adaptive step size* algorithm. This allows it to take larger steps in regions where the functions are changing slowly and smaller steps where there is rapid change. This gives more accurate and quicker solutions. For a thorough discussion of the common ODE solver algorithms, see Press et al.[21]

FORCE-CLOSED, ONE-MASS VALVE TRAIN MODEL To put a higher-order ODE into the state space configuration is a simple process. We will illustrate it first for the force-closed, single-mass system with its return spring at the end effector as shown in Figure 10-3a (p. 267). Its equation 10.1b is repeated here, rearranged and renumbered.

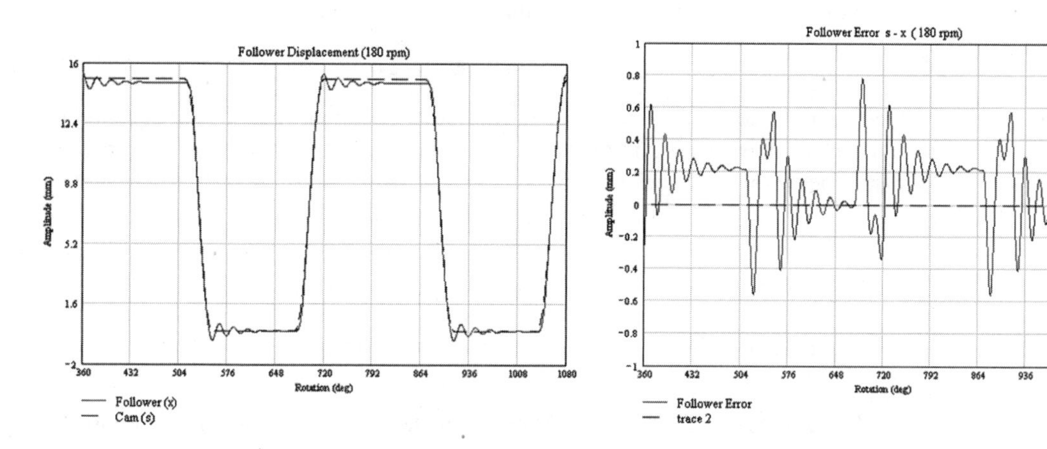

(*a*) Cam input *s* and follower displacement *x*

(*b*) Follower error $s - x$

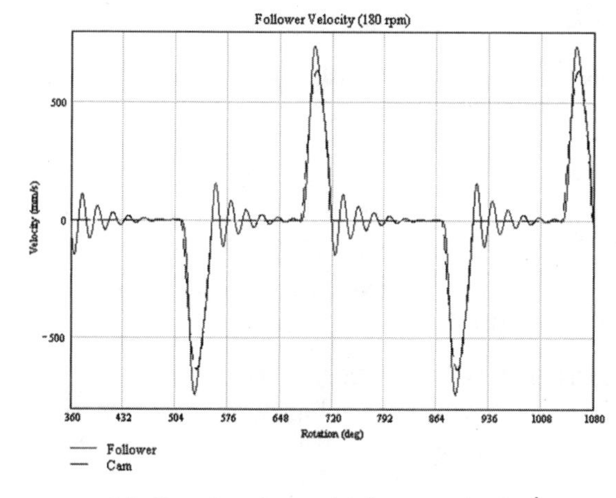

(*c*) Cam input *v* and follower velocity $\dot{x}$

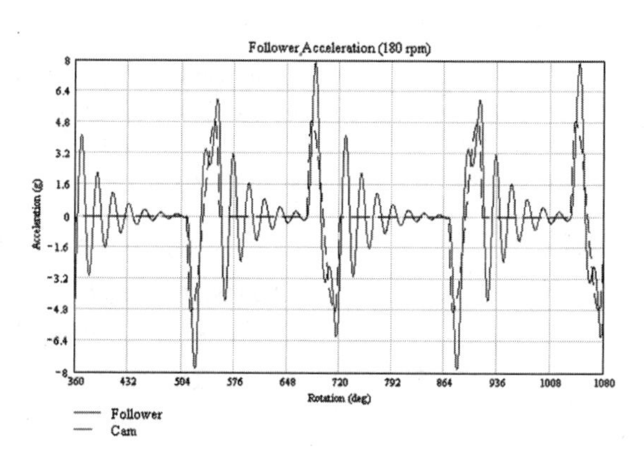

(*d*) Cam input *a* and follower acceleration $\ddot{x}$

FIGURE 10-16

Mathcad solution to a cam-follower system modeled with equation 10.1 *(Courtesy of Corey Maynard, of The Gillette Co.)*

$$\ddot{x} = \omega_1^2 s + 2\zeta_1\omega_1\dot{s} - 2\zeta_2\omega_2\dot{x} - \omega_2^2 x \tag{10.14a}$$

First, create as many dummy variables as the order of the equation, here two, say y_1 and y_2. Set one dummy variable to equal the independent variable, and one to equal its first derivative. (If the order were higher than 2, or there were more variables, then additional dummy variables would be needed).

$$y_1 = x \qquad\qquad y_2 = \dot{x} \tag{10.14b}$$

Now substitute the dummy variables into the original equation 10.14a to get:

$$\dot{y}_2 = \omega_1^2 s + 2\zeta_1\omega_1\dot{s} - 2\zeta_2\omega_2 y_2 - \omega_2^2 y_1 \tag{10.14c}$$

This has reduced equation 10.14a to a first-order equation. The second equation is obtained by combining the dummy variable equations 10.14b. Together these comprise the state-space form of the system equations and can be solved simultaneously for y_1 and y_2.

$$\begin{aligned} \dot{y}_1 &= y_2 \\ \dot{y}_2 &= \omega_1^2 s + 2\zeta_1\omega_1\dot{s} - 2\zeta_2\omega_2 y_2 - \omega_2^2 y_1 \end{aligned} \tag{10.14d}$$

The results of their simultaneous solution are then back-substituted in the original equation 10.14a according to equation 10.14b and $\ddot{x}$ computed.

FORM-CLOSED, ONE-MASS MODEL The form-closed system has no return spring and was modeled in Figure 10-8a (p.276). Its system equation 10.7b is repeated here, rearranged and renumbered.

$$\ddot{x} = \omega_n^2 s + 2\zeta\omega_n\dot{s} - 2\zeta\omega_n\dot{x} - \omega_n^2 x \tag{10.15a}$$

Establish two dummy variables as in equation 10.14b and create two coupled first-order state space ODEs for solution.

$$\begin{aligned} \dot{y}_1 &= y_2 \\ \dot{y}_2 &= \omega_n^2 s + 2\zeta\omega_n\dot{s} - 2\zeta\omega_n y_2 - \omega_n^2 y_1 \end{aligned} \tag{10.15b}$$

The results of their simultaneous solution are then back-substituted in the original equation 10.15a according to equation 10.14b and $\ddot{x}$ computed.

FORCE-CLOSED, ONE-MASS INDUSTRIAL CAM MODEL The one-mass, force-closed system with its return spring at the cam was modeled in Figure 10-8b. Its system equation 10.8b for the in-contact condition is repeated here, rearranged and renumbered.

$$\ddot{x} = \omega_2^2 s + 2\zeta_2\omega_2\dot{s} - 2\zeta_2\omega_2\dot{x} - \omega_2^2 x \tag{10.16a}$$

Establish two dummy variables as in equation 10.14b and create two coupled first-order state space ODEs for solution.

$$\begin{aligned} \dot{y}_1 &= y_2 \\ \dot{y}_2 &= \omega_2^2 s + 2\zeta_2\omega_2\dot{s} - 2\zeta_2\omega_2 y_2 - \omega_2^2 y_1 \end{aligned} \tag{10.16b}$$

The results of their simultaneous solution are then back-substituted in the original equation 10.16a according to equation 10.14b and $\ddot{x}$ computed.

FORCE-CLOSED, TWO-MASS INDUSTRIAL CAM MODEL The two-mass force-closed system with its return spring at the cam was modeled in Figure 10-11 (p. 279). Its system equation 10.12a for the in-contact condition is repeated and renumbered here.

$$\ddot{x} = \frac{c_2}{m_2}\dot{z} + \frac{k_2}{m_2}z - \frac{c_2}{m_2}\dot{x} - \frac{k_2}{m_2}x \tag{10.17a}$$

Establish two dummy variables as in equation 10.14b and create two coupled first-order state space ODEs for solution.

$$\begin{aligned} \dot{y}_1 &= y_2 \\ \dot{y}_2 &= \frac{c_2}{m_2}\dot{z} + \frac{k_2}{m_2}z - \frac{c_2}{m_2}y_2 - \frac{k_2}{m_2}y_1 \end{aligned} \tag{10.17b}$$

As described in the previous section, this equation only covers the case of contact between cam and follower (stage 1). It is solved assuming that the displacement z of m_1 equals cam displacement s for a given time step. Then the contact force F_c is calculated from equation 10.12b using the values of x and its derivatives found from equations 10.17b. If F_c is positive, then equations 10.17b are used again at the next time step. If F_c becomes negative, then separation has occurred and the solution must switch to stage 2, which requires simultaneous solution of equations 10.12a and 10.12b for x and z with F_c set to zero and the initial conditions set to the current values of position and velocity.

$$\begin{aligned} \ddot{x} &= \frac{c_2}{m_2}\dot{z} + \frac{k_2}{m_2}z - \frac{c_2}{m_2}\dot{x} - \frac{k_2}{m_2}x \\ \ddot{z} &= -\frac{c_1 + c_2}{m_1}\dot{z} - \frac{k_1 + k_2}{m_1}z + \frac{c_2}{m_1}\dot{x} + \frac{k_2}{m_1}x \end{aligned} \tag{10.17c}$$

These two second-order ODEs require four dummy variables.

$$y_1 = x \qquad y_2 = \dot{x} \qquad y_3 = z \qquad y_4 = \dot{z} \tag{10.17d}$$

Substitution of equation 10.17d into 10.17c leads to the following set of state space equations.

$$\begin{aligned} \dot{y}_1 &= y_2 \\ \dot{y}_2 &= \frac{c_2}{m_2}y_4 + \frac{k_2}{m_2}y_3 - \frac{c_2}{m_2}y_2 - \frac{k_2}{m_2}y_1 \\ \dot{y}_3 &= y_4 \\ \dot{y}_4 &= -\frac{c_1 + c_2}{m_1}y_4 - \frac{k_1 + k_2}{m_1}y_3 + \frac{c_2}{m_1}y_2 + \frac{k_2}{m_1}y_1 \end{aligned} \tag{10.17e}$$

These equations are solved, and their values of x and z used in equation 10.12b, to determine F_c. As long as F_c continues to be negative, this stage 2 solution is continued. When F_c goes positive, the solution is switched back to equations 10.17b. Another possibility is to compare the calculated value of z to the cam displacement s at each time step in stage 2 to determine when z again becomes equal to or less than s, indicating that contact has been reestablished.

IMPLEMENTATION This state-space approach has been implemented in program DYNACAM and can be solved for any of these five models (equations 10.5c-d, 10.8, 10.14, 10.15, 10.16, or 10.17).* It uses a fourth-order Runge-Kutta algorithm with adaptive step size control. A typical solution screen (for equation 10.14) is shown in Figure

* For the two-mass cam models as defined in equations 10.5c-d, and 10.17, program Dynacam only solves equations 10.17b and calculates the dynamic force from equation 10.12b. If the force goes negative, it records the angle at which that occurred and thereafter ceases to calculate the force. The system parameters can be changed to eliminate follower jump and the system dynamics recalculated.

10-17. The system parameters (m, k_1, k_2, ζ_1, ζ_2) are supplied by the user. The natural frequencies ω_1 and ω_2 are calculated from the m-k data to be used in the solution. The user can also specify the desired start and end times for the calculation and the initial value of displacement (initial velocity is assumed to be zero). The starting time step, end time, minimum step size, and accuracy are also user controllable, though good results will be obtained by accepting their default values. The end time is required to be at least two cam rotation periods in order to let the starting transients die out. The data plotted from DYNACAM's plot routines shows one cam revolution cycle, beginning with the second revolution, in order to eliminate the transient effects.

The example shown in Figure 10-17 is of a double-dwell cam having a modified-trapezoidal acceleration rise of 1 in over 90°, a 90° dwell, a 90° modified-sine acceleration fall, and a 90° dwell. Camshaft speed is 180 rpm. Compare the simulated acceleration of the modified trapezoidal rise with that of Figure 9-10 (p. 232), which shows a measurement of the acceleration of a similar cam-follower program. They are quite similar.

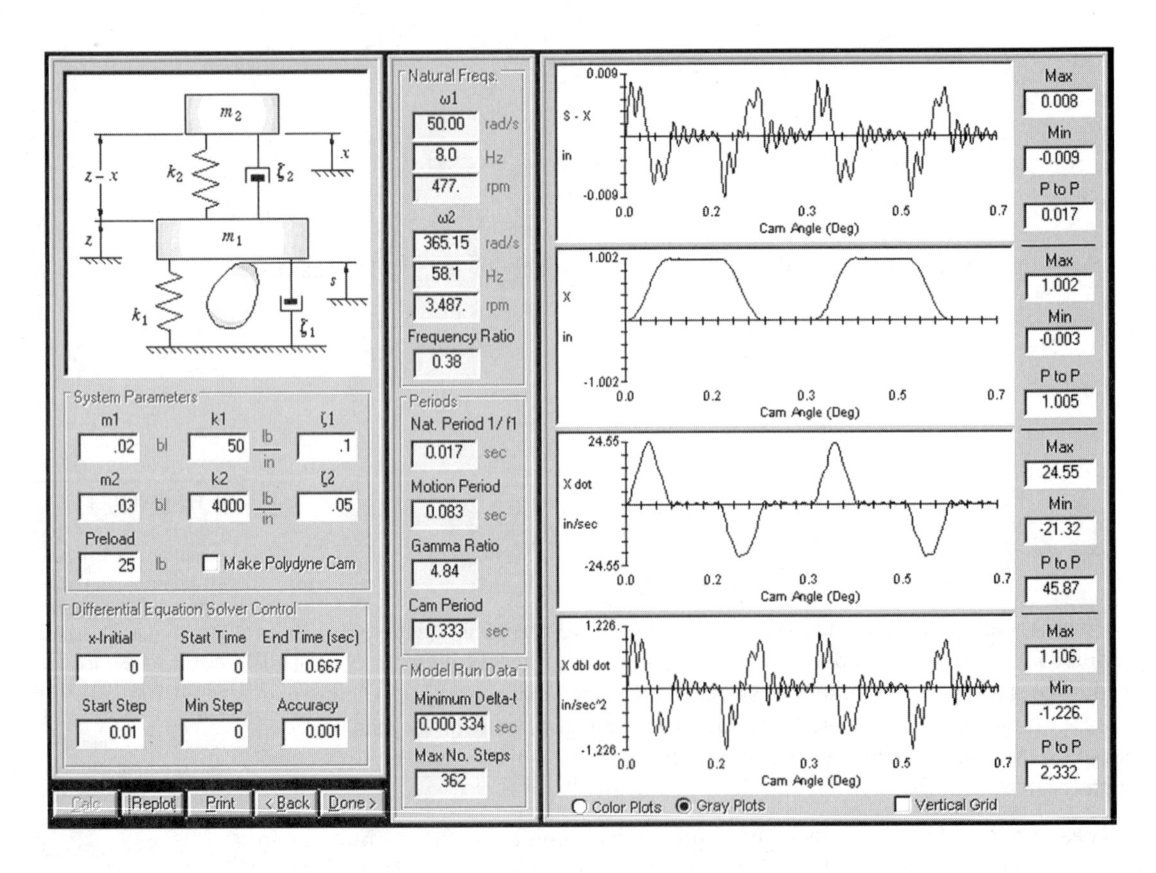

FIGURE 10-17

Dynamic model input and calculation in program DYNACAM

10.8 POLYDYNE CAM FUNCTIONS

The term polydyne is a contraction of "polynomial" and "dynamic." It was coined by Thoren, Engemann, and Stoddart in 1953[12] to describe a cam design method first proposed and implemented by Dudley in 1948,[13] who was the first to use a dynamic model of a cam-follower system such as those described in the previous sections of this chapter to determine a cam profile that would, in effect, compensate for the dynamic vibration of the follower train, at least at one particular cam speed. Dudley used a simple SDOF model that did not include damping, as shown in Figure 10-18. The equation of motion for this system is:

$$\begin{aligned} m\ddot{x} &= k_1(s-x) - k_2 x \\ &= k_1 s - (k_1 + k_2)x \end{aligned} \tag{10.18a}$$

This equation relates the cam displacement s to the follower displacement x. In the previous sections, we defined the cam displacement s and solved for x. However, the equation works in the other direction as well. We can define the desired follower motion x and its derivatives and compute the cam displacement s needed to obtain that displacement with the assumed spring rates and known mass of the system. Solving equation 10.18a for s gives:

$$s = \frac{m}{k_1}\ddot{x} + \frac{k_1 + k_2}{k_1}x \tag{10.18b}$$

The acceleration $\ddot{x}$ has units of length/sec^2. It will be useful to convert it to an angle base (in degrees) rather than a time base and to introduce the camshaft angular velocity N in rpm.

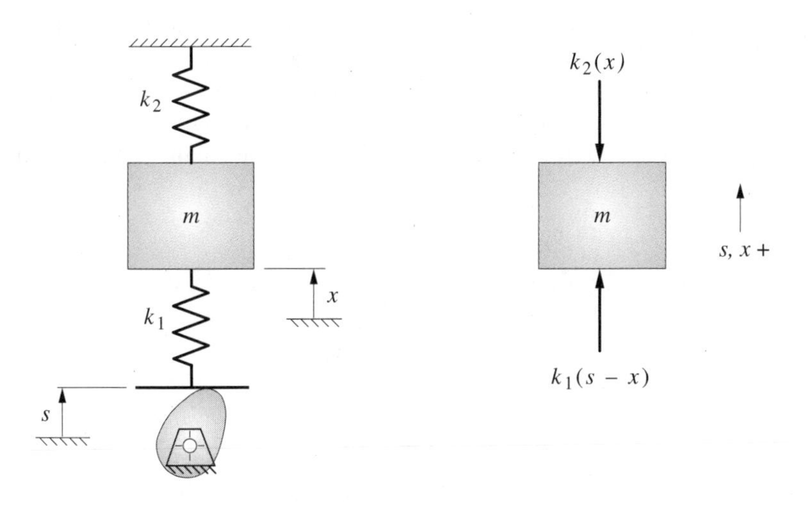

FIGURE 10-18

Dudley's SDOF model and FBD of a cam-follower system for a polydyne cam (13)

10

$$\frac{\text{length}}{\sec^2} = \frac{\text{length}}{\deg^2}\left(\frac{360^2 \deg^2}{\text{rev}^2}\right)\left(\frac{N^2 \text{rev}^2}{60^2 \sec^2}\right)$$

$$\ddot{x} = \frac{d^2x}{dt^2} = 36N^2 \frac{d^2x}{d\theta^2} \tag{10.19}$$

Substitute equation 10.19 in 10.18b.

$$s = \left(36N^2 \frac{m}{k_1}\right)\frac{d^2x}{d\theta^2} + \left(\frac{k_1 + k_2}{k_1}\right)x \tag{10.20a}$$

Differentiate with respect to θ to get the polydyne cam velocity function.

$$\dot{s} = \left(36N^2 \frac{m}{k_1}\right)\frac{d^3x}{d\theta^3} + \left(\frac{k_1 + k_2}{k_1}\right)\frac{dx}{d\theta} \tag{10.20b}$$

Differentiate again with respect to θ to get the polydyne cam acceleration function.

$$\ddot{s} = \left(36N^2 \frac{m}{k_1}\right)\frac{d^4x}{d\theta^4} + \left(\frac{k_1 + k_2}{k_1}\right)\frac{d^2x}{d\theta^2} \tag{10.20c}$$

Note that the cam acceleration function involves the fourth derivative of the selected follower displacement function, which we call "ping." This means that any function selected for the follower displacement must be continuous through at least four derivatives, velocity, acceleration, jerk, and ping. Since polynomial functions allow control of continuity at the ends of the segment to any derivative, they provide a useful means to this end. Spline functions can also satisfy these requirements.

Dudley's polydyne model of Figure 10-18 is for a system with the return spring acting at the end effector. For an industrial cam that has the return spring acting on the cam shoe, the model of Figure 10-8 is more appropriate. If we assume that contact is maintained between cam and follower and that damping is negligible, we can write the system equation as a modification of equation 10.7, leaving out the damping terms.

$$\ddot{x} + \frac{k}{m}x = \frac{k}{m}s \tag{10.21a}$$

Rearrange to solve for s:

$$s = \frac{m}{k}\ddot{x} + x \tag{10.21b}$$

Substitute equation 10.19 in 10.21b

$$s = 36N^2 \frac{m}{k}\frac{d^2x}{d\theta^2} + x \tag{10.22a}$$

Differentiate with respect to θ to get the polydyne cam velocity function.

$$\dot{s} = 36N^2 \frac{m}{k}\frac{d^3x}{d\theta^3} + \frac{dx}{d\theta} \tag{10.22b}$$

Differentiate again with respect to θ to get the polydyne cam acceleration function.

$$\ddot{s} = 36N^2 \frac{m}{k}\frac{d^4x}{d\theta^4} + \frac{d^2x}{d\theta^2} \tag{10.22c}$$

Note the presence of cam speed N in equations 10.20 and 10.22. For any selected follower function x and its derivatives, the cam profile, as defined by the function s (equation 10.20a or 10.22a) will be different for any value of N selected. This means that the dynamic behavior of the system can be optimized (i.e, vibration minimized) only for one operating speed. If the system will be operated at a single speed, this is not a problem. If not, then one speed in its range will have to be selected for the cam profile definition and the system will have some vibration when operated at any other speed. When applied to engine valve trains that obviously operate over a wide range of speeds, a speed close to the highest expected operating speed is typically used in equations 10.20. Fawcett and Fawcett[20] show that a polydyne cam will always give lower vibration than a non-polydyne cam having the same follower program when its speed is greater than 0.707N. They also state that:

> *By a suitable choice of N, the speeds at which a polydyne cam is inferior may be relegated to the lower part of the speed range of the mechanism where (the inherently lower, ed.) vibration amplitudes are not a problem.*

Given an effective dynamic model of the system as described by equation 10.18a, the problem devolves to identifying one or more suitable polynomial functions for x that provide sufficient control of continuity and that also yield acceptable peak values of velocity and acceleration of the follower for any given set of kinematic constraints on the follower motion. However, Dudley points out that:

> *With a flexible valve linkage, one must accept higher accelerations at the valve (follower) than would occur with a rigid connecting system. There is a corollary: If a cam is designed on the assumption of a rigid system, the actual limiting speed will be appreciably lower than the theoretical value.*

The derivations of these polydyne polynomials are long and involved and will not be reproduced here. The interested reader can explore them in the original references given. Dudley addressed the problem of valve cam design for the overhead valve-pushrod cam-follower system as shown in Figure 10-1, which is a relatively soft system and can have significant vibration problems. In valve trains, it also is necessary to maximize the area under the valve lift curve since it directly affects engine breathing. Dudley included this constraint in his development of the equations. He also attempted to minimize peak accelerations and to have an acceleration curve not dissimilar in shape to those that had proven successful in the past Some trial and error was involved, particularly in respect to choice of powers to be used in the x equation. He determined that a polynomial equation of this form was suitable.

$$s = h\left[1 + C_2\left(\frac{\theta}{\beta}\right)^2 + C_p\left(\frac{\theta}{\beta}\right)^p + C_q\left(\frac{\theta}{\beta}\right)^{p+2} + C_r\left(\frac{\theta}{\beta}\right)^{p+4}\right] \tag{10.23a}$$

where the constants are defined by:

10

$$C_2 = \frac{-6p^2 - 24p}{6p^2 - 8p - 8} \tag{10.23b}$$

$$C_p = \frac{p^3 + 7p^2 + 14p + 8}{6p^2 - 8p - 8} \tag{10.23c}$$

$$C_q = \frac{-2p^3 - 4p^2 + 16p}{6p^2 - 8p - 8} \tag{10.23d}$$

$$C_r = \frac{p^3 - 3p^2 + 2p}{6p^2 - 8p - 8} \tag{10.23e}$$

He comments that "retaining the power 2 provides nearly constant acceleration in the center of the curve, and increasing the exponents of the other three terms has the effect of moving the inflection point further out." This leads to a set of polynomials with exponents of 2-4-6-8, 2-6-8-10, 2-8-10-12, 2-10-12-14, and 2-12-14-16. Of these, Dudley claims "the 2-10-12-14 is probably the best." Its equation is:

$$s = \frac{h}{64}\left[64 - 105\left(\frac{\theta}{\beta}\right)^2 + 231\left(\frac{\theta}{\beta}\right)^{10} - 280\left(\frac{\theta}{\beta}\right)^{12} + 90\left(\frac{\theta}{\beta}\right)^{14}\right] \tag{10.24}$$

10

The lift curve and its first four derivatives are shown in Figure 10-19. Equations 10.23 define only the fall portion of the RFD function. The rise is a mirror image of the fall, switched left to right. To calculate the rise, the independent variable θ/β must be run backward from 1 to 0, and then the velocity and jerk functions must be negated as well. Note in Figure 10-19 that the jerk is discontinuous at the dwells, violating the rule stated by Dudley for functions to be used for polynomial cams (see the discussion of equation 10.20 on p. 292).

Thoren, Engemann, and Stoddart also developed other polynomials for the pushrod overhead valve train that used higher powers than Dudley's and that are continuous in all derivatives through ping.[12] They define a 6-term polynomial of the form

$$s = h\left[1 + C_2\left(\frac{\theta}{\beta}\right)^2 + C_p\left(\frac{\theta}{\beta}\right)^p + C_q\left(\frac{\theta}{\beta}\right)^q + C_r\left(\frac{\theta}{\beta}\right)^r + C_s\left(\frac{\theta}{\beta}\right)^s\right] \tag{10.25a}$$

where the constants are defined by:

$$C_2 = \frac{-pqrs}{(p-2)(q-2)(r-2)(s-2)} \tag{10.25b}$$

$$C_p = \frac{2qrs}{(p-2)(q-p)(r-p)(s-p)} \tag{10.25c}$$

$$C_q = \frac{-2prs}{(q-2)(q-p)(r-q)(s-q)} \tag{10.25d}$$

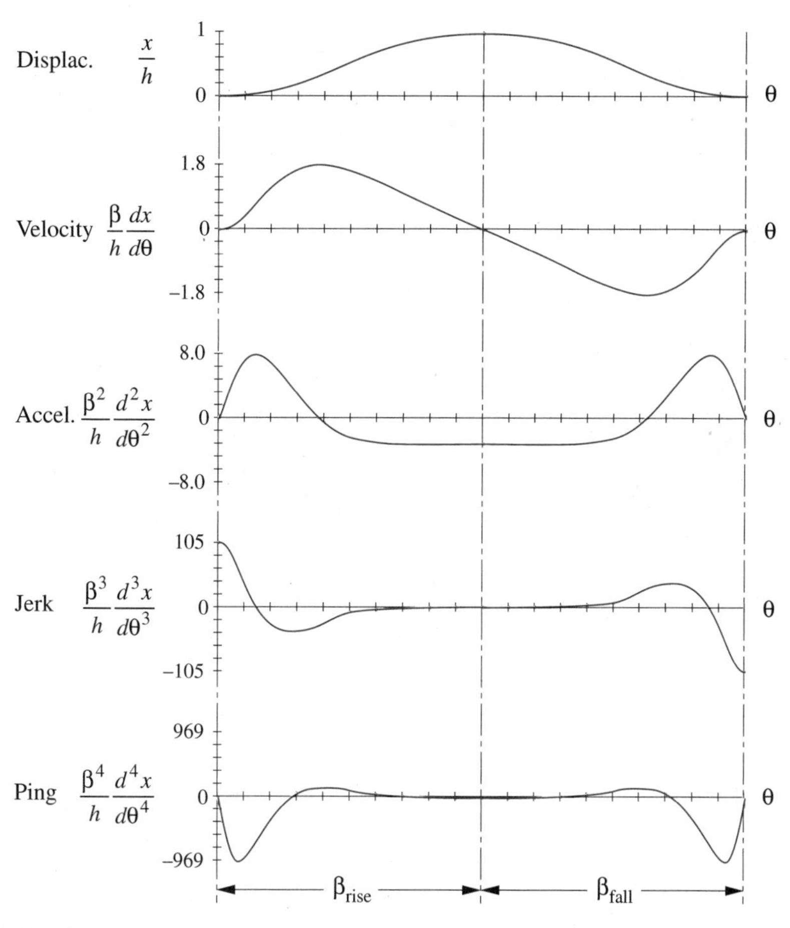

FIGURE 10-19

A 2-10-12-14 polynomial rise-return follower function for a polydyne valve cam

$$C_r = \frac{2pqs}{(r-2)(r-p)(r-q)(s-r)} \tag{10.25e}$$

$$C_s = \frac{-2pqr}{(s-2)(s-p)(s-q)(s-r)} \tag{10.25f}$$

They used evenly spaced exponent sets, i.e, $q - p = r - q = s - r$. Figure 10-20 shows the follower functions for a 2-10-20-30-40 polynomial, and Figure 10-21 shows a 2-14-26-38-50 polynomial. These plots are normalized to show dimensionless factors for their peak values. Note that the higher-order polynomial (Figure 10-21) has significantly higher peak acceleration than the one in Figure 10-20 (13.2 vs. 10.4), but also has a smaller negative acceleration (–2.8 vs. –3.1) because of the steeper and shorter positive acceleration pulses. Less negative acceleration requires less spring force to keep the follower in contact with the cam, which can be an advantage.

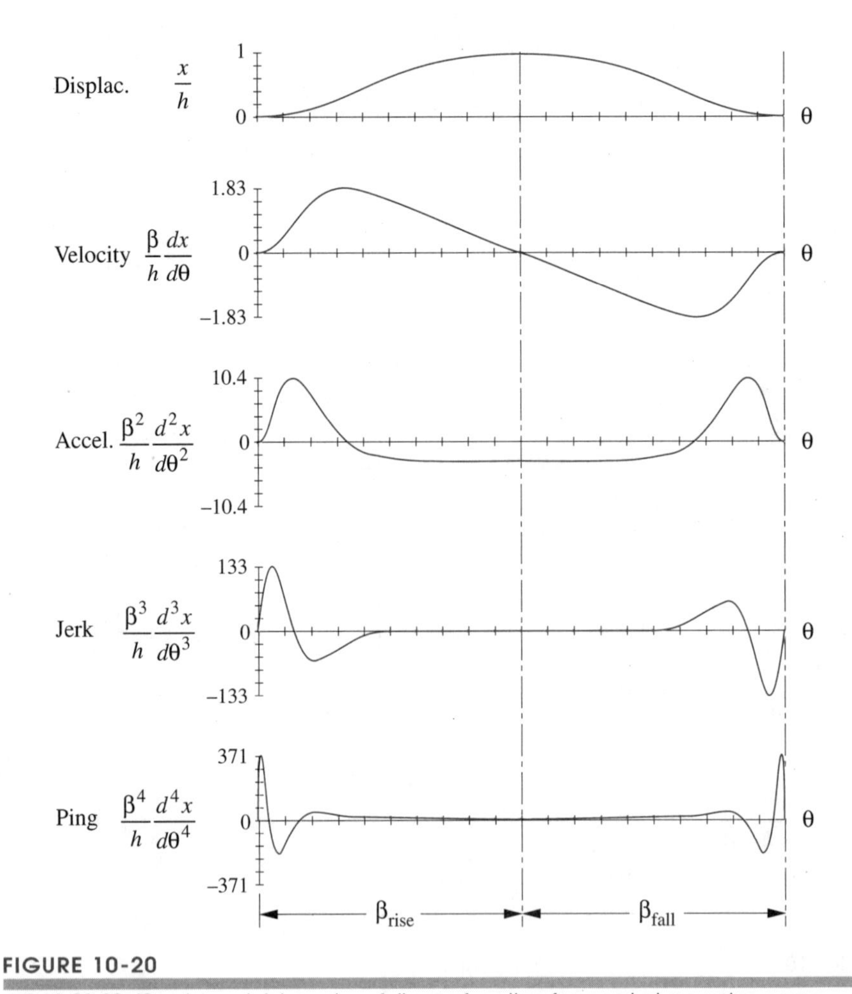

FIGURE 10-20

A 2-10-20-30-40 polynomial rise-return follower function for a polydyne valve cam

10

Figure 10-22 (p. 298) shows both the cam and the follower curves for displacement, velocity, and acceleration for a 2-10-20-30-40 polynomial on the rise and a 2-10-16-24-30 polynomial on the fall from [16]. Note that the cam displacement s is larger than the valve displacement x over the entire motion, indicating that the "spring" in the system is "wound up" during the entire event. Recognize that the cam is cut to the solid curve, but the follower takes the dotted path. The velocity of the valve is greater than that of the cam function at its peaks. The acceleration curves show the greatest differences. Note how the cam acceleration curve dips to nearly zero at the beginning of the rise to compensate for the tendency of the valve acceleration to overshoot. A similar but smaller difference is seen near the end of the fall.

Stoddart [14],15,[16] continued Dudley's work on polydyne cams for valve trains and developed a characteristic relationship of vibration amplitude versus cam speed as shown in Figure 10-23 (p. 299). The actual amplitude A of this curve will vary with the partic-

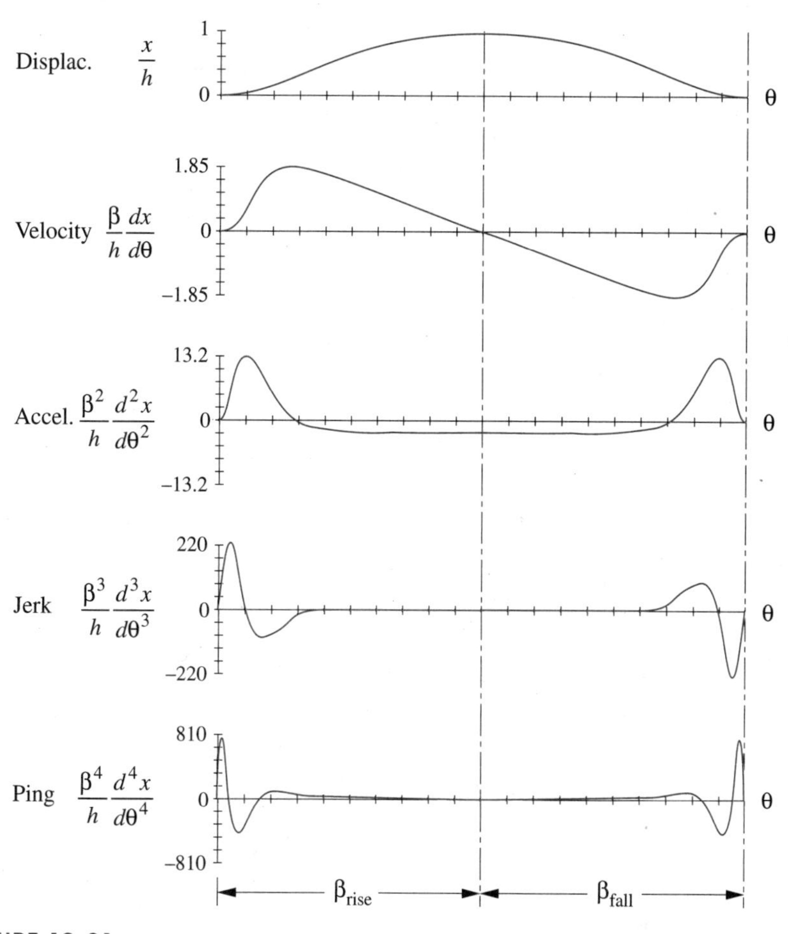

FIGURE 10-21

A 2-14-26-38-50 polynomial rise-return follower function for a polydyne valve cam

ular follower train, but the shape will be similar in all cases. The sign of A is not of interest since the direction of vibration is not relevant, only its magnitude. Note that the vibration amplitude is zero at the design speed N, peaks at $0.707N$, and reaches the same level A at $1.1N$, increasing rapidly at higher speeds.

Polydyne functions were used extensively for automotive valve cams in the U. S. in the 1950's and 1960's, but they have not been used in these applications in recent years. Instead, spline functions are now more commonly used for valve trains. Figure 10-24 shows a modern spline-function design for a single overhead camshaft (SOHC) valve train.[17] Note the asymmetry of the acceleration function, a design freedom allowed by the use of spline functions. This design uses a quartic spline for displacement, but is designed from the acceleration using designer-shaped quadratic splines.[18] The step functions in the acceleration function before and after the lift event are the opening and closing ramps used to wind up the system compliance.[17]

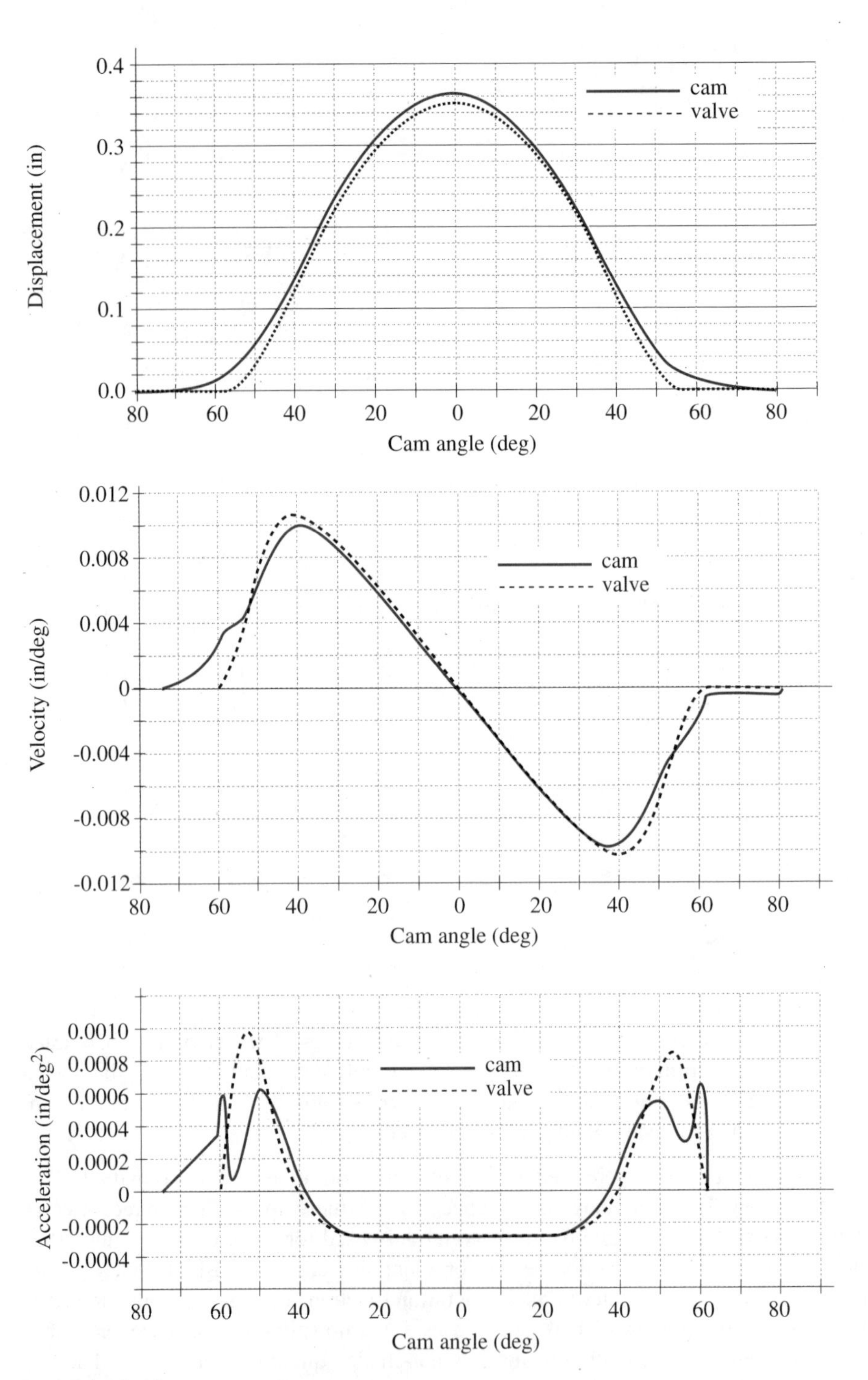

FIGURE 10-22

Cam and follower (valve) motions with a polydyne cam [14, 15, 16]

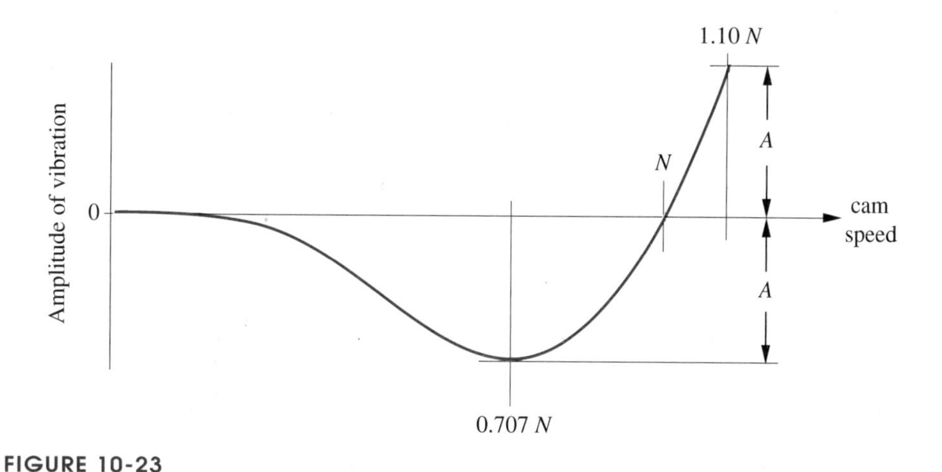

FIGURE 10-23

Characteristic relationship of vibration with cam speed for polydyne cams (14, 15, 16)

The abandonment of polydyne cams by most automobile engine manufacturers was due mainly to the inherent limitation of the polydyne method to one cam speed. Also, the current trend in engine design is to overhead camshaft valve trains as shown in Figure 10-25 (p. 301), rather than the pushrod-overhead valve train type shown in Figure 10-1 (p. 265). The former have much shorter and stiffer follower trains than the latter and so do not exhibit as much variation in dynamic motion between cam and valve. Current valve cam design uses very sophisticated multi-DOF dynamic models of the valve train (analogous to Seidlitz's model) that contain accurate representations of the hydraulic lash adjuster's nonlinear behavior as well as multi-DOF models of the valve spring. Computer programs written specifically to simulate the dynamic behavior of valve trains are commercially available.*

Double-Dwell Polydyne Curves

Polydyne cams were first developed for the single-dwell automobile engine valve train but others have adapted the technique to double-dwell machine cams as well. Stoddart[14] shows how to develop double-dwell polynomials suitable for a polydyne application but does not appear to provide a viable solution in reference [14]. Peisekah[19] derives a useful solution to this problem. He uses the same model as Dudley (Figure 10-18 on p. 291) and its system equation 10.18 (p. 291) to define follower motion. The polydyne cam displacement function and its derivatives are as shown in equation 10.20 (p. 292). As these equations show, the chosen follower function $x(\theta)$ should be continuous through the fourth derivative of displacement or ping. For a double-dwell motion with a total rise of h and a normalized independent variable $u = \theta/\beta$, this requires the minimum set of boundary conditions (BCs):

$$\begin{aligned}
&\text{when } u = 0, \quad && x = 0 \quad x' = 0 \quad x'' = 0 \quad x''' = 0 \quad x'''' = 0 \\
&\text{when } u = 1, \quad && x = h \quad x' = 0 \quad x'' = 0 \quad x''' = 0 \quad x'''' = 0
\end{aligned} \tag{10.26}$$

* e.g., *Valdyne*, available from Ricardo, Bridge Works, Shoreham-by-Sea, West Sussex BN43 5FG England U.K.

10

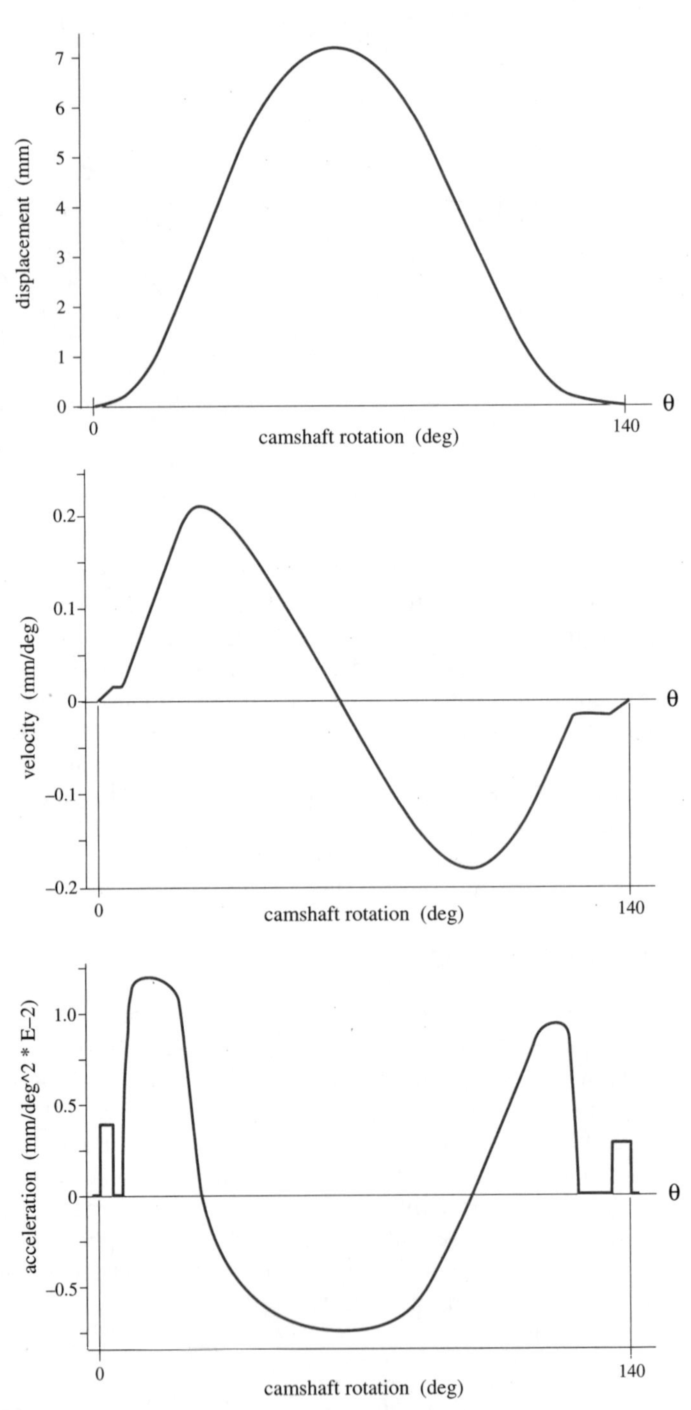

FIGURE 10-24

Spline functions for follower motion of a SOHC valve train (17)

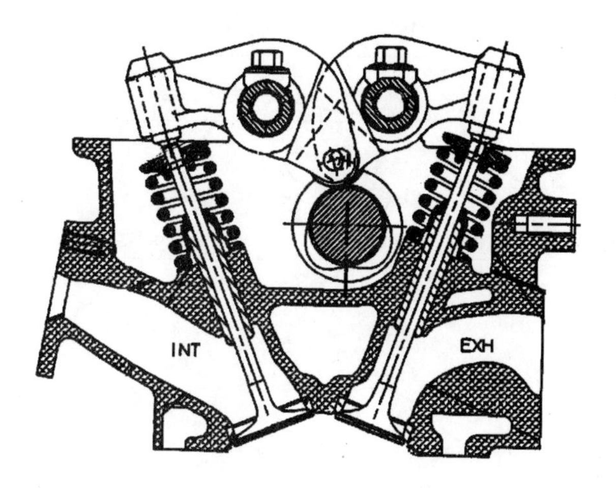

FIGURE 10-25

Single overhead camshaft valve train (Courtesy of DaimlerChrysler, Auburn Hills, MI)

These 10 BCs will yield the 9th-degree polynomial

$$x = h\left[126u^5 - 420u^6 + 540u^7 - 315u^8 + 70u^9\right] \tag{10.27}$$

Unfortunately, this function's second derivative has a maximum acceleration factor of 9.37 h/β^2, which is rather high compared to the standard double-dwell functions shown in Table 3-2 (p. 50).

Peisekah[19] added two additional BCs to derive an 11th-degree polynomial as shown in Figure 10-26. One of the additional BCs was used to force symmetry by making $x(0.5) = 0.5h$. The remaining BC was used as a free variable, and iterated to minimize the value of peak acceleration. The resulting function is:

$$x = h\begin{bmatrix} 336u^5 - 1890u^6 + 4740u^7 - 6615u^8 \\ +5320u^9 - 2310u^{10} + 420u^{11} \end{bmatrix} \tag{10.28a}$$

$$x' = h\begin{bmatrix} 1680u^4 - 11\,340u^5 + 33\,180u^6 - 52\,920u^7 \\ +47\,880u^8 - 23\,100u^9 + 4620u^{10} \end{bmatrix} \tag{10.28b}$$

$$x'' = h\begin{bmatrix} 6720u^3 - 56\,700u^4 + 199\,080u^5 - 370\,440u^6 \\ +383\,040u^7 - 207\,900u^8 + 46\,200u^9 \end{bmatrix} \tag{10.28c}$$

$$x''' = \frac{h}{\beta^3}\begin{bmatrix} 20\,160u^2 - 226\,800u^3 + 995\,400u^4 - 2\,222\,640u^5 \\ +2\,681\,280u^6 - 1\,663\,200u^7 + 415\,800u^8 \end{bmatrix} \tag{10.28d}$$

This polynomial function has a maximum theoretical acceleration factor of 7.91 h/β^2, which, though still high compared to the functions in Table 3-2 (p. 50), may be ac-

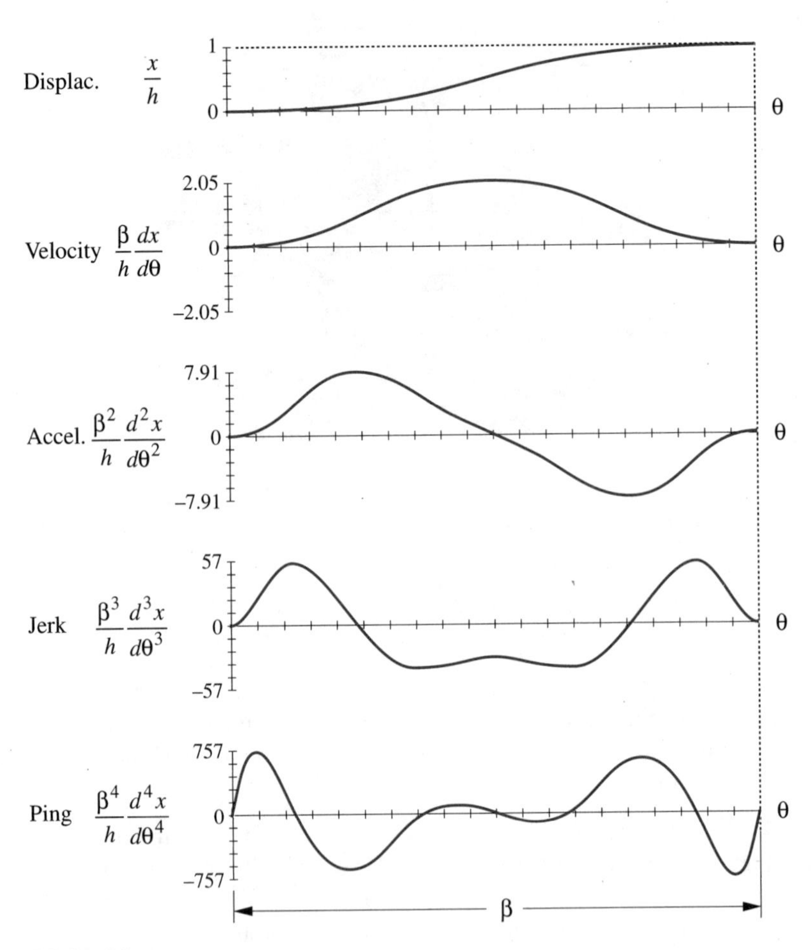

FIGURE 10-26

A Peisekah 5-6-7-8-9-10-11 polynomial rise-return follower function for a polydyne cam[19]

ceptable when the reduction of vibration at the design speed achieved by the polydyne method is taken into account as is shown in the following example.

EXAMPLE 10-1

Designing a Polydyne Double-Dwell Cam With Closure Spring at End Effector.

Problem: Consider the following statement of a double-dwell motion problem:

Rise	1 in over 100°
Dwell	for 80°
Fall	1 in over 100°
Dwell	for 80°
Speed	425 rpm (7.083 Hz)
m_{eff}	0.0104 blob* (4 lb)

* See Section 1.6, p. 10

k_1	1 000 lb/in for the linkage
k_2	50 lb/in at the end effector
Preload	25 lb
ζ_1, ζ_2	0.05

Assumptions: The joint closure spring is at the end effector, so a SDOF, one-mass model as shown in Figure 10-2a (p. 266) will be used.

Solution:

1 A polydyne function theoretically requires continuity through ping, so a Peisekah polynomial will be used as defined in equations 10.28. Calculate the theoretical kinematic *S V A J P* functions for this design. The peak kinematic acceleration is 5143 in/sec^2.

2 Create the differential equation of the system using the relevant effective mass and spring constants based on equation 10.13b (p. 282).

$$\ddot{x} = \omega_1^2 s + 2\zeta_1\omega_1\dot{s} - 2\zeta_2\omega_2\dot{x} - \omega_2^2 x$$

$$\ddot{x} = \frac{k_1}{m}s + 2\zeta_1\sqrt{\frac{k_1}{m}}\dot{s} - 2\zeta_2\sqrt{\frac{k_1+k_2}{m}}\dot{x} - \frac{k_1+k_2}{m}x$$

$$\ddot{x} = \frac{1\,000}{0.0104}s + 2(0.05)\sqrt{\frac{1\,000}{0.0104}}\dot{s} - 2(0.05)\sqrt{\frac{1\,050}{0.0104}}\dot{x} - \frac{1\,050}{0.0104}x \tag{a}$$

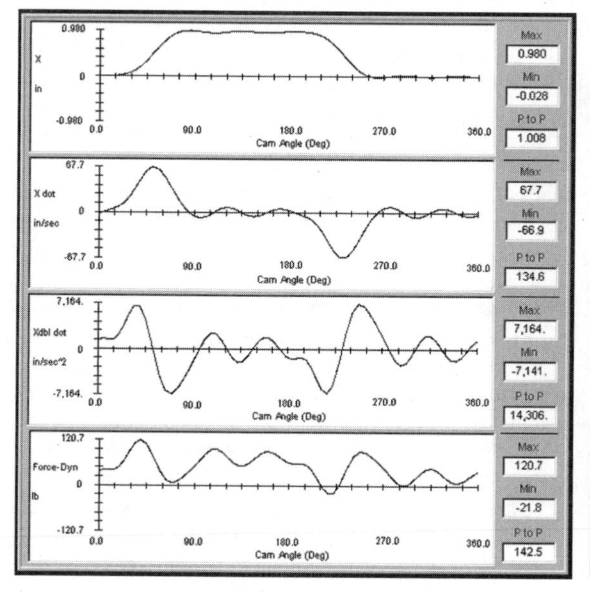

(*a*) Dynamic response of a nonpolydyne cam

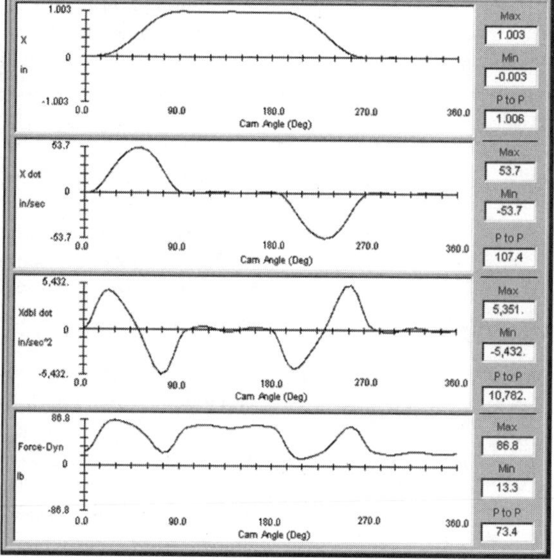

(*b*) Dynamic response of a polydyne cam

FIGURE 10-27

Dynamic response of double-dwell, 11th-degree polynomial functions, non-polydyne and polydyne cams

3 Apply the Runge-Kutta algorithm to solve this ODE for x and its derivatives, and solve for the dynamic force. The results are shown in Figure 10-27a. This non-polydyne cam undershoots its peak kinematic displacement by 0.020 in, overshoots its kinematic velocity and acceleration, and shows follower jump at 210°. The peak dynamic acceleration is 7138 in/sec^2, 39% larger than the kinematic acceleration.

4 To determine the polydyne cam function needed to create the desired motion at the follower, substitute the kinematic displacement and its derivatives through ping (as defined by equations 10.28) into equations 10.20 for x and its derivatives, then solve for s and its first two derivatives. Figure 10-27b shows the dynamic displacement, velocity, acceleration, and force of this polydyne cam. This polydyne version of the same cam program shows well-controlled motion with a displacement error of only 0.003 in and only small errors in velocity and acceleration. These errors can be due to the lack of a damping term in equation 10.20. The peak follower acceleration is reduced to 5431 in/sec^2 , 6% larger than kinematic, and there is no follower jump.

5 Figure 10-28 shows the normalized kinematic displacement, velocity, and acceleration functions superposed on the corresponding dynamic responses of the follower rise for both the non-polydyne and polydyne cams. Figure 10-28a shows the dynamic displacement of the non-polydyne follower and the cam shoe kinematic displacement (labeled *cam*). The follower lags the cam shoe for the first half of the rise and falls short of the full kinematic displacement at the dwell. Figure 10-28b shows the difference in the polydyne cam shoe displacement and the follower displacement. Note that the polydyne cam shoe displacement is greater than the follower displacement over the entire rise to compensate for the dynamic deflection of the follower train.

10

6 Figure 10-28c shows the normalized kinematic velocity as imparted to the cam shoe and the resulting non-polydyne follower velocity that substantially overshoots the peak kinematic value. Figure 10-28d shows the polydyne cam shoe velocity, which initially leads the follower, and then lags it between about 25% and 75% of the rise period as the stored energy in the elastic follower train speeds up the follower mass. The follower velocity is nearly kinematic for the polydyne case.

7 Figure 10-28e shows the kinematic acceleration of the cam shoe and the resulting dynamic response of the follower for the non-polydyne cam. There is significant error coming off the previous dwell and a large overshoot in the follower acceleration response. In Figure 10-28f , the polydyne cam shoe acceleration function is distinctly different from that of the follower in order to compensate for system elasticity. The sudden drop in cam shoe acceleration during the start of the rise serves to slow the follower and control overshoot.

8 Note the "role reversal" of cam and follower between the polydyne and non-polydyne cam systems. The non-polydyne system has an "ideal" kinematic cam shoe motion but a distorted follower motion. The polydyne cam system deliberately distorts the cam shoe motion to compensate for system elasticity and mass, resulting in a follower motion that is closer to the kinematic ideal at the design speed. Vibration of the follower of a polydyne cam will be close to zero at the design speed as long as there is sufficient follower spring force to keep it in contact. The trade-off can be a cam contour with larger pressure angle, smaller radius of curvature, thus higher stresses, and even undercutting in severe cases.

9 It is reasonable to ask whether anything has been gained by using this polydyne approach over a more conventional one of, say, a modified trapezoidal (MT) acceleration motion program. After all, the MT will have a much lower theoretical kinematic acceleration

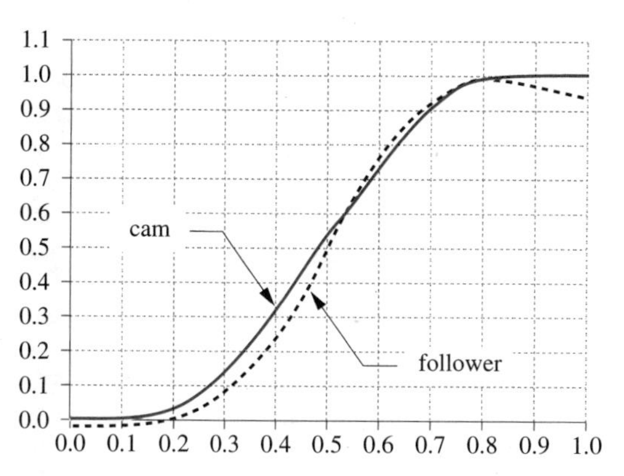

(*a*) Displacement of nonpolydyne cam and follower

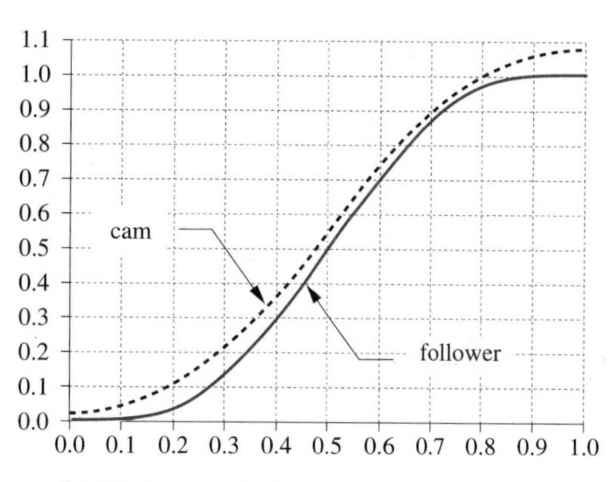

(*b*) Displacement of polydyne cam and follower

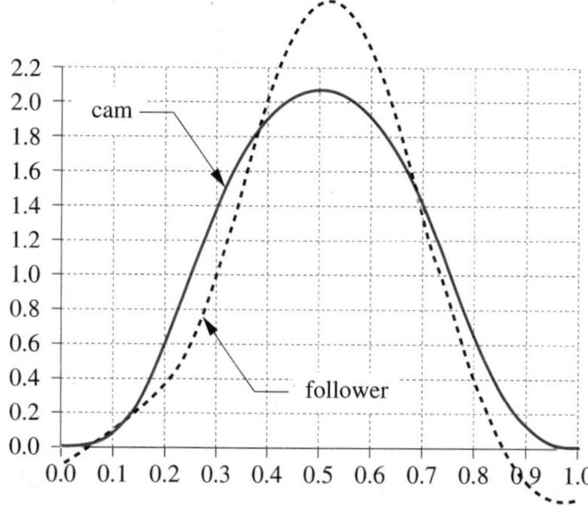

(*c*) Velocity factor for nonpolydyne cam and follower

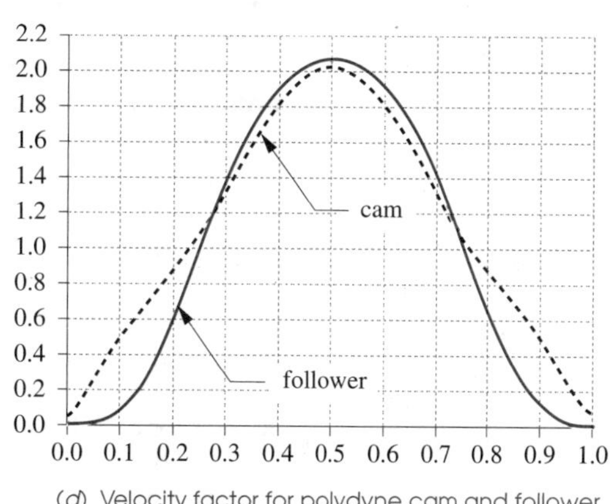

(*d*) Velocity factor for polydyne cam and follower

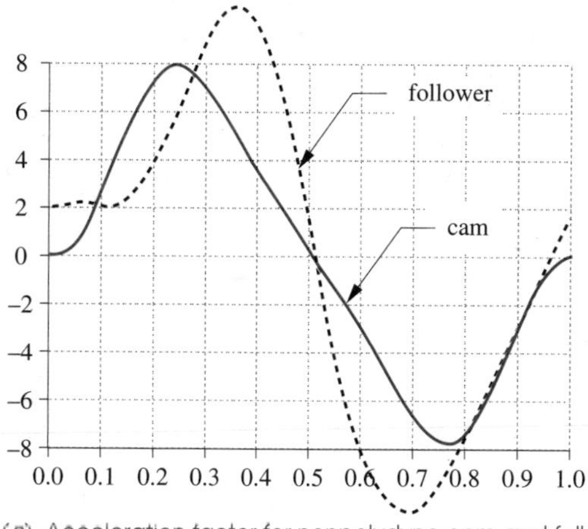

(*e*) Acceleration factor for nonpolydyne cam and follower

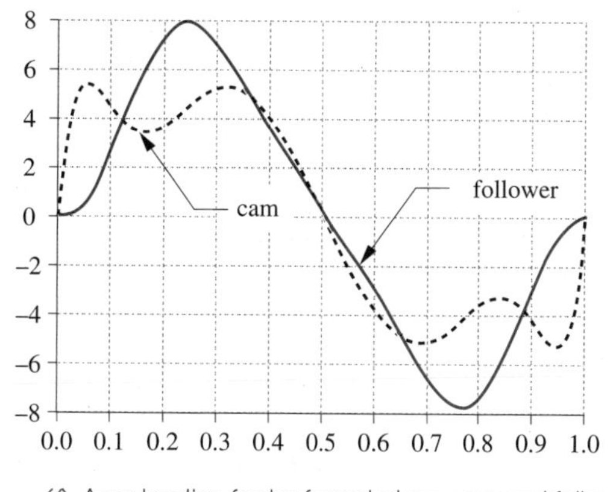

(*f*) Acceleration factor for polydyne cam and follower

FIGURE 10-28

Normalized double-dwell, 11th-degree polynomial functions, non-polydyne and polydyne cams

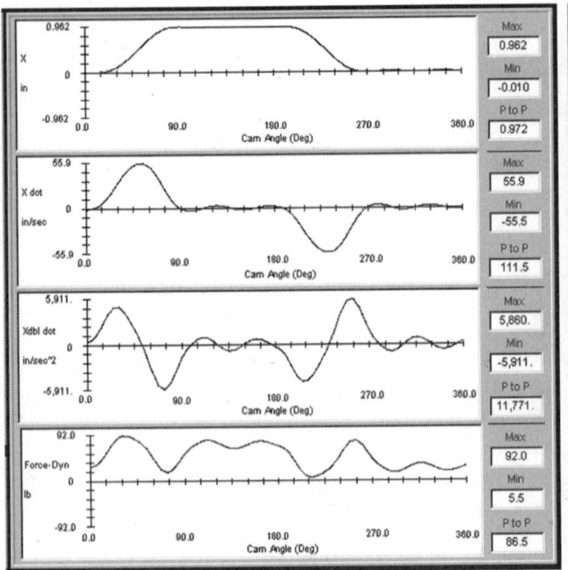

(a) Dynamic response of a modified trapezoid cam

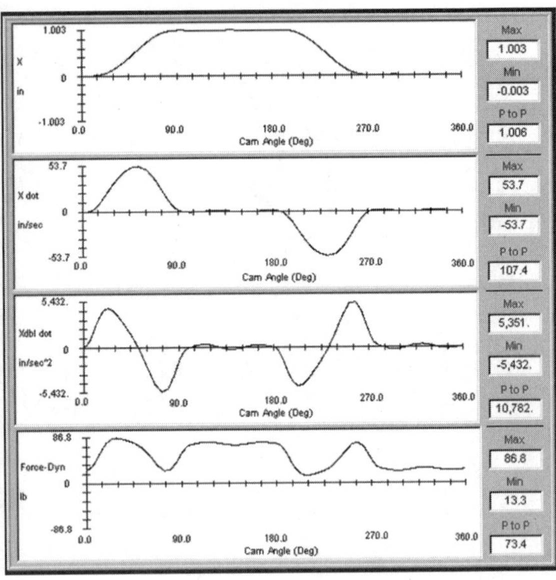

(b) Dynamic response of a Peisekah polydyne cam

FIGURE 10-29

Dynamic response of double-dwell, non-polydyne modified trapezoid and 11th-degree polynomial polydyne cams

peak value (3 179 in/sec^2) than the Peisekah polynomial used here (4.88 versus 7.91 acceleration factor). Figure 10-29 shows the dynamic displacement, velocity, acceleration, and force for an MT cam with the same lift and mass property specifications. It also shows the dynamic response of the polydyne cam from Figure 10-27b for comparison. Though the MT cam does not exhibit any follower jump (it comes close), it has a higher peak acceleration (5 955 versus 5 431 in/sec^2) than the polydyne cam despite the lower theoretical acceleration of the MT. Thus, the real peak acceleration of the MT cam with this follower system is 87% higher than the MT kinematic acceleration and 10% higher than the polydyne cam. The peak dynamic force of the MT cam (92 lb) is also higher than the polydyne cam (87 lb) by 6%. Thus, the polydyne cam is a dynamically superior solution despite its 11th-degree polynomial function's theoretical disadvantages in terms of its higher kinematic peak values. Also note the lack of vibration of the polydyne cam during the dwells as compared to the MT. Overall, the polydyne cam is a better solution.

EXAMPLE 10-2

Designing a Polydyne Double-Dwell Cam With Closure Spring at Cam Follower.

Problem: Consider the following statement of a double-dwell motion problem:

Rise 1 in over 90°
Dwell for 90°
Fall 1 in over 90°

Dwell	for 90°
Speed	180 rpm (3 Hz)
m_{eff}	0.03 bl (11.6 lb)
k_1	1 000 lb/in for the linkage
k_2	50 lb/in at the end effector
Preload	30 lb
ζ_1, ζ_2	0.05, 0.10, respectively

Assumptions: The joint closure spring is at the cam follower, so a SDOF, one-mass model as shown in Figure 10-8b (p. 276) will be used.

Solution:

1 Figure 10-30 shows the difference between displacement, velocity, and acceleration of the follower for a non-polydyne and a polydyne cam with the same motion program, as calculated in program Dynacam. The motion is RDFD, 90-90-90-90 deg, with 1-in rise and fall at 180 rpm. A Peisekah 11th-degree polynomial is used on both rise and fall. The one-mass industrial cam dynamic model of Figure 10-8 is used with a return spring rate of 50 lb/in and a 30-lb preload. The stiffness of the follower train is 1 000 lb/in and the mass of the follower is 0.03 bl. Damping is between 0.05 and 0.1 of critical.

2 Figure 10-30a shows the non-polydyne solution. Note the significant error in the follower's velocity and acceleration functions. The displacement overshoots by 0.034 in and undershoots by 0.036 in. The peak follower acceleration is 1 450 in/sec^2.

3 Figure 10-30b shows the same data for a polydyne cam using the same Peisekah 11th-degree polynomial and equations 10.21. The velocity and acceleration functions now look like the theoretical functions, with the peak acceleration reduced to 1 121 in/sec^2. This is 1.5% lower than the designed kinematic acceleration of 1 138 in/sec^2, within the numerical error expected from the simulation. The displacement error is reduced to +0.005 / –0.004 in. Some of this error is due to the absence of a damping term in equations 10.21.

4 Figure 10-31 shows the kinetostatic and dynamic follower force functions superposed for each of the cases, non-polydyne and polydyne. In Figure 10-31a, the dynamic force is grossly different than the kinetostatic force and is close to zero at one point, indicating incipient jump. The polydyne cam contour has reduced the dynamic force significantly in Figure 10-31b. It is now close to the kinetostatic ideal and would be exactly equal if the damping were included in the polydyne equations and the dynamic model was exact.

10.9 SPLINEDYNE CAM FUNCTIONS

In Chapters 5 and 6 we showed how spline functions, particularly B-splines, can provide superior solutions to motion control problems than polynomials can in some cases. It would seem logical that splines then might offer some advantages to polydyne-type cam designs as well. The polydyne approach requires a function for the follower motion that has continuity through the 4th derivative of displacement, or ping. This is easy to accomplish with B-splines since they have the ability to control the unwanted excursions that are typical of high-order polynomials. Moreover, manipulation of knot locations can reduce peak velocity and acceleration. Thus, we introduce the concept of a splinedyne cam, a combination of spline functions for the motion and the dynamics of the follower train.

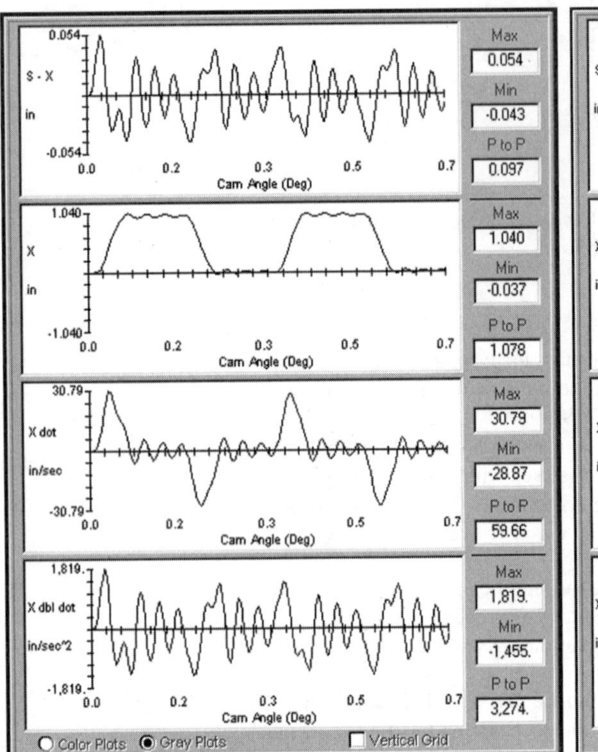

(a) Non-polydyne dynamic response

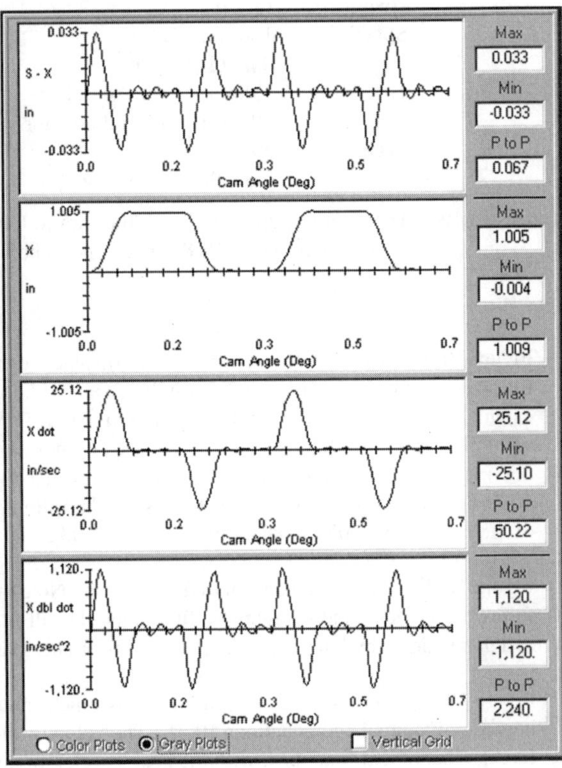

(b) Polydyne dynamic response

FIGURE 10-30

Non-polydyne and polydyne responses of a follower train to the same cam motion program

Figure 10-32 (p. 310) shows a 6th-order B-spline for a double-dwell rise motion with ten boundary conditions that make the displacement and its first four derivatives (through ping) zero at the beginning of the rise. At the end of the rise the displacement is equal to the total lift h, and its first four derivatives are zero. Choosing a 6th-order (5th-degree, quintic) spline for this motion leaves four internal knots available for manipulation and still provides continuity through ping. The functions in the figure were calculated with a camshaft speed of 1 rad/sec, a rise of 1 unit, and a period of 1 radian. This makes their peak values nondimensional, thus providing factors comparable to those of Table 3-2 (p. 50).

Pulling pairs of knots close to the beginning and end of the interval has the effect of forcing the displacement curve to rise (and fall) more rapidly at the extremes and so lowers the velocity and acceleration peaks (at the expense of peak jerk). The functions shown in Figure 10-32 have knots at points approximately 7% and 8% of the period in from each end, respectively. This gives a peak velocity factor of 2.1 and a peak acceleration factor of 7.1. These are better than the Peisekah polynomial of the previous sec-

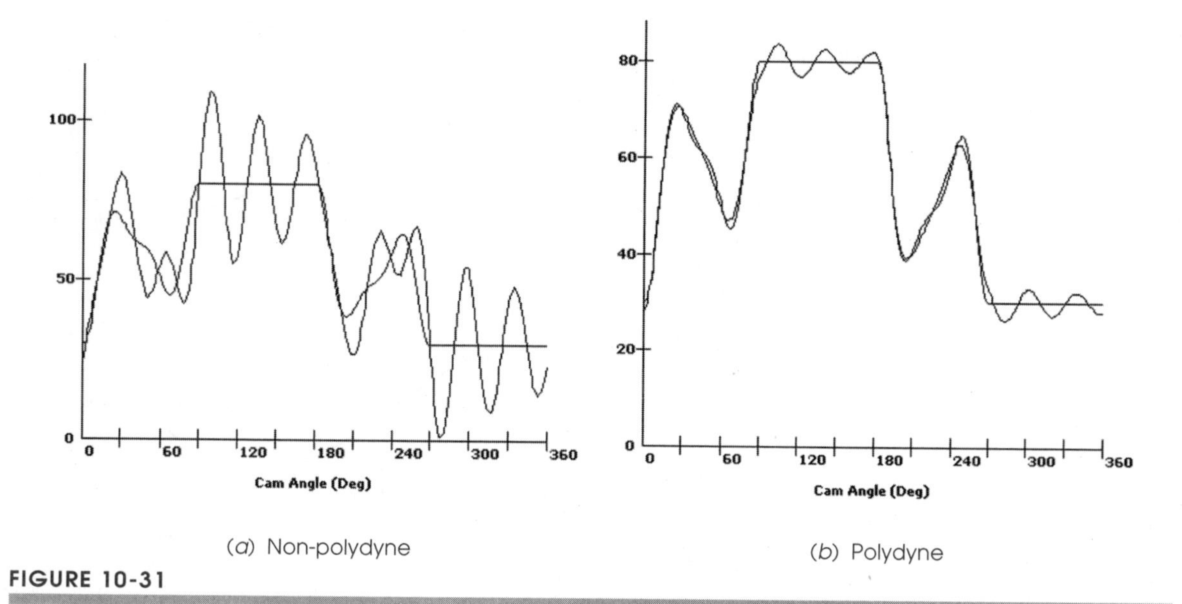

(a) Non-polydyne (b) Polydyne

FIGURE 10-31

Kinetostatic and dynamic force for a polydyne and non-polydyne cam made to the same motion program

tion, and their peak values are not much higher than the classic cycloidal motion. But, because of their 4th-order continuity, these functions also have the ability to be modified according to the system's dynamics to provide a splinedyne, low-vibration motion at one design speed by using equations 10.20 or 10.22 (p. 292) to define the cam shape based on these follower motion functions.

Figure 10-33 (p. 311),shows an 8th-order B-spline for a double-dwell rise motion with the same ten boundary conditions as used in the functions of Figure 10-32. Now there are only two interior knots available for manipulation. These are placed 3.5% in from each end of the interval. Note that the functions through jerk and ping are much smoother than the 6th-order B-spline of Figure 10-32 due to the higher-order continuity of this spline. The penalty is higher peak factors, now 2.23 for velocity and 7.78 for acceleration, but comparable to the Peisekah 11th-degree polynomial.

EXAMPLE 10-3

Designing a Splinedyne Double-Dwell Cam With Spring at Cam Follower.

Problem: Consider the same double-dwell motion problem as in Example 10-2:

Rise	1 in over 90°
Dwell	for 90°
Fall	1 in over 90°
Dwell	for 90°
Speed	180 rpm (3 Hz)
m_{eff}	0.03 bl (11.6 lb)

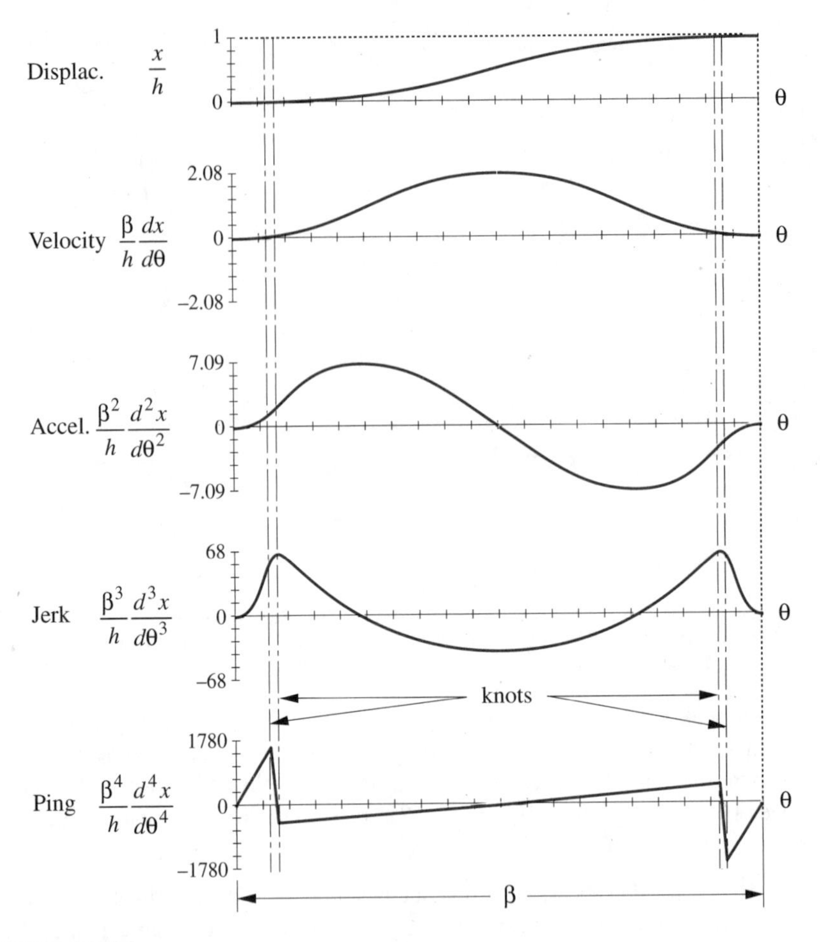

FIGURE 10-32

6th-order spline function for a splinedyne cam, knots at 7%, 8%, 92%, and 93% of period

k_1	1 000 lb/in for the linkage
k_2	50 lb/in at the end effector
Preload	30 lb
ζ_1, ζ_2	0.05, 0.10, respectively

Assumptions: The joint closure spring is at the cam follower, so a SDOF, one-mass model as shown in Figure 10-8b (p. 276) will be used.

Solution:

1 Figure 10-34 (p. 312) shows the difference between displacement, velocity, and acceleration of the follower for a non-splinedyne and a splinedyne cam with the same motion program, as calculated in program DYNACAM. The motion is RDFD, 90-90-90-90 deg, with 1-in rise and fall at 180 rpm. The 8th order B-spline of Figure 10-33 is used on both rise and fall. The one-mass dynamic model of Figure 10-8 is used with a return spring rate

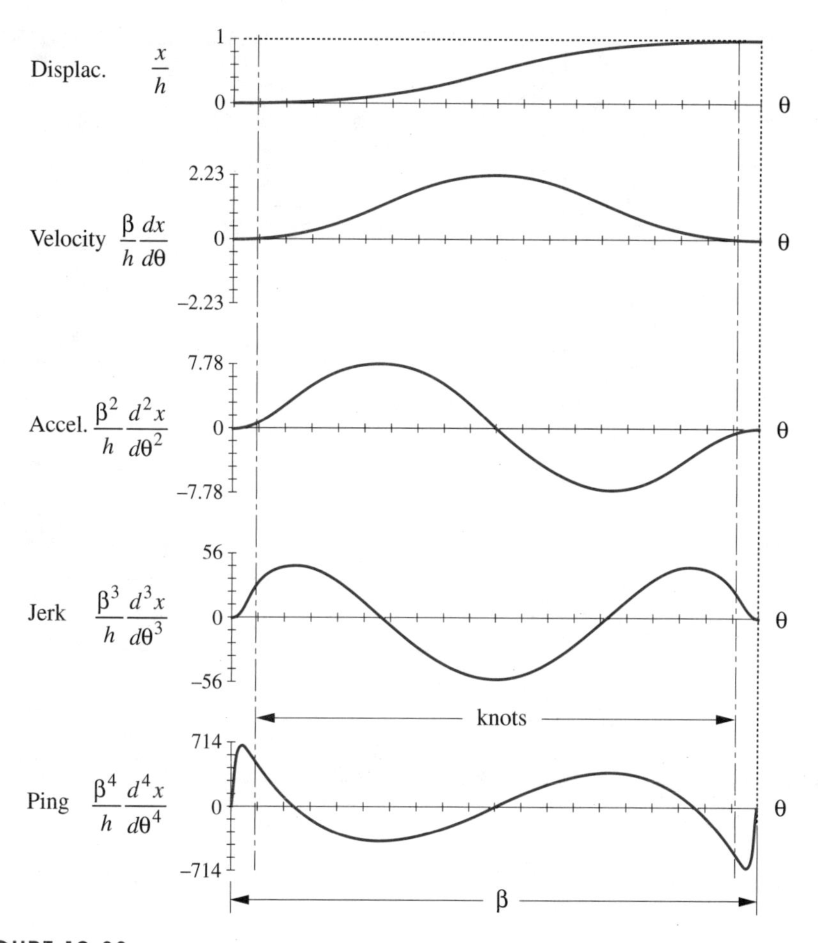

FIGURE 10-33

8th-order spline function for a splinedyne cam, knots at 3.5% and 96.5% of period

10

of 50 lb/in and a 30-lb preload. The stiffness of the follower train is 1 000 lb/in and the mass of the follower is 0.03 bl. Damping is between 0.05 and 0.1 of critical. All these data are identical to those used for the Peisekah polynomial of Example 10-2.

2 Figure 10-34a shows the non-splinedyne solution. There is less error in the follower's velocity and acceleration functions than for the polynomial solution, but they are nevertheless oscillatory. The displacement overshoots by 0.012 in and undershoots by 0.012 in. The peak follower acceleration is 1 600 in/sec^2.

3 Figure 10-34b shows the same data for the splinedyne cam. The velocity and acceleration functions now look very much like the theoretical functions, with the peak acceleration reduced to 1 166 in/sec^2. This is only 4% higher than the designed kinematic acceleration of 1 121 in/sec^2, within the numerical error expected from the simulation using equation

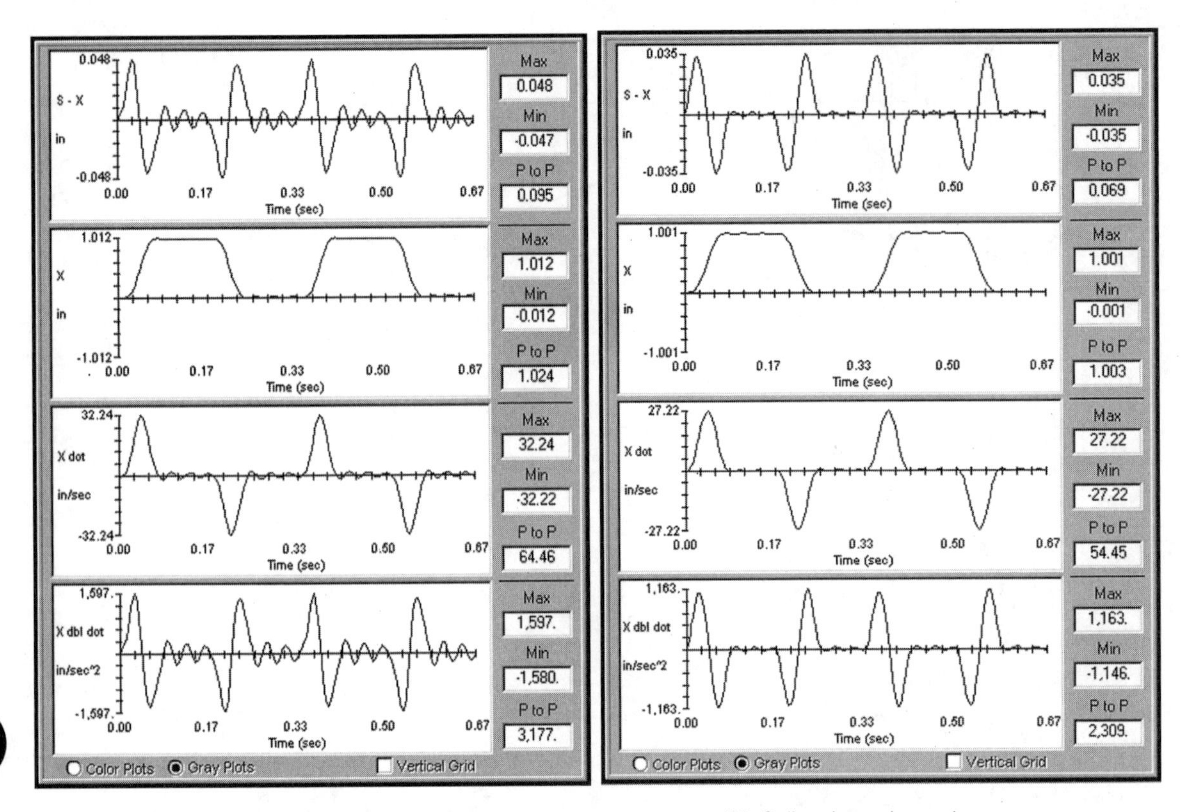

(*a*) Non-splnedyne dynamic response (*b*) Splinedyne dynamic response

FIGURE 10-34

Non-splinedyne and splinedyne responses to the same cam motion program

10.21. The displacement error is reduced to ±0.001 in, a better result than the polydyne cam.

4 Figure 10-35 shows the kinetostatic and dynamic follower force functions superposed for each of the cases, non-splinedyne and splinedyne. In Figure 10-35a, the dynamic force is much different than the kinetostatic force, though not as bad as that of the Peisekah polynomial of Figure 10-31. The splinedyne cam contour in Figure 10-35b has reduced the dynamic force significantly. It is now very close to the kinetostatic ideal. These figures show the superiority of the splinedyne cam over the polydyne cam for reducing dynamic oscillations at any one design speed. Either is significantly better than a cam design that does not account for the dynamics of the follower system, however.

10.10 REFERENCES

1 **Barkan, P.** (1953). "Calculation of High Speed Valve Motion with a Flexible Overhead Linkage." SAE Transactions, 61, pp. 687-700.

10

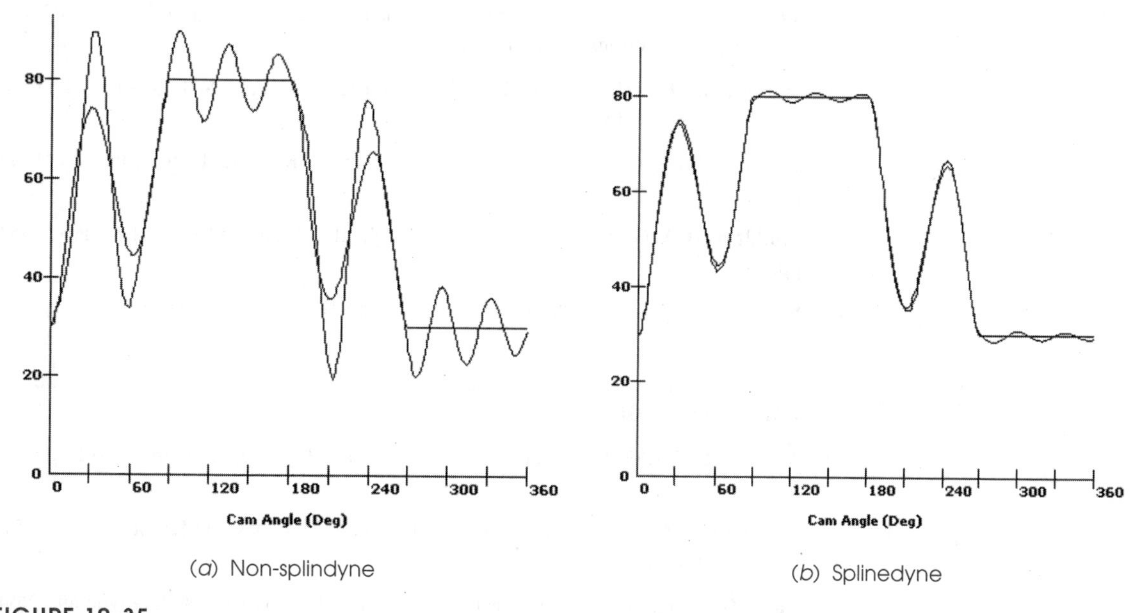

FIGURE 10-35

Kinetostatic and dynamic force for a splinedyne and non-splinedyne cam made to the same motion program

2 **Koster, M. P.** (1974). *Vibrations of Cam Mechanisms*. Phillips Technical Library Series, Macmillan Press Ltd.: London.

3 **Akiba, K., and H. Sakai.** (1981). "A Comprehensive Simulation of High Speed Driven Valve Trains." 810865, SAE.

4 **Dresner, T. L., and P. Barkan.** (1995). "New Methods for the Dynamic Analysis of Flexible Single-Input and Multi-Input Cam-Follower Systems." Journal of Mechanical Design, 117(1), p. 151.

5 **Koster, M. P.** (1974). *Vibrations of Cam Mechanisms*. Phillips Technical Library Series, Macmillan Press Ltd.: London, p. 172.

6 **Seidlitz, S.** (1989). "Valve Train Dynamics - A Computer Study." 890620, SAE.

7 **Ibid**, p. 11.

8 **Pisano, A. P.** (1981). "The Analytical Development and Experimental Verification of a Predictive Model of a High-Speed Cam-Follower System." Ph.D. Thesis, Columbia University.

9 **Ibid**, p. 127.

10 **Wiederrich, J. L.** (1973). "Design of Cam Profiles for Systems with High Inertial Loading." Ph.D. Thesis, Stanford University.

11 **Norton, R. L.** (2000). Machine Design: An Integrated Approach. 2ed, Prentice-Hall: Upper Saddle River, NJ, pp. 111-116.

12 **Thoren, T. R., et al.** (1952). "Cam Design as Related to Valve Train Dynamics." SAE Quarterly Transactions, 6(1), pp. 1-14.

13 **Dudley, W. M.** (1948). "New Methods in Valve Cam Design." SAE Quarterly Transactions, 2(1), pp. 19-33.

14 **Stoddart, D. A.** (1953). "Polydyne Cam Design-I." Machine Design, January, 1953, pp. 121-135.

15 **Stoddart, D. A.** (1953). "Polydyne Cam Design-II." Machine Design, February, 1953, pp. 146-154.

16 **Stoddart, D. A.** (1953). "Polydyne Cam Design-III." Machine Design, March, 1953, pp. 149-164.

17 **Norton, R. L., et al.** (1999). "Effect of Valve-Cam Ramps on Valve Train Dynamics." SAE 1999-01-0801.

18 **Mosier, R. G.** (2000). "Modern Cam Design." Int. J. Vehicle Design, 23(1/2), pp. 38-55.

19 **Peisekah, E. E.** (1966). "Improving the Polydyne Cam Design Method." Russian Engineering Journal, XLVI(12), pp. 25-27.

20 **Fawcett, G. F., and J. N. Fawcett**, eds. (1978). "Comparison of Polydyne and Non-Polydyne Cams." Cams and Cam Mechanisms, Jones, J. R., ed., Institution of Mechanical Engineers: London.

21 **Press, W. H., et al.** (1986). *Numerical Recipes: The Art of Scientific Computing*. Cambridge University Press: Cambridge, pp. 547-577.

Chapter 11

RESIDUAL VIBRATIONS IN CAM-FOLLOWER SYSTEMS

11.0 INTRODUCTION

The dynamic models of cam-follower systems in Chapter 10 demonstrate that the actual motion of any real follower system consists of the motion requested by the cam function plus some dynamic oscillation superposed on the output. This chapter will explore the phenomena of vibration, particularly residual vibrations, in these systems and discuss methods to minimize them.

11.1 RESIDUAL VIBRATION

Consider a double-dwell cam design. Most such applications in production machinery are designed to move the follower to a precise position and then hold it there, absolutely still and in an accurate, repeatable location so that the production operation can be done on the workpiece. In some machines, an end position accuracy of 0.001 or 0.0005 in is required. If the actual motion of the follower during the rise (or fall) does not exactly match the theoretical cam function (assuming no follower bounce), it might not matter as long as the endpoint is reached at the right time and position. However, if the follower is vibrating during the rise because of the dynamics of the system, it will continue to vibrate freely during the dwell. It is this residual vibration that causes problems in respect to workpiece accuracy. Errors in workpiece position during pickup of new parts from feed-rails, insertion of new parts to the assembly, the welding of components, and even automatic inspection operations, etc., can lead directly to an increase in scrap rate. The first three listed cause real scrap. The fourth can cause false scrap by fooling the inspection algorithm that expects a stationary workpiece.

Each rise or fall cycle provides an additional excitation to the system, ringing it again and precipitating residual, free vibrations during the dwells. Damping in the system will cause these free vibrations to die out. But as was demonstrated in earlier chapters, these systems are typically very underdamped with damping ratios in the range of 0.05 to 0.15, at most. Thus, if the dwell is short, the system may still be vibrating when the next rise/fall event arrives. This situation can lead to severe vibration if the phase of the residual vibrations with respect to the forced vibration of the rise/fall event are such as to add rather than cancel. Koster[1] reports from his extensive testing that situations in which the dwell time was of the same order as the rise time appeared to allow the residual vibrations to damp out before the end of the dwell, even with the typically low damping present. However, the wise designer will avoid dwells so short as to precipitate this reinforcement of vibration. This can be checked by using the logarithmic decrement technique described in Section 9.3 to determine a suitable damping ratio for the system and then calculating the length of dwell needed at design speed to allow die-out of the residual vibration.

11.2 RESIDUAL VIBRATION OF DOUBLE-DWELL FUNCTIONS

One strategy to reduce the residual vibration problem is to choose a function for the rise (or fall) that generates lower levels of residual vibration than others. Neklutin was the first to suggest this approach.[2] It is possible to compare the residual vibration levels of rise/fall functions on the assumption of zero damping in the system. This greatly simplifies the mathematics with no loss of generality as the goal is to find functions with inherently lower residual vibration. The presence of damping will have the same lowering effect on any one of them and be all to the good. It also simplifies the mathematics to consider a form-closed cam joint to eliminate the physical follower spring. This does not reduce generality; the inclusion of a force-closed spring merely changes the effective spring rate of the (nonjumping) system and thus alters its first natural frequency.

Mercer and Holowenko[3] were the first to derive the equations to generalize Neklutin's analysis technique and we will follow their approach. Figure 11-1 shows the simple model of an undamped, force-closed system. All the follower mass is lumped as m and the effective spring rate of the follower train is k. The cam input displacement is $s(t)$ and the follower motion is $x(t)$. The relative displacement between cam and follower (across the "spring" of the follower train) that we are interested in is $y(t)$. The equation of motion is:

$$m\ddot{x} = -ky \tag{11.1a}$$

The absolute position x of the follower mass is:

$$x = s + y \tag{11.1b}$$

Substitute equation 11.1b in equation 11.1a to eliminate x.

$$m(\ddot{s} + \ddot{y}) = -ky$$

$$\ddot{y} + \frac{k}{m}y = -m\ddot{s} \tag{11.1c}$$

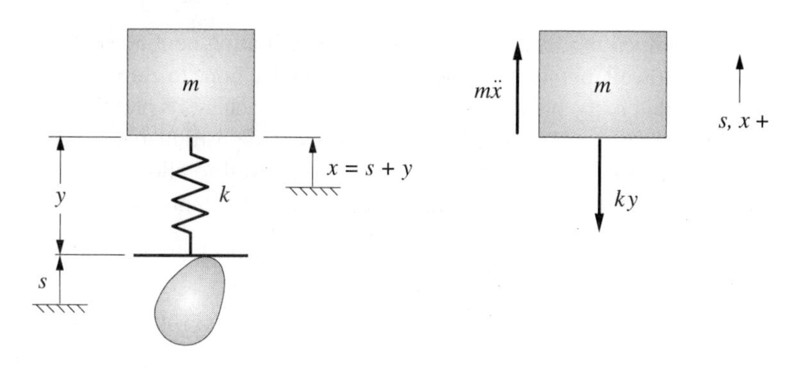

FIGURE 11-1

Mercer and Holowenko's model of an undamped, form-closed cam-follower system

For initial conditions of $y = \dot{y} = 0$ at $t = 0$, the solution of equation 11.1c using Duhamel's integral is

$$y = \frac{1}{\omega_n}\left[\left(\sin\omega_n t \int_0^t -\ddot{s}\cos\omega_n \tau \, d\tau\right) - \left(\cos\omega_n t \int_0^t -\ddot{s}\sin\omega_n \tau \, d\tau\right)\right] \tag{11.2a}$$

where τ is a dummy integration variable. Note that the residual vibration is a function of only the acceleration of the cam shoe $\ddot{s}$ and the circular natural frequency of the system $\omega_n^2 = k/m$.

The solution of equation 11.2a requires evaluating the two definite integrals:

$$A = \frac{1}{\omega_n}\int_0^t \ddot{s}\sin\omega_n \tau \, d\tau$$
$$B = -\frac{1}{\omega_n}\int_0^t \ddot{s}\cos\omega_n \tau \, d\tau \tag{11.2b}$$

in order to obtain a trigonometric result for y as:

$$y = A\cos\omega_n t + B\sin\omega_n t \tag{11.2c}$$

Note that A and B are functions of ω_n and t, so the residual displacement of the follower will vary with both time and natural frequency during the rise when $\ddot{s}$ is nonzero. However, during the dwell period, $\ddot{s} = 0$; then A and B are independent of t. Thus, at the beginning of the dwell, the follower starts a free vibration at its natural frequency with an amplitude of:

$$y_{max} = \sqrt{A^2 + B^2} \tag{11.3a}$$

seeded by the accumulation of the effects of acceleration over the entire rise interval t. For a given cam function acceleration and rise time, this residual vibration amplitude is a function only of the follower natural frequency.

11

This undamped model assumes that the free vibration continues at this amplitude throughout the dwell, but in reality, it will attenuate (we hope to zero), before the start of the next cycle due to the presence of damping. Nevertheless, its maximum amplitude at the beginning of the dwell may be the culprit in errors of position that increase scrap rate. If one delays the workpiece event until later in the dwell to allow some attenuation, it will ultimately cost in terms of throughput by effectively lengthening the cycle time.

Vibration amplitude is not a direct analog of follower force but can be made so by introducing follower stiffness.

$$F_{max} = k y_{max} = k\sqrt{A^2 + B^2} \tag{11.3b}$$

This can be divided by the peak theoretical inertial force on a rigid follower

$$\frac{F_{max}}{m\ddot{s}_{max}} = \frac{k\sqrt{A^2 + B^2}}{m\ddot{s}_{max}} = \frac{\omega_n^2\sqrt{A^2 + B^2}}{\ddot{s}_{max}} = \frac{F_R}{\ddot{s}_{max}}$$

$$F_R = \omega_n^2\sqrt{A^2 + B^2} \tag{11.3c}$$

to create a dimensionless force ratio F_R, the ratio of actual to theoretical cam follower force for a peak cam acceleration of unity. So to determine the actual peak force for any cam, multiply F_R for that system by the actual peak acceleration during the rise.

To compute the value of F_R for any given system requires an estimate of the system's fundamental natural frequency ω_n. It also requires evaluation of the integrals in equation 11.2b to obtain A and B. These definite integrals are best evaluated numerically using, e.g., Simpson's rule. They each contain the cam acceleration function multiplied by a trigonometric function (sin or cos) of the natural frequency of the system multiplied by time. Since the peak value of acceleration increases nonlinearly with reduction in rise time, this integral will increase with shorter rise times, giving larger absolute F_R for the same system ω_n.

To see the effect of changing ω_n for a given rise time, it is easiest to normalize the calculation. Define the rise time period as T_r and, for the following analysis consider it to be unity. The period of the natural frequency of the system is

$$T_n = \frac{1}{\omega_n} \tag{11.4a}$$

Define a "frequency" ratio:

$$\lambda = \frac{T_r}{T_n} \tag{11.4b}$$

When $\lambda = 1$, the system is in resonance, and we should expect a large response. The larger λ becomes, the smaller F_R will be. Note that this ratio has a reciprocal meaning compared to the ratio of forcing frequency to natural frequency ω_f / ω_n of Section 9.2 (p. 224) and Figure 9-6 (p. 225) where ω_f is the rotational forcing frequency of the camshaft. But a difference is that λ involves the "frequency" ($1/T_r$) of just the rise time rather than of the entire cam cycle. Also note that λ is the reciprocal of Koster's[1] factor τ, also introduced in Section 9.2. The largest possible λ ratio (smallest τ) is desirable.

The value of F_R versus λ was calculated for a variety of standard double-dwell acceleration functions over a range $0 \le \lambda \le 20$ with the peak acceleration normalized to 1 and a rise time of 1. The results are shown in Figure 11-2, plotted at a uniform scale, and arranged in order of decreasing average magnitude of their residual vibration spectra.

The parabolic displacement (constant acceleration) function that was shown in Chapter 2 to be an unacceptable choice confirms that conclusion with a residual vibration spectrum that remains at 4 times the nominal acceleration at all odd multiples of the rise time. It also has zeros at all even harmonics, since it is an odd function. So, if you could guarantee that the system would operate at a speed that had a rise time that was an even submultiple of the natural frequency, you could expect zero residual vibration. But move away from those ratios and you will have the largest possible residual vibration amplitudes of any double-dwell function. Parabolic motion is still unacceptable.

The simple harmonic motion function was also shown in Chapter 2 to be unacceptable for the double-dwell case. This is confirmed by its residual vibration spectrum, which remains at a constant peak level (at odd harmonics) of about twice the nominal level.

Next is the modified trapezoid acceleration whose magnitude decays fairly rapidly with frequency ratio, but even system natural frequencies up to 5 times that of the rise time show a gain over the nominal value. At frequency ratios greater than about 9, it has very low residual vibration. The modified sine is lower in magnitude than the modified trapezoid at low frequency ratios and decays exponentially to similarly low values above $\lambda = 9$.

The 3-4-5 polynomial displacement function has lower residual vibration magnitudes at all but the resonant frequency (at which you absolutely do not want to operate anyway) than either the modified trapezoid or the modified sine. Of the six classical functions shown in this figure, the cycloidal displacement (sine acceleration) shows the lowest residual vibration across the spectrum. This should not be surprising since it has the lowest acceleration frequency content of this set of functions, being a pure sinusoid.

Note that all of these functions exhibit "holes in the spectrum" (zeros) at particular frequency ratios. This is typical, and for symmetrical functions (those that pass through $h/2$ with $v = \max$ and $a = 0$ at $T_r/2$), the zeros will occur at even frequency ratios. Non-symmetrical functions will not have true zeros in the residual vibration spectrum, but will come to low values at similar intervals. So, if there is any flexibility in machine speed, it may be possible to adjust the operating speed to take advantage of these "holes."*

11.3 DOUBLE-DWELL FUNCTIONS FOR LOW RESIDUAL VIBRATION

Once the relationship between the harmonic content of the cam acceleration function and the natural frequencies of the follower system was recognized, many researchers devised cam functions to reduce the harmonic content. One method was to use a Fourier series representation of the function to control the harmonic content. The first to propose such an approach was Freudenstein,[4] who derived several double-dwell functions that were intended to reduce the F_R ratio across the frequency ratio spectrum. He limited the highest term in the Fourier series approximation of the acceleration to respectively, n = 1, 3, 5, 7, and 9 harmonics. When n = 1, this technique yields the cycloidal function. One of

* For a practical example of this, see the second case study in Chapter 17 (p. 511).

11

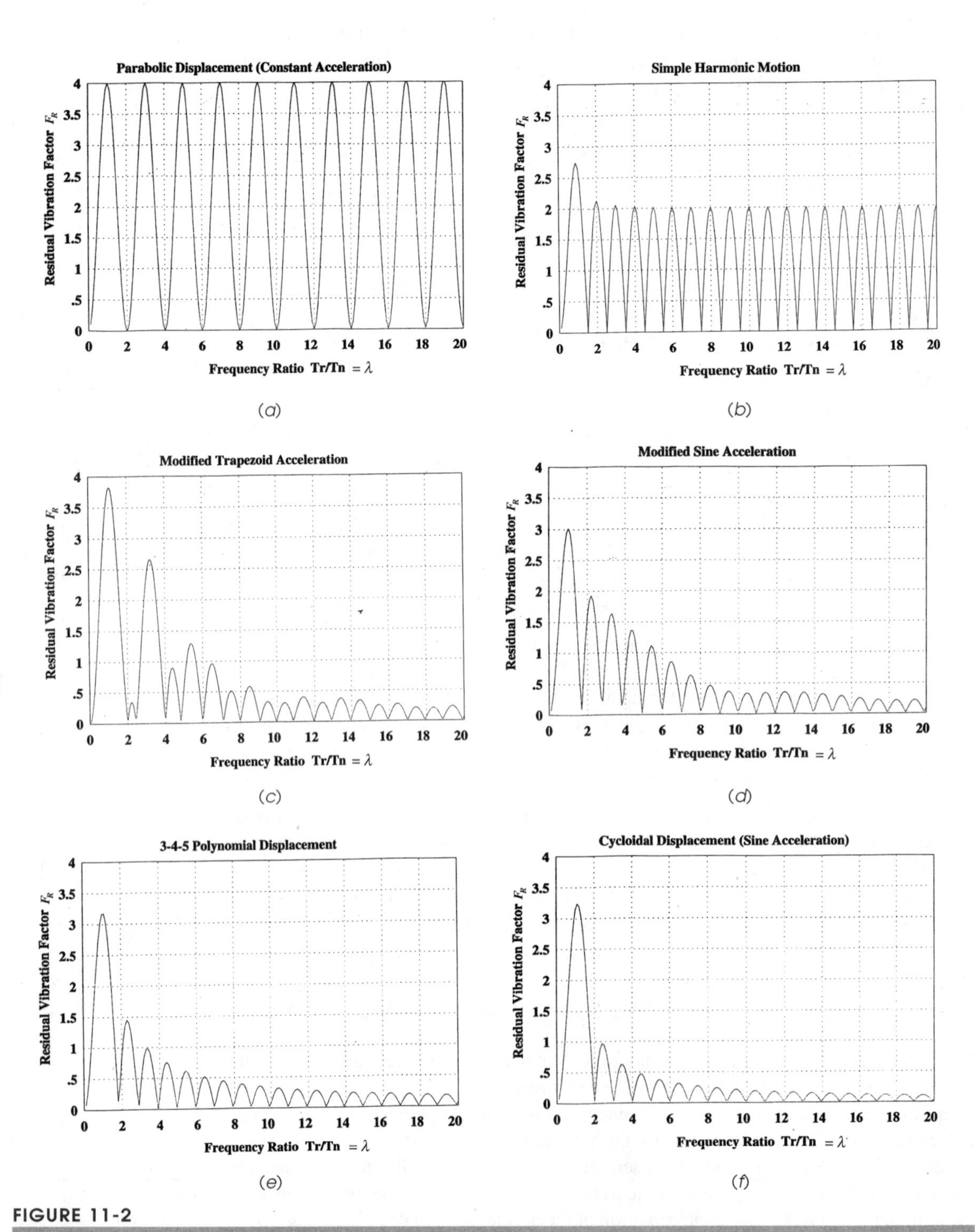

FIGURE 11-2

Dimensionless force ratio F_R for residual vibration levels in six standard double-dwell cam functions

the others, the so-called Freudenstein 1-3 that retains only the 1st and 3rd harmonics will be presented here.

Freudenstein 1-3 Fourier Series (Harmonic) Function

The equations for this function are:

$$s = h\left\{\frac{\theta}{\beta} - \frac{1}{2\pi}\left[\frac{27}{28}\sin\left(2\pi\frac{\theta}{\beta}\right) + \frac{1}{84}\sin\left(6\pi\frac{\theta}{\beta}\right)\right]\right\} \tag{11.5a}$$

$$v = \frac{h}{\beta}\left[1 - \frac{27}{28}\cos\left(2\pi\frac{\theta}{\beta}\right) - \frac{1}{28}\cos\left(6\pi\frac{\theta}{\beta}\right)\right] \tag{11.5b}$$

$$a = 2\pi\frac{h}{\beta^2}\left[\frac{27}{28}\sin\left(2\pi\frac{\theta}{\beta}\right) + \frac{3}{28}\sin\left(6\pi\frac{\theta}{\beta}\right)\right] \tag{11.5c}$$

$$j = 4\pi^2\frac{h}{\beta^3}\left[\frac{27}{28}\cos\left(2\pi\frac{\theta}{\beta}\right) + \frac{9}{28}\cos\left(6\pi\frac{\theta}{\beta}\right)\right] \tag{11.5d}$$

This function has a peak velocity factor $C_v = 2.0$ and a peak acceleration factor $C_a = 5.39$. Its *s, v, a, j*, and *p* diagrams are shown in Figure 11-3. Note the similarity of its acceleration waveshape to that of the modified trapezoid function of Figure 3-5 (p. 35).

Gutman F-3 Fourier Series (Harmonic) Function

Gutman[5] also proposed using a truncated Fourier series to derive double-dwell functions while controlling their harmonic content. He started with the parabolic displacement (constant acceleration) function because of its low theoretical peak acceleration. He recognized that its spectrum contained high harmonic content, and so modified it by truncating its Fourier series approximation to the first two terms, calling this the Gutman F-3 function. Its equations are:

$$s = h\left\{\frac{\theta}{\beta} - \frac{1}{2\pi}\left[\frac{15}{16}\sin\left(2\pi\frac{\theta}{\beta}\right) + \frac{1}{48}\sin\left(6\pi\frac{\theta}{\beta}\right)\right]\right\} \tag{11.6a}$$

$$v = \frac{h}{\beta}\left[1 - \frac{15}{16}\cos\left(2\pi\frac{\theta}{\beta}\right) - \frac{1}{16}\cos\left(6\pi\frac{\theta}{\beta}\right)\right] \tag{11.6b}$$

$$a = 2\pi\frac{h}{\beta^2}\left[\frac{15}{16}\sin\left(2\pi\frac{\theta}{\beta}\right) + \frac{3}{16}\sin\left(6\pi\frac{\theta}{\beta}\right)\right] \tag{11.6c}$$

$$j = 4\pi^2\frac{h}{\beta^3}\left[\frac{15}{16}\cos\left(2\pi\frac{\theta}{\beta}\right) + \frac{9}{16}\cos\left(6\pi\frac{\theta}{\beta}\right)\right] \tag{11.6d}$$

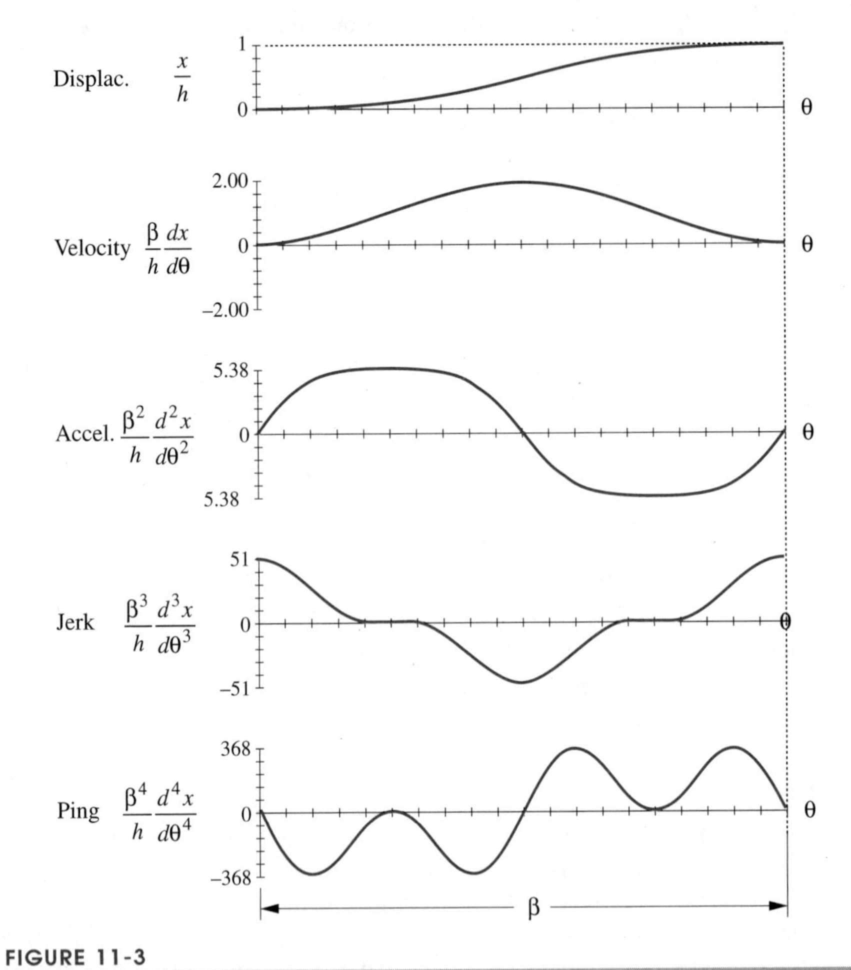

FIGURE 11-3

The Freudenstein 1-3 harmonic function for double-dwell motion

Its *s*, *v*, *a*, *j*, and *p* diagrams are shown in Figure 11-4. Note the similarity to the Freudenstein 1-3 function. The Gutman F-3 function has a peak velocity factor $C_v = 2.0$ and a peak acceleration factor $C_a = 5.15$.

Berzak-Freudenstein Polynomials

Kanzaki and Itao[6] developed functions to minimize residual vibrations at multiple speeds in high-speed printing equipment. Berzak and Freudenstein[7] optimized Kanzaki and Itao's approach to minimize peak velocity, peak acceleration, and residual vibrations in the general double-dwell case.

Berzak and Freudenstein used a simple SDOF dynamic model that contains one damper as shown in Figure 11-5. The nondimensionalized equation of motion for this model is derived in [7] as:

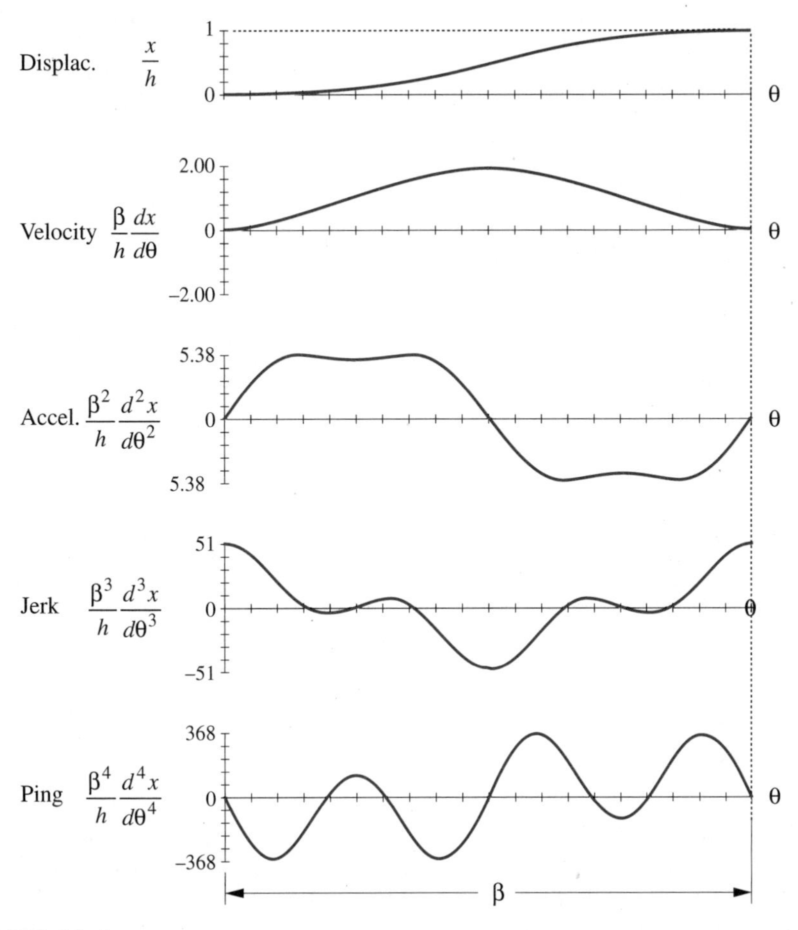

FIGURE 11-4

The Gutman F3 harmonic function for double-dwell motion

$$\ddot{X}+2\zeta(2\pi\lambda)\dot{X}+(2\pi\lambda)^2 X=(2\pi\lambda)^2 S \tag{11.7}$$

where:

$X=x/h$ $h=$ maximum follower displacement (length)

$S=s/h_c$ $h_c=$ maximum cam displacement (length)

$\zeta=$ damping ratio $=c/(2m\omega_n)$

$\omega_n=\sqrt{(k_1+k_2)/m}$

$t=$ time (sec)

$T_n=1/w_n$ fundamental period of vibration of follower system (sec)

$T_r=$ rise (or fall) time of cam (sec)

$\lambda=T_r/T_n$ ratio of rise time to fundamental vibration period

$\tau=t/T_r$ dimensionless time

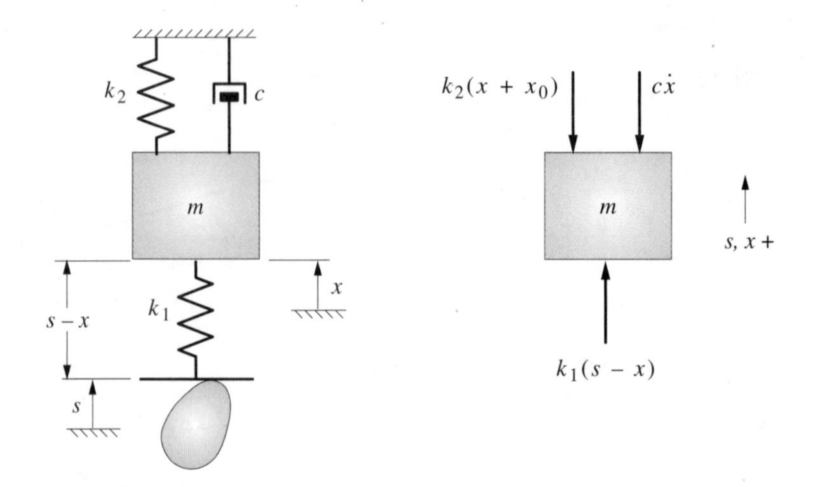

FIGURE 11-5

Berzak and Freudenstein's SDOF dynamic model and free body diagram

In this nondimensional form of the system equation, the rise is from 0 to 1 and the range of the independent variable τ is also 0 to 1. The dot denotes differentiation with respect to dimensionless time τ.

11

Using this model they derived five polynomial functions to give low residual vibrations in the double dwell rise or fall case. Three of these are 3-5-6-7 polynomials and two are 3-4-5-6-7 polynomials. Of these, the two 3-4-5-6-7 polynomial functions have the best properties. We will call them, respectively, the Berzak-D and Berzak-E polynomials. Both have the general form:

$$x = h\left[C_3\left(\frac{\theta}{\beta}\right)^3 + C_4\left(\frac{\theta}{\beta}\right)^4 + C_5\left(\frac{\theta}{\beta}\right)^5 + C_6\left(\frac{\theta}{\beta}\right)^6 + C_7\left(\frac{\theta}{\beta}\right)^7\right] \tag{11.8a}$$

$$\dot{x} = \frac{h}{\beta}\left[3C_3\left(\frac{\theta}{\beta}\right)^2 + 4C_4\left(\frac{\theta}{\beta}\right)^3 + 5C_5\left(\frac{\theta}{\beta}\right)^4 + 6C_6\left(\frac{\theta}{\beta}\right)^5 + 7C_7\left(\frac{\theta}{\beta}\right)^6\right] \tag{11.8b}$$

$$\ddot{x} = \frac{h}{\beta^2}\left[6C_3\left(\frac{\theta}{\beta}\right) + 12C_4\left(\frac{\theta}{\beta}\right)^2 + 20C_5\left(\frac{\theta}{\beta}\right)^3 + 30C_6\left(\frac{\theta}{\beta}\right)^4 + 42C_7\left(\frac{\theta}{\beta}\right)^5\right] \tag{11.8c}$$

$$\dddot{x} = \frac{h}{\beta^3}\left[6C_3 + 24C_4\left(\frac{\theta}{\beta}\right) + 60C_5\left(\frac{\theta}{\beta}\right)^2 + 120C_6\left(\frac{\theta}{\beta}\right)^3 + 210C_7\left(\frac{\theta}{\beta}\right)^4\right] \tag{11.8d}$$

Figure 11-6 (p. 326) shows the velocity, acceleration and jerk functions for these two polynomials. The coefficients $C_3 - C_7$ are shown in Table 11-1, which also shows the peak velocity, acceleration, and jerk factors for these functions. Compare these values to those for the standard double-dwell functions in Table 3-2 (p. 50). These two functions compare quite favorably with the modified sine, 3-4-5 polynomial, and cycloidal functions on the basis of theoretical peak velocity and acceleration. The Berzak-D polynomial has slightly lower peak velocity and acceleration than the 3-4-5 polynomial and is almost as low on both counts as the modified sine.

In addition, these functions offer the bonus of vibration cancellation at particular ratios of rise time to follower fundamental period as shown in Table 11-1 and Figure 11-7. At other frequency ratios their residual vibration magnitudes are low, as shown in Table 11-1 and Figure 11-7 (p. 327), which shows a plot of the absolute value of the residual vibration as a function of λ. There are multiple zeros. The first occur at $\lambda = 2.85$ for the Berzak-D and at 3.075 for the Berzak-E. These functions will only cancel vibrations completely if the follower system happens to operate at one of the values of λ that corresponds to a zero. Lambda is the ratio of cam rise time and follower system fundamental period. From a design standpoint, one may not have a great deal of control over rise time if it is dictated by the machine's timing diagram and speed. If there is any design flexibility in that regard, then altering the cam rise time to achieve the desired ratio versus follower period may be appropriate. It also may be possible to alter the natural frequency of the follower train to achieve the desired ratio λ for a given rise time. The natural frequency of the follower system is determined entirely by the masses and spring rates of the system as shown in the models of this chapter. Designing the parts' geometries to maximize stiffness with minimum mass is the correct strategy to increase ω_n and reduce T_n. Reducing ω_n to increase T_n is simple—just add mass, but watch the stresses—they will increase due to the larger inertial loading. This same design strategy can be applied to any cam function based on its residual vibration spectrum, as shown in Figures 11-2 (p. 320) and 11-8 (p. 328).

Residual Vibration Spectra for "Low Vibration" Functions

Figure 11-8 (p. 328) shows F_R as a function of frequency ratio for the Freudenstein 1-3, Gutman F-3, Berzak-D, Berzak-E, the 4-5-6-7 polynomial, and the 11th-degree Peisekah polynomial from Chapter 10. All are plotted at a uniform scale. Chen[8] did a similar analysis and comparison of residual vibration spectra of a number of double-dwell functions and reached similar conclusions to those presented here. He included other "low vibration" functions in addition to those shown here, such as the Baranyi,[9] and A.R.

TABLE 11-1 Cofficients for Berzak Double-Dwell Polynomials

Type	Coefficients					Peak Velocity Factor	Peak Accel. Factor	Peak Jerk Factor	Peak Residual Vibration	Residual Vibration is Zero at λ =
	C_3	C_4	C_5	C_6	C_7					
D	12.10	−25.50	24.90	−14.70	4.20	1.8	5.6	73	0.027	2.850
E	5.35	8.20	−35.74	32.46	−9.27	2.0	6.5	41	0.014	3.075

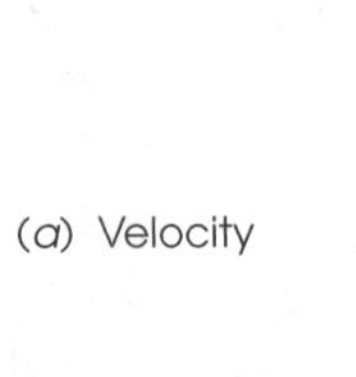

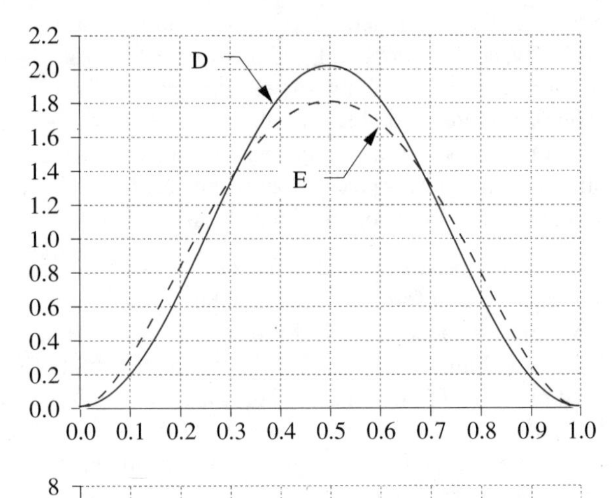

(b) Acceleration

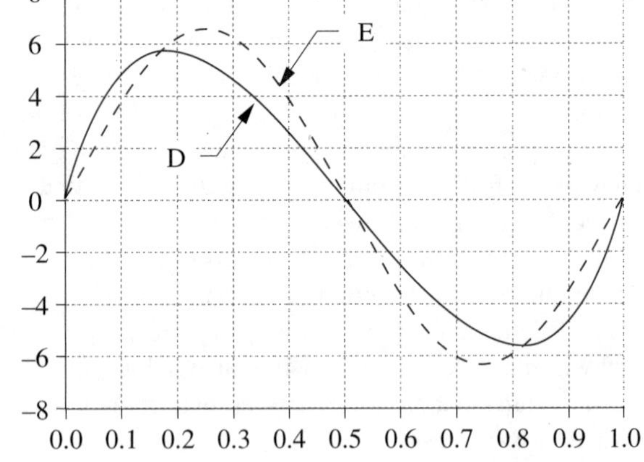

(c) Jerk

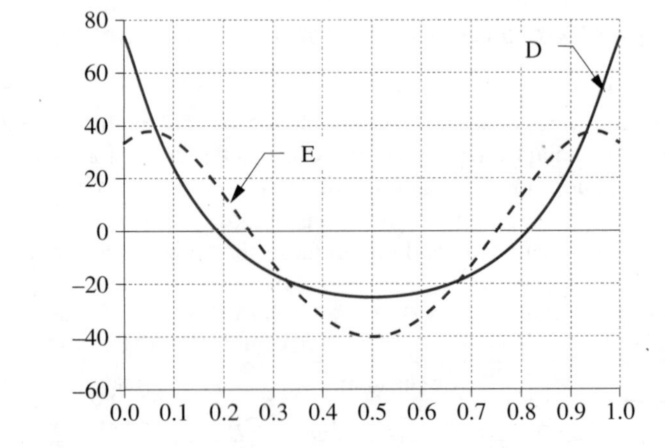

FIGURE 11-6

Berzak-D and Berzak-E polynomials for double-dwell motions

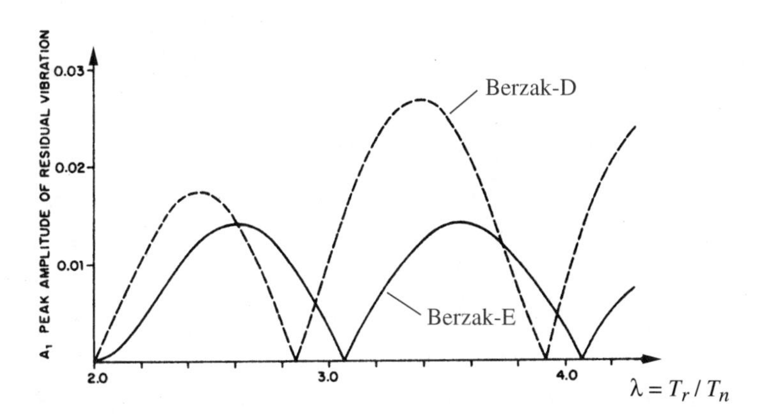

FIGURE 11-7

Residual vibration levels in the Berzak-D and Berzak-E double-dwell polynomial functions (7)

Johnson 15th degree polynomial.[10] These are not superior to those shown here and so are omitted. See Chen[8] for more information.

Table 11-2 (p. 329) shows the 12 double dwell functions whose residual vibration spectra are depicted in Figures 11-2 (p. 320) and 11-8 (p. 328) along with the peak, average, rms, and C_a • rms products for their residual vibration spectra. These are calculated over a bandwidth of frequency ratio from 2 to 20 on the premise that any reasonable design of a cam follower system should be operating at a rise time whose period is (substantially) more than twice the period of the fundamental frequency of the follower system. Therefore, the magnitudes of the residual vibration spectrum below $\lambda = 2$ are irrelevant. The rms averages of the residual vibration spectra from $2 \le \lambda \le 20$ are multiplied by the theoretical acceleration factors C_a of each function to put them on a comparable scale. Table 11-3 (p. 329) shows the same type of data for a frequency ratio bandwidth of 3 to 20, and Table 11-4 (p. 330) shows data for a frequency ratio bandwidth of 4 to 20. In each table, the data are arranged in order of decreasing rms average magnitude of their residual vibration spectra times their C_a factors. These tables can be used to select an appropriate double-dwell function for low-residual vibration, based on the particular ratio of follower-system natural frequency period and rise time.

These rankings are quite interesting. Not surprisingly, the parabolic displacement (constant acceleration) is the worst, and it is taken as 100% in each bandwidth. The simple harmonic motion (SHM) function is nearly as bad, again confirming the tenets of Chapter 2 that infinite jerk is a bad idea. The modified trapezoid, long a mainstay of industrial cams, is the noisiest of the acceptable functions in the lower two bandwidths shown (Tables 11-2 and 11-3 on p. 329). When Neklutin[2]* devised this function in the 1940s to replace the then standard parabolic motion, it was a revolutionary improvement. As with most technology, it has since been surpassed by better solutions. Schmidt's modified sinc function is marginally better than the modified trapezoid in terms of average residual vibration levels when frequency ratios below 4 are included, and continues to be a popular choice (along with the modified trapezoid). However, if low vibration is the goal, the functions at the bottom of the lists in Tables 11-2 to 11-4 (pp. 329-330) de-

* Neklutin's seminal paper of a half-century ago is an absolute gem. He was a pioneer in cam design and anticipated just about everything relevant to cams that has since been discovered.

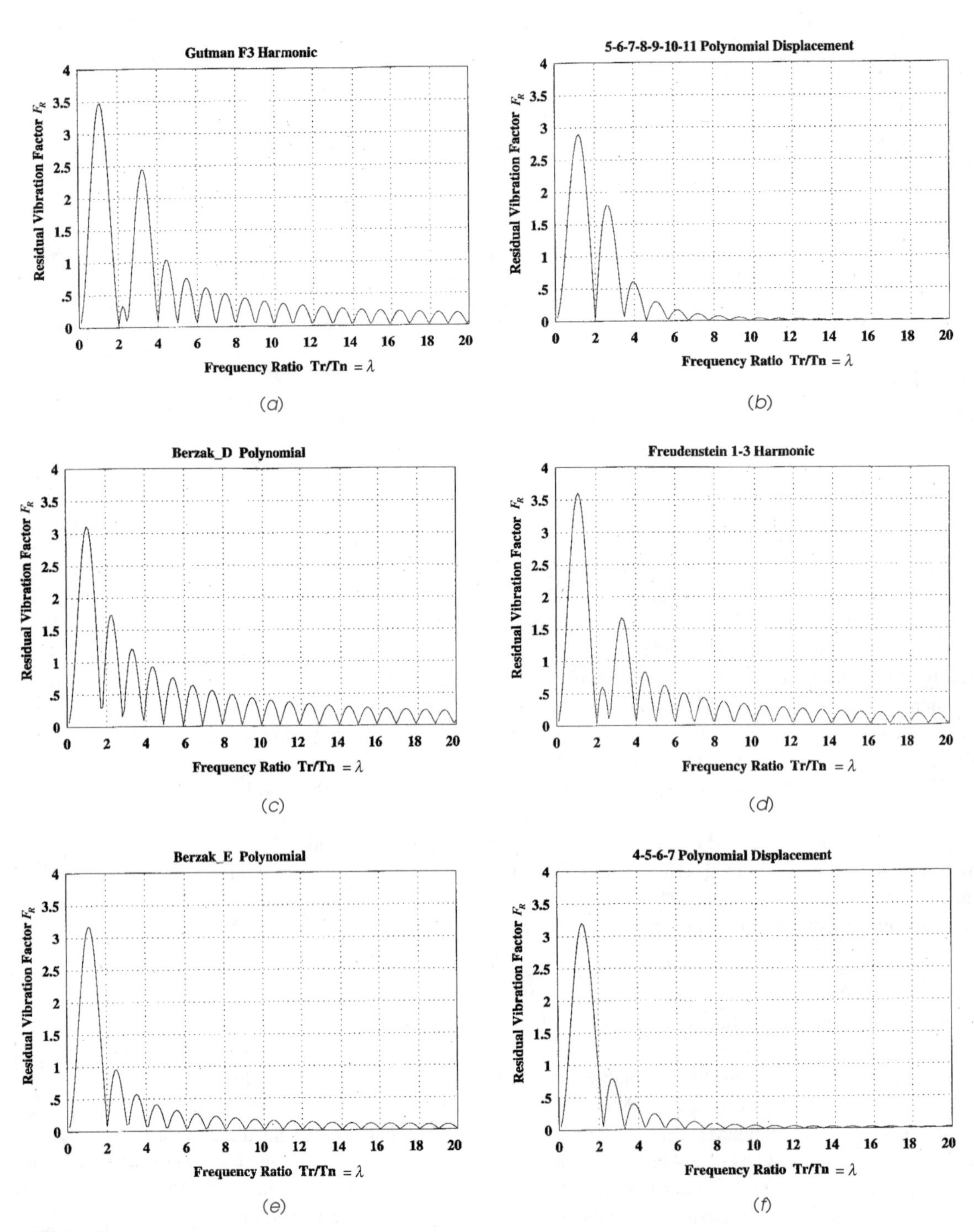

FIGURE 11-8

Dimensionless force ratio F_R for residual vibration levels in six "low-vibration" double-dwell cam functions

TABLE 11-2 Residual Vibrations Over Frequency Ratios of $2 \leq \lambda \leq 20$

Function	C_a	peak	mean	rms	C_a • rms	%
Parabolic	4.00	4.03	2.012	0.183	0.732	100.0%
SHM	4.94	2.13	1.283	0.107	0.528	72.1%
Mod trap	4.89	2.65	0.412	0.049	0.240	32.7%
Mod Sine	5.53	1.93	0.406	0.043	0.238	32.5%
Gutman F3	5.15	2.56	0.373	0.045	0.232	31.7%
11-deg poly	7.91	1.80	0.147	0.028	0.221	30.3%
Berzak-D	5.60	1.73	0.349	0.036	0.202	27.5%
345 poly	5.78	1.45	0.283	0.029	0.168	22.9%
Freudenstein 1-3	5.39	1.67	0.280	0.031	0.167	22.8%
Cycloidal	6.28	0.96	0.175	0.018	0.113	15.4%
Berzak-E	6.50	0.96	0.153	0.017	0.111	15.1%
4567 poly	7.53	0.83	0.088	0.013	0.098	13.4%

serve another look, though two of them (cycloidal and 4-5-6-7 polynomial) have long had bad press due to their relatively high C_a values and their perceived difficulties of manufacture. (Manufacturing issues are addressed in Chapter 14.)

Because in any cam-follower system vibrations increase the dynamic forces beyond the theoretical ones predicted by the classical C_a factors, it seems more sensible to choose a cam function based on inclusion of vibratory effects. Then, the cycloidal, Berzak-E, Peisekah 11th-degree polynomial, and 4-5-6-7 polynomial functions begin to look like good solutions. Of course, one cannot determine if any cam function is truly

TABLE 11-3 Residual Vibrations Over Frequency Ratios of $3 \leq \lambda \leq 20$

Function	C_a	peak	mean	rms	Ca • rms	%
Parabolic	4.00	4.03	2.025	0.189	0.756	100.0%
SHM	4.94	2.06	1.280	0.110	0.543	71.8%
Mod trap	4.89	2.65	0.398	0.048	0.235	31.0%
Gutman F3	5.15	2.56	0.359	0.045	0.232	30.7%
Mod Sine	5.53	1.63	0.359	0.038	0.210	27.8%
Berzak-D	5.60	1.21	0.303	0.030	0.168	22.2%
Freudenstein 1-3	5.39	1.67	0.271	0.031	0.167	22.1%
345 poly	5.78	0.98	0.244	0.024	0.139	18.3%
11-deg poly	7.91	1.32	0.081	0.015	0.119	15.7%
4567 poly	7.53	0.83	0.088	0.013	0.098	12.9%
Cycloidal	6.28	0.62	0.150	0.015	0.094	12.5%
Berzak-E	6.50	0.58	0.125	0.013	0.085	11.2%

TABLE 11-4 Residual Vibrations Over Frequency Ratios of $4 \leq \lambda \leq 20$

Function	C_a	peak	mean	rms	C_a • rms	%
Parabolic	4.00	4.03	2.014	0.195	0.780	100.0%
SHM	4.94	2.03	1.280	0.113	0.558	71.5%
Mod Sine	5.53	1.36	0.315	0.034	0.188	24.1%
Mod trap	4.89	1.28	0.304	0.032	0.156	20.1%
Berzak-D	5.60	0.93	0.273	0.026	0.146	18.7%
Gutman F3	5.15	1.09	0.265	0.027	0.139	17.8%
345 poly	5.78	0.76	0.220	0.021	0.121	15.6%
Freudenstein 1-3	5.39	0.83	0.212	0.021	0.113	14.5%
Cycloidal	6.28	0.47	0.134	0.013	0.082	10.5%
Berzak-E	6.50	0.41	0.109	0.011	0.072	9.2%
11-deg poly	7.91	0.60	0.050	0.009	0.071	9.1%
4567 poly	7.53	0.31	0.045	0.006	0.045	5.8%

viable until a stress analysis of the cam and follower (and related parts) has been completed. Chapter 12 discusses this topic. However, any stress analysis is only as good as the accuracy of the estimates of forces acting on the elements. Vibrations need to be taken into account when determining the dynamic forces on a cam-follower joint, as they can significantly increase force and stress levels.

11

As an example, Figure 11-9 shows the acceleration of a follower as measured in a running assembly machine. This machine has a large number of similar cam-follower trains. This cam is a modified trapezoid double-dwell. Its theoretical waveform is superposed on the measured one. The vibration during the rise and fall events is clearly visible, more than doubling the theoretical peaks. During the dwell between the rise and fall, the residual vibrations can be clearly seen and appear to damp out before the fall motion begins. Note that there is a low level of continual residual vibration in the machine. This is due to other mechanisms within the machine and other machines in the same plant contributing to a general excitation of all machine structures on the same floor. This is typical in any factory. In fact, the background level of vibration in Figure 11-9 is lower than normal since these data were taken during the debug run of the machine when the factory was not in full operation and the machine was running at less than design speed. In the middle of the high dwell, a second impulse can be seen re-exciting the follower. This disturbance is from the start of the conveyor's index motion, transmitted through the chassis to all stations' follower trains. At the beginning of the trace, a larger disturbance can be seen during the low dwell that is the end of index.

Note the difference between the vibratory noise in the first half of the rise event (positive acceleration) and the second half (negative acceleration). The positive pulse shows a classical impulse response of the follower train, decaying exponentially. The second half is more ragged. This is due to interaction between the vibration during the positive pulse that has not fully died out before the imposition of a second step in jerk that begins the transition from positive peak to negative peak acceleration. This new event rings the

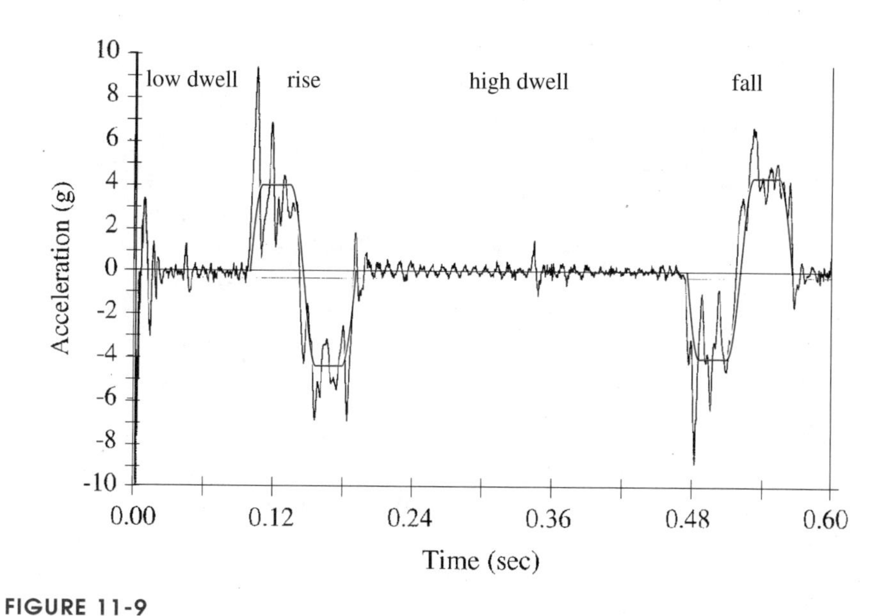

FIGURE 11-9

Acceleration of a modified trapezoid cam follower motion—machine running at 100 rpm

system and its vibrations convolve with whatever vibration is still present of those that were initiated at the start of the rise. The phase relationship of the old and new vibration events causes reinforcement and cancellation, resulting in the distorted vibration of the negative pulse of acceleration. In like fashion, disturbances from external sources such as those seen here during the dwells, also can reinforce the response of the follower system when it is excited by the rise event. This cam was later changed to a polynomial design to reduce its level of residual vibration.

Actual Cam Performance Compared to Theoretical Performance

The relative peak accelerations of several common double-dwell cam functions were discussed in Chapter 3 and summarized in Table 3-2 (p. 50). That discussion also emphasized the importance of a smooth jerk function for minimizing vibrations. The theoretical differences between peak accelerations of different cam functions will be altered by the presence of vibratory noise in the actual acceleration waveforms. This noise will be due in part to errors introduced in the manufacturing process, as discussed above, but there will also be inherent differences due to the degree to which the cam function excites vibrations in the cam-follower train. These vibrations will be heavily influenced by the structural dynamic characteristics of the follower train itself. In general, a very stiff follower train will vibrate less than a flexible one, but the presence of harmonic frequencies of significant magnitude in the cam function that are near the natural frequencies of the follower train will exacerbate the problem.*

Figure 11-10 shows the actual acceleration waveforms of four common double-dwell cam programs—**modified trapezoidal**, **modified sine**, **cycloidal**, and **4-5-6-7 polynomial**—that were ground on the same four-dwell cam. These waveforms were

* For a practical example of this see the second case study in Chapter 17 (p. 511).

11

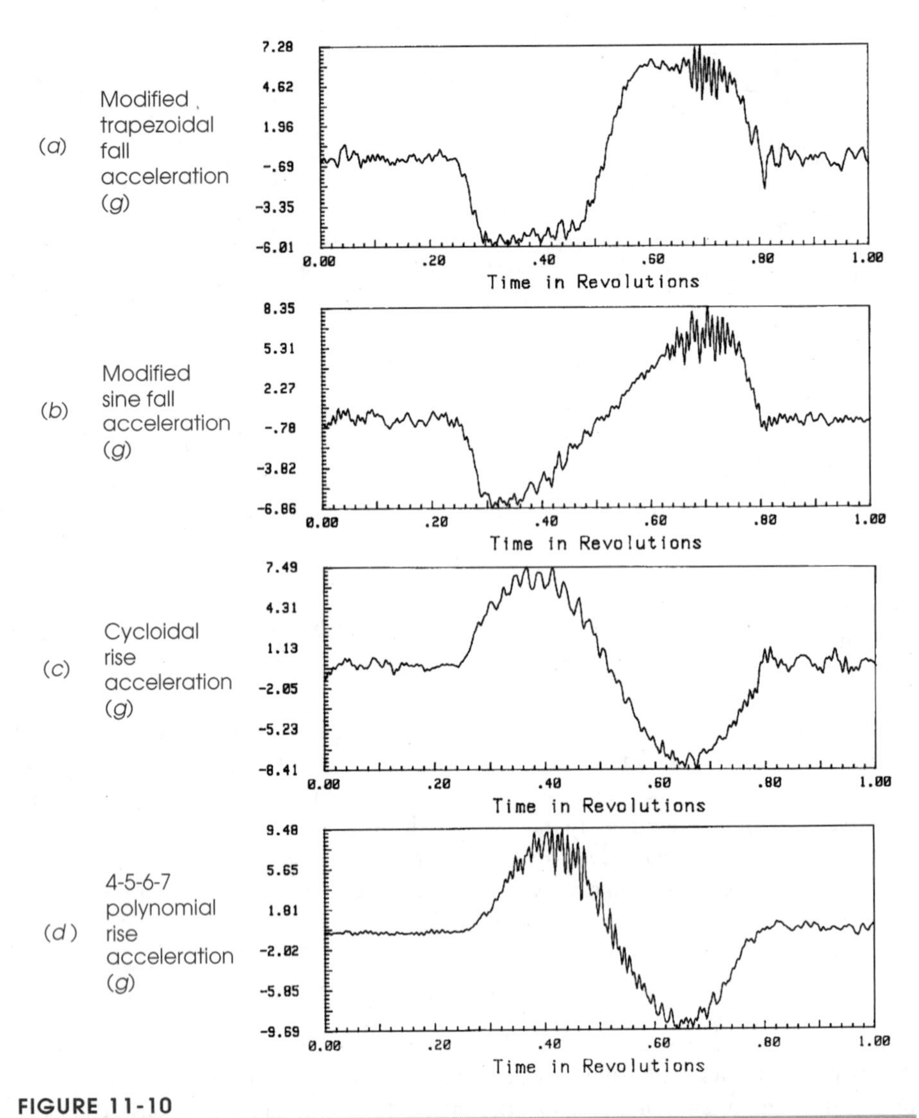

FIGURE 11-10

Actual measured follower acceleration curves of four, double-dwell programs

measured with an accelerometer mounted on the stiff follower train of the cam dynamic test fixture (CDTF) specially designed for low vibration and noise[11] (shown in Figure 14-6, p. 436). These test cams were made by a specialty cam manufacturer using linear interpolation continuous numerical control (CNC)* with a 1/4 degree digitizing increment on very high quality machining and grinding equipment. After hardening, the cam contour was ground to a surface finish with an average roughness (R_a) of 0.125 microns. Comparing these functions on the same cam, made on the same production machinery, and run against the same follower system gives a means to measure their actual differences in relative accelerations.

* See Chapter 14

TABLE 11-5 Actual Peak Acceleration Compared to the Theoretical
for Various Double-Dwell Cam Programs (g)

Cam Program	Theor. A_{pp}, g	Actual A_{pp}, g	Ratio A_{mf} (Eq. 11.9)	Theor.* Accel. Factor	Actual Accel. Factor
Mod trap accel	11.78	13.29	1.13	4.888 h/β^2	5.514 h/β^2
Mod sine accel	13.32	15.21	1.14	5.528 h/β^2	6.313 h/β^2
Cycloidal disp	15.14	15.90	1.05	6.283 h/β^2	6.597 h/β^2
4-5-6-7 poly disp	18.14	19.17	1.06	7.526 h/β^2	7.954 h/β^2

* From Table 3-2 (p. 50)

All four acceleration waveforms have some amount of vibratory noise present that is seen as oscillations or ripples on the curves. Compare these curves to their exact theoretical equivalents in Figure 3-16 (p. 50). Table 11-5 compares the expected theoretical peak-to-peak acceleration $A_{pp_{theoretical}}$ for each cam (column 2) with its actual, measured peak-to-peak acceleration $A_{pp_{actual}}$ (column 3). The error is expressed in column 4 as an acceleration multiplier factor A_{mf} defined as:

$$A_{mf} = \frac{A_{pp_{actual}}}{A_{pp_{theoretical}}} \tag{11.9}$$

Column five in Table 11-5 shows the maximum acceleration factors for these four functions taken from Table 3-2 (p. 50). The last column shows the actual maximum acceleration values based on these test data and is the product of columns four and five.

These values of actual maximum acceleration show smaller differences between the functions than are predicted by their theoretical waveforms. The modified trapezoid, which has the lowest theoretical acceleration, has a 13% noise penalty due to vibration. The modified sine has a 14% noise penalty, while the cycloidal and 4-5-6-7 polynomial functions have only 5-to-6% noise. This is due to the fact that the last two functions have smoother jerk waveforms than the first two. The cycloidal waveform, with its cosine jerk function, is a good choice for high-speed operations. Its acceleration is actually only about 19% greater than the modified trapezoid's and 5% more than the modified sine, rather than the 28% and 14% differentials predicted by the theoretical peak values.

11.4 REFERENCES

1 **Koster, M. P.** (1974). *Vibrations of Cam Mechanisms*. Phillips Technical Library Series, Macmillan Press Ltd.: London.

2 **Neklutin, C. N.** (1952). "Designing Cams for Controlled Inertia and Vibration". *Machine Design*, June, 1952, pp. 143-160.

3 **Mercer, S., and A. R. Holowenko**. (1958). "Dynamic Characteristics of Cam Forms Calculated by the Digital Computer." *Trans ASME*, **80**(8), pp. 1695-1705.

4 **Freudenstein, F.** (1960). "On the Dynamics of High-Speed Cam Profiles." *Int. J. Mech. Sci.*, **1**, pp. 342 - 349.

5 **Gutman, A. S.** (1961). "To Avoid Vibration, Try This New Cam Profile". *Product Engineering*, December 25, 1961.

6 **Kanzaki, K., and K. Itao**. (1972). "Polydyne Cam Mechanisms for Typehead Positioning." *ASME Journal of Engineering for Industry*, **94**(1), pp. 250 - 254.

7 **Berzak, N., and F. Freudenstein**. (1979). "Optimization Criteria in Polydyne Cam Design." *Proc. of Fifth World Congress on Theory of Machines and Mechanisms*, pp. 1303-1310.

8 **Chen, F. Y.** (1981). "Assessment of the Dynamic Quality of a Class of Dwell-Rise-Dwell Cams." *Journal of Mechanical Design*, **103**, pp. 793 - 802.

9 **Baranyi, S. J.** (1970). "Multiple Harmonic Cam Profiles" 70-Mech-59, ASME.

10 **Johnson, A. R.** (1965). "Motion Control for a Series System of N-Degrees of Freedom Using Numerically Derived and Evaluated Equations." *Journal of Engineering for Industry*, **87 Series B**, pp. 191-204.

11 **Norton, R. L.** (1988). "Effect of Manufacturing Method on Dynamic Performance of Double Dwell Cams." *Mechanism and Machine Theory*, **23**(3), pp. 191-208.

Chapter 12

FAILURE OF CAM SYSTEMS—STRESS, WEAR, CORROSION

12.0 INTRODUCTION

There are only three ways in which parts or systems can "fail": *obsolescence, breakage,* or *wearing out*. Most systems are subject to all three types of possible failure. Failure by obsolescence is somewhat arbitrary. Failure by breakage is often sudden and may be permanent. We will not address failure by breakage as it applies to general machine elements such as follower arms, bearings, etc. That topic is well covered in many other references such as [1]. We will address only the failure mechanisms to which the cam-follower interface are typically subject. These fall under the general rubric of wear.

Failure by "wearing out" is generally a gradual process and is sometimes repairable. Ultimately, any system that does not fall victim to one of the other two modes of failure will inevitably wear out if kept in service long enough. Wear is the final mode of failure, that nothing escapes. Thus, we should realize that we cannot design to avoid all types of wear completely, only to postpone them.

Wear is a broad term that encompasses many types of failures, all of which involve changes to the **surface** of the part. Some of these so-called *wear mechanisms* are still not completely understood, and rival theories exist in some cases. Most experts describe five general categories of wear: **adhesive wear**, **abrasive wear**, **erosion, corrosion wear**, and **surface fatigue**. In general, the (noncorrosive) wear rate is inversely proportional to hardness.

The following sections discuss these topics in detail. In addition, there are other types of surface failure that do not fit neatly into one of the five categories or that can fit into more than one. **Corrosion fatigue** has aspects of the last two categories as does **fretting corrosion**. For simplicity, we will discuss these hybrids in concert with one of the five main categories listed above. Table 12-0 shows the variables used in this chapter.

Table 12-0 Variables Used in This Chapter

Symbol	Variable	ips units	SI units
a	half-width of contact patch major axis	in	m
A_a	apparent area of contact	in^2	m^2
A_r	real area of contact	in^2	m^2
B	geometry factor	1/in	1/m
b	half-width of contact patch minor axis	in	m
d	depth of wear	in	m
E	Young's modulus	psi	Pa
F, P	force or load	lb	N
f	friction force	lb	N
f_{max}	maximum tangential force	lb	N
K	wear coefficient	none	none
l	length of linear contact	in	m
L	length of cylindrical contact	in	m
m_1, m_2	material constants	1/psi	m^2/N
H	penetration hardness	psi	kg/mm^2
N	number of cycles	none	none
N_f	safety factor in surface fatigue	none	none
p	pressure in contact patch	psi	Pa
p_{avg}	avg pressure in contact patch	psi	Pa
p_{max}	max pressure in contact patch	psi	Pa
R_1, R_2	radii of curvature	in	m
S_{us}	ultimate shear strength	psi	Pa
S_{ut}	ultimate tensile strength	psi	Pa
S_y	yield strength	psi	Pa
S_{yc}	yield strength in compression	psi	Pa
V	volume	in^3	m^3
x, y, z	generalized length variables	in	m
μ	coefficient of friction	none	none
ν	Poisson's ratio	none	none
σ	normal stress	psi	Pa
σ_1	principal stress	psi	Pa
σ_2	principal stress	psi	Pa
σ_3	principal stress	psi	Pa
τ	shear stress	psi	Pa
τ_{13}	maximum shear stress	psi	Pa
τ_{21}	principal shear stress	psi	Pa
τ_{32}	principal shear stress	psi	Pa

Portions of this chapter were adapted from R. L. Norton, *Machine Design: An Integrated Approach*, 2ed, Prentice-Hall, 2000, with permission.

Failure from wear usually involves the loss of some material from the surf
solid parts in the system. The wear motions of interest are sliding, rolling, or so
bination of both. It only requires the loss of a very small volume of material t
the entire system nonfunctional. Rabinowicz[2] estimates that a 4 000-lb auto
when completely "worn out," will have lost only a few ounces of metal from its w
surfaces. Moreover, these damaged surfaces will not be visible without extensive
sembly, so it is often difficult to monitor and anticipate the effects of wear before
occurs.

12.1 SURFACE GEOMETRY

Before discussing the types of wear mechanisms in detail, it will be useful to define
characteristics of an engineering surface that are relevant to these processes. (Mate
strength and hardness will also be factors in wear.) Most solid surfaces that are subj
to wear in machinery will be either machined, ground, or EDM'd, though some will
as-cast or as-forged. In any case, the surface will have some degree of roughness that
concomitant with its finishing process. Its degree of roughness or smoothness will hav
an effect on both the type and degree of wear that it will experience.

Even an apparently smooth surface will have microscopic irregularities. These can be measured by any of several methods. A profilometer passes a lightly loaded, hard (e.g., diamond) stylus over the surface at controlled (low) velocity and records its undulations. The stylus has a very small (about 0.5 μm) radius tip that acts, in effect, as a low-pass filter, since contours smaller than its radius are not sensed. Nevertheless, it gives a reasonably accurate profile of the surface with a resolution of 0.125 μm or better. Figure 12-1 shows the profiles and SEM* photographs (100X) of both (*a*) ground and (*b*) machined surfaces of hardened steel cams. The profiles were measured with a Hommel T-20 profilometer that digitizes 8 000 data points over the sample length (here 2.5 mm). The microscopic "mountain peaks" on the surfaces are called **asperities**.

From these profiles a number of statistical measures may be calculated. ISO defines at least 19 such parameters. Some of them are shown in Figure 12-2 along with their mathematical definitions. Perhaps the most commonly used parameters are R_a, which is the average of the absolute values of the measured points, and R_q, which is their rms average. These are very similar in both value and meaning. Unfortunately, many engineers specify only one of these two parameters, neither of which tells enough about the surface. For example, the two surfaces shown in Figures 12-3*a* and *b* have the same R_a and R_q values, but are clearly different in nature. One has predominantly positive, and the other predominantly negative, features. These two surfaces will react quite differently to sliding or rolling against another surface.

In order to differentiate these surfaces that have identical R_a or R_q values, other parameters should be calculated. Skewness S_k is a measure of the shape or symmetry of the amplitude distribution of the surface rougness contour. A negative value of S_k indicates that the surface has a predominance of valleys (Figure 12-3*a*) and a positive S_k defines a predominance of peaks (Figure 12-3*b*). Many other parameters can be com-

* Scanning Electron Microscope

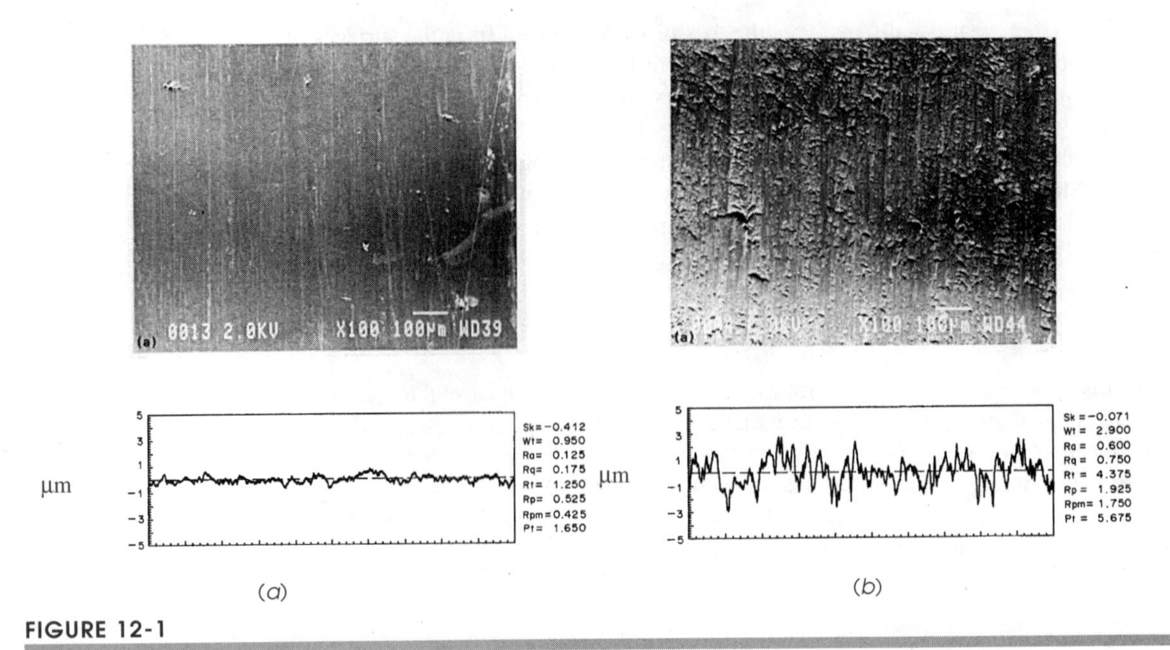

FIGURE 12-1

SEM surface-replica photographs (100x) and profiles of (*a*) ground and (*b*) milled cam surfaces

puted (see Figure 12-2). For example, R_t defines the largest peak-to-valley dimension in the sample length, R_p the largest peak height above the mean line, and R_{pm} the average of the five largest peak heights. All the roughness measurements are calculated from an electronically filtered measurement that zeros out any slow-changing waves in the surface. An average line is computed from which all peak/valley measurements are then made. In addition to these roughness measurements (denoted by R), the waviness W_t of the surface can also be computed. The W_t computation filters out the high-frequency contours and preserves the long-period undulations of the raw surface measurement. If you want to completely characterize the surface-finish condition, note that using only R_a or R_q is not sufficient. A cam surface finish specification should, at a minimum, include limits for R_a, R_t, R_{pm}, S_k, and W_t.

12

12.2 MATING SURFACES

When two surfaces are pressed together under load, their apparent area of contact A_a is easily calculated from geometry, but their real area of contact A_r is affected by the asperities present on their surfaces and is more difficult to accurately determine. Figure 12-4 shows two parts in contact. The tops of the asperities will initially contact the mating part and the initial area of contact will be extremely small. The resulting stresses in the asperities will be very high and can easily exceed the compressive yield strength of the material. As the mating force is increased, the asperity tips will yield and spread until their combined area is sufficient to reduce the average stress to a sustainable level, i.e., some *compressive penetration strength* of the weaker material.

R_a (CLA) (AA) **Arithmetic mean roughness value** DIN 4768 DIN 4762 ISO 4287/1

The arithmetic average value of filtered roughness profile determined from deviations about the centre line within the evaluation length l_m.

$$R_a = \frac{1}{l_m} \int_{x=0}^{x=l_m} |y| dx$$

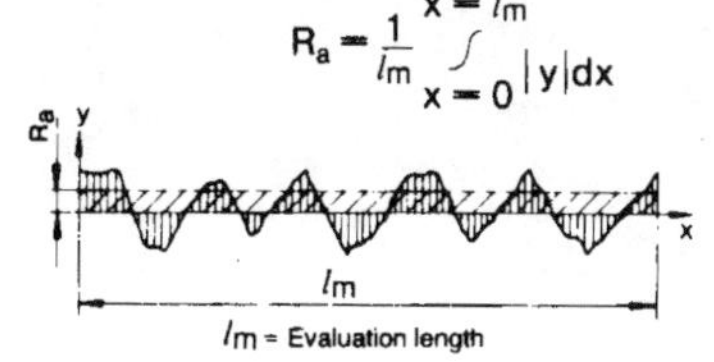

R_q (RMS) **Root mean square roughness value** DIN 4762 ISO 4287/1

The RMS value obtained from the deviations of the filtered roughness profile over the evaluation length l_m.

$$R_q = \sqrt{\frac{1}{l_m} \int_0^{l_m} y^2(x)\, dx}$$

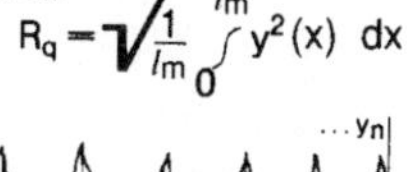

R_{pm} **Mean peak height value above the mean line** DIN 4762

The arithmetic average value of the five single highest peaks above the mean line $R_{p1} - R_{p5}$, similar to the R_z (DIN) definition specified in DIN 4768.
The five highest peaks are determined from the "centre line" of the filtered roughness profile each from a single sampling length l_e.

$$R_{pm} = \frac{1}{5} \cdot (R_{p1} + R_{p2} + \ldots + R_{p5})$$

R_p **Single highest peak above mean line** DIN 4762 ISO 4287 / 1

The value of the highest single peak above the centre line of the filtered profile as obtained from R_{pm}.

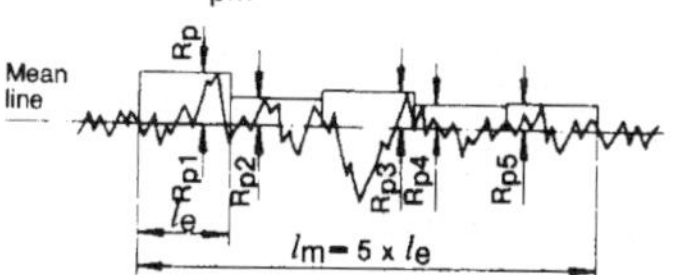

S_k **Skewness of the profile**

A measure of the shape or symmetry of the amplitude distribution curve obtained from the filtered roughness profile. A negative skewness would represent good bearing properties.

Amplitude-distribution curve DIN 4762 ISO 4287 / 1

$$S_k = \frac{1}{R_q^3} \cdot \frac{1}{n} \sum_{i=1}^{i=n} (y_i - \bar{y})^3$$

A graph of the frequency in % of profile amplitudes.

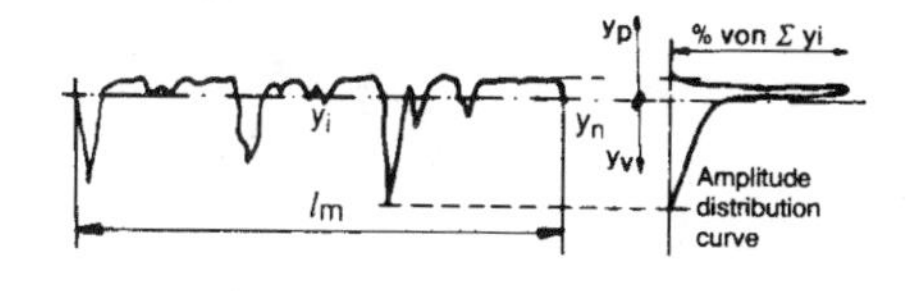

R_t (R_h) (R_d) **Maximum peak-to-valley height** DIN 4762 (1960) since 1978 it is R_{max}.

The maximum peak-to-valley height of the filtered profile over the evaluation length l_m irrespective of the sampling lengths l_e.

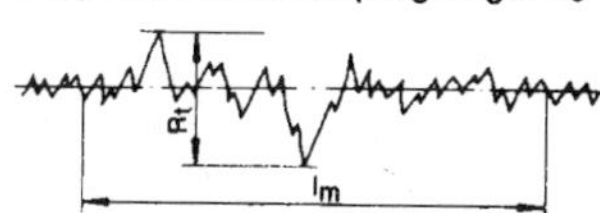

W_t **Waviness depth** DIN 4774

The maximum peak-to-valley height of levelled waviness profile (roughness eliminated) within the evaluation length l_m.

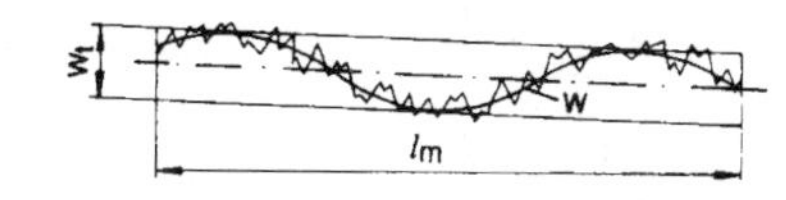

FIGURE 12-2

DIN and ISO surface roughness, waviness, and skewness parameter definitions (*Courtesy of Hommel America Inc.*)

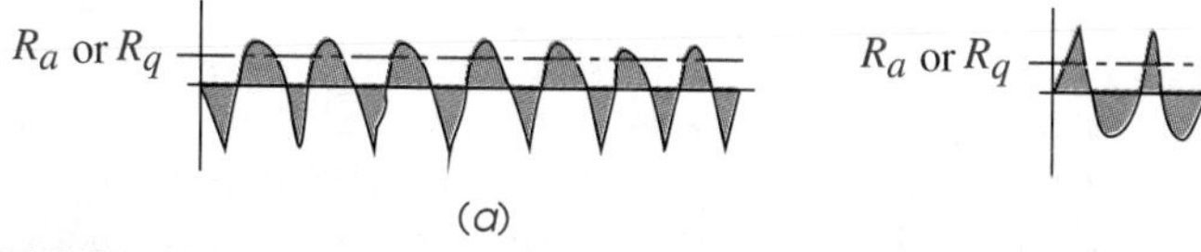

FIGURE 12-3

Different surface contours can have the same R_a or R_q values

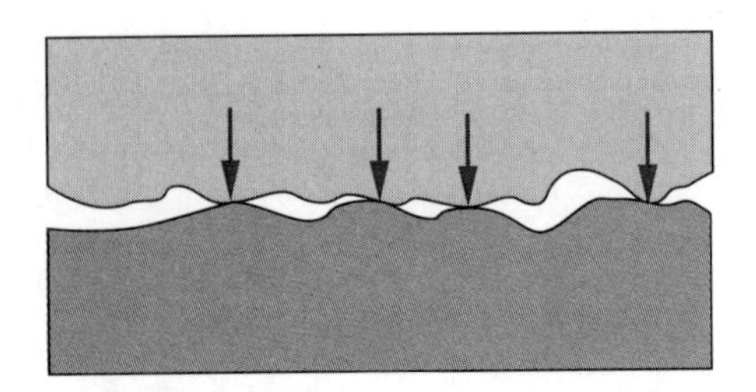

FIGURE 12-4

The actual contact between two surfaces is only at the asperity tips

We can get a measure of a material's compressive penetration strength from conventional hardness tests (Brinell, Rockwell, etc.), that force a very smooth stylus into the material and deform (yield) the material to the stylus' shape. The penetration strength S_p is easily calculated from these test data and tends to be of the order of 3 times the compressive yield strength S_{yc} of most materials.[3]

The real area of contact can then be estimated from

$$A_r \cong \frac{F}{S_p} \cong \frac{F}{3S_{yc}} \tag{12.1}$$

where F is the force applied normal to the surface and the strengths are as defined in the above paragraph, taken for the weaker of the two materials. *Note that the contact area for a material of particular strength under a given load will be the same regardless of the apparent area of the mating surfaces.*

12.3 ADHESIVE WEAR

When (clean) surfaces such as those shown in Figure 12-1 (p. 338) are pressed against one another under load, some of the asperities in contact will tend to adhere to one another due to the attractive forces between the surface atoms of the two materials.[4] As sliding between the surfaces is introduced, these adhesions are broken, either along the original interface or along a new plane through the material of the asperity peak. In the latter case, a piece of part *A* is transferred to part *B*, causing surface disruption and damage. Sometimes, a particle of one material will be broken free and become debris in the interface, which can then scratch the surface and plough furrows in both parts. This damage is sometimes called **scoring** or **scuffing*** of the surface. Figure 12-5 shows an example of a shaft failed by adhesive wear in the absence of adequate lubricant.[5]

The original adhesion theory postulated that all asperity contacts would result in yielding and adhesion due to the high stresses present. It is now believed that in most cases of contact, especially with repeated rubbing, only a small fraction of the asperity contacts actually result in yielding and adhesion; elastic deformations of the asperities also play a significant role in the tractive forces (friction) developed at the interface.[6]

* Note that scuffing is often associated with gear teeth, which typically experience a combination of rolling and sliding.

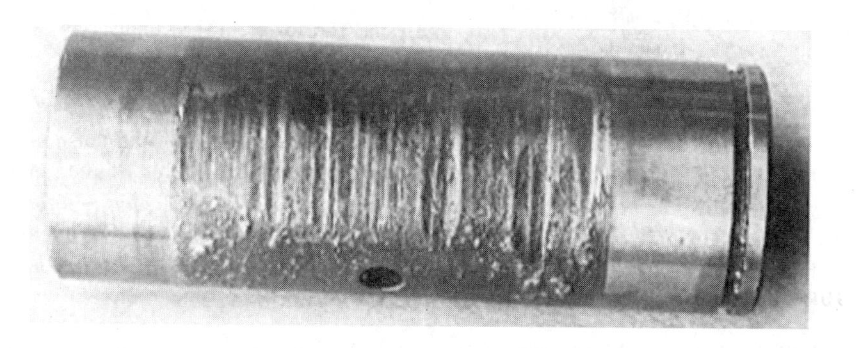

FIGURE 12-5

Adhesive wear on a shaft. (5)

CONTAMINANTS Adhesive bonding at the asperities can only occur if the material is clean and free of contaminants. Contaminants can take the form of oxides, skin oils from human handling, atmospheric moisture, etc. Contaminants in this context also include materials deliberately introduced to the interface such as coatings or lubricants. In fact, one of the chief functions of a lubricant is to prevent these adhesions and thus reduce friction and surface damage. A lubricant film effectively isolates the two materials and can prevent adhesion even between identical materials.

SURFACE FINISH It is not necessary for the surfaces to be "rough" for this adhesive-wear mechanism to operate. The fine-ground finish of the part in Figure 12-1*a* is seen to have as many asperities available for this process as the rougher milled surface in Figure 12-1*b* (p. 338).

12

COLD-WELDING If the mating materials are metals, and are extremely clean, the adhesive forces will be high and the sliding friction can generate enough localized heat to weld the asperities together. If the clean metal surfaces are also finished to a low roughness value (i.e., polished), and then rubbed together (with sufficient force), they can cold-weld (seize) with a bond virtually as strong as the parent metal. This process is enhanced if done in a vacuum, as the absence of air eliminates contamination from surface oxidation.

GALLING This describes a situation of incomplete cold-welding where, for whatever reason (usually contamination), the parts do not completely weld together. But, portions of the surfaces do adhere, causing material to be transferred from one part to the other in large streaks visible to the naked eye. Galling generally ruins the surfaces in one pass.

These factors explain the reasons for what is common knowledge among machinists and experienced engineers: *the same material should generally not be run against itself*. There are some exceptions to this rule, notably for hardened steel on hardened steel, but other combinations such as aluminum on aluminum *must* be avoided.

12.4 ABRASIVE WEAR

Abrasion occurs in two modes, referred to as the *two-body* and *three-body* abrasive wear processes.[7] **Two-body abrasion** refers to a *hard, rough material sliding against a softer one*. The hard surface digs into and removes material from the softer one. An example is a file used to contour a metal part. **Three-body abrasion** refers to the *introduction of hard particles between two sliding surfaces, at least one of which is softer than the particles*. The hard particles abrade material from one or both surfaces. Lapping and polishing are in this category. **Abrasion** is then *a material removal process in which the affected surfaces lose mass at some controlled or uncontrolled rate*.

UNCONTROLLED ABRASION Machine parts that operate in clean environments can be designed to minimize or eliminate abrasive wear through proper selection of materials and finishes. Smooth, hard materials will tend to not abrade soft ones in two-body contact. Smooth finishes minimize abrasion at the outset and, unless hard particulate contaminants are later introduced to the interface in service, that situation should continue.

CONTROLLED ABRASION In addition to designing systems to avoid abrasion, engineers also design them to *create* controlled abrasive wear. Controlled abrasion is widely used in manufacturing processes. Two-body **grinding** is perhaps the most common example, in which abrasive media such as silicon carbide (Carborundum) are forced against the part under high sliding velocities to remove material and control size and finish. Cam surfaces are ground in this manner.

Abrasion-Resistant Materials

Some engineering materials are better suited to abrasive-wear applications than others, based largely on their hardness. However, with hardness usually comes brittleness, and thus their resistance to impact or fatigue loads can be less than optimum. Table 12-1 shows the hardness of some materials that are suitable for abrasive-wear applications.

COATINGS Some ceramic materials can be plasma-sprayed onto metal substrates to provide a hard facing that also has high corrosion and chemical resistance. Some of these plasma-sprayed coatings are quite rough upon application (like severely orange-peeled paint) and thus must be diamond ground to obtain a finish suitable for a sliding joint. Some of these coatings are also very brittle and may chip from the substrate if overstressed either mechanically or thermally. Other hard coatings are available for steel that take on, or even improve, the surface finish of the base material. Most of these are proprietary formulas that are kept as trade secrets. Some trade names are: *Armaloy, Diamond Black, Dicronite, Titancote, Titanium Nitriding*, and others.*

* For more information on coatings, contact: *The American Electroplaters and Surface Finishers Society* at http://www.aesf.org/ or *The National Asociation of Metal Finishers* at http://www.namf.org/.

12.5 CORROSION WEAR

Corrosion occurs in normal environments with virtually all materials except those termed noble, i.e., gold, platinum, etc. The most common form of corrosion is oxidation. Most metals react with the oxygen in air or water to form oxides. Elevated temperatures greatly increase the rate of all chemical reactions.

TABLE 12-1 Materials Resistant to Abrasion

Material	Hardness (kg/mm^2)	Relative Wear
Tungsten carbide (sintered)	1 400-1 800	0.5-5
High-chromium white cast iron		5-10
Tool steel	700-1 000	20-30
Bearing steel	700-950	
Chromium (electroplated)	900	
Carburized steel	900	20-30
Nitrided steel	900-1 250	20-30
Pearlitic white iron		25-50
Austenitic manganese steel		30-50
Pearlitic low-alloy (0.7%C) steel	480	30-60
Pearlitic unalloyed (0.7%C) steel	300	50-70
As-rolled or normalized low-carbon (0.2%C) steel		100 (reference)

Sources: E. Rabinowicz, *Friction and Wear of Materials,* 1965, reprinted by permission of John Wiley & Sons, Inc., New York
T. E. Norman, *Abrasive Wear of Metals*, in *Handbook of Mechanical Wear*, C. Lipson, ed., U. Mich. Press, 1961.

Corrosion wear adds to the chemically corrosive environment a mechanical disruption of the surface layer due to a sliding or rolling contact of two bodies. This surface contact can act to break up the oxide (or other) film and expose new substrate to the reactive elements, thus increasing the rate of corrosion. If the products of the chemical reaction are hard and brittle (as with oxides), flakes of this layer can become loose particles in the interface and contribute to other forms of wear such as abrasion.

Some reaction products of metals such as metallic chlorides, phosphates, and sulfides are softer than the metal substrate and are also not brittle. These corrosion products can act as beneficial contaminants to reduce adhesive wear by blocking the adhesion of the metal asperities. This is the reason for adding compounds containing chlorine, sulphur, and other reactive agents to create EP (extreme pressure) oils. The strategy is to trade a slow rate of corrosive wear for a more rapid and damaging rate of adhesive wear on metal surfaces such as gear teeth and cams, which can have poor lubrication due to their nonconforming geometry.

Corrosion Fatigue

The phenomenon variously called corrosion fatigue or stress corrosion is not yet fully understood, but the empirical evidence of its result is strong and unequivocal. When a part is stressed in the presence of a corrosive environment, the corrosion process is accelerated and failure occurs more rapidly than would be expected from either the stress state alone or the corrosion process alone.

Static stresses are sufficient to accelerate the corrosion process. The combination of stress and corrosive environment has a synergistic effect and the material corrodes more rapidly than if unstressed. This combined condition of static stress and corrosion is termed **stress corrosion**. If the part is *cyclically stressed in a corrosive environment*, a crack will grow more rapidly than from either factor alone. This is called **corrosion fatigue**. While the frequency of stress cycling (as opposed to the number of cycles) appears to have no detrimental effect on crack growth in a noncorrosive environment, in the presence of corrosive environments it does. Lower cyclic frequencies allow the environment more time to act on the stressed crack tip while it is held open under tensile stress, and this substantially increases the rate of crack growth.

Fretting Corrosion

When two metal surfaces are in intimate contact, such as press-fit or clamped, one would expect no severe corrosion to occur at the interface, especially if in air. However, these kinds of contacts are subject to a phenomenon called **fretting corrosion** (or **fretting**) that can cause significant loss of material from the interface. Even though no gross sliding motions are possible in these situations, even small deflections (of the order of thousandths of an inch) are enough to cause fretting. Vibrations are another possible source of small fretting motions.

The fretting mechanism is believed to be some combination of abrasion, adhesion, and corrosion.[8] Free surfaces will oxidize in air, but the rate will slow as the oxides formed on the surface gradually block the substrate from the atmosphere. Some metals actually self-limit their oxidation if left undisturbed. The presence of vibrations or repeated mechanical deflections tends to disturb the oxide layer, scraping it loose and exposing new base metal to oxygen. This promotes adhesion of the "cleaned" metal asperities between the parts and also provides abrasive media in the form of hard oxide particles in the interface for three-body abrasion. All of these mechanisms tend to slowly reduce the solid volume of the materials and produce a "dust" or "powder" of abraded/oxidized material. Over time, significant dimensional loss can occur at the interface. In other cases, the result can be only a minor discoloration of the surfaces or adhesion similar to galling. All this from a joint that has no designed-in relative motion and was probably thought of by the designer as rigid and static! Of course, nothing is truly rigid, and fretting is evidence that microscopic motions are enough to cause wear. Figure 12-6 shows fretting on a shaft where a hub was press-fitted.[5] Fretting is sometimes encountered on roller-follower studs that have no relative motion versus the follower arm.

Some techniques that have proven to reduce fretting are the reduction of deflections (i.e., stiffer designs or tighter clamping) and the introduction of dry or fluid lubricants to the joint to act as an oxygen barrier and friction reducer. The introduction of a gasket, especially one with substantial elasticity (such as rubber) to absorb the vibrations has been shown to help. Harder and smoother surfaces on the metal parts are more resistant to abrasion and will reduce fretting damage. Corrosion-resistant platings such as chromium are sometimes used. The best method (impractical in most instances) is to eliminate the oxygen by operating in a vacuum or inert-gas atmosphere.

FIGURE 12-6

Fretting wear on a shaft beneath a press-fit hub (5)

12.6 STRESS

Stress is defined as force per unit area with units of psi or MPa. In a part subjected to some forces, stress is generally distributed as a continuously varying function within the continuum of material. Every infinitesimal element of the material can conceivably experience different stresses at the same time. Thus, we must look at stresses as acting on vanishingly small elements within the part. These infinitesimal elements are typically modeled as cubes, shown in Figure 12-7. The stress components are considered to be acting on the faces of these cubes in two different manners. **Normal stresses** act perpendicular (i.e., normal) to the face of the cube and tend to either pull it out (tensile normal stress) or push it in (compressive normal stress). **Shear stresses** act parallel to the faces of the cubes, in pairs (couples) on opposite faces, which tends to distort the cube into a rhomboidal shape. This is analogous to grabbing both pieces of bread of a peanut-butter sandwich and sliding them in opposite directions. The peanut butter will be sheared as a result. These normal and shear components of stress acting on an infinitesimal element make up the terms of a **tensor**.*

Stress is a tensor of order two† and thus requires nine values or components to describe it in three dimensions. The 3-D stress tensor can be expressed as the matrix:

$$\begin{bmatrix} \sigma_{xx} & \tau_{xy} & \tau_{xz} \\ \tau_{yx} & \sigma_{yy} & \tau_{yz} \\ \tau_{zx} & \tau_{zy} & \sigma_{zz} \end{bmatrix} \tag{12.2a}$$

where each stress component contains three elements, a magnitude (either σ or τ), the direction of a normal to the reference surface (first subscript), and a direction of action (second subscript). We will use σ to refer to normal stresses and τ for shear stresses.

* For a discussion of tensor notation, see C. R., Wylie and L. C. Barrett, *Advanced Engineering Mathematics*, 6th ed., McGraw-Hill, New York, 1995.

† Equation 12.2a is more correctly a tensor for rectilinear Cartesian coordinates. The more general tensor notation for curvilinear coordinate systems will not be used here.

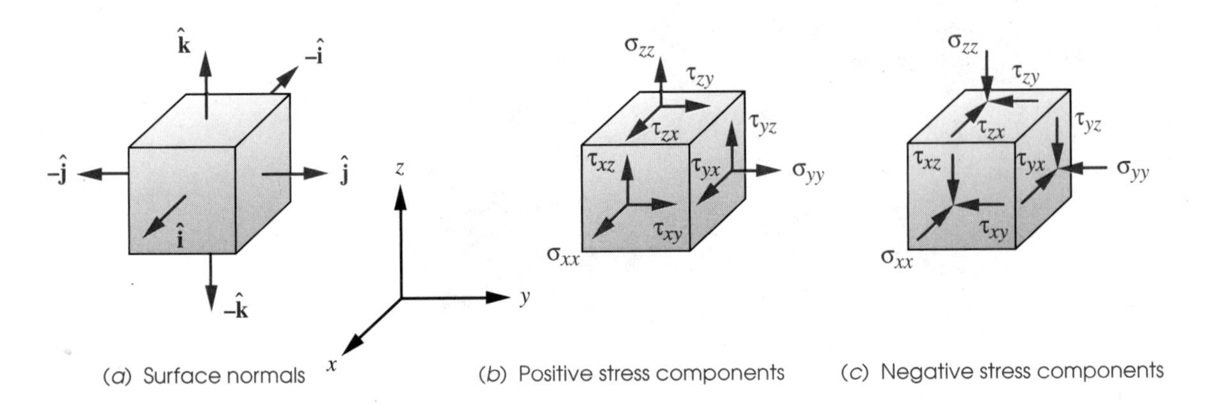

FIGURE 12-7

The stress cube, its surface normals, and its stress components

Many elements in machinery are subjected to three-dimensional stress states and thus require the stress tensor of equation 12.2a. There are some special cases, however, which can be treated as two-dimensional stress states. The stress tensor for 2-D is

$$\begin{bmatrix} \sigma_{xx} & \tau_{xy} \\ \tau_{yx} & \sigma_{yy} \end{bmatrix} \tag{12.2b}$$

Figure 12-7 shows an infinitesimal cube of material taken from within the material continuum of a part that is subjected to some 3-D stresses. The faces of this infinitesimal cube are made parallel to a set of *xyz* axes taken in some convenient orientation. The orientation of each face is defined by its surface normal vector* as shown in Figure 12-7*a*. The *x* face has its surface normal parallel to the *x* axis, etc. Note that there are thus two *x* faces, two *y* faces, and two *z* faces, one of each being positive and one negative as defined by the sense of its surface normal vector.

The nine stress components are shown acting on the surfaces of this infinitesimal element in Figure 12-7*b* and *c*. The components σ_{xx}, σ_{yy}, and σ_{zz} are the normal stresses, so called because they act, respectively, in directions normal to the *x, y,* and *z* surfaces of the cube. The components τ_{xy} and τ_{xz}, for example, are shear stresses that act on the *x* face and whose directions of action are parallel to the *y* and *z* axes, respectively. The sign of any one of these components is defined as positive if the signs of its surface normal and its stress direction are the same, and as negative if they are different. Thus the components shown in Figure 12-7*b* are all positive because they are acting on the positive faces of the cube and their directions are also positive. The components shown in Figure 12-7*c* are all negative because they are acting on the positive faces of the cube and their directions are negative. This sign convention makes tensile normal stresses positive and compressive normal stresses negative.

For the 2-D case, only one face of the stress cube may be drawn. If the *x* and *y* directions are retained and *z* eliminated, we look normal to the *xy* plane of the cube of Fig-

* A surface normal vector is defined as "growing out of the surface of the solid in a direction normal to that surface." Its sign is defined as the sense of this surface normal vector in the local coordinate system.

ure 12-7 and see the stresses shown in Figure 12-8, acting on the unseen faces of the cube. The reader should confirm that the stress components shown in Figure 12-8 are all positive by the sign convention stated above.

Note that the definition of the dual subscript notation given above is consistent when applied to the normal stresses. For example, the normal stress σ_{xx} acts on the x face and is also in the x direction. Since the subscripts are simply repeated for normal stresses, it is common to eliminate one of them and refer to the normal components simply as σ_x, σ_y, and σ_z. Both subscripts are needed to define the shear-stress components and they will be retained. It can also be shown[2] that the stress tensor is symmetric, which means that

$$\begin{aligned} \tau_{xy} &= \tau_{yx} \\ \tau_{yz} &= \tau_{zy} \\ \tau_{zx} &= \tau_{xz} \end{aligned} \tag{12.3}$$

This reduces the number of stress components to be calculated.

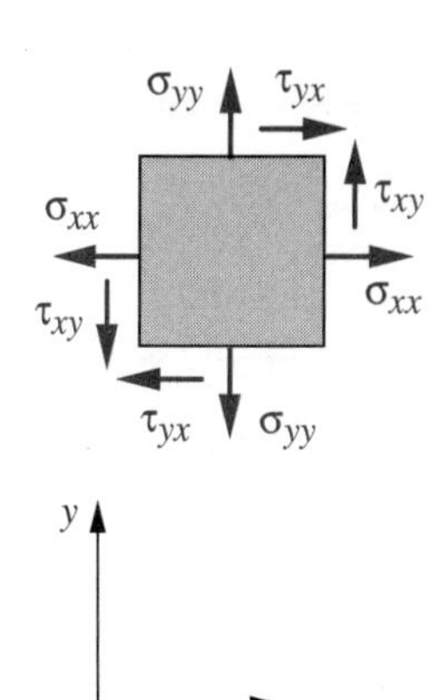

FIGURE 12-8

Two-dimensional stress element

12.7 STRAIN

Stress and strain are linearly related by Hooke's law in the elastic region of most engineering materials as discussed in Chapter 2. Strain is also a second-order tensor and can be expressed for the 3-D case as

$$\begin{bmatrix} \varepsilon_{xx} & \varepsilon_{xy} & \varepsilon_{xz} \\ \varepsilon_{yx} & \varepsilon_{yy} & \varepsilon_{yz} \\ \varepsilon_{zx} & \varepsilon_{zy} & \varepsilon_{zz} \end{bmatrix} \tag{12.4a}$$

and for the 2-D case as

$$\begin{bmatrix} \varepsilon_{xx} & \varepsilon_{xy} \\ \varepsilon_{yx} & \varepsilon_{yy} \end{bmatrix} \tag{12.4b}$$

where ε represents either a normal or a shear strain, the two being differentiated by their subscripts. We will also simplify the repeated subscripts for normal strains to ε_x, ε_y, and ε_z for convenience while retaining the dual subscripts to identify shear strains. The same symmetric relationships shown for shear stress components in equation 12.3 also apply to the strain components.

12.8 PRINCIPAL STRESSES

The axis systems taken in Figures 12-7 and 12-8 are arbitrary and are usually chosen for convenience in computing the applied stresses. For any particular combination of applied stresses, there will be a continuous distribution of the stress field around any point analyzed. The normal and shear stresses at the point will vary with direction in any coordinate system chosen. There will always be planes on which the shear-stress components are zero. The normal stresses acting on these planes are called the principal stresses. The planes on which these principal stresses act are called the **principal planes**. The directions of the surface normals to the principal planes are called the **principal axes**, and the normal stresses acting in those directions are the **principal normal stresses**. There will also be another set of mutually perpendicular axes along which the shear stresses will be maximal. The **principal shear stresses** act on a set of planes that are at 45° angles to the planes of the principal normal stresses. The principal planes and principal stresses for the 2-D case of Figure 12-8 are shown in Figure 12-9.

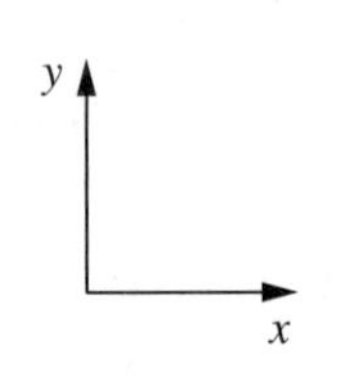

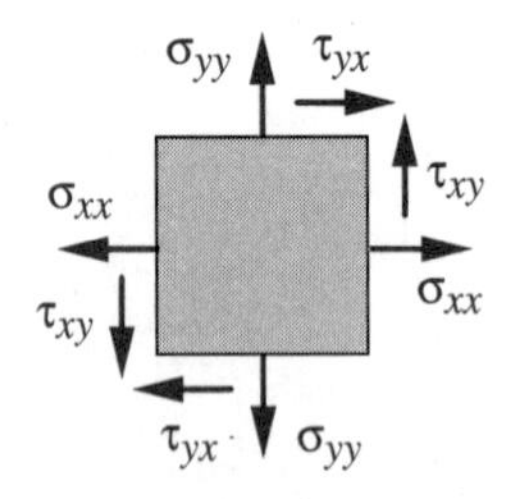

(a) Applied stresses

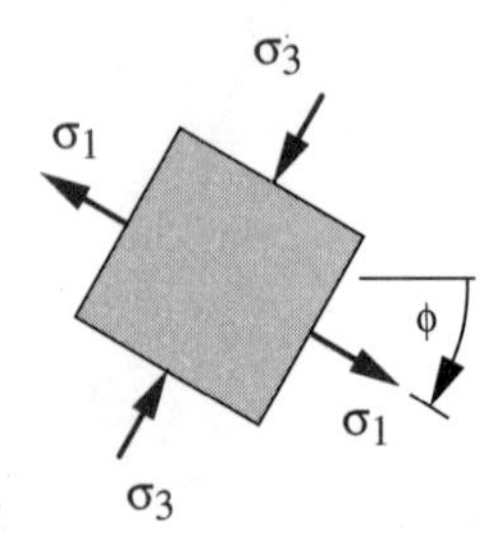

(b) Principal normal stresses

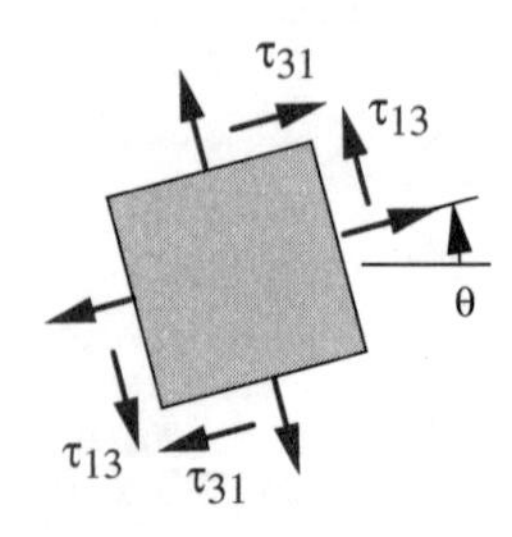

(c) Principal shear stresses

FIGURE 12-9

Principal stresses on a two-dimensional stress element

Since, from an engineering standpoint, we are most concerned with designing our machine parts so that they will not fail, and since failure will occur if the stress at any point exceeds some safe value, we need to find the largest stresses (both normal and shear) that occur anywhere in the continuum of material that makes up our machine part. We may be less concerned with the directions of those stresses than with their magnitudes as long as the material can be considered to be at least macroscopically isotropic, thus having strength properties that are uniform in all directions. Most metals and many other engineering materials meet these criteria, although wood and composite materials are notable exceptions.

The expression relating the applied stresses to the principal stresses is

$$\begin{bmatrix} \sigma_x - \sigma & \tau_{xy} & \tau_{xz} \\ \tau_{yx} & \sigma_y - \sigma & \tau_{yz} \\ \tau_{zx} & \tau_{zy} & \sigma_z - \sigma \end{bmatrix} \begin{bmatrix} n_x \\ n_y \\ n_z \end{bmatrix} = \begin{bmatrix} 0 \\ 0 \\ 0 \end{bmatrix} \tag{12.5a}$$

where σ is the principal stress magnitude and n_x, n_y, and n_z are the direction cosines of the unit vector **n**, which is normal to the principal plane:

$$\begin{aligned} \hat{\mathbf{n}} \cdot \hat{\mathbf{n}} &= 1 \\ \hat{\mathbf{n}} &= n_x \hat{\mathbf{i}} + n_y \hat{\mathbf{j}} + n_z \hat{\mathbf{k}} \end{aligned} \tag{12.5b}$$

For the solution of equation 12.5a to exist, the determinant of the coefficient matrix must be zero. Expanding this determinant and setting it to zero, we obtain

$$\sigma^3 - C_2\sigma^2 - C_1\sigma - C_0 = 0 \tag{12.5c}$$

where

$$\begin{aligned} C_2 &= \sigma_x + \sigma_y + \sigma_z \\ C_1 &= \tau_{xy}^2 + \tau_{yz}^2 + \tau_{zx}^2 - \sigma_x\sigma_y - \sigma_y\sigma_z - \sigma_z\sigma_x \\ C_0 &= \sigma_x\sigma_y\sigma_z + 2\tau_{xy}\tau_{yz}\tau_{zx} - \sigma_x\tau_{yz}^2 - \sigma_y\tau_{zx}^2 - \sigma_z\tau_{xy}^2 \end{aligned}$$

Equation 12.5c is a cubic polynomial in σ. The coefficients C_0, C_1, and C_2 are called the tensor invariants because they have the same values regardless of the initial choice of xyz axes in which the applied stresses were measured or calculated. The units of C_2 are psi (MPa), of C_1 psi^2 (MPa^2), and of C_0 psi^3 (MPa^3). The three principal (normal) stresses σ_1, σ_2, σ_3 are the three roots of this cubic polynomial The roots of this polynomial are always real[2] and are usually ordered such that $\sigma_1 > \sigma_2 > \sigma_3$. If needed, the directions of the principal stress vectors can be found by substituting each root of equation 12.5c into 12.5a and solving for n_x, n_y, and n_z for each of the three principal stresses. The directions of the three principal stresses are mutually orthogonal.

The principal shear stresses can be found from the values of the principal normal stresses using

$$\tau_{13} = \frac{|\sigma_1 - \sigma_3|}{2}$$
$$\tau_{21} = \frac{|\sigma_2 - \sigma_1|}{2} \tag{12.6}$$
$$\tau_{32} = \frac{|\sigma_3 - \sigma_2|}{2}$$

If the principal normal stresses have been ordered as shown above, then $\tau_{max} = \tau_{13}$. The directions of the planes of the principal shear stresses are 45° from those of the principal normal stresses and are also mutually orthogonal.

The solution of equation 12.5c for its three roots can be done trigonometrically by Viete's method* or using an iterative root-finding algorithm. The file Stres_3D.tk provided on CD-ROM solves equation 12.5c and finds the three principal stress roots by Viete's method and orders them by the above convention. Stres_3D.tk also computes the stress function (Eq. 12.5c) for a list of user-defined values of σ and then plots that function. The root crossings can be seen in Figure 12-10, which shows the stress function for an arbitrary set of applied stresses plotted over a range of values of σ that includes all three roots.

* See *Numerical Recipes* by W. H. Press et al., Cambridge Univ. Press, 1986, p. 146, or *Standard Mathematical Tables*, CRC Press, 22d ed., 1974, p. 104, or any collection of standard mathematical formulas.

For the special case of a two-dimensional stress state, the equations 12.5c for principal stress reduce to†

$$\sigma_a, \sigma_b = \frac{\sigma_x + \sigma_y}{2} \pm \sqrt{\left(\frac{\sigma_x - \sigma_y}{2}\right)^2 + \tau_{xy}^2}$$
$$\sigma_c = 0 \tag{12.7a}$$

The two nonzero roots calculated from equation 12.7a are temporarily labeled σ_a and σ_b, and the third root σ_c is always zero in the 2-D case. Depending on their resulting values, the three roots are then labeled according to the convention: *algebraically largest* = σ_1,

† Equations 12.7 can also be used when one principal stress is nonzero but is directed along one of the axes of the xyz coordinate system chosen for calculation. The stress cube of Figure 12-7 is then rotated about one principal axis to determine the angles of the other two principal planes.

algebraically smallest = σ_3, and *other* = σ_2, Using equation 12.7a to solve the example shown in Figure 12-10 would yield values of $\sigma_1 = \sigma_a$, $\sigma_3 = \sigma_b$, and $\sigma_2 = \sigma_c = 0$ as labeled in the figure.* Of course, equation 12.5c for the 3-D case can still be used to solve any two-dimensional case. One of the three principal stresses found will then be zero. The example in Figure 12-10 is of a two-dimensional case solved with equation 12.5c. Note the root at $\sigma = 0$.

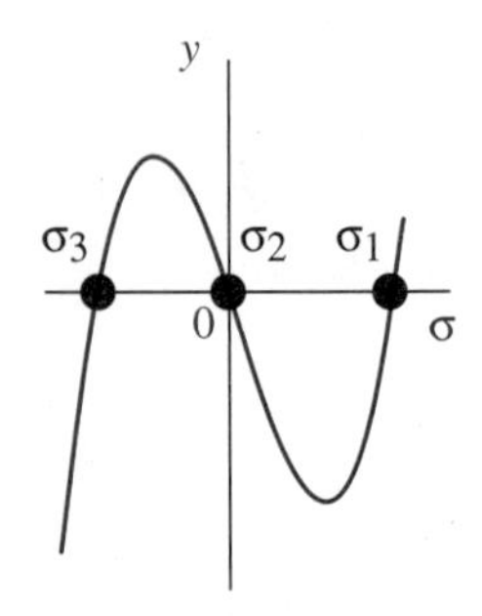

FIGURE 12-10

The three roots of the stress function for a plane stress case

Once the three principal stresses are found and ordered as described above, the maximum shear stress is found from equation 12.6:

$$\tau_{max} = \tau_{13} = \frac{|\sigma_1 - \sigma_3|}{2} \tag{12.7b}$$

12.9 PLANE STRESS AND PLANE STRAIN

The general state of stress and strain is three-dimensional but there exist particular geometric configurations that can be treated differently.

Plane Stress

The two-dimensional, or biaxial, stress state is also called plane stress. **Plane stress** requires that one principal stress be zero. This condition is common in some applications. For example, a thin plate or shell may also have a state of plane stress away from its boundaries or points of attachment. These cases can be treated with the simpler approach of equations 12.7.

Plane Strain

There are principal strains associated with the principal stresses. If one of the principal strains (say ε_3) is zero, and if the remaining strains are independent of the dimension along its principal axis, $\mathbf{n}_3$, it is called **plane strain**. This condition occurs in particular geometries. For example, if a long, solid, prismatic bar is loaded only in the transverse direction, regions within the bar that are distant from any end constraints will see essentially zero strain in the direction along the axis of the bar and be in plane strain. (However, the stress is not zero in the zero-strain direction.) A long, hydraulic dam can be considered to have a plane strain condition in regions well removed from its ends or base at which it is attached to surrounding structures.

* If the 3-D numbering convention is strictly observed in the 2-D case, then sometimes the two nonzero principal stresses will turn out to be σ_1 and σ_3 if they are opposite in sign. Other times they will be σ_1 and σ_2 when they are both positive and the smallest (σ_3) is zero. A third possibility is that both nonzero principal stresses are negative (compressive) and the algebraically largest of the set (σ_1) is then zero. Equation 12.7a arbitrarily calls the two nonzero 2-D principal stresses σ_a and σ_b with the remaining one (σ_c) reserved for the zero member of the trio. Application of the standard convention can result in σ_a and σ_b being called any of the possible combinations σ_1 and σ_2, σ_1 and σ_3, or σ_2 and σ_3 depending on their relative values.

EXAMPLE 12-1

Determining 3-D Principal Stresses Using Analytical Methods.

Given: A triaxial stress element as shown in Figure 12-7 (p. 346) has $\sigma_x = 40\,000$, $\sigma_y = -20\,000$, $\sigma_z = -10\,000$, $\tau_{xy} = 5\,000$, $\tau_{yz} = -1\,500$, and $\tau_{zx} = 2\,500$ psi.

Problem: Find the principal stresses using a numerical method.

Solution:

1 Calculate the tensor invariants C_0, C_1, and C_2 from equation 12.5c.

$$C_2 = \sigma_x + \sigma_y + \sigma_z = 40\,000 - 20\,000 - 10\,000 = 10\,000 \tag{a}$$

$$\begin{aligned} C_1 &= \tau_{xy}^2 + \tau_{yz}^2 + \tau_{zx}^2 - \sigma_x\sigma_y - \sigma_y\sigma_z - \sigma_z\sigma_x \\ &= (5\,000)^2 + (-1\,500)^2 + (2\,500)^2 - (40\,000)(-20\,000) \\ &\quad -(-20\,000)(-10\,000) - (-10\,000)(40\,000) = 10.335E8 \end{aligned} \tag{b}$$

$$\begin{aligned} C_0 &= \sigma_x\sigma_y\sigma_z + 2\tau_{xy}\tau_{yz}\tau_{zx} - \sigma_x\tau_{yz}^2 - \sigma_y\tau_{zx}^2 - \sigma_z\tau_{xy}^2 \\ &= 40\,000(-20\,000)(-10\,000) + 2(5\,000)(-1\,500)(2\,500) \\ &\quad -40\,000(-1\,500)^2 - (-20\,000)(2\,500)^2 \\ &\quad -(-10\,000)(5\,000)^2 = 8.248E12 \end{aligned} \tag{c}$$

2 Substitute the invariants into equation 12.5c (p. 348) and solve for its three roots using Viete's or a numerical method.

$$\sigma^3 - C_2\sigma^2 - C_1\sigma - C_0 = 0$$

$$\sigma^3 - 10\,000\,\sigma^2 - 10.335E8\,\sigma - 8.248E12 = 0 \tag{d}$$

$$\sigma_1 = 40\,525 \qquad \sigma_2 = -9\,838 \qquad \sigma_3 = -20\,687$$

3 The principal shear stresses can now be found from equation 12.6 (p. 349).

$$\begin{aligned} \tau_{13} &= \frac{|\sigma_1 - \sigma_3|}{2} = \frac{|40\,525 - (-20\,687)|}{2} = 30\,606 \\ \tau_{21} &= \frac{|\sigma_2 - \sigma_1|}{2} = \frac{|-9\,838 - 40\,525|}{2} = 25\,182 \\ \tau_{32} &= \frac{|\sigma_3 - \sigma_2|}{2} = \frac{|-20\,687 - (-9\,838)|}{2} = 5\,425 \end{aligned} \tag{e}$$

12.10 APPLIED VERSUS PRINCIPAL STRESSES

We now want to summarize the differences between the stresses **applied to an element** and the principal stresses that may occur on other planes as a result of the applied stresses. The **applied stresses** are the nine *components of the stress tensor* (equation 12.5a, p. 348) that result from whatever loads are applied to the particular geometry of the object as defined in a coordinate system chosen for convenience. The **principal stresses** are the three *principal normal stresses* and the three *principal shear stresses* defined in Section 12-8. Of course, many of the applied-stress terms may be zero in a given case. For example, in a tensile-test specimen the only nonzero applied stress is the σ_x term in equation 12.5a (p. 348), which is unidirectional and normal. There are no applied shear stresses on the surfaces normal to the force axis in pure tensile loading. However, the principal stresses are **both normal and shear.**

In a tensile-test specimen, the applied stress is pure tensile and the maximum principal normal stress is equal to it in magnitude and direction. But a principal shear stress of half the magnitude of the applied tensile stress acts on a plane 45° from the plane of the principal normal stress. Thus, the principal shear stresses will typically be nonzero even in the absence of any applied shear stress. This fact is important to an understanding of why parts fail. The most difficult task for the machine designer is to correctly determine the locations, types, and magnitudes of all the applied stresses acting on the part. The calculation of the principal stresses is then *pro forma* using equations 12.5 to 12.7 (pp. 348-350).

12.11 SURFACE FATIGUE

All the surface-failure modes discussed above apply to situations in which the relative motions between the surfaces are essentially pure sliding, such as a cam running against a flat-faced follower. When two surfaces are in **pure rolling** contact, or are primarily rolling in combination with a small percentage of sliding, a different surface failure mechanism comes into play, called **surface fatigue**. Many applications of this condition exist such as cams with roller followers, ball and roller bearings, nip rolls, and spur or helical gear tooth contact. All except the gear teeth and nip rolls typically have essentially pure rolling with only about 1% sliding.

The stresses introduced in two materials contacting at a rolling interface are highly dependent on the geometry of the surfaces in contact as well as on the loading and material properties. The general case allows any three-dimensional geometry on each contacting member and, as would be expected, its calculation is the most complex. Two special-geometry cases are of practical interest and are also somewhat simpler to analyze. These are *sphere-on-sphere* and *cylinder-on-cylinder*. In all cases, the radii of curvature of the mating surfaces will be significant factors. By varying the radii of curvature of one mating surface, these special cases can be extended to include the sub-cases of *sphere-on-plane*, *sphere-in-cup, cylinder-on-plane*, and *cylinder-in-trough*. It is only necessary to make the radii of curvature of one element infinite to obtain a plane, and negative radii of curvature define a concave cup, or concave trough surface. For example, some ball bearings can be modeled as *sphere-on-plane* and some roller bearings and cylindrical cam followers as *cylinder-in-trough*.

As a ball passes over another surface, the theoretical contact patch is a point of zero dimension. A roller against a cylindrical or flat surface theoretically contacts along a line of zero width. Since the area of each of these theoretical contact geometries is zero, any applied force will then create an infinite stress. We know that this cannot be true, as the materials would instantly fail. In fact, the materials must deflect to create sufficient contact area to support the load at some finite stress. This deflection creates a semi-ellipsoidal pressure distribution over the contact patch. In the general case, the contact patch is elliptical as shown in Figure 12-11*a*. Spheres will have a circular contact patch, and cylinders create a rectangular contact patch as shown in Figure 12-11*b*.

Consider the case of a spherical ball rolling in a straight line against a flat surface with no slip, and under a constant normal load. If the load is such as to stress the material only below its yield point, the deflection in the contact patch will be elastic and the surface will return to its original curved geometry after passing through contact. The same spot on the ball will contact the surface again on each succeeding revolution. The resulting stresses in the contact patch are called **contact stresses or Hertzian stresses**. The contact stresses in this small volume of the ball are **repeated** at the ball's rotation frequency. This creates a fatigue-loading situation that eventually leads to a **surface-fatigue failure**.

This repeated loading is similar to a tensile fatigue-loading case. The significant difference in this case is that the principal contact stresses at the center of the contact patch are all compressive, not tensile. Fatigue failures are considered to be initiated by shear stress and continued to failure by tensile stress. There is also a shear stress associated with these compressive contact stresses, and it is believed to be the cause of crack formation after many stress-cycles. Crack growth then eventually results in failure by **pitting**—*the fracture and dislodgment of small pieces of material from the surface.* Once the surface begins to pit, its surface finish is compromised and it rapidly proceeds to failure by **spalling**—*the loss of large pieces of the surface.* Figure 12-12 shows some examples of pitted and spalled surfaces.

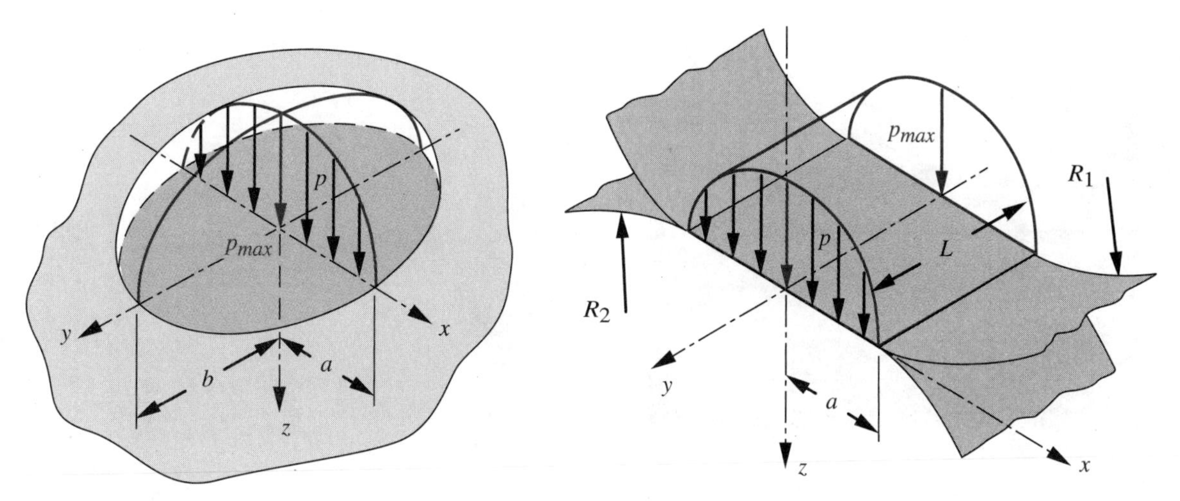

(*a*) Ellipsoidal pressure distribution in general contact—for spherical contact $a = b$

(*b*) Ellipsoidal-prism pressure distribution in cylindrical contact

FIGURE 12-11

Pressure distributions and contact zones of spherical, cylindrical, and general Hertzian contact

If the load is large enough to raise the contact stress above the material's compressive yield strength, then the contact-patch deflection will create a permanent flat on the ball. This condition is sometimes called **false brinelling**, because it has a similar appearance to the indentation made to test a material's Brinell hardness. Such a flat on even one of its balls (or rollers) makes a ball (or roller) bearing useless.

We will now investigate the contact-patch geometries, pressure distributions, stresses, and deformations in rolling contacts starting with the relatively simple geometry of *sphere-on-sphere*, next dealing with the *cylinder-on-cylinder* case, and finally discussing the *general* case. Derivation of the equations for these cases are among the more complex sets of examples from the theory of elasticity. The equations for the area of contact, deformation, pressure distribution, and contact stress on the centerline of two bodies with static loading were originally derived by Hertz in 1881,[9] an English translation of which can be found in [10]. Many others have since added to the understanding of this problem.[11],[12],[13],[14]

12.12 SPHERICAL CONTACT

Cross sections of two spheres in contact are shown in Figure 12-13. The dotted lines indicate the possibilities that one is a flat plane or a concave cup. The difference is only in the magnitude or the sign of its radius of curvature (convex +, concave –). Figure 12-11*a* (p. 353) shows the general semi-ellipsoidal pressure distribution over the contact patch. For a sphere-on-sphere, it will be a hemisphere with a circular contact patch ($a = b$).

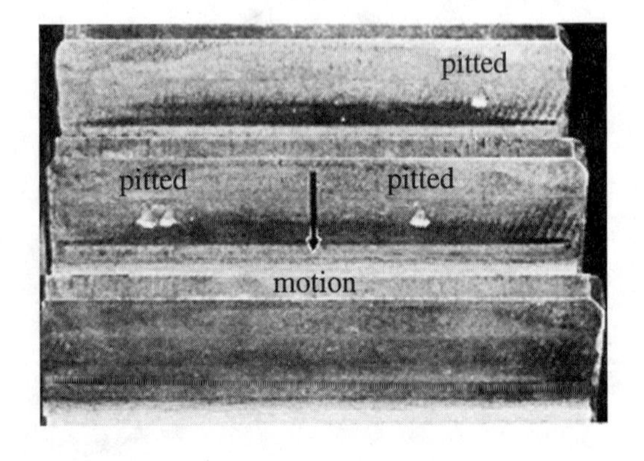

(a) Mild pitting on gear teeth

(b) Severe pitting, spalling, and disintegration of gear teeth

FIGURE 12-12

Examples of surfaces failed by pitting and spalling due to surface fatigue (Source: J. D. Graham, *Pitting of Gear Teeth*, in *Handbook of Mechanical Wear*, C. Lipson, ed., U. Mich. Press, 1961, pp. 138, 143, with permission)

Contact Pressure and Contact Patch in Spherical Contact

The contact pressure is a maximum p_{max} at the center and zero at the edge. The total applied load F on the contact patch is equal to the volume of the hemisphere:

$$F = \frac{2}{3}\pi a^2 p_{max} \tag{12.8a}$$

where a is the half-width (radius) of the contact patch. This can be solved for the maximum pressure:

$$p_{max} = \frac{3}{2}\frac{F}{\pi a^2} \tag{12.8b}$$

The average pressure on the contact patch is the applied force divided by its area:

$$p_{avg} = \frac{F}{area} = \frac{F}{\pi a^2} \tag{12.8c}$$

and substituting equation 12.8c in 12.8b gives:

$$p_{max} = \frac{3}{2} p_{avg} \tag{12.8d}$$

We now define *material constants* for the two spheres

$$m_1 = \frac{1 - \nu_1^2}{E_1} \qquad m_2 = \frac{1 - \nu_2^2}{E_2} \tag{12.9a}$$

where E_1, E_2 and ν_1, ν_2 are the Young's moduli and Poisson's ratios for the materials of sphere 1 and sphere 2, respectively.

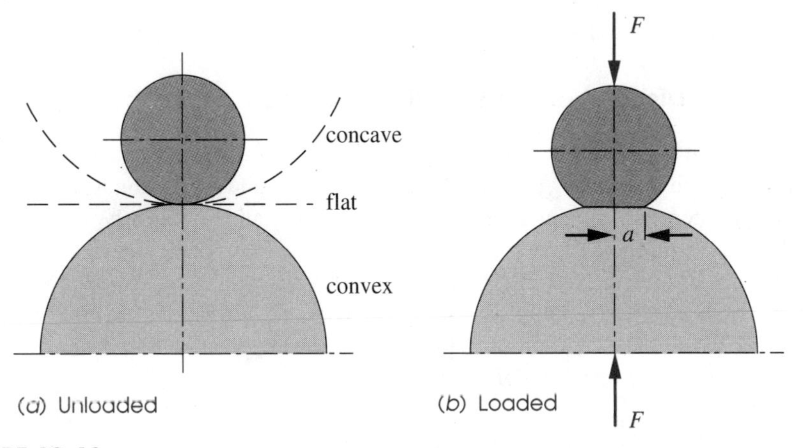

FIGURE 12-13
Contact zone of two spheres or cylinders

The dimensions of the contact area are typically very small compared to the radii of curvature of the bodies, which allows the radii to be considered constant over the contact area despite the small deformations occurring there. We can define a *geometry constant* that depends only on the radii R_1 and R_2 of the two spheres,

$$B = \frac{1}{2}\left(\frac{1}{R_1} + \frac{1}{R_2}\right) \tag{12.9b}$$

To account for the case of a sphere-on-plane, R_2 becomes infinite, making $1/R_2$ zero. For a sphere-in-cup, R_2 becomes negative. (See Figure 12-13, p. 355.) Otherwise R_2 is finite and positive, as is R_1.

The contact-patch radius a is then found from

$$a = \frac{\pi}{4} p_{max} \frac{m_1 + m_2}{B} \tag{12.9c}$$

Substitute equation 12.8b in 12.9c:

$$a = \frac{\pi}{4}\left(\frac{3}{2}\frac{F}{\pi a^2}\right)\frac{m_1 + m_2}{B}$$

$$a = \sqrt[3]{0.375 \frac{m_1 + m_2}{B} F} \tag{12.9d}$$

The pressure distribution within the hemisphere is

$$p = p_{max}\sqrt{1 - \frac{x^2}{a^2} - \frac{y^2}{a^2}} \tag{12.10}$$

We can normalize the pressure p to the magnitude of p_{avg} and the patch dimension x or y to the patch radius a and then plot the normalized pressure distribution across the patch, which will be an ellipse as shown in Figure 12-14.

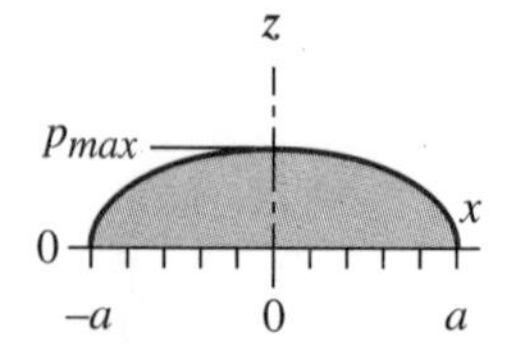

FIGURE 12-14

Pressure distribution across contact patch

Static Stress Distributions in Spherical Contact

The pressure on the contact patch creates a three-dimensional stress state in the material. The three applied stresses σ_x, σ_y, and σ_z are compressive and are maximal at the sphere's surface in the center of the patch. They diminish rapidly and nonlinearly with depth and with distance from the axis of contact. They are called **Hertzian stresses** in honor of their original discoverer. A complete derivation of these equations can be found in reference 16. Note that these applied stresses in the x, y, and z directions are also the principal stresses in this case. If we look at these stresses as they vary along the z axis (with z increasing into the material) we find

$$\sigma_z = p_{max}\left[-1+\frac{z^3}{\left(a^2+z^2\right)^{3/2}}\right] \tag{12.11a}$$

$$\sigma_x = \sigma_y = \frac{p_{max}}{2}\left[-(1+2\nu)+2(1+\nu)\left(\frac{z}{\sqrt{a^2+z^2}}\right)-\left(\frac{z}{\sqrt{a^2+z^2}}\right)^3\right] \tag{12.11b}$$

Poisson's ratio is taken for the sphere of interest in this calculation. These normal (and principal) stresses are maximal at the surface, where $z = 0$:

$$\sigma_{z_{max}} = -p_{max} \tag{12.11c}$$

$$\sigma_{x_{max}} = \sigma_{y_{max}} = -\frac{1+2\nu}{2}p_{max} \tag{12.11d}$$

There is also a principal shear stress induced from these principal normal stresses:

$$\tau_{13} = \frac{p_{max}}{2}\left[\frac{(1-2\nu)}{2}+(1+\nu)\left(\frac{z}{\sqrt{a^2+z^2}}\right)-\frac{3}{2}\left(\frac{z}{\sqrt{a^2+z^2}}\right)^3\right] \tag{12.12a}$$

which is not maximum at the surface but rather at a small distance $z_{@\tau_{max}}$ below the surface.

$$\tau_{13_{max}} = \frac{p_{max}}{2}\left[\frac{(1-2\nu)}{2}+\frac{2}{9}(1+\nu)\sqrt{2(1+\nu)}\right] \tag{12.12b}$$

$$z_{@\tau_{max}} = a\sqrt{\frac{2+2\nu}{7-2\nu}} \tag{12.12c}$$

Figure 12-15 shows a plot of the principal normal and maximum shear stresses as a function of depth z along a radius of the sphere. The stresses are normalized to the maximum pressure p_{max}, and the depth is normalized to the half-width a of the contact patch. This plot provides a dimensionless picture of the stress distribution on the centerline under a spherical contact. Note that all the stresses have diminished to less than10% of p_{max} within $z = 5a$. The subsurface location of the maximum shear stress can also be seen. If both materials are steel, it occurs at a depth of about $0.63a$ and its magnitude is about 0.34 p_{max}. The shear stress is about 0.11 p_{max} at the surface on the z axis.

The subsurface location of the maximum shear stress is believed by some to be a significant factor in surface-fatigue failure. The theory says cracks that begin below the surface eventually grow to the point that the material above the crack breaks out to form a pit as shown in Figure 12-12 (p. 354).

12

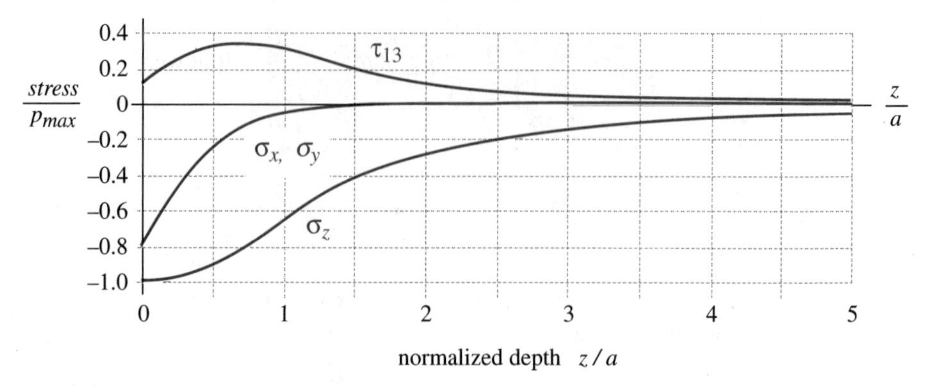

FIGURE 12-15

Normalized stress distribution along *z* axis in spherical contact; *xyz* stresses are principal

Figure 12-16 shows a photoelastic model of the contact stresses in a cam immediately beneath a loaded roller follower.[15] Experimental photoelastic stress analysis uses a physical model of the part to be analyzed made from a transparent plastic material (Lexan in this example) that shows fringes of constant stress magnitude when loaded and viewed in polarized light. The maximum shear stress can be clearly seen a small distance into the cam directly under the follower. While this is a cylindrical rather than a spherical contact, their stress distributions along the centerline are similar, as will be seen in the next section.

When we move off the centerline of the contact patch on the surface of the sphere, the stresses diminish. At the edge of the patch the radial stress σ_z is zero, but there is a condition of pure shear stress with the magnitude:

$$\tau_{xy} = \frac{1-2\nu}{3} p_{max} \tag{12.13a}$$

The two nonzero principal stresses will be $\pm\,\tau_{xy}$, which means that there is also a tensile stress at that point of

$$\sigma_{1_{edge}} = \frac{1-2\nu}{3} p_{max} \tag{12.13b}$$

12.13 CYLINDRICAL CONTACT

Cylindrical contact is common in machinery. A non-crowned cylindrical roller follower running against a face cam is one example. Roller bearings are another application. The mating cylinders can be both convex, one convex and one concave (cylinder-in-trough), or, in the limit, a cylinder-on-plane. In all such contacts there is the possibility of sliding as well as rolling at the interface. The presence of tangential sliding forces has a significant effect on the stresses compared to pure rolling. We will first consider the case of two cylinders in pure rolling and later introduce a sliding component.

FIGURE 12-16

Photoelastic analysis of contact stresses under a cam-follower (Source: V. S. Mahkijani, *Study of Contact Stresses as Developed on a Radial Cam Using Photoelastic Model and Finite Element Analysis.* M.S. Thesis, Worcester Polytechnic Institute, 1984) [15]

Contact Pressure and Contact Patch in Parallel Cylindrical Contact

When two cylinders roll together, their contact patch will be rectangular as shown in Figure 12-11*b* (p. 353). The pressure distribution will be a semi-elliptical prism of half-width a. The contact zone will look as shown in Figure 12-13 (p. 355). The contact pressure is a maximum p_{max} at the center and zero at the edges as shown in Figure 12-14 (p. 357). The applied load F on the contact patch is equal to the volume of the half-prism:

$$F = \frac{1}{2} \pi a L p_{max} \tag{12.14a}$$

where F is the total applied load and L is the length of contact along the cylinder axis. This can be solved for the maximum pressure:

$$p_{max} = \frac{2F}{\pi a L} \tag{12.14b}$$

The average pressure is the applied force divided by the contact-patch area:

$$p_{avg} = \frac{F}{area} = \frac{F}{2aL} \tag{12.14c}$$

Substituting equation 12.14c in 12.14b gives

$$p_{max} = \frac{4}{\pi} P_{avg} \cong 1.273 P_{avg} \tag{12.14d}$$

We now define a cylindrical geometry constant that depends on the radii R_1 and R_2 of the two cylinders, (Note that it is the same as equation 12.9b (p. 356) for spheres.)

$$B = \frac{1}{2}\left(\frac{1}{R_1} + \frac{1}{R_2}\right) \tag{12.15a}$$

To account for the case of a cylinder-on-plane, R_2 becomes infinite, making $1/R_2$ zero. For a cylinder-in-trough, R_2 becomes negative. Otherwise R_2 is finite and positive, as is R_1. The contact-patch half-width a is then found from

$$a = \sqrt{\frac{2}{\pi}\frac{m_1 + m_2}{B}\frac{F}{L}} \tag{12.15b}$$

where m_1 and m_2 are material constants as defined in equation 12.9a (p. 355).

The pressure distribution within the semi-elliptical prism is

$$p = p_{max}\sqrt{1 - \frac{x^2}{a^2}} \tag{12.16}$$

which is an ellipse, as shown in Figure 12-11 (p. 353).

Static Stress Distributions in Parallel Cylindrical Contact

Hertzian stress analysis is for static loading, but is also applied to pure rolling contact. The stress distributions within the material are similar to those shown in Figure 12-15 (p. 358) for the sphere-on-sphere case. Two cases are possible: *plane stress,* where the cylinders are very short axially as in some cam roller-followers, and *plane strain,* where the cylinders are long axially such as in squeeze-rollers. In the plane-stress case, one of the principal stresses is zero. In plane strain, all three principal stresses may be nonzero.

Figure 12-17 shows the principal and maximum shear stress distributions across the patch width at the surface and along the z axis (where they are largest) for two cylinders in static or pure rolling contact. The normal stresses are all compressive and are maximal at the surface. They diminish rapidly with depth into the material and also diminish away from the centerline, as shown in the figure.

At the surface on the centerline, the maximum applied normal stresses are

$$\begin{aligned} \sigma_x &= \sigma_z = -p_{max} \\ \sigma_y &= -2\nu p_{max} \end{aligned} \tag{12.17a}$$

These stresses are principal, since there is no applied shear stress. The maximum shear stress τ_{13} on the z axis that results from the combination of stresses is beneath the surface as it was in the spherical-contact case. For two steel cylinders in static contact, the peak value and location of the maximum shear stress are[16]

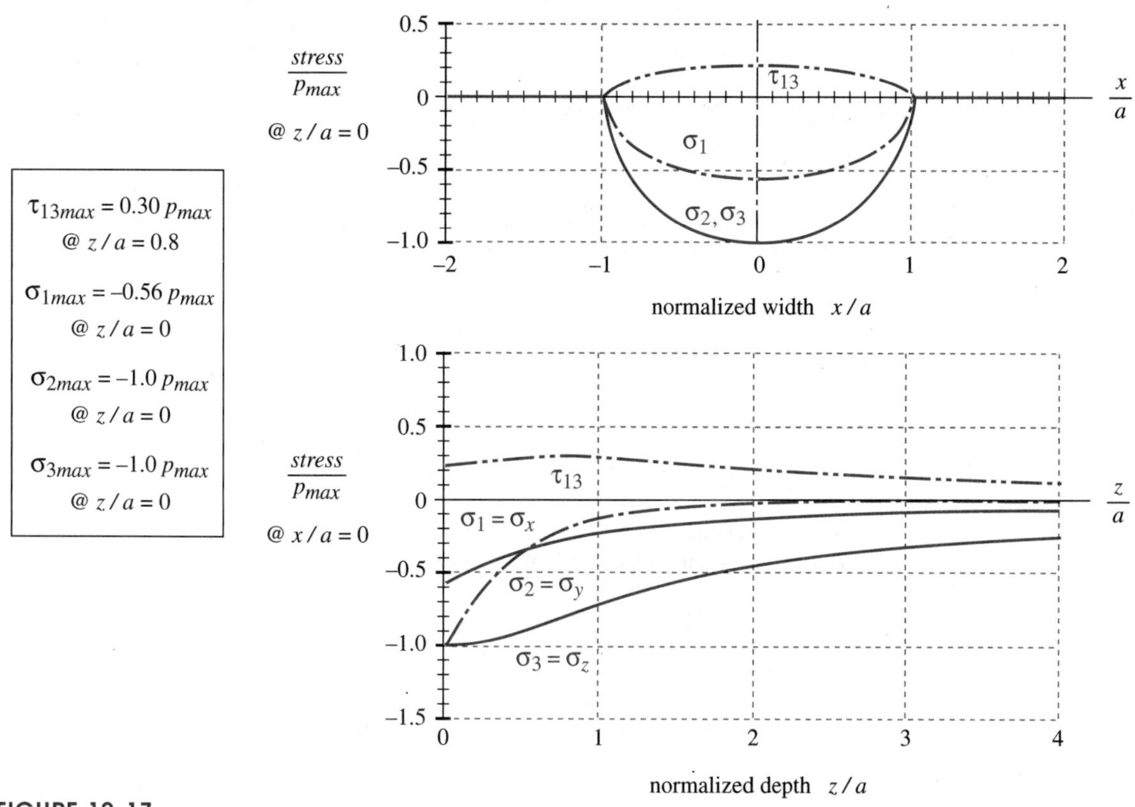

FIGURE 12-17

Principal, maximum shear, and von Mises stress distributions for steel cylinders in static loading or pure rolling

$$\tau_{13_{max}} = 0.304\,p_{max}$$
$$z_{@\tau_{max}} = 0.786a \tag{12.17b}$$

However, note in Figure 12-17 that, on the z axis, the shear stress is not zero but is $0.22\ p_{max}$ at the surface and does not vary greatly over the depth $0 < z < 2a$.

EXAMPLE 12-2

Stresses in Cylindrical Contact.

Problem: A cylindrical roller follower runs against a radial cam. What is the size of the contact patch between cam and roller and what are the stresses at the point of minimum radius of curvature? What is the depth of the maximum shear stress?

Given: The minimum positive radius of curvature of the cam is 1.25 in. The roller diameter is 1 in. The cam is 0.875-in thick and the roller is 1 in long. Both parts are steel. The radial load at that point is 500 lb.

Assumptions: The roller axis is exactly parallel to the cam surface.

Solution:

1 First determine the size of the contact patch, for which the geometry constant and material constants are found from equations 12.15a (p. 360) and 12.9a (p. 355).

$$B=\frac{1}{2}\left(\frac{1}{R_1}+\frac{1}{R_2}\right)=\frac{1}{2}\left(\frac{1}{1.25}+\frac{1}{0.5}\right)=1.4 \tag{a}$$

$$m_1=m_2=\frac{1-\nu_1^2}{E_1}=\frac{1-0.28^2}{3E7}=3.072E-8 \tag{b}$$

Note both materials are the same in this example. The material and geometry constants can now be used in equation 12.15b (p. 360).

$$a=\sqrt{\frac{2}{\pi}\frac{m_1+m_2}{B}\frac{F}{L}}=\sqrt{\left(\frac{2}{\pi}\right)\frac{2(3.072E-8)}{1.4}\left(\frac{500}{0.875}\right)}=0.004 \text{ in} \tag{c}$$

where a is the half-width of the contact patch. The rectangular contact-patch area is

$$area=2aL=2(0.004)(0.875)=0.007 \text{ in}^2 \tag{d}$$

2 The average and maximum contact pressure are found from equations 12.14b and c (p. 359).

$$p_{avg}=\frac{F}{area}=\frac{500}{0.007}=71\,507 \text{ psi} \tag{e}$$

$$p_{max}=\frac{2F}{\pi aL}=\frac{2(500)}{\pi(0.004)(0.875)}=90\,946 \text{ psi} \tag{f}$$

3 The maximum normal stresses in the center of the contact patch at the surface are then found using equations 12.17a (p. 360).

$$\sigma_{z_{max}}=\sigma_{x_{max}}=-p_{max}=-90\,946 \text{ psi} \tag{g}$$

$$\sigma_{y_{max}}=-2\nu p_{max}=-2(0.28)(90\,946)=-50\,930 \text{ psi} \tag{h}$$

4 The maximum shear stress and its location (depth) are found from equations 12.17b (p. 361).

$$\tau_{13_{max}} = 0.304 p_{max} = 0.304(90\,946) = 27\,648 \text{ psi}$$
$$z_{@\tau_{max}} = 0.786a = 0.786(0.004) = 0.003 \text{ in} \qquad \text{(i)}$$

5 All the stresses found exist on the z axis and the normal stresses are principal. These stresses apply to the cam and roller follower, as both are steel.

12.14 GENERAL CONTACT

When the geometry of the two contacting bodies is allowed to have any general curvature, the contact patch is an ellipse and the pressure distribution is a semi-ellipsoid, as shown in Figure 12-11*a* (p. 353). Even the most general curvature can be represented as a radius of curvature over a small angle with minimal error. The size of the contact patch for most practical materials in these applications is so small that this approximation is reasonable. Thus the compound curvature of each body is represented by two mutually orthogonal radii of curvature at the contact point.

Contact Pressure and Contact Patch in General Contact

The contact pressure is a maximum p_{max} at the center and zero at the edge. The total applied load F on the contact patch is equal to the volume of the semi-ellipsoid:

$$F = \frac{2}{3}\pi a b p_{max} \qquad (12.18a)$$

where a is the half-width of the major axis and b the half-width of the minor axis of the contact-patch ellipse. This can be solved for the maximum pressure:

$$p_{max} = \frac{3}{2}\frac{F}{\pi a b} \qquad (12.18b)$$

The average pressure on the contact patch is the applied force divided by its area:

$$p_{avg} = \frac{F}{area} = \frac{F}{\pi a b} \qquad (12.18c)$$

and substituting equation 12.18c in 12.18b gives

$$p_{max} = \frac{3}{2} P_{avg} \qquad (12.18d)$$

We must define three geometry constants that depend on the radii of curvature of the two bodies,

$$A = \frac{1}{2}\left(\frac{1}{R_1} + \frac{1}{R_1'} + \frac{1}{R_2} + \frac{1}{R_2'}\right) \qquad (12.19a)$$

$$B = \frac{1}{2}\left[\left(\frac{1}{R_1} - \frac{1}{R_1'}\right)^2 + \left(\frac{1}{R_2} - \frac{1}{R_2'}\right)^2 + 2\left(\frac{1}{R_1} - \frac{1}{R_1'}\right)\left(\frac{1}{R_2} - \frac{1}{R_2'}\right)\cos 2\theta\right]^{\frac{1}{2}} \tag{12.19b}$$

$$\phi = \cos^{-1}\left(\frac{B}{A}\right) \tag{12.19c}$$

where R_1 and R_1' are the two radii of curvature* of body 1, R_2 and R_2' are the radii* of body 2, and θ is the angle between the planes containing R_1 and R_2.

The contact-patch dimensions a and b are then found from

$$a = k_a \sqrt[3]{\frac{3F(m_1 + m_2)}{4A}} \qquad b = k_b \sqrt[3]{\frac{3F(m_1 + m_2)}{4A}} \tag{12.19d}$$

where m_1 and m_2 are material constants as defined in equation 12.9a (p. 355) and the values of k_a and k_b are from Table 12-2 corresponding to the value of ϕ from equation 12.19c.

The pressure distribution within the semi-ellipsoid is

$$p = p_{max}\sqrt{1 - \frac{x^2}{a^2} - \frac{y^2}{b^2}} \tag{12.20}$$

which is an ellipse as shown in Figure 12-11 (p. 353).

Stress Distributions in General Contact

The stress distributions within the material are similar to those shown in Figure 12-17 for the cylinder-on-cylinder case. The normal stresses are all compressive and are maximal at the surface. They diminish rapidly with depth into the material and away from the centerline. At the surface on the centerline, the maximum normal stresses are:[16]

$$\sigma_x = -\left[2\nu + (1 - 2\nu)\frac{b}{a + b}\right]p_{max}$$

$$\sigma_y = -\left[2\nu + (1 - 2\nu)\frac{a}{a + b}\right]p_{max} \tag{12.21a}$$

$$\sigma_z = -p_{max}$$

* Measured in mutually perpendicular planes

12

TABLE 12-2 Factors For Use in Equation 12.19d

ϕ	0	10	20	30	35	40	45	50	55	60	65	70	75	80	85	90
k_a	∞	6.612	3.778	2.731	2.397	2.136	1.926	1.754	1.611	1.486	1.378	1.284	1.202	1.128	1.061	1
k_b	0	0.319	0.408	0.493	0.530	0.567	0.604	0.641	0.678	0.717	0.759	0.802	0.846	0.893	0.944	1

Sources: H. Hertz, *Contact of Elastic Solids*, in *Miscellaneous Papers*, P. Lenard, ed. Macmillan & Co. Ltd.: London, 1896, pp. 146-162.
H. L. Whittemore and S. N. Petrenko, *Natl. Bur. Std. Tech. Paper* 201, 1921.

$$k_3 = \frac{b}{a} \qquad k_4 = \frac{1}{a}\sqrt{a^2 - b^2} \tag{12.21b}$$

These applied stresses are also the principal stresses. The maximum shear stress at the surface associated with these stresses can be found from equation 12.6. The largest shear stress occurs slightly below the surface, with that distance dependent on the ratio of the semiaxes of the contact ellipse. For $b/a = 1.0$, the largest shear stress occurs at $z = 0.63a$, and for $b/a = 0.34$ at $z = 0.24a$. Its peak magnitude is approximately $0.34\,p_{max}$.[16]

At the ends of the major axis of the contact ellipse, the shear stress at the surface is

$$\tau_{xz} = (1 - 2\nu)\frac{k_3}{k_4^2}\left(\frac{1}{k_4}\tanh^{-1} k_4 - 1\right) p_{max} \tag{12.21c}$$

At the ends of the minor axis of the contact ellipse, the shear stress at the surface is

$$\tau_{xz} = (1 - 2\nu)\frac{k_3}{k_4^2}\left[1 - \frac{k_3}{k_4}\tan^{-1}\left(\frac{k_4}{k_3}\right)\right] p_{max} \tag{12.21d}$$

The location of the largest surface shear stress will vary with the ellipse ratio k_3. For some cases it is as shown in equation 12.21c, but in others it moves to the center of the ellipse and is found from the principal stresses in equation 12.21a, using equation 12.6 (p. 349).

EXAMPLE 12-3

Stresses in a Crowned Cam Follower.

Problem: A crowned cam roller-follower has a gentle radius transverse to its rolling direction to eliminate the need for critical alignment of its axis with that of the cam. The cam's radius of curvature and dynamic load vary around its circumference. What is the size of the contact patch between cam and follower and what are the worst-case stresses?

Given: The roller radius is 1 in with a 20-in crown radius at 90° to the roller radius. The cam's radius of curvature at the point of maximum force is 3.46 in and it is flat axially. (The minimum radius of curvature is 1.72 in.) The rotational axes of the cam and roller are parallel, which makes the angle between the two bodies zero. The maximum force is 250 lb, normal to the contact plane.

Assumptions: Materials are steel. The relative motion is rolling with <1% sliding.

Solution:

1 Find the material constants from equation 12.9a.

$$m_1 = m_2 = \frac{1 - \nu_1^2}{E_1} = \frac{1 - 0.28^2}{3E7} = 3.072E - 8 \tag{a}$$

2 Two geometry constants are needed from equations 12.19a and b.

$$A=\frac{1}{2}\left(\frac{1}{R_1}+\frac{1}{R_1'}+\frac{1}{R_2}+\frac{1}{R_2'}\right)=\frac{1}{2}\left(\frac{1}{1}+\frac{1}{20}+\frac{1}{3.46}+\frac{1}{\infty}\right)=0.669\,5 \tag{b}$$

$$B=\frac{1}{2}\left[\left(\frac{1}{R_1}-\frac{1}{R_1'}\right)^2+\left(\frac{1}{R_2}-\frac{1}{R_2'}\right)^2+2\left(\frac{1}{R_1}-\frac{1}{R_1'}\right)\left(\frac{1}{R_2}-\frac{1}{R_2'}\right)\cos 2\theta\right]^{\frac{1}{2}}$$

$$B=\frac{1}{2}\left[\left(\frac{1}{1}-\frac{1}{20}\right)^2+\left(\frac{1}{3.46}-\frac{1}{\infty}\right)^2+2\left(\frac{1}{1}-\frac{1}{20}\right)\left(\frac{1}{3.46}-\frac{1}{\infty}\right)\cos 2(0)\right]^{\frac{1}{2}}=0.619\,5 \tag{c}$$

The angle ϕ is found from their ratio (equation 12.19c),

$$\phi=\cos^{-1}\left(\frac{B}{A}\right)=\cos^{-1}\left(\frac{0.6195}{0.6695}\right)=22.3^\circ \tag{d}$$

and used in Table 12-2 (p. 364) to find the factors k_a and k_b. Cubic interpolation* for k_a and linear interpolation* for k_b gives

$$k_a=3.444 \qquad\qquad k_b=0.427 \tag{e}$$

3 The material and geometry constants can now be used in equation 12.19d (p. 364).

$$a=k_a\sqrt[3]{\frac{3F(m_1+m_2)}{4A}}=3.444\sqrt[3]{\frac{3(250)2(3.072E-8)}{4(0.6695)}}=0.088\,9$$

$$b=k_b\sqrt[3]{\frac{3F(m_1+m_2)}{4A}}=0.427\sqrt[3]{\frac{3(250)2(3.072E-8)}{4(0.6695)}}=0.011\,0 \tag{f}$$

where a is the half-width of the major axis, and b is the half-width of the minor axis of the contact patch. The contact-patch area is then:

$$area=\pi ab=\pi(0.0889)(0.011)=0.0031\text{ in}^2 \tag{g}$$

4 The average and maximum contact pressure can be found from equations 12.18b and c (p. 363).

$$p_{avg}=\frac{F}{area}=\frac{250}{0.0031}=81119\text{ psi} \tag{h}$$

12

* The different interpolation methods are used to best fit the functions, one of which is linear and the other nonlinear. Plot the values in Table 12-2 to see this.

$$p_{max} = \frac{3}{2} p_{avg} = \frac{3}{2}(81119) = 121\,679 \text{ psi} \tag{i}$$

5 The maximum normal stresses in the center of the contact patch at the surface are then found using equations 12.21a (p. 365).

$$\begin{aligned}
\sigma_x &= -\left[2\nu + (1-2\nu)\frac{b}{a+b}\right] p_{max} \\
&= -\left[2(.28) + (1-2(.28))\frac{0.011}{0.0889+0.011}\right] 121\,679 = -74\,051 \text{ psi} \\
\sigma_y &= -\left[2\nu + (1-2\nu)\frac{a}{a+b}\right] p_{max} \\
&= -\left\{2(.28) + [1-2(.28)]\frac{0.0889}{0.0889+0.011}\right\} 121\,679 = -115\,678 \text{ psi} \\
\sigma_z &= -p_{max} = -121\,679 \text{ psi}
\end{aligned} \tag{j}$$

These stresses are principal: $\sigma_1 = \sigma_x$, $\sigma_2 = \sigma_y$, $\sigma_3 = \sigma_z$. The maximum shear stress associated with them at the surface will be (from equation 12.6, p. 349):

$$\tau_{13} = \left|\frac{\sigma_1 - \sigma_3}{2}\right| = \left|\frac{-74\,051 + 121\,679}{2}\right| = 23\,814 \text{ psi (at surface)} \tag{k}$$

6 The largest shear stress under the surface on the z axis is approximately:

$$\tau_{13} \cong 0.34 p_{max} = 0.34(121\,679) \cong 41\,000 \text{ psi (below surface)} \tag{l}$$

7 All the stresses found so far exist on the centerline of the patch. At the edge of the patch, at the surface, there will also be a shear stress. Two constants are found from equation 12.21b (p. 365) for this calculation.

$$\begin{aligned}
k_3 &= \frac{b}{a} = \frac{0.0110}{0.0889} = 0.124 \\
k_4 &= \frac{1}{a}\sqrt{a^2 - b^2} = \frac{1}{0.0889}\sqrt{0.0889^2 - 0.0110^2} = 0.992
\end{aligned} \tag{m}$$

These constants are used in equations 12.21c and *d* (p. 365) to find the shear stresses on the surface at the ends of the major and minor axes.

$$\begin{aligned}
\tau_{xz} &= (1-2\nu)\frac{k_3}{k_4^2}\left(\frac{1}{k_4}\tanh^{-1} k_4 - 1\right) p_{max} \\
\tau_{xz} &= (1-0.56)\frac{0.124}{(0.992)^2}\left(\frac{1}{0.992}\tanh^{-1} 0.992 - 1\right) 121\,679 = 12\,131 \text{ psi}
\end{aligned} \tag{n}$$

$$\tau_{xz} = (1-2\nu)\frac{k_3}{k_4^2}\left[1 - \frac{k_3}{k_4}\tan^{-1}\left(\frac{k_4}{k_3}\right)\right]p_{max}$$

$$\tau_{xz} = (1-0.56)\frac{0.124}{(0.992)^2}\left[1 - \frac{0.124}{0.992}\tan^{-1}\left(\frac{0.992}{0.124}\right)\right]121\,679 = 5\,528 \text{ psi} \tag{o}$$

12.15 DYNAMIC CONTACT STRESSES

The equations presented above for contact stresses assume that the load is pure rolling. When rolling and sliding are both present, the stress field is distorted by the tangential loading. Figure 12-18 shows a photoelastic study of a cam-follower pair[15] loaded *(a)* statically and *(b)* dynamically with sliding. The distortion of the stress field from the sliding motion can be seen in part *b*. This is a combination of rolling contact with relatively low-velocity sliding. Increased sliding causes more distortion of the stress field.

Effect of a Sliding Component on Contact Stresses

Smith and Lui[14] analyzed the case of parallel rollers in combined rolling and sliding, and developed the equations for the stress distribution beneath the contact point. The sliding (frictional) load has a significant effect on the stress field. The stresses can be expressed as separate components, one set due to the normal load on the rolls (denoted by a subscript *n*) and the other set due to the tangential friction force (denoted by a subscript *t*). These are then combined to obtain the complete stress situation. The stress field can be two-dimensional in a very short roll, such as a thin plate cam or thin gear, assumed to be in plane stress. If the rolls are long axially, then a plane strain condition will exist in regions away from the ends, giving a three-dimensional stress state.

The contact geometry is as shown in Figure 12-11*b* (p. 353) with the *x* axis aligned to the direction of motion, the *z* axis radial to the rollers, and the *y* axis axial to the rollers. The stresses due to the normal loading p_{max} are:

$$\begin{aligned}
\sigma_{x_n} &= -\frac{z}{\pi}\left[\frac{a^2+2x^2+2z^2}{a}\alpha - \frac{2\pi}{a} - 3x\beta\right]p_{max} \\
\sigma_{z_n} &= -\frac{z}{\pi}\left[a\beta - x\alpha\right]p_{max} \\
\tau_{xz_n} &= -\frac{1}{\pi}z^2\alpha p_{max}
\end{aligned} \tag{12.22a}$$

and those due to the frictional unit force f_{max} are:

12

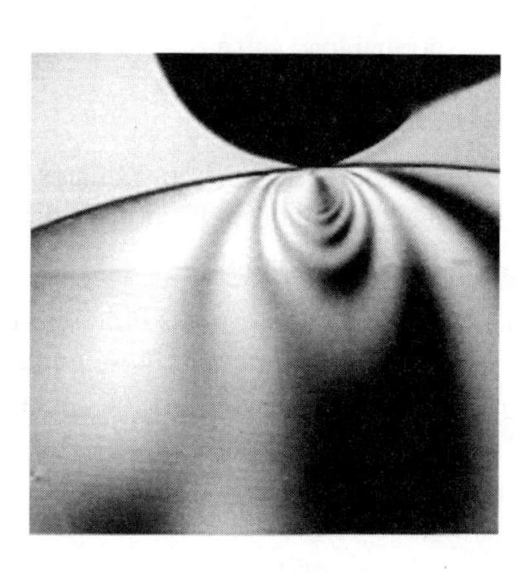

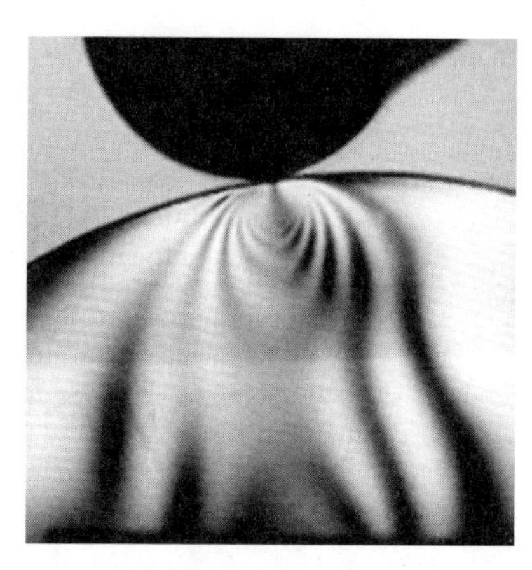

(a) Static loading (b) Dynamic loading

FIGURE 12-18

Photoelastic study of stresses for two cylinders in contact in (a) static and (b) dynamic pure rolling loading (Source: V. S. Mahkijani, *Study of Contact Stresses as Developed on a Radial Cam Using Photoelastic Model and Finite Element Analysis*, M.S. Thesis, Worcester Polytechnic Institute, 1984)

$$\sigma_{x_t} = -\frac{1}{\pi}\left[\left(2x^2 - 2a^2 - 3z^2\right)\alpha + 2\pi\frac{x}{a} + 2\left(a^2 - x^2 - z^2\right)\frac{x}{a}\beta\right] f_{max}$$

$$\sigma_{z_t} = -\frac{1}{\pi} z^2 \alpha f_{max} \tag{12.22b}$$

$$\tau_{xz_t} = -\frac{1}{\pi}\left[\left(a^2 + 2x^2 + 2z^2\right)\frac{z}{a}\beta - 2\pi\frac{z}{a} - 3xz\alpha\right] f_{max}$$

12

where the factors α and β are given by

$$\alpha = \frac{\pi}{k_1}\frac{1-\sqrt{\dfrac{k_2}{k_1}}}{\sqrt{\dfrac{k_2}{k_1}}\sqrt{2\sqrt{\dfrac{k_2}{k_1}}+\left(\dfrac{k_1+k_2-4a^2}{k_1}\right)}} \tag{12.22c}$$

$$\beta = \frac{\pi}{k_1}\frac{1+\sqrt{\dfrac{k_2}{k_1}}}{\sqrt{\dfrac{k_2}{k_1}}\sqrt{2\sqrt{\dfrac{k_2}{k_1}}+\left(\dfrac{k_1+k_2-4a^2}{k_1}\right)}} \tag{12.22d}$$

$$k_1 = (a+x)^2 + z^2 \qquad k_2 = (a-x)^2 + z^2 \tag{12.22e}$$

The tangential unit force f_{max} is found from the normal load and a coefficient of friction μ.

$$f_{max} = \mu p_{max} \tag{12.22f}$$

The independent variables in these equations are then the coordinates x, z in the cross section of the roller, referenced to the contact point, the half-width a of the contact patch, and the maximum normal load p_{max} at the contact point.

Equations 12.22 (pp. 369–370) define the behavior of the stress functions below the surface, but when $z = 0$, the factors α and β become infinite and these equations fail. Other forms are needed to account for the stresses on the surface of the contact patch.

When $z = 0$

$$\begin{aligned} &\text{if } |x| \le a \text{ then } \sigma_{x_n} = -p_{max}\sqrt{1-\frac{x^2}{a^2}} \text{ else } \sigma_{x_n} = 0 \\ &\sigma_{z_n} = \sigma_{x_n} \\ &\tau_{xz_n} = 0 \end{aligned} \tag{12.23a}$$

$$\begin{aligned} &\text{if } x \ge a \text{ then } \sigma_{x_t} = -2 f_{max}\left(\frac{x}{a} - \sqrt{\frac{x^2}{a^2} - 1}\right) \\ &\text{if } x \le -a \text{ then } \sigma_{x_t} = -2 f_{max}\left(\frac{x}{a} + \sqrt{\frac{x^2}{a^2} - 1}\right) \\ &\text{if } |x| \le a \text{ then } \sigma_{x_t} = -2 f_{max}\frac{x}{a} \end{aligned} \tag{12.23b}$$

$$\begin{aligned} &\sigma_{z_t} = 0 \\ &\text{if } |x| \le a \text{ then } \tau_{xz_t} = -f_{max}\sqrt{1-\frac{x^2}{a^2}} \text{ else } \tau_{xz_t} = 0 \end{aligned} \tag{12.23c}$$

The total stress on each Cartesian plane is found by superposing the components due to the normal and tangential loads:

$$\begin{aligned} \sigma_x &= \sigma_{x_n} + \sigma_{x_t} \\ \sigma_z &= \sigma_{z_n} + \sigma_{z_t} \\ \tau_{xz} &= \tau_{xz_n} + \tau_{xz_t} \end{aligned} \tag{12.24a}$$

For short rollers in plane stress, σ_y is zero, but if the rollers are long axially, then a plane strain condition will exist away from the ends and the stress in the y direction will be:

$$\sigma_y = \nu\left(\sigma_x + \sigma_z\right) \tag{12.24b}$$

where ν is Poisson's ratio.

These stresses are maximum at the surface and decrease with depth. Except at very low ratios of tangential force to normal force (< about 1/9)[14],[17] the maximum shear stress occurs at the surface as well, unlike the pure rolling case. A computer program was written to evaluate equations 12.22 and 12.23 (pp. 369–370) for the conditions at the surface and plot them. The stresses are all normalized to the maximum normal load p_{max} and the locations normalized to the patch half-width a. A coefficient of friction of 0.33 and steel rollers with $\nu = 0.28$ were assumed for the examples. The magnitudes and shapes of the stress distributions will be a function of these factors.

Figure 12-19*a* shows the *x* direction stresses at the surface, which are due to the normal and tangential loads, and also shows their sum from the first of equations 12.24a. Note that the stress component σ_{x_t} due to the tangential force, is tensile from the contact point to and beyond the trailing edge of the contact patch. This should not be surprising, as one can picture that the tangential force is attempting to pile up material in front of the contact point and stretch it behind that point, just as a carpet bunches up in front of anything you try to slide across it. The stress component σ_{x_n} due to the normal force is compressive everywhere. However, the sum of the two σ_x components has a significant normalized tensile value of twice the coefficient of friction (here 0.66 p_{max}) and a compressive peak of about –1.2 p_{max}. Figure 12-19*b* shows all of the applied stresses in *x, y*, and *z* directions across the surface of the contact zone. Note that the stress fields on the surface extend beyond the contact zone when a tangential force is present, unlike the situation in pure rolling where they extend beyond the contact zone only under the surface. (See Figure 12-17 on p. 361)

Figure 12-20 shows the principal and maximum shear stresses for the plane strain, applied stress state in Figure 12-19. Note that the magnitude of the largest compressive principal stress is about 1.38 p_{max} and the largest tensile principal stress is 0.66 p_{max} at the trailing edge of the contact patch. The presence of an applied tangential shear stress in this example increases the peak compressive stress by 40% over a pure rolling case and introduces a tensile stress in the material. The principal shear stress reaches a peak value of 0.40 p_{max} at $x / a = 0.4$. All the stresses shown in Figures 12-19 and 12-20 are at the surface of the rollers.

Beneath the surface, the magnitudes of the compressive stresses due to the normal load reduce. However, the shear stress τ_{xz_n} due to the normal loading increases with depth, becoming a maximum beneath the surface at $z = 0.5a$, as shown in Figure 12-21. Note the sign reversal at the midpoint of the contact zone. There are fully reversed shear-stress components acting on each differential element of material as it passes through the contact zone. The peak-to-peak range of this fully reversed shear stress in the *xz* plane is greater in magnitude than the range of the maximum shear stress and is considered by some to be responsible for subsurface pitting failures.[13]

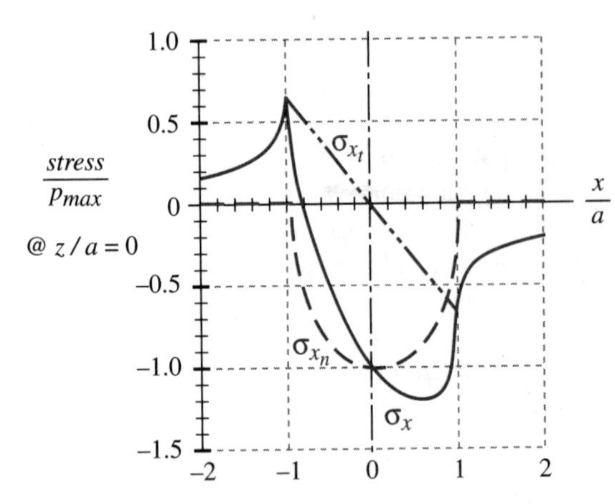

(a) Normal and tangential components of σ_x

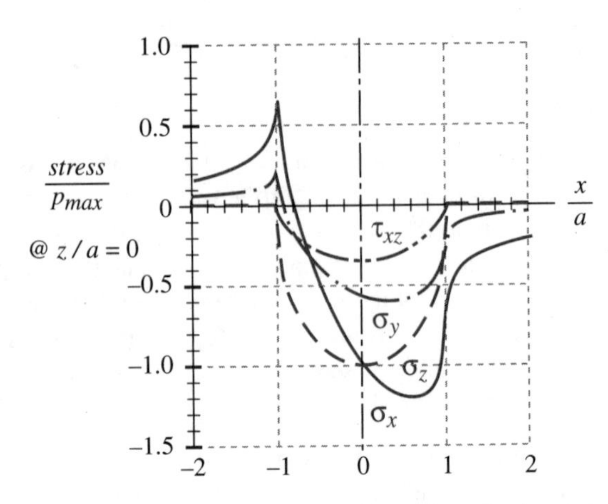

(b) All applied stresses at surface of contact patch

FIGURE 12-19

Applied tangential, normal, and shear stresses at surface for cylinders in combined rolling and sliding with $\mu = 0.33$

Figure 12-22 plots the principal and maximum shear stresses (calculated for $\mu = 0.33$ and a plane strain condition) versus the normalized depth z / a taken at the $x / a = 0.3$ plane (where the principal stresses are maximum as shown in Figure 12-20). All the stresses are maximum at the surface. The principal stresses diminish rapidly with depth, but the shear stress remains nearly constant over the first $1a$ of depth.

At the surface, the maximum shear stress is relatively uniform across the patch width with a peak of 0.4 at $x / a = 0.4$ when $\mu = 0.33$, as shown in Figure 12-20. This τ_{max} peak location moves versus the patch centerline with increasing depth, but its magnitude var-

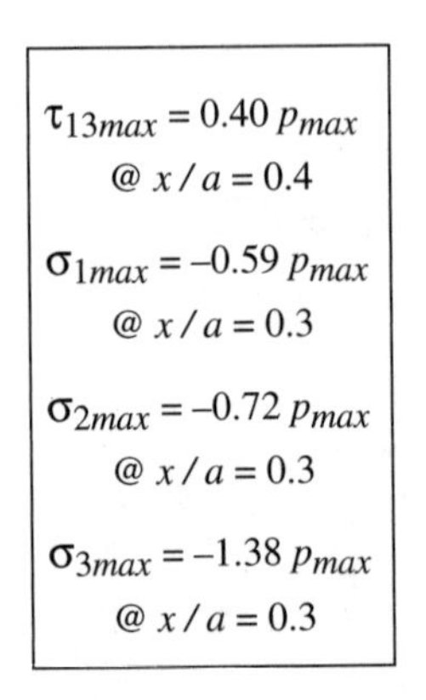

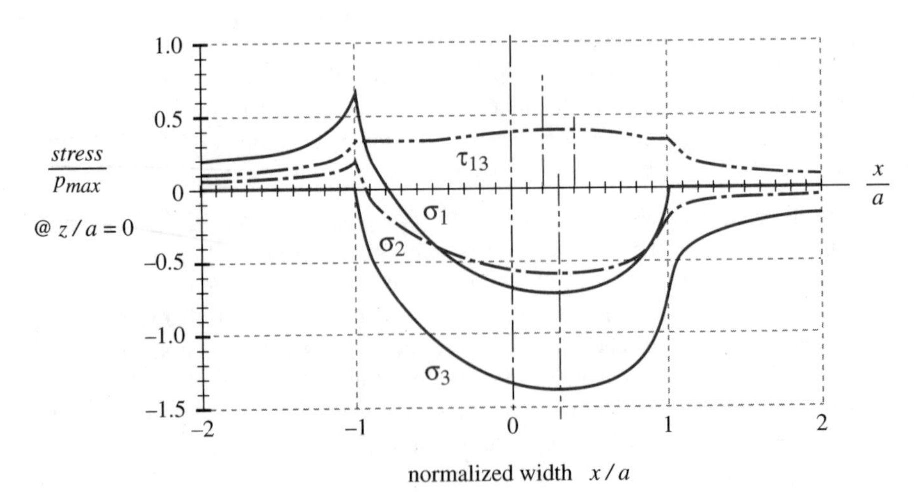

FIGURE 12-20

Principal stresses across contact zone at surface for cylinders in combined rolling and sliding with $\mu = 0.33$

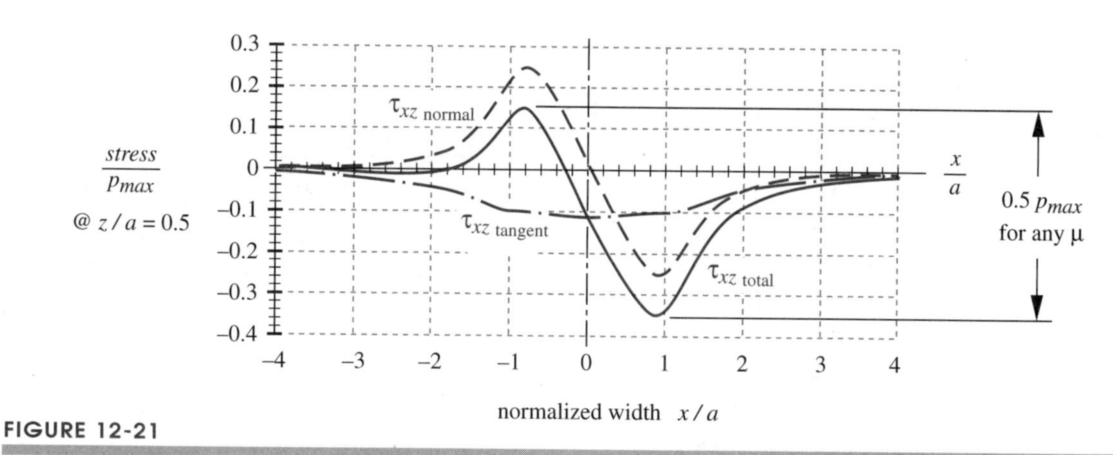

FIGURE 12-21

Shear stresses below surface at $z/a = 0.5$ for cylinders in combined rolling and sliding with $\mu = 0.33$

ies only slightly with depth. Figure 12-23 plots the largest peak value of the shear stress τ_{13} occurring at any value of x across the patch zone, and so is a composite plot of the peak shear stress value in each z plane. For $0 < \mu < 0.5$ the peak value remains within 60–80% of its largest value over the first a of depth and is still 58–70% of its peak value at $z/a = 2.0$. As the coefficient of friction is increased to 0.5 or greater, the normalized maximum shear-stress value becomes equal to μ and is constant across the contact-patch surface.

The limited variation of τ_{max} over small z depths may explain why some pitting failures appear to start at the surface and some below it. With a relatively uniform-magnitude maximum shear stress over the entire near-surface region, any inclusion in that region of the material creates a stress concentration and serves as a crack initiation point. The fact that the peak value of the maximum shear stress occurs at slightly different

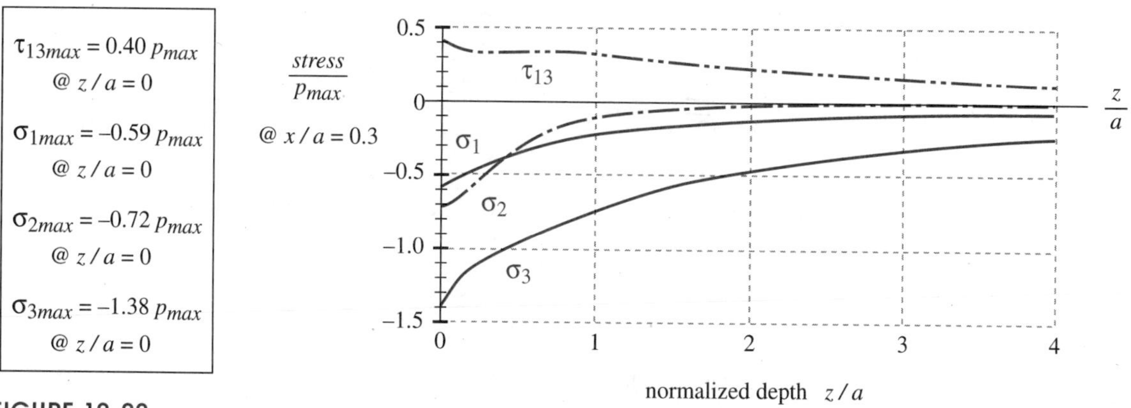

FIGURE 12-22

Principal stresses below surface at $x/a = 0.3$ for cylinders in combined rolling and sliding with $\mu = 0.33$

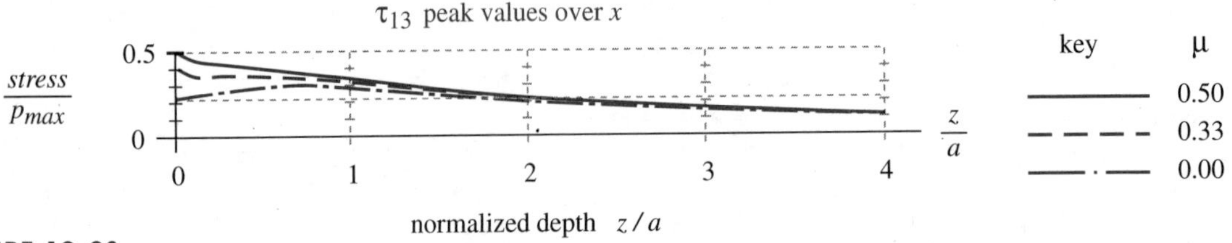

FIGURE 12-23

Peak values of maximum shear stress at all values of x / a for cylinders in combined rolling and sliding with $0 \le \mu \le 0.5$

transverse locations at different depths within the contact zone is irrelevant, since an inclusion at any particular depth will pass through that location once per revolution and be exposed to the peak stress value.

EXAMPLE 12-4

Stresses in Combined Rolling and Sliding of Cylinders.

Problem: A radial track cam and cylindrical roller follower have a combination of rolling and sliding. Find the maximum tensile, compressive, and shear stresses in the cam and roller.

Given: The roller radius is 1.25 and the minimum radius of curvature of the cam is 2.5 in. The cam is 0.875 in thick and the roller is 1-in-long. The force is 500 lb, normal to the contact plane.

Assumptions: The roller axis is exactly parallel to the cam track surface. Both materials are steel. The coefficient of friction is 0.33.

Solution:

1 The contact-patch geometry is found in the same way as was done in Example 12-3. Find the material constants from equation 12.9a (p. 355).

$$m_1 = m_2 = \frac{1 - \nu_1^2}{E_1} = \frac{1 - 0.28^2}{3E7} = 3.072E-8 \tag{a}$$

The geometry constant is found from equation 12.15a (p. 360)

$$B = \frac{1}{2}\left(\frac{1}{R_1} + \frac{1}{R_2}\right) = \frac{1}{2}\left(\frac{1}{1.25} + \frac{1}{2.5}\right) = 0.600 \tag{b}$$

and the patch half-width from equation 12.15b (p. 360).

12

$$a = \sqrt{\frac{2}{\pi}\frac{m_1 + m_2}{B}\frac{F}{L}} = \sqrt{\left(\frac{2}{\pi}\right)\frac{2(3.072E-8)}{0.600}\left(\frac{500}{0.875}\right)} = 0.006\,103 \text{ in} \tag{c}$$

where a is the half-width of the contact patch. The rectangular contact-patch area is then

$$area = 2aL = 2(0.006\,103)(0.875) = 0.010\,681 \text{ in}^2 \tag{d}$$

2 The average and maximum contact pressure can now be found from equations 12.14b and c (p. 359).

$$p_{avg} = \frac{F}{area} = \frac{500}{0.010\,681} = 46\,812 \text{ psi} \tag{e}$$

$$p_{max} = \frac{2F}{\pi a L} = \frac{2(500)}{\pi(0.006\,103)0.875} = 59\,603 \text{ psi} \tag{f}$$

The tangential pressure is found from equation 12.22*f* (p. 370):

$$f_{max} = \mu\, p_{max} = 0.33(59\,603) = 19\,669 \text{ psi} \tag{g}$$

3 With $\mu = 0.33$, the principal stresses in the contact zone will be maximal on the surface ($z = 0$) at $x = 0.3a$ from the centerline as shown in Figures 12-20 (p. 364) and 12-22 (p. 369). The applied stress components are found from equation 12.23a (p. 370) for the normal force and equation 12.23b (p. 370) for the tangential force.

$$\sigma_{x_n} = -p_{max}\sqrt{1 - \frac{x^2}{a^2}} = -59\,603\sqrt{1-0.3^2} = -56\,858 \text{ psi}$$

$$\sigma_{x_t} = -2 f_{max}\frac{x}{a} = -2(19\,669)(0.3) = -11\,801 \text{ psi} \tag{h}$$

$$\sigma_{z_n} = -p_{max}\sqrt{1 - \frac{x^2}{a^2}} = -59\,603\sqrt{1-0.3^2} = -56\,858 \text{ psi}$$

$$\sigma_{z_t} = 0 \qquad \tau_{xz_n} = 0 \tag{i}$$

$$\tau_{xz_t} = -f_{max}\sqrt{1 - \frac{x^2}{a^2}} = -19\,669\sqrt{1-0.3^2} = -18\,763 \text{ psi} \tag{j}$$

4 Equations 12.24a and b (p. 371) can now be solved for the total applied stresses along the x, y, and z axes.

$$\sigma_x = \sigma_{x_n} + \sigma_{x_t} = -56\,858 - 11\,801 = -68\,659 \text{ psi} \tag{k}$$

$$\sigma_z = \sigma_{z_n} + \sigma_{z_t} = -56\,858 + 0 = -56\,858 \text{ psi} \tag{l}$$

$$\tau_{xz} = \tau_{xz_n} + \tau_{xz_t} = 0 - 18\,763 = -18\,763 \text{ psi} \tag{m}$$

5 Since the roller is short, we expect a plane stress condition to exist. The stress in the third dimension is then:

$$\sigma_y = 0 \tag{n}$$

6 Unlike the pure rolling case, these stresses are not principal because of the applied shear stress. The principal stresses are found from equations 12.5 (p. 348) using a cubic root-finding solution.

$$\begin{aligned}\sigma_1 &= 0\\ \sigma_2 &= -43\,090 \text{ psi}\\ \sigma_3 &= -82\,428 \text{ psi}\end{aligned} \tag{o}$$

The maximum shear stress is found from the principal stresses using equation 12.6 (p. 349).

$$\tau_{13} = \frac{|\sigma_1 - \sigma_3|}{2} = \frac{|0 + 82\,428|}{2} = 41\,214 \text{ psi} \tag{p}$$

7 The principal stresses are maximum at the surface as seen in Figures 12-20 and 12-22.

12.16 SURFACE FATIGUE FAILURE MODELS—DYNAMIC CONTACT

There is still some disagreement among experts as to the actual mechanism of failure that results in pitting and spalling of surfaces. The possibility of having a maximum shear stress at a subsurface location (in pure rolling) has led some to conclude that pits begin at or near that location. Others have concluded that pitting begins at the surface. It is possible that both mechanisms are at work in these cases, since failure initiation usually begins at an imperfection, which may be on or below the surface. Figure 12-24 shows both surface and subsurface cracks in a case-hardened steel roll subjected to heavy rolling loads.[18]

An extensive experimental study of pitting under rolling contact was done by Way[19] in 1935. Over 80 tests were made with contacting, pure rolling, parallel rollers of different materials, lubricants, and loads, run for up to 18 million cycles, though most samples failed between 0.5*E*6 and 1.5*E*6 cycles. The samples were monitored for the appearance of minute surface cracks, which inevitably presaged a pitting failure within less than about 100 000 additional cycles in the presence of a lubricant.

Harder and smoother surfaces better resisted pitting failure. Highly polished samples did not fail in over 12*E*6 cycles. Nitrided rolls with very hard cases on a soft core

FIGURE 12-24

Photomicrograph (100x) of surface and subsurface cracks in a carburized and hardened roll (HRC 52-58) subjected to a heavy rolling load (Source: J. D. Graham, *Pitting of Gear Teeth*, in C. Lipson, *Handbook of Mechanical Wear*, U. Mich. Press, 1961, p.137.)

were longer-lived than other materials tested. **No pitting occurred on any samples in the absence of a lubricant** even though dry-running produced surface cracks. The cracked parts would continue to run dry with no failure for as many as $5E6$ cycles until some lubricant was added. Then the surface cracks would rapidly enlarge and turn to pits of a characteristic arrowhead shape within 100 000 additional cycles.

The suggested explanation for the deleterious effect of the lubricant was that once suitably oriented surface cracks form, they are pumped full of oil on approaching the roll-nip, and then are pressed closed within the roll-nip, pressurizing the fluid trapped in the crack. The fluid pressure creates tensile stress at the crack tip, causing rapid crack growth and subsequent break-out of a pit. Higher-viscosity lubricants did not eliminate metal-to-metal contact but did delay the pitting failure, indicating that the fluid must be able to readily enter the crack to do the damage.

Way reached a number of conclusions regarding how to design rollers to delay surface fatigue failure.[19]

1 Use no oil (though he was quick to point out that this is not a practical solution, as it promotes other types of wear as discussed in previous sections).

2 Increase the viscosity of the lubricant.

3 Polish the surfaces (though this is expensive to do).

4 Increase the surface hardness (preferably on a softer, tough core).

No conclusions were drawn with respect to the reasons for the initiation of the initial cracks on the surface. Though, with pure rolling, the shear stresses are not maximal

TABLE 12-3 Modes of Surface Failure and Their Causes

Mode of Failure	Factors That Promote Occurrence
Inclusion origin	Frequency and severity of oxide or other hard inclusions.
Geometric stress concentration	End-of-contact geometry. Misalignment and deflections. Possible lubricant-film thickness effects.
Point-surface origin (PSO)	Low lubricant viscosity. Thin elastohydrodynamic film compared to asperities on contact surfaces. Tangential forces and/or gross sliding.
Peeling (superficial pitting)	Low lubricant viscosity. Frequent asperities in surface finish exceeding elastohydrodynamic-film thickness. Loss of elastohydrodynamic pressure due to side leakage or scratches in contact surface.
Subcase fatigue (on carburized components)	Low core hardness. Thin case depth relative to radius of curvature for elements in contact.

Source: W. E. Littmann and R. L. Widner, Propagation of Contact Fatigue from Surface and Subsurface Origins, *J. Basic Eng. Trans. ASME*, vol. 88, pp. 624–636, 1966.

at the surface, they **are** nonzero there at some locations (see Figures 12-12, p. 354 and 12-17, p. 361).

Littman and Widner[20] performed an extensive analytical and experimental study on contact fatigue in 1966 and describe five different modes of failure in rolling contact. These are listed in Table 12-3 along with some factors that promote their occurrence. Some of these modes address the crack initiation issue and others the crack propagation issue. We will briefly discuss each in the order listed.

Inclusion Origin This describes a mechanism for crack initiation. It is assumed that the crack originates in a shear-stress field at a subsurface or surface location containing a small inclusion of "foreign" matter. The most commonly identified inclusions are oxides of the material that formed during processing and were captured within it. These are typically hard and irregular in shape and create stress concentration. Several researchers[21],[22],[23] have published photomicrographs of (or otherwise identified) subsurface cracks starting at oxide inclusions. "These oxide inclusions are often present as stringers or elongated aggregates of particles . . ., which provide a much greater chance for a point of high stress concentration to be in an unfavorable position with respect to the applied stress."[24] The propagation of a crack from the inclusion may remain subsurface, or break out to the surface. In the latter case it provides a site for hydraulic pressure propagation, as described above. In either case, it ultimately results in pitting or spalling.

Geometric Stress Concentration (GSC) This mechanism can act on a surface when, for example, one contacting part is shorter axially than the other (common with cam-follower joints and roller bearings). The ends of the shorter roller create line-contact stress concentration in the mating roller as shown in Figure 12-25*a,* and pitting or spalling will likely occur at that location. This is one reason for using crowned rollers, which have a large crown radius of curvature in the *yz* plane in addition to their roller radius in the *xz* plane. If the contact load is predictable, the crown radius can be sized to provide a more uniform stress distribution across the axial length of the contact area due to the deflections of the rollers, as shown in Figure 12-25*b*. However, at lighter loads, there will be reduced contact area and thus higher stresses at the center, and at higher-than-design loads the stress concentrations at the ends will return. A partial crown can be used as shown in Figure 12-25*c*, but may cause some stress concentration at the transition from straight to crown. Reusner[25] has shown that a logarithmic curve on the crown, as shown in Figure 12-25*d* ,will give a more uniform stress distribution under varied load levels.

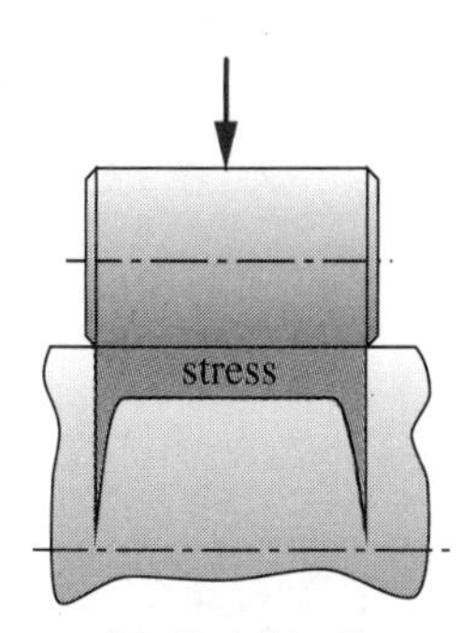

(*a*) Straight roller

POINT-SURFACE ORIGIN (PSO) This phenomenon is described by Way and discussed above. Littman et al.[20] consider PSO to be more a manner of crack propagation than crack initiation and suggest that an inclusion at or near the surface may be responsible for starting the crack. Handling nicks or dents can also provide a crack nucleus on the surface. Once present, and if pointing in the right direction to capture oil, the crack rapidly propagates to failure. Once spalling starts, the debris can create new nicks to serve as additional crack sites.

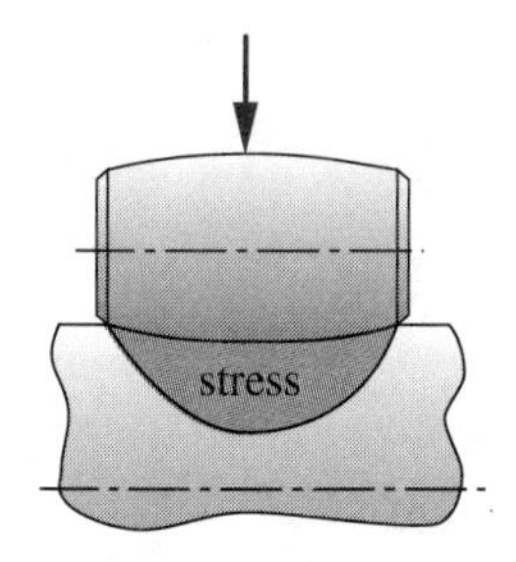

(*b*) Crowned roller

PEELING This refers to a situation in which the fatigue cracks are at shallow depth and extend over a large area such that the surface "peels" away from the substrate. Rough surfaces exacerbate peeling if the surface asperities are larger than the lubricant film thickness.

SUBCASE FATIGUE Also called *case crushing*, this occurs only on case-hardened parts and is more likely if the case is so thin that the subsurface stresses extend into the softer, weaker core material. The fatigue crack starts below the case and eventually causes the case to either collapse into the failed subsurface material or break out in pits or spalls.

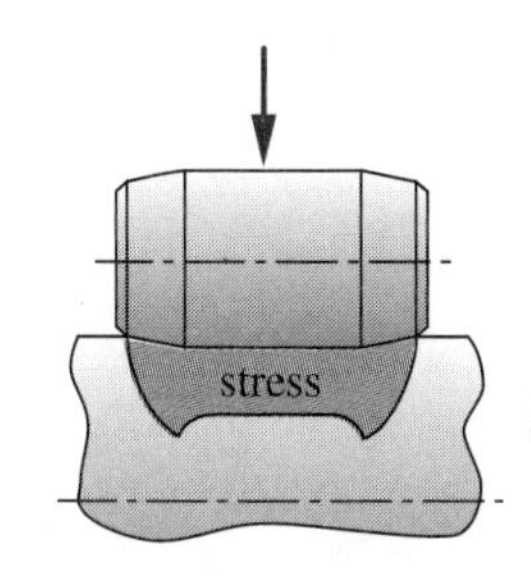

(*c*) Part-crown roller

Whatever the detailed cause of the start of a crack, once started the outcome is predictable. So, the designer needs to take all possible precautions to improve the part's resistance to pitting as well as to all other wear modes. The summary section to this chapter will attempt to set some guidelines to this end.

12.17 SURFACE FATIGUE STRENGTH

Repeated, time-varying loads tend to fail parts at lower stress levels than the material can stand in static load applications. The concept of **surface fatigue strength** is similar to that of bending- and axial-fatigue strength[2] except for one main difference. While steels and a few other materials loaded in bending or axial fatigue show an endurance limit, *no materials* show an equivalent property when loaded in surface fatigue. Thus, we should expect that our machine, though carefully designed to be safe against all other forms of failure, will eventually succumb to surface fatigue if so loaded for enough cycles.

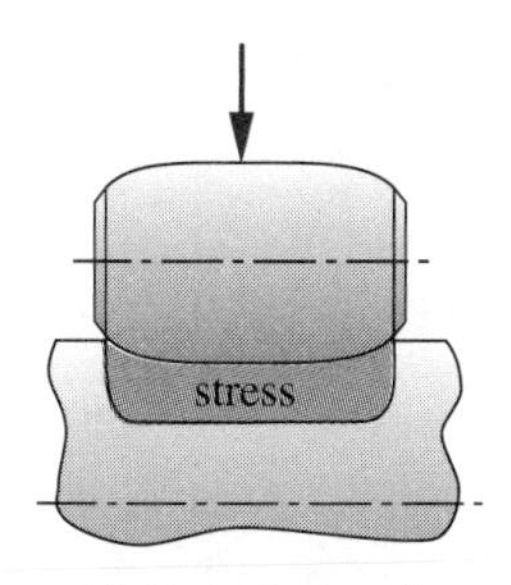

(*d*) Logarithmic roller

FIGURE 12-25

Stress concentrations beneath variously shaped rollers

Morrison[26] and Cram[27] report separately on an experimental study of the surface-fatigue strength of materials done at USM Corp. from 1932 to 1956. Four wear-testing machines were operated **24 hours per day for 24 years** to gather surface fatigue strength data on cast iron, steel, bronze, aluminum, and nonmetallic materials. Their tests included rollers in pure rolling as well as rolling plus varying percentages of sliding. Most of their roll/slide data are done at 9% sliding, since that simulates the average conditions on spur and helical gear teeth. The percent sliding figure is defined as the relative sliding velocity between the rollers or gear teeth divided by the pitch-line velocity of the interface.

Previous sections have shown the complexity of the stress state that exists in the surface and subsurface regions of the contact zones of mating cylinders, spheres, or other bodies. The discussion of crack initiation mechanisms above indicates that the location of an incipient crack is quite unpredictable, given the random distribution of inclusions in the material. Therefore it is more difficult to accurately predict the condition of stress at an expected point of failure in a contact zone than was the case in designing a cantilever beam, for example. This dilemma is resolved by using one, easily calculated contact-zone stress as a *reference value* to compare to material strengths. The one chosen is the largest negative (compressive) principal contact stress. In a pure rolling case, its magnitude will be equal to the applied maximum contact pressure P_{max}. But it will be greater than that value if sliding is present.

To develop allowable surface fatigue strengths, the material is typically run under controlled loading conditions (i.e., controlled p_{max}) and the number of cycles to failure recorded and reported along with other loading factors such as percent sliding, lubrication, body geometry, etc. This "virtual strength" can be compared to the peak magnitude of compressive stress in other applications having similar loading factors. Thus the reported surface fatigue strength has only an indirect relationship to the actual stresses that may have been present in the test piece and in your similarly loaded part, since the Hertzian stress equations are only valid for static loading.

12

The expression for the normal, compressive Hertzian static stress in cylindrical contact is found by combining equations 12.14b (p. 359) and 12.17a (p. 360):

$$\sigma_z = -p_{max} = \frac{2F}{\pi a L} \tag{12.25a}$$

Substitute the expression for a from equation 12.15b (p. 360), square both sides, and simplify:

$$\sigma_z^2 = \frac{2}{\pi}\frac{F}{L}\frac{B}{(m_1 + m_2)} \tag{12.25b}$$

Rearrange to solve for the load F,

$$F = \sigma_z^2 \frac{\pi L(m_1 + m_2)}{2B} \tag{12.25c}$$

and collect terms in a constant K,

$$F = \frac{KL}{2B} \tag{12.25d}$$

where

$$K = \pi(m_1 + m_2)\sigma_z^2 \tag{12.25e}$$

This factor K is termed an *experimental load factor* and is used to determine the safe endurance load F at a specified number of cycles or the number of cycles that can be expected before failure occurs at a given load.

Table 12-4 shows experimentally determined load factors, K, fatigue strengths, S_c, and strength factors for a number of materials running either against themselves or against hardened tool steel.[26] See the original reference for a complete listing, as some materials were omitted here due to lack of space. Two different loading modes are also addressed in separate sections of the table: pure rolling, and rolling with 9% sliding. The first column of the table defines the material. In each section, the next two columns give the K value and the surface fatigue strength at $1E8$ cycles as tested. The next two columns contain strength factors λ and ζ, which represent the slope and intercept of the *S-N* diagram (on log-log coordinates) for the surface fatigue strength of the material as determined by regression on large amounts of test data. These factors can be used in the equation of the statistically fitted *S-N* line to find the expected cycle life N for the applied stress level.

$$\log_{10} K = \frac{\zeta - \log_{10} N}{\lambda} \tag{12.26}$$

The K values in Table 12-4 can be used directly in equation 12.25d to calculate an allowable load F for the selected material at $1E8$ cycles of stress. For other desired design cycle lives, first calculate the largest negative (compressive) radial stress for your design from the appropriate equations as defined in the preceding sections. Then calculate K from equation 12.25e and use it and the values of λ and ζ from Table 12-4 to find the value of N for the application from equation 12.26. Since there is no endurance limit for surface fatigue loading, we can expect pitting to begin after approximately N stress cycles at the level of nominal stress contained in your calculated K factor.

Alternatively, a desired number of cycles N can be chosen and an allowable design-stress level σ_z for a chosen material computed from equations 12.25e and 12.26. A safety factor can be applied either by selecting a material with a longer cycle life than required for the application or by sizing the parts to have a stress level below the calculated allowable stress level for a necessary number of cycles.

The strength values in Table 12-4 were obtained using rollers in contact, lubricated with a light mineral oil of 280-320 SSU at 100°F. The researchers report that "an orderly transition occurs from pitting fatigue to abrasive wear as percent sliding is increased." Pitting failures were observed under as high as 300% sliding on some cast irons, and abrasive wear was seen at as low as 9% sliding on hardened steels under high stress. They also note that the addition of oxide coatings, fortified (EP) lubricants, or lead as an alloy-

TABLE 12-4 Surface Fatigue Strength Data for Various Materials
Part 1: Materials running against an HRC 60-62 tool-steel roller

#	Material	Pure Rolling				Rolling & 9% Sliding			
		K psi	Sc @ 1E8 cycles, psi	λ	ζ	K psi	Sc @ 1E8 cycles, psi	λ	ζ
1	1020 steel, carburized, 0.045 in min depth HRC 50-60	12 700	256 000	7.39	38.33	10 400	99 000	13.20	61.06
2	1020 steel, HB 130-150	—	—	—	—	1 720	94 000	4.78	23.45
3	1117 steel, HB 130-150	1 500	89 000	4.21	21.41	1 150	77 000	3.63	19.12
4	X1340 steel, induction hardened, 0.045 in min depth HRC 45-58	10 000	227 000	6.56	34.24	8 200	206 000	8.51	41.31
5	4150 steel, h-t, HB 270-300, flash-chrome plated	6 060	177 000	11.18	50.29	—	—	—	—
6	4150 steel, h-t, HB 270-300, phosphate coated	9 000	216 000	8.80	42.81	6 260	180 000	11.56	51.92
7	4150 cast steel, h-t, HB 270-300	—	—	—	—	2 850	121 000	17.86	69.72
8	4340 steel, induction hardened, 0.045 in min depth HRC 50-58	13 000	259 000	14.15	66.22	9 000	216 000	14.02	63.44
9	4340 steel, h-t, HB 270-300	—	—	—	—	5 500	169 000	18.05	75.55
10	6150 steel, HB 300-320	1 170	78 000	3.10	17.51	—	—	—	—
11	6150 steel, HB 270-300	—	—	—	—	1 820	97 000	8.30	35.06
12	18% Ni maraging tool steel, air hardened, HRC 48-50	—	—	—	—	4 300	146 000	3.90	22.18
13	Gray iron, Cl. 20, HB 140-160	790	49 000	3.83	19.09	740	47 000	4.09	19.72
14	Gray iron, Cl. 30, HB 200-220	1 120	63 000	4.24	20.92	—	—	—	—
15	Gray iron, Cl. 30, h-t (austempered) HB 255-300, phosphate-coated	2 920	102 000	5.52	27.11	2 510	94 000	6.01	28.44
16	Gray iron, Cl. 35, HB 225-255	2 000	86 000	11.62	46.35	1 900	84 000	8.39	35.51
17	Gray iron, Cl. 45, HB 220-240	—	—	—	—	1 070	65 000	3.77	19.41
18	Nodular iron, Gr. 80-60-03, h-t HB 207-241	2 100	96 000	10.09	41.53	1 960	93 000	5.56	26.31
19	Nodular iron, Gr. 100-70-03, h-t HB 240-260	—	—	—	—	3 570	122 000	13.04	54.33
20	Nickel bronze, HB 80-90	1 390	73 000	6.01	26.89	—	—	—	—
21	SAE 65 phosphor-bronze sand casting, HB 65-75	730	52 000	2.84	16.13	350	36 000	2.39	14.08
22	SAE 660 cont-cast bronze, HB 75-80	—	—	—	—	320	33 000	1.94	12.87
23	Aluminum bronze	2 500	98 000	5.87	27.97	—	—	—	—
24	Zinc die-casting, HB 70	250	28 000	3.07	15.35	220	26 000	3.11	15.29
25	Acetal resin	620	—	—	—	580	—	—	—
26	Polyurethane rubber	240	—	—	—	—	—	—	—

TABLE 12-4 Surface Fatigue Strength Data for Various Materials
Part 2: Materials running against the same material

#	Material	Pure Rolling				Rolling & 9% Sliding			
		K psi	Sc @ 1E8 cycles, psi	λ	ζ	K psi	Sc @ 1E8 cycles, psi	λ	ζ
27	1020 steel, HB 130-170, and same but phosphate coated	2 900	122 000	7.84	35.17	1 450	87 000	6.38	28.23
28	1144 steel CD steel, HB 260-290, (stress-proof)	—	—	—	—	2 290	109 000	4.10	21.79
29	4150 steel, h-t, HB 270-300, and same but phosphate coated	6 770	187 000	10.46	48.09	2 320	110 000	9.58	40.24
30	4150 leaded steel, phosphate coated, h-t, HB 270-300	—	—	—	—	3 050	125 000	6.63	31.1
31	4340 steel, h-t, HB 320-340, and same but phosphate coated	10 300	230 000	18.13	80.74	5 200	164 000	26.19	105.31
32	Gray iron, Cl. 20, HB 130-180	960	45 000	3.05	17.10	920	43 900	3.55	18.52
33	Gray iron, Cl. 30, h-t (austempered) HB 270-290	3 800	102 000	7.25	33.97	3 500	97 000	7.87	35.90
34	Nodular iron, Gr. 80-60-03, h-t HB 207-241	3 500	117 000	4.69	24.65	1 750	82 000	4.18	21.56
35	Meehanite, HB 190-240	1 600	80 000	4.77	23.27	1 450	76 500	4.94	23.64
36	6061-T6 aluminum, hard anodized coating	350	—	10.27	34.15	260	—	5.02	20.12
37	HK31XA-T6 magnesium, HAE coating	175	—	6.46	22.53	275	—	11.07	35.02

Source: R. A. Morrison, "Load/Life Curves for Gear and Cam Materials," *Machine Design,* vol. 40, pp. 102–108, Aug. 1, 1968, A Penton Publication, Cleveland, Ohio, with the publisher's permission.

ing element all reduced tangential stress levels and increased fatigue life or allowable % sliding. The addition of phosphate coatings to the surfaces reduced sparking and flashing of lubricant, reduced the friction coefficient, and also increased fatigue life. They saw evidence of pitting starting both at the surface under high % sliding and below the surface in pure rolling or low-percent-sliding situations.[26] Increased sliding percentages reduce fatigue life but not linearly. Figure 12-26 shows some *S-N* curves (from reference 26) for three materials with various percentages of sliding.

The speed of stress cycling only affected nonmetallic materials, wherein friction heat blistered or yielded the material. A material's stiffness is a factor, however. Lower-modulus materials reduce the contact stress because their larger deflections increase the contact-patch area. Cast iron on cast iron had longer life than cast iron on hardened steel. The free graphite in cast iron also makes it a good choice in contact situations, as it acts to retard adhesion as well as being a dry lubricant, though the lower grades of CI have strengths too low to be useful in this situation. Nodular iron in its harder forms may be a better choice. Hardness of a material was not found to correlate closely with its surface endurance. Some softer steels performed better than some harder ones.[26]

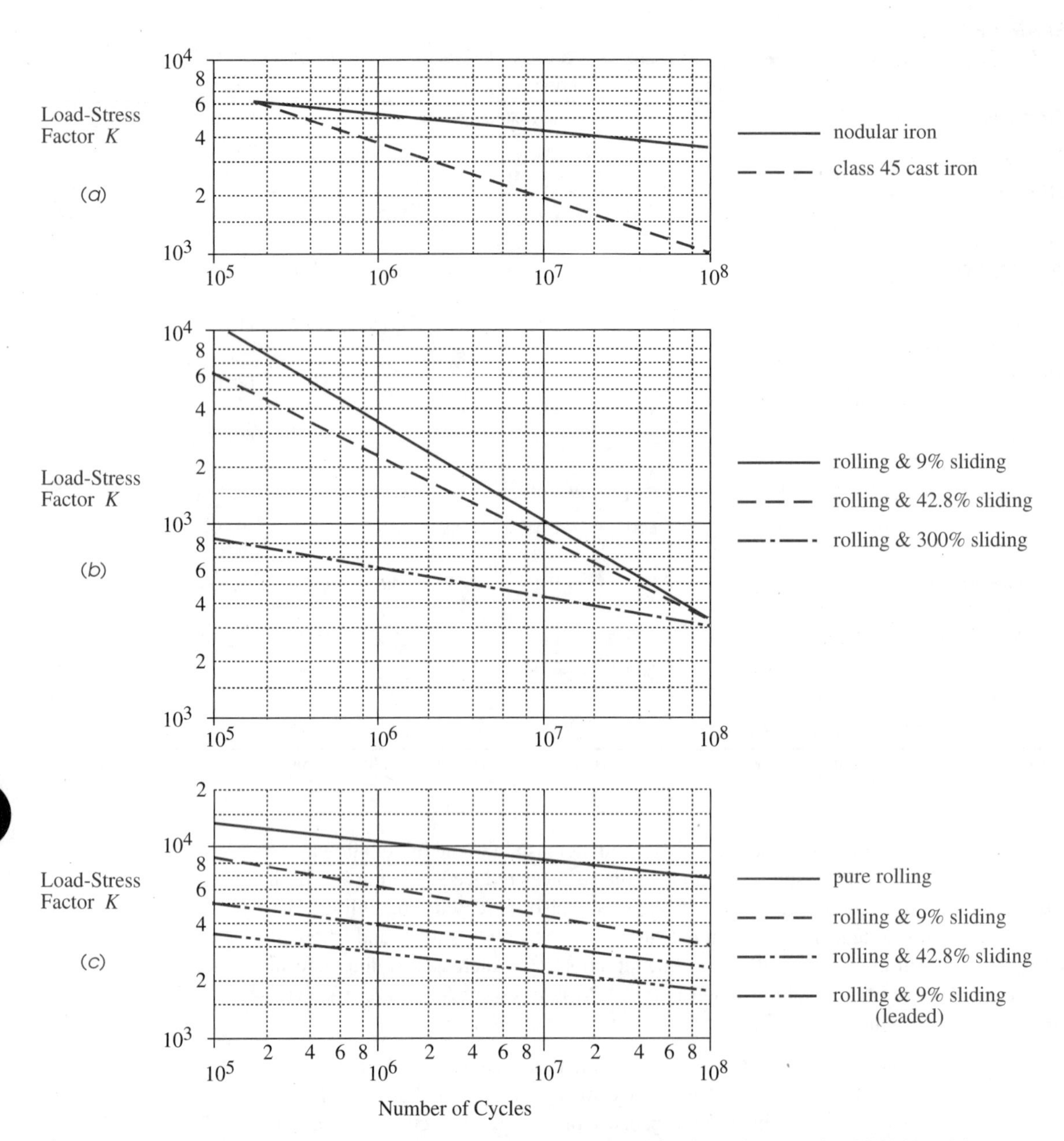

Typical curves showing load-life relationships for common gear and cam materials. Curves in (*a*) are for 100-70-30 nodular iron (HB 240-260) and class 45 gray cast iron (HB 220-240), both materials running on carbon tool steels (HRC 60-62). Curves in (*b*) are for continuous-cast bronze running on hardened steel. Curves in (*c*) are for heat-treated 4150 steel running against the same material, but phosphate coated. In all charts, 9% sliding velocity is 54 fpm; 42.8% sliding velocity is 221 fpm.

FIGURE 12-26

Load-life curves for some combinations of materials in combined rolling and sliding Source: R. A. Morrison, "Load/Life Curves for Gear and Cam Materials," *Machine Design*, vol. 40, pp. 102–108, Aug. 1, 1968, A Penton Publication, Cleveland, Ohio, with permission

EXAMPLE 12-5

Finding the Safety Factor in Surface Fatigue.

Problem: Choose a material to provide 10 years of life for the cam and roller follower in Example 12-4.

Given: The stresses are as shown in Example 12-4. The roller follower is turning at 4 000 rpm.

Assumptions: There is 9% sliding combined with rolling. The roller follower is made from HRC 60-62 tool steel. The cam can be of any suitable material from Table 12-4. The machine will operate 3 shifts/day for 345 days/year.

Solution:

1 Calculate the required cycle life from the given data:

$$cycles = 4\,000 \frac{rev}{min} \cdot 60 \frac{min}{hr} \cdot 24 \frac{hr}{day} \cdot 345 \frac{day}{yr} \cdot 10 \text{ yr} = 2.0E10 \tag{a}$$

2 The maximum normal stress calculated in Example 12-4 is 56 858 psi compressive. Its K factor can be calculated from equation 12.25d (p. 381). The previously calculated material constants m_1 and m_2 are needed:

$$m_1 = m_2 = \frac{1 - \nu_1^2}{E_1} = \frac{1 - 0.28^2}{3E7} = 3.072E-8 \tag{b}$$

$$K = \pi(m_1 + m_2)\sigma_z^2 = 2\pi(3.072E-8)(56\,858)^2 = 624 \tag{c}$$

3 A trial material must be selected from Table 12-4 (pp. 382–383). With a K this low, virtually any of the steels or ductile irons can probably be used. We will try the Nodular iron, Gr. 80-60-03, h-t HB 207-241 (#18 in part 1 of the table), since it is running against a hardened tool-steel roller. The slope and intercept factors of this steel for rolling with 9% sliding are

$$\lambda = 5.56 \qquad \zeta = 26.31 \tag{d}$$

4 These are used in equation 12.26 (p. 381) along with the value of K from equation (c) above to find the number of cycles that can be expected at this load before pitting begins.

$$\begin{aligned} \log_{10} K &= \frac{\zeta - \log_{10} N_{life}}{\lambda} \\ \log_{10} N_{life} &= \zeta - \lambda \log_{10} K = 26.31 - 5.56 \log_{10}(624) \\ N_{life} &= 10^{[26.31 - 5.56 \log_{10}(624)]} = 5.9E10 \end{aligned} \tag{e}$$

5 A safety factor against pitting can now be calculated from the ratio of the projected cycle life and the desired number of cycles.

$$N_f = \frac{N_{life}}{cycles} = \frac{5.9E10}{2.0E10} = 3.0 \qquad \text{(f)}$$

12.18 ROLLER FOLLOWERS

Cam roller followers are essentially rolling element bearings made in special configurations to suit the application. Stud and cam yoke type followers are shown in Figure 12-27. These are available with needle, caged-roller, ball, and plain bushing bearings. The most commonly used are the needle and caged-roller types. The outside diameter of the bearing's outer race runs against the cam surface. Cam followers are typically made of tool steel, hardened to Rockwell C60-62 (HRC60-62). The cam surface should be equally hard or as hard as its material will permit.

Two types of outer surface configuration are available, cylindrical and crowned. The cylindrical type is intended to provide line contact with the cam surface. To achieve this condition requires that care be taken to guarantee that the roller axis is parallel to the cam surface at assembly. If errors or inaccuracies skew the roller axis, then it may only contact the cam along one edge. It will then have a much smaller contact patch than intended and the stresses will be much higher, leading to early failure. This condition can also be caused by excessive dynamic deflections in the links that carry the roller. The crowned roller has a large radius in the plane of the roller axis, causing its contact patch with a flat cam to be elliptical as shown in Figure 12-11*a* (p. 353). This reduces the theoretical contact patch area, but in the face of any misalignment or deflection of the follower, the contact patch and thus the stresses will remain essentially the same as designed. The crowned follower should be used in any situation where good parallel alignment of roller and cam cannot be guaranteed or where the cam surface is not flat in a direction parallel to the roller axis.

(*a*) Stud type with roller bearing

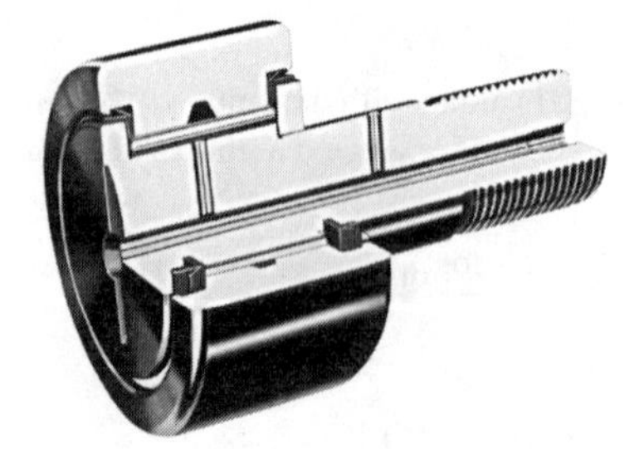

(*b*) Stud type with needle bearing

(*c*) Yoke type with needle bearing

FIGURE 12-27

Commercial cam rollers

A low coefficient of friction in the bearing promotes rolling at the cam-follower interface. If the torque due to the friction force at the cam-follower interface exceeds the friction torque within the bearing, it will roll; otherwise it will slide on the cam. In addition to calculating the surface stresses at the cam-follower interface, it is necessary to determine if the rolling element bearing will have sufficient life at the design loads. The Anti-Friction Bearing Manufacturers Association (AFBMA), an industry-sponsored organization, has developed test procedures and design data to allow this task to be easily done.

Types of Rolling-Element Bearings

Rolling-element bearings can be grouped into two broad categories, ball and roller bearings, both of which have many variants within these divisions.

BALL BEARINGS These capture a number of hardened and ground steel spheres between two raceways, an inner and outer race for radial bearings, or top and bottom races for thrust bearings. A retainer (also called a cage or separator) is used to keep the balls properly spaced around the raceways. Ball bearings can support combined radial and thrust loads to varying degrees depending on their design and construction. Ball bearing roller followers are available with shields to keep out foreign matter and seals to retain factory-applied lubricant.

ROLLER BEARINGS These use straight, tapered, or contoured rollers running between raceways. In general, roller bearings can support larger static and dynamic (shock) loads than ball bearings because of their line contact. Unless the rollers are tapered or contoured, they can support a load in only one direction, either radial or thrust according to the bearing design. Figure 12-27*a* shows a cam follower with straight, cylindrical roller bearings designed to support only radial loads. A needle bearing cam follower, as shown in Figures 12-27*b* and *c*, uses small-diameter rollers without retainers (meaning the rollers can rub on one another). Its advantages are higher load capacity due to the full complement of rollers and its compact radial dimension.*

12.19 FAILURE OF ROLLING-ELEMENT BEARINGS

If sufficient, clean lubricant is provided, failure in rolling bearings will be by surface fatigue. Failure is considered to occur when either raceway or balls (rollers) exhibit the first pit. Typically the raceway will fail first. The bearing will give an audible indication that pitting has begun by emitting noise and vibration. It can be run beyond this point, but as the surface continues to deteriorate, the noise and vibration will increase, eventually resulting in spalling or fracture of the rolling elements and possible jamming and damage to other connected elements.

Any large sample of bearings will exhibit wide variations in life among its members. The failures do not distribute statistically in a symmetrical Gaussian manner, but rather according to a Weibull distribution, which is skewed. Bearings are typically rated based on the life, stated in revolutions (or in hours of operation at the design speed), that 90% of a random sample of bearings of that size can be expected to reach or exceed at their design load. In other words, 10% of the batch can be expected to fail at that load before

* Experience has shown that despite the higher load capacity of uncaged needle bearings, the roller rubbing combined with their lower capacity for grease storage compared to caged roller bearings often leads to shorter life and higher failure rates for needle roller bearings.

the design life is reached. This is called the L_{10} life.* For critical applications, a smaller failure percentage can be designed for, but most manufacturers have standardized on the L_{10} life as a means of defining the load-life characteristic of a bearing. The rolling-bearing selection process largely involves using this parameter to obtain whatever life is desired under the anticipated loading or overloading conditions expected in service.

Figure 12-28 shows a curve of bearing failure and survival percentages as a function of relative fatigue life. The L_{10} life is taken as the reference. The curve is relatively linear to 50% failure, which occurs at a life 5 times that of the reference. In other words, it should take 5 times as long for 50% of the bearings to fail as it does for 10% to do so. After that point the curve becomes quite nonlinear, showing that it will take about 10 times as long to fail 80% of the bearings as to fail 10%. At 20 times the L_{10} life, there are still a few percent of the original bearings running.

12.20 SELECTION OF ROLLING-ELEMENT BEARINGS

Once a bearing type suited to the application is chosen based on considerations discussed above, selection of an appropriate-size bearing depends on the magnitudes of applied static and dynamic loads and the desired fatigue life.

Basic Dynamic Load Rating C

Extensive testing by bearing manufacturers, based on well-established theory, has shown that the fatigue life L of rolling bearings is inversely proportional to the third power of the load for ball bearings, and to the 10/3 power for roller bearings. These relationships can be expressed as

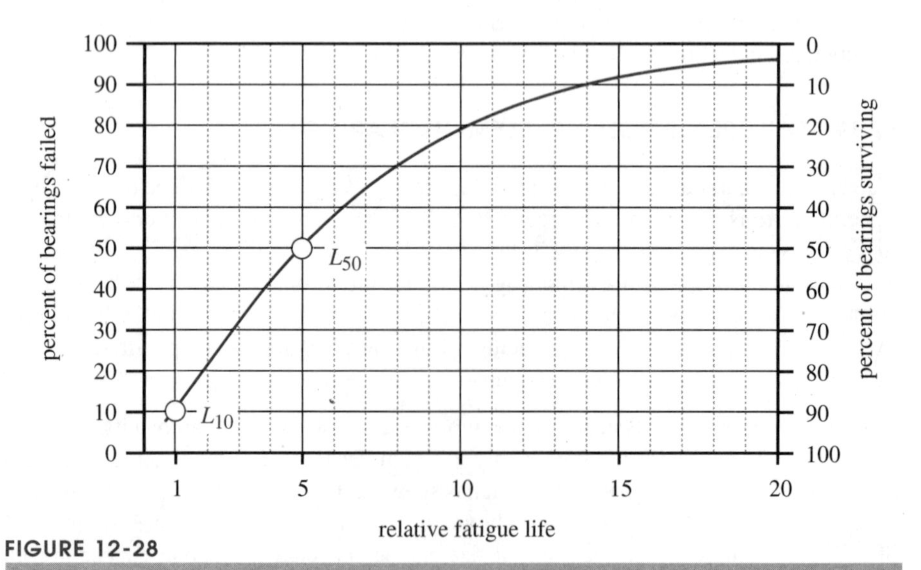

FIGURE 12-28

Typical life distribution in rolling-element bearings *(Adapted from SKF USA Inc.)*

* Some bearing manufacturers refer to this as the B_{90} or C_{90} life, referring to the survival of 90% of the bearings rather than the failure of 10%.

ball bearings:

$$L = \left(\frac{C}{P}\right)^3 \tag{12.27a}$$

roller bearings:

$$L = \left(\frac{C}{P}\right)^{10/3} \tag{12.27b}$$

where L is fatigue life expressed in millions of revolutions, P is the applied load,* and C is the *basic dynamic load rating* for the particular bearing that is defined by the manufacturer and published for each bearing in the bearing catalogs. The ***basic dynamic load rating*** C is defined as *the load that will give a life of 1 million revolutions of the inner race.* This load C is typically larger than any practical load that one would subject the particular bearing to, because the desired life is usually much higher than 1 million revolutions. In fact, some bearings will fail statically if actually subjected to a load equal to C. It is simply a reference value that allows bearing life to be predicted at any level of actual applied load. Figure 12-29 shows a page from a cam-follower bearing manufacturer's catalog that specifies the value of C for each bearing.

* Note that even a constant external load applied to a rotating bearing creates dynamic loads in the bearing elements in the same manner that a constant moment on a rotating shaft causes dynamic stresses, because any one point on a ball, roller, or raceway sees the load come and go as the bearing rotates.

BRG. NO.	ROLLER DIA. (RD) +.000 -.001	ROLLER WIDTH (W) +.000 -.005	STUD DIA. (SD) +.001 -.000	STUD L'GTH. (SL)	MIN. THR'D L'GTH. (TL)	FINE THR'DS.	OIL HOLE		LUB. FIT-TING SIZE (F)	MIN. BOSS DIA.	HOUSING BORE DIA. +.0002 -.0003	*RECOM. CLAMPING TORQUE LBS.-IN.	MAX. STATIC CAPACITY LBS.	BASIC DYN. RATING LBS.
							HOLE CENTER (HC)	HOLE DIA. (HD)						
CCFH-1/2-SB	.500	.375	.250	5/8	1/4	1/4-28	-	-	-	13/32	.2503	35	1580	680
CCFH-9/16-SB	.5625	.375	.250	5/8	1/4	1/4-28	-	-	-	13/32	.2503	35	1580	680
CCFH-5/8-SB	.625	.4375	.3125	3/4	5/16	5/16-24	-	-	-	15/32	.3128	90	2480	955
CCFH-11/16-SB	.6875	.4375	.3125	3/4	5/16	5/16-24	-	-	-	15/32	.3128	90	2480	955
CCFH-3/4-SB	.750	.500	.4375	7/8	3/8	7/16-20	1/4	3/32	3/16	39/64	.4378	250	4130	1660
CCFH-7/8-SB	.875	.500	.4375	7/8	3/8	7/16-20	1/4	3/32	3/16	39/64	.4378	250	4130	1660
CCFH-1-SB	1.000	.625	.625	1	1/2	5/8-18	1/4	3/32	3/16	25/32	.6253	650	6120	2225
CCFH-1 1/8-SB	1.125	.625	.625	1	1/2	5/8-18	1/4	3/32	3/16	25/32	.6253	650	6120	2225
CCFH-1 1/4-SB	1.250	.750	.750	1 1/4	5/8	3/4-16	5/16	3/32	3/16	63/64	.7503	1250	8500	3930
CCFH-1 3/8-SB	1.375	.750	.750	1 1/4	5/8	3/4-16	5/16	3/32	3/16	63/64	.7503	1250	8500	3930
CCFH-1 1/2-SB	1.500	.875	.875	1 1/2	3/4	7/8-14	3/8	3/32	3/16	1 3/32	.8753	1500	11280	4840
CCFH-1 5/8-SB	1.625	.875	.875	1 1/2	3/4	7/8-14	3/8	3/32	3/16	1 3/32	.8753	1500	11280	4840
CCFH-1 3/4-SB	1.750	1.000	1.000	1 3/4	7/8	1-14	7/16	3/32	3/16	1 1/4	1.0003	2250	15840	6385
CCFH-1 7/8-SB	1.875	1.000	1.000	1 3/4	7/8	1-14	7/16	3/32	3/16	1 1/4	1.0003	2250	15840	6385
CCFH-2-SB	2.000	1.250	1.125	2	1	1 1/8-12	1/2	1/8	3/16	1 13/32	1.1253	2800	21140	8090
CCFH-2 1/4-SB	2.250	1.250	1.125	2	1	1 1/8-12	1/2	1/8	3/16	1 13/32	1.1253	2800	21140	8090
CCFH-2 1/2-SB	2.500	1.500	1.250	2 1/4	1 1/8	1 1/4-12	9/16	1/8	3/16	1 11/16	1.2503	3450	32900	11720
CCFH-2 3/4-SB	2.750	1.500	1.250	2 1/4	1 1/8	1 1/4-12	9/16	1/8	3/16	1 11/16	1.2503	3450	32900	11720
CCFH-3-SB	3.000	1.750	1.500	2 1/2	1 1/4	1 1/2-12	5/8	1/8	① 1/4	2 1/8	1.5003	5000	49820	15720
CCFH-3 1/4-SB	3.250	1.750	1.500	2 1/2	1 1/4	1 1/2-12	5/8	1/8	① 1/4	2 1/8	1.5003	5000	49820	15720
CCFH-3 1/2-SB	3.500	2.000	1.750	2 3/4	1 3/8	1 3/4-12	11/16	1/8	① 1/4	2 7/16	1.7503	5000	63250	22800
CCFH-4-SB	4.000	2.250	2.000	3 1/2	1 1/2	2-12	3/4	1/8	① 1/4	2 51/64	2.0003	5000	89540	29985

FIGURE 12-29

Dimensions and load ratings for CCFH-SB series Camrol cam follower bearings *(Courtesy of McGill Precision Bearings, Valparaiso, IN.)*

Basic Static Load Rating C_0

Permanent deformations on rollers or balls can occur at even light loads because of the very high stresses within the small contact area. The limit on static loading in a bearing is defined as the load that will produce a total permanent deformation in the raceway and rolling element at any contact point of 0.0001 times the diameter *d* of the rolling element. Larger deformations will cause increased vibration and noise, and can lead to premature fatigue failure. The stresses required to cause this $0.0001d$ static deformation in bearing steel are quite high, ranging from about 4 GPa (580 kpsi) in roller bearings to 4.6 GPa (667 kpsi) in ball bearings. Bearing manufacturers publish a basic static load rating C_0 for each bearing, calculated according to AFBMA standards. This loading can sometimes be exceeded without failure, especially if rotating speeds are low, which avoids vibration problems. It usually takes a load of $8C_0$ or larger to fracture a bearing. Figure 12-29 shows a page from a bearing manufacturer's catalog that specifies the value of C_0 for each bearing.

Calculation Procedures

Equations 12.27 can be solved for any situation in which either the applied load or a desired fatigue life is known. Usually, the radial loads acting on the follower bearing will be known from a load analysis of the design. A bearing manufacturer's catalog should then be consulted, a trial bearing (or bearings) selected, and the values of *C*, and C_0 extracted. The load *P* and basic dynamic load rating *C* are used in equations 12.27 to find the predicted fatigue life *L*.

Alternatively, equations 12.27 can be solved for the value of dynamic load factor *C* required to achieve a desired life *L*. The bearing catalogs can then be consulted to find a suitably sized bearing with the necessary *C* value. In either case, the static load should also be compared to the static load factor C_0 for the chosen bearing to guard against excessive deformations.

EXAMPLE 12-6

Selection of Cam-Roller for a Designed Cam.

Problem: Select a suitable cam follower for a cam with the following loading.

Given: The maximum dynamic load on the cam is 400 lb. The camshaft speed is 18.85 rad/sec (180 rpm or 3 rps). Minimum radius of curvature of the cam is 1.7 in.

Assumptions: Thrust loads are negligible. The machine operates 24 hrs/day, 365 days/year.

Find: The projected bearing L_{10} fatigue life in years.

Solution:

1 From Figure 12-29, choose a 1.5-in-dia #CCFH- 1 1/2-SB cam follower, which is 1/2.26 the radius of the cam's minimum ρ. This is an acceptable ratio. Its dynamic load rating factor is C = 4 840 lb.

2 The static load rating C_0 = 11 280 lb. The static applied load of 400 lb is obviously well below the bearing's static rating.

3 Calculate the projected life using equation 12.27b, since it is a roller bearing.

$$L = \left(\frac{C}{P}\right)^{\frac{10}{3}} = \left(\frac{4\,840}{400}\right)^{\frac{10}{3}} = 4.067E3 \text{ millions of revs } = 4E9 \text{ revs} \tag{a}$$

4 Calculate the projected life in years

$$4E9 \text{ revs}\left(\frac{1\text{sec}}{3\text{ rev}}\right)\left(\frac{1\text{ hr}}{3600\text{ sec}}\right)\left(\frac{1\text{ day}}{24\text{ hr}}\right)\left(\frac{1\text{ year}}{365\text{ day}}\right) = 43 \text{ years} \tag{b}$$

5 This bearing is obviously very lightly loaded and should outlast the machine.

12.21 REFERENCES

1 **R. L. Norton**. (2000). *Machine Design: An Integrated Approach*, 2ed. Prentice-Hall.

2 **E. Rabinowicz**. (1965). *Friction and Wear of Materials*. John Wiley & Sons: New York, pp. 110.

3 *Ibid*., pp. 21, 33.

4 *Ibid*., p. 125.

5 **D. J. Wulpi**. (1990). *Understanding How Components Fail*. American Society for Metals: Metals Park, OH.

6 **J. F. Archard**. (1980). "Wear Theory and Mechanisms," in *Wear Control Handbook*, M. B. Peterson and W. O. Winer, eds., McGraw-Hill: New York, pp. 35-80.

7 **J. T. Burwell**. (1957). Survey of Possible Wear Mechanisms. *Wear*, **1**: pp. 119-141.

8 **J. R. McDowell**. (1961). Fretting and Fretting Corrosion, in *Handbook of Mechanical Wear*, C. Lipson and L. V. Colwell, ed. Univ. of Mich. Press: Ann Arbor. pp. 236-251.

9 **H. Hertz**. (1881). On the Contact of Elastic Solids. *J. Math*., **92**: pp. 156-171, (in German).

10 **H. Hertz**. (1896). Contact of Elastic Solids, in *Miscellaneous Papers*, P. Lenard, ed. Macmillan & Co. Ltd.: London. pp. 146-162.

11 **H. L. Whittemore and S. N. Petrenko**. (1921). *Friction and Carrying Capacity of Ball and Roller Bearings*, Technical Paper 201, National Bureau of Standards, Washington, D.C.

12 **H. R. Thomas and V. A. Hoersch**. (1930). *Stresses Due to the Pressure of One Elastic Solid upon Another*, Bulletin 212, U. Illinois Engineering Experiment Station, Champaign, IL, July 15.

13 **E. I. Radzimovsky**. (1953). *Stress Distribution and Strength Condition of Two Rolling Cylinders*, Bulletin 408, U. Illinois Engineering Experiment Station, Champaign, Ill., Feb.

14 **J. O. Smith and C. K. Lui**. (1953). Stresses Due to Tangential and Normal Loads on an Elastic Solid with Application to Some Contact Stress Problems. *J. Appl. Mech. Trans. ASME*, **75**: pp. 157-166.

15 **V. S. Mahkijani**. (1984). *Study of Contact Stresses as Developed on a Radial Cam using Photoelastic Model and Finite Element Analysis*. M.S. Thesis, Worcester Polytechnic Institute.

16 **S. P. Timoshenko and J. N. Goodier**. (1970). *Theory of Elasticity*, 3rd ed., McGraw-Hill: New York, pp. 403-419.

17 **J. Poritsky**. (1950). Stress and Deformations due to Tangential and Normal Loads on an Elastic Solid with Applications to Contact of Gears and Locomotive Wheels, *J. Appl. Mech.* Trans ASME, **72**: p. 191.

18 **E. Buckingham and G. J. Talbourdet**. (1950). *Recent Roll Tests on Endurance Limits of Materials.* in Mechanical Wear Symposium. ASM.

19 **S. Way**. (1935). Pitting Due to Rolling Contact. *J. Appl. Mech. Trans. ASME*, **57**: p. A49-58.

20 **W. E. Littman and R. L. Widner**. (1966). Propagation of Contact Fatigue from Surface and Subsurface Origins, *J. Basic Eng. Trans. ASME*, **88**: pp. 624-636.

21 **H. Styri**. (1951). *Fatigue Strength of Ball Bearing Races and Heat-Treated 52100 Steel Specimens.* Proceedings ASTM, **51**: p. 682.

22 **T. L. Carter, et al.** (1958). Investigation of Factors Governing Fatigue Life with the Rolling Contact Fatigue Spin Rig, *Trans. ASLE,* **1**: p. 23.

23 **H. Hubbell and P. K. Pearson**. (1959). *Nonmetallic Inclusions and Fatigue under Very High Stress Conditions, in Quality Requirements of Super Duty Steels.* AIME Interscience Publishers: p. 143.

24 **W. E. Littmann and R. L. Widner**. (1966). Propagation of Contact Fatigue from Surface and Subsurface Origins, *J. Basic Eng. Trans. ASME*, **88**: p. 626.

25 **H. Reusner**. (1987). The Logarithmic Roller Profile—the Key to Superior Performance of Cylindrical and Tapered Roller Bearings, *Ball Bearing Journal*, SKF, **230** , June.

26 **R. A. Morrison**. (1968). "Load/Life Curves for Gear and Cam Materials." *Machine Design*, v. 40, pp. 102-108, Aug. 1.

27 **W. D. Cram**. (1961). Experimental Load-Stress Factors, in *Handbook of Mechanical Wear*, C. Lipson and L. V. Colwell, eds., Univ. of Mich. Press: Ann Arbor. pp. 56-91.

Chapter 13

CAM PROFILE DETERMINATION

13.0 INTRODUCTION

Once the cam follower motion functions have been defined, the follower train's dynamic behavior modeled, its stresses computed, and the design deemed satisfactory on all counts, the next step is to create the cam profile for manufacture. The procedure used for this depends on the geometry of the cam follower. A roller follower requires a different approach than does a flat-faced follower. Manufacturing issues such as the cutter or grinder diameter may require the calculation of a so-called cutter-compensation path. The type of machine to be used may also influence the choice of profile definition in either polar or Cartesian coordinates. This chapter addresses the procedures for determining the geometry of the cam envelope, as defined either at the centerline of a roller follower or on the surface of the cam for a flat-faced follower. Manufacturing issues will be discussed in the next chapter.

Other factors that affect the geometry of the cam contour are the type of cam, either radial or barrel, the motion of the follower, either translating or rotating, and the presence of a linkage in the follower train. The designed kinematic/dynamic motion of the follower should ideally be imparted to the end effector of the follower train. If there is a multibar linkage between the cam and end effector, then its geometry should be taken into account. Note that, if properly done, the dynamic modeling of the follower train should have taken the dynamic characteristics (mass, compliance) of any follower linkage into account as described in Chapters 8 through 10. Now any follower linkage's basic geometry also needs to be accounted for in creating the cam profile for maximum accuracy. Note that for simple fourbar follower linkages with small angular strokes and near parallelogram geometry, ignoring the linkage geometry contribution will make little difference in the cam contour. However more complex linkages or ones with substantial angular motions will distort the end effector motion if it is not accounted for in the contour calculation.

Another issue affecting cam contour definition would be a requirement for a conjugate cam set. Conjugate cams were defined in Chapter 1 (p. 5) and comprise a pair of cams on a common shaft that drive a pair of followers arranged on a common lever or slider that maintain a fixed distance or angle between the (flat) follower surfaces or (roller) centerlines. This requires that two conjugate cam surfaces (or roller centerline) paths be defined.

Inversion

The conventional approach to cam contour definition is to invert the problem. Though in most cases the cam rotates and the follower train moves with respect to a fixed pivot or slide track, for contour specification the follower train and its ground plane are considered to rotate about the cam centerline with the cam held stationary. The follower train must of course "rotate" around the cam in the opposite direction to that of the cam's actual rotation, i.e., making it an inversion of the actual motion. For the case where the cam is in fact stationary and the follower train actually rotates around it, no inversion is necessary to define the cam contour.

It is usually left to the cam manufacturer to correctly determine the direction in which to run the cutter around the cam blank, based on information provided on the cam drawing. Thus it is important to completely specify the cam on the drawing in respect to its rotation direction versus hub up/down, the location of zero rotation angle and keyway, the roller diameter (if any), and to show its timing diagram on the drawing. The actual cam contour coordinates are seldom provided on the drawing nowadays since they are invariably computer generated and typically are supplied to the manufacturer as a text file or as a CAD file.

Digitization Increment

While all of the mathematical definitions of follower motion discussed in previous chapters result in continuous (or analog) functions for all parameters, the realities of manufacturing dictate that we can only specify a discrete and finite number of coordinates for the cam contour. Thus we must decide on the frequency at which we will specify the cam coordinate data. Various manufacturing techniques to be discussed in the next chapter can be used to improve the accuracy of this discrete estimation of the actual cam functions, but at this stage, the designer must choose an increment at which to sample the continuous mathematical functions.

For small diameter cams, the sampling frequency can be lower (i.e., larger steps) than for large diameter cams. The tangent of 1° is 0.017. So sampling the cam displacement function at 1° intervals will give data points spaced about 0.017 in (0.44 mm) apart on a cam of 1-in radius, or 0.034 in (0.88 mm) apart on a 2-in radius cam, etc. Typical digitization increments used in cam manufacture range from 1° down to 1/8° or 1/10°* with the choice based on cam size in order to make the distance between data points reasonable for manufacturing accuracy. As noted above, in the next chapter we will discuss methods to improve the accuracy of cam contour manufacture in the case of a sparse set of contour data.

* Program DYNACAM allows increments of 1, 1/2, 1/4, and 1/8°.

13.1 RADIAL CAMS WITH ROLLER FOLLOWERS

The conventional approach for the definition of a roller-follower cam contour is to define the path of the centerline of the follower around the cam. If the manufacturer were to use a cutter or grinding wheel of the same diameter as the roller follower, then following the path of the roller centerline would generate the proper cam surface contour. Since it is not always convenient or even possible to use a cutting/grinding tool of the same diameter as the roller, it may be necessary for the designer or manufacturer to modify the centerline path by offsetting it along the normal vector to the cam surface at each defined position around the cam. This process is called cutter compensation and results in a definition of the path of the centerline of the cutter that is a "normal offset" path to that of the roller follower. The cam surface coordinates are also on a normal offset path to that of the roller centerline, being effectively the path of a cutter of zero radius.

Offset Translating Roller Follower

The general case for a translating roller follower allows for the axis of the follower motion to be offset from the cam centerline. A zero-offset follower is a special case in which the follower axis extended passes through the cam centerline. We will derive the equations for the general case. They can be used for the special case by setting the offset to zero. Follower offset is sometimes also referred to as follower eccentricity.

Figure 13-1 shows the geometry for this case. The cam is shown at its $\theta = 0$ position and the follower is rotated around the cam in the opposite direction to cam rotation. The follower is shown in two positions in the figure, at zero cam angle and at an arbitrary cam angle θ. Since the prime circle radius R_p will have been chosen prior to this stage of cam design in order to control pressure angle and radius of curvature (including the effects of any follower offset), it will be used unchanged in this calculation. (See Chapter 7 for a discussion of pressure angle and radius of curvature.)

ROLLER CENTERLINE COORDINATES The cam coordinates depend in combination upon the rotation direction of the cam and on the direction of the follower offset. Figure 13-1 shows a global *XY* coordinate system rooted at the cam centerline with the *X* axis oriented at cam zero. Cam rotation is defined by the right-hand-rule as positive counterclockwise (CCW) and negative clockwise (CW).

The initial follower displacement d at $\theta = 0$ (labeled A in the figure) is determined by the offset e and the prime circle radius R_p according to

$$d = \sqrt{R_p^2 - e^2} \tag{13.1a}$$

Note that the offset e is defined as positive (leading) if it shifts the follower opposite to the direction of the tangential velocity of the cam surface and negative (trailing) if it shifts the follower in the same direction as the tangential velocity of the cam surface.

The initial offset angle γ of the follower with respect to the cam angle θ is

$$\gamma(0) = \tan^{-1}\left(\frac{e}{d}\right) \tag{13.1b}$$

from triangle OAD.

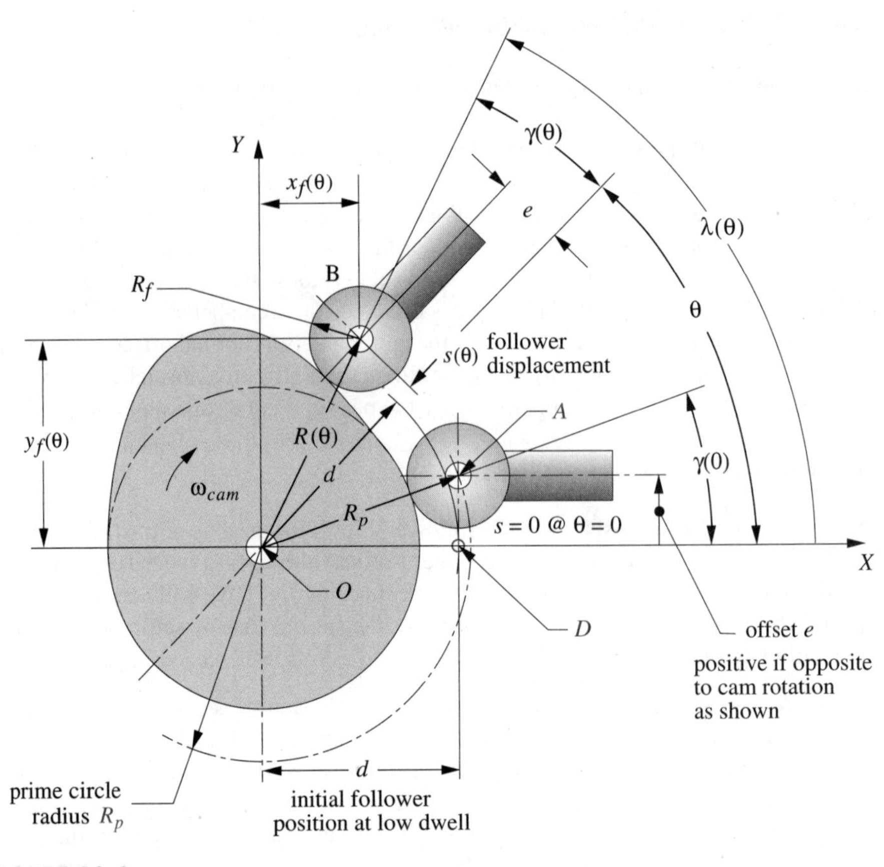

FIGURE 13-1

Geometry for calculation of the cam contour for an offset translating follower

13

Note that angle $\gamma(\theta)$ will vary at each position of the follower around the cam as defined in equation 13.2b. The initial position d of the follower from equation 13.1a will be used to calculate the coordinates of the follower at each succeeding position as the cam angle θ is varied in the chosen increments (in the opposite direction to cam rotation).

At the later (arbitrary) position shown (labeled B in the figure), when the cam has rotated through angle θ, the follower has risen above its initial position d by the displacement $s(\theta)$. Its new radius $R(\theta)$, offset angle $\gamma(\theta)$, and angle to the follower $\lambda(\theta)$ are now:

$$R(\theta) = \sqrt{[d + s(\theta)]^2 + e^2} \tag{13.2a}$$

$$\gamma(\theta) = \tan^{-1}\left(\frac{e}{(d + s(\theta))}\right) \tag{13.2b}$$

$$\lambda(\theta) = \gamma(\theta) + \theta \tag{13.2c}$$

Equations 13.2a and 13.2c will generate the polar coordinates of the follower centerline path around the cam. Equation 13.2c is valid for both directions of rotation of the cam. To create the Cartesian coordinates of the cam simply requires a transformation.

$$\begin{aligned} x_f(\theta) &= \left[R(\theta)\right]\left\{\cos\left[\lambda(\theta)\right]\right\} \\ y_f(\theta) &= -\operatorname{sgn}(\omega)\left[R(\theta)\right]\left\{\sin\left[\lambda(\theta)\right]\right\} \end{aligned} \tag{13.3}$$

where ω is the (constant) angular velocity of the cam and its sign (*sgn*) obeys the right-hand-rule. The negative sign on the *sgn* function reflects the inversion of the follower around the cam in the opposite direction to cam rotation.

Note that for a cam with a nonzero follower offset, this approach to cam contour generation results in an initial position of the follower that is less than the prime radius. This fact must be taken into account in the design of the follower system to ensure that the follower's end effector is in the proper assembly location at the low dwell or bottom point of the cam motion.

CAM SURFACE COORDINATES Once the roller centerline coordinates have been calculated, they can be used to compute the coordinates of the cam surface for either an open (external or internal surface) radial cam, or for a closed-track cam with both external and internal surfaces. Note that the cam surface coordinates cannot be determined by simply subtracting or adding the roller radius from the roller center location except on a dwell. The contact point between roller and cam will lead the line that connects the cam center and roller center on a rise, and lag it on a fall. This can be seen in Figure 13-1 at position *B* where the follower is on a rise. The cam-roller contact point will lie along the normal to the path of the roller centerline (pitch curve), called the common normal. The common normal will only intersect the cam center in a dwell.

Figure 13-2 shows an offset translating roller follower in an arbitrary (inverted) position on a rise. The cam angle is θ and the pressure angle at this position is $\phi(\theta)$. The roller centerline is at coordinates x_f, y_f as defined in equation 13.3. The point of contact x_s, y_s between cam and roller lies along the common normal defined by angle $\phi(\theta)$.

The angle of the common normal in the *XY* coordinate system for any position θ is

$$\begin{aligned} \sigma(\theta) &= -\operatorname{sgn}(\omega)\left[\theta - \phi(\theta)\right] \\ &= \operatorname{sgn}(\omega)\left[\phi(\theta) - \theta\right] \end{aligned} \tag{13.4a}$$

where θ is a positively increasing cam angle between 0 and 2π, $\phi(\theta)$ is the signed pressure angle as calculated by equation 7.3d (p. 156). Multiplying by the sign of the camshaft angular velocity ω accounts for CW or CCW cam rotation (right-hand rule).

The cam surface coordinates for a translating follower are then defined by:

$$\begin{aligned} x_s(\theta) &= x_f(\theta) \pm R_f \cos\left[\sigma(\theta)\right] \\ y_s(\theta) &= y_f(\theta) \pm R_f \sin\left[\sigma(\theta)\right] \end{aligned} \tag{13.4b}$$

The plus sign gives the outer envelope and the minus sign gives the inner envelope.

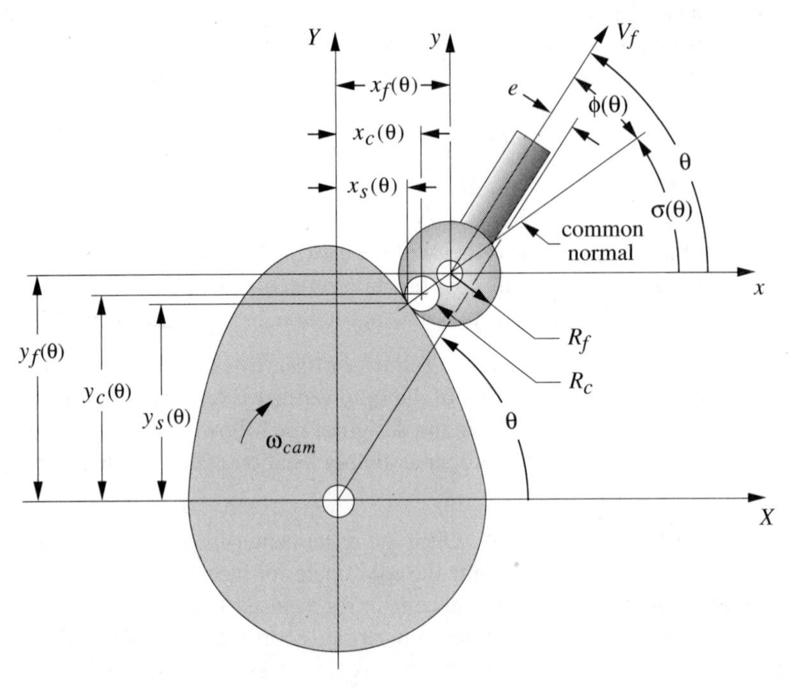

FIGURE 13-2

Geometry for the cam surface and cutter coordinates for an offset translating follower

13

Cam Cutter Coordinates In the case that the cam roller diameter is inappropriate for a milling cutter or grinding wheel, or if the designer provides only cam surface coordinates to the manufacturer, then a so-called **cutter compensation** calculation must be done to generate a set of coordinates at the locus of the center of the chosen cutter diameter. These points must lie along the common normal for each position of the roller around the cam, i.e., on the line connecting the roller centerline and the contact point between cam and follower.

Figure 13-2 also shows a small diameter cutter of radius R_c in contact with the cam. The coordinates of its center, x_c, y_c can be found in similar fashion to the surface coordinates.

$$x_c(\theta) = x_f(\theta) \pm (R_f - R_c)\cos[\sigma(\theta)]$$
$$y_c(\theta) = y_f(\theta) \pm (R_f - R_c)\sin[\sigma(\theta)] \tag{13.4c}$$

Equation 13.4c is valid for cutters either larger or smaller in diameter than the roller. Be sure to check the cutter or grinding wheel radius against the minimum (positive and negative) radii of curvature of the cam to avoid undercutting. (See Chapter 7.)

Any of the Cartesian coordinates calculated can of course be converted to polar (R,θ) coordinates by the Pythagorean formula and a two-argument arctangent function.

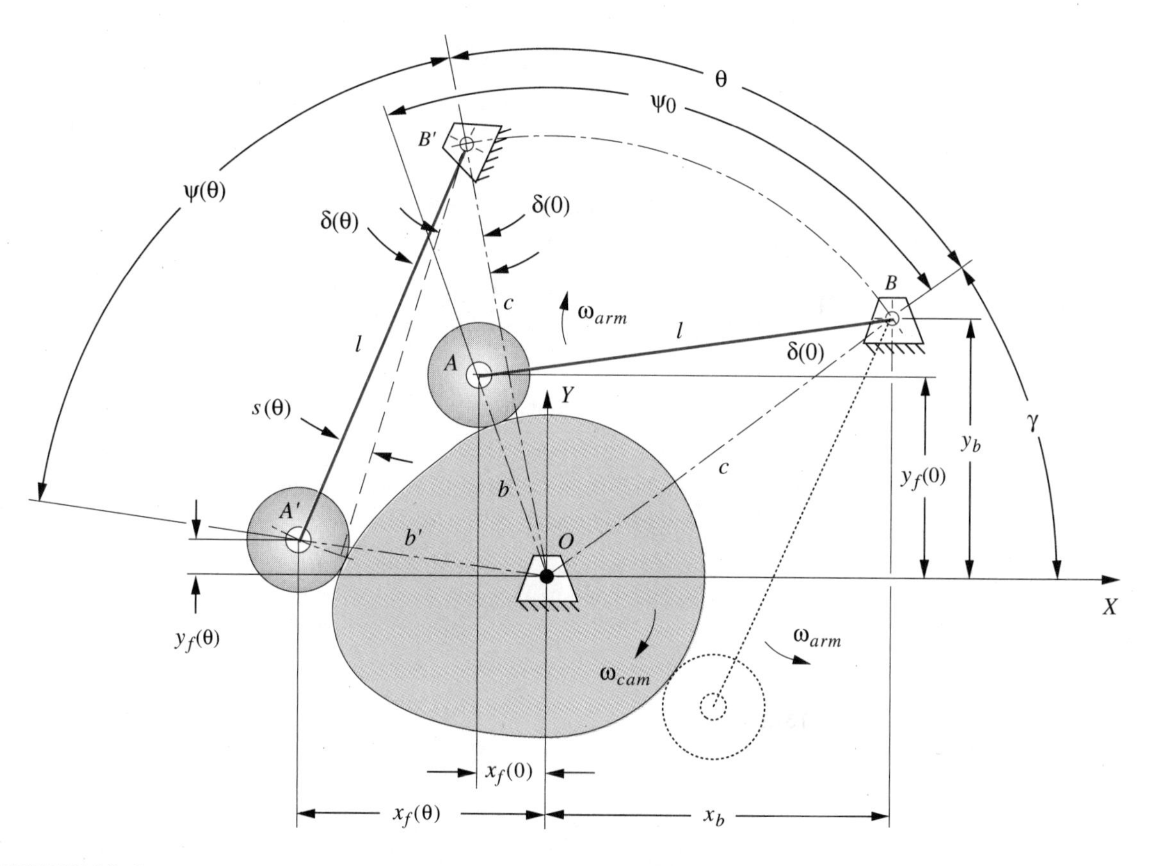

FIGURE 13-3

Geometry for calculation of the cam contour for an oscillating roller follower at the roller centerline (1)

13

Oscillating Roller Follower

The geometry of a radial cam with an oscillating arm roller follower is depicted in Figure 13-3,[1] which shows two positions of the follower arm *BA* being rotated around a "stationary" cam in the typical inversion of motion for analysis purposes. (Usually, the follower arm pivot *B* remains stationary and the cam rotates.) The initial arm position *BA* becomes *B´A´* at a later time after the cam has rotated through the angle θ.

ROLLER CENTERLINE COORDINATES The coordinates of the follower arm pivot x_b, y_b in the global *XY* coordinate system as well as the follower arm radius l and direction of arm rotation must be known *a priori*. For this analysis, it is assumed that at cam angle $\theta = 0$, the follower is on the low dwell or low point of the cam, and thus the radius to the cam follower (shown as b in Figure 13-3) is equal to the prime circle radius R_p. For any other situation, the value of b should be the radius to the roller follower centerline at $\theta = 0$. Follower motion on a rise is then a positively increasing angle $s(\theta)$, starting at zero. The length of center distance c from cam center to arm pivot and its angle γ at that position can be found from trigonometry as:

$$c = \sqrt{x_b^2 + y_b^2} \tag{13.5a}$$

$$\gamma = \tan^{-1}\left(\frac{y_b}{x_b}\right) \tag{13.5b}$$

being careful to compute a two-argument arctangent in order to accommodate the possibility of angle values in all four quadrants.*

The initial angle of the follower arm $\delta(0)$ with respect to the line of centers c can be found from the law of cosines for the case of 3 known sides.

$$\delta(0) = \cos^{-1}\left(\frac{c^2 + l^2 - b^2}{2cl}\right) \tag{13.6a}$$

Note that this initial angle $\delta(0)$ is valid for all values of cam angle θ. The arm rotation angle $s(\theta)$ will be added to it at each position to calculate a new roller displacement $\delta(\theta)$.

The initial angle $\psi(0)$ between the line of centers OB and the follower vector OA can also be found from the law of cosines.

$$\psi(0) = \cos^{-1}\left(\frac{c^2 + b^2 - l^2}{2cb}\right) \tag{13.6b}$$

The coordinates of the initial follower position at $\theta = 0$ can now be found from

$$\begin{aligned} x_f(0) &= b\cos\left[\gamma - \operatorname{sgn}(\omega_{arm})\psi(0)\right] \\ y_f(0) &= b\sin\left[\gamma - \operatorname{sgn}(\omega_{arm})\psi(0)\right] \end{aligned} \tag{13.7}$$

where the sign of the arm rotation ω_{arm} accounts for the two possible contact points on the cam for any arm pivot location. Figure 13-3 shows the case for CW rotation of the arm during a rise from low dwell (or point) to high dwell (or point) on the cam. If the arm rotated CCW, it would contact the underside of the cam as is shown dotted in the figure.

The coordinates of each successive follower position can be found by calculating the new angle $\psi(\theta)$ that results as the follower arm moves. From the triangle $OB'A'$ in Figure 13-3, using the law of cosines,

$$b' = \sqrt{c^2 + l^2 - 2cl\cos\delta(\theta)} \tag{13.8a}$$

$$\psi(\theta) = \cos^{-1}\left(\frac{c^2 + b'^2 - l^2}{2cb'}\right) \tag{13.8b}$$

where

$$\delta(\theta) = \delta(0) + s(\theta) \tag{13.8c}$$

---

* A two-argument arctangent function must be used to obtain angles in all four quadrants. The single argument arctangent function found in most calculators and computer programming languages returns angle values in only the first and fourth quadrants. You can calculate your own two-argument arctangent function very easily by testing the sign of the x component of the arguments and, if x is negative, adding π radians or 180° to the result obtained from the available single-argument arctangent function.

For example (in Fortran or Visual Basic):

```
FUNCTION Atan2( x, y )
IF x = 0 THEN
   Atan2 = SGN(y) * 1.571
   RETURN
END IF
IF x <> 0 THEN Q = y / x
Temp = ATAN(Q)
IF x < 0 THEN
   Atan2 = Temp + 3.14159
  ELSE
   Atan2 = Temp
END IF
END
```

The above code assumes that the language used has a built-in single-argument arctangent function called ATAN(x) which returns an angle between ± π/2 radians when given a signed argument representing the value of the tangent of that angle and a function SGN(x) that returns the sign of its argument.

and $s(\theta)$ is the arm rotation angle at each cam position θ. The follower centerline coordinates for each cam angle θ are then:

$$x_f(\theta) = b\cos\left[\gamma - \text{sgn}(\omega_{cam})\theta - \text{sgn}(\omega_{arm})\psi(\theta)\right]$$
$$y_f(\theta) = b\sin\left[\gamma - \text{sgn}(\omega_{cam})\theta - \text{sgn}(\omega_{arm})\psi(\theta)\right] \qquad (13.9)$$

Note that there are four cases, one for each permutation of the directions of rotation of cam and follower arm. These are accounted for by applying the algebraic signs of their respective rotations (according to the right-hand rule) in equations 13.7 and 13.9. The negative signs on the *sgn* functions reflect the inversion of the follower around the cam in the opposite direction to cam rotation. So, if the cam is in fact stationary, equations 13.9 need to use a virtual cam rotation that is opposite to the actual follower rotation.

CAM SURFACE COORDINATES Figure 13-4 shows an oscillating roller follower in an arbitrary (inverted) position on a rise. The initial position of the follower at $\theta = 0$ is shown dotted. The cam angle is θ and the pressure angle at this position is $\phi(\theta)$. The roller centerline is at coordinates x_f, y_f as defined in equations 13.9. The point of contact x_s, y_s between cam and a roller of radius R_f lies along the common normal defined by angle $\phi(\theta)$. The angle of the follower arm at any cam angle θ is

$$\lambda(\theta) = \theta + \gamma - \delta(\theta) \qquad (13.10a)$$

where $\delta(\theta)$ is defined in equation 13.8c, and γ in equation 13.5b.

The angle of the common normal in the XY coordinate system for any position is

$$\sigma(\theta) = \lambda(\theta) + \frac{\pi}{2} + \phi(\theta) \qquad (13.10b)$$

where $\phi(\theta)$ is the signed pressure angle of the follower.

The coordinates of the cam surface for an oscillating roller follower can be found from:

$$x_s(\theta) = x_f(\theta) \pm \text{sgn}(\omega_{arm}) R_f \cos\left[\sigma(\theta)\right]$$
$$y_s(\theta) = y_f(\theta) \pm \text{sgn}(\omega_{arm}) R_f \sin\left[\sigma(\theta)\right] \qquad (13.10c)$$

The plus sign gives the outer envelope and the minus sign gives the inner envelope.

CAM CUTTER COORDINATES The coordinates of a cam cutter or grinding wheel of radius R_c for an oscillating roller follower can be found from:

$$x_c(\theta) = x_f(\theta) \pm \text{sgn}(\omega_{arm})(R_f - R_c)\cos\left[\sigma(\theta)\right]$$
$$y_c(\theta) = y_f(\theta) \pm \text{sgn}(\omega_{arm})(R_f - R_c)\sin\left[\sigma(\theta)\right] \qquad (13.10d)$$

The plus sign gives the outer envelope and the minus sign gives the inner envelope. Be sure to check the cutter or grinding wheel radius against the minimum (positive and negative) radii of curvature of the cam to avoid undercutting. (See Chapter 7.)

13

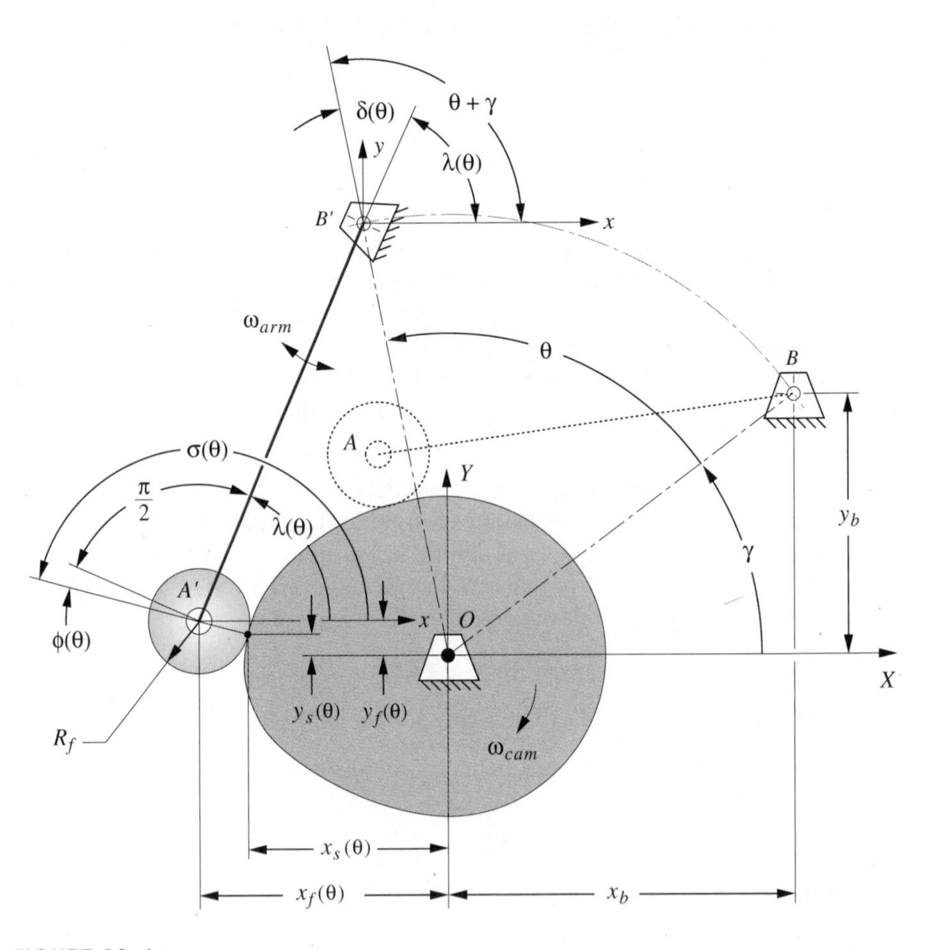

13

FIGURE 13-4

Geometry to calculate the cam contour for an oscillating roller follower

Any of the Cartesian coordinates calculated can of course be converted to polar (R,θ) coordinates by the Pythagorean formula and a two-argument arctangent function.

13.2 RADIAL CAMS WITH FLAT-FACED FOLLOWERS

For a flat-faced follower cam, the coordinates of the physical cam surface must be provided to the machinist as there is no pitch curve to work to. A flat follower can be either translating or oscillating. Oscillating flat-faced followers are seldom used because a roller on an oscillating arm provides lower friction, lower wear rates, ease of replacement, and does not require that the cam's radius of curvature be always positive. Radial cams with translating flat-faced followers are sometimes used, especially in automotive valve trains.

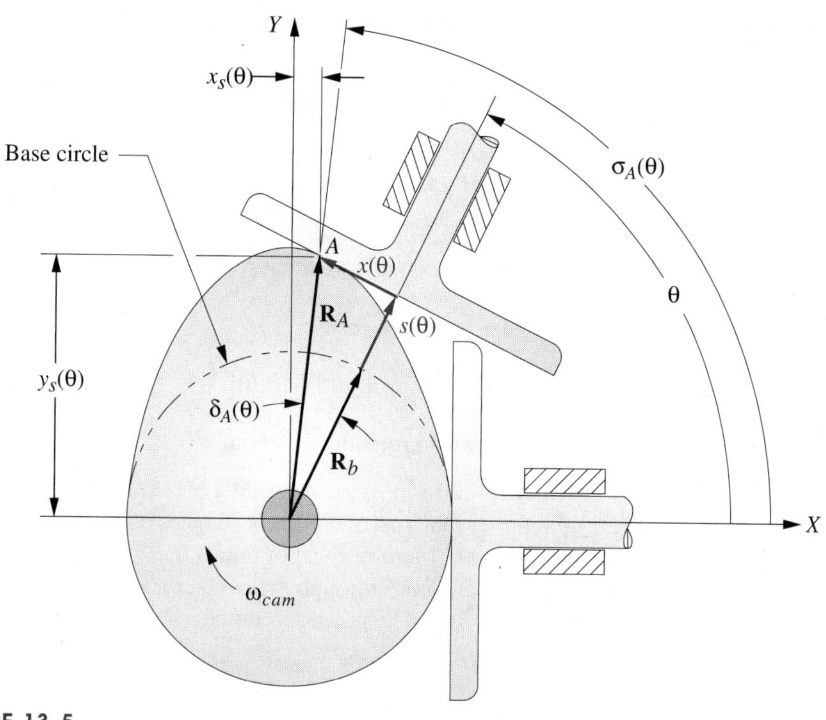

FIGURE 13-5

Geometry for derivation of the surface contour of a radial cam with a flat-faced follower

Radial Cams with Translating Flat-Faced Followers

CAM SURFACE COORDINATES Figure 13-5 shows the geometry of a translating flat-faced follower and cam in two positions, the initial position at $\theta = 0$ and an inverted position at an arbitrary value of cam angle θ. The follower motion is directed along the X axis. For any other follower direction, set up an axis system orthogonal to that direction as shown and transform the resulting coordinates into the global system when done. The y component of the cam surface contour for any cam angle θ, is defined as

$$y(\theta) = R_b + s(\theta) \tag{13.11a}$$

where R_b is the base circle radius and $s(\theta)$ is the displacement of the follower.

It was shown in Chapter 7 that an interesting relationship exists between the follower velocity and the lateral (i.e., orthogonal to follower motion) distance $x(\theta)$ to the contact point A between the cam and flat follower. These parameters are numerically equal if the follower velocity $v(\theta)$ is expressed in units of length/radian as was shown in equation 7.16 (p. 171), expanded here for your convenience.

$$x(\theta) = \frac{V(\theta)}{\omega}\,\frac{\text{in/sec}}{\text{rad/sec}} = v(\theta)\ \text{in/rad} \tag{13.11b}$$

13

The polar coordinates of the contact point A for any θ are:

$$R_A(\theta) = \sqrt{\left[R_b + s(\theta)\right]^2 + \left[x(\theta)\right]^2}$$
$$\sigma_A(\theta) = \theta + \tan^{-1}\left(\frac{x(\theta)}{\left[R_b + s(\theta)\right]}\right) \tag{13.12a}$$

The Cartesian coordinates of the contact point are then:

$$x_s(\theta) = \left[R_A(\theta)\right]\cos\left[\sigma(\theta)\right]$$
$$y_s(\theta) = -\operatorname{sgn}(\omega)\left[R_A(\theta)\right]\sin\left[\sigma(\theta)\right] \tag{13.12b}$$

where the *sgn* function accounts for cam rotation direction.

Cam Cutter Coordinates To machine or grind a cam for the flat-faced follower shown in Figure 13-5 requires that a set of cutter coordinates be created on the path $P(\theta)$ of the centerline of the cutter or grinding wheel of radius R_c. The cutter centerline must lie on the common normal that passes through the contact point orthogonal to the follower face as shown in Figure 13-6. The polar coordinates R_p, δ_p to the center p of the cutter/grinding wheel at any cam angle θ are:

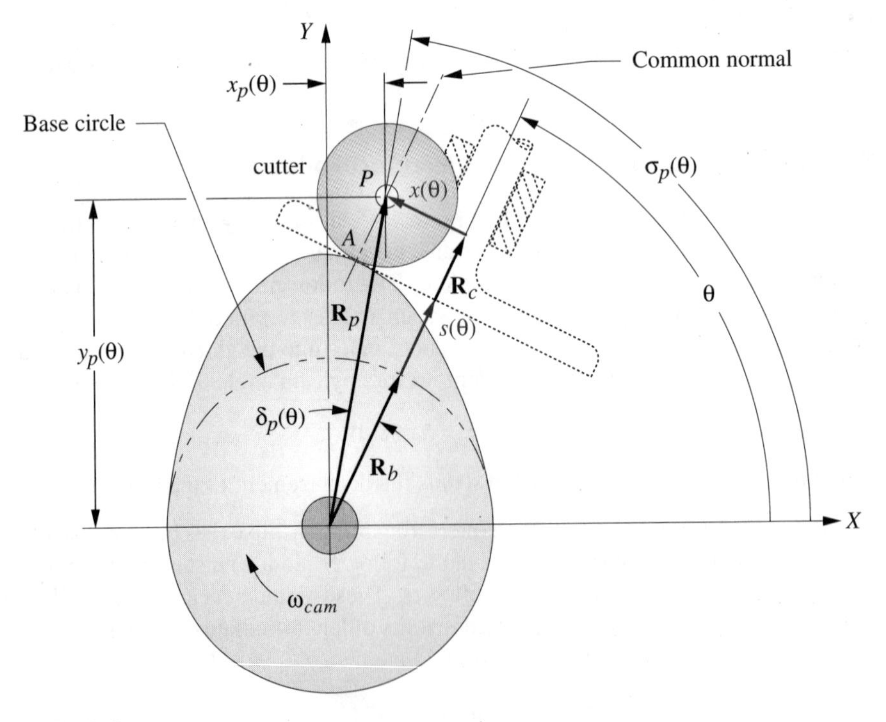

FIGURE 13-6

Geometry for derivation of the cutter coordinates of a radial cam with flat-faced follower

$$R_p(\theta) = \sqrt{\left[R_b + s(\theta) + R_c\right]^2 + \left[x(\theta)\right]^2}$$

$$\delta_p(\theta) = \tan^{-1}\left[\frac{x(\theta)}{R_b + s(\theta) + R_c}\right] \qquad (13.13a)$$

$$\sigma_p(\theta) = \delta_p(\theta) + \theta$$

where R_b is the base circle of the cam.

The Cartesian coordinates of the cutter centerline are:

$$x_p(\theta) = R_p(\theta)\cos\left[\sigma_p(\theta)\right]$$

$$y_p(\theta) = -\text{sgn}(\omega)\,R_p(\theta)\sin\left[\sigma_p(\theta)\right] \qquad (13.13b)$$

where the *sgn* function accounts for cam rotation direction.

Radial Cams with Oscillating Flat-Faced Followers

Figure 13-7 shows a flat-faced oscillating arm follower in two positions around a "stationary" radial cam in the typical inversion of motion for analysis purposes. The initial arm position *BA* becomes *B′A′* at a later time after the cam has rotated through the angle θ. The arm rotation can be either CCW or CW, positioning it either above or below the cam for any arm pivot location. The coordinates of the follower arm pivot x_b, y_b in the global *XY* coordinate system and the direction of arm rotation must be known *a priori*. The offset *h* of the follower surface from the arm centerline is positive if directed toward the cam center and negative otherwise.

For this analysis, it is assumed that at cam angle $\theta = 0$, the follower is on the low dwell or low point of the cam, and thus the radius to the cam follower shown as *b* in Figure 13-7 is equal to the base circle radius R_b when $\theta = 0$. Follower motion on a rise is then a positively increasing angle $s(\theta)$, starting at zero. The length of center distance *c* from cam center to arm pivot and its angle γ at that position are defined in equations 13.5 (p. 400). Note that triangle *OAB* is a right triangle for all positions of the follower. The initial angle of the follower arm on the low dwell is:

$$\delta_0 = \sin^{-1}\left(\frac{R_b + h}{c}\right) \qquad (13.14a)$$

and the distance to the initial contact point is

$$l_0 = c\cos\delta_0 \qquad (13.14b)$$

The angle of the initial position of the follower arm in the local *xy* coordinate system is

$$\psi_0 = \gamma + \pi + \text{sgn}(\omega_{arm})\delta_0 \qquad (13.14c)$$

At any arbitrary value of cam angle θ, the angle of the (noninverted) follower arm in the local *xy* coordinate system will be:

$$\psi(\theta) = \gamma + \pi + \text{sgn}(\omega_{arm})\left[\delta_0 + s(\theta)\right] \qquad (13.15a)$$

13

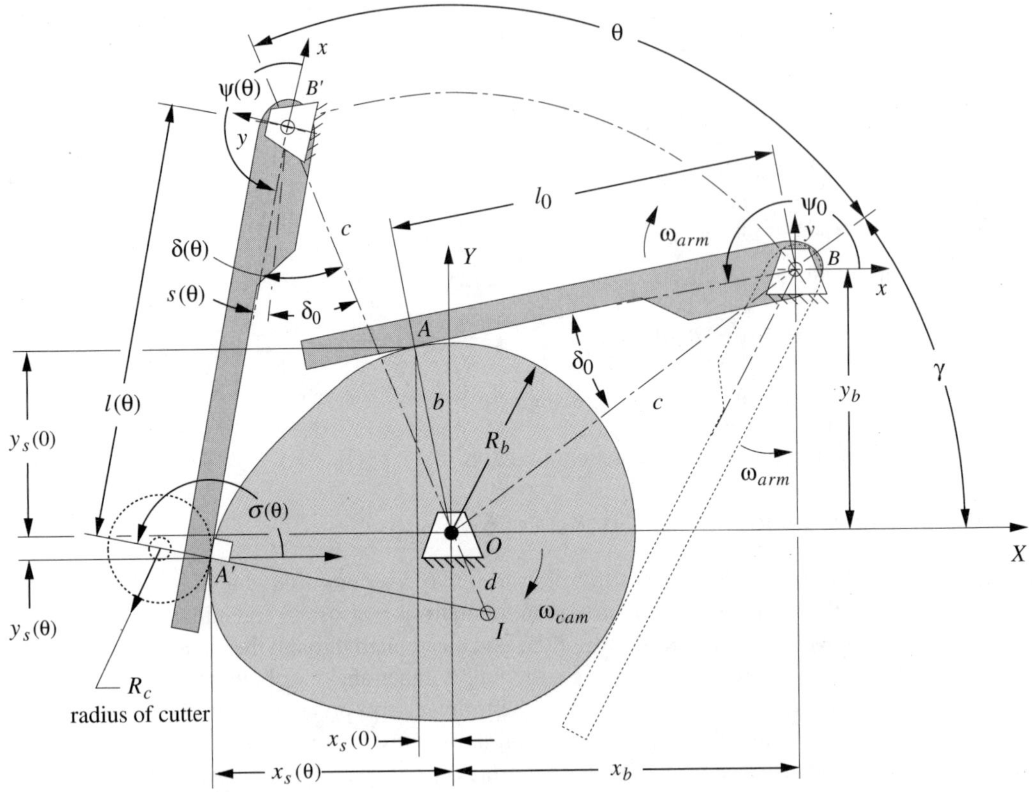

FIGURE 13-7

Geometry for calculation of the cam contour for an oscillating flat-faced follower

where the *sgn* function accounts for direction of arm rotation.

Distance d to instant center I is a function of follower velocity v in rad/rad.

$$d = c\frac{v}{1-v} \tag{13.15b}$$

and the distance $l(\theta)$ to the contact point at cam angle θ in the local xy coordinate system is:

$$\begin{aligned} l(\theta) &= (c+d)\cos\left[\delta_0 + s(\theta)\right] \\ &= c\frac{1}{1-v}\cos\left[\delta_0 + s(\theta)\right] \end{aligned} \tag{13.15c}$$

and the coordinates of the contact point in the local xy coordinate system are:

$$\begin{aligned} x_s(\theta) &= l(\theta)\cos\left[\psi(\theta)\right] \\ y_s(\theta) &= l(\theta)\sin\left[\psi(\theta)\right] \end{aligned} \tag{13.15d}$$

These coordinates can be transformed to the global *XY* system by a translation to the global origin and a rotation through the angle $-sgn(\omega)\cdot\theta$ about the global origin, in that order.

CAM CUTTER COORDINATES The cutter centerline must lie on the common normal that passes through the contact point orthogonal to the follower face as shown in Figure 13-7. The angle of the common normal at any cam angle θ is

$$\sigma(\theta) = \theta + \psi(\theta)\gamma + \text{sgn}(\omega_{arm})\pi/2 \tag{13.16a}$$

The Cartesian coordinates to the center of a cutter of radius R_c at any cam angle θ are

$$\begin{aligned} x_c(\theta) &= x_s(\theta) + R_c(\theta)\cos[\sigma(\theta)] \\ y_c(\theta) &= y_s(\theta) + R_c(\theta)\sin[\sigma(\theta)] \end{aligned} \tag{13.16b}$$

13.3 BARREL CAMS WITH ROLLER FOLLOWERS

A barrel or axial cam is shown in Figure 1-8 (p. 7). The follower motion is along the axis of cam rotation rather than radial. Creating the cam profile for a barrel cam is a somewhat simpler geometric exercise than for the radial cam, especially for a translating follower. It is not practical to use a flat-faced follower on an axial cam as there will always be regions of negative radius of curvature regardless of the barrel diameter. Thus a roller follower is required, especially if it is a form-closed cam, which is common for barrel cams.

Barrel Cam With Translating Roller Follower

To create the profile of a barrel cam, it is only necessary to wrap the *s*-diagram around the circumference of a cylinder of the chosen radius. Unlike the radial cam, there is no distortion of the *s*-curve other than a scaling of its length (but not its height) to match the cylinder circumference. A prime cylinder radius R_p to the centerline of the follower's length is chosen and it determines the cam diameter, pressure angles, and radii of curvature as described in Chapter 7. This gives the most accurate path for a crowned roller follower, which are recommended in this application to minimize slip between roller and cam that must occur with a cylindrical roller.

ROLLER CENTERLINE COORDINATES Figure 13-8 shows the developed surface of a cylindrical (barrel) cam with the *s*-curve applied to it. The translating roller follower is shown in two locations, at $\theta = 0$ and at an arbitrary cam angle θ. Follower motion in this case is in the *z* direction, parallel to the cam axis. The z_0 position is taken at the low dwell or low point of the displacement. The *z*-coordinates of the roller centerline are identical to those of the displacement or *s*-curve. The independent parameter is the cam angle θ. The three-dimensional cylindrical coordinates R, λ, z of the roller centerline are:

$$\begin{aligned} R(\theta) &= R_p \\ \lambda(\theta) &= \theta \\ z(\theta) &= s(\theta) \end{aligned} \tag{13.17}$$

Cylindrical coordinates are preferred for machining a barrel cam.

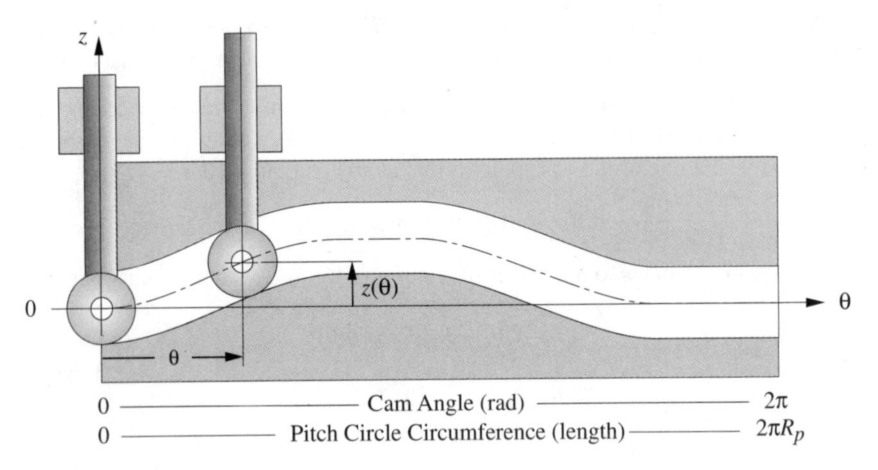

FIGURE 13-8

Developed surface of a barrel cam with translating follower at pitch radius

Equation 13.17 defines the nominal track centerline coordinates at the pitch cylinder radius R_p. These data should be used for cam manufacture. Barrel cams should ideally be manufactured with cutters or grinders that are essentially equal in diameter to the roller follower because of the change in geometry that occurs from the outside to the inside of the cam track as the barrel radius changes.

CAM SURFACE COORDINATES Figure 13-9 shows a translating roller in an arbitrary position along the developed surface of the cam pitch circle. The pressure angle $\phi(\theta)$ at each cam angle θ and at the pitch cylinder radius defines the common normal along which the two contact points lie. The contact points at the pitch cylinder can be defined in cylindrical coordinates by:

$$\begin{aligned} z_{s_{1,2}}(\theta) &= z(\theta) \pm \operatorname{sgn}(\omega) R_f \cos\left[\phi\left(\theta, R_p\right)\right] \\ \delta_{s_{1,2}}(\theta) &= \lambda(\theta) \mp \operatorname{sgn}(\omega) \frac{R_f}{R_p} \sin\left[\phi\left(\theta, R_p\right)\right] \end{aligned} \tag{13.18}$$

where the $sgn(\omega)$ function accounts for cam rotation direction and the sign of the pressure angle accounts for rise or fall. The upper sign on the *sgn* function gives the top surface and the lower sign gives the bottom surface of the track. All angles are in radians.

Equation 13.18 defines the cam surface coordinates only at the pitch cylinder radius R_p. If you want to compute the surface coordinates at another radius, then the pressure angles must also be recalculated for use in equation 13.18 as they are a function of cam radius. While it is certainly possible to calculate multiple sets of cam surface contour coordinates for a family of cylinder radii across the cam track, and then use them to cut the surface, it may not be practical to do so due to the large number of data and the complications of NC programming associated with the task. Therefore, it is recommended that barrel cams be cut to the path of the roller centerline, using a tool of the same diameter as the roller.

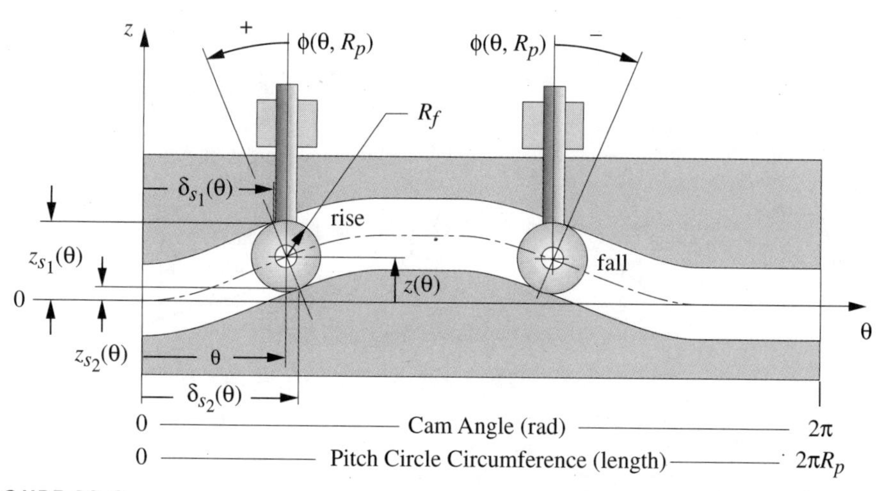

FIGURE 13-9

Surface coordinates of a barrel cam with translating follower

CAM CUTTER COORDINATES Cutter compensation should not, in general, be used to cut a barrel cam for the same reasons as described in the preceding paragraph regarding the effect of cylinder radius on the pressure angle and the cam contour. The cutter coordinates are different for each cylinder radius used in their calculation. However, in some circumstances, it may be possible to use cutter compensation on a barrel cam with minimal error. If the ratio of track depth to prime cylinder radius is small, i.e., a shallow track groove on a large diameter cam, then the change in cam surface or cutter coordinates across the track will be small. Also, if a crowned roller follower, centered at the pitch cylinder radius, is used, then small contour errors at other track radii may not significantly affect follower motion. For such circumstances, the following equation will compute the cutter coordinates at any chosen cylinder radius R_p.

$$
\begin{aligned}
z_{s_{1,2}}(\theta) &= z(\theta) \pm \operatorname{sgn}(\omega)\left(R_f - R_c\right)\cos\left[\phi\left(\theta, R_p\right)\right] \\
\delta_{s_{1,2}}(\theta) &= \lambda(\theta) \mp \operatorname{sgn}(\omega)\frac{R_f - R_c}{R_p}\sin\left[\phi\left(\theta, R_p\right)\right]
\end{aligned}
\tag{13.19}
$$

Barrel Cam With Oscillating Roller Follower

Figure 13-10a shows a barrel cam with oscillating roller follower. The follower arm has radius R_a and pivots about point P that lies in and defines the datum plane, which is orthogonal to the cam rotation axis z and through the arm pivot. The initial position of the roller with respect to the datum plane is defined by the x and z offsets, x_0, $z(0)$, which can each be either positive or negative. Note that x_0 is a constant while z is a function of θ. Depending on CW or CCW arm rotation ω_{arm}, on the rise in combination with x_0, $z(0)$, the roller's arc motion may or may not cross the datum plane or cam axis. For best balance of forces and pressure angle, good design suggests that the arm's motion should be symmetrical about the datum plane and cam axis, as shown in Figure 13-10a.

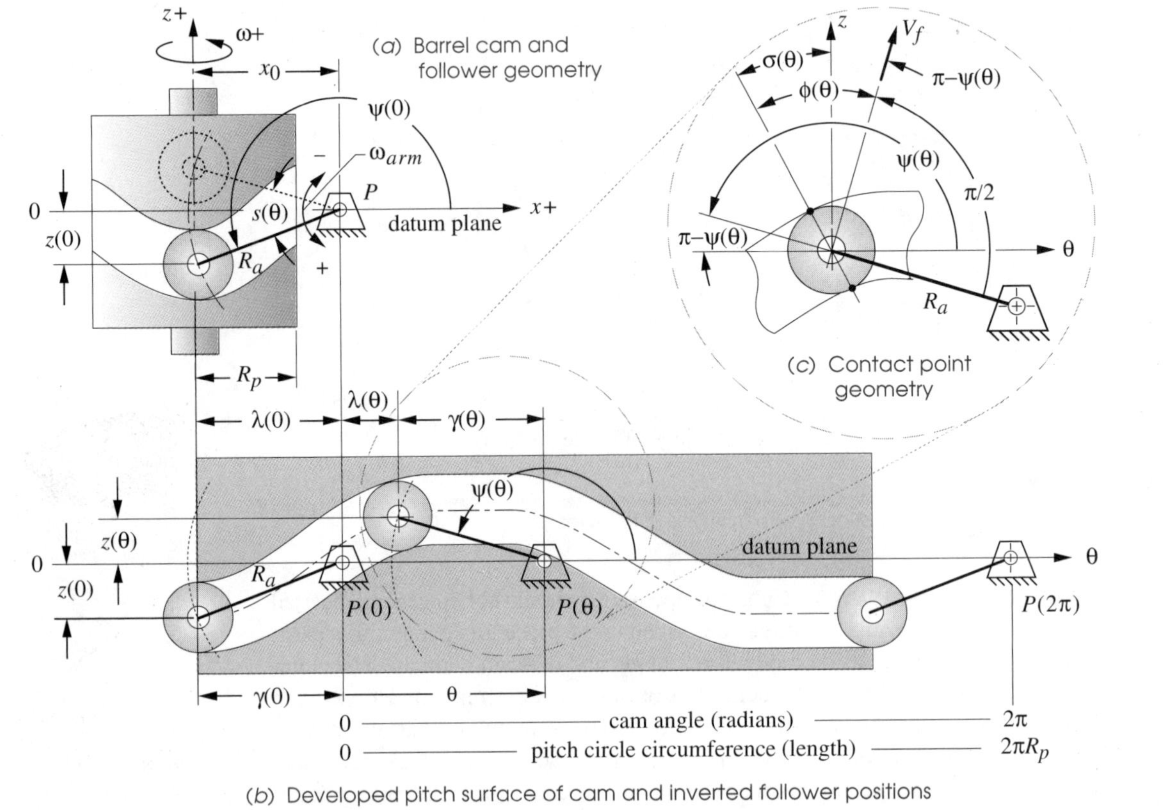

(a) Barrel cam and follower geometry

(c) Contact point geometry

(b) Developed pitch surface of cam and inverted follower positions

FIGURE 13-10

Geometry of a barrel cam with oscillating roller follower and its developed surface at the pitch radius

For any oscillating follower, the displacement $s(\theta)$ is defined in relative angular measure as a positively increasing angle starting from zero at the minimum displacement position (typically the low dwell). The initial angle of the arm $\psi(0)$ at $s(0)$ with respect to the follower datum plane is defined as

$$\psi(0) = \sin^{-1}\left(\frac{z(0)}{R_a}\right) \qquad x_0 \le 0$$

$$\psi(0) = \pi - \sin^{-1}\left(\frac{z(0)}{R_a}\right) \qquad x_0 > 0 \tag{13.20a}$$

Figure 13-10b shows the developed pitch surface of the cam and several inverted positions of the follower. The length of the developed surface is the circumference of the prime cylinder $2\pi R_p$. Note that this is dimensioned from the initial to final positions of the arm pivot P. Point $P(0)$ is offset a distance along the surface equal to the arc subtended by the horizontal component of the initial position of the follower at radius R_p, expressed in terms of cam angle as:

$$\gamma(0)=\frac{R_a}{R_p}\cos\left[\psi(0)\right] \quad \text{radians} \tag{13.20b}$$

At an arbitrary cam angle θ, the inverted arm's pivot location is at $P(\theta)$ and the angle of the arm with respect to the datum plane is:

$$\psi(\theta)=\psi(0)+\operatorname{sgn}\left(\omega_{arm}\right)s(\theta) \tag{13.20c}$$

where the *sgn* function accounts for direction of arm rotation on the rise.

The cam angle subtended by the arc length corresponding to the horizontal component of the roller position at cam angle θ is:

$$\gamma(\theta)=\frac{R_a}{R_p}\cos\left[\psi(\theta)\right] \quad \text{radians} \tag{13.20d}$$

The location of the roller centerline on the pitch cylinder at cam angle θ is:

$$\begin{gathered} z(\theta)=R_a\sin\left[\psi(\theta)\right] \\ \lambda(\theta)=\theta+\operatorname{sgn}(\omega)\gamma(\theta) \end{gathered} \tag{13.20e}$$

where the $sgn(\omega)$ function accounts for direction of cam rotation.

Note that the above derivation does not account for error associated with the arc motion of the oscillating roller that results from its contact point moving out of the plane of the axis of the cam during its sweep and effectively increasing the cam radius at which the center of the roller contacts the slot. If the arm radius is reasonably long in respect to the lift of the cam and has a small angular excursion, and/or the pitch cylinder is large, then this error may be small. If these conditions do not obtain, then the nominal pitch cylinder radius R_p needs to be corrected for each position of the roller follower as:

$$R_p' = \sqrt{R_p^2+\left\{R_a\cos\left[\psi(\theta)\right]+x_0\right\}^2} \tag{13.20f}$$

13

Substitute R_p' for R_p in equations 13.20b and d.

Cam Surface Coordinates The same arguments given for the barrel cam with translating roller follower in respect to manufacturing apply here as well. Only the roller centerline coordinates at the pitch cylinder should be used to cut the cam, and a cutter/grinder of essentially the same diameter as the roller ideally should be used. Cutter compensation does not give correct results for a barrel cam except at one pitch cylinder radius. Therefore the surface coordinates should not be used for manufacturing as they, by definition, require cutter compensation also be used. The following equations for cam surface coordinates at the pitch cylinder radius R_p are nevertheless provided.

Figure 13-10c shows an oscillating roller in an arbitrary position along the developed surface of the cam pitch cylinder. The pressure angle $\phi(\theta)$ at each cam angle θ defines the common normal along which the two contact points lie with respect to the direction of the roller velocity V_f. We need to resolve the pressure angle at radius R_p into the z-θ coordinate system to find the deviation angle of the contact point $\sigma(\theta)$:

$$\begin{aligned}\sigma(\theta) &= \phi\left(\theta, R_p\right)-\left[\pi-\psi(\theta)\right] \\ &= \phi\left(\theta R_p\right)+\psi(\theta)-\pi\end{aligned} \tag{13.21a}$$

The contact points can then be defined in cylindrical coordinates by:

$$\begin{aligned}z_{s_{1,2}}(\theta) &= z(\theta) \pm \operatorname{sgn}(\omega) R_f \cos\left[\sigma(\theta)\right] \\ \delta_{s_{1,2}}(\theta) &= \lambda(\theta) \mp \operatorname{sgn}(\omega)\frac{R_f}{R_p}\sin\left[\sigma(\theta)\right]\end{aligned} \tag{13.21b}$$

where the $sgn(\omega)$ function accounts for cam rotation direction and the sign of the pressure angle accounts for rise or fall. The upper sign on the *sgn* function gives the top surface and the lower sign gives the bottom surface of the track for the case of the arm pivot leading the roller follower as shown in Figure 13.10. The signs are reversed for the case of the arm pivot trailing the roller follower. All angles are in radians.

CAM CUTTER COORDINATES As was pointed out in the discussion of barrel cams with translating followers, cutter compensation should not, in general, be used to cut a barrel cam. The cutter coordinates will be different for each cylinder radius used in their calculation. However, if the ratio of track depth to prime cylinder radius is small, i.e., a shallow track groove on a large diameter cam, then the change in cam surface or cutter coordinates across the track will be small. Also, if a crowned roller follower, centered at the pitch cylinder radius, is used, then contour errors at other track radii may not significantly affect follower motion. For such circumstances, the following equation will compute the cutter coordinates for an oscillating follower on a barrel cam at any chosen cylinder radius R_p.

$$\begin{aligned}z_{s_{1,2}}(\theta) &= z(\theta) \mp \operatorname{sgn}(\omega)\left(R_f - R_c\right)\cos\left[\sigma(\theta)\right] \\ \delta_{s_{1,2}}(\theta) &= \lambda(\theta) \pm \operatorname{sgn}(\omega)\frac{R_f - R_c}{R_p}\sin\left[\sigma(\theta)\right]\end{aligned} \tag{13.22}$$

13

13.4 LINEAR CAMS WITH ROLLER FOLLOWERS

A linear cam is one that either has rectilinear motion with respect to a stationary follower assembly, or one that is stationary with respect to a rectilinearly translating follower assembly that passes over it. In the first case, the cam must be oscillated back and forth past the follower assembly and thus has a time-varying velocity that complicates its dynamic analysis. In the second case, a follower assembly passing over a stationary cam will often have constant velocity in the direction of the cam axis. An example would be a machine in which the workpieces are carried on a constant speed conveyor past one or more stationary cams that impart transverse motion to the assemblies as they pass over the cam(s).

The contour of either type of linear cam can be calculated by the same methods as described in the previous section for barrel cams. The developed barrel cam used for that analysis is in fact a "linear" cam. A true linear cam will of course not have a prime circle or prime cylinder radius, as it is not "wrapped around" anything. Its contour at the roller centerline is in fact just that of the displacement diagram. It will have a specified length over which the cam follower will travel, which we will call the *prime length* L_p.

Equations 13.17 through 13.22 can be used to calculate a linear cam profile if a simple trick is used to convert the linear cam to a pseudo-barrel cam for computation purposes. Given a desired prime length L_p of cam, calculate an effective prime cylinder radius R_p that would have a circumference equal to that prime length:

$$R_p = \frac{L_p}{2\pi} \tag{13.23}$$

Then the barrel cam equations can be used and any angular results (θ, λ) converted to linear measure using R_p. Note also that since the linear cam is effectively a barrel cam of infinite radius, there is no contour error across its thickness due to using R_p in equations 13.18, 13.19, 13.21, and 13.22, and cutter compensation can be used with no error.

If the linear cam is stationary, and the follower passes by at constant velocity, then the dynamics of the follower can be calculated as if a pseudo barrel cam as well by creating a pseudo cam angular velocity that will give the actual linear velocity at the chosen effective prime cylinder radius.

If the linear cam is actually oscillated back and forth past the follower with some linkage- or cam-driven velocity pattern, then the follower dynamics can be handled by calculating the linear cam follower's kinematics and dynamics for a pseudo-barrel cam running at 1 rad/sec, and then combining the results of this analysis with that of the linear cam's motion due to its driver mechanism.

13.5 CONJUGATE CAMS

Conjugate cams are a pair of cams that impart simultaneous motions to two coupled followers. The followers can be attached to the same translating carriage as shown in Figure 13-11 or to the same oscillating arm as shown in Figure 13-12. These figures both show roller followers on radial cams, but the same result can also be obtained with flat-faced followers on radial cams and with roller followers on barrel cams. In the case of translating followers, the distance between the two roller centers or between the two flat faces is a constant value, and for an oscillating arm set is a constant angle.

The principal advantage of conjugate cams is that they do not require any closure spring. Together they create a form-closed cam-follower system. A track cam also provides form-closure, but has the problem of backlash due to the need for clearance in the track. A pair of conjugate cams with fixed follower spacing cannot practically be run at a fixed clearance of zero, and so will also have backlash. However, they offer the possibility of spring loading the followers together against the cams to take up the backlash and accommodate tolerance variations in the cam contour. Because the dynamic deflection of this backlash spring is extremely small (being only the variation in tolerance), it can be a very high rate spring to minimize its vibration effects, and be given sufficient preload to avoid separation.

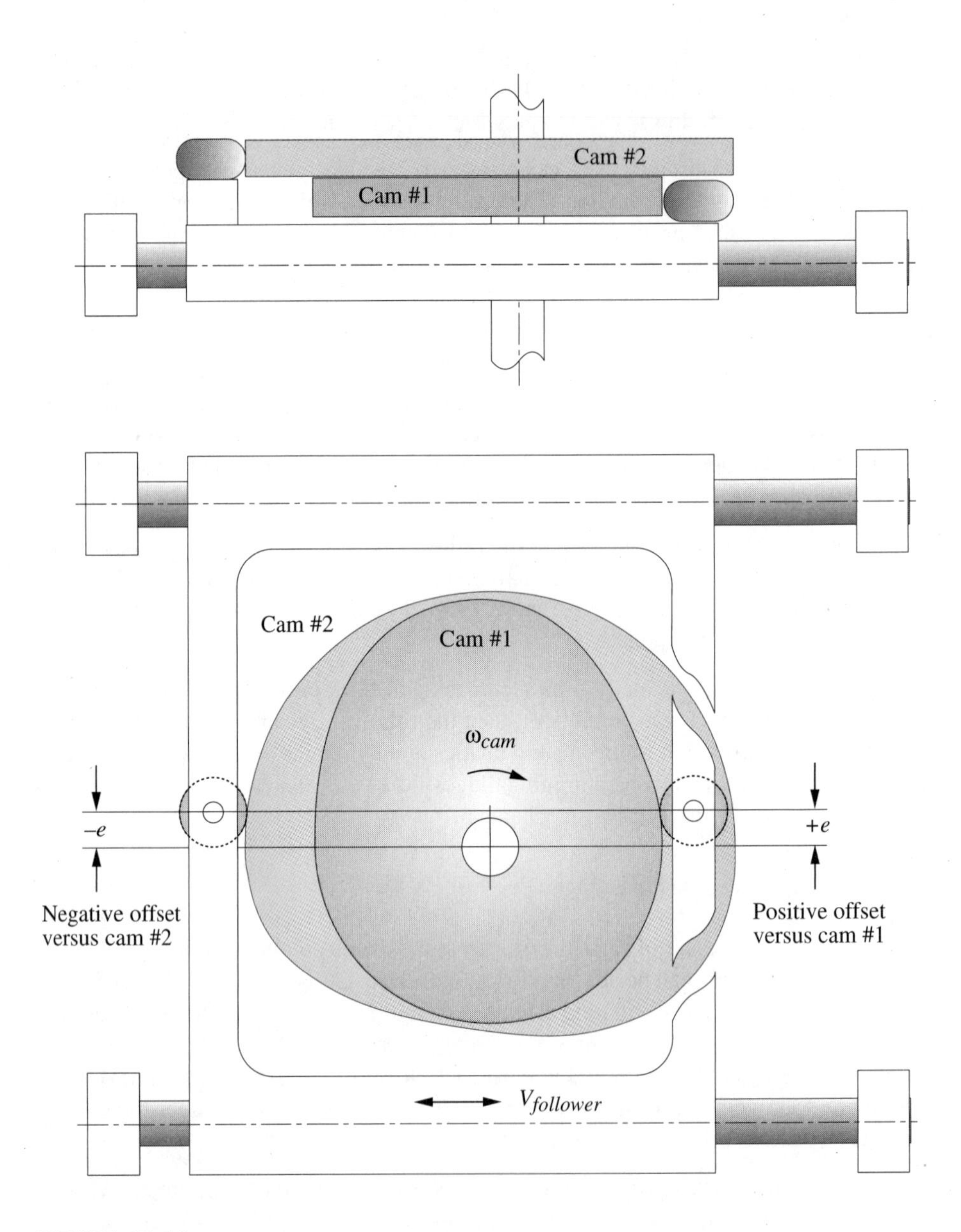

FIGURE 13-11

Conjugate cams with offset inline translating followers

The principal disadvantage of conjugate cams is their added cost and complexity, and so they are not very often used in production machinery. Internal combustion (IC) engine valve trains also seldom use conjugate cam systems, reserving them for racing applications where the cost and complexity are justified by the need for higher engine

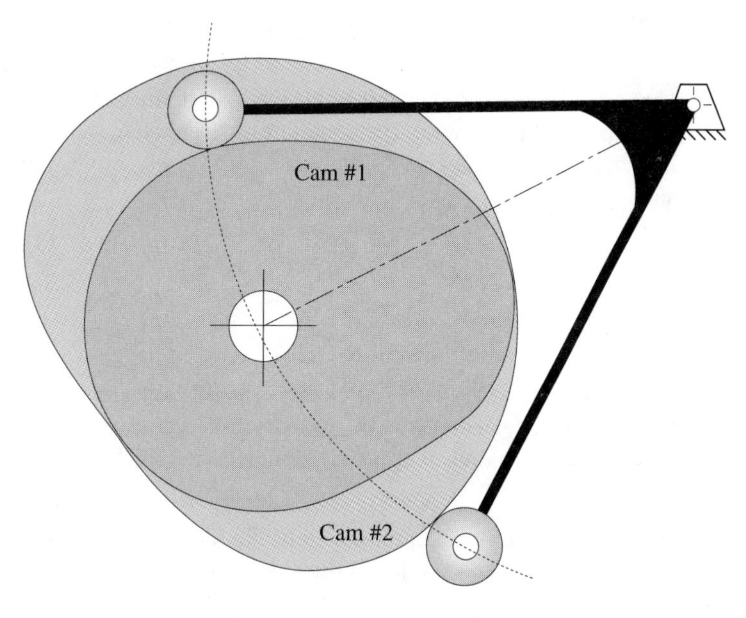

FIGURE 13-12

Conjugate cams with an oscillating arm and roller followers

speeds, which are otherwise limited by valve float with spring-closed cam-followers. These are called *desmodromic* systems in IC engine valve train applications.*

Designing Conjugate Cams

Consider one of the pair of conjugate cams to be the primary cam, and design it for the desired follower motion as for any other nonconjugate cam set. In effect, one only has to design the $s\ v\ a\ j$ functions for one cam of the conjugate pair as both must have the same effective cam-follower motion program. The conjugate cam is created later at the stage of cam contour generation by inverting the s-diagram and appropriately phase-shifting the start of the cam contour versus the keyway for a translating follower or negating the rotation direction of the conjugate's follower arm as explained below.

Conjugate Radial Cams With Translating Followers

Figure 13-11 shows a pair of rise-fall-dwell conjugate cams driving a set of translating roller followers. These could as well be flat-faced followers and the following discussion would still apply (with the exception of an offset). The two rollers are shown attached to a platform that straddles the cam and is guided in linear bearings not shown. Many other arrangements are possible and this diagram should be considered schematic.

The conjugate cam pair is shown rotating CW on a common shaft. Consider cam #1 to be the primary cam. Its follower has a positive offset as defined in Section 13.1 because it is opposite to the direction of the cam's tangential velocity. Note that for inline

* e.g., the Ducati "Desmo" motorcycle engine.

13

offset followers as shown, the offset of the conjugate cam will be of opposite sign to that of the primary cam.

The contour of cam #1 is created by one of the methods defined in an earlier section of this chapter for its type of follower. The contour of its conjugate, cam #2, is formed by inverting the *s*-diagram used for cam #1, since as cam #1 rises, cam #2 must fall with the identical motion program.

$$\begin{aligned} s_{conjugate} &= h - s_{primary} \\ v_{conjugate} &= -v_{primary} \\ a_{conjugate} &= -a_{primary} \end{aligned} \tag{13.24a}$$

where h is the total lift of the follower. The velocity, acceleration, and jerk of the conjugate follower must all be negated versus the primary follower as well. These factors affect the pressure angle and radius of curvature calculations, which must be redone for the conjugate cam.

The start of the contour of cam #2 must also be phase-shifted 180° with respect to cam #1 as the conjugate's follower motion is flipped 180° versus the primary follower.

$$\theta_{conjugate} = \theta_{primary} + 180° \tag{13.24b}$$

The sign of the offset e (see Figure 13-1, p. 396) must also be inverted if the followers are to be in line. (If not, then one roller will be above and one below the cam center.)

$$e_{conjugate} = -e_{primary} \tag{13.24c}$$

The prime or base circle diameters do not have to be the same for both cams. These can be adjusted to tailor the pressure angles and radii of curvature independently for each cam in the conjugate pair if desired. A different prime or base circle between the conjugates will only affect the spacing between the two followers. The two cams in Figure 13-11 have the same prime circle radii. Their contours are not tangent due to the opposite offsets.

Conjugate Radial Cams With Oscillating Followers

Figure 13-12 shows a pair of conjugate cams operating an oscillating arm with two rollers. The following discussion applies equally to the case of a pair of flat-faced arms. In both cases the angle between the two arms is fixed and can be chosen by the designer. Actually the designer's choice of prime or base circle radius and roller radius for each cam in combination with the common pivot location will define the included angle between the two branches of the follower arm. The two rollers must obviously be in different planes to contact their respective cams.

Consider cam #1 to be the primary cam. Its contour is defined in the same manner as described in previous sections of this chapter for the follower style chosen. The contour of its conjugate, cam #2, is formed by inverting the *s*-diagram used for cam #1 according to equation 13.24a. Using either the same or different values for the second arm radius, cam prime/base circle, and roller radius as used for cam #1, but using the same arm pivot location, the contour of cam #2 is calculated for opposite rotation of the arm

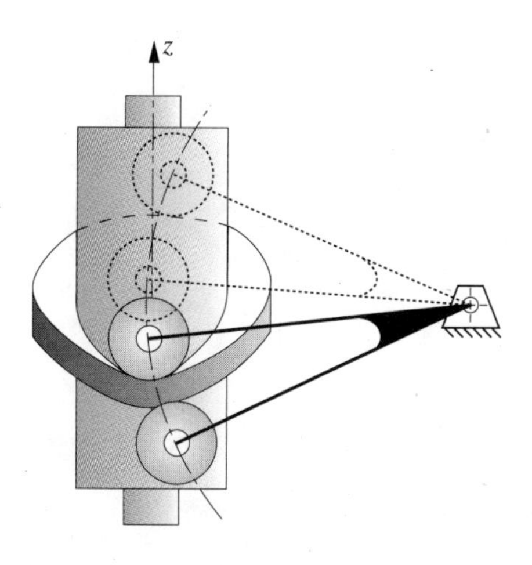

FIGURE 13-13

Ribbed barrel cam with oscillating roller follower

than used for cam #1. This is because arm rotation was defined as CW or CCW based on its direction of rotation during the rise event, i.e., away from the cam pivot. Since one branch of the arm must fall as the other rises, their rotation directions are reversed for the contour calculation with this approach even though both branches of the arm in fact rotate in the same direction at any instant. The phase shift of the second cam is taken care of by the above considerations and offset is not a factor with oscillating followers, so equations 13.24b and c are not needed in this case.

Conjugate Axial Ribbed Cams With Oscillating Followers

An alternative to the tracked barrel cam shown in Figure 13-10 is the ribbed barrel cam of Figure 13-13. This provides a rib rather than a groove in the cam and uses a pair of roller followers that straddle the rib on a common arm. This geometry is somewhat more difficult to machine than a groove or track cam, but offers the advantage of spring loading the two rollers together to pinch the rib and eliminate backlash. Tapered rollers are sometimes used to match the changing tangential velocity along the cam's radius. The rib can also be cut with nonparallel sides in the radial direction, making it thicker at the base than at the tip. This allows tapered rollers to be adjusted radially toward the cam axis to take up backlash due to wear.

13

The two rib surfaces are conjugates of one another and the same approach used for radial conjugate cams works here as well. Considering one of the surfaces as the primary cam, its contour is defined as for one side of a barrel groove according to equations 13.19 and 13.20 (pp. 409-411). The contour of the conjugate surface is then found by changing the value of z_0 in equation 13.20a to suit the initial position of the second roller and inverting the displacement function according to equation 13.24a. The proper sign must be used in equation 13.23b to generate the desired surface of each side of the rib.

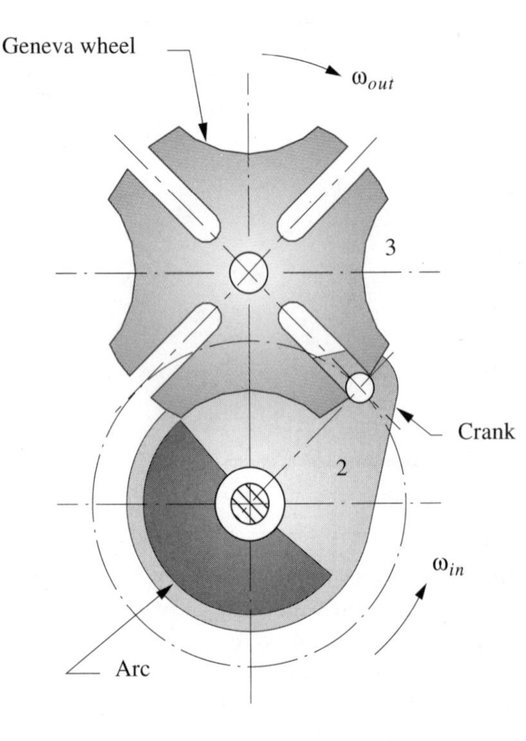

(a) Four-stop Geneva mechanism

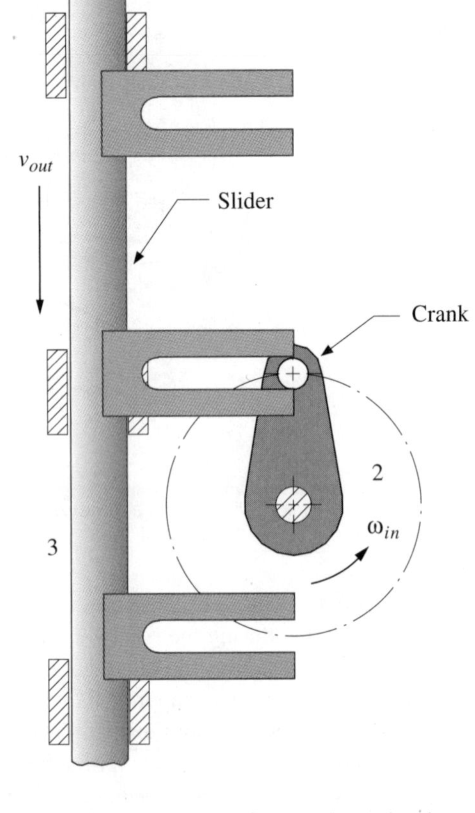

(b) Linear intermittent motion "Geneva" mechanism

FIGURE 13-14

Rotary and linear intermittent motion mechanisms

Indexing Cams

It is common in production machinery to require that a dial or conveyor belt/chain have intermittent rotary motion. Typically the product is indexed (moved) from station to station, then stopped for a portion of the cycle while an operation is performed on it, then indexed to the next station. The product stations can be arranged around the periphery of a rotating dial or along the length of a belt or chain conveyor. In either case, the input motion to the dial or belt pulley/chain sprocket is rise-dwell-rise-dwell, etc.

This motion is accomplished with some sort of indexing mechanism. Perhaps the best known of these is the Geneva wheel. Figure 13-14 shows a four-stop example of a Geneva wheel and a linear motion variant. It is a form of cam follower mechanism but has some disadvantages such as high acceleration, infinite jerk at the points where the pin enters a slot, and high wear in the slots.

A more commonly-used approach is the cam indexer shown in Figure 13-15. Originally developed by the Ferguson Company, similar devices are built by other companies such as Camco Inc. These rotary indexers can be made with three or more dwells, called stops, per revolution. The cam is of the ribbed barrel type with an interrupted rib. It is essentially a worm thread with a customized, nonconstant lead over a revolution—its lead function being the desired cam motion program of rise and dwell. Modified sine rise functions are often used. One revolution of the cam input axis constitutes one cycle of the machine, i.e., a rise (index) and a dwell. There are multiple roller followers (as many as the required number of stops) arranged radially around the output shaft such that at least one pair is always engaged with the cam rib. The roller pairs enter and leave the interrupted rib as the follower turns. The arrangement shown in Figure 13-15 is a six-stop indexer.

The principal difference between an axial cam with an oscillating roller follower and an indexer is that the follower "arm" of the indexer has unidirectional rotary motion rather than oscillatory rotary motion. In addition, the follower-cam surface geometry of the indexer is more complicated than that of the conventional ribbed barrel cam because of the orientation of the rollers with respect to the rotating follower arm. The roller axis of a conventional barrel cam is normal to the radius of the arm and to the cam axis as shown in Figure 13-13 (p. 417). The roller axis of an indexer is radial to the follower arm and intersects the cam axis, but it is only normal to the cam axis in one position as can be seen in Figure 13-15. This results in a complicated three-dimensional geometry of the rib. The pair of indexing cam surfaces are conjugate but their contours are more complicated and require a different approach than outlined in the previous section on ribbed cams with oscillating followers. The mathematics of this application will not be addressed here. The cam contour is most conveniently developed for manufacture with a solids modeling CAD system. Some information on the geometry of indexing cams can be found in references [2] and [3] noted at the end of this chapter.

13.6 CAM-LINKAGE COMBINATIONS

Many industrial (and automotive) cam-follower systems use linkages as follower trains to transfer the cam motion program to the end effector. These linkages can be fourbar, sixbar, or multibar linkages as described in reference 4. If the cam motion program is of the critical end position (CEP) type, for which only the dwell positions are critical and for which the position accuracy during the rise and fall is not critical, then the distortion of the motion program due to linkage geometry may be of little concern provided that the extreme input follower link angular positions have been adjusted for linkage geometry to match the desired extreme end effector locations. Also, if the total angular displacements of the cam follower arm and the other rotating links in the train are small, then the arc error will be small, and the linkage geometry's effect on the motion may be ignored with little resulting position error, provided that the total excursion at the cam is set to yield the desired total motion at the end effector. On the other hand, if the angular excursions of the follower links are substantial or if there are critical intermediate positions that must be accurately met during rise or fall, then the linkage's distortion of the cam motion program should be accounted for in the calculation of the cam contour.

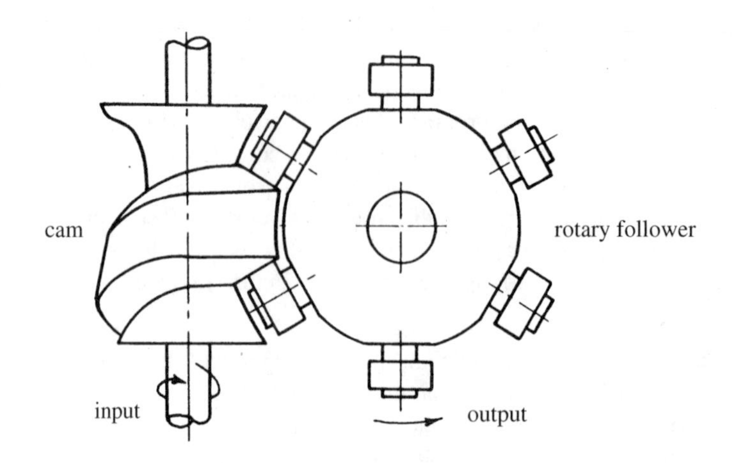

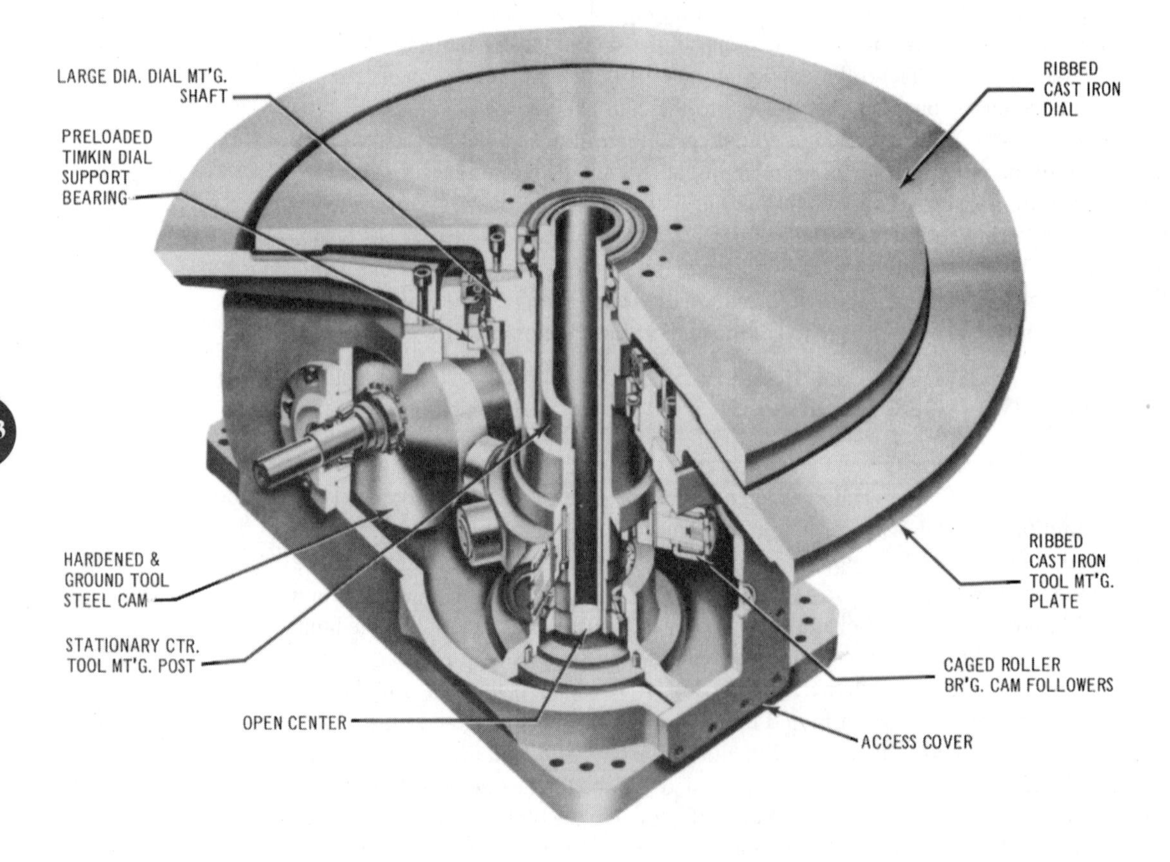

FIGURE 13-15

Multistop rotary indexer. *(Courtesy of The Ferguson Co., St. Louis, MO)*

Linkage analysis is complex and is beyond the scope of this work. The reader is directed to reference 4 for a complete treatment of the subject.

Modifying the Cam Contour for Follower Linkage Geometry

The designed follower motion should ideally be applied to the end effector and the cam follower link's motion back-calculated to account for the linkage geometry between the two. To do this only requires that the (assumed constant) linkage angular displacement ratio λ for the particular follower linkage be known. Then the cam-follower displacement function can be modified before applying it to the appropriate cam contour calculation according to:

$$s_{in} = \lambda s_{out} \tag{13.25a}$$

where s_{in} is the input function to be used to determine the cam contour, and s_{out} is the desired displacement function. In general, the linkage angular displacement ratio λ is defined as the ratio of the angular displacement of the output link δ_{out} divided by the angular displacement of the input link δ_{in}.

$$\lambda = \frac{\delta_{out}}{\delta_{in}} \tag{13.25b}$$

Equations for the calculation of the linkage ratio λ for a variety of linkages can be found in [4]. Note that if a polydyne or splinedyne function was calculated for the cam motion, it should be used as the s_{out} function. Demonstration versions of programs FOURBAR, SIXBAR, and SLIDER are provided on CD-ROM. These will calculate the geometry and ratios of linkages of those configurations for use with cam-follower systems.

13.7 SHIFTING THE CAM CONTOUR TO MACHINE ZERO

It often happens that the machine zero position on the timing diagram occurs within a rise or fall event for a particular cam. It is not practical to define a cam motion starting within a motion event. It is only practical to start the cam motion calculation at the beginning of a rise or fall, or within a dwell. Thus the "cam zero" position may necessarily differ from machine zero. This difference only affects the phase of the cam contour information, in effect its orientation vis-a-vis the cam's keyway location. Nevertheless, this is a critical relationship that if misapplied can result in disastrous results when the machine is run.

We will define a "start angle" θ_{start} for the cam displacement function that is the angle on the timing diagram that the designer chooses to make the "cam zero" position for definition of the start point of the cam motion program, being one of: the start or end of a rise, the start or end of a fall, the start of a dwell, or any point within a dwell. Then:

$$\theta_{cam} - \theta_{start} = \theta_{machine} \tag{13.26}$$

The cam is initially defined with respect to the chosen cam angle. When the cam contour is created, it can either be referenced to the cam angle, or to the machine zero angle by the phase shift defined in equation 13.26, as desired. There are reasons for doing either or both of these in any given case. If the desire is to compare the cam profiles of multiple cams on the same camshaft, then it makes sense to reference them all to a common angular position, and machine zero is the logical choice. This choice is also appropriate for the purpose of importing a set of cam contour coordinates into a CAD model of the machine in order to see if it appears correct in respect to the phasing and displacement of rise, fall, and dwell events. To shift the cam contour coordinates back to machine zero simply requires a coordinate transformation. If the cam contour is defined in polar or cylindrical coordinates, its radius at any point R remains the same and the angle shifts according to equation 13.26. If the contour is in Cartesian coordinates, R is found from the Pythagorean formula

$$R = \sqrt{\left[x(\theta_{cam})\right]^2 + \left[y(\theta_{cam})\right]^2} \tag{13.27a}$$

and the transformation is:

$$\begin{aligned} x_{\theta_{machine}} &= R\cos(\theta_{machine}) \\ y_{\theta_{machine}} &= R\sin(\theta_{machine}) \end{aligned} \tag{13.27b}$$

However, for cam manufacturing purposes, which will be discussed in more detail in the next chapter, it is not a good idea to start the cam contour data within a rise or fall event. It is better to begin cutting the cam at the start of a motion, at the end of a motion, or somewhere within a dwell event. Thus, for manufacturing, it is better practice to reference the cam contour data to cam zero rather than to machine zero if the latter occurs within a motion event. The keyway must be properly located with respect to machine zero in any event.

13.8 REFERENCES

1 **Kloomok, M.**, and **R. V. Muffley**. (1955). "Plate Cam Design-Pressure Angle Analysis". *Product Engineering*, v. 26, May, 1955, pp. 155-171.

2 **Gonzales-Palacios, M. A.**, and **J. Angeles**. (1993). *Cam Synthesis*. Kluwer Academic Publishers: Dordrecht.

3 **Angeles, J.**, and **C. S. Lopez-Cajun**. (1991). *Optimization of Cam Mechanisms*. Kluwer Academic Publishers: Dordrecht.

4 **Norton, R. L.** (2000). *Design of Machinery*, 2ed. McGraw-Hill, New York, NY.

Chapter 14

CAM MATERIALS AND MANUFACTURING

14.0 INTRODUCTION

The preceding chapters illustrate that there are a number of factors to consider when designing a cam. A great deal of care in design is necessary to obtain a good compromise of all factors, some of which conflict. Once the cam design is complete a whole new set of considerations must be dealt with that involve manufacturing the cam. After all, if your design cannot be successfully machined in metal in a way that reasonably represents the theoretical functions chosen, their benefits will not be realized. Unlike linkages, which are relatively easy to make, cams are a challenge to manufacture properly.

Cams are usually made from strong, hard materials such as **medium to high carbon steels** (case- or through-hardened) or **cast ductile iron** or **grey cast iron** (case-hardened). Cams for low loads and speeds or marine applications are sometimes made of bronze or stainless steel. Even plastic cams are used in such applications as washing machine timers where the cam is merely tripping a switch at the right time. We will concentrate on the higher load-speed situations here for which steel or cast/ductile iron are the only practical choices. These materials range from fairly difficult to very difficult to machine depending on the alloy. At a minimum, a reasonably accurate milling machine is needed to make a cam. A computer controlled machining center is far preferable and is most often the choice for serious cam production.

Cams are typically milled with rotating cutters that in effect "tear" the metal away leaving a less than perfectly smooth surface at a microscopic level as can be seen in Figure 12-1b (p. 338). For a better finish and better geometric accuracy, the cam can be ground after milling away most of the unneeded material. (See Figure 12-1a, p. 338.) Heat treatment is usually necessary to get sufficient hardness to prevent rapid wear. Tool

Portions of this chapter are adapted from R. L. Norton, *Machine Design: An Integrated Approach*, 2ed., Prentice-Hall, 2000, with permission.

steel cams are typically hardened to about 60-62HRC, ductile iron cams to at least 50-55HRC. Heat treatment introduces some geometric distortion. Grinding is usually done after heat treatment to correct the contour as well as to improve the finish.* The grinding step can nearly double the cost of an already expensive part, so it may sometimes be skipped in order to save money. A hardened but unground cam will have some heat distortion error despite accurate milling before hardening.

14.1 CAM MATERIALS

Table 12-4 in Chapter 12 (p. 382) contains surface fatigue strength data for a number of materials, many of which are suitable for cams. Other material properties are provided in Appendix B. The most commonly used cam materials are cast iron and steel. Gray cast iron can be used when loads are relatively low and cost is important. Cast ductile or nodular iron has higher strength than gray cast iron. Steels offer higher strengths and hardness than any of the cast irons. Bronze or stainless steel is sometimes used for applications in marine or corrosive environments.

Cast Irons

Cast irons constitute a whole family of materials. Their main advantages are relatively low cost and ease of fabrication. Some are weak in tension compared to steels but, like most cast materials, have good compressive strength. Their densities are slightly lower than steel at about 0.25 lb/in^3 (6 920 kg/m^3). Most cast irons do not exhibit a linear stress-strain relationship below the elastic limit; they do not obey Hooke's law. Their modulus of elasticity E is estimated by drawing a line from the origin through a point on the curve at 1/4 the ultimate tensile strength, and is in the range of 14–25 Mpsi (97–172 MPa). Cast iron's chemical composition differs from steel principally in its higher carbon content, being between 2 and 4.5% versus less than 1% for steel. The large amount of carbon, present in some cast irons as graphite, makes some of these alloys easy to pour as a casting liquid and also easy to machine as a solid. The most common means of fabrication is sand casting with subsequent machining operations. Cast irons are not easily welded, however.

GRAY CAST IRON This is the most commonly used form of cast iron. Its graphite flakes give it its gray appearance and name. The ASTM grades gray cast iron into seven classes based on the minimum tensile strength in kpsi. Class 20 has a minimum tensile strength of 20 kpsi (138 MPa). The class numbers of 20, 25, 30, 35, 40, 50, and 60 then represent the tensile strength in kpsi. Cost increases with increasing tensile strength. This alloy is easy to pour, easy to machine, and offers good acoustical damping. This makes it the popular choice for machine frames, engine blocks, brake rotors and drums, etc. The graphite flakes also give it good lubricity and wear resistance. Its relatively low tensile strength recommends against its use in situations where large bending or fatigue loads are present, though it is sometimes used in low-cost engine crankshafts, low-cost cams and low-cost IC engine camshafts e.g., lawn-mower engines. It runs reasonably well against steel if lubricated.

* Some automotive camshafts are ground soft, then hardened and polished to remove the carburization film from the hardening process. The reason for this is to avoid the possibility of "burning" the surface during grinding that would locally anneal the material and lead to premature surface fatigue failure.

NODULAR (DUCTILE) CAST IRON This has the highest tensile strength of the cast irons, ranging from about 70 to 135 kpsi (480 to 930 MPa). The name *nodular* comes from the fact that its graphite particles are spheroidal in shape. Ductile cast iron has a higher modulus of elasticity [about 25 Mpsi (172 GPa)] than gray cast iron and exhibits a linear stress-strain curve. It is tougher, stronger, more ductile, and less porous than gray cast iron. It is the cast iron of choice for fatigue-loaded parts such as crankshafts, pistons, cams, and cam-follower arms.

Wrought Steels

The term "wrought" refers to all processes that manipulate the shape of the material without melting it. Hot rolling and cold rolling are the two most common methods used, though many variants exist, such as wire drawing, deep drawing, extrusion, and cold-heading. The common denominator is a deliberate yielding of the material to change its shape either at room or at elevated temperatures.

HOT-ROLLED STEEL This is produced by forcing hot billets of steel through sets of rollers or dies that progressively change their shape into I-beams, channel sections, angle irons, flats, squares, rounds, tubes, sheets, plates, etc. The surface finish of hot-rolled shapes is rough due to oxidation at the elevated temperatures. The mechanical properties are also relatively low because the material ends up in an annealed or normalized state, unless deliberately heat-treated later. This is the typical choice for low-carbon structural steel members used for building- and machine-frame construction. Hot-rolled material is also used for machine parts that will be subjected to extensive machining (gears, cams, etc.) where the initial finish of the stock is irrelevant and uniform, non-cold-worked material properties are desired in advance of a planned heat treatment. A wide variety of alloys and carbon contents are available in hot-rolled form.

COLD-ROLLED STEEL Produced from billets or hot-rolled shapes, the shape of cold-rolled steel is brought to final form and size by rolling between hardened steel rollers or drawing through dies at room temperature. The rolls or dies burnish the surface and cold work the material, increasing its strength and reducing its ductility. The result is a material with good surface finish and accurate dimensions compared to hot-rolled material. Its strength and hardness are increased at the expense of significant built-in strains, which can later be released during machining, welding, or heat treating, then causing distortion. Cold-rolled shapes commonly available are sheets, strips, plates, round and rectangular bars, tubes, etc. Cold rolled steel is not recommended for cams. Hot rolled steel is preferred.

Forged Steel

Large cams or complex shapes such as IC engine camshafts are often formed by hot forging a steel billet to an approximate shape for later machining. If sufficient quantity is required to offset the cost of forging dies, significant savings of machining time can be realized over starting each cam with a billet. Also, the strength of the forged part, especially against fatigue loading can be superior to that of a cam made from billet.

Sintered Metals

Sintered materials start with a powder of the desired alloy that is compacted into the desired shape in dies. The resulting "green" has little strength until it is fired at high temperature, creating a part with good mechanical properties, especially in compression. Advantages of this process include the ability to accurately form complex shapes with good surface finish and little waste. Less machining is required than with either cast or forged parts. The mechanical properties of the material can be closely controlled for a particular application. The sintered parts can also be heat treated.

Steel Numbering Systems

Several steel numbering systems are in general use. The ASTM, AISI, and SAE* have devised codes to define the alloying elements and carbon content of steels. Table 14-1 lists some of the AISI/SAE designations for commonly used steel alloys. The first two digits indicate the principal alloying elements. The last two digits indicate the amount of carbon present, expressed in hundredths of a percent. ASTM and the SAE have developed a Unified Numbering System for all metal alloys, which uses the prefix UNS followed by a letter and a 5-digit number. The letter defines the alloy category, *F* for cast iron, *G* for carbon and low-alloy steels, *K* for special-purpose steels, *S* for stainless steels, and *T* for tool steels. For the *G* series, the numbers are the same as the AISI/SAE designations in Table 14-1 with a trailing zero added. For example, SAE 4340 becomes UNS G43400.

* ASTM is the American Society for Testing and Materials; AISI is the American Iron and Steel Institute; and SAE is the Society of Automotive Engineers. AISI and SAE both use the same designations for steels.

PLAIN CARBON STEEL This steel is designated by a first digit of 1 and a second digit of 0, since no alloys other than carbon are present. The low-carbon steels are those numbered AISI 1005 to 1030, medium-carbon from 1035 to 1055, and high-carbon from 1060 to 1095. The AISI 11xx series adds sulphur, principally to improve machinability. These are called free-machining steels and are not considered alloy steels as the sulphur does not improve its mechanical properties and also makes it brittle. The ultimate tensile strength of plain carbon steel can vary from about 60 to 150 kpsi (414 to 1 034 MPa) depending on heat treatment.

ALLOY STEELS These have various elements added in small quantities to improve the material's strength, hardenability, temperature resistance, corrosion resistance, and other properties. Any level of carbon can be combined with these alloying elements. Chromium is added to improve strength, ductility, toughness, wear resistance, and hardenability. Nickel is added to improve strength without loss of ductility, and it also enhances case hardenability. Molybdenum, used in combination with nickel and/or chromium, adds hardness, reduces brittleness, and increases toughness. Many other alloys in various combinations, as shown in Table 14-1, are used to achieve specific properties. Specialty steel manufacturers are the best source of information and assistance for the engineer trying to find the best material for any application. The ultimate tensile strength of alloy steels can vary from about 80 to 300 kpsi (550 to 2 070 MPa), depending on its alloying elements and heat treatment. Appendix B contains tables of mechanical property data for a selection of carbon and alloy steels. Figure 14-1 shows approximate ultimate tensile strengths of some normalized carbon and alloy steels.

AISI #
4340
4140
1095
6150
9255
3140
1050
1040
1030
1118
1020
1015
Kpsi 0 100 200
MPa 0 700 1 400
tensile strength

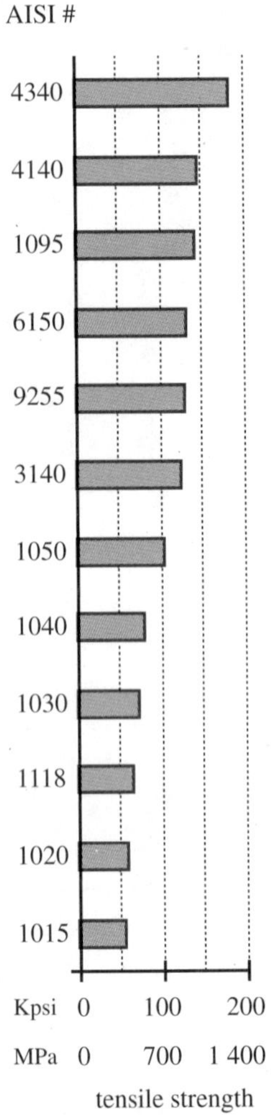

FIGURE 14-1

Approximate ultimate tensile strengths ofsome normalized steels

TABLE 14-1 AISI/SAE Designations of Steel Alloys

A partial list - other alloys are available - consult the manufacturers

Type	AISI/SAE Series	Principal Alloying Elements
Carbon Steels		
Plain	10xx	Carbon
Free-cutting	11xx	Carbon plus Sulphur (resulphurized)
Alloy Steels		
Manganese	13xx	1.75% Manganese
	15xx	1.00 to 1.65% Manganese
Nickel	23xx	3.50% Nickel
	25xx	5.00% Nickel
Nickel-Chrome	31xx	1.25% Nickel and 0.65 or 0.80% Chromium
	33xx	3.50% Nickel and 1.55% Chromium
Molybdenum	40xx	0.25% Molybdenum
	44xx	0.40 or 0.52% Molybdenum
Chrome-Moly	41xx	0.95% Chromium and 0.20% Molybdenum
Nickel-Chrome-Moly	43xx	1.82% Nickel, 0.50 or 0.80% Chromium, and 0.25% Molybdenum
	47xx	1.45% Nickel, 0.45% Chromium, and 0.20 or 0.35% Molybdenum
Nickel-Moly	46xx	0.82 or 1.82% Nickel and 0.25% Molybdenum
	48xx	3.50% Nickel and 0.25% Molybdenum
Chrome	50xx	0.27 to 0.65% Chromium
	51xx	0.80 to 1.05% Chromium
	52xx	1.45% Chromium
Chrome-Vanadium	61xx	0.60 to 0.95% Chromium and 0.10 to 0.15% Vanadium minimum

TOOL STEELS These are medium- to high-carbon alloy steels especially formulated to give very high hardness in combination with wear resistance and sufficient toughness to resist the shock loads experienced in service as cutting tools, dies and molds. There is a very large variety of tool steels available. AISI uses a different coding system for them consisting of a single letter followed by a one or two-digit number. The letter indicates the principal quenching medium, e.g., O-oil, W-water, A-air, and the number defines their position in the series. Because of their high strength and high hardenability, tool steels are commonly used for cams that see high stresses or high wear. Type A-2 tool steel is often used for cams. Type A-10 tool steel offers the advantage of low heat treat distortion and higher wear resistance, but is more expensive and slightly more difficult to machine. A partial list of AISI designated tool steels commonly used for cams with their relevant properties and features is shown in Table 14-2.

TABLE 14-2 Some Tool Steels Commonly Used for Cams (W1 shown for reference only)

AISI #	Features	Wear Resistance	Toughness	Distortion	Mach-inability	Quench Medium	Hardness HRC
W1	General purpose tool steel	Med+	Good	High	100	Water	64
A2	High carbon, medium chrome, low distortion, fairly good machining	Med+	Med+	Low-	65	Air	62
A6	Medium alloy, low temperature hardening, good free machining characteristics	Med	Med+	Low	85	Air	60
A8	Maximum toughness, good abrasion resistance	Med	High	Low	85	Air	61
A10	Excellent wear, good toughness and machinability, minimum heat-treat distortion, low temperature hardening, contains free graphite	High+	High	Lowest	90	Air	59/61
D2	High carbon, high chrome, good compressive strength	High+	Med	Very low	65	Air-oil	62
D3	High carbon, high chrome, good cutting edge	High+	Low	Low-	45	Air-oil	62
D5	Nondeforming, cobalt for extra wear resistance and anti-galling	High	Low	Low	60	Air	64
D7	Super abrasion resistance	Highest	Poor	Very low	35	Air	64

14.2 HARDNESS

The hardness of a material can be an indicator of its resistance to wear (but is not a guarantee of wear resistance). The strengths of some materials such as steels are also closely correlated to their hardness. Various treatments are applied to steels and other metals to increase hardness and strength. These are discussed below.

14

Hardness is most often measured on one of three scales: *Brinell, Rockwell,* or *Vickers.* These hardness tests all involve the forced impression of a small probe into the surface of the material being tested. The **Brinell test** uses a 10-mm hardened steel or tungsten-carbide ball impressed with either a 500- or 3000-kg load depending on the range of hardness of the material. The diameter of the resulting indent is measured under a microscope and used to calculate the Brinell hardness number, which has the units of kg_f/mm^2. The **Vickers test** uses a diamond-pyramid indenter and measures the width of the indent under the microscope. The **Rockwell test** uses a 1/16-in ball or a 120° cone-shaped diamond indenter and measures the depth of penetration. For cast and ductile irons, the Brinell test should be used rather than the Rockwell test as the latter gives erroneous readings due to the amount of free carbon in those materials.

Hardness is indicated by a number followed by the letter H, followed by letter(s) to identify the method used, e.g., 375 HB or 396 HV. Several lettered scales (A, B, C, D,

F, G, N, T) are used for materials in different Rockwell hardness ranges. It is necessary to specify both the letter and number of a Rockwell reading, such as 60 HRC. In the case of the Rockwell N scale, a narrow-cone-angle diamond indenter is used with loads of 15, 30, or 40 kg and the specification must include the load used as well as the letter specification, e.g., 84.6 HR15N. This Rockwell N scale is typically used to measure the "superficial" hardness of thin or case-hardened materials. The smaller load and narrow-angle N-tip give a shallow penetration that measures the hardness of the case without including effects of the soft core.

All these tests are nondestructive in the sense that the sample remains intact. However, the indentation can present a problem if the surface finish is critical or if the section is thin, so they are actually considered destructive tests. The Vickers test has the advantage of having only one test setup for all materials. Both the Brinell and Rockwell tests require selection of the tip size or indentation load, or both, to match the material tested. The Rockwell test is favored for its lack of operator error, since no microscope reading is required and the indentation tends to be smaller, particularly if the N-tip is used. But, the Brinell hardness number provides a very convenient way to quickly estimate the ultimate tensile strength (S_{ut}) of the material from the relationship

$$\begin{aligned} S_{ut} &\cong 500 H_B \pm 30 H_B \quad \text{psi} \\ S_{ut} &\cong 3.45 H_B \pm 0.2 H_B \quad \text{MPa} \end{aligned} \tag{14.1}$$

where H_B = the Brinell hardness number. This gives a convenient way to obtain a rough experimental measure of the strength of any low- or medium-strength carbon or alloy steel sample, even one that has already been placed in service and cannot be truly destructively tested.

Microhardness tests use a low force on a small diamond indenter and can provide a profile of microhardness as a function of depth across a sectioned sample. The hardness is computed on an absolute scale by dividing the applied force by the area of the indent. The units of **absolute hardness** are kg_f/mm^2. Brinell and Vickers hardness numbers also have these hardness units, though the values measured on the same sample can differ with each method. For example, a Brinell hardness of 500 HB is about the same as a Rockwell C hardness of 52 HRC and an absolute hardness of 600 kg_f/mm^2. Note that these scales are not linearly related, so conversion is difficult. Table 14-3 shows approximate conversions between the Brinell, Vickers, and Rockwell B and C hardness scales for steels and their approximate equivalent ultimate tensile strengths.

14.3 HEAT TREATMENT

The steel heat-treatment process is quite complicated and is dealt with in detail in materials texts such as those listed in the bibliography at the end of this chapter. The reader is directed to such references for a more complete discussion. Only a brief review of some of the salient points is provided here.

The hardness and other characteristics of many steels and some nonferrous metals can be changed by heat treatment. Steel is an alloy of iron and carbon. The weight per-

Table 14-3 Approximate Equivalent Hardness Numbers and Ultimate Tensile Strengths for Steels

Brinell	Vickers	Rockwell		Ultimate, σ_u	
HB	*HV*	*HRB*	*HRC*	MPa	ksi
627	667	—	58.7	2393	347
578	615	—	56.0	2158	313
534	569	—	53.5	1986	288
495	528	—	51.0	1813	263
461	491	—	48.5	1669	242
429	455	—	45.7	1517	220
401	425	—	43.1	1393	202
375	396	—	40.4	1267	184
341	360	—	36.6	1131	164
311	328	—	33.1	1027	149
277	292	—	28.8	924	134
241	253	100	22.8	800	116
217	228	96.4	—	724	105
197	207	92.8	—	655	95
179	188	89.0	—	600	87
159	167	83.9	—	538	78
143	150	78.6	—	490	71
131	137	74.2	—	448	65
116	122	67.6	—	400	58

Note: Load 3000 kg for HB.

Source: N. E. Dowling, (1993). *Mechanical Behavior of Materials*, Prentice Hall, Englewood Cliffs, N.J., , Table 5-10, p.185, with permission.

cent of carbon present affects the alloy's ability to be heat-treated. A low-carbon steel will have about 0.03 to 0.30% of carbon, a medium-carbon steel about 0.35 to 0.55%, and a high-carbon steel about 0.60 to 1.50%. (Cast irons will have greater than 2% carbon.) Hardenability of steel increases with carbon content. Low-carbon steel has too little carbon for effective through-hardening, so other surface-hardening methods must be used (see below). Medium- and high-carbon steels can be through-hardened by appropriate heat treatment. The depth of hardening will vary with alloy content.

QUENCHING To harden a medium- or high-carbon steel, the part is heated above a critical temperature [about 1 400°F (760°C)], allowed to equilibrate for some time, and then suddenly cooled to room temperature by immersion in a water or oil bath. The rapid cooling creates a supersaturated solution of iron and carbon called martensite, which is extremely hard and much stronger than the original soft material. Unfortunately, it is also very brittle. In effect, we have traded off the steel's ductility for its increased strength. The rapid cooling also introduces strains to the part. While the increased strength is desirable, the severe brittleness of a fully quenched steel usually makes it unusable without tempering.

TEMPERING Subsequent to quenching, the same part can be reheated to a lower temperature [400–1300°F (200–700°C)], heat-soaked, and then allowed to cool slowly. This will cause some of the martensite to convert to ferrite and cementite, which reduces the strength somewhat but restores some ductility. A great deal of flexibility is possible in terms of tailoring the resulting combination of properties by varying time and temperature during the tempering process.

ANNEALING The quenching and tempering process is reversible by annealing. The part is heated above the critical temperature (as for quenching), but now allowed to cool slowly to room temperature. This restores the solution conditions and mechanical properties of the unhardened alloy. Annealing is often used even if no hardening has been previously done in order to eliminate any residual stresses and strains introduced by the forces applied in forming the part. It effectively puts the part back into a "relaxed" and soft state, restoring its original stress-strain curve.

NORMALIZING Many tables of commercial steel data indicate that the steel has been normalized after rolling or forming into its stock shape. Normalizing is similar to annealing, but involves a shorter soak time at elevated temperature and a more rapid cooling rate. The result is a somewhat stronger and harder steel than a fully annealed one, but one that is closer to the annealed condition than to any tempered condition.

Surface (Case) Hardening

When a part is large or thick, it is difficult to obtain uniform hardness within its interior by through-hardening. An alternative is to harden only the surface, leaving the core soft. This also avoids much of the distortion associated with quenching a large, through-heated part. If the steel has sufficient carbon content, its surface can be heated, quenched, and tempered as would be done for through-hardening. For low-carbon (mild) steels, other techniques are needed to obtain a hardened condition. These involve heating the part in a special atmosphere rich in either carbon, nitrogen, or both and then quenching it, a process called *carburizing*, *nitriding*, or *cyaniding*. In all situations, the result is a hard surface (i.e., *case*) on a soft core, referred to as being **case-hardened.**

Note that for cast and ductile iron cams that are to be case-hardened, it is important to first normalize them to make sure that there is enough carbon in solution with the ferrite. For a flame or induction hardened cast or ductile iron cam, the surface will not come up to the proper hardness if it was not normalized to a proper base hardness.

14

CARBURIZING This process heats low-carbon steel in a carbon monoxide gas atmosphere, causing the surface to take up carbon in solution. **Nitriding** heats low-carbon steel in a nitrogen-gas atmosphere and forms hard iron nitrides in the surface layers. **Ion nitriding** bombards the surface with ions of nitrogen to achieve a similar effect. **Cyaniding** heats the part in a cyanide salt bath at about 1 500°F (800°C), and the low-carbon steel takes up both carbides and nitrides from the salt.

For medium- and high-carbon steels, no artificial atmosphere is needed, as the steel has sufficient carbon for hardening. Two methods are in common use. **Flame hardening** passes an oxyacetylene flame over the surface to be hardened and follows it with a water jet or an oil-water emulsion for quenching. This results in a somewhat deeper hardened case than obtainable from the artificial-atmosphere methods. **Induction hardening** uses electric coils to rapidly heat the part surface, which is then quenched before the core can get hot.

CASE-HARDENING By any appropriate method, this is a very desirable hardening treatment for many applications. It is often advantageous to retain the full ductility (and thus the toughness) of the core material for better energy absorption capacity, while also obtaining high hardness on the surface in order to reduce wear and increase surface strength. Large machine parts such as cams and gears are examples of elements that can benefit more from case hardening than from through-hardening, as heat distortion is minimized and the tough, ductile core can better absorb impact energy. Note that distortion can be a particular problem in a case-hardened cam depending on cam shape and case depth. It is possible to grind through the hard case if excessive distortion forces too much metal removal in order to achieve the required finished contour.

14.4 CAM MANUFACTURING METHODS

There are several methods of cam manufacture that have been used over the years as shown in Table 14-4. Of these, the first two are less common in current practice, and the last listed is becoming less common, except perhaps in high volume applications such as automotive camshaft grinding. Historically, cam manufacturing was fairly crude in the first half of the 20th century, with the exception of automotive engine applications where volume demands required some degree of automation and precision. For low-volume cams, the "old-fashioned" method was to transfer the dimensional information from the cam blueprint manually onto a "blued" cam blank using a scriber, scale, and dividers at perhaps 5° increments, and then mill the contour "by eye" to the scribed outline, finishing with a hand filing operation to smooth the surface.* This did not make a highly accurate cam, but for that matter, the mathematical definition of the cam contour was equally crude by our present standards, often consisting of circle arcs and straight-line segments, or perhaps a constant acceleration or simple-harmonic motion that could be geometrically constructed by a draftsman.

* As an even cruder example of "old-time" cam manufacturing methods, an engineer of the author's acquaintance tells of his experience in the 1960s as a new apprentice machinist at a well-known manufacturing company, being given the task of making a cam. He was told "Go get the blueprint (of the cam) and paste it onto the steel." He then was instructed to cut out the contour of the pasted-on drawing with a bandsaw, leaving a little stock outside the cam outline so that it could be belt sanded down to the blueprint line. Voila! A cam!

Geometric Generation

Geometric generation refers to the continuous "sweeping out" of a surface as in turning a cylinder on a lathe. This is perhaps the ideal way to make a cam because it creates a truly continuous surface with an accuracy limited only by the quality of the machine and

TABLE 14-4 Common Cam Manufacturing Methods

1	Geometric Generation
2	Manual or **N**umerical **C**ontrol (**NC**) machining to cam coordinates (plunge-cutting)
3	Continuous **N**umerical **C**ontrol (**CNC**) with **L**inear **I**nterpolation (**LI**)
4	Continuous **N**umerical **C**ontrol (**CNC**) with **C**ircular **I**nterpolation (**CI**)
5	**A**nalog **D**uplication (**AD**) of a master cam

tools used. Unfortunately there are very few types of cams that can be made by this method. The most obvious one is the eccentric cam (Figure 2-5, p. 24) which can be turned and ground on a lathe. A cycloid can also be geometrically generated. Very few other curves can. The presence of dwells makes it extremely difficult to apply this method. Thus, it is seldom used for cams. However, when it can be, as in the case of the eccentric cam of Figure 2-5, the resulting acceleration, though not perfect, is very close to the theoretical cosine wave as seen in Figure 14-2. This eccentric cam was made by turning and grinding on a high-quality lathe. This is essentially the best that can be obtained in cam manufacture. Note that the displacement function is virtually perfect. The errors are only visible in the more sensitive acceleration function measurement.

Manual or NC Machining to Cam Coordinates (Plunge-Cutting)

Computer-aided manufacturing (CAM) has become the virtual standard for high accuracy machining in the United States. Numerical control (NC) machinery comes in many types. Lathes, milling machines, grinders, etc., are all available with on-board computers which control either the position of the workpiece, the tool, or both. The simplest type of NC machine moves the tool (or workpiece) to a specified x, y location and then drives the tool (say a drill) down through the workpiece to make a hole. This process is repeated as many times as necessary to create the part. This simple process is referred to as NC to distinguish it from continuous numerical control (CNC).

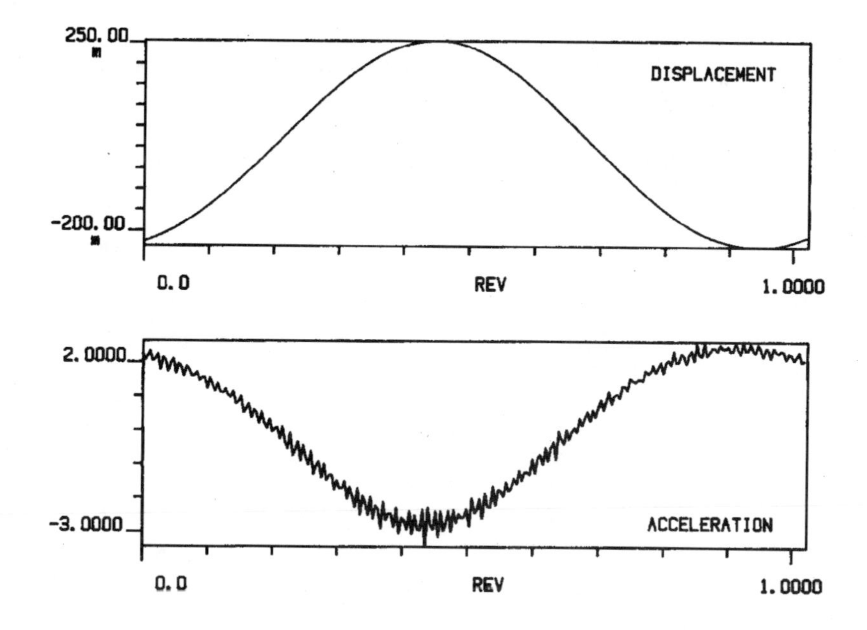

FIGURE 14-2

Measured follower displacement and acceleration of a geometrically generated ground eccentric cam

This NC process has been used for cam manufacture, and even for master cams as described below. However, its use has declined in favor of CNC as also described below. NC is, in fact, merely a computerized version of the old manual method of cam milling, which is often called **plunge-cutting** to refer to plunging the spinning milling cutter down through the workpiece. This is not the best way to machine a cam because it leaves "scallops" on the surface as shown in Figure 14-3, due to the fact that the machinist can only *plunge* at a discrete number of positions around the cam. In effect, the displacement function that we developed has to be "discretized" or sampled at some finite number of places around the cam. For plunge cutting, practicality limits this digitizing process to increments of about 1/2 to 1 degree. With an NC process, the increment might be reduced to 1/4 or 1/10 degree. At some point "diminishing returns" will set in as the machine's ability to resolve positions spaced too closely will limit the accuracy. Standard milling machines can be expected to give accuracies in the 0.001 inch tolerance range. Tooling-quality machining centers, jig borers, and grinders can be as much as 2 to 20 times that accurate [0.000 5-in down to 0.000 050-in (50 millionths) tolerance].

The scallops that are left on the cam after plunge-cutting have to be removed by hand-dressing with files and grindstones. This obviously introduces more error. Even if the bottoms of the scallops at the sample increments were exactly correct, all points between are subject to the vagaries of hand work. With this manufacturing method, the chance of exactly achieving the designed $s\ v\ a\ j$ functions, especially the higher derivatives, is slight.

Continuous Numerical Control with Linear Interpolation

In a CNC machine, the tool is in constant contact with the workpiece, always cutting, while the computer controls the movement of the workpiece from position to position as stored in its memory. This is a **continuous cutting process** as opposed to the discrete one of NC. However, the cam displacement function must still be discretized or sampled at some angular increment. Common increments are 1/4, 1/2, and 1 degree. Since the machine only has information about the x, y locations of these 360, 720, or 1440

14

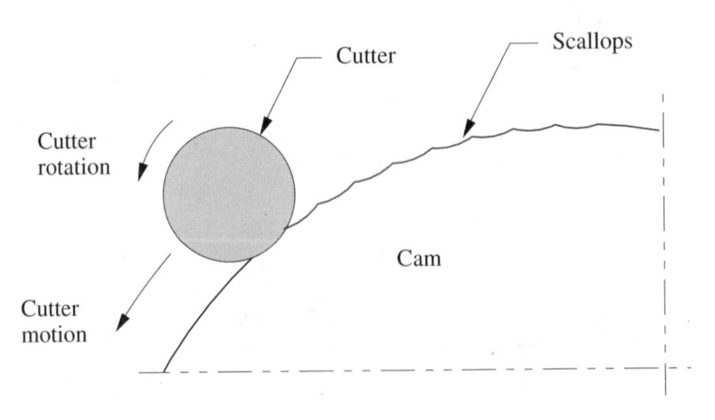

FIGURE 14-3

Plunge-cutting a cam leaves scallops on its surface

points around the cam, it has to figure out how to get from one point to another while cutting. The most common method used to "fill in" the missing data is **linear interpolation** (LI). The machine's computer calculates the straight line between each pair of data points and then drives the cutter (or workpiece) so as to stay as close to that line as it can. If it could do this perfectly (which it can't), we would get a piecewise continuous first-order approximation to the cam contour as shown in Figure 14-4, which depicts a portion of the rise of a multi-dwell cam. This would introduce slope (velocity) discontinuities which will create theoretically infinite pulses of acceleration. We would be back to our "bad cam by naive designer" of Example 2-1 (p. 20) having infinite acceleration regardless of the actual motion function selected.

An improvement can be made in this linear interpolation scheme by fitting a cubic spline curve to the cam coordinate data and then resampling the spline approximation at finer spacing down to the milling machine's resolution. This denser data set is then used to drive the cutter that still must traverse approximate straight-line paths between the closely spaced data points. The curve fitting and resampling process will be discussed in a later section of this chapter.

With sufficiently small interpolation increments, the dynamics of the cutting process, which are a function of speeds, feeds, tool sharpness, tool chatter, deflection of the spindle, etc., will act to limit the formation of distinct "flats" that would give the derivatives shown in Figure 2-3 (p. 20). Rather, the kind of acceleration curve that actually results from a cam which was milled (but not ground) on a very high-quality CNC machining center, using 1-degree linear interpolation, is as shown in Figure 14-5. The program is a simple harmonic eccentric with no dwells. The dynamic curves were measured with instrumentation on the roller follower while the cam was running at 600 rpm on a custom-designed cam dynamic test fixture (CDTF) as shown in Figure 14-6.[2]* The actual displacement is quite true to the theoretical, but the acceleration has a significant

* This research at WPI was supported by the National Science Foundation under grants MEA-82-10865 and MSM-85-12913, and by AMP Incorporated, Harrisburg, Pa.

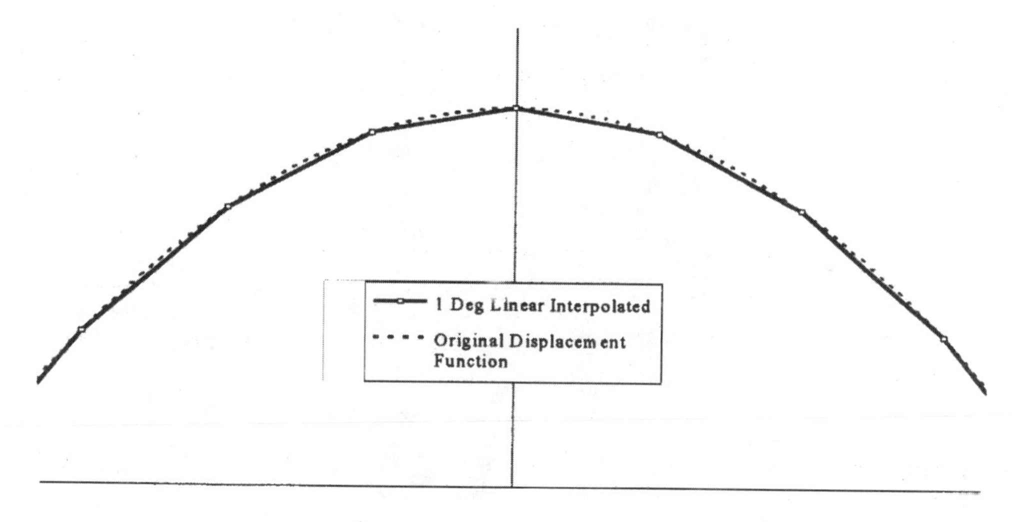

FIGURE 14-4

Cam contour as designed and as made with 1° linear interpolation CNC[1]

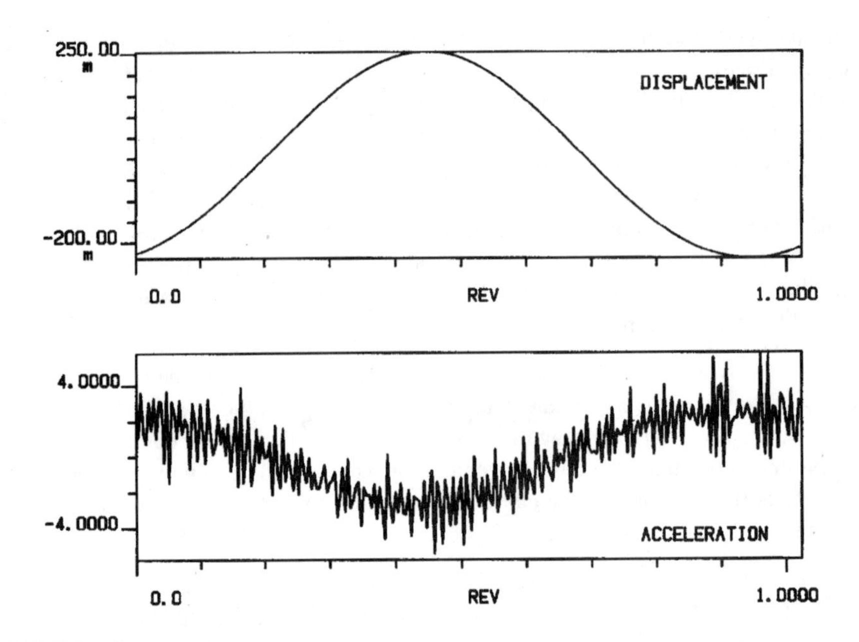

FIGURE 14-5

Displacement and acceleration of eccentric cam made with 1° linear interpolation CNC

amount of vibratory noise present which distorts the function from its theoretical cosine waveshape. The acceleration is shown in g's. Compare the peak-to-peak values in Figure 14-4 (8 g) with the same cam design made with geometric generation in Figure 14-2

FIGURE 14-6

Cam Dynamic Test Fixture (CDTF) used for testing of cam follower dynamics at WPI

(5 *g*). The error is of the order of 3 *g*'s on a base of 5 *g*'s. These errors in the acceleration are due to a combination of manufacturing factors as described above. It was found that the use of 1/4-, 1/2- or 1-degree digitization increments on this size cam* made no statistically significant difference in the fidelity of the actual acceleration function to its theoretical waveform using linear interpolation.[2]

The physical fidelity of the cam surface of the same 1-degree linear interpolated eccentric cam to a geometrically generated (turned and ground) reference cam can be seen in Figure 14-7 which is an enlarged section of a portion of the cams as measured to 0.0005-in accuracy. The axes are calibrated in inches. These figures show that linear interpolated CNC is a reasonably accurate method of cam manufacture.

Continuous Numerical Control with Circular Interpolation

This process is similar to CNC with linear interpolation except that a **circular interpolation** (CI) algorithm is used between data points. Most CNC and NC machines have a built-in algorithm to generate and cut circle arcs quickly and efficiently, as this is a common requirement in normal machining needed for the many instances of internal or external arcuate features on machine parts. Using this circle-arc generation capability for cam contour cutting potentially allows a sparser (thus smaller) database in the machine, which can be an advantage. The computer tries to fit a circle arc to as many adjacent data points as possible without exceeding a user-selected error band around the actual displacement function, as shown in Figure 14-8, which depicts a portion of the rise of a

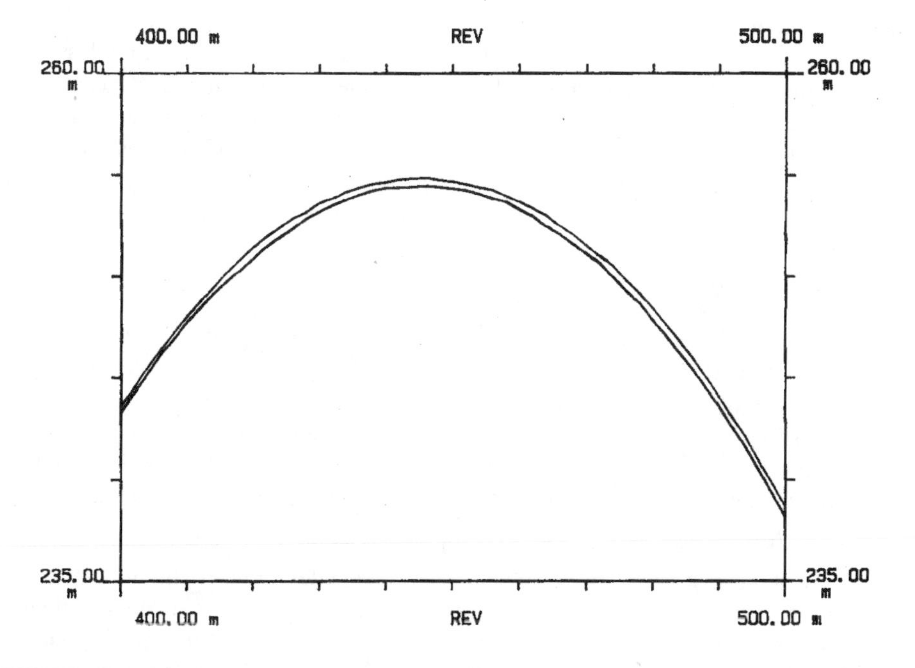

FIGURE 14-7

Contours of 2 cams: one turned and ground; one milled with 1° linear interpolation CNC

* These cams were about 8 in (200 mm) in diameter. If the cam diameter is larger, then smaller angular increments of digitization will be needed as the distance along the pitch curve between data points for any angular increment increases linearly with prime circle diameter.

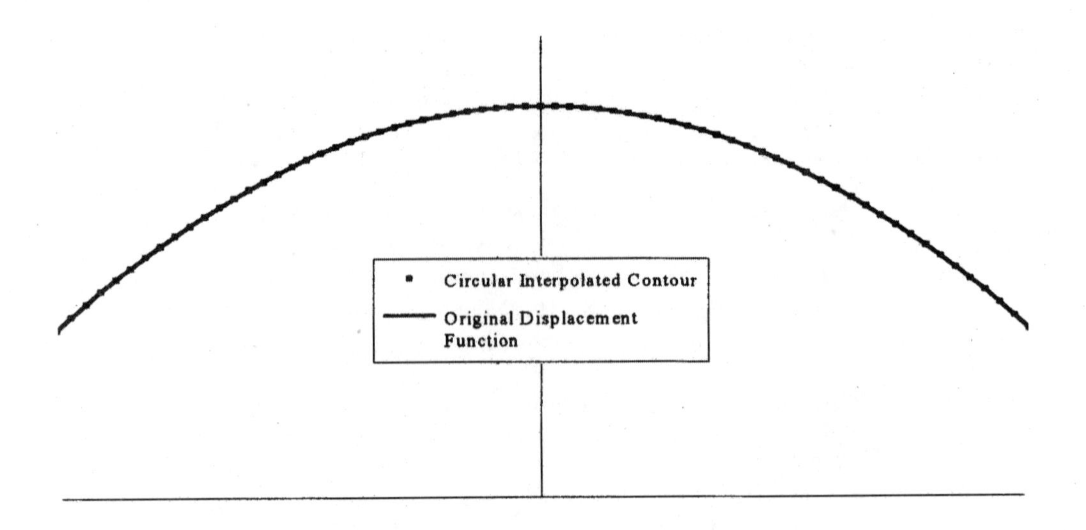

FIGURE 14-8
Cam contour as designed and as made with circular interpolation CNC[1]

multi-dwell cam. This reduces the (variable) number of fitted data points for any one arc segment to a radius, its two center coordinates and arc angle.

Figure 14-9 shows the same cam design as in Figures 14-2 and 14-5, made on the same machining center from the same bar of tool steel, but with circular interpolation (CI). The error in acceleration is more than the turned-ground (TG) cam, but less than the linear interpolated (LI) one. Figure 14-10 shows the cam contour of the CI cam compared to the TG cam. Based on dynamic performance the circular interpolated cam has lower acceleration error than the linear interpolated cam and the difference is statistically significant. The "blip" in the middle of the period is due to the slight ridge formed at the point where the cutter starts and stops its continuous sweep around the cam contour.

Analog Duplication

The last method listed in Table 14-4, analog duplication, involves the creation of a master cam, sometimes at larger than full scale, that is subsequently used in a cam duplicating machine to turn out large quantities of the finished cams. Some automotive camshafts are still made by this method, though CNC methods are replacing the automotive cam duplicating machines. Analog duplication can be the most economical method where production quantities are high enough to warrant the cost of making a master cam.

A cam duplicating machine has two spindles. The master cam is mounted on one, and the workpiece is placed on the other. A dummy cutter mounted on one crank of a pantograph linkage is used as a follower against the master. The actual cutter is mounted on the other crank of the pantograph linkage. A multiplication ratio can be introduced between the dummy follower and the cutter to size the finished cam versus the master. This allows master cams of larger size to be used, which increases accuracy. As the mas-

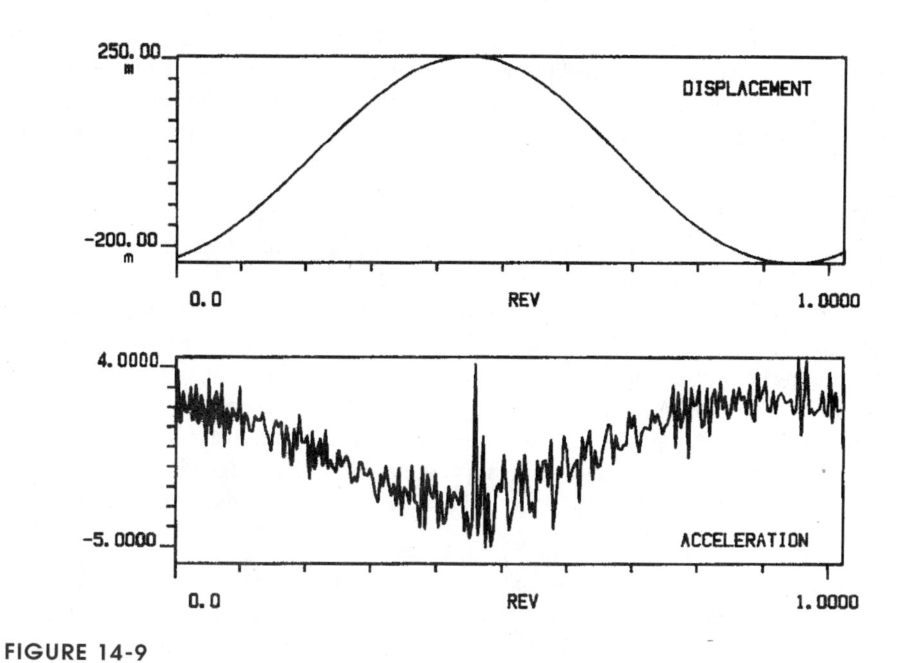

FIGURE 14-9

Displacement and acceleration of eccentric cam made with circular interpolation CNC

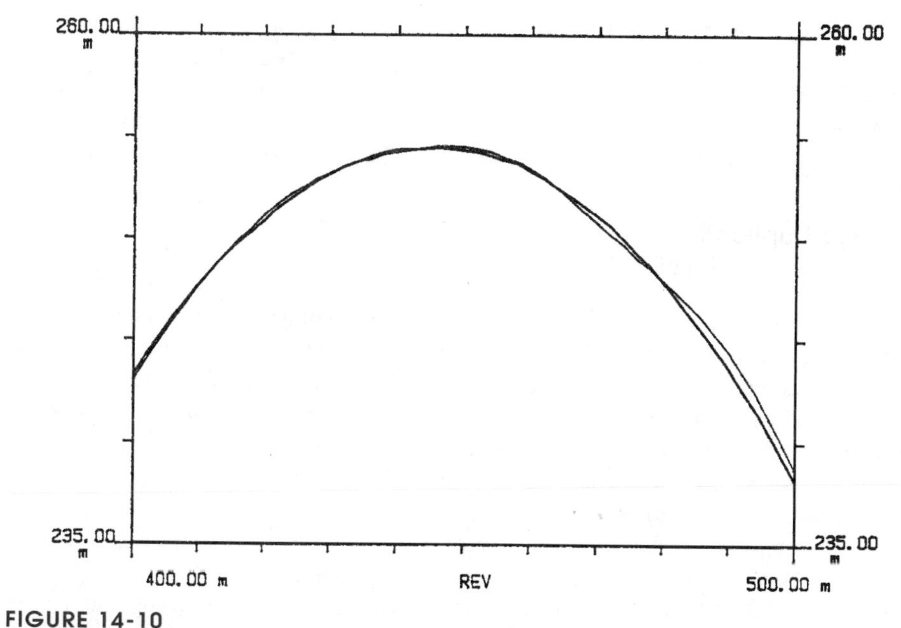

FIGURE 14-10

Contours of 2 cams: one turned and ground; one milled with circular interpolation CNC

ter and slave slowly turn together, synchronous and in phase, the dummy cutter follows the master's contour and the workpiece is cut to match. This process can be done with either milling (cutting) or grinding of the cam surface. Typically the cam is rough cut first, then heat treated and reground to finished size. Some cams are left as-milled with no post heat-treat grinding.

The analog duplication method can obviously create a cam that is only as good as the master cam at best. Some errors will be introduced in the duplicating process due to deflections of the tool or machine parts, but the quality of the master cam ultimately limits the quality of the finished cams. The master cam is typically made by one of the other methods listed in Table 14-4, each of which has its limitations. The master cam may require some hand-dressing with files or polishing with hand grindstones or emery cloth to smooth its surface. A plunge-cut cam requires a lot of hand-dressing, the CNC cams less so. If hand-dressing is done, it will result in a very smooth surface, but the chances that the resulting contour is an accurate representation of the designed *s v a j* functions, especially the higher derivatives, is slim. Thus the finished cam may not be an accurate representation of the design.

Figure 14-11 shows the same cam design as in Figures 14-2, 14-4 and 14-7, made from the same bar of steel, but analog duplicated (milled) from a hand-dressed, plunge-cut master on a Rowbottom duplicating cam machine. This represents the worst case in terms of manufacturing error. The error in acceleration is more than any of the other cams. Note the peak-to-peak acceleration is now 20g including the noise from the manufacturing errors. This is four times the level of the turned-ground cam of Figure 14-2.

Figure 14-12 shows the cam contour of the analog milled cam compared to the reference turned-ground cam. It is much less accurate than either of the CNC versions. Based on dynamic performance, the analog milled cam from a hand-dressed, plunge-cut master has a higher acceleration error than any other cam tested and the difference is statistically significant. Even an analog ground cam had greater error in contour and acceleration than any of the CNC cams. If the master cam were made by a more accurate method, the accuracy of the production cam could be better, but would still be potentially inferior to one made with direct CNC.

14.5 CUTTING THE CAM

In this discussion we will assume that a CNC method will be used for cutting the cam, either with linear or circular interpolation. The cam contour data for a roller follower cam is usually provided to the shop as coordinates of the roller centerline path. The discretization increment is usually chosen by the designer and may or may not be optimal from a manufacturing standpoint. If the shop were to use a cutter of exactly the roller diameter moved along the centerline path, then it would theoretically generate the correct surface contour. However, it is often not convenient to use a cutter of exactly the same diameter as the roller follower. Moreover, multiple passes around the cam are necessary to remove all the material from the blank, and so multiple NC paths must be generated. Standard NC programming software* in the hands of a trained programmer will take care of this task and its details will not be addressed here.

* e.g., Mastercam, Smartcam, Surfcam, and others.

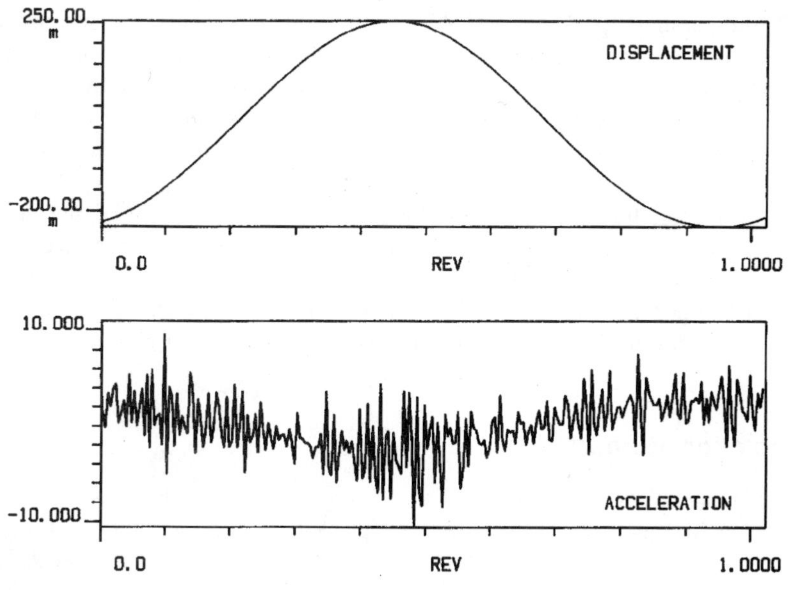

FIGURE 14-11

Displacement and acceleration of eccentric cam made with analog duplication of a plunge-cut and hand-dressed master cam

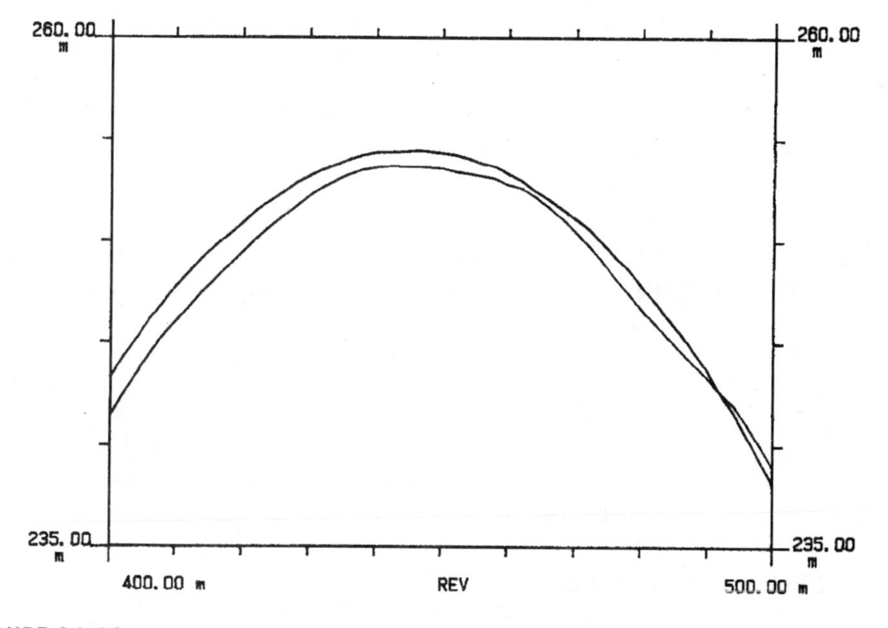

FIGURE 14-12

Contours of 2 cams: one turned and ground; one milled with analog duplication of a plunge-cut and hand-dressed master cam

NC programming packages and NC machines themselves usually contain a cutter compensation feature. Cutter compensation was defined and its equations derived in Chapter 13. Essentially, a new set of cutter path coordinates is generated as a normal offset curve from the given path. This not only allows a cutter or grinder of arbitrary diameter (larger or smaller than the roller, but smaller than the minimum negative radius of curvature of the cam surface) to be used, but also allows for tool wear or wheel dressing to be accounted for between passes around the workpiece. The tool diameter can be measured after each complete traverse around the cam and a new compensated tool path calculated for the next pass. For a flat-follower cam, there is no roller path, only a cam surface definition, so cutter compensation must be used to create the cutter path. Note that cutter compensation is not recommended for barrel cams for reasons described in Chapter 13.

Interpolation Method

The default interpolation method among most CNC shops and cam manufacturers is linear interpolation. Testing has shown that while acceptable for cam manufacturing when properly done, linear interpolation can be less dynamically accurate than circular interpolation.[2]

Circular interpolation was devised in the early years of CNC machining to solve a problem that technology has since made disappear. In the 1970's, NC machines were controlled by paper tape readers. The contour data was encoded as punched holes in a strip of paper tape, essentially a continuous IBM card. The paper tape was fed into the milling machine, which read enough data to get its next cutter location and then moved the cutter accordingly. If one chose to define a cam at say 0.25° increments (1440 *x, y* data pairs) to get the desired accuracy, there would be too much data for the machine to handle. It would become "tape bound," meaning that it took longer to read the data than to cut metal. This not only slowed production to a crawl, but also made an inferior cam. If the cutter stops moving (i.e., feeding) at any point during its traverse, the still spinning cutter digs a small "divot" or mark into the surface as the dynamic deflection of the spindle caused by the feed force relaxes. The cutter needs to move continuously, and preferably at a constant feed rate, to get the best finish and accuracy.

The General Electric Co., which pioneered much of CNC machine controller technology, invented circular interpolation in order to compress large data sets so the machines would not be tape bound. By mathematically fitting a circle arc to as many data points as possible within a user specified least squares error band, a large number of *x, y* data can be reduced to a smaller set of arc lengths and center coordinates. Present day CNC machines are fitted with dedicated computers that have no trouble generating cutter coordinate data faster than the machine can cut metal. Nevertheless, some specialty cam manufacturers still use circular interpolation as it can give better results and is less sensitive to the cam digitization increment than linear interpolation.

Digitization Increment

The angular increment at which the cam contour function was digitized by the designer may or may not be optimal from a manufacturing standpoint if linear interpolation is used. The tangent of 1° is 0.017, which means for every inch (25.4 mm) of cam radius, the spacing between data sampled at 1° is 0.017 in or about 0.5 mm. This can make noticeable flats on the surface of a cam of large radius. Even if the designer chose a smaller digitization increment than 1°, it is likely that the distance between data points on the cam surface will be greater than the resolution of the milling machine or machining center. If one wants to take full advantage of the resolution accuracy of the machine, then it will be necessary to resample the data.

Resampling the Data

Most commercial Computer Aided Manufacturing (CAM) software provides a resampling feature. Typically, the software takes a digitized representation of a continuous surface, say a cam contour calculated at every half degree, and fits a spline to the data to get a continuous function that approximates the original surface. Then it resamples the spline at a user specified digitization increment that can result in an average spacing of data as small as the machine's encoder resolution, typically about 0.001 in. Some resampling software allows the setting of a desired error band and resamples at varying angular step size to maintain contour accuracy within the specified band. Though there will be some small numerical sampling error with any of these methods, the surface can now be cut with vanishingly small linear interpolation flats.

The type of spline available for curve fitting will vary with the particular CAM package. Some seem to be limited to using cubic natural splines, which are a piecewise continuous function that has only displacement and velocity continuity at the piecewise junctions. Other, more powerful CAM packages, typically those within a solids modeling CAD package, allow some choice of spline order up to at least quintic. These have continuity through acceleration and are to be preferred. Nevertheless, even the relatively crude cubic spline fit is a better bet than using a sparse digitization of the original cam contour directly with linear interpolation in the machine. Though this has not been tested, it seems likely that a splined and finely resampled, then linearly interpolated cam would give dynamic results comparable to, if not better than, circular interpolation.

A recent experiment at The Gillette Company has demonstrated the efficacy and fidelity of this approach to cam manufacture. A radial cam was designed in DYNACAM and its follower centerline path calculated at 0.25° increments. The path data was exported to a disk file and imported into the *Unigraphics* solids modeling program. A quintic spline was then fitted to the discrete follower path in Unigraphics. The splined follower path was used to drive the follower linkage within the mechanisms package in Unigraphics to generate dynamic information and check for possible interferences.

A second quintic spline was created as a normal offset curve at one follower radius from the spline that approximated the follower centerline path This generated a continuous quintic spline function for the cam surface and allowed a visual check of the cam contour within the CAD model of the follower assembly. The surface-spline model was then transmitted electronically to the shop where the Unigraphics CAM package was used to generate the NC tool path information for a chosen cutter radius offset from the cam surface as another quintic spline. This CNC information was used to manufacture the cam.

The finished cam was measured on a 12-millionths-resolution coordinate measuring machine (CMM) at 0.001 in increments and the resulting data brought back into Un-

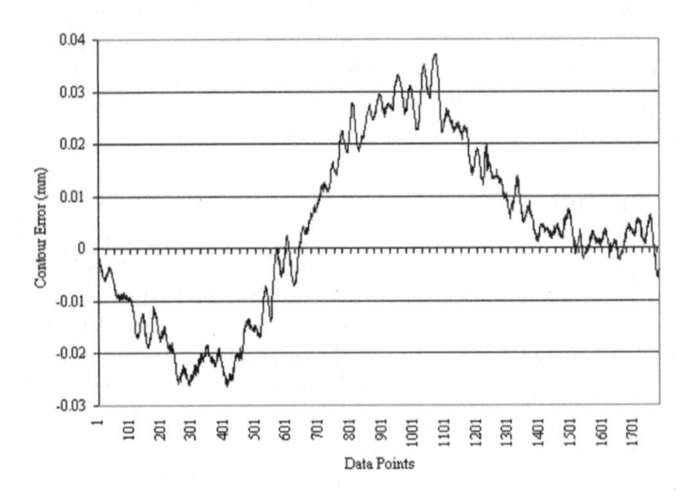

FIGURE 14-13

Difference between a designed cam contour as quintic-splined in a CAD system and the CMM measurement of the actual cam surface ground with CNC to that spline path digitized at the machine's resolution *(Courtesy of The Gillette Company, Boston, MA)*

igraphics where it was compared to the original surface spline. The surface contour was within ±0.04 mm (±0.0015 in) at all points, as shown in Figure 14-13. Though this just confirms that the finished cam accurately represents the splined approximation to the original designed function, the error between those two curves is less than the machining tolerance. This approach represents the current state of the art in cam manufacturing and is a far cry from the hand filing of scallops that was standard practice 30-40 years ago.

Pythagorean Hodographs

The purpose of this section is to inform the reader about problems inherent in machining a cam surface and of the most recently proposed methods for dealing with them. A new approach to the generation of cam contours and their CNC tool path information has recently been proposed in the form of Pythagorean hodographs (PH), which are a form of spline. The name comes from the Pythagorean formula for a right triangle and the term hodograph, which means the derivative curve of a curve. Computer numerical control machining requires the conversion of mathematical descriptions of cam contour tool paths (such as a NURBs description of the cam profile) into instructions to servomechanisms that control the cutting/grinding machine. Furthermore, all of the conversions must be done in real time, while the machine is operating. One difficulty is that the cam designer has usually specified the cam shape in terms of some parameter that made perfect sense when defining follower motion, but that has no direct geometric meaning to the CNC machinery and no direct time equivalent to its operating speed. For example, equation 13.4c (p. 398) gives the cam cutter center coordinates for the offset translating roller follower in terms of θ. The angle θ, however, has geometric meaning only for a cutter of a radius equal to that of the follower, which is not necessarily true in all cases.

The engineer concerned with manufacturing the actual cam would be well advised to consider θ as just an arbitrary parameter. Ideally, the spacing of the sampled points along the cutter path should be equal in terms of distance along the curved path as opposed to equal in terms of, say, cam angle. Then the feed rate will be constant along the path, giving better accuracy and finish. Pythagorean hodograph splines have this property which makes them well suited to the task.

Suppose that a milling machine is arranged such that the cam blank rotates about the center of a rotary table and the cutter or grinding wheel moves radially toward or away from the center of the cam blank in order to shape the profile. The cutter motion must be synchronized with cam blank rotation. The machine has to know when to move, whether to move toward the center or away from the center, and how far and how fast to move. These decisions can be made by the machine's computer using the *specified feed rate*, the speed at which it is cutting/grinding the profile.

In creating the cam profile, the cam design engineer seldom thinks about the arc length of the cam profile or about the speed that a circular follower would move over the cam surface when the cam blank is rotating at a constant rpm on the cutting machine's rotary table. (Note that similar issues apply with an x-y coordinate based CNC machine.) To the manufacturing engineer, these are the two most important properties of the cam profile. For a fixed rotational speed of the cam blank, the actual feed rate will vary as the cutter/grinding wheel moves over the surface. Setting the cam rotational speed constant and letting the feed rate vary is not acceptable because the cutter feed mechanism will move too quickly over relatively gently-curved parts of the profile and spend undue time at sections of the profile with small radii of curvature.

One common method used is to fix the feed rate at a constant value and let the cam blank rotation speed vary. A better method is to vary the feed rate according to the profile radius of curvature, which also leads to a varying cam blank rotational speed. Both of these methods, coupled with the lack of a meaningful computational parameter, can involve difficult mathematics and, because of the real-time computation constraints, can result in crude approximations. We will not go into detail here about the mathematics involved, but will just give the reader some idea of the task and then introduce an ingenious solution to many of these problems—the Pythagorean hodograph.

Let us assume that the feed rate V is to be held constant and denote the profile arc length by l and also denote the parameter used to describe the profile with θ. We cannot be sure that the cam engineer specified a parameter to guarantee that the time derivative of the parameter and the cam blank rotational speed satisfy a simple relationship, but we will assume that is true so as not to further complicate what follows. We have then

$$V = \frac{d\theta}{dt}\frac{dl}{d\theta} \qquad \text{so} \qquad \frac{d\theta}{dt} = V \Big/ \frac{dl}{d\theta} \tag{14.2}$$

Using the parametric equations that define the cam profile, we have then

$$\frac{d\theta}{dt} = V \Big/ \sqrt{\left(\frac{dx}{d\theta}\right)^2 + \left(\frac{dy}{d\theta}\right)^2} \tag{14.3}$$

With this formula, the cam blank rotational speed can be found as a function of time. However, the machine's controller must also know the position of its cutter center as a function of time. This is done by solving the differential equation 14.3 for fixed increments of time

$$\Delta t, 2\Delta t, 3\Delta t, \ldots, N\Delta t \tag{14.4}$$

to get the parameter values at those times

$$\theta_1, \theta_2, \ldots, \theta_N \tag{14.5}$$

and then finding the cutter center locations at those values of the parameter

$$x(\theta_j), y(\theta_j) \qquad j = 1, 2, \ldots, N \tag{14.6}$$

called *interpolation points*.

All of these expressions are generally quite complicated. The original equations for cutter center location are offset curves from the cam profile and, as is well known to engineers who concern themselves with geometry, they can be much more mathematically complex than the profile equations. Usually approximations are used so as to solve the differential equation 14.3 on the fly.

The square root expression in equation 14.3 is particularly troublesome. It would be of great benefit if somehow that square root were a simple polynomial. Suppose that it were. Then not only would it be easy to compute the cam blank rotational speed as a function of time, but the interpolation points would be solutions of the equations

$$l(\theta_k) - kV\Delta t = 0 \qquad k = 1, 2, \ldots, N \tag{14.7}$$

where the left hand side is a low degree polynomial. There is a large and dependable body of methods for finding the roots of low degree polynomials.

The idea behind the Pythagorean hodograph is to express the cutter center equations, such as equation 13.4c (p. 398) with a spline whose polynomial pieces are constructed such that the square roots of the sums of the squares of their derivatives are also polynomials.

Here is how to make such a polynomial. Let $u(\theta)$ and $v(\theta)$ be polynomials. Define

$$\begin{aligned} \frac{dx}{d\theta} &= u^2(\theta) - v^2(\theta) \\ \frac{dy}{d\theta} &= 2u(\theta)v(\theta) \end{aligned} \tag{14.8}$$

Then we have that

$$\begin{aligned} \sqrt{\left(\frac{dx}{d\theta}\right)^2 + \left(\frac{dy}{d\theta}\right)^2} &= \sqrt{u^4(\theta) - 2u^2(\theta)v^2(\theta) + v^4(\theta) + 4u^2(\theta)v^2(\theta)} \\ &= \sqrt{u^4(\theta) + 2u^2(\theta)v^2(\theta) + 4u^2(\theta)v^2(\theta)} \\ &= u^2(\theta) + v^2(\theta) \end{aligned} \tag{14.9}$$

which is another polynomial.

The arc length is therefore also a polynomial since it is the integral of a polynomial, and the radius of curvature at any point is a polynomial divided by a polynomial. The radius of curvature is in fact

$$\rho(\theta) = \frac{\left[u^2(\theta) + v^2(\theta)\right]^2}{u(\theta)\frac{dv}{d\theta} - v(\theta)\frac{du}{d\theta}} \tag{14.10}$$

Thus if the feed rate is not constant but is proportional to the radius of curvature, the equations get only a little more difficult, but are still quite tractable. Furthermore, in all cases much more accurate values of the profile can be determined and the resulting cam will be more faithful to the designer's intent.

This idea is relatively new. Usually a Pythagorean hodograph quintic spline is constructed to describe the position of the center of the cutter as it traverses the cam surface. More details on how this is done can be found in the papers listed in the bibliography to this chapter.

14.6 MANUFACTURING METHODS

Several methods of manufacture can be used effectively for cams, sometimes in combination. Cam blanks can be made by sand casting in appropriate materials, forged to near net shape, or cut from billet. Cost, required material properties, and quantities needed usually dictate the choice. Many industrial cams are made from cast (gray or ductile) iron blanks. Many others are machined from billet when wrought steel is needed and quantities are small. If quantities are high, as in the automotive industry, the camshaft blanks are often cast or forged to near net shape. Some automotive camshafts are also made from sintered metals. Referred to as assembled camshafts, the cam lobes are sintered to near net shape and then attached by crimping to a ductile steel tube that serves as the camshaft. Cost and weight savings are achieved over other methods of camshaft production and the sintered cam lobe shape requires less grinding than a forged camshaft due to the closer tolerances of sintering. Sintered metals' material properties are also easily tailored to an application.

Finishing Processes

The details of cam bore, keyway, and contours other than the actual cam surface typically are just machined by conventional milling, turning, and broaching methods. Milling is the most common process for cam surface contour generation. The cam surface will be milled (usually with CNC) and possibly ground after heat treatment. Grinding improves surface finish, corrects distortion due to heat treatment, and provides closer tolerance than milling. Nevertheless, some industrial cams are used as milled to save cost. As pointed out in an earlier section of this chapter, a cam milled on a high quality machining center can provide satisfactory performance in many applications. Where speeds are high and tolerances tight, grinding is usually necessary. Automotive cams typically are ground directly from their near-net-shape forged, cast, or sintered blanks with no intermediate contour milling operation. Cams made from billet or rough castings will first be contour milled to near-net shape when soft and then finish ground after heat treatment.

Wire EDM (electrical discharge machining) is another option for radial cam contour generation. The workpiece is immersed in an oil bath and a thin wire is run through the material while a large electrical current is passed through the wire. The arc produced burns its way through the material leaving a cut approximately as wide as the wire diameter. The wire's path motion is controlled by CNC.

Wire EDM has several advantages over conventional milling and grinding methods. First, full-hard materials can be cut by EDM, so the cam can be hardened before any surface contour generation is done. (Other details such as bore, keyway etc., can be done before hardening or can also be cut with EDM.) No distortion is introduced in the EDM process, nor is material hardness compromised. It leaves an excellent surface finish, better than milled, but not quite as good as ground. Only one pass around the contour is needed regardless of the amount of material being removed from the blank. Close tolerance can be maintained, akin to grinding processes. The only disadvantage of this process is its relatively slow speed. It can require several hours to traverse.a cam contour. The equipment is expensive, thus making its use expensive, though the elimination of secondary grinding operations after milling can actually make wire EDM an economic way to make hardened radial cams. Because of their geometry, barrel and radial track cams are not suitable for manufacturing by wire EDM.

Polishing Processes

Automotive camshaft lobes are often subjected to a polishing operation after grinding. This is accomplished with fine grit carborundum cloth held against a spinning cam lobe. The abrasive follows the existing cam contour and does not change it significantly. Its purpose is to burnish the surface to a better (smoother) finish and, in the case of a cam that was heat treated after grinding, to remove the resulting carburization layer. Cams for industrial machines are typically ground after heat treat and are not generally polished after grinding. Some burnishing of the cam surface occurs during initial run-in in any case. Figure 14-14 shows the change in the surface of a milled cam over the first 500 min of running under load. The sharp tops of the surface asperities are worn off and most of the change occurs in the first 20 min of run-in.[3] A ground cam shows less burnishing, but not zero. See Figure 12-2 (p. 339) for definitions of the surface finish parameters shown in Figure 14-14. Note their significant decrease in value with run-in.

14.7 SURFACE COATINGS

A number of commercial surface coating processes are available and are sometimes used on cam surfaces to enhance wear resistance. These range from chemical surface conversions that add no thickness to ceramic coatings that do increase dimensions slightly. Chrome plating is sometimes used on roller followers to obtain increased wear life. The chrome plate provides a hardness of 70HRC and chrome runs well against steel, reducing adhesive wear. Benefits of surface coatings generally include increased hardness, reduced friction, and increased resistance to abrasion and adhesion. The manufacturers of these coatings and processes provide engineering assistance in selecting the right treatment for each application. Some coating sources are provided in the bibliography of this chapter.

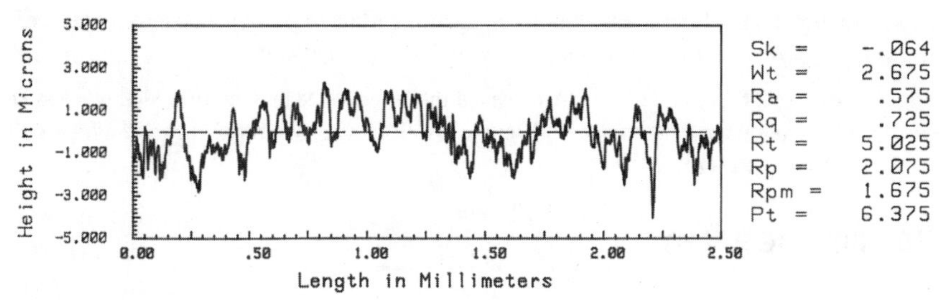

(*a*) Cam surface profile before running

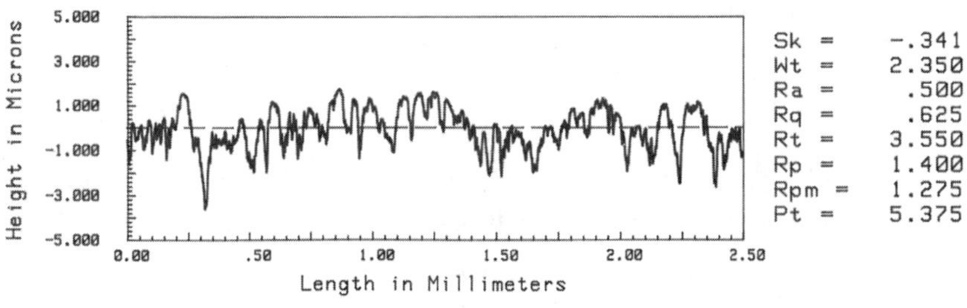

(*b*) Cam surface in same location after 500 minutes of running

FIGURE 14-14

Change in a milled cam surface profile and parameters with initial run-in[3]

14.8 MEASURING THE CAM

In the past, it was quite difficult to accurately measure an entire cam surface to determine whether it matched the design specifications. Typically, a quality control check of a finished cam would consist of measuring the dimensions of bore, keyway, low dwell radius, high dwell radius, and perhaps a single or at most a few points on the rise and fall, often only the mid-rise and mid-fall points at best.

The development of digital coordinate measuring machines (CMM) in recent years has made it possible, and even relatively simple (though not inexpensive), to accurately measure the entire surface contour of a cam and compare it to the design specifications, assuming that the designed cam contour is available in a database form for comparison to the measured data. A description of this process is given in an earlier section of this chapter (p. 443) and the result shown in Figure 14-13 (p. 444). This is the recommended procedure for cam quality control along with conventional surface metrology to classify the surface finish. The relatively high cost of CMM equipment is still a barrier to the general use of this approach to cam quality control. That situation may change in the future however.

14.9 REFERENCES

1 **Casseres, M. G.** (1994). "An Experimental Investigation of the Effect of Manufacturing Methods and Displacement Functions on the Dynamic Performance of Quadruple Dwell Plate Cams," M. S. Thesis, Worcester Polytechnic Institute, Worcester, MA.

2 **Norton, R. L.** (1988). "Effect of Manufacturing Method on Dynamic Performance of Cams." *Mechanism and Machine Theory*, **23**(3), pp. 191-208.

3 **Norton, R. L., et al.** (1988). "Analysis of the Effect of Manufacturing Methods and Heat Treatment on the Performance of Double Dwell Cams." *Mechanism and Machine Theory*, **23**(6), pp. 461-473.

14.10 BIBLIOGRAPHY

For information on material properties see:

H. E. Boyer and T. L. Gall, ed. (1985). *Metals Handbook.* Vol. 1. American Society for Metals: Metals Park, Ohio.

For information on surface coatings see:

Balzers Tool Coating Inc. 1-800-435-5010 www.btc.balzers.com

Surface Conversion Technologies, Inc., Cumming, GA 30130. 404-889-6240

For information on Pythagorean hodographs see

Farouki, R. T., et al. (1998). "Design of Rational Cam Profiles with Pythagorean Hodograph Curves." *Mechanism and Machine Theory*, 33(6), pp. 669-682.

Farouki, R. T. (1996), The elastic bending energy of Pythagorean-hodograph curves. *Computer Aided Geometric Design* 13 (April), 227-241.

Farouki, R. T., and C. A. Neff. (1995). "Hermite Interpolation by Pythagorean Hodographic Quintics." *Mathematics of Computation*, 64(212), pp. 1589-1609.

Farouki, R. T. (1992), Pythagorean-hodograph curves in practical use, in: R.E. Barnhill, ed. *Geometry Processing for Design and Manufacturing*, SIAM, Philadelphia, 3-33.

Farouki, R. T., and C. A. Neff. (1990). "Algebraic Properties of Plane Offset Curves." *Computer Aided Geometric Design*, 7, pp. 101-127.

Farouki, R. T., and C. A. Neff. (1990). "Analytic Properties of Plane Offset Curves." *Computer Aided Geometric Design*, 7, pp. 83-99.

Farouki, R. T. and V.T.Rajan. (1987)., On the numerical condition of polynomials in Bernstein form, *Computer Aided Geometric Design* 4, 191-216.

Farouki, R. T. and V.T.Rajan. (1988), Algorithms for polynomials in Bernstein form, *Computer Aided Geometric Design* 5, 1-26.

Farouki, R. T., and T. Sakkalis. (1990). "Pythagorean Hodographs." *IBM J. Res. Develop.*, 34(5), pp. 736-752.

Farouki, R. T., and S. Shah. (1996). "Real Time CNC Interpolators for Pythagorean-Hodograph Curves." *Computer Aided Geometric Design*, 13, pp. 583-600.

B. K. Fussell, C. Ersoy, and R.B. Jerard. (1992), Computer generated CNC machining rates, *Proceedings of the Japan/USA Symposium on flexible Automation*, Vol. 1, ASME, 377-384.

H. Pottmann. (1995). Curve design with rational Pythagorean-hodograph curves, *Advances in Computational Mathematics.* 3, 147-170.

Chapter 15

LUBRICATION OF CAM SYSTEMS

15.0 INTRODUCTION

We use the term **bearing** here in its most general sense. Whenever two parts have relative motion, they constitute a bearing by definition, regardless of their shape or configuration. Usually, lubrication is needed in any bearing to reduce friction and remove heat. Bearings may roll or slide or do both simultaneously. Thus, a cam-follower joint fits this definition of bearing, whether a flat-faced follower sliding against a cam or a roller follower rolling against a cam with a small amount of slip. The design and/or selection of bearings to support the camshaft will not be addressed here. See reference [3] for information on that topic.

Lubrication theory for surfaces in relative motion is extremely complex mathematically. Solutions to the partial differential equations that govern the behavior are based on simplifying assumptions that yield only approximate solutions. This chapter does not attempt to present a complete discussion or explanation of all the complicated phenomena of dynamic lubrication, as that is far beyond the scope of this text. Rather an introductory discussion of a few of the common cases encountered in machine design is presented. Boundary, hydrostatic, hydrodynamic, and elastohydrodynamic lubrication are introduced and described, and the theory for the last two conditions is discussed without presentation of complete derivations of the governing equations, due to space limitations.

Such topics as squeeze-film theory are not addressed at all, nor are the issues of lubricant supply to, and heat transfer from, the cam-follower joint. Derivations of the governing equations are presented in some of the referenced works. Reference 2 provides an excellent introduction to lubrication theory with minimal mathematics, and reference 3 is a very complete, up-to-date, and mathematically rigorous treatment of the subject.

In this chapter, we present a simple and reasonably accurate approach to lubrication theory and its application to nonconforming contacts such as cam-follower joints. This

TABLE 15-0 Variables Used in This Chapter

Symbol	Variable	ips units	SI units
A	area	in^2	m^2
a, b	half-width, major, minor, of contact patch	in	m
C	basic dynamic load rating	lb	N
C_0	basic static load rating	lb	N
c_d, c_r	diametral and radial clearance	in	m
d	diameter	in	m
E'	effective Young's modulus	psi	Pa
F	force (with various subscripts)	lb	N
f	friction force	lb	N
h	lubricant film thickness	in	m
K_ε	dimensionless parameter	none	none
l	length	in	m
L	fatigue life of rolling bearings	10^6 revs	10^6 revs
n'	angular velocity	rps	rps
O_N	Ocvirk number	none	none
P	force or load	lb	N
p	pressure	psi	N/m^2
r	radius	in	m
R'	effective radius	in	m
S	Sommerfeld number	none	none
T	torque	lb-in	N-m
U	linear velocity	in/sec	m/sec
X, Y	radial and axial force factors	none	none
α	pressure-viscosity exponent	in^2/lb	m^2/N
ε	eccentricity ratio	none	none
ε_x	empirical eccentricity ratio	none	none
ϕ	angle to resultant force	rad	rad
Φ	power	hp	watts
η	absolute viscosity	reyn	cP
Λ	specific film thickness	none	none
μ	coefficient of friction	none	none
ν	Poisson's ratio	none	none
θ_{max}	angle to maximum pressure	rad	rad
ρ	mass density	$blob/in^3$	kg/mm^3
τ	shear stress (with various subscripts)	psi	Pa
υ	kinematic viscosity	in^2/sec	cS
ω	angular velocity	rad/sec	rad/sec

Portions of this chapter were adapted from R. L. Norton, *Machine Design: An Integrated Approach*, 2ed, Prentice-Hall, 2000, with permission.

first requires a discussion of lubrication in conforming contacts such as journal (sleeve) bearings because that is the basis for the theory which is extended to the nonconforming cam-follower joint.

15.1 LUBRICANTS

Introduction of a lubricant to a sliding interface has several beneficial effects on the friction coefficient. Lubricants may be gaseous, liquid, or solid. Liquid or solid lubricants share the properties of low shear strength and high compressive strength. A liquid lubricant such as petroleum oil is essentially incompressible at the levels of compressive stress encountered in bearings, but it readily shears. Thus, it becomes the weakest material in the interface, and its low shear strength reduces the coefficient of friction. Lubricants can also act as contaminants to the metal surfaces and coat them with monolayers of molecules that inhibit adhesion.

Liquid lubricants are the most commonly used and mineral oils the most common liquid. Greases are oils mixed with soaps to form a thicker, stickier lubricant used where liquids cannot be supplied to or retained on the surfaces. Solid lubricants are used in situations where liquids either cannot be kept on the surfaces or lack some required property such as high-temperature resistance. Gaseous lubricants are used in special situations such as air bearings to obtain extremely low friction. Lubricants, especially liquids, also remove heat from the interface. Lower bearing temperatures reduce surface interactions and wear.

LIQUID LUBRICANTS These are largely petroleum-based or synthetic oils, though water is sometimes used as a lubricant in aqueous environments. Many commercial lubricant oils are mixed with various additives that react with the metals to form monolayer contaminants. So-called EP (*Extreme Pressure*) lubricants add fatty acids or other compounds to the oil that attack the metal chemically and form a contaminant layer that protects and reduces friction even when the oil film is squeezed out of the interface by high contact loads. Oils are classified by their viscosity as well as by the presence of additives for EP applications. Table 15-1 shows some common liquid lubricants, their properties, and typical uses. Lubricant manufacturers should be consulted for particular applications.

SOLID-FILM LUBRICANTS These are of two types: materials that exhibit low shear stress, such as graphite and molybdenum disulfide, which are added to the interface, and coatings such as phosphates, oxides, or sulfides that are caused to form on the material surfaces. The graphite and MoS_2 materials are typically supplied in powder form and can be carried to the interface in a binder of petroleum grease or other material. These dry lubricants have the advantage of low friction and high-temperature resistance, though the latter may be limited by the choice of binder. Coatings such as phosphates or oxides can be chemically or electrochemically deposited. These coatings are thin and tend to wear through in a short time. The EP additives in some oils provide a continuous renewal of sulfide or other chemically induced coatings. Table 15-2 shows some common solid-film lubricants, their properties, and typical uses.

TABLE 15-1 Types of Liquid Lubricants

Type	Properties	Typical Uses
Petroleum oils (mineral oils)	Basic lubrication ability fair, but additives produce great improvement. Poor lubrication action at high temperatures	Very wide and general
Polyglycols	Quite good lubricants, do not form sludge on oxidizing	Brake fluid
Silicones	Poor lubrication ability, especially against steel. Good thermal stability	Rubber seals. Mechanical dampers
Chlorofluorocarbons	Good lubricants, good thermal stability	Oxygen compressors. Chemical processing equipment
Polyphenyl ethers	Very wide liquid range. Excellent thermal stability. Fair lubricating ability	High-temperature sliding systems
Phosphate esters	Good lubricants—EP action	Hydraulic fluid + lubricant
Dibasic esters	Good lubricating properties. Can stand higher temperatures than mineral oils.	Jet engines

Source: E. Rabinowicz, *Friction and Wear of Materials*, 1965, reprinted by permission of John Wiley & Sons, Inc.

15.2 VISCOSITY

Viscosity is a measure of a fluid's resistance to shear. Viscosity varies inversely with temperature and directly with pressure, both in a nonlinear fashion. It can be expressed either as an **absolute viscosity** η or as **kinematic viscosity** υ. They are related as

$$\eta = \upsilon\rho \tag{15.1}$$

where ρ is the mass density of the fluid. The units of absolute viscosity η are either lb-sec/in^2 (reyn) in the English system or Pa-s in *SI* units. These are often expressed as μreyn and mPa-s to better suit their typical magnitudes. A centipoise (cP) is 1 mPa-s. Typical absolute viscosity values at 20°C (68°F) are 0.0179 cP (0.0026 μreyn) for air, 1.0 cP (0.145 μreyn) for water, and 393 cP (57μreyn) for SAE 30 engine oil. Oils operating in hot bearings typically have viscosities in the 1 to 5 μreyn range. The term viscosity used without modifiers implies absolute viscosity.

KINEMATIC VISCOSITY is measured in a *viscometer,* which may be either rotational or capillary. A capillary viscometer measures the rate of flow of the fluid through a capillary tube at a particular temperature, typically 40 or 100°C. A rotational viscometer measures the torque and speed of rotation of a vertical shaft or cone running inside a bearing with its concentric annulus filled with the test fluid at the test temperature. The *SI* units of kinematic viscosity are cm^2/sec (stoke) and the English units are in^2/sec. Stokes are quite large, so centistokes (cS) are often used.

15

TABLE 15-2 Types of Solid Film Lubricants

Type	Properties	Typical Uses
Graphite and/or MoS_2 + binder	Best general-purpose lubricants. Low friction (0.12-0.06) reasonably long life (@ 10^4–10^6 cycles)	Locks and other intermittent mechanisms
Teflon + binder	Life not quite as long as previous type, but resistance to some liquids better	As above
Rubbed graphite or MoS_2 film	Friction very low (0.10–0.04), but life quite short (10^2–10^4 cycles)	Deep drawing and other metalworking
Soft metal (lead, indium, cadmium)	Friction higher (0.30–0.15) and life not as long as resin-bonded types	Running-in protection (temporary)
Phosphate, anodized film. Other chemical coatings	Friction high (@ 0.20). Galling preventatives leave "spongy" surface layer.	Undercoating for resin-bonded film

Source: E. Rabinowicz, *Friction and Wear of Materials*, 1965, reprinted by permission of John Wiley & Sons, Inc.

Absolute Viscosity is needed for calculation of lubricant pressures and flows within bearings. It is determined from the measured kinematic viscosity and the density of the fluid at the test temperature. Figure 15-1 shows a plot of the variation of absolute viscosity with temperature for a number of common petroleum oils, designated by their ISO numbers and by SAE numbers on both the engine-oil and gear-oil scales.

15.3 TYPES OF LUBRICATION

Three general types of lubrication can occur in bearings: **full-film, mixed film**, and **boundary lubrication**. Full-film lubrication describes a situation in which the bearing surfaces are fully separated by a film of lubricant, eliminating any contact. Full-film lubrication can be **hydrostatic**, **hydrodynamic**, or **elastohydrodynamic**, each discussed below. Boundary lubrication describes a situation where, for reasons of geometry, surface roughness, excessive load, or lack of sufficient lubricant, the bearing surfaces physically contact, and adhesive or abrasive wear may occur. Mixed-film lubrication describes a combination of partial lubricant film plus some asperity contact between the surfaces.

Figure 15-2 shows a curve depicting the relationship between friction and the relative sliding speed in a bearing. At slow speeds, boundary lubrication occurs with concomitant high friction. As the sliding speed is increased beyond point A, a hydrodynamic fluid film begins to form, reducing asperity contact and friction in the mixed-film regime. At higher speeds, a full film is formed at point B, separating the surfaces com-

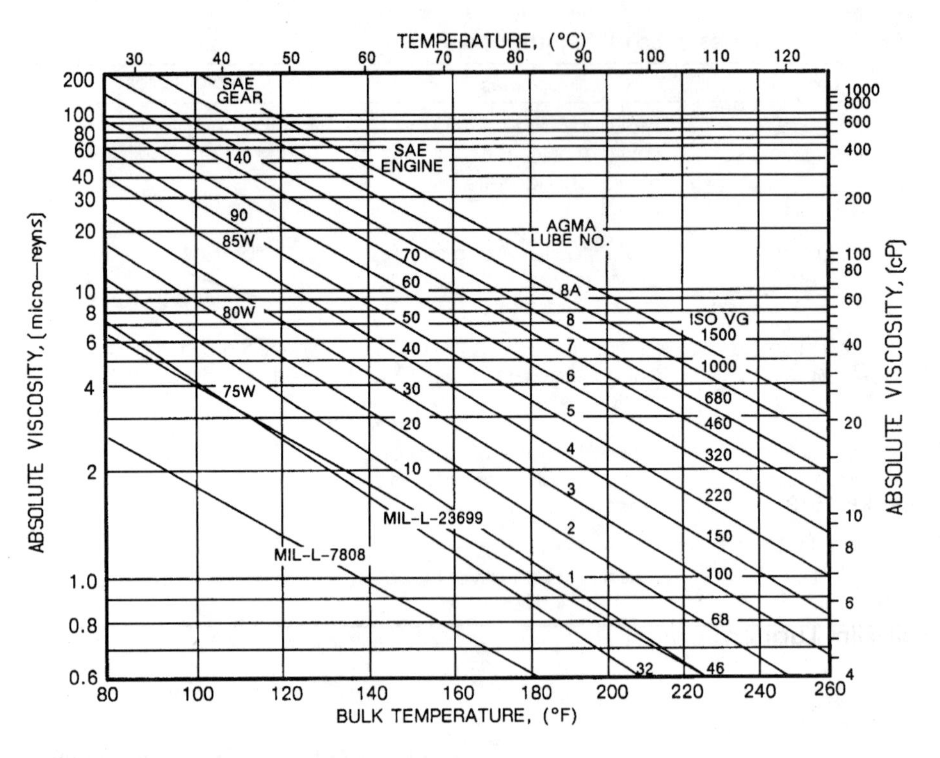

FIGURE 15-1

Absolute viscosity versus temperature of petroleum lubricating oils in ISO viscosity grades (Source: Extracted from AGMA Standard 2001-B88, *Fundamental Rating Factors and Calculation Methods for Involute Spur and Helical Gear Teeth* with the permission of the publisher, American Gear Manufacturers Association, 1500 King St., Suite 201, Alexandria, Va., 22314)

pletely with reduced friction. (This is the same phenomenon that causes automobile tires to *aquaplane* on wet roads. If the relative velocity of the tire versus the wet road exceeds a certain value, the tire motion pumps a film of water into the interface, lifting the tire off the road. The tire's coefficient of friction is drastically reduced, and the sudden loss of traction can cause a dangerous skid.) At still higher speeds the viscous losses in the sheared lubricant increase friction.

In rotating journal (sleeve) bearings, all three of these regimes will be experienced during start-up and shutdown. As the shaft begins to turn, it will be in boundary lubrication. If its top speed is sufficient, it will pass through the mixed regime and reach the desired full-film regime where wear is reduced virtually to zero if the lubricant is kept clean and not overheated. We will briefly discuss the conditions that determine these lubrication states, and then explore a few of them in somewhat greater detail.

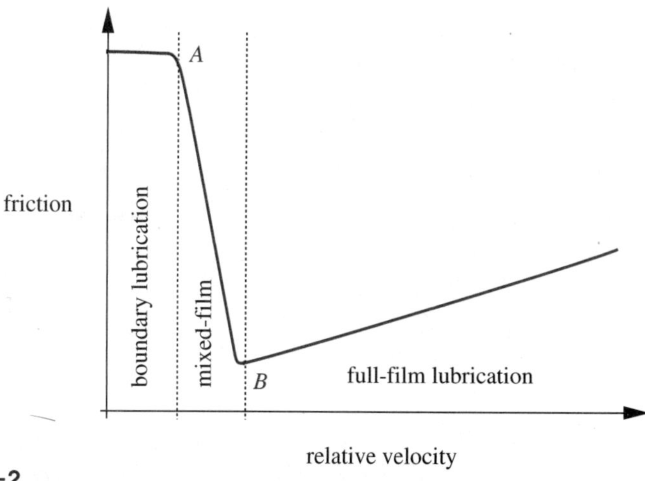

FIGURE 15-2

Change in friction with relative velocity in a sliding bearing

Full-Film Lubrication

Three mechanisms can create full-film lubrication: **hydrostatic**, **hydrodynamic**, and **elastohydrodynamic** lubrication.

HYDROSTATIC LUBRICATION refers to the continuous supply of a flow of lubricant (typically an oil) to the sliding interface at some elevated hydrostatic pressure ($\approx 10^2$–10^4 psi). This requires a reservoir (sump) to store, a pump to pressurize, and plumbing to distribute the lubricant. When properly done, with appropriate bearing clearances, this approach can eliminate all metal-to-metal contact at the interface during sliding. The surfaces are separated by a film of lubricant, that, if kept clean and free of contaminants, reduces wear rates to virtually zero. At zero relative velocity, the friction is essentially zero. With relative velocity, the coefficient of friction in a hydrostatically lubricated interface is about 0.002 to 0.010. This is also the principle of a so-called air bearing, used on "air pallets" to lift (thrust) a load from a surface, allowing it to be moved sideways with very little effort. Hovercraft operate on a similar principle. Water is sometimes used in hydrostatic bearings. Denver's Mile High Stadium has a 21 000-seat grandstand that slides back on hydrostatic water films to convert the stadium from baseball to football.[1] Hydrostatic thrust bearings are more common than radial ones.

HYDRODYNAMIC LUBRICATION refers to the supply of sufficient lubricant (typically an oil) to the sliding interface to allow the relative velocity of the mating surfaces to pump the lubricant within the gap and separate the surfaces on a dynamic film of liquid. This technique is most effective in journal bearings, where the shaft and bearing create a thin annulus within their clearance that can trap the lubricant and allow the shaft to pump it around the annulus. A leakage path exists at the ends, so a continuous supply of oil must be provided to replace the losses. This supply may be either gravity fed or

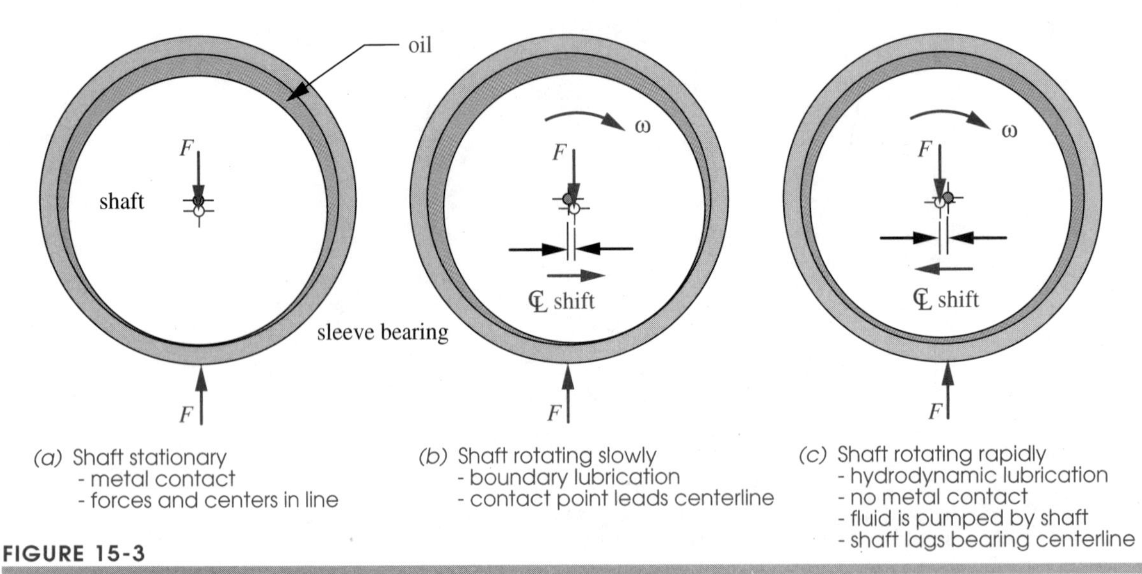

FIGURE 15-3

Boundary and hydrodynamic lubrication conditions in a sleeve bearing—clearance and motions exaggerated

pressure fed. This is the system used to lubricate the crankshaft and camshaft bearings in an internal-combustion engine. Filtered oil is pumped to the bearings under relatively low pressure to replenish the oil lost through the bearing ends, but the condition within the bearing is hydrodynamic, creating much higher pressures to support the bearing loads.

In a hydrodynamic sleeve bearing at rest, the shaft or journal sits in contact with the bottom of the bearing, as shown in Figure 15-3*a*. As it begins to rotate, the shaft centerline shifts eccentrically within the bearing and the shaft acts as a pump to pull the film of oil clinging to its surface around with it, as shown in Figure 15-3*b*. (The "outer side" of the oil film is stuck to the stationary bearing.) A flow is set up within the small thickness of the oil film. With sufficient relative velocity, the shaft "climbs up" on a wedge of pumped oil and ceases to have metal-to-metal contact with the bearing, as shown in Figure 15-3*c*.

Thus, a hydrodynamically lubricated bearing touches its surfaces together only when stopped or when rotating below its "aquaplane speed." This means that adhesive wear can occur only during the transients of start-up and shutdown. As long as sufficient lubricant and velocity are present to allow hydrodynamic lifting of the shaft off the bearing at its operating speed, there is essentially no adhesive wear. This greatly increases wear life over that of a continuous-contact situation. As with hydrostatic lubrication, the oil must be kept free of contaminants to preclude other forms of wear such as abrasion. The coefficient of friction in a hydrodynamically lubricated interface is about 0.002 to 0.010.

ELASTOHYDRODYNAMIC LUBRICATION When the contacting surfaces are nonconforming, as with the gear teeth or cam and follower shown in Figure 15-4, then it is

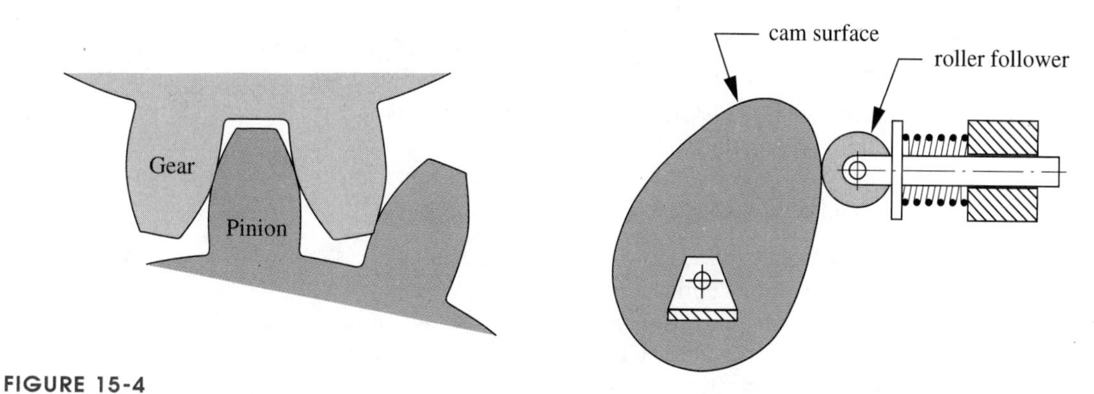

FIGURE 15-4

Open joints that can have EHD, mixed, or boundary lubrication

more difficult to form a full film of lubricant, since the nonconforming surfaces tend to expel rather than entrap the fluid. At low speeds these joints will be in boundary lubrication, and high wear rates can result with possible scuffing and scoring. The load creates a contact patch from the elastic deflections of the surfaces, as discussed in Chapter 12. This small contact patch can provide enough of a flat surface to allow a full hydrodynamic film to form if the relative sliding velocity is high enough (see Figure 15-2). This condition is termed **elastohydrodynamic lubrication** (EHD), as it depends on the elastic deflections of the surfaces and the fact that the high pressures (100 to 500 kpsi) within the contact zone greatly increase the viscosity of the fluid. (In contrast, the film pressure in conforming bearings is only several thousand psi and the change in viscosity due to this pressure is small enough to ignore.)

Gear teeth can operate in any of the three conditions depicted in Figure 15-2. Boundary lubrication occurs in start-stop operation and if prolonged will cause severe wear. Cam-follower joints can also experience any of the regimes in Figure 15-2, but are more likely to be in a boundary-lubricated mode at locations of small radius of curvature on the cam. Rolling-element bearings can see any of the three regimes as well.

The most important parameter that determines which situation occurs in nonconforming contacts is the ratio of the oil-film thickness to the surface roughness. To get full-film lubrication and avoid asperity contact, the rms average surface roughness (R_q)* needs to be no more than about 1/2 to 1/3 of the oil-film thickness. An EHD full-film thickness is normally of the order of 1 μm. At very high loads, or low speeds, the EHD film thickness may become too small to separate the surface asperities and mixed-film or boundary lubrication conditions may recur. Factors that have the most effect in creating EHD conditions are increased relative velocity, increased lubricant viscosity, and increased radius of curvature at the contact. Reduction in unit load and reduced stiffness of the material have less effect.[6]

* See Section 12.1 and Figure 12-2 on p. 339 for a discussion of surface finish and a definition of R_q.

15

Boundary Lubrication

Boundary lubrication refers to situations in which some combination of the geometry of the interface, high load levels, low velocity, or insufficient lubricant quantity preclude the initiation of a hydrodynamic condition. The properties of the contacting surfaces and lubricant properties, other than its bulk viscosity, determine friction and wear in this situation. Viscosity of the lubricant is not a factor. Note in Figure 15-2 that friction is independent of velocity in boundary lubrication.

Boundary lubrication implies that there is always some metal-to-metal contact in the interface. If the lubricant film is not thick enough to "bury" the asperities on the surfaces, this will be true. Rough surfaces could cause this condition. If the relative velocity or the supply of lubricant to a hydrodynamic interface is reduced, it will revert to a boundary-lubrication condition. Surfaces such as gear teeth and cam/follower interfaces (see Figure 15-4) that do not envelop each other can be in a boundary-lubrication mode if EHD conditions do not prevail. Ball and roller bearings can also operate in boundary-lubrication mode if the combination of speeds and loads does not allow EHD to occur.

Boundary lubrication is a less desirable condition than the other types described above because it allows the surface asperities to contact and wear rapidly. It is sometimes unavoidable as in the examples of cams, gears, and rolling-element bearings cited. The EP lubricants mentioned above were created for these boundary-lubrication applications, especially for hypoid gears, which experience both high sliding velocities and high loads. The coefficient of friction in a boundary-lubricated sliding interface depends on the materials used as well as on the lubricant, but ranges from about 0.05 to 0.15, with most being about 0.10.

15.4 MATERIAL COMBINATIONS IN CAM-FOLLOWER JOINTS

Bearings used to support rotating shafts offer a variety of suitable materials, as discussed in [3]. The cam-follower joint typically has such high local stresses that the only practical materials for both cam and follower are hardenable ferrous alloys such as steel and cast ductile iron. For some applications in corrosive environments such as seawater, bronzes may be the only suitable materials and the design may have to be such as to reduce stresses to levels tolerable to these weaker materials.

For lightly loaded cams,* nonmetallic materials of some types offer the possibility of dry running if they have sufficient lubricity. Some thermoplastics such as nylon, acetal, and filled Teflon offer a low coefficient of friction μ against any metal but have low strengths and low melt temperatures that, when combined with their poor heat conduction, severely limits the loads and speeds of operation that they can sustain. Teflon has a very low μ (approaching rolling values), but requires fillers to raise its strength to usable levels. Inorganic fillers such as talc or glass fiber add significant strength and stiffness to any of the thermoplastics but at the cost of a higher μ and increased abrasiveness. Graphite and MoS_2 powder are also used as fillers and these add lubricity as well as strength and temperature resistance. Some mixtures of polymers such as acetal-teflon are also offered.

* Examples are cams used in toys and in household appliances such as sewing machines, and dishwasher or washing machine timers.

15.5 HYDRODYNAMIC LUBRICATION THEORY

Consider the sleeve bearing shown in Figure 15-3 (p. 458). Figure 15-5*a* shows a similar journal and bearing, but concentric and with the axis vertical. The diametral clearance c_d between journal and bearing is very small, typically about one-thousandth of the diameter. We can model this as two flat plates because the gap *h* is so small compared to the radius of curvature. Figure 15-5*b* shows two such plates separated by an oil film with a gap of dimension *h*. If the plates are parallel, the oil film will not support a transverse load. This is also true of a concentric journal and bearing. A concentric horizontal journal will become eccentric from the weight of the shaft, as in Figure 15-3. If the axis is vertical, as in Figure 15-5*a*, the journal can spin concentric with the bearing, since there is no transverse gravity force.

Petroff's Equation for No-Load Torque

If we hold the lower plate of Figure 15-5*b* stationary and move the upper plate to the right with a velocity *U*, the fluid between the plates will be sheared in the same manner as in the concentric gap of Figure 15-5*a*. The fluid wets and adheres to both plates, making its velocity zero at the stationary plate and *U* at the moving plate. Figure 15-5*c* shows a differential element of fluid in the gap. The velocity gradient causes the angular distortion β. In the limit, as Δt approaches 0, $\beta = dx / dy$. The shear stress τ_x acting on a differential element of fluid in the gap is proportional to the shear rate:

$$\tau_x = \eta \frac{d\beta}{dt} = \eta \frac{d}{dt}\frac{dx}{dy} = \eta \frac{d}{dy}\frac{dx}{dt} = \eta \frac{du}{dy} \tag{15.2a}$$

and the constant of proportionality is the viscosity η. In a film of constant thickness *h*, the velocity gradient $du / dy = U / h$ and is constant. The force to shear the entire film is

$$F = A\tau_x = \eta A \frac{U}{h} \tag{15.2b}$$

where *A* is the area of the plate.

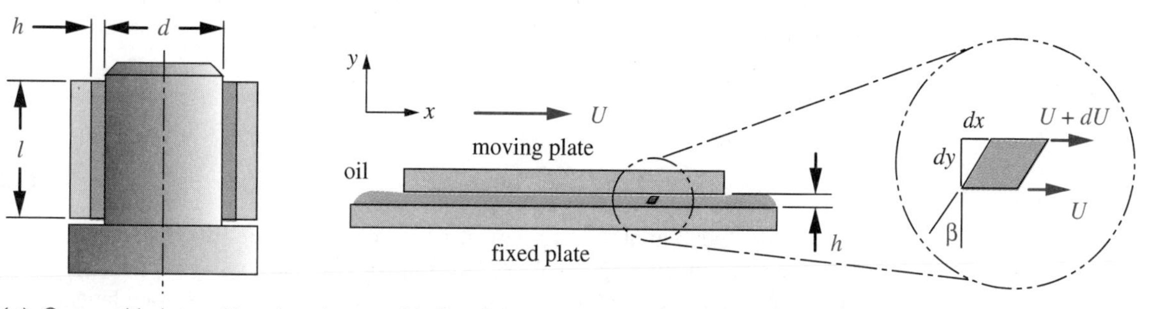

(*a*) Concentric journal in a bearing (*b*) Parallel plates shearing an oil film (*c*) Differential element in shear

FIGURE 15-5

An oil film sheared between two parallel surfaces cannot support a transverse load (clearances exaggerated)

For the concentric journal and bearing of Figure 15-5*a*, let the gap $h = c_d / 2$ where c_d is the diametral clearance. The velocity is $U = \pi d n'$ where n' is revolutions per second, and the shear area is $A = \pi d l$. The torque T_0 required to shear the film is then

$$T_0 = \frac{d}{2} F = \frac{d}{2} \eta A \frac{U}{h} = \frac{d}{2} \eta \pi d l \frac{\pi d n'}{c_d / 2}$$

$$T_0 = \eta \frac{\pi^2 d^3 l n'}{c_d} \tag{15.2c}$$

This is *Petroff's equation* for the no-load torque in a fluid film.

Reynolds' Equation for Eccentric Journal Bearings

To support a transverse load, the plates of Figure 15-5*b* must be nonparallel. If we rotate the lower plate of Figure 15-5*b* slightly counterclockwise and move the upper plate to the right with a velocity U, the fluid between the plates will be carried into the decreasing gap as shown in Figure 15-6*a*, developing a pressure that will support a transverse load P. The angle between the plates is analogous to the varying clearance due to the eccentricity e of the journal and bearing in Figure 15-6*b*.* When a transverse load is applied to a journal, it must assume an eccentricity with respect to the bearing in order to form a changing gap to support the load by developing pressure in the film.†

Figure 15-6*b* shows a greatly exaggerated eccentricity e and gap h for a journal bearing. The eccentricity e is measured from the center of the bearing O_b to the center of the journal O_j. The zero-to-π axis for the independent variable θ is established along the line O_bO_j as shown in Figure 15-6*b*. The maximum possible value of e is $c_r = c_d / 2$, where c_r is the radial clearance. The eccentricity can be converted to a dimensionless eccentricity ratio ε:

$$\varepsilon = \frac{e}{c_r} \tag{15.3}$$

which varies from 0 at no load to 1 at maximum load when the journal contacts the bearing. An approximate expression for the film thickness h as a function of θ is

$$h = c_r (1 + \varepsilon \cos\theta) \tag{15.4a}$$

The film thickness h is maximum at $\theta = 0$ and minimum at $\theta = \pi$, found from

$$h_{min} = c_r (1 - \varepsilon) \qquad h_{max} = c_r (1 + \varepsilon) \tag{15.4b}$$

Consider the journal bearing shown in Figure 15-7. In the analysis that follows, the gap is given by equation 15.4a. We can take the origin of an xy coordinate system at any point on the circumference of the bearing, such as O. The x axis is then tangent to the bearing, the y axis is through the bearing center O_b, and the z axis (not shown) is parallel to the axis of the bearing. Generally, the bearing is stationary and only the journal rotates, but in some cases the reverse may be true, or both may rotate, as in the planet shaft

* The amount of angle needed to create a supporting force is surprisingly small. For example, in a bearing of about 32-mm-dia, the circumference is 100 mm. A typical entrance gap h_{max} might be 25 μm (0.001 in), and the exit gap h_{min}, 12.5 μm (0.0005 in). The slope is then 0.0125 / 100 or about 7 / 1000 of a degree (26 seconds of arc). This is equivalent to about a 3-cm rise over a 300-foot-long football field.

† In England, in the 1880s, Beauchamp Tower was experimentally investigating the friction in hydrodynamically lubricated bearings for the railroad industry (though the term hydrodynamic and its theory were only then about to be discovered). His results showed much lower friction coefficients than expected. He drilled a radial hole through the bearing in order to add oil while running, but was surprised to find that oil flowed out of the hole when the shaft turned. He corked the hole, but the cork was expelled. He plugged the hole with wood, but that also popped out. When he put a pressure gage in the hole, he measured pressures well above the average pressure expected from a calculation of load / area. He then mapped the pressure distribution over 180° of the bearing and discovered the now familiar pressure distribution whose average

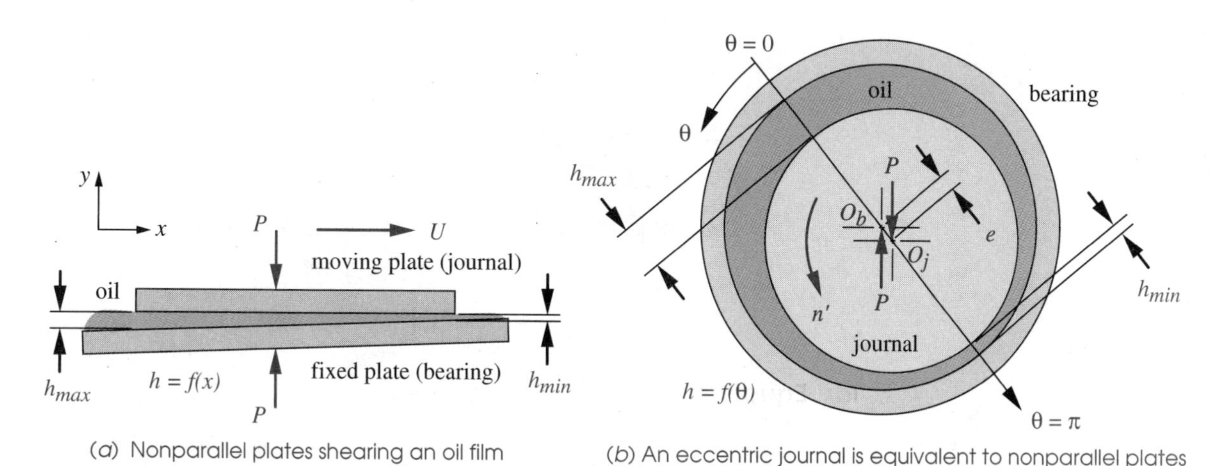

(a) Nonparallel plates shearing an oil film

(b) An eccentric journal is equivalent to nonparallel plates

FIGURE 15-6

An oil film sheared between nonparallel surfaces can support a transverse load

of an epicyclic gear train. Thus we show a tangential velocity U_1 for the bearing as well as a tangential velocity T_2 for the journal. Note that the directions (angles) of the vectors U_1 and T_2 are not the same due to the eccentricity. The tangential velocity T_2 of the journal can be resolved into components in the x and y directions as U_2 and V_2, respectively. The angle between T_2 and U_2 is so small that its cosine is essentially 1 and so we can set $U_2 \cong T_2$. The component V_2 in the y direction is due to the closing (or opening) of the gap h as the journal rotates and is $V_2 = \partial h / \partial x$.

value is load / area. On learning of this discovery, Osborne Reynolds set out to develop the mathematical theory to explain it, publishing the results in 1886.[12]

Using the above assumptions, we can write Reynolds' equation§ relating the changing gap thickness h, the relative velocities between the journal and bearing V_2 and $U_1 - U_2$, and the pressure in the fluid p as a function of the two dimensions x and z, assuming that the journal and bearing are parallel in the z direction and the viscosity η is constant,

§ For a derivation of Reynolds' equation, see reference 2, 3, 4, or 10.

$$\frac{1}{6\eta}\left[\frac{\partial}{\partial x}\left(h^3\frac{\partial p}{\partial x}\right)+\frac{\partial}{\partial z}\left(h^3\frac{\partial p}{\partial z}\right)\right]=\left(U_1-U_2\right)\frac{\partial h}{\partial x}+2V_2$$

$$=\left(U_1-U_2\right)\frac{\partial h}{\partial x}+2U_2\frac{\partial h}{\partial x}=\left(U_1+U_2\right)\frac{\partial h}{\partial x}=U\frac{\partial h}{\partial x} \tag{15.5a}$$

where $U = U_1 + U_2$.

Equation 15.5a does not have a closed-form solution but can be solved numerically. Raimondi and Boyd did so in 1958 and provide a large number of design charts for its application to finite-length bearings.[4] Reynolds solved a simplified version in series form (in 1886)[5] by assuming that the bearing is infinitely long in the z direction, which makes the flow zero and the pressure distribution over that direction constant, and thus makes the term $\partial p / \partial z = 0$. With this simplification, the Reynolds' equation becomes:

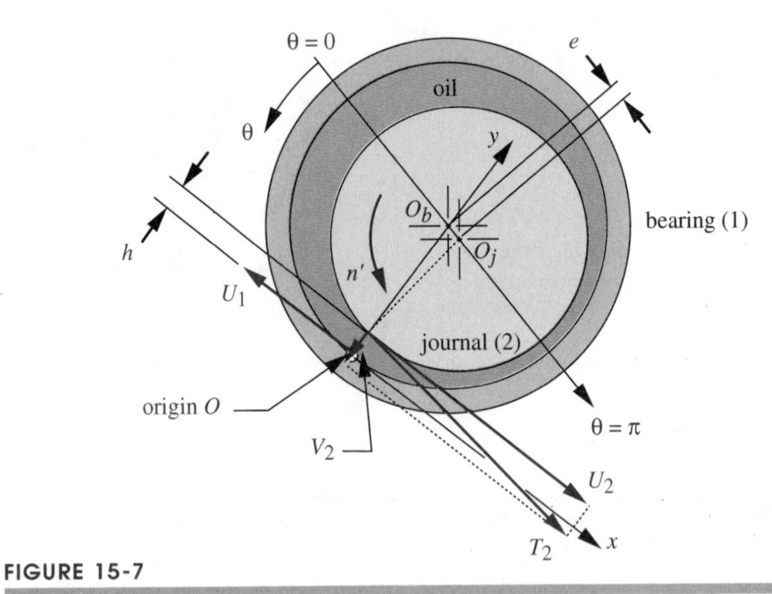

FIGURE 15-7

Velocity components in an eccentric journal bearing

$$\frac{\partial}{\partial x}\left(h^3 \frac{\partial p}{\partial x}\right) = 6\eta U \frac{\partial h}{\partial x} \tag{15.5b}$$

In 1904, A. Sommerfeld found a closed-form solution for the infinitely-long-bearing equation 15.5b as

$$p = \frac{\eta U r}{c_r^2}\left[\frac{6\varepsilon(\sin\theta)(2+\varepsilon\cos\theta)}{\left(2+\varepsilon^2\right)(1+\varepsilon\cos\theta)^2}\right] + p_0 \tag{15.6}$$

which gives the pressure p in the lubricant film as a function of angular position θ around the bearing for particular dimensions of journal radius r, radial clearance c_r, eccentricity ratio ε, surface velocity U, and viscosity η. The term p_0 accounts for any supply pressure at the otherwise zero-pressure position at $\theta = 0$. Equation 15.6 is referred to as the *Sommerfeld solution* or the *long-bearing solution.*

15

Long bearings are not often used in modern machinery for several reasons. Small shaft deflections or misalignment can reduce the radial clearance to zero in a long bearing, and packaging considerations often require short bearings. Typical l / d ratios of modern bearings are in the range of 1/4 to 2. The long-bearing (Sommerfeld) solution assumes no end leakage of oil from the bearing, but at these small l / d ratios, end leakage can be a significant factor. Ocvirk and DuBois[6],[7],[8],[9] solved a form of Reynolds' equation for a short bearing that includes the end-leakage term:

$$\frac{\partial}{\partial z}\left(h^3 \frac{\partial p}{\partial z}\right) = 6\eta U \frac{\partial h}{\partial x} \tag{15.7a}$$

This form neglects the term that accounts for the circumferential flow of oil around the bearing on the premise that it will be small in comparison to the flow in the z direction (leakage) in a short bearing. Equation 15.7a can be integrated to give an expression for pressure in the oil film as a function of both θ and z:

$$p = \frac{\eta U}{r c_r^2}\left(\frac{l^2}{4} - z^2\right)\frac{3\varepsilon \sin\theta}{(1+\varepsilon\cos\theta)^3} \tag{15.7b}$$

Equation 15.7b is known as the *Ocvirk solution* or the *short-bearing solution*. It is typically evaluated for $\theta = 0$ to π, with the pressure assumed to be zero over the other half of the circumference.

15.6 NONCONFORMING CONTACTS

Nonconforming contacts such as cam-follower joints, gear teeth, and rolling-element bearings (balls, rollers) can operate in boundary, mixed, or elastohydrodynamic (EHD) modes of lubrication. The principal factor that determines which of these situations will occur is the specific film thickness Λ, which is defined as the minimum film thickness at the patch center divided by the composite rms surface roughness of the two surfaces.

$$\Lambda = h_c \Big/ \sqrt{R_{q1}^2 + R_{q2}^2} \tag{15.8a}$$

where h_c is the film thickness of the lubricant at the center of the contact patch and R_{q1} and R_{q2} are the rms average roughnesses of the two contacting surfaces. The denominator of equation 15.8a is termed the composite surface roughness. (See Section 12.1 on p. 337 for a discussion of surface roughness.) The film thickness at the center of the contact patch can be related to the minimum film thickness h_{min} at the trailing edge of contact by

$$h_c \cong \frac{4}{3} h_{min} \tag{15.8b}$$

Figure 15-8*a* shows the experimentally measured frequency of asperity contact within an EHD gap as a function of specific film thickness.[10] When $\Lambda < 1$, the surfaces are in continuous metal-to-metal contact, i.e, in boundary lubrication. When $\Lambda > 3$ to 4, there is essentially no asperity contact. Between these values there is some combination of partial EHD and boundary-lubrication conditions. A majority of Hertzian contacts in gears, cams, and rolling-element bearings operate in this partial EHD (mixed lubrication) region of Figure 15-2 (p. 457).[11] From Figure 15-8*a* we can conclude that Λ needs to be > 1 for partial EHD to begin[11] and > 3 to 4 for full-film EHD.[2],[11] Effective partial EHD conditions begin at about $\Lambda = 2$ and if $\Lambda < 1.5$, it indicates an effective boundary-lubrication condition in which significant asperity contact occurs.[11]

Figure 15-8*b* shows the effect of specific film thickness on fatigue life of a rolling bearing.[12] The ordinate defines a ratio of expected life over the predicted catalog life for a bearing. This plot also shows the desirability of maintaining $\Lambda > 1.2$ in order to obtain the catalog life. A small increase in Λ from 1.2 to about 2 can double the fatigue life. Further increases in Λ have less dramatic effect on life and may cause higher friction due to viscous-drag losses if a heavier oil is used to obtain the greater Λ.

Surface roughness is fairly easy to measure and control. The lubricant film thickness is more difficult to predict. Chapter 12 discusses calculation of the Hertzian pressure in surface contact and shows that the pressures in the contact zone between stiff materials in nonconforming (theoretical point or line) contact are extremely high, commonly as much as 80 to 500 kpsi (0.5 to 3 GPa) if both materials are steel. It was once believed that lubricants could not withstand these pressures and thus could not separate the metal surfaces. It is now known that viscosity is an exponential function of pressure, and at typical contact pressures, oil can become essentially as stiff as the metals it separates. Figure 15-9 shows the viscosity-pressure relationship for several common lubricants on a semilog plot. The curve for mineral oils can be approximated by:

$$\eta = \eta_0 e^{\alpha p} \tag{15.9a}$$

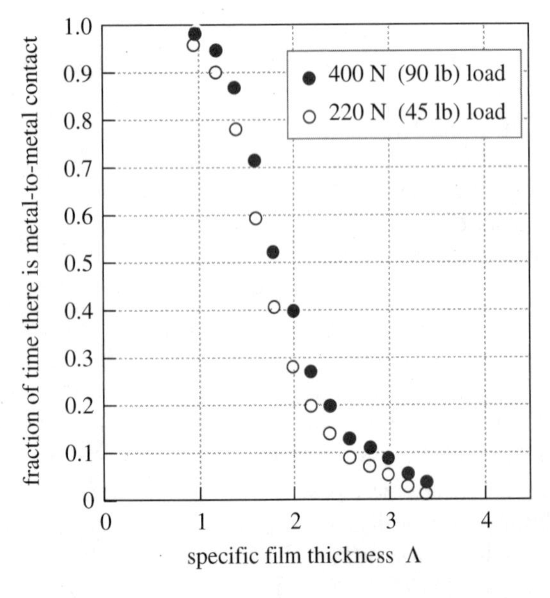

(*a*) Penetration of EHD film by surface asperities (10)

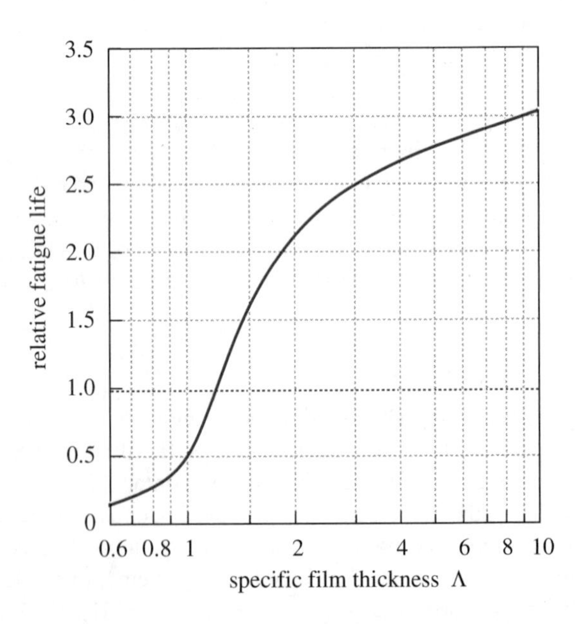

(*b*) Effect of film thickness on fatigue life (13)

FIGURE 15-8

Effect of specific film thickness Λ on the asperity contacts and fatigue life

15

where η_0 is the absolute viscosity (reyn) at atmospheric pressure and p is the pressure (psi). An approximate expression for the pressure-viscosity exponent α for mineral oils is (with units of υ_0 = in²/s, η_0 = reyn, and ρ = lb-sec²/in⁴):*[13]

$$\alpha \cong 7.74E-4\left(\frac{\upsilon_0}{10^4}\right)^{0.163} \cong 7.74E-4\left(\frac{\eta_0}{\rho\left(10^4\right)}\right)^{0.163} \tag{15.9b}$$

Cylindrical Contact Dowson and Higginson[14],[15] determined a formula for the minimum film thickness in an EHD contact between cylindrical rollers as*

$$h_{min} = 2.65R'(\alpha E')^{0.54}\left(\frac{\eta_0 U}{E'R'}\right)^{0.7}\left(\frac{P}{lE'R'}\right)^{-0.13} \tag{15.10}$$

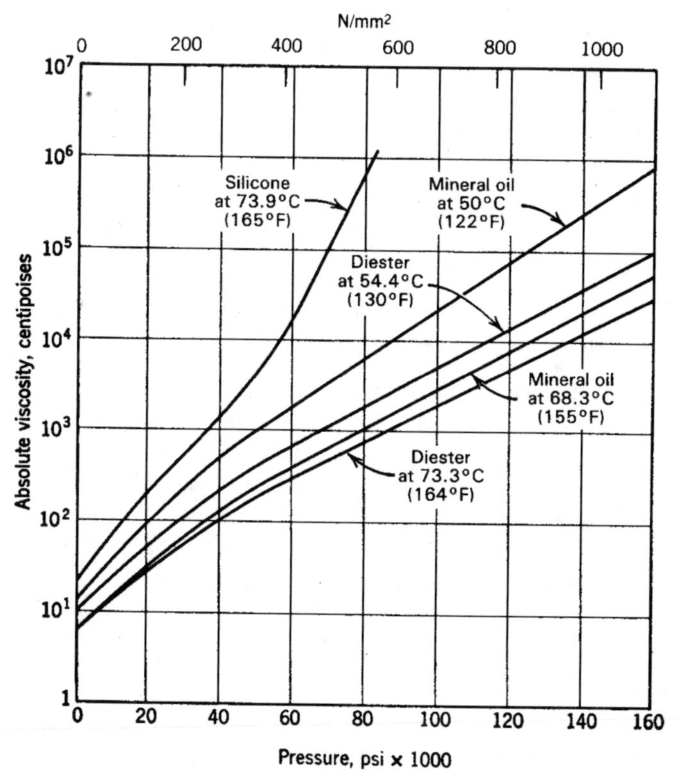

FIGURE 15-9

Absolute viscosity versus pressure of various lubricating oils (Source: ASME Research Committee on Lubrication,"Pressure Viscosity Report–Vol. 11," 1953.)

* Equations 15.9b and 15.10 contain empirical constants that are correct only when all variables are expressed in U.S. customary units. These equations cannot be used with *SI* units or any other non-U.S. units.

where P is the transverse load (lb), l is the length of axial contact (in), U = average velocity $(U_1 + U_2) / 2$ (in/s), η_0 is the absolute lubricant viscosity (reyn) at atmospheric pressure and operating temperature, and α is the pressure-viscosity exponent for the particular lubricant from equation 15.9b. The effective radius R' is defined as

$$\frac{1}{R'} = \frac{1}{R_{1_x}} + \frac{1}{R_{2_x}} \tag{15.11a}$$

where R_{1x} and R_{2x} are the radii of the contacting surfaces in the direction of rolling. The effective modulus is defined as

$$E' = \frac{2}{m_1 + m_2} = \frac{2}{\frac{1-\nu_1^2}{E_1} + \frac{1-\nu_2^2}{E_2}} \tag{15.11b}$$

where E_1, E_2 are Young's moduli, and ν_1, ν_2 are Poisson's ratio for each material.

General Contact In general point contact, the contact patch is an ellipse as discussed in Chapter 12. The contact ellipse is defined by its major and minor half-axis dimensions, a and b, respectively. Contact between two spheres, or between a sphere and a flat plate, will have a circular contact patch, which is a special case of elliptical contact wherein $a = b$. Hamrock and Dowson[16] developed an equation for the minimum film thickness in generalized point contact as

$$h_{min} = 3.63R'(\alpha E')^{0.49}\left(\frac{\eta_0 U}{E'R'}\right)^{0.68}\left(1 - e^{-0.68\psi}\right)\left[\frac{P}{E'(R')^2}\right]^{-0.073} \tag{15.12}$$

where U is average velocity = $(U_1 + U_2) / 2$, and ψ is the ellipticity ratio of the contact patch a / b (see Section 12.11 on p. 352).

In both of these equations the film thickness is most dependent on speed and lubricant viscosity, but is relatively insensitive to load. Figure 15-10 shows pressure-distribution and film-thickness plots for light and heavy load conditions in an EHD contact between steel rollers lubricated with mineral oil.[17] Note that the fluid pressure is the same as the dry Hertzian contact pressure except for the pressure spike that occurs as the film thickness contracts near the exit. Except for that local contraction, the film thickness is essentially constant throughout the contact patch.

Equations 15.10, 15.11, and 15.12 allow a minimum film thickness to be calculated for a nonconforming contact joint such as a pair of gear teeth, cam-follower, or rolling-element bearing. The specific film thickness from equation 15.8 will indicate whether EHD or boundary lubrication can be expected in the contact. An oil with EP additives is needed if EHD is not present.

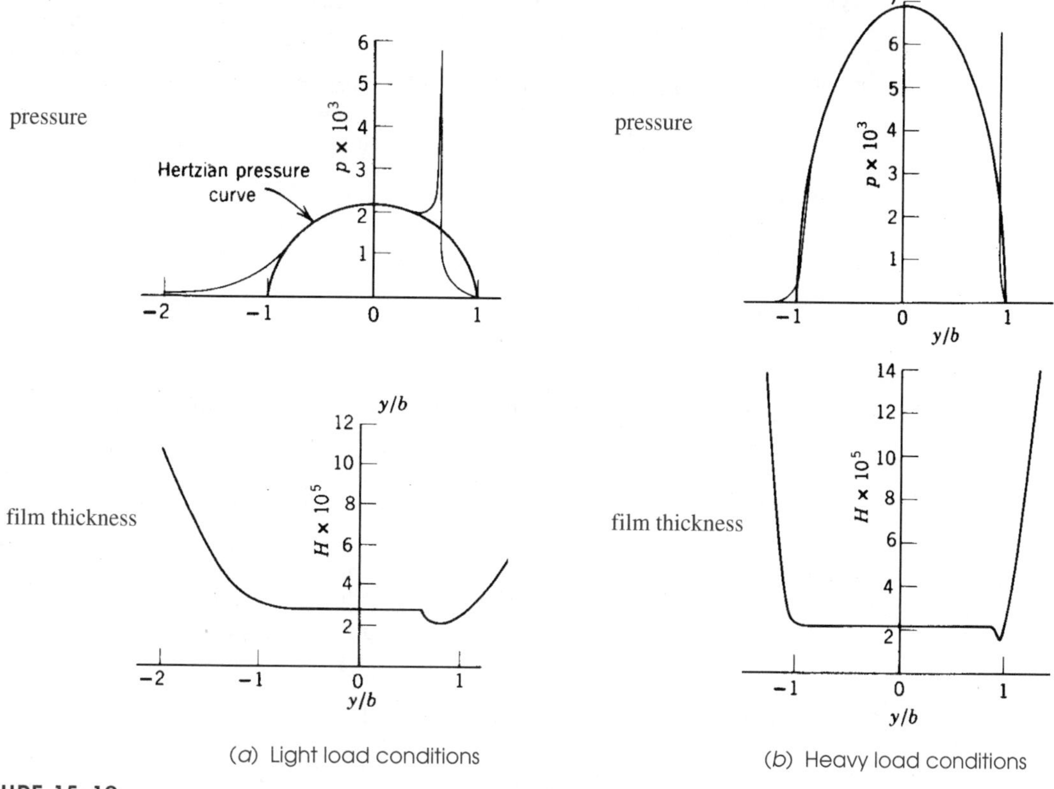

FIGURE 15-10

Pressure distribution and film thickness in an EHD joint (Source: D. Dowson and G. Higginson, "The Effect of Material Properties on the Lubrication of Elastic Rollers," *J. Mech. Eng. Sci.*, vol. 2, no. 3, 1960, with permission)

EXAMPLE 15-1

Lubrication in a Crowned Cam-Follower Interface.

Problem: A cam-follower system was analyzed for contact-patch geometry and contact stresses in Example 12-3 (p. 365). Determine the film-thickness parameter and the lubrication condition for a ground roller running against both a ground cam and a milled cam.

Given: The roller follower radius is 1 in with a 20-in crown radius at 90° to the roller radius with an rms surface roughness of $R_q = 7$ μin. The cam's minimum radius of curvature is 1.72 inches, in the direction of rolling. It is flat axially. It forms an elliptical contact patch with the cam. The half-dimensions of this ellipse are $a = 0.0889$ in and $b = 0.0110$ in. The cam angular velocity is 18.85 rad/sec and the radius to its surface at the point of minimum radius of curvature is 3.92 in. The bulk oil temperature is 180°F. The ground cam has an rms surface roughness of $R_q = 7$ μin and the milled cam has $R_q = 30$ μin.

Assumptions: Try an ISO VG 460 oil with an assumed specific gravity of 0.9. The roller has 1% slip versus the cam.

Find: The specific film thickness and lubrication condition for the assumed lubricant and the viscosity of lubricant required to obtain effective partial or full EHD conditions for each cam, if possible.

Solution:

1 Figure 15-1 (p. 456) gives the viscosity of an ISO VG 460 oil as about 6.5 μreyn at 180°F.

2 Find the mass density ρ of the oil from the given specific gravity SG of the oil and the weight density of water.

$$\rho = SG\frac{\gamma}{g} = 0.9\left(0.036\,11\frac{\text{lb}}{\text{in}^3} \Big/ 386\frac{\text{in}}{\text{sec}^2}\right) = 84.2E-6\frac{\text{lb}-\text{sec}^2}{\text{in}^4} \text{ or } \frac{\text{blob}}{\text{in}^3} \tag{a}$$

3 Find the approximate pressure-viscosity exponent α from equation 15.9b (p. 467).

$$\alpha \cong 7.74E-4\left(\frac{\eta_0}{\rho(10^4)}\right)^{0.163} \cong 7.74E-4\left(\frac{6.5E-6}{84.2E-6(10^4)}\right)^{0.163} = 1.136E-4 \tag{b}$$

4 Find the effective radius from equation 15.11a (p. 468).

$$\frac{1}{R'} = \frac{1}{R_{1_x}} + \frac{1}{R_{2_x}} = \frac{1}{1} + \frac{1}{1.720} \qquad R' = 0.632 \text{ in} \tag{c}$$

5 Find the effective modulus of elasticity from equation 15.11b (p. 468).

$$E' = \frac{2}{\frac{1-\nu_1^2}{E_1} + \frac{1-\nu_2^2}{E_2}} = \frac{2}{\frac{1-0.28^2}{3E7} + \frac{1-0.28^2}{3E7}} = 3.255E7 \tag{d}$$

6 Find the average velocity U as in equation 15.12. The roller has 99% of the cam velocity.

$$U_2 = r\omega = 3.92 \text{ in } (18.85 \text{ rad/sec}) = 73.892 \text{ in/sec}$$
$$U_1 = 0.99U_2 = 0.99(73.892) = 73.153 \text{ in/sec}$$
$$U = (U_1 + U_2)/2 = (73.892 + 73.153)/2 = 73.523 \text{ in/sec} \tag{e}$$

7 Find the ellipticity ratio = major / minor axis. The minor axis is in the direction of rolling in this case.

$$\psi = a/b = 0.0889/0.0110 = 8.082 \tag{f}$$

8 Find the minimum film thickness from equation 15.12 (p. 468).

$$h_{min} = 3.63R'(\alpha E')^{0.49}\left(\frac{\eta_0 U}{E'R'}\right)^{0.68}\left(1-e^{-0.68\psi}\right)\left[\frac{P}{E'(R')^2}\right]^{-0.073}$$

$$= 3.63(0.632)[1.136E-4(3.255E7)]^{0.49}\left[\frac{(6.5E-6)(73.523)}{(3.255E7)(0.632)}\right]^{0.68}$$

$$\cdot\left[1-e^{-0.68(8.082)}\right]\left[\frac{250}{3.255E7(0.632)^2}\right]^{-0.073} = 16.6\ \mu\text{in} \qquad \text{(g)}$$

9 Convert this minimum value at the exit to an approximate thickness at the center of the contact patch with equation 15.8b (p. 465).

$$h_c \cong \frac{4}{3}h_{min} = \frac{4}{3}(16.6) = 22.2\ \ \mu\text{in} \qquad \text{(h)}$$

10 The specific film thickness values for each cam can now be found from equation 15.8a (p. 465):

$$\text{ground cam:}\quad \Lambda = h_c\Big/\sqrt{R_{q1}^2 + R_{q2}^2} = 22.2\Big/\sqrt{7^2+7^2} = 2.24$$

$$\text{milled cam:}\quad \Lambda = h_c\Big/\sqrt{R_{q1}^2 + R_{q2}^2} = 22.2\Big/\sqrt{7^2+30^2} = 0.72 \qquad \text{(i)}$$

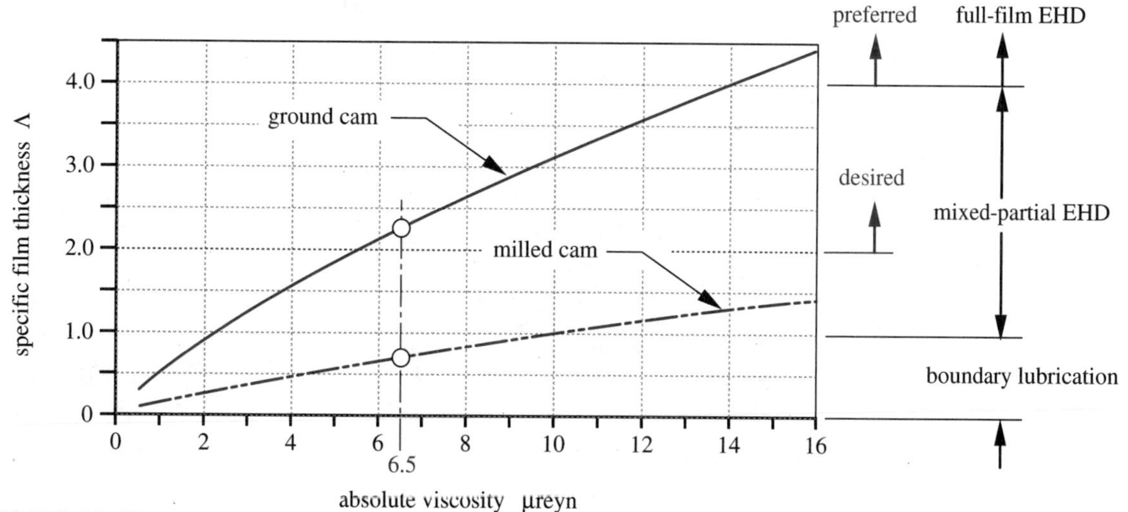

FIGURE 15-11

Variation of specific film thickness Λ with lubricant viscosity η_0 in example 15-1

15

which indicates that the milled cam is in boundary lubrication and the ground cam is in partial EHD with the specified oil. These are common conditions for ground or milled cams, respectively, running against a ground roller-follower.

11 To determine what viscosity of oil would be needed to put each system into partial or full EHD condition, the model was solved for a range of possible η_0 values from 0.5 to 16 μreyn as shown in Figure 15-1 (p. 456) for 180°F. A plot of the results in Figure 15-11 shows that an oil with $\eta_0 \geq 14$ μreyn is needed to put this ground cam into full EHD lubrication, and that an oil with $\eta_0 > 10$ μreyn will get this milled cam to $\Lambda > 1$ and into the low end of the mixed-partial EHD lubrication regime. But, none of the oils shown in Figure 15-1 (p. 456) can provide $\Lambda > 4$ for full-film EHD with the milled cam.

15.7 CAM LUBRICATION

The best form of lubrication for a cam-follower interface is an oil bath. Only by flooding the interface with copious amounts of oil of appropriate viscosity is it possible to obtain elastohydrodynamic lubrication conditions. Even then, the lubrication regime may be mixed or even boundary depending on the velocity, surface finishes, and geometry at the joint. Automotive camshafts and valve trains are enclosed within the engine and flooded with filtered engine oil pumped at low pressure over the cam-follower interface.

The industrial cam and follower is typically enclosed in a cam box that is partially filled with a static charge of oil. The box should never be completely filled with oil as that will cause severe "windage" losses from the drag of the oil on the moving cam and follower surfaces. This will heat the oil excessively, expand it, pressurize the box and blow lubricant past the seals. As long as the cam dips into the static oil reservoir, its motion will carry sufficient lubricant to the joint. Depending on the cam box's orienta-

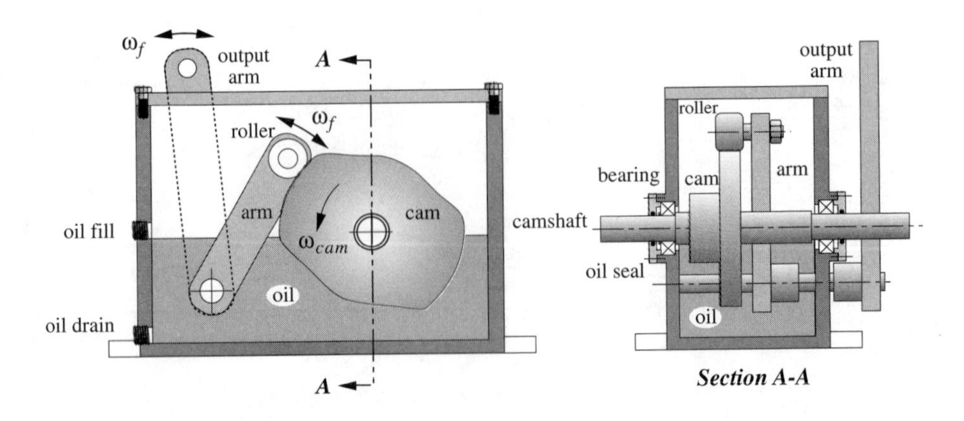

FIGURE 15-12

Cam box (schematic)

tion in the gravity field, the level of oil may be different to ensure that it reaches all bearings and other moving parts.

Figure 15-12 shows one possible arrangement for a cam box enclosure. The camshaft protrudes from two sides of the box to allow it to be mounted in line with other cam boxes to form a line shaft. Seals are provided around the camshaft to prevent oil leaks. The camshaft is supported in ball bearings capable of supporting both radial and thrust loads, though the thrust load in both directions (if any) is taken to ground across only one of the bearings. The other is left free to adjust itself axially to changes in shaft length due to thermal expansion. The second camshaft bearing can be a roller type rather than a ball type as it supports no thrust load. The oscillating follower arm is pivoted on a shaft within the cam box that also protrudes through the box wall for attachment of an external lever to drive the follower train. This allows the rocker shaft to be easily sealed against oil leakage. Rolling element bearings can be used on the rocker shaft, but this is not an ideal application for them as the shaft does not make a full revolution. Plain bronze bushings could be used instead if desired. If the cam requires a translating follower, then a linear oil seal is fitted to the follower shaft where it exits the box. Whether rotary or linear, o-rings can be used as oil seals. Many other types of oil seals are commercially available as well.

On some industrial machines, it is not practical to enclose all the cams and followers in oil tight boxes. These machines often have open line shafts to which split cams are clamped at intervals along the camshaft length. These cams receive only topical grease lubrication at maintenance intervals. Grease on a cam surface can, at best, only achieve boundary lubrication conditions, meaning that there is significant metal-to-metal contact—elastohydrodynamic conditions are not achievable with grease lubrication. In a dirty environment (which, practically speaking, means anything short of a "cleanroom," since the air in any factory is full of particulate matter), grease lubrication on open cams is at once a blessing and a curse. Over time, the exposed grease traps and entrains dust and dirt particles from the air, eventually making it into a form of "grinding compound." Unless maintenance is thorough and frequent about cleaning and regreasing, cams in this condition can be expected to show significant wear over time.

Other industrial applications may require that cams be run unlubricated if there is any possibility that lubricant tossed from the parts might contaminate a product such as medical or food items. A better solution is to enclose them in well-sealed cam boxes with proper lubricant therein, and/or arrange to shield the product from possible contamination from the mechanisms.

15.8 REFERENCES

1 **Elwell, R. C.** (1983). "Hydrostatic Lubrication," in *Handbook of Lubrication*, E. R. Booser, ed., CRC Press: Boca Raton Fla., p. 105.

2 **Radovich, J. L.** (1983). "Gears," in *Handbook of Lubrication*, E. R. Booser, ed., CRC Press: Boca Raton, Fla., p. 544.

3 **Norton, R. L.** (2000). *Machine Design: An Integrated Approach*, 2ed, Prentice-Hall.

4 **Raimondi, A. A.,** and **J. Boyd**. (1958) "A Solution for the Finite Journal Bearing and its Application to Analysis and Design–Parts I, II, and III", *Trans. Am. Soc. Lubrication Engineers,* 1(1): pp. 159-209.

5 **Reynolds, O.** (1886). "On the Theory of Lubrication and its Application to Mr. Beauchamp Tower's Experiments". *Phil. Trans. Roy. Soc.* (London), 177: pp. 157-234.

6 **Dubois G. B.**, and **F. W. Ocvirk**. (1955). "The Short Bearing Approximation for Full Journal Bearings". *Trans. ASME*, 77: p. 1173-1178.

7 **Dubois, G. B., F. W. Ocvirk,** and **R. L. Wehe**. (1955). *Experimental Investigation of Eccentricity Ratio, Friction, and Oil Flow of Long and Short Journal Bearings–With Load Number Charts*, TN3491, NACA.

8 **Ocvirk, F. W.**, *Short Bearing Approximation for Full Journal Bearings*, TN2808, NACA.

9 **Dubois G. B.,** and **F. W. Ocvirk**. (1952). *Analytical Derivation and Experimental Evaluation of Short Bearing Approximation for Full Journal Bearings*, TN1157, NACA, 1953.

10 **Tallian, T. E.** (1964). "Lubricant Films in Rolling Contact of Rough Surfaces," *ASLE Trans.,* 7(2): pp. 109-126.

11 **Cheng, H. S.** (1983). "Elastohydrodynamic Lubrication," in *Handbook of Lubrication*, E. R. Booser, ed., CRC Press: Boca Raton Fla., pp. 155-160.

12 **Bamberger, E. N.**, et al. (1971). "Life Adjustment factors for Ball and Roller Bearings," *ASME Engineering Design Guide.*

13 ASME Research Committee on Lubrication. (1953). "Pressure-Viscosity Report—Vol. 11," ASME.

14 **Dowson D.** and **G. Higginson**. (1959). "A Numerical Solution to the Elastohydrodynamic Problem." *J. Mech. Eng. Sci.*, 1(1): p. 6.

15 **Dowson D.** and **G. Higginson**. (1961). " New Roller Bearing Lubrication Formula," *Engineering*, 192: p. 158-9.

16 **Hamrock B.** and **D. Dowson.** (1977). "Isothermal Elastohydrodynamic Lubrication of Point Contacts—Part III—Fully Flooded Results," *ASME J. Lubr. Technol.*, 99: pp. 264-276.

17 **Dowson D.** and **G. Higginson**. (1968). *Proceedings of Institution of Mechanical Engineers*, 182 (Part 3A): pp. 151-167.

Chapter 16

MEASURING CAM-FOLLOWER PERFORMANCE

16.0 INTRODUCTION

"The proof of the pudding is in the eating," as the saying goes, and the proof of a cam-follower system's performance is in the testing. The dynamic performance of internal combustion (IC) engine cam-follower systems is usually measured in some fashion during their development, but this is less often done for cam-follower systems that are used in industrial machines. When it is done in either venue, the results seldom bear very close resemblance to the theoretical follower motions as designed. (Though, if the system was dynamically modeled, the test data should look similar to the model's simulated results.) There are many reasons for the disparities between theory and experiment, some of which were mentioned in Chapter 14 on cam manufacturing. Other issues will be discussed here.

Significant improvements in measurement technology have occurred in recent years. Dynamic phenomena such as angular position, follower displacement, velocity, acceleration, force, and strain can be measured with very good accuracy, high bandwidth, and excellent repeatability at relatively low cost. This is mainly due to advances in microprocessor and integrated circuit technology that have significantly affected transducer design. The high speed computational capability of modern microcomputers also makes it possible to do extensive analysis of dynamic measurements in real time and use the results for machine or process control as well as for off-line analysis.

16.1 TRANSDUCERS

Many types of transducers are available for the measurement of dynamic parameters relevant to cam-follower performance. Some of them will be described briefly here.

Angular Position Transducers

The angular position of a shaft often needs to be measured with high accuracy, especially for servomotor control. Many types of shaft encoders are available for this purpose. The most accurate and popular devices are optical shaft encoders (Figure 16-1) that use a transparent disk fitted with equispaced opaque radial lines that interrupt a light beam as the disk turns. These devices are powered by a low-voltage DC source (5 -15V) that drives a LED or other light source and a photodiode detector positioned to straddle the transparent disk. As the disk (which is mechanically coupled to the shaft) turns, its opaque lines interrupt the light beam. The pulses from the interrupted beam are conditioned by self-contained circuitry for output as TTL level (0-5V) square or sine waves.

Both *absolute* and *incremental* types of encoders are available. An absolute encoder will provide the shaft's absolute angular position and retain it through a power-down cycle, but an incremental encoder only gives relative position and loses its information when de-powered. Incremental encoders are more common and are less expensive than absolute encoders.

Typically a shaft encoder will output at least two pulse trains at the frequency of the line spacing *n* with a 90° phase shift between them (referred to as being in *quadrature**) to allow detection of motion direction. Also, at least one additional pulse is typically provided once per disk revolution to serve as a trigger and to define the home position of

* See G. S. Gordon, "Understanding Quadrature" at http://www.gpi-encoders.com/technical_articles.htm#"Understanding%20Quadrature"

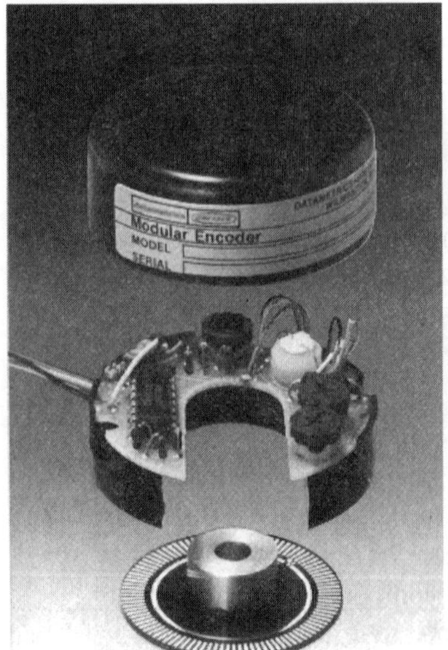

(*a*) Encoder disk and circuitry

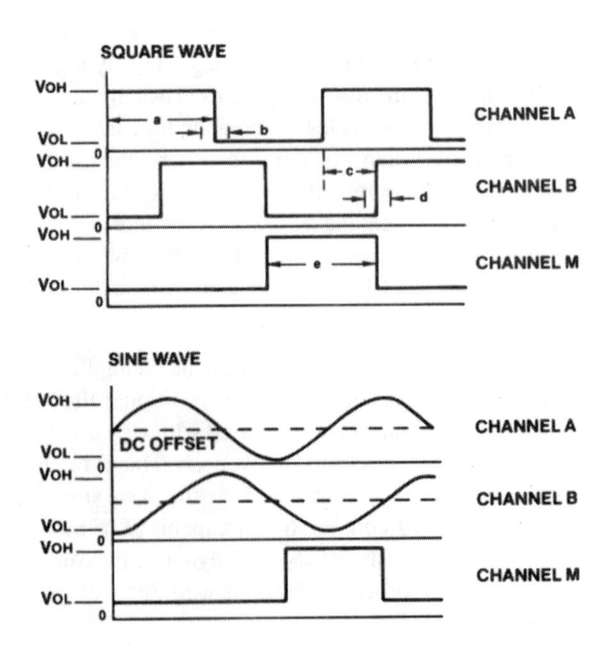

(*b*) Two forms of quadrature encoder output

FIGURE 16-1

Optical shaft encoder and typical output signals

the shaft. When ordering an encoder, the number of lines n can be specified within a wide range to give anywhere from a few hundred to tens of thousands of counts per revolution.

Optical shaft encoders are highly accurate, extremely repeatable, quite reliable, and relatively inexpensive. They provide an excellent way to measure shaft angular position and velocity. Their main limitation is that they generally must be directly coupled mechanically to the open end of, or around the shaft for best accuracy. Sometimes the shaft end or the shaft surface space is not accessible for this purpose. The encoder's case must also be mechanically attached to ground and be well aligned to the shaft.

Displacement Transducers

Several types of devices are available to measure the linear displacement of a cam-follower. We will discuss only two, one contacting device (LVDT) and one noncontacting device (proximity probe).

LVDT The *Linear Variable Differential Transformer*, or LVDT, (Figure 16-2) provides a reliable, high accuracy measurement of instantaneous displacement, but requires that its moving core be attached to the follower train. This is seldom a serious limitation unless the cam follower train is inaccessible or in a hostile/inaccessible environment, such as the combustion chamber of an IC engine.

Figure 16-3 shows a diagram of a cross-section of an LVDT. It is essentially a set of three stationary coils of wire, a primary and two secondaries, through which a magneti-

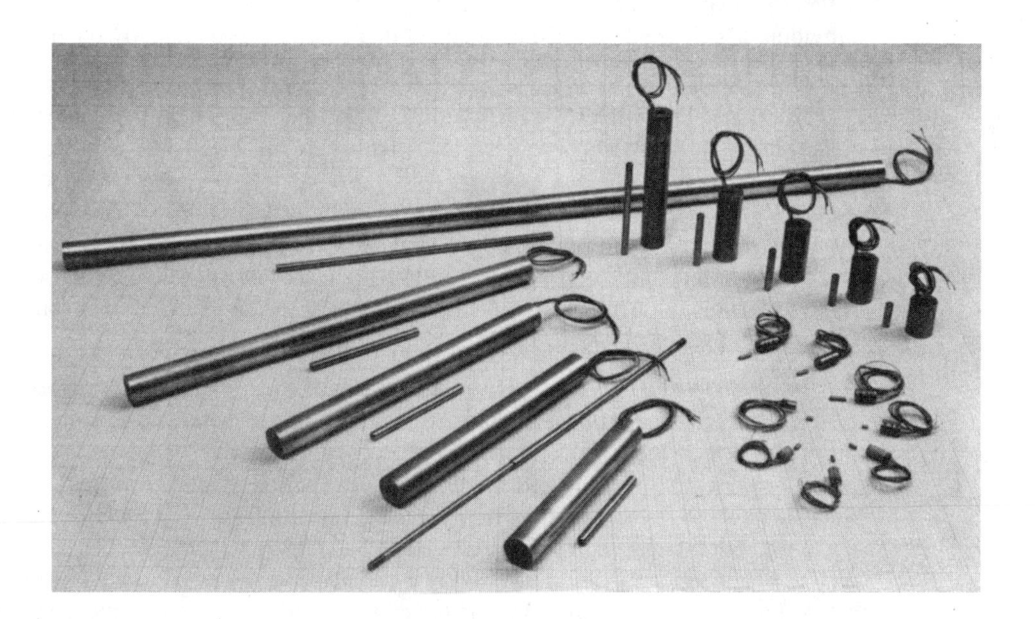

FIGURE 16-2

Linear Variable Differential Transformers (LVDT) of various strokes *(Courtesy of Novatronics Inc., www.novatronics.com)*

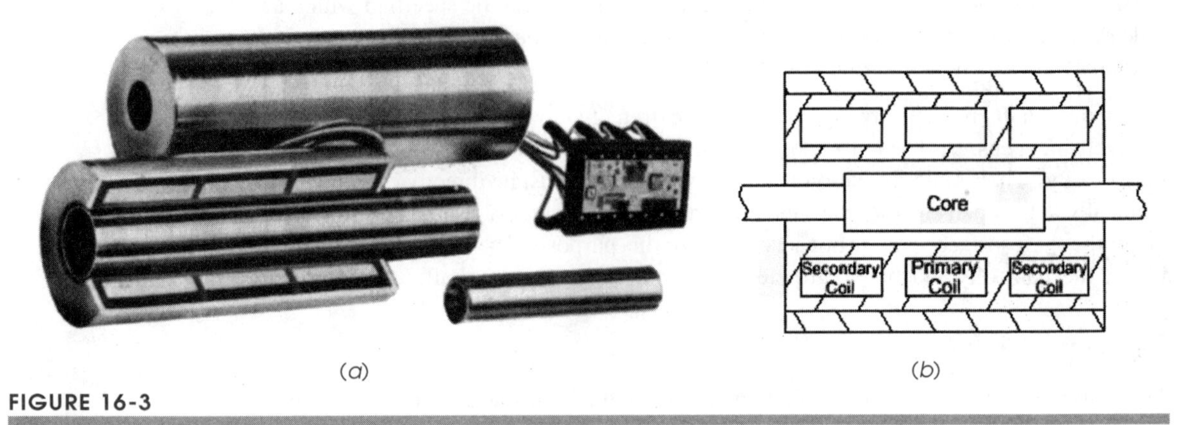

FIGURE 16-3

Cutaway (*a*) and schematic (*b*) of an LVDT *(Courtesy of Schaevitz Div. of Lucas Contol Systems, Hampton, VA 23666)*

cally permeable core is moved. This core is attached to the object whose displacement is being measured. The primary coil is energized by a relatively high frequency (2.5 kHz) alternating current supplied by built-in circuitry. The LVDT output is the differential AC voltage between the two secondary coils, which is a function of core position.

Many LVDTs contain signal conditioning circuitry that generates the AC driving signal from a DC input voltage and converts the AC output signal to a DC voltage that is linearly proportional to core position. These are referred to as *DC in-DC out* devices and are quite convenient for cam work, requiring only a DC power supply for excitation and providing a well-conditioned output signal that can be viewed directly on an oscilloscope.

LVDT resolution is theoretically infinitesimal, but is limited to the order of millionths of an inch by its signal conditioning circuitry . Available in different lengths, they can measure motions continuously over distances of up to several feet with linearity errors of less than 0.25%. Repeatability is essentially exact. Output is absolute and no accuracy is lost through a power cycle. There is very low friction between core and coils and the mass of the core is typically small, giving reasonable dynamic response. Their response bandwidth is limited to about 1 kHz, which can be a limitation in some cam-follower applications.

PROXIMITY PROBES are noncontacting devices that use eddy current principles to detect the presence of an electrically conductive object. Their main advantage is that they require no attachment to the moving object being measured. Their main disadvantage for cam work is their limited stroke capability, typically a small fraction of an inch. Some are able to sense position with reasonable linearity at up to 0.5-in distance from the object, but these are about 1-in diameter and so require a flat target surface of about that size on the moving object. The eddy currents sense any metal in the vicinity, so a protrusion or odd shape on the moving target will compromise measurement accuracy.

Proximity probes are frequently used to measure valve motion on a firing IC engine by arranging them to detect the position of the retainer washer on the back end of the valve spring (thus keeping the probe away from combustion events). However, the limited range of a proximity probe small enough to detect this narrow target only allows measurement of a few mm of motion near the closed valve position. Nevertheless, this gives good information regarding valve bounce on closure. We have successfully used a long-stroke, large diameter proximity probe to measure the entire range of valve motion in a nonfiring, motored IC engine by detecting valve head displacement in an empty combustion chamber.[1]

Velocity Transducers

Velocity transducers are similar to LVDTs. Passing a ferrous core through a coil generates a current in the coil that is proportional to velocity. A linear velocity transducer (LVT) is based on this principle. Like the LVDT, the LVT requires that its core be attached to the moving element. The accuracy and repeatability are similar to that of the LVDT, but the bandwidth is about 2 kHz. Some LVTs are manufactured in the same housing as the manufacturer's LVDTs and so can be interchanged within a given test setup. This was done in the automotive valve train experiment described in [2].

Strain Transducers

Strain gages of either the resistance or semiconductor type are widely used in engineering. These are small, light, thin-film transducers that can be cemented with epoxy to the surface of a part and provide a linear measure of strain at that location when the part is loaded. The electrical resistance of the gage varies in proportion to the change in strain. The gages are connected into a Wheatstone bridge configuration that is excited with a low voltage DC current. The change in resistance within the bridge is measured as the part is loaded and provides an analog to strain. Strain gages are inexpensive, relatively easy to apply (though some skill and training are needed to do so), and they add negligible mass to the part or assembly.

Within the elastic range of a material's strength, strain is proportional to deflection, stress is proportional to strain, and force has a definable relationship to stress that depends on the loading and geometry of the part. Strain gages can be arranged on the part to selectively measure strains due to bending stress, torsional stress, axial stress, or some combination thereof in any given case. Thus, a strain measurement can be used as an analog of the force or deflection that is applied to the part or assembly.

Some disadvantages of strain gages compared to other methods for force measurement are their relative fragility, limited reliability, and their sensitivity to temperature variation. Under dynamic loading, the bond between gage and substrate can become broken as can the delicate wires that connect the gage to the bridge amplifier. Also, their frequency bandwidth is usually limited by the bridge amplifier to about 1 to 2 kHz, and this can be a disadvantage in some vibration measurements.

Force Transducers

LOAD CELLS Force can be measured by strain-gage-based devices as described above. Load cells typically are constructed by arranging strain gages on a structure of simple geometry, such as a beam, which can then be attached as a unit to the assembly in such a way that the force to be measured passes through the load cell. Proper design of the load cell can provide excellent linearity over its usable load range. These devices have the same bandwidth limitations as the strain gage, but without the advantage of low mass.

PIEZOELECTRIC FORCE SENSORS contain thin piezoelectric quartz crystals that generate a voltage proportional to any applied load. In order to measure both tensile and compressive force, the crystals are preloaded (squeezed) between two steel platens that become the threaded mounting surfaces, as shown in Figure 16-4. When fastened in series with a load path, tensile forces applied across the platen faces unload the crystals and generate a negative voltage. Compressive forces generate a positive voltage. Linear load ranges of up to 1 000 lb are commonly available. Their cost is typically several hundred dollars apiece.

An excitation to the crystals must be provided from either a charge-mode amplifier or a current source depending on the presence or absence of circuitry internal to the force sensor. The most convenient sensor type for cam work is the ICP®* or LIVM®† type that has the charge amplifier built into the transducer as microcircuitry and requires only a current source for excitation. Many measuring instruments such as *Dynamic Signal Analyzers* (see below) have a built-in ICP® current source for this purpose. If the transducer will be subjected to temperatures high enough to damage its internal microcircuits, or if more flexibility in terms of sensitivity adjustment is needed in a laboratory setting, then the charge-mode type that has no internal circuitry should be used instead.

Advantages of the piezoelectric force sensor compared to the strain gage are its ruggedness, simplicity of installation (two screws or studs), excellent linearity, high accuracy, good repeatability, temperature insensitivity, cross-axis insensitivity, and relatively high bandwidth of about 2 Hz to 5 kHz. Disadvantages are that it must be "designed in" to the assembly (as opposed to being "stuck on") and it adds some mass. It is essentially a dynamic force transducer, in that the crystals cannot hold a static measurement. The crystal's charge bleeds off fairly rapidly. Thus, if you need to measure static force level and hold the reading, then the strain-gage based load cell is needed.

* Integrated Circuit Piezoelectronic

† Low Impedance Voltage Mode

(a) Transducer, impact cap and 1/4-28 stud

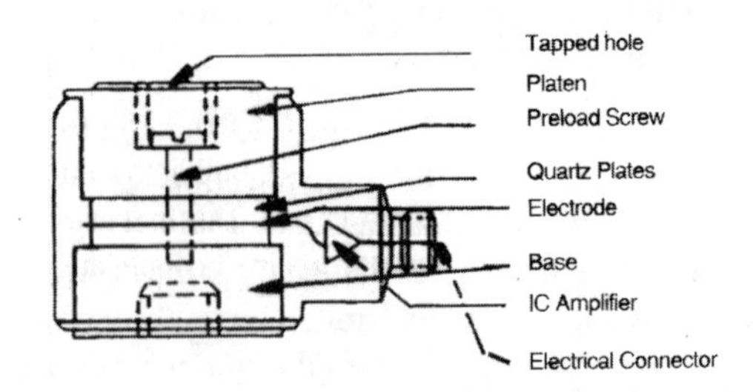

(b) Cross section of transducer

FIGURE 16-4

Piezoelectric force transducer *(Courtesy of Dytran Instruments Inc., Chatsworth, CA 91311)*

Acceleration Transducers

For cam work, the best acceleration transducer is the piezoelectric accelerometer, as shown in Figure 16-5. It is essentially a piezoelectric force transducer to which a seismic mass has been added. So instead of two mounting surfaces that provide a push-pull force across the quartz crystals, they are sandwiched and preloaded between a single mounting surface and a seismic mass. When subjected to an acceleration in the direction of crystal compression/tension, the seismic mass provides an inertial force to excite the crystals. The result is a voltage proportional to (signed) acceleration.

These devices have the same set of advantages as the piezoelectric force sensor on which they are based (see above). The shear mode type (Figure 16-5b) is more sensitive than the compression type (Figure 16-5a); however, their sensitivity is directly proportional to the size of the seismic mass (more mass = more force per unit acceleration). So sensitivity costs in terms of increased accelerometer size and mass, and reduced bandwidth. Mass loading of the system being measured can be a concern, especially with light structures, because adding mass changes the system's natural frequencies. Small accelerometers with sensitivities of about 100 g/V weigh less than 10 grams, have bandwidths of up to 12 kHz, and a usable linear range of 500g. Larger accelerometers with sensitivities of 2 g/V weigh several tens of grams, have bandwidths of about 5-8 kHz, and a usable linear range of 10g. Many other configurations are available as well.

As with their force sensor cousins, these transducers are sensitive along only one axis. To measure simultaneous 3-axis acceleration requires three accelerometers. Triaxial configurations on a common mounting block are available. Attachment of the transducer is best accomplished with threaded fasteners as it is necessary to have tight coupling to the assembly being measured. Most piezoelectric accelerometers are provided with tapped holes or studs on their mounting surface for this purpose (usually a

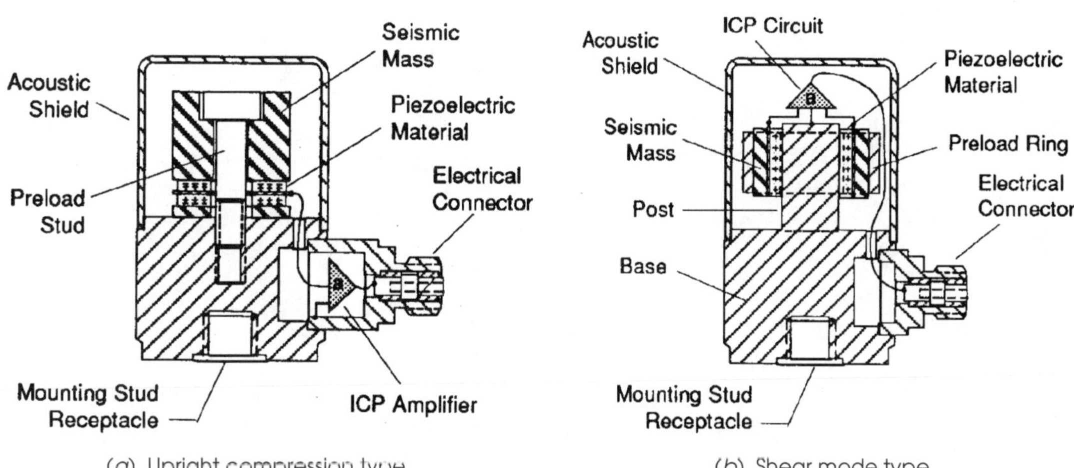

(a) Upright compression type

(b) Shear mode type

FIGURE 16-5

Piezoelectric accelerometers *(Courtesy of PCB Piezotronics, Depew, NY 14043)*

16

#10-32 thread in the U.S.). Magnets can be used to attach accelerometers to ferrous materials, but very small accelerometers have to be cemented to the part. For low shock applications on any material, one can often use beeswax to attach an accelerometer. It is surprising that beeswax works so well. Figure 16-6 shows the mounted natural frequencies and relative transmission bandwidths of mechanical, cement, beeswax, and magnetic connections of accelerometers to a part. In general, one should not make measurements at frequencies higher than about 1/3 of the mounted natural frequency of the transducer. Mechanical connection provides the highest bandwidth, beeswax or cement the next best, and magnetic coupling has the lowest usable bandwidth of these four options.

Vibration Measurement

Vibration measurement has a long history. The methods used have changed as technology provided new devices. In the early days of vibration measurement, displacement

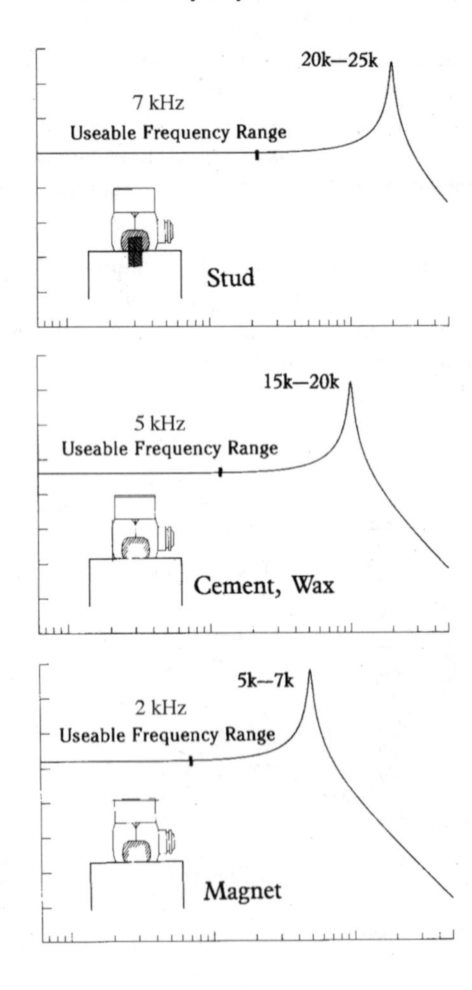

FIGURE 16-6

Dynamic bandwidths of various accelerometer mountings[3]

measurement methods (largely mechanical in nature) were all that were available. Thus, most early vibration analysis dealt with displacement output. This, of course, provided a direct correlation with the mathematics of vibration analysis that solves for displacement as a function of force input as described in Chapter 10.

Historically, velocity transducers were developed next and they soon supplanted displacement methods for vibration measurement. Some vibration analysts still prefer to use velocity measurements because the product of force and velocity is power, and that provides some insight into the system's dynamic behavior.

However, now that inexpensive and accurate accelerometers are generally available, they provide the best means for the measurement of vibration. The best argument for using acceleration for vibration measurement is that the higher the derivative measured, the more dynamic information will be available mathematically. Also, the bandwidth of accelerometers is significantly larger than that of either velocity or displacement transducers. Moreover, if one can only measure one parameter, it is better to measure the highest derivative possible. Then, if desired, the lower derivatives can be determined by numerical integration. Mathematically, numerical integration is a smoothing operation. Conversely, measuring a lower derivative and attempting to numerically differentiate it to create its derivatives can lead to numerically noisy results. Using smoothing algorithms as described in [1] can improve the situation, but will nevertheless limit the bandwidth of the simulated higher derivatives as compared to their direct measurement.

16.2 EXPERIMENTAL CAM-FOLLOWER MEASUREMENTS

Based on the preceding discussion, it should be clear that the primary interest should be to measure acceleration of the follower train, preferably as close to the end effector as possible. Luckily, this is quite easy to do in most cases, especially in industrial machinery. In this instance, the environment is favorable (as compared to a combustion chamber in a firing IC engine, for example) in that it is usually at room temperature with plenty of line current available for powering measurement equipment.

Most industrial machinery was designed without any consideration for the need or desire to measure its dynamic performance. Nevertheless, it is often possible to convince the manufacturing supervisor to allow the placement of an accelerometer on a cam-follower lever with a magnet or beeswax, provided that one swears that it will not interfere with the daily production quota.* The main difficulty with this procedure is protecting the relatively delicate wire to the accelerometer from being chewed up in the machinery.

It is more difficult to arrange the installation of LVDT or LVT sensors on existing production machinery as this requires attaching the core to the moving lever of interest and some bracketry to mount the transducer body to ground. At a minimum, this requires a tapped hole to attach the core to the lever. If to do this requires interrupting production, it will be an uphill battle. On the other hand, the "midnight acceleration thief" with magnet or beeswax can be in and gone in a matter of minutes before the production monitors even know what happened. All joking aside, the information obtained from acceleration measurements in industrial machinery is of much greater value than that which

* Woe to you if it does! You will henceforth be *persona non grata*.

might have been obtained from more logistically and technically difficult measurements of the lower derivatives.

16.3 DATA ANALYSIS

Once the choice of transducers is made and the system has been fitted with them, data can be gathered. Transducers typically convert the phenomena measured to a time-varying electrical signal whose voltage is linearly proportional to the measured parameter. These are analog signals that can be directly displayed on an oscilloscope. For a quick observation of system performance, this may be adequate, but most likely you will want to save the data to make further comparisons and calculations with it. This usually requires that the analog signal be converted to digital form for storage within a computer. In fact, modern oscilloscopes are digital devices that convert the analog input signal to digital form before displaying it.

Analog to Digital Conversion

Digitization of an analog signal presents some challenges that, if not properly understood, can result in erroneous information being extracted from the data. A digital representation of a continuous function is only defined at a set of discrete values of the independent variable (usually time). The sampling rate (samples/sec, or Hz) is defined as the rate at which the value of the continuous analog signal is sampled. Sample period (sec) is the reciprocal of sampling rate. Once digitized, all information between the discrete sampling times has been lost. Thus, the chosen sampling rate can have a significant effect on the result.

NYQUIST THEOREM Nyquist discovered that the minimum sampling rate that will not lose significant frequency information is exactly twice the highest frequency present in the original signal. Sampling at any higher rate is called oversampling and will, up to some limit determined by numeric roundoff errors, potentially improve a digitally sampled waveform's fidelity to the original analog signal in a time-based presentation. In terms of fidelity of the frequency content of the signal, however, exact Nyquist sampling is theoretically adequate. Nevertheless, it is common to oversample, especially if an accurate waveshape in the digital time domain is desired. Digital frequency analyzers that make no attempt to display an accurate time-based signal typically sample at about 1.3X the Nyquist rate, while digital oscilloscopes sample at much higher factors to create an accurate representation of the signal in the time domain.

ALIASING describes the result of undersampling a signal in violation of the Nyquist rule. In such cases, frequencies actually present in the signal that are above the Nyquist frequency will be "aliased" down to a lower frequency and appear in the result. There is no way to determine that these aliased signals are imposters, and erroneous conclusions may be engendered by their presence. Aliasing **will occur** during the digitization of an analog signal **any time** the sampling rate is less than twice the frequency of any information present in the signal. It is important to realize that the process of digitization is what potentially causes aliasing. It has nothing to do with whether the resulting signal is looked at in the time or frequency domains or in both.*

* You have undoubtedly seen the effects of aliasing when watching an old western movie or a modern TV commercial for automobiles with fancy spoked wheels. Sometimes the spoked wheels on vehicles moving forward appear to be turning backward. This is aliasing caused by the digitization frequency (the film's frame/sec rate) being less than twice the frequency of the spokes' motion. As the stagecoach slows to a stop, you will see the spokes suddenly switch from apparent backwards rotation to forward rotation when the wheel's spoke-pass frequency drops below the Nyquist frequency for the film's frame/sec rate. Note that the aliasing in this example creates clearly erroneous information about wheel motion. While our experience tells us that it is incorrect in this case, one will likely never know that their signal measured from a machine contains aliased information as there may be no experiential reference for comparison.

So how do we guard against aliasing? If we know with certainty what the highest frequency content of our yet to be measured signal is, then we can use a sampling rate high enough to avoid aliasing. However, it is rare to have that knowledge a priori; if we did, we probably wouldn't need to measure the signal in the first place. Given the uncertainty of our knowledge of the signal, the best bet is to filter it in analog form before we sample it.

ANTI-ALIAS FILTERING refers to the passing of the measured analog signal (i.e., the transducer's raw output signal) through a low-pass filter whose cutoff frequency is at the Nyquist frequency for our sample rate (i.e., 2X the sampling rate). This cuts out any frequency content that would otherwise alias down into our digitized signal. Note that it is not sufficient to first sample (digitize) the signal and then digitally filter the result. By then it is too late. Any aliasing will have already occurred in the digitization process.

Analog anti-alias filters need to be of high quality to avoid corrupting the data, and so tend to be expensive. Moreover, if you are taking multiple channels of simultaneous data and plan to do some computations using more than one of these signals as described in the following sections, then it becomes important that all the anti-alias filters used for simultaneous sampling (one per channel) be matched for both amplitude and phase. This makes them even more expensive.

An ideal filter (which does not exist) would pass all frequencies between the cutoff frequencies unchanged (a gain of 1) and then abruptly attenuate all frequencies outside the cutoff frequencies to zero, as shown in Figure 16-7.* A real filter has a "rolloff" zone within whose frequencies the signal is partially attenuated as also shown in Figure 16-7. For accurate digital signal processing work, we must avoid using data that is affected by

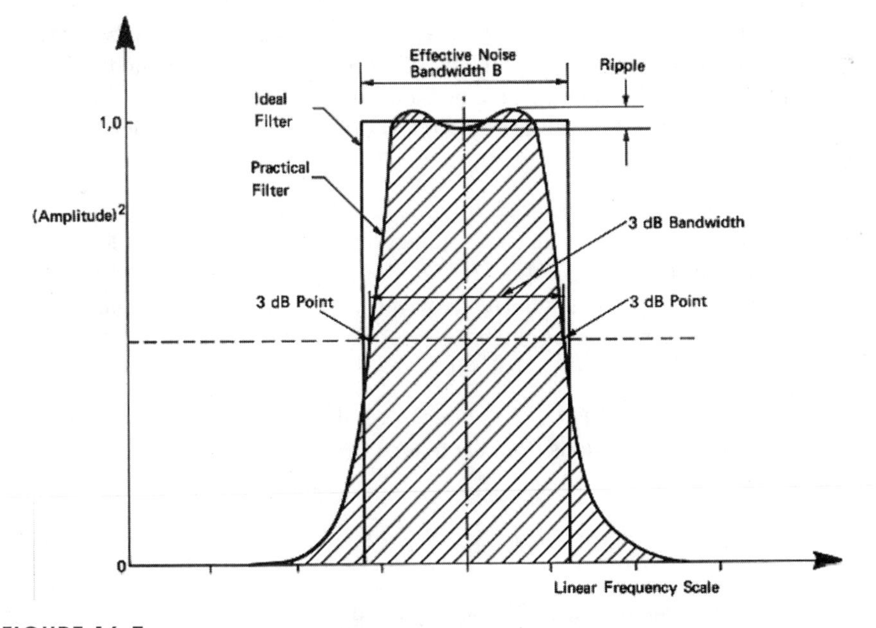

FIGURE 16-7
Ideal versus real filter characteristics[4]

* Note that Figure 16-7 shows a "band-pass" filter designed to attenuate signals both above and below its two cutoff frequencies. An anti-alias filter is a low-pass filter that is essentially the right half of the filter shown in the figure. The center of the band pass filter shown would then be at zero frequency and it would pass all frequencies up to its high cutoff. One can also make a high-pass filter which would be the left half of the band-pass filter shown that attenuated all frequencies below a low cutoff.

anti-alias filter rolloff. The best available analog low-pass filters have a rolloff zone that extends about 30% beyond the frequency at which they begin to attenuate the signal. This is the reason digital signal analyzers that use anti-alias filters (as all should) sample at about 1.3X the Nyquist rate. They then "throw away" (i.e., don't display) the frequency band that was affected by filter rolloff.

Note that when using the "do-it-yourself" approach to digital signal processing of sticking an inexpensive analog-to-digital (A/D) board in a PC and hooking up a transducer or two, you typically will have no anti-alias protection unless you supply it externally. Some higher-end A/D boards are now available with on-board anti-alias filters. Without them, you are taking a risk that your data may be corrupted by aliasing.

Spectrum Analysis

While it is instructive to look at your measured data in the time domain, often more can be learned about the system's dynamic behavior by transforming the time data to the frequency domain. Fourier published the mathematics for the series and transform that bear his name in 1822.

THE FOURIER SERIES provides a means to break down any periodic function into an infinite series of sine and cosine terms

$$y(t)=\frac{a_0}{2}+\sum_{n-1}^{\infty}\left[a_n\cos(2\pi nf_0t)+b_n\sin(2\pi nf_0t)\right] \tag{16.1}$$

Each succeeding pair of sine and cosine terms has a frequency that is an integer multiple of the first, or fundamental, frequency. These terms are called the harmonics of the function. The original function can be approximated by summing a finite number of these terms as was shown in Figures 3-19 through 3-21 (pp. 54-55). For a periodic (i.e., repeating) function, continuous in the time domain, such as a cam-follower acceleration function, the Fourier series can be used to find its discrete set of harmonics in the frequency domain.

Figure 16-8 shows such a function, simplified to contain only two harmonics. Note that the time and frequency domains are nothing more than alternate ways to view the same data. Theoretically, converting a function from the time domain to the frequency domain or vice-versa does not alter the function. Of course, summing of a finite number of terms of an infinite series leaves out the effects of the excluded terms. Also, numeric roundoff errors will always take a small toll.

16

THE FOURIER TRANSFORM or Fourier integral, provides a means to transform continuous, nonperiodic functions from the time to the frequency domain.

$$H(f)=\int_{-\infty}^{\infty}h(t)e^{-j2\pi ft}\,dt \tag{16.2}$$

The inverse Fourier transform allows conversion in the opposite direction.

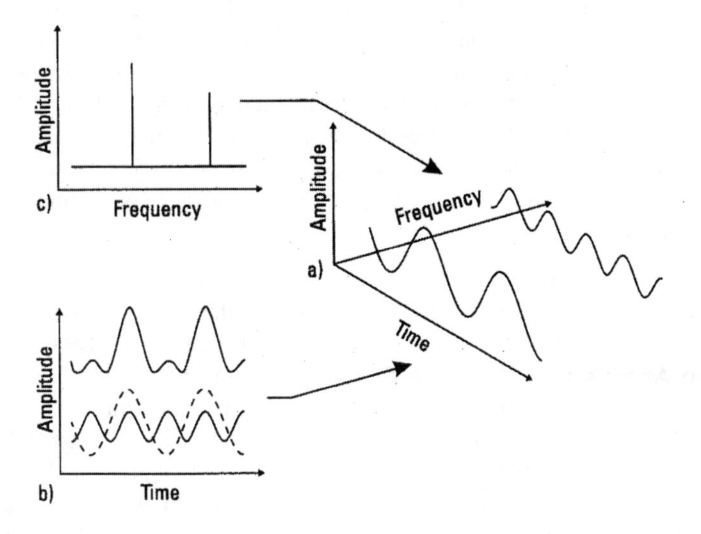

FIGURE 16-8

Relationship between the time and frequency domains[5]

$$h(t) = \int_{-\infty}^{\infty} H(f) e^{j2\pi ft} \, df \tag{16.3}$$

Theoretically these mathematical operations are exact and loss-free. Note that the integrations of these transforms are from $-\infty$ to $+\infty$ in both time and frequency. Obviously the practical computation of these functions will introduce some roundoff error.

Both the Fourier series and Fourier transform require analytic functions for which an algebraic expression is known. This is useful as a mathematical exercise, but a different approach is needed for experimentally measured functions in digital form.

THE DISCRETE FOURIER TRANSFORM or DFT, is defined as

$$G\left(\frac{n}{NT}\right) = \sum_{k=0}^{N-1} g(kT) e^{-j2\pi nk/N} \qquad n = 0,\ 1,\ \ldots,\ N-1 \tag{16.4}$$

and provides a means to transform a discrete function from the time to frequency domain and vice-versa. The result will be a discrete function as well. The **Fast Fourier Transform**, or FFT, is a variant on the DFT; in fact it is just an efficient calculation algorithm that runs faster than the DFT algorithm. The main difference is that the FFT requires that the discrete function supplied for transformation contain a number of data that is a power of 2. But the DFT algorithm can operate on an array of any number of data. Many equation solver packages and spreadsheets contain a DFT or FFT algorithm. The computer code to compute it is also published in many references. Program DYNACAM contains a DFT algorithm that will calculate the Fourier transform of any set of data that has been calculated within the program.

WINDOWING Like the analytical Fourier series, the DFT expects that the function to be transformed is periodic and exists from $-\infty$ to $+\infty$ in time. We cannot give it an infinite amount of data, so after sampling and digitizing the measured function, we must *window* it to "cut out" a finite set of data for computation of the DFT. For the DFT calculation to give correct results requires that the piece of the function "windowed" be an integer multiple of its period, i.e., exactly N cycles of the periodic function. In general, we do not know the period of the measured function in advance of calculating the DFT, so this presents a dilemma.

It is resolved in general by first multiplying the digitized data by one of several *window functions* that have the property of zero value at each end and a positive value in the center, as shown in Figure 16-9. By forcing the sampled data to zero at each end, it becomes "pseudo periodic" and thus satisfies that requirement of the DFT. The particular window function is selected based on its effects on the resulting transform. Some window functions, such as the *Hanning,* will give good frequency resolution at the expense of some amplitude error and others, such as the *Flat Top**, will give good amplitude accuracy at the expense of frequency resolution. All nonuniform window functions introduce some distortion to the data, but this is a useful trade-off against the worse distortion (called leakage) that would occur if the data were nonperiodic in the window.

There are many other window functions available, but their discussion is beyond the scope of this brief introduction to the topic of dynamic signal analysis. The interested reader should consult the references listed in the bibliography of this chapter for more complete information. Let it simply be said here that, in general, it is necessary to use a

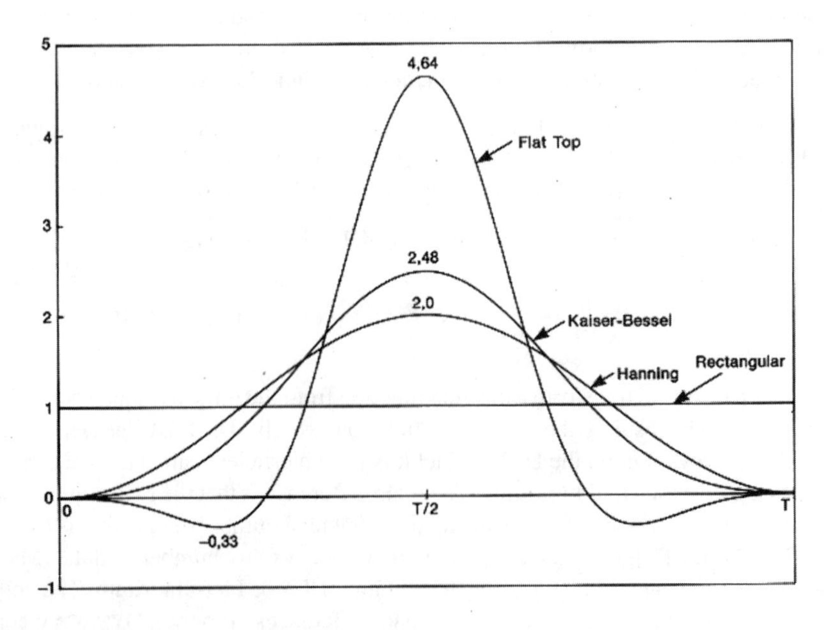

FIGURE 16-9

Shapes of several common window functions in the time domain [6]

16

* The "flat-top" window function shown in the figure is clearly not very flat on top. Its name comes from its shape in the frequency domain where it does have a flat top with very little ripple on it, thus giving accurate amplitude of the transformed spectrum.

suitable window function to get good results from a DFT or FFT calculation. Keep this in mind if attempting the "do-it-yourself" approach to this exercise, as noted above, since commercial A/D boards and their software seldom provide window functions. Another trap for the unwary experimenter.

FREQUENCY VS. ORDER DOMAINS There is another (and better) way around this windowing problem for situations involving rotating machinery, and cams are usually part of a rotating machine. The DFT mathematically requires a sample set that is "periodic in the window." When sampling is being done "against the clock," i.e., each sample taken after a discrete time step or "clock tick" in the computer, then there will be no particular relationship between the time length of the sample window and the period of rotation of the machine. Thus the chance of the sampled data being an integer multiple of a rotation period is about nil. This standard approach to sampling (i.e., against the clock) gives results in the *frequency domain* after the Fourier transform is done. The units of time are *seconds* and the units of frequency are *Hz* in this case.

An alternate approach for rotating machinery is to use a shaft encoder with, say, 2048 counts per revolution, to drive the sampling algorithm. Now one sample is taken every 2048th of a revolution regardless of shaft speed, rather than at fixed time steps, making the window contain exactly one period of data in 2048 samples that can be fed to an FFT. This is referred to either as *external sampling* or as *order analysis*. The first name refers to the fact that the sampling "clock" (the shaft encoder) is now external to the computer. The second name refers to the units of the resulting Fourier transform that is now in *orders* or *harmonics* of the fundamental frequency of shaft rotation. The time data now has units of *revolutions* rather than seconds and the frequency information is in the *order domain*. To convert revolutions to a time base, simply divide by shaft speed in rps. To convert orders to Hz, multiply by shaft speed in rps.

There are several advantages to using order analysis for rotating machinery measurements. First, the measurement is insensitive to variations in shaft speed, so data can be taken during runup and rundown transients without smearing the spectra, and speed variations during steady state operation will not affect data accuracy. Second, properly done, the data is inherently periodic in the window, and so there is no need to multiply by a special window function. The resulting spectrum is then as undistorted as is possible in this business. Third, by triggering each data set at the same point of shaft rotation, time averaging can be used to minimize the effects of random noise in the data, resulting in the cleanest possible time and spectral information.* Order analysis is recommended for measurements in rotating machinery whenever possible. It requires, at a minimum, a reliable once-per-revolution pulse from the drive shaft,† and is best done with a train of pulses (preferably a power of 2 for the FFT) per shaft revolution from a shaft encoder.

* Note that triggering and time averaging can also be used with clock-based measurements as well as with externally sampled ones, giving the same advantages of random noise reduction. However, a window function will then be needed to force pseudo-periodicity in the sample window.

† If a once-per-revolution pulse is all that is available, then an electronic device called a ratio synthesizer will also be needed to lock on to the frequency of the pulse and generate a string of synchronous pulses of some power of 2 (say 2048) between the once-per-revolution pulses to use as the external "clock." Some dynamic signal analyzers have this feature built in. However, the response time of the ratio synthesizer limits the rate at which the sampler can follow transient changes in shaft speed, which is not a problem with a directly coupled N counts per revolution encoder.

Forms of Spectra

There are several forms in which a spectrum can be displayed. A "pure" Fourier spectrum is a complex quantity sometimes called a linear spectrum:

$$X(f) = \Im[x(t)] = [a_i + jb_i]_{i=1}^{n} \tag{16.5}$$

This can be displayed in Cartesian form as real and imaginary components or in polar form as magnitude and phase.

An autospectrum, or power spectrum, is a real quantity calculated as the product of the complex conjugates of a single linear spectrum.

$$\begin{aligned} G_{xx}(f) &= X(f)\cdot X^*(f) \\ &= (a+jb)(a-jb) \\ &= a^2-(j^2)(b^2) \\ &= a^2-(-1)(b^2) \\ G_{xx}(f) &= a^2+b^2 \end{aligned} \tag{16.6}$$

where the * denotes a complex conjugate. Its units are volts^2, so it is proportional to power (V^2/R) in an electrical circuit, accounting for the name power spectrum. Note that calculating a power spectrum loses all phase information in the linear spectrum.

The cross power spectrum (or simply cross spectrum) computes the elements common to two spectra and is useful to determine the degree to which one part of a system is influencing another. The cross spectrum of Y versus X is:

$$\begin{aligned} & & G_{yx}(f) &= Y(f)\cdot X^*(f) \\ &\text{let:} & X(f) &= a+jb \\ &\text{and:} & Y(f) &= c+jd \\ &\text{then:} & G_{yx}(f) &= Y(f)\cdot X^*(f) \\ & & &= (c+jd)(a-jb) \\ & & &= ac+jad-jbc+bd \\ & & G_{yx}(f) &= (ac+bd)+j(ad-bc) \end{aligned} \tag{16.7a}$$

and the cross spectrum of X versus Y is:

$$\begin{aligned} G_{xy}(f) &= X(f)\cdot Y^*(f) \\ &= (a+jb)(c-jd) \\ &= ac+jbc-jad+bd \\ &= (ac+bd)+j(bc-ad) \\ G_{xy}(f) &= (ac+bd)-j(ad-bc) = G^*_{yx}(f) \end{aligned} \tag{16.7b}$$

These are complex quantities. The two constituents $X(f)$ and $Y(f)$ must be measured simultaneously.

16

Modal Domain

The modal behavior of a structure refers to the spatial distribution of its vibratory behavior. The structural natural frequencies of a solid part will have the same values of frequency at all points within the structure. However, the amplitude of vibration will vary with location over the part's geometry. Consider a simply-supported beam of uniform cross section subjected to unidirectional transverse vibration. The amplitude of its first natural frequency will be maximal midway between the supports, zero at the supports, and somewhere between those values everywhere else along the beam. The curve formed by the deflected beam at its peak vibrational excursion is termed its mode shape for that frequency. Each of its higher natural frequencies will have a unique mode shape and a number of nodes, or zero crossovers, equal to the frequency number plus 1 for this beam. For example, the second natural frequency of this beam will have nodes at the supports and in the center with peak amplitudes at the 1/4 points. In two dimensions, the mode shape becomes a surface. The modal behavior of the structure of a cam-follower system is of interest in terms of analyzing and ultimately minimizing the vibration of the follower system when subjected to cam-induced motion.

Frequency Response Functions (FRF)

The **transfer function** of a linear, time invariant system is defined as the quotient of the Laplace transforms of the system's output and input functions.

$$L(f)=\frac{\ell(output)}{\ell(input)} \tag{16.8}$$

The transfer function (TF) is a physical property of a linear system and defines its behavior (output) in response to any input function. If the transfer function of a system can be defined or measured, then its response to any proposed input can be predicted in advance of its application.

We are dealing with Fourier transforms rather than Laplace transforms here and there is an analogous quantity called the frequency response function (FRF) defined as the quotient of the Fourier transforms of the system's output and input functions.

$$H(f)=\frac{\Im(output)}{\Im(input)} \tag{16.9}$$

The FRF is a subset of the TF and lacks only the damping information present in the TF.* To experimentally determine the FRF of a system requires a simultaneous measurement of a suitable input excitation function as applied to the system and the resulting system output response. Transforming both measurements to the frequency domain and forming their quotient gives the FRF. This measurement proves very useful, as will be seen in the next chapter.

The FRF is the Fourier transform of the *impulse response* of the system, i.e., the system's output in response to an impulse excitation, effectively a hammer blow that causes the system to "ring" (like a bell) at all its natural frequencies.

* See reference 7 for a complete explanation.

Dynamic Signal Analyzers

Dynamic signal analyzers (DSA) are designed to analyze the behavior of physical systems (typically mechanical systems) in the frequency and order domains. They have at least two input channels, the minimum needed to measure the FRF of a system. A good quality DSA will have amplitude and phase matched analog anti-alias filters on both channels through which the analog input signals are passed before being digitized by a high resolution (12 or 16-bit) A/D converter for each channel. A good DSA will do true simultaneous sampling of both channels rather than multiplexed sequential sampling so as to guarantee no phase shift between two simultaneous signals used to form an FRF. (Note that most PC-based A/D cards do only multiplexed sequential sampling among channels.)

After digitizing, the time-domain signals will be transformed to the frequency domain with an FFT algorithm hard-coded in a chip for fastest processing speed. The sampling rate of a typical DSA is about 265 ksamples/sec, which after applying the Nyquist rule and allowing for the rolloff of the anti-alias filters allows a maximum bandwidth of about 100 kHz to be displayed. This is more than ample for mechanical systems that seldom display frequency content at even that high a level. For cam-follower measurements using accelerometers or other transducers as described above, one seldom has valid information in the measured signals above 5 to 10 kHz and often has much less. For this reason, DSAs typically provide a variety of selectable measurement bandwidths from milli-Hz to the maximum bandwidth of the analyzer.

Memory size in a DSA's computer limits the resolution possible within the frequency band measured. A common arrangement provides for storage of 2048 digitized points in the time domain signal, though some DSAs use 4096 or more points. The FFT computes a symmetrical two-sided spectrum containing negative and positive frequencies. The negative frequency information is redundant and so is discarded. This means that 2048 data points in time can at best yield 1024 spectral lines in the positive frequency regime. If the anti-alias filter were ideal, these could all be used, but filter rolloff compromises the amplitude of the high end of the spectrum. For this reason, only 800 spectral lines are typically displayed from a measurement of 2048 points of time data. Obviously, increasing the time data to 4096 points will provide 1600 lines, etc. The limitation on this situation is essentially economic. More memory and faster processing speed in the computer will allow higher resolution without compromising its ability to capture rapidly changing signals.

Since the spectral resolution is practically limited to some number of lines (800, 1600, etc.), the delta frequency between lines will be a function of the maximum frequency (bandwidth) selected. To obtain maximum frequency resolution, the measurement bandwidth should be set as close to the maximum usable bandwidth of the transducer as possible, or even lower if the frequencies of interest allow.

Measuring the FRF

The tools needed for measurement of the FRF of a system are an LVDT, LVT, or accelerometer to measure the output term in equation 16.9 and a force transducer to measure the input term in that equation. For cam-follower systems, the best combination is a piezoelectric accelerometer and a piezoelectric force transducer. Some means to excite the structure with a controlled force is also needed. The two most practical means available for this are an instrumented hammer or a shaker. Either will work and each has its advantages and disadvantages.

INSTRUMENTED HAMMER An instrumented hammer has a built-in piezoelectric transducer to measure the force of the hammer blow. It also has interchangeable, conical hammer tips of different stiffness. The stiffer the tip, the higher the frequency content that can be imparted to the structure. For a steel structure, an aluminum tip will give good results to several thousand kHz. Hard and soft plastic tips limit the input frequency to lower bandwidth. Because an impulse in time has a frequency transform that is theoretically uniform in magnitude over all frequencies, it is ideal for uniformly exciting all the natural frequencies in a structure. Some practice is required to become proficient at making "clean" hammer hits that closely resemble a pure impulse. The advantages of hammer testing are its short setup time, ease of use, and flexibility in terms of input location for modal testing. Its disadvantages are poor repeatability, danger of "dirty hits" if not skilled and careful, and the tediousness of doing multiple hits for averaging.

SHAKER An electromagnetic shaker is essentially a loudspeaker whose core is attached by a long thin rod, called a "stinger" to the structure to be tested. A force transducer is placed in series with the rod to measure the input force. The shaker body must be suspended or otherwise supported so as to be mechanically decoupled from the tested structure so that the shaker energy travels only through the stinger to the structure. The shaker is excited by one of several functions that contain a broad range of spectral energy at uniform amplitude. One such function is pseudorandom noise. Another is a burst sine wave sweep. Most DSAs contain signal generators that provide these functions. The output is measured by placing an accelerometer on the structure at one or more points of interest. If the purpose is to measure the FRF of the cam-follower system as it functions in service, then it is best to apply the input force to the follower train as close to the cam-follower interface as possible and measure the output acceleration as close to the end effector as possible. If a modal survey is desired, then either the accelerometer will have to be moved to a different location for each measurement or the input force moved in similar fashion, or both. Advantages are good repeatability and ease of averaging. Disadvantages are fussy and difficult setup requirements to isolate the shaker.

MEASUREMENT A DSA is also most useful for an FRF measurement as it is essentially designed for that purpose. The input force is input to one channel and the output acceleration to the other. The proper window functions are chosen, the desired bandwidth set, and some number of averages selected. Averaging of multiple signals is required for the computation of the *coherence function* that gives a quality control measure of the validity of the FRF measurement. If the coherence is one, then all of the output is due to the supplied input (hammer or shaker). If it is zero, then none of the output is due to the input. Instead, the structure was excited by some other uncontrolled source of energy, e.g., the floor, a passing truck, etc. One hopes to see coherence values greater than 0.9 at the natural frequencies. The DSA calculates the FRF according to equation 16.9 and displays both it and its coherence. The results can usually be extracted to a storage medium, such as a floppy disk, for off-line computation.

Once a valid estimate of a system's FRF is obtained, we have a tool to predict its behavior in response to any input (e.g., a cam function) that we may wish to mathematically apply. Figure 16-10 shows the FRF of the combination of follower arm, camshaft, cam, and frame of the cam dynamic test fixture of Figure 14-6 (p. 436) in the direction of follower motion as measured with a shaker applying random noise to the cam follower in place on the cam and an accelerometer attached to the follower arm.[7] The machine was of course not running at the time. Several natural frequencies are evident in this FRF that spans a bandwidth of 3200 Hz. Each peak depicts a structural natural frequency (SNF) of some element of the system. Among those seen in Figure 16-10 are the follower arm in bending, the camshaft in bending, the machine's case in various modes, and even the metal frame used to mount the shaker. Each of these was separately measured in a series of tests that involved partial disassembly of the apparatus. The FRF in the figure is of the fully assembled machine, ready to run.

The "Q" of a System

Electrical engineers refer to the "Q" of a circuit by which they mean the degree of damping in the system. Q stands for "quality" and a high quality or "high Q" circuit is one with low damping, thus low losses. Mechanical engineers have adopted the term to apply to "mechanical circuits" such as cam-follower systems. Mechanical and electrical systems are directly analogous as pointed out in Chapter 8 (see Section 8.10 and Tables 8-1 through 8-3). The level of damping in a system can be discerned qualitatively from its FRF. Sharp, narrow peaks at its natural frequencies indicate low damping or high Q. Broad, rounded, and gently sloping peaks indicate greater damping and low Q.

Note that systems of the type used for cam-follower trains and their supporting structures, typically do not have any dampers built in. The only damping usually present is

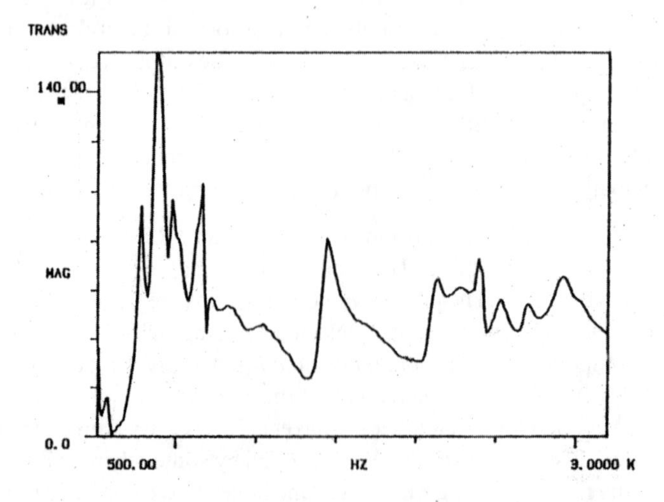

FIGURE 16-10

Frequency response function (FRF) of the Cam Dynamic Test Fixture (CDTF)[7]

16

structural damping, i.e., damping within the materials themselves. Some materials such as wood have relatively good internal damping. However the metals of which the usual cam-follower train are made have low internal damping. Steel has very low internal damping, aluminum and cast iron a little more, but none have enough to make it a low Q system. Thus these systems are typically very high Q, as is the one shown in Figure 16-10. See also Figures 9-9 (p. 230) and 9-10 (p. 232). High Q has a significant effect on the system's response to the excitation typically provided by cams as will be demonstrated in the next chapter.

Convolution and Deconvolution

When an input function of time such as a cam displacement is applied to a physical system such as a follower train, the input function $x(t)$ is said to be convolved with the impulse response $h(t)$ of the system. The output $y(t)$ is the convolution (*) of the two functions $x(t)$ and $h(t)$ defined mathematically as

$$y(t) = \int_{-\infty}^{\infty} x(\tau)h(t-\tau)d\tau = x(t) * h(t) \tag{16.10}$$

where τ is a dummy integration variable. Its execution can be described in 4 steps:

1 Folding: Take the mirror image of $h(\tau)$ about the ordinate axis.

2 Displacement: Shift $h(-\tau)$ by the amount t.

3 Multiplication: Multiply the shifted function $h(t-\tau)$ by $x(\tau)$.

4 Integration: The area under the product of $h(t-\tau)$ and $x(\tau)$ is $y(t)$.

CONVOLUTION THEOREM The convolution theorem relates convolution in the time and frequency domains. Stated simply, convolution in one domain equals multiplication in the other.

$$x(t) * h(t) = X(f) \cdot H(f) \tag{16.11a}$$

$$X(f) * H(f) = x(t) \cdot h(t) \tag{16.11b}$$

where the operator * denotes convolution and · denotes multiplication. $X(f)$ and $H(f)$ are the Fourier transforms of $x(t)$ and $h(t)$, respectively.

The significance of this relationship is that a complicated mathematical operation (convolution) in the time domain can be accomplished by first Fourier transforming the functions to the frequency domain and then performing a simple operation: multiplication. Thus, if we know the FRF of the system, we know $H(f)$; they are one and the same. If we transform the designed cam function $x(t)$ to $X(f)$, their multiplication yields $Y(f)$, the system response to that cam input function. Inverse Fourier transformation of $Y(f)$ gives $y(t)$, the system impulse response in the time domain. Note in equations 16.11 that the convolution theorem is reciprocal. We can go in either direction. Figure 16-11 shows this theorem applied to a dynamic system. The system is shown as a "black box" having a transfer function in FRF form of $H(f)$ and its inverse transform, the system impulse response $h(t)$. An input function $x(t)$ or $X(f)$ is applied and an output $y(t)$ or $Y(f)$ results.

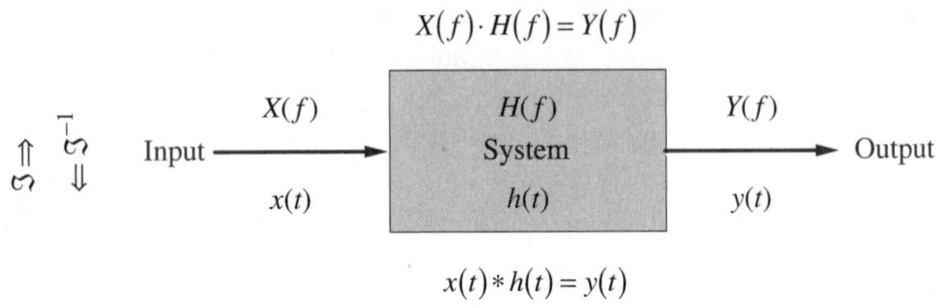

$$X(f) \cdot H(f) = Y(f)$$

Input — $X(f)$, $x(t)$ → $H(f)$ System $h(t)$ → $Y(f)$, $y(t)$ — Output

$$x(t) * h(t) = y(t)$$

FIGURE 16-11

The convolution theorem as applied to a dynamic system

DECONVOLUTION The real power of the convolution theorem lies in its inverse, deconvolution. When two signals become convolved, it is difficult to separate them into their original components within the same domain in which they were convolved. However, transformation to the other domain allows deconvolution to be done by a division operation. Why would one want to, you say? Suppose that you wish to determine the nature of the input function (i.e., the cam) in a machine for which you have no up-to-date engineering information. It is a simple matter to measure the output function acceleration on the end effector in the time domain and transform it to the frequency domain. It is also possible (though not always simple) to get a frequency domain measurement of the FRF of the follower system. We now have two of the three terms of equation 16.9 and can find the remaining one (the input function) by complex division in the frequency domain. Inverse Fourier transformation of this quotient back to the time domain gives the input function in time. Any of the three components can be found by this method when the other two are known or have been measured.

Figure 16-12 shows the results of such an operation. In this case, it was desired to determine the effect that surface finish and cam manufacturing technique had on cam performance.[7] The cam was run in the CDTF of Figure 14-6 (p. 436) and the output acceleration of the follower measured, transformed to the frequency domain and its complex (linear) spectrum stored. Its autospectrum is shown in Figure 16-12a. Note that this contains the input function from the cam convolved with the impulse response of the CDTF. The FRF of the CDTF was separately measured and is shown in Figure 16-12b for a bandwidth equivalent to that of the output acceleration measurement (about 800 Hz). Figure 16-12c shows the cam input function as deconvolved by dividing part (a) by part (b) of the figure. Part (c) represents the frequency spectrum of the cam itself, devoid of the effects of the follower system in which it was tested. The cam's time function could be obtained by inverse Fourier transformation of the spectrum in part (c).

Deconvolution provides a way to mathematically "take apart" a system and determine the contributions to the output of each element or subsystem and how they modify the input function. This is an extremely powerful diagnostic tool when carefully and properly applied. The greatest difficulty lies in getting "clean" measurements of the FRFs of the various parts of a machine, especially in the typically noisy environment of a factory. It can even be difficult to do under carefully controlled laboratory conditions.

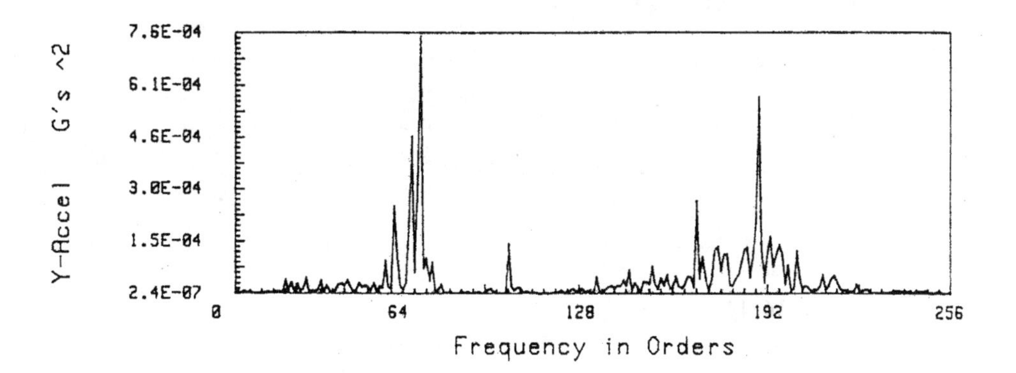

(*a*) Output acceleration of follower train

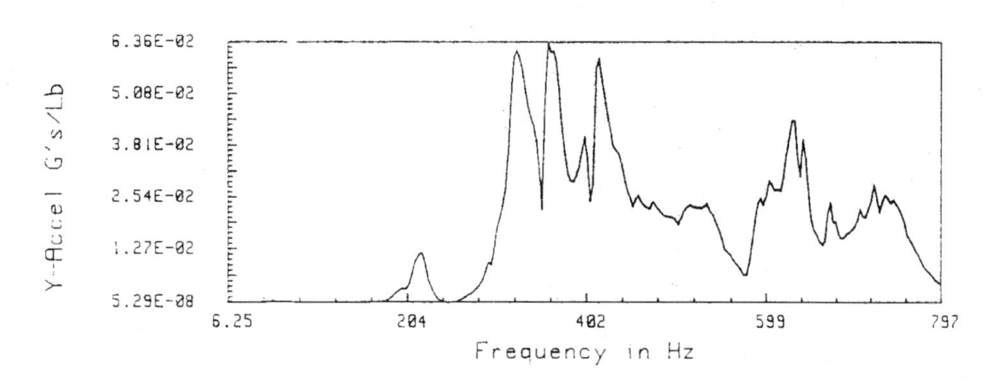

(*b*) Frequency response function (FRF) of follower train

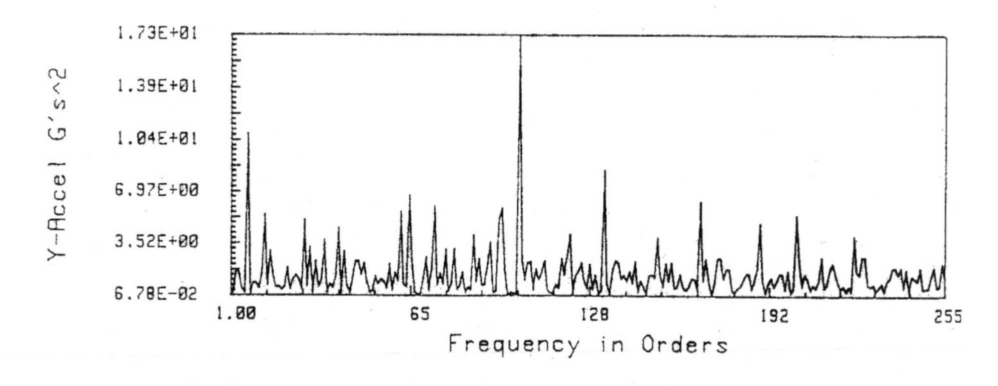

(*c*) Cam input function deconvolved from (*a*)

FIGURE 16-12

A cam input function deconvolved from the output acceleration using the FRF of the follower train [7]

16.4 REFERENCES

1 **Johnson, B., D. Retzke, D. Dewitt, R. L. Norton, and J. R. Hall**. (2001). "Approximation of IC Engine Valve Acceleration from Proximity Probe Displacement Data," SAE Paper 2001-01-0075, SAE International Convention, Detroit, MI.

2 **Norton, R. L., R. L. Stene, J. Westbrook III, and D. Eovaldi**. (1999). "Analyzing Vibrations in an IC Engine Valve Train." Transactions of the SAE. .

3 **Hewlett Packard**. (1986). *The Fundamentals of Modal Testing*, Hewlett Packard Application Note 243-3, p. 23.

4 **Gade, S. and H. Herlufsen**. (1987). "Use of Weighting Functions in DFT/FFT Analysis (Part I)" Bruel & Kjaer Technical Review, No. 3, p. 5.

5 **Hewlett Packard**. (1986). *The Fundamentals of Signal Analysis*. Hewlett Packard Application Note 243, p. 7.

6 **Gade, S. and H. Herlufsen**. (1987). "Use of Weighting Functions in DFT/FFT Analysis (Part I)," Bruel & Kjaer Technical Review, No. 3, p. 7.

7 **Petit, A. E.** (1987). "An Analytical and Experimental Investigation of Cam Shaft Vibration," M. S. Thesis, Worcester Polytechnic Institute.

16.5 BIBLIOGRAPHY

Information on Dynamic Signal Analysis

Brigham, E. O. (1988) *The Fast Fourier Transform and its Applications*. Prentice-Hall.

Ramirez, R. W. (1985) *The FFT, Fundamentals and Concepts*. Prentice-Hall.

Wright, C. P. (1995) *Applied Measurement Engineering*. Prentice-Hall.

Goldman, S. (1999) *Vibration Spectrum Analysis,* 2ed. Industrial Press.

Information on shaft encoders

http://www.gpi-encoders.com/rotinc.htm

http://www.leinelinde.se/

http://www.beiied.com/

Information on LVDTs and LVTs

http://www.macrosensors.com/

http://www.schaevitz.com/

Information on Strain Gages

http://www.entran.com

http://www.blh.de/

Practical Strain Gage Measurements, Hewlett Packard Application Note 290-1, 1999.

Information on Accelerometers

http://www.dytran.com/

http://www.pcb.com/

http://www.endevco.com/

Chapter 17

CASE STUDIES

17.0 INTRODUCTION

This chapter presents two case studies of cam-follower systems taken from the author's consulting experience that proved to be both interesting and challenging. One is an automotive valve train and the other is an application from industrial machinery.

17.1 ANALYZING VIBRATIONS IN AN IC ENGINE VALVE TRAIN*

This study was done to determine what information could be obtained from the analysis of experimentally measured acceleration, velocity, and displacement of the valve in a modern overhead camshaft (OHC) valve train. The measured data were compared to the theoretical cam functions as designed. This information has been reported previously in [1]. Modern OHC IC engine valve trains are designed to operate at high speed for 300 million camshaft revolutions or more with very little maintenance. The displacement functions used for valve motion in the DaimlerChrysler 2.0L inline 4-cylinder, 16-valve engine are "shape-preserving" splines[2]. The acceleration function over the valve motion event shown in Figure 17-1 is a quadratic spline, making the velocity a cubic spline, and the displacement a quartic spline.[3] The two "square waves" of acceleration that precede and follow the motion event are the "ramps" of constant acceleration. The ramp that precedes motion is intended to wind up the compliance in the follower train (principally in the hydraulic lash adjuster) prior to the actual opening of the valve so that the timing of valve opening will be accurate. The ramp following the motion event is intended to control the impact velocity of the valve against the valve seat on closing. Some impact is desired to insure that any particles remaining from combustion are crushed and a good seal is maintained. A poorly sealed exhaust valve will not only compromise performance, but will quickly burn up if it cannot transfer its heat through the valve seat to the cooling fluid every cycle. Studies have demonstrated that these constant acceleration ramps are less than optimal due to their high harmonic content. Other possible shapes are suggested in [4].

* Excerpted with permission from SAE paper number 980570 © 1998 Society of Automotive Engineers, Inc.

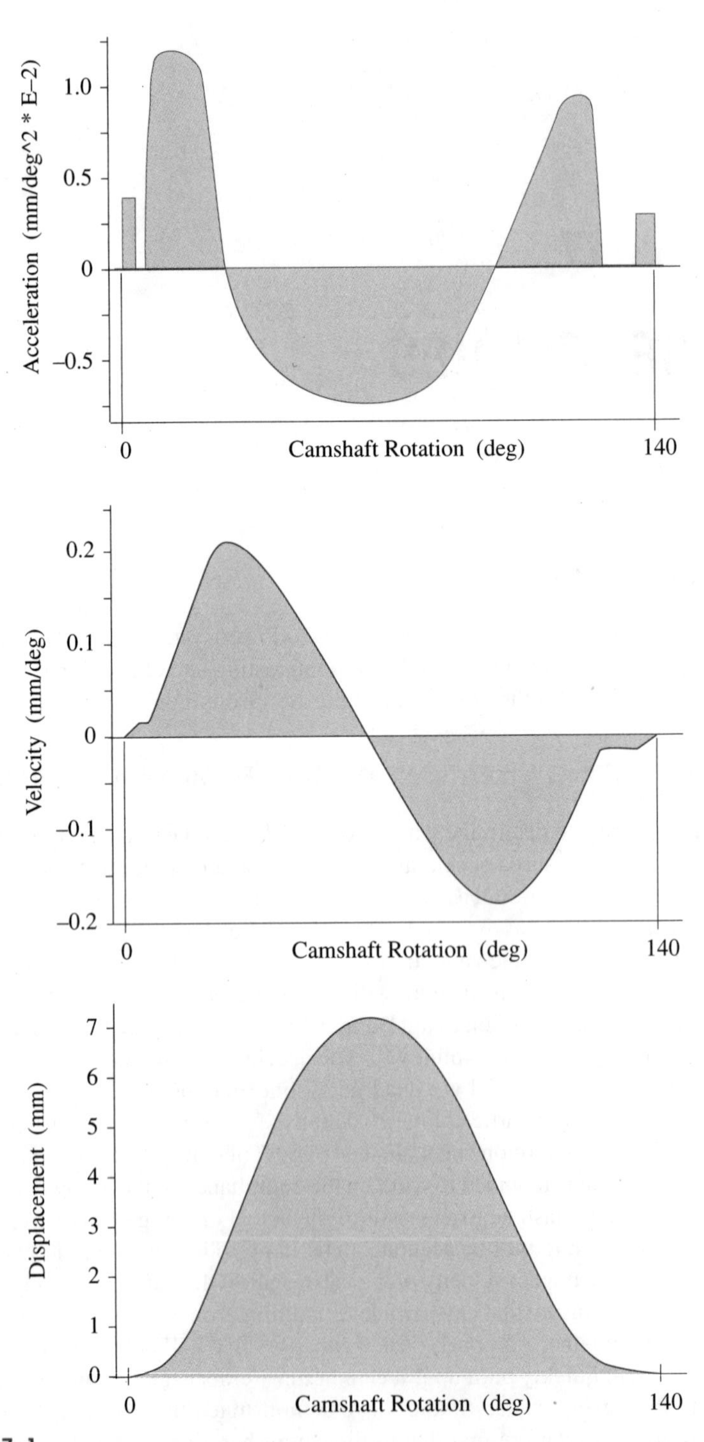

FIGURE 17-1

Valve motion functions as designed [1]

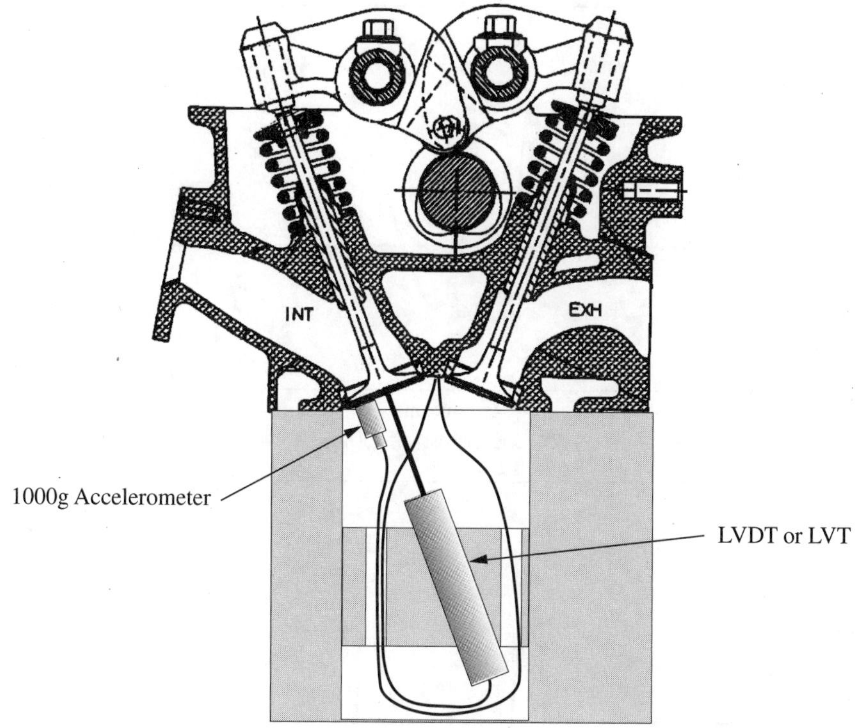

1000g Accelerometer

LVDT or LVT

INT

EXH

FIGURE 17-2

Test setup for measurement of valve dynamics[1]

The test protocol involved instrumenting one valve in an engine that was driven by a *motoring fixture* in which the engine, stripped of pistons and connecting rods, was rotated (motored) by an electric motor at speeds consistent with a firing engine. The empty combustion chamber was then available for the fitting of instrumentation to measure valve motion, as shown in Figure 17-2. All valves but the one being measured were removed to eliminate crosstalk. Two transducers were simultaneously attached (mechanically, by threading) to the head of one intake valve, either an accelerometer and an LVDT, or the accelerometer and an LVT. Mass was removed from the valve to compensate for the added mass of the transducers. Lubricating oil and heated cooling water were supplied to the normal engine passages during the tests. The motoring fixture was capable of running the crankshaft to and beyond the engine's redline speed of 6800 crankshaft rpm with accurate speed control.

For this test, the engine was motored at constant speeds of 500 to 3500 camshaft rpm in 500 rpm increments. At each speed the two transducers mounted (either displacement and acceleration, or velocity and acceleration) were monitored and recorded in both time and frequency domains (either Hz or order in various tests) in an HP35665A DSA. Representative plots of measured time-domain valve acceleration compared to theoretical acceleration for three camshaft speeds (500, 2000, and 3500 rpm) are shown in Figure 17-3,

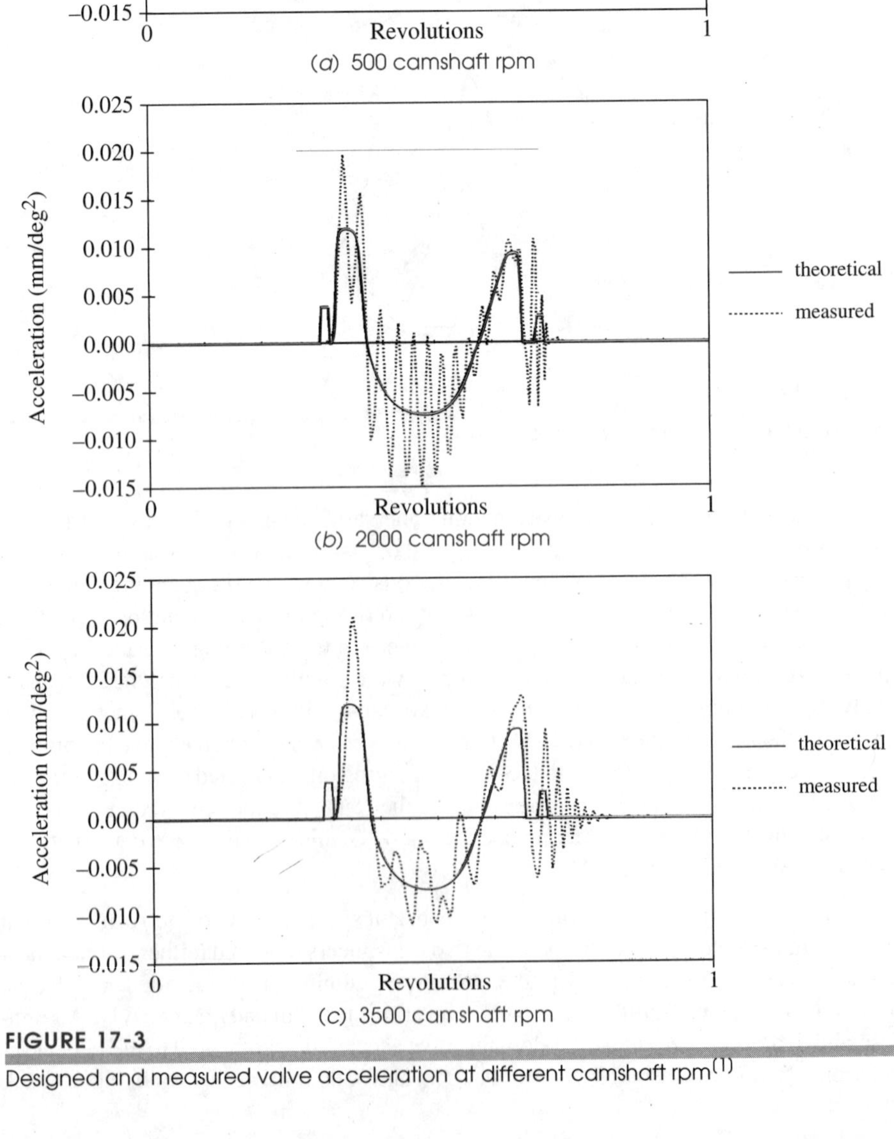

FIGURE 17-3

Designed and measured valve acceleration at different camshaft rpm[1]

all plotted to the same scale. The acceleration curves have been normalized to shaft speed in units of mm/deg^2 for direct comparison of the effects of rotational speed on vibration. Note that at 500 camshaft rpm (1000 crankshaft rpm or idle speed) the measured acceleration is quite close to the theoretical with a small amount of overshoot on the opening pulse and some vibration during the negative acceleration. Ringing due to impact at valve closure is also seen. At 3500 camshaft rpm, or redline speed for this engine, there is significant overshoot on the opening acceleration pulse, almost doubling the theoretical peak acceleration to about 1000g in this test. Significant oscillations seen during the negative acceleration indicate vibratory response of the follower train. At 2000 rpm, the vibration during the negative acceleration phase is significantly larger than at either 500 or 3500 rpm. This is due to an interaction between the harmonic content of the cam function and the structural natural frequency of the valve spring. This will be discussed in detail below.

Figure 17-4 shows the displacement and velocity of the valve as measured at 3500 camshaft rpm compared to its theoretical curves. The corresponding acceleration measurement is shown in Figure 17-3c. Note that the actual displacement very closely follows the theoretical even at this high speed. No valve toss is seen at the peak and no bounce is seen at closing. The follower train is in good dynamic control. The measured velocity lags the theoretical at opening, an effect of the compliance of the follower train, and shows some vibratory behavior during the negative acceleration phase. In general, the velocity is well behaved. Note how the acceleration measurements (Figure 17-3) give more dynamic information than do either the velocity or displacement. The reasons for this were discussed in Chapter 16. Further evidence for this can be found in the spectra of these functions.

Figure 17-5a shows the Fourier order spectra for the theoretical and measured displacement functions to 25 orders. Note that there is little frequency content beyond about 10 orders. Note also the very good correlation between theoretical and measured displacement spectra with only slight gains in the harmonic magnitudes. The velocity spectrum measured at 3500 rpm is shown in Figure 17-5b. Compare its harmonic content to that of the displacement. The velocity shows activity out to about 24 orders.

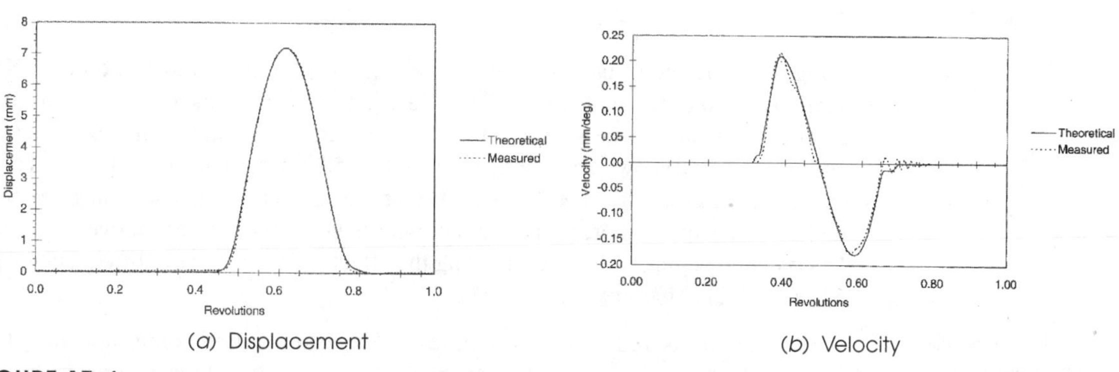

FIGURE 17-4

Designed and measured valve displacement and velocity at 3500 camshaft rpm[1]

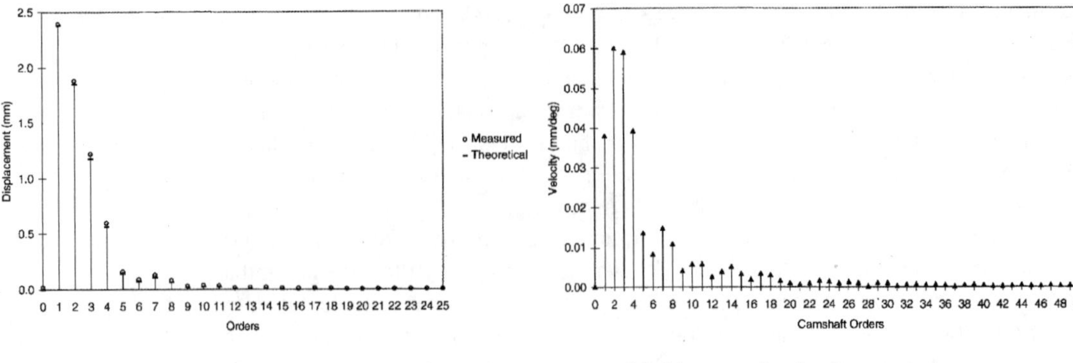

(*a*) Theoretical and measured displacement spectra

(*b*) Measured velocity spectrum

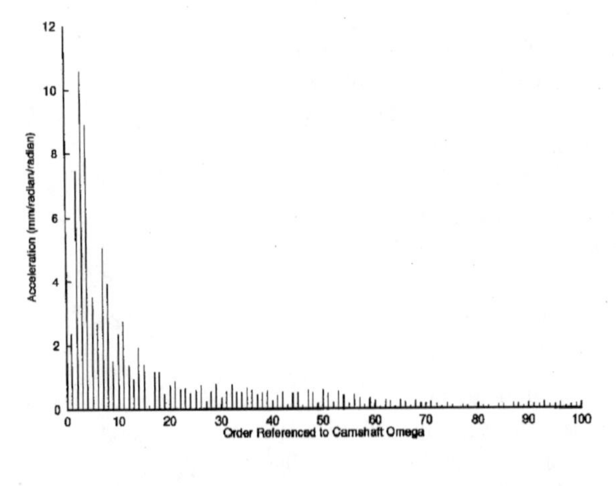

(*c*) Theoretical acceleration spectrum to 100 orders

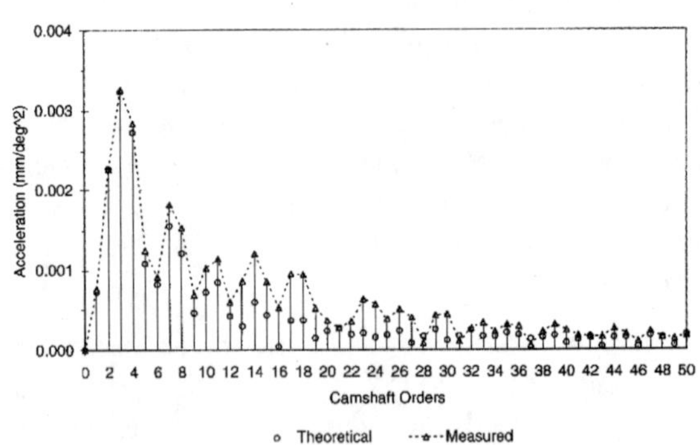

(*d*) Theoretical and measured acceleration spectra to 50 orders

FIGURE 17-5

Theoretical and measured order spectra of cam functions at 3500 rpm[1]

Figure 17-5c shows the theoretical acceleration spectrum out to 100 orders. Note the activity to 50 orders and more. Clearly, acceleration functions have more harmonic excitation available* to energize structural natural frequencies than do the lower derivatives. Figure 17-5d shows a comparison of the first 50 orders of the theoretical and measured acceleration spectra at 3500 rpm. At that speed this is a frequency range of 58 to 2900 Hz. The measured order spectrum closely matches the theoretical over the first 6 orders but then begins to deviate, showing the effects of system natural frequencies excited by the higher harmonics.

All of the measured spectra in Figure 17-5 are shown in the order domain. This makes it convenient to compare them to the theoretical spectra of the cam functions that are defined as harmonics of whatever fundamental frequency the cam may be driven at.

* Due to the fact that the acceleration spectrum is the displacement spectrum multiplied by ω_2.

But for comparison of measured spectra to structural natural frequencies of the follower system, it may be more convenient to generate spectra in the frequency domain as shown in Figure 17-6a. This shows the measured valve acceleration spectrum at 3500 rpm on a frequency axis from 0 to 3.2 kHz. In this form, it can be directly compared to the FRF of any part (or of the whole) of the follower train that has been independently measured. Figure 17-6b shows such a measurement. The valve spring is the "softest" mechanical element in this system and so has the lowest natural frequencies and largest deflections of any of the other parts (rocker arm, valve, etc.) The spring was removed from the assembly and placed in a specially designed, low mass fixture that compressed it to the mid-range of its working length. Spring and fixture were mounted to a shaker and excited with a broadband periodic chirp function. The resulting spring FRF function is shown in Figure 17-6b. It shows natural frequencies at about 648, 808, 984, 1010, 1270, 1340, 1400, and 1750 Hz. The lowest of these is the first longitudinal mode. Others are higher longitudinal, transverse, and torsional modes.

Note that the need for a fixture to hold the spring having boundary constraints consistent with its assembled condition necessarily alters the resulting FRF and so the result can only be considered an approximation to the true function. It is actually quite challenging and sometimes very difficult to get good quality FRF measurements of parts with realistic boundary condition because the test apparatus always alters the system in some way. Finite element analysis (FEA) can also be used to obtain an estimate of the natural frequencies of any part. It is a difficult task to establish realistic boundary conditions within the FEA model that accurately represent the real constraints on the part. Errors in boundary conditions can make large differences in the computed natural frequencies. The best approach is usually a combination of FEA and experiment, each used to corroborate the other.

Figure 17-6a represents the output acceleration function of the system. Figure 17-6b represents the FRF of a portion of the system. Together, these constitute two terms of equation 16.9 (p. 491). The third term, essentially the input function, can be found by deconvolution, i.e., dividing the spectrum in part (*a*) by the FRF of part (*b*) to obtain the deconvolved spectrum shown in Figure 17-6c. The result shows the effects of the remaining elements in the follower train on its vibratory behavior in the absence of the spring. This process can be repeated, dividing out the FRF of each element in turn (rocker arm, valve, etc.) until all that is left is the input function of the cam. In the process the individual contributions of each element to vibration can be discerned.

Figure17-6c approximates the forcing function on the system, but it is still colored by the presence of the FRFs of other valve train elements such as the rocker arm, rocker shaft, etc. Only the effects of the spring have been approximately eliminated in this computation. The harmonic pattern of the cam function can be seen up to about 500 Hz, which is 9th order at this speed. Significant activity is seen at 576 Hz and 1.52 kHz which appear to be natural frequencies of other structural elements in the system since they do not order track with changing rpm. Note that the 576 Hz frequency cannot be attributed to the valve spring as its contribution has been removed from these data by the deconvolution process. Moreover, the spring's lowest natural frequency is 648 Hz not 576 Hz. Separate FRF hammer tests of the camshaft and the rocker arm showed the pres-

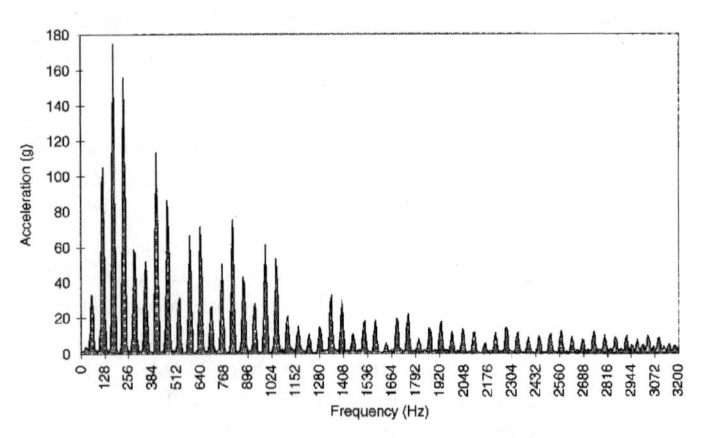

(a) Output acceleration spectrum in frequency domain

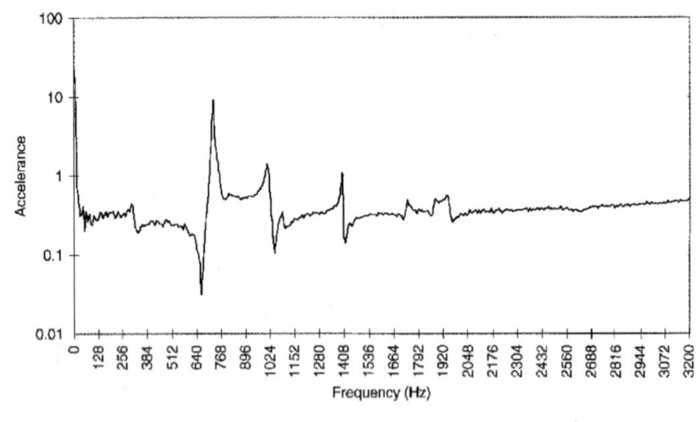

(b) FRF of valve spring

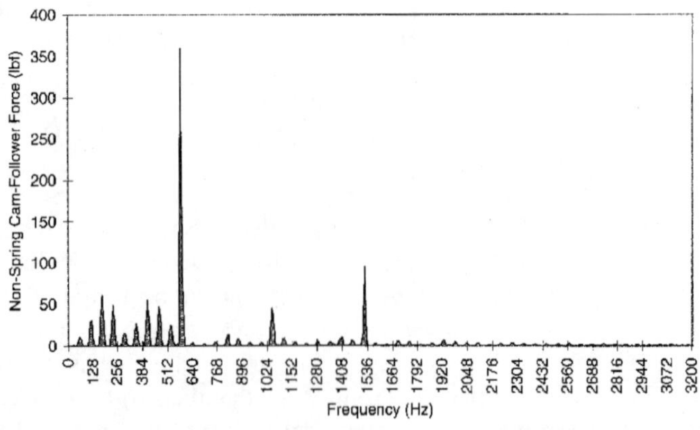

(c) Spectrum of acceleration with spring deconvolved

FIGURE 17-6

Output spectrum, FRF, and their deconvolution[1]

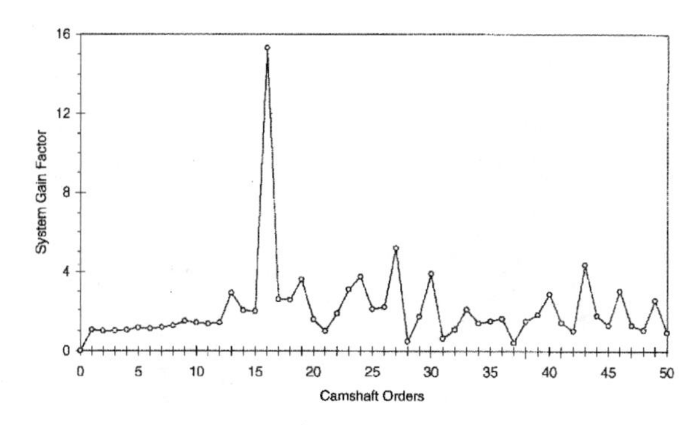

FIGURE 17-7

Deconvolution of the theoretical acceleration spectrum from the output at 3500 rpm[1]

ence of natural frequencies at about 540 Hz and 1.48 kHz in the camshaft and 1.49 kHz in the rocker arm, which roughly correspond to the frequencies seen in this deconvolution.

Just as an FRF can be deconvolved from an output function to reveal the underlying input function, if available, a spectrum of an input function can be deconvolved from an output function to reveal the system FRF. In this case, we have the theoretical spectrum of the cam acceleration (Figure 17-5c) in the order domain and we have the output acceleration measured in the order domain at various speeds (Figure 17-3). Figure 17-7 shows the result of deconvolving the theoretical acceleration spectrum (Figure 17-5c truncated to 50 orders) from the output acceleration at 3500 rpm (Figure 17-3c). Here the most prominent feature is a 16th-order spike which order tracks, indicating that it is due to a rotating element in the system.

Figure 17-8a shows a waterfall plot of the measured acceleration frequency spectra over a fixed bandwidth of 3.2 kHz as the rpm is increased from 500 to 3500. Here, vibrations of structural components, such as the spring, appear always at the same frequencies and the cam acceleration orders will sweep across the spectrum with changes in speed. So, if the effects of the structural components are of prime interest, this approach has an advantage. In this case, the first three harmonics of the valve spring are in the 500 to 1500 Hz range.

At 500 rpm, the cam acceleration harmonic information ends at about 167 Hz (20th order) and beyond that frequency, little energy is present. At 2000 rpm, 20th order is 667 Hz and the cam order spectrum can be seen below that value in Figure 17-8a. The spring activity is quite apparent within its bandwidth as the energy level is now sufficient to excite it. There is a valley in the spectrum around 700 Hz between the cam and spring activity that have not yet coupled at this speed. At 3000 rpm, significant coupling between the cam and spring's harmonics occurs. The cam's 20th order is now at 1000 Hz and well into the spring's natural frequency regime. The coupling between spring and cam is evident at the higher shaft speeds as the cam spectral activity sweeps up in frequency.

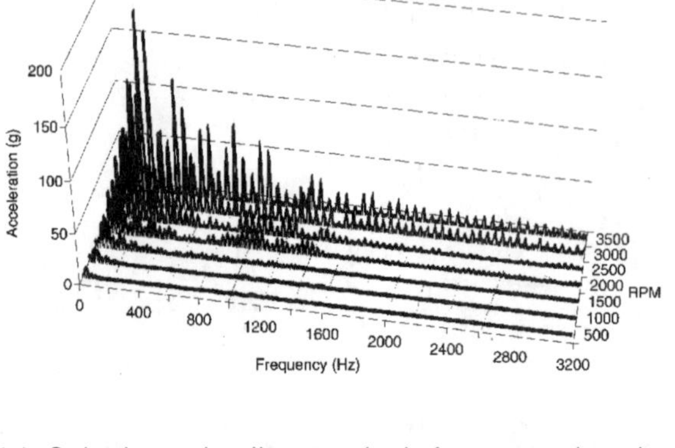

(*a*) Output acceleration spectra in frequency domain

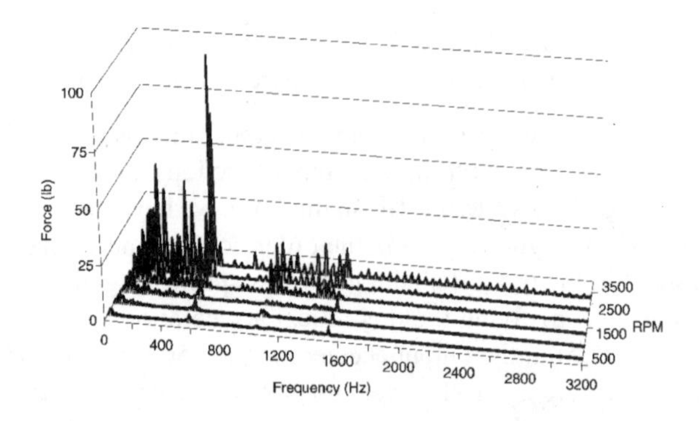

(*b*) Acceleration spectra after deconvolving spring FRF

FIGURE 17-8

Waterfall plots of acceleration frequency domain spectra as a function of rpm [1]

Figure 17-8b shows a waterfall plot of the deconvolution of the spring FRF from the acceleration spectra over the range of tested camshaft speeds. A division of an output function in acceleration units by an FRF in accelerance units yields an input function in units of force, i.e., the input forcing function to the system. However, since we are only deconvolving a portion of the system's complete FRF in this computation, the resulting units should not be considered to have correct magnitudes, but rather be used for relative comparison only.

In these deconvolved spectra, the cam frequencies are still present as they should be. However, two frequencies show quite clearly in Figure 17-8b and are present in all other data as well, indicating their structural nature. The spike at 576 Hz is quite strong and couples with cam excitation at the higher speeds. The spike at 1.52 kHz is as strong as the one at 576 Hz in the 500 rpm measurement but does not grow in magnitude with speed as much as the lower frequency one. This is probably due to the lower level of excitation available from the cam at the higher frequency that corresponds to 26th order at 3500 rpm. The cam has very little energy at that harmonic.

Figure 17-9a shows a waterfall plot of the acceleration order spectra as the camshaft rpm is increased from 500 to 3500 rpm. At 500 rpm, there is little activity because 50 orders at that speed is only 417 Hz, below the spring's first natural frequency. At 2000 rpm, a great deal of activity is seen in the regime between 20 orders (667 Hz) and 50 orders (1667 Hz). This is the range of valve spring natural frequencies as measured (see Figure17-6b). At 3000 rpm, the first spring frequency occurs at about 13th order where the cam spectrum is still active. This creates coupling between the cam harmonics and the spring, resulting in larger vibrations.

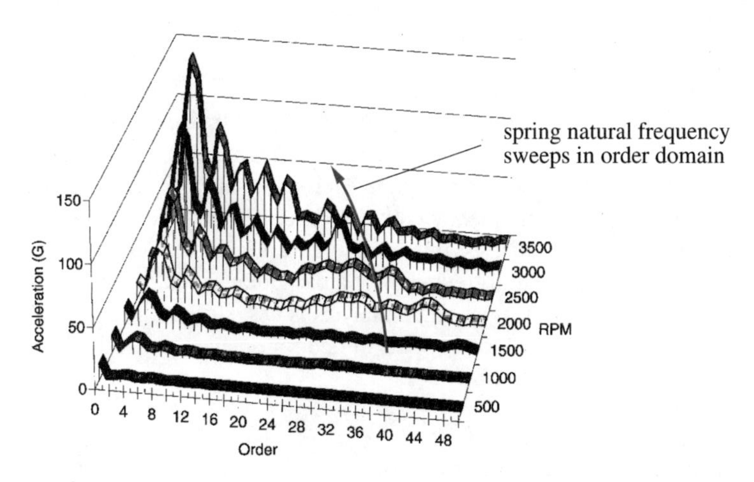

(a) Output acceleration spectra in order domain

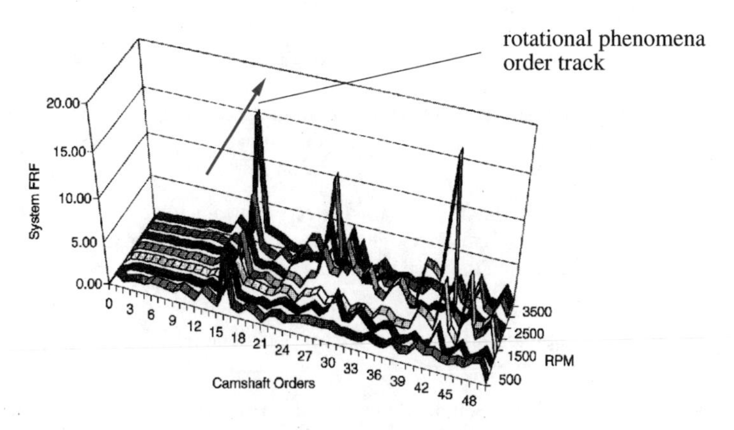

(b) Acceleration spectra after deconvolving input order spectrum

FIGURE 17-9

Waterfall plots of acceleration order domain spectrra as a function of rpm[1]

17

Note that the energy contained in the cam harmonics (to about 20th order) increases with the square of the speed, but remains in "order" because it is rotational in nature. But, information related to structural vibration, which is fixed in frequency, sweeps across the order spectrum as speed increases. For example, the hump at 40th order in the 2000 rpm spectrum corresponds to about 1333 Hz, which is close to one of the spring's natural frequencies. At 2500 rpm, this frequency shows up at 32nd order, at 3000 rpm about 27th order, and at 3500 rpm about 23rd order. The first natural frequency of the spring (648 Hz) can be seen coupling with the cam theoretical spectrum at about the 15th order at 2500 rpm, 13th order at 3000 rpm, and 11th order at 3500 rpm.

The cam input function was deconvolved from the measured acceleration spectra for each tested speed by dividing each 50-order output acceleration spectrum by the first 50 orders of the cam input order spectrum. The results are shown as a waterfall plot in Figure 17-9b. This purports to be the FRF of the entire cam-follower system sans cam lobe. Camshaft runout or imbalance will order track. Structural elements such as the spring or rocker arm will frequency track.

A distinct 16th order spike is apparent in Figure 17-9b and it order tracks. Spring activity can be seen sweeping across the order spectrum in an apparently haphazard way from high order at low speed to mid order at high speed. The delta Hz between orders increases with each test speed and ranges from 8.3 Hz at 500 rpm to 58.3 Hz at 3500 rpm. The "giant steps" in frequency between orders at the higher speeds cause some spring frequency spectral data to be missed in Figure 17-9b. Unless an order happens to coincide with a spring natural frequency at a particular speed, it will not appear in the order spectrum. Not surprisingly, order analysis is best suited for detecting rotating phenomena, and frequency analysis is better suited for detecting structural phenomena.

Conclusions

This study was undertaken primarily to demonstrate some measurement and analysis techniques that can be quite useful in the design of high-speed rotating machinery such as the valve-train system.

1 Displacement measurement will always be of value to determine the presence of system separation and to accurately measure valve lift at various cam speeds. However, in general, measurement of displacement in high-speed, underdamped mechanical systems is of little dynamic value because displacement functions and transducers typically have bandwidths too limited for effective vibration detection.

2 Acceleration measurement is the most effective means to determine the vibratory response of this type of mechanical system due to its combination of high bandwidth in both the theoretical acceleration function and in the typical piezoelectric accelerometer.

3 The most effective way to analyze the dynamic behavior of valve trains requires both time and frequency domain information. Either one alone tells only part of the story. Both should be utilized.

4 In analyzing rotating machinery, there are distinct advantages to be gained by looking at frequency-domain data as both frequency (Hz) spectra and as order spectra. Rotating machinery is unique in that it can possess high energy levels due to the rotating elements and also high energy levels due to vibrating structural components such as springs. Each is revealed most clearly in its own subdomain of orders or Hertz.

5 The value of the FRF in analyzing the dynamic behavior of a system cannot be overstated. With modern analysis equipment it is relatively straightforward to measure many systems' FRF. However, it can be difficult to get good quality FRF measurements in cases where the system is mechanically complex or noisy. If obtainable, they will provide very useful information.

6 The power of the deconvolution technique for "taking apart" measurements of dynamic systems to reveal hidden flaws or behaviors has been demonstrated.

17.2 ANALYZING VIBRATIONS IN CAM-DRIVEN AUTOMATED ASSEMBLY MACHINERY*

Many of the consumer products that are available in our economy are produced at high speed on fully automated machinery, often with minimal human intervention. Production rates as high as several hundred parts per minute (PPM) are now fairly common. Most of the operations on the product as it passes through the machine are performed by cam-follower systems, often driving linkages that take the motion from the cam to the tooling at the end effector where an operation is performed on the product. These machines divide into two general categories, intermittent motion and continuous motion. Intermittent motion, or indexing, machines, carry the product through the cycle on a timing belt, timing chain, a series of dials, or other mechanism. The parts to be assembled are typically fed by vibratory bowl feeders, blown through tubes, or fed from preloaded magazines and loaded into nests attached to the belt, chain, or dial. The cam driven tooling escapes the parts from the feeder mechanisms or slices them from a magazine and places them in the nest. Successive parts in the assembly are added, typically one per station, until the assembly is complete and the finished product is off-loaded, also by cam-driven linkages.

In an indexing machine, the belt, chain, or dial can be moved between stations by a cam of the type shown in Figure 13-15 (p. 419) and held stationary in dwell while the tooling cams perform their tasks at each station. In continuous motion machines, the belt, chain, or dials never stop, and parts must be fed and operations performed on the product while in motion. Much higher throughput speeds are possible using continuous motion as there is no time wasted in accelerating and decelerating the nests. But, this adds complexity to the tooling mechanisms that now must either chase the moving product or ride with it on the conveyor or dial. Rotary dials offer some advantage over linear belts and chains in continuous motion machines because tooling can be mounted to the dial and the parts passed tangentially from dial to dial or from conveyor to dial. With tooling mounted to a continuously rotating dial, the cams that activate the tools can be stationary while the followers pass over them and operate on the product captured in the dial-mounted nests. While the parts are on the dial there is, of course, no relative motion between them and the tools except for the programmed operational motions desired.

The machine in this case study was a continuous motion, dial type machine with multiple tools per dial. The product passed through the machine on carriers that were transferred tangentially from dial to dial. There were a number of dials, some of which served as idlers to pass the product carriers between tooling dials. Each dial was driven by its own servomotor through a gear reduction unit. The cams to activate the tooling

* This case study is presented with permission of The Gillette Company, Boston, Mass.

were stationary beneath or above the dial and the followers on the dials' multiple tools ran around the cams.

The author was asked to investigate the reason for severe vibrations present in some dials as the machine was brought on line with its speed being periodically increased in arbitrary (but regular) increments on the way to its design speed of several hundred PPM. At about half the design speed, the vibrations had become so severe that some were wondering if a practical limit had been reached and if it would be possible to reach the design speed.

No provision in the machine's design had been made for the inclusion of instrumentation to measure cam system performance (i.e., vibration). Serendipitously, a location was found where vibrations in the reaction torque from the cam motions in any dial could be conveniently measured with an accelerometer. Figure 17-10 shows a schematic cross section of a dial drive assembly. By design, the motor reaction torque was taken back to the bed plate of the machine through a flanged, cantilevered, standoff to which the motor and gearbox were mounted. An accelerometer placed near one corner of its flange measured the vibrations induced by the action of the cam-follower systems in the dial

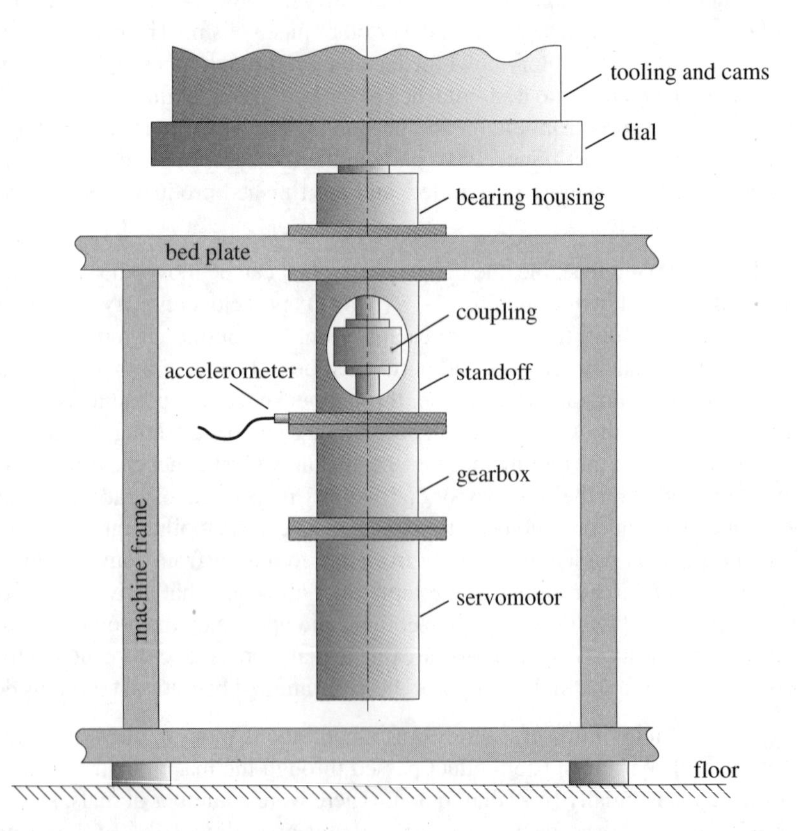

FIGURE 17-10

Schematic of dial drive assembly

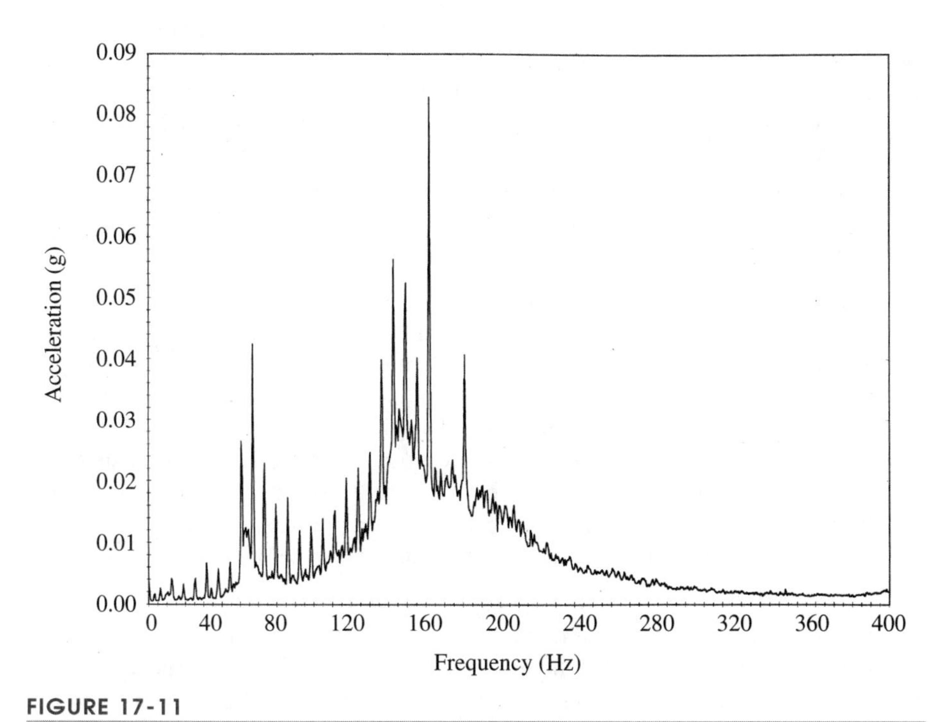

FIGURE 17-11

Autospectrum of output acceleration with machine running

above. Figure 17-11 shows a typical 400-Hz bandwidth acceleration spectrum as measured at that point. The signal level was quite low and, of course, was contaminated with energy from all the other dials in the machine as well as by the general noise present in the factory floor. Nevertheless, the energy from the dial to which the accelerometer was attached proved to be 2-3 orders of magnitude above background noise levels. Also, large numbers of averages (50-100) were taken to improve the signal to noise ratio.

Figure 17-11 shows some very interesting clues to the machine's behavior. There is an underlying "mountainous" shape across the spectrum out of which a series of uniformly spaced "spikes" grow. The underlying shape is indicative of the transfer function or FRF of the machine at this location and the spikes are the collective cam/dial harmonics superposed on it. This spectrum represents the output function in equation 16.9 and so contains the convolution of the machine's FRF at this location and the input function of the collective set of cams that drive the tooling in this dial (there were typically several cams per dial). If we could somehow get an independent measure of the FRF of the dial at this point, then deconvolution would reveal the spectrum of the cam set's input function.

Several attempts were made to measure the FRF of the dial with no success. Hammer tests proved futile, since the energy available from the hammer hit was swamped by the background noise in the factory (which for some reason could not be shut down to do this test), and the resulting coherence was unacceptable. The machine was too large to effectively use a shaker approach. However, the level of background noise in the factory floor proved to be a boon as well as a bane in that it was sufficiently random (given the hundreds of asynchronous machines running at any time) to excite the machine when not running and give a reasonable estimate of its FRF at the measuring point. Thus the "FRF" was measured by shutting the particular machine down and recording the acceleration due to the "random" noise imparted to the machine from the factory floor. In effect, we used the factory floor as the "shaker" to excite the machine's natural frequencies with (very) pseudo-random noise. Nevertheless, it worked well enough.

Figure 17-12 shows the machine's "FRF" measured by this method at one location. It is not a true FRF because it is in autospectrum form and thus has no phase information. Nevertheless it turns out to be useful. There are several natural frequencies apparent in the figure; the 10 Hz peak is the shaft's first torsional natural frequency, confirmed by independent calculation. There are peaks at about 27, 39, 44, 47, 50, 55, 62, 70, 140, and 197 Hz. These are natural frequencies of various mechanical elements or substructures in the system. There are also large peaks at exactly 60 and 120 Hz. These are artifacts from the line frequency of the electrical current supplied to the factory by the power company.* These were later notch filtered out of the data before doing any computations.

Another thing that is apparent from the "FRF" in Figure 17-12 is the high "Q" of its natural frequencies. There is very little internal damping. This is no surprise since the machine is made almost exclusively of steel, bolted or welded together with no vibration absorption material used anywhere because of the high accuracy of assembly need-

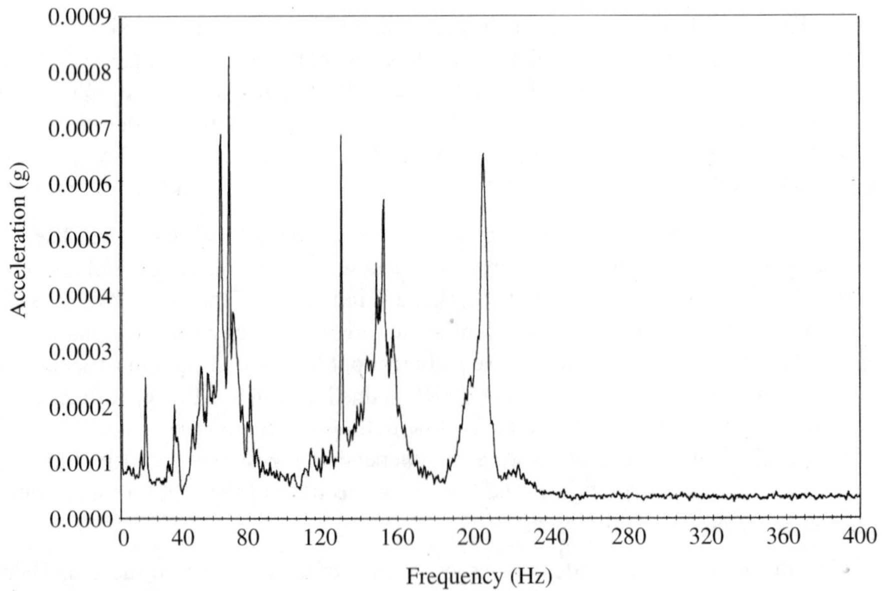

FIGURE 17-12

Autospectrum of background acceleration with machine off

* The same test done on the same type of machine in Europe showed these artifacts at 50 and 100 Hz due to the line frequency there.

ed. The high Q of the system coupled with the inherently sharp nature of the cam's forcing function harmonics makes the machine's vibratory response very sensitive to speed as will be described below.

The next step was to deconvolve the FRF from the measured output to get an estimate of the input function to this dial by dividing the data in Figure 17-11 by that of Figure 17-12. The result is shown in Figure 17-13. Since we are dividing acceleration by acceleration in this case (due to our unconventional measure of the FRF) rather than dividing acceleration by accelerance, the result in Figure 17-13 is expressed as a gain ratio which represents the distribution of spectral energy in the collective set of cams and other rotating elements such as gear trains in this dial system. Note the sharp, evenly-spaced spikes. These are the rotating system's harmonics.

The data in Figure 17-13 can be easily converted to the order domain because the speed at which the machine was run to generate the data from Figure 17-11 used to calculate Figure 17-13 is known. Figure 17-14 shows the harmonics, or Fourier coefficients, of the rotating elements (cams, gears, bearings, motor, etc.) in this dial obtained by selecting the values from the data in Figure 17-13 that correspond to integer multiples of the fundamental shaft speed. With this information, we now have a good empirical measure of the forcing function for this dial. Note the high harmonic content. The largest harmonic is the 27th and there is measurable activity to 50 harmonics, or orders. This means that any natural frequencies present in the system are subjected to excitation when

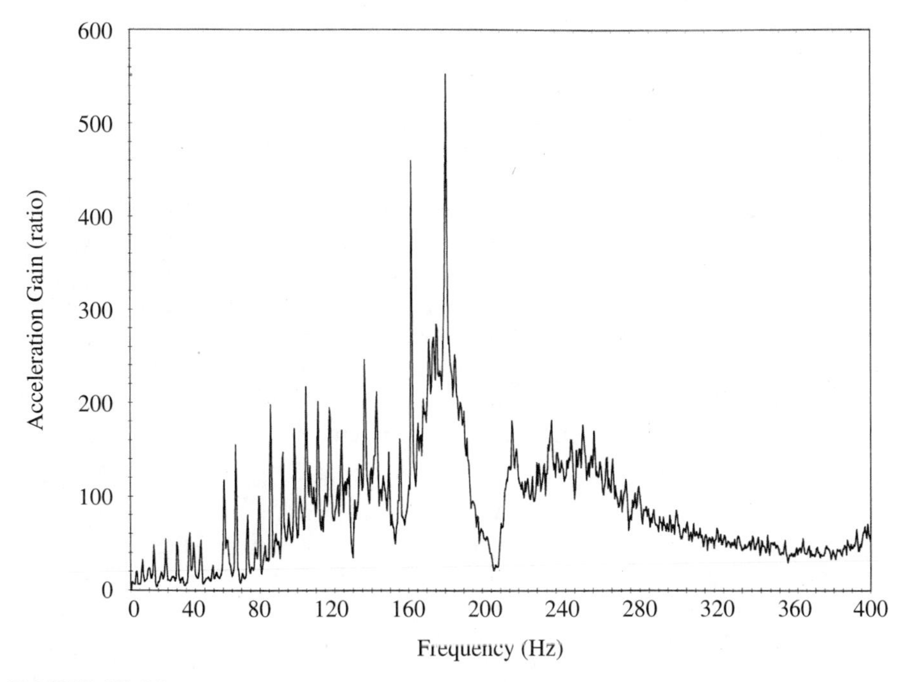

FIGURE 17-13

Input forcing function from all cams in dial shown as acceleration gain over background

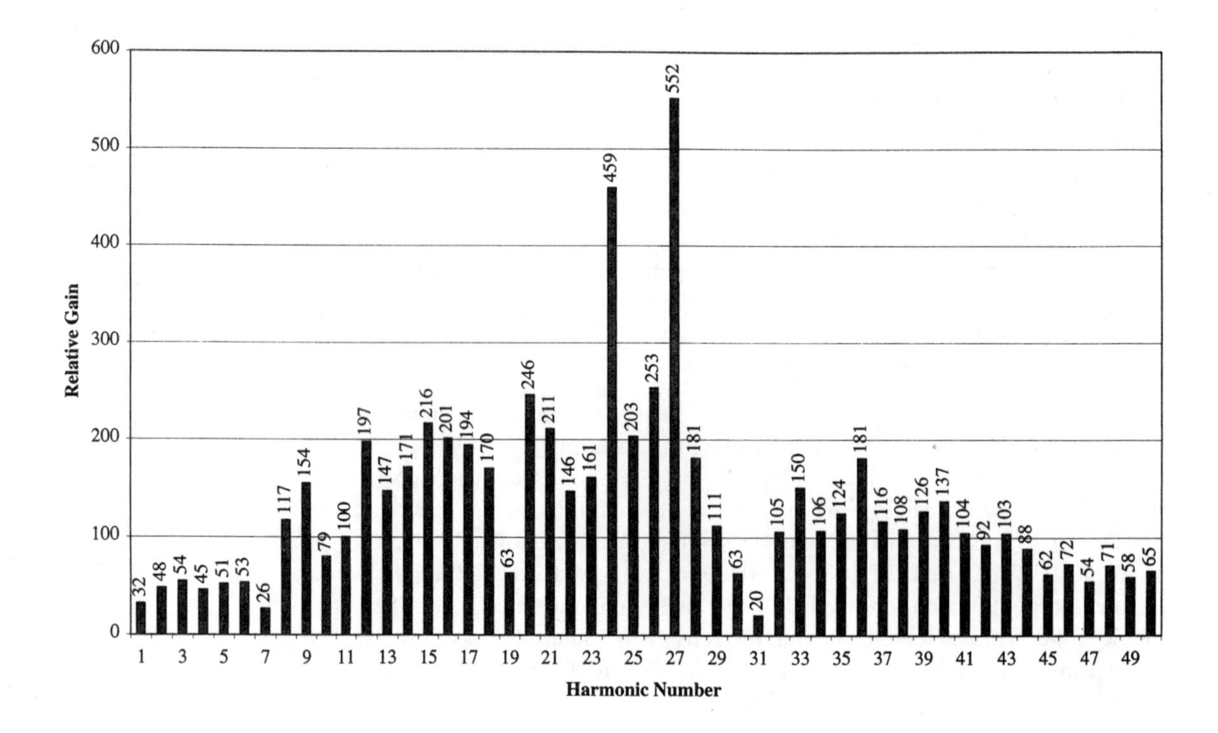

FIGURE 17-14

Fourier coefficients of forcing function

any of these harmonics of the rotating elements happens to land on top of them at a given machine speed.

At this juncture, we have an experimentally derived measure of the forcing function for this dial (Figure 17-14) that is independent of machine speed because it is in terms of harmonics or orders. We also have a crude but effective measure of the natural frequencies of this dial (Figure 17-12). These constitute two of the terms in equation 16.9, namely the input function and the FRF at that location. Their multiplication gives a predicted output of the system for any machine speed we choose. This provides a model of the system and its particular forcing function with which we can simulate the behavior of the machine over any range of operating speeds desired.

A computer program was written to "run" the machine over a range of speeds of interest, ranging from the speed at which it was tested up to or beyond the desired design speed. This program computed the "hits" between the cam harmonics (Figure 17-14) and the natural frequencies (Figure 17-12) at each speed and weighted them according to two gain factors: the ordinal value in Figure 17-14 for the particular harmonic that "hit" a natural frequency at that speed, and the magnitude of the natural frequency that was hit.

This process can be depicted in a Campbell diagram as shown in Figure 17-15. A Campbell diagram plots frequency against frequency, here machine rpm against FRF fre-

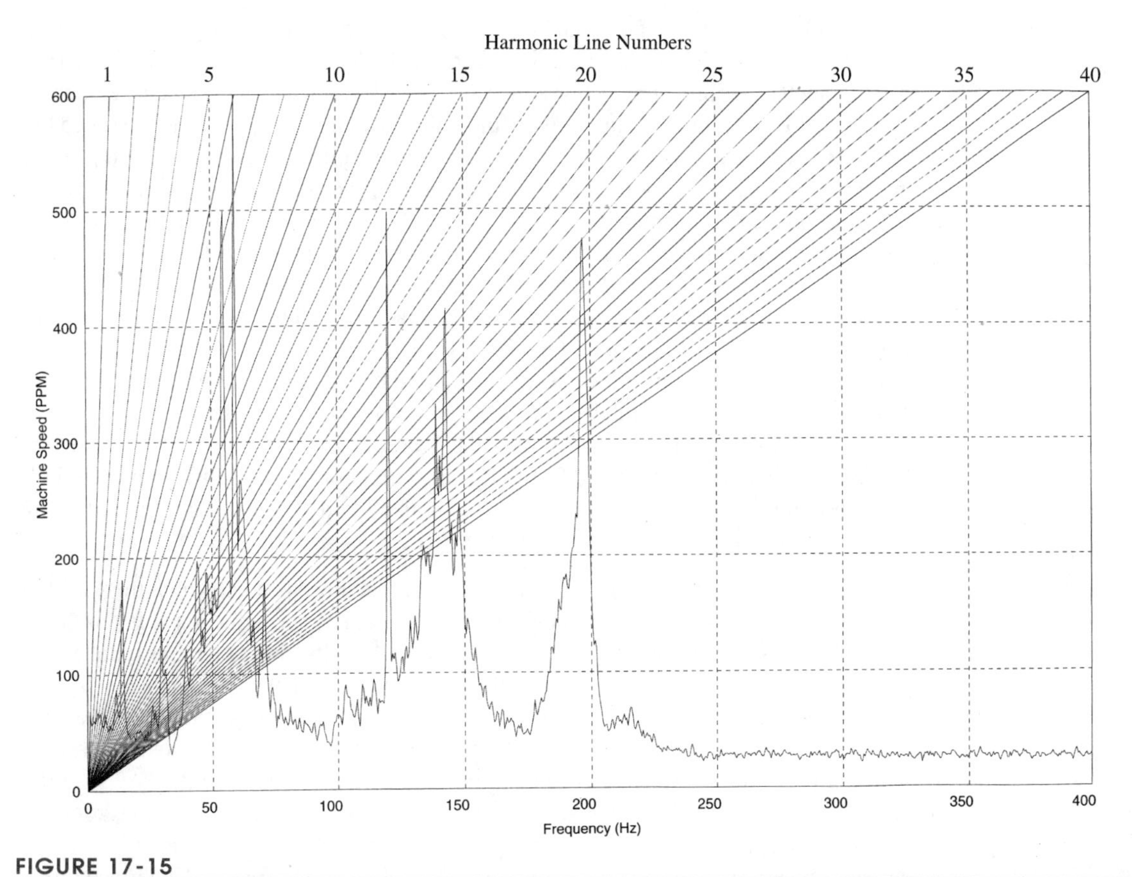

FIGURE 17-15

Campbell diagram showing the intersections of harmonic lines and natural frequencies

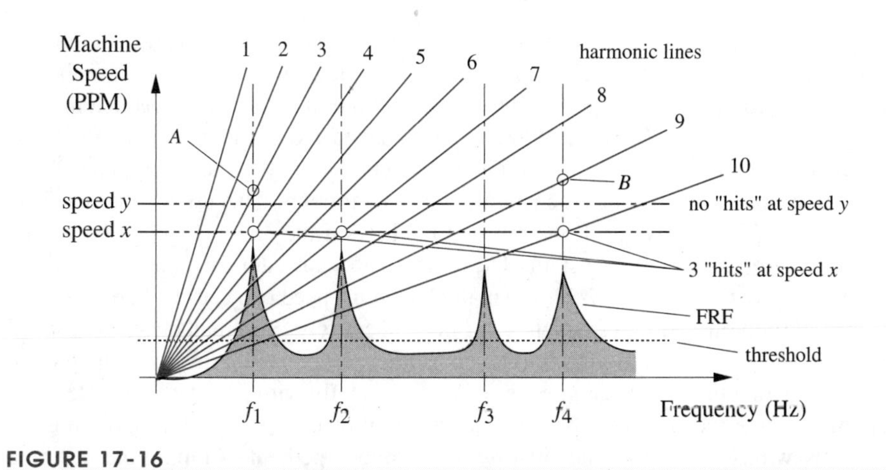

FIGURE 17-16

Interpreting the Campbell diagram

quency in Hz. The "fan" of diagonal lines depicts the harmonics of the forcing function. These overlay the FRF of the system. Figure 17-16 (p. 517) explains how to interpret the Campbell diagram. Draw a horizontal line across the diagram at any chosen operating speed. Look for intersections among three things along that speed line: the speed line itself, any harmonic line, and any natural frequency spike (extended to the top of the chart). At any such 3-way intersection, that harmonic of the cam will be exciting that natural frequency at that machine speed. The level of excitation will be a function of the product of the magnitude of the natural frequency spike from Figure 17-12 and the magnitude of the harmonic from Figure 17-14 that are coupling at that speed. In Figure 17-16, if the machine is run at speed *x*, then there are three "hits" of harmonic lines with natural frequencies. Harmonic 4 hits natural frequency f_1, harmonic 7 hits f_2 and harmonic 10 hits f_4. If the machine speed is increased to speed *y*, then there are no hits between any harmonic lines and the 4 natural frequencies shown. If the speed is increased above *y*, at some point harmonic 3 will strike f_1 at *A* and at a slightly higher speed, harmonic 9 will strike f_4 at *B*, and so forth. It is also necessary to establish some threshold as a minimum above which hits will be recognized and below which ignored. This threshold should be above the valleys in the FRF.

While it is theoretically possible to find some machine speed at which there will be no intersections between any harmonic lines and any natural frequencies, given the 50 harmonics shown in Figure 17-14 and the number of peaks in the FRF in Figure 17-12, it is quite unlikely in this case. What will occur is that some machine speeds will have fewer and some will have more intersections between harmonics and natural frequencies. Also, the collective energy or sum of all the "hits" between harmonics and natural frequencies will vary with speed. The computer simulation model essentially tests every speed (in 1 PPM increments) between selected minimum and maximum values for these intersections, applies the appropriate weighting factors for harmonic and natural frequency magnitudes, and sums the collective hits for each machine speed. This procedure is repeated with data from each of the dials measured in order to obtain an overall map of the relative vibration that can be expected from the coincidences of harmonics and natural frequencies at each simulated machine speed.

Figure 17-17 shows such a map for this machine, calculated for a large number of dials, essentially the entire machine, involving multiple dozens of cams. Note the extreme sensitivity to machine speed of the overall vibration level. Look, for example, at the transition between 570, 571, and 572 PPM. From 570 to 571 PPM the vibration level on an arbitrary scale drops from 2 to about 0.2, a factor of 10. One additional PPM of speed jumps it back up to about 1, changing it by a factor of 5. This is exquisite sensitivity to speed. Why is this so? Look again at the FRF of the dial in Figure 17-12. The peaks are sharp; it is a high-Q system. The harmonics of any cam are inherently sharp as shown in Figures 17-13 and 17-14. A small change in speed can cause a harmonic line either to sit right on a natural frequency or fall off it entirely. So, small changes in speed make big changes in overall vibration levels. Other tests done on this machine proved conclusively that these vibration levels correlated directly with product scrap rate. The mapping of the good and bad vibratory speeds has allowed these machines to run more efficiently with lower scrap rates just by the simple expedient of tuning the machine speed to one that shows low overall vibration levels.

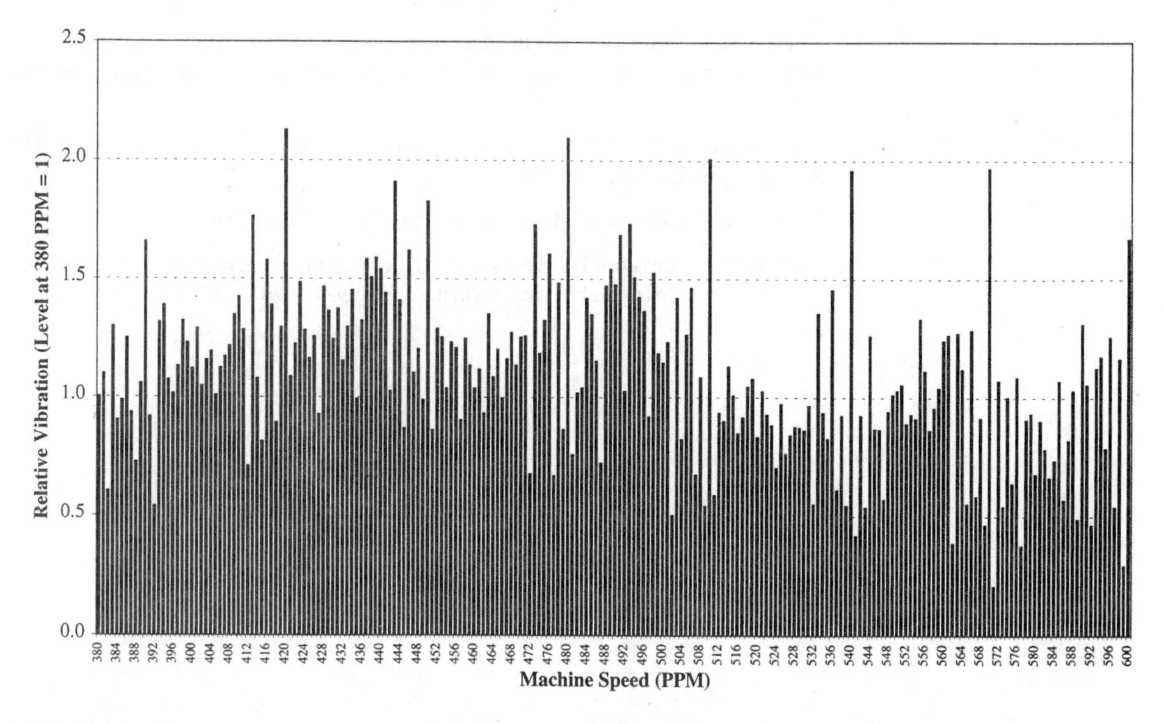

FIGURE 17-17

Overall vibration levels in a machine as a function of operating speed

There are several of these machines in production. While theoretically identical, there are individual differences that make the vibration/speed map of Figure 17-17 slightly different for each machine. Also, whenever engineering upgrades are applied that change the mass or stiffness of any components, or change cam designs, a remapping of the machine's good and bad speeds is necessary. This study and analysis has also been performed on indexing type machines with similar results, leading to the conclusion that this may be a universal phenomenon in cam-driven machinery.

Conclusions

1 Cams are known typically to have significant harmonic content in their acceleration functions. The more cams that a machine has, the more dense will be the overall harmonic excitation.

2 Industrial machinery is usually structurally high-Q, having very low damping.

3 The combination of 1 and 2 can create a situation in which the overall vibration levels in a machine can be extremely sensitive to operating speed. A speed change as small as a fraction of an rpm has been shown to have a measurable effect on vibration levels.

4 Using conventional frequency domain measurement techniques and deconvolution, it is possible to map the "good" and "bad" speeds of operation of a particular machine.

17.3 REFERENCES

1 **Norton, R. L., Stene, R., Westbrook III, J. W.,** and **Eovaldi, D.** (1998). "Analyzing Vibrations in an IC Engine Valve Train." 980570, SAE International Symposium, Detroit, MI.

2 **Schumaker, L.** (1983). "On Shape Preserving Quadratic Spline Interpolation." *SIAM J. Numer. Anal.*, 20(4), 854-864.

3 **Mosier, R. G.** (2000). "Modern Cam Design." *Int. J. Vehicle Design*, 23(1/2), 38-55.

4 **Norton, R. L., Eovaldi, D., Westbrook III, J. W.**, and **Stene, R. L.** (1999). "Effect of Valve-Cam Ramps on Valve Train Dynamics." 1999-01-0801, SAE International Symposium, Detroit, MI.

Chapter 18

CAM DESIGN GUIDELINES

18.0 INTRODUCTION

This brief chapter attempts to put in summary form some of the lessons "learned the hard way" over 40 years of cam and machine design experience. Some myths and old shibboleths are also debunked. It is the author's impression that there is still a lot of misinformation and misunderstanding in the machine design world about the design of cam follower systems. It is hoped that these general guidelines will be of some help in improving the situation.

18.1 PRACTICAL DESIGN CONSIDERATIONS

The cam designer is often faced with many confusing decisions, especially at an early stage of the design process. Many early decisions, often made somewhat arbitrarily and without much thought, can have significant and costly consequences later in the design. The following is a discussion of some of the trade-offs involved with such decisions in the hope that it will provide the cam designer with some guidance in making these decisions.

Translating or Oscillating Follower?

There are many cases, especially early in a design, when either translating or rotating motion could be accommodated as output from the cam, though in other situations, the follower motion and geometry is dictated to the designer. If some design freedom is allowed, and straight-line motion is specified, the designer should consider the possibility of using an approximate straight line motion, which is often adequate and can be obtained from a large-radius rocker follower. The rocker or oscillating follower has advantages over the translating follower when a roller is used. A round-cross-section translating follower slide is free to rotate about its axis of translation and needs to have some antirotation guiding provided (such as a keyway or second slide) to prevent *z*-axis misalignment of the roller follower with the cam. Many commercially available, nonrotating slide assemblies are now available, often fitted with ball bearings, and these provide a good way to deal with this issue. However, an oscillating follower arm will keep the roller follower aligned in the same plane as the cam with no guiding other than its own pivot.

Also, the pivot friction in an oscillating arm follower has a small moment arm compared to the moment of the force from the cam acting on the follower arm at the roller. Conversely, the friction force on a translating follower has a one-to-one geometric relationship with the cam force. This can have a larger parasitic effect on the system.

Another advantage of oscillating arm followers over translating followers is their ability to accommodate somewhat larger pressure angles since there is less tendency to jam the follower than with a sliding joint. This allows a smaller cam diameter, all else equal. However, due to their low friction, ball bearing slides used for translating followers can also allow somewhat larger maximum pressure angles than the traditional 30° limitation.

Translating flat-faced followers are often deliberately arranged with their axis slightly out of the plane of the cam in order to create a rotation about their own axis due to the frictional moment resulting from the offset. The flat follower will then *precess* around its own axis and distribute the wear over its entire face surface. This is common practice with automotive valve cams that use flat-faced followers or "tappets."

Force or Form-Closed?

A form-closed (track or groove) cam or conjugate cams are more expensive to make than a force-closed (open) cam simply because there are two surfaces to machine and grind. Also, heat treating will often distort the track of a form-closed cam, narrowing or widening it such that the roller follower will not fit properly. This virtually requires post heat-treat grinding for track cams in order to resize the slot. An open (force-closed) cam will also distort on heat-treating, but can still be usable without grinding.

FOLLOWER JUMP The principal advantage of a form-closed (track) cam is that it does not need a return spring, and thus can be run at higher speeds than a force-closed cam whose spring and follower mass will go into resonance at some speed, causing potentially destructive follower jump. High-speed automobile and motorcycle racing engines often use form-closed (desmodromic) valve cam trains to allow higher engine rpm without incurring valve "float," or **follower jump**.

CROSSOVER SHOCK Though the lack of a return spring can be an advantage, it comes, as usual, with a trade-off. In a form-closed track cam there will be **crossover shock** each time the acceleration changes sign. Crossover shock describes the impact force that occurs when the follower suddenly jumps from one side of the track to the other as the dynamic force (*ma*) reverses sign. There is no flexible spring in this system to absorb the force reversal as in the force-closed case. The high impact forces at crossover cause noise, high stresses, and local wear. Also, the roller follower has to reverse rotation direction at each crossover which causes sliding and accelerates follower wear.

Studies in our laboratory have shown that roller followers running against a well-lubricated open radial cam can have slip rates of less than 1%.[1] Crossover shock can significantly increase this slippage on single-roller form-closed (track) cams.* Dual-roller conjugate cams do not have this roller-rotation-reversal problem.

Designers sometimes add a spring to the follower train of a form closed track cam, often at the end effector, in order to bias the roller to one side of the track and eliminate crossover shock. This arrangement effectively becomes a force-closed cam with a "belt and suspenders" insurance policy in the form of the other side of the track, that will take over and guarantee that the follower provides the required motion, even if dynamic forces overcome the spring and "jump" the roller off the spring-loaded side of the track, or when a jam in the tooling tends to hang up the follower. Then the other side of the track will take over and drive the follower through at the expense of some impact force.

Another common way to deal with the crossover shock problem is to use a ribbed cam rather than a track cam as shown on this book's cover and in Figure 1-9 (p. 8) and also schematically in Figure 13-15 (p. 419). These cams use a pair of rollers that straddle the rib and can (optionally) be spring-loaded together to pinch the rib in order to eliminate backlash. This spring needs to have a high rate to avoid resonance problems. Such an arrangement has the added benefit of providing wear compensation. Another way to achieve the same effect is to make a track cam's slot wide enough to accommodate two rollers that are spring loaded apart to take up any backlash.

LOCKOUT If it is necessary to lock out the follower motion, then a force-closed cam is mandatory and an air cylinder will be needed as the closure spring. Moreover, the cam and follower must be arranged so that the cam rise actually returns the tooling to the rest position so that pulling the follower off the cam at the high dwell (or high point) holds the tooling out of the nest. In other words, you want the air cylinder spring to drive the tooling in and the cam to pull it out.

Radial or Barrel Cam?

This choice is largely dictated by the overall geometry of the machine for which the cam is being designed. If the follower must move parallel to the camshaft axis, then a barrel cam is dictated. If there is no such constraint, a radial cam is probably a better choice simply because it is a less complicated, thus less expensive, cam to manufacture.

Roller or Flat-Faced Follower?

The roller follower is a better choice from a cam design standpoint simply because it accepts negative radius of curvature on the cam. This allows more variety in the cam program. Also, for any production quantities, the roller follower has the advantage of being available from several manufacturers in any quantity from one to a million. For low quantities, it is not usually economical to design and build your own custom follower. In addition, replacement roller followers can be obtained from suppliers on short notice when repairs are needed. Also, they are not particularly expensive, even in small quantities.

Perhaps the largest users of flat-faced followers are automobile engine makers. Their quantities are high enough to allow any custom design they desire. Flat followers

* Hollis [2] reports that high-speed video of a track cam with roller follower in one case showed that at 1000 cam rpm the roller was always sliding. It never was able to get its rotational velocity up to a level to match the cam surface speed before the crossover event demanded that it reverse direction.

can be made or purchased economically in large quantity and can be less expensive than a roller follower in that case. Also with engine valve cams, a flat follower can save space over a roller. Nevertheless, many manufacturers have switched to roller followers in automobile engine valve trains to reduce friction and improve fuel economy. Most new automotive IC engines designed in the U.S. in recent years have used roller followers for those reasons. Diesel engines have long used roller followers (tappets) as have racers who "hop-up" engines for high performance.

Cams used in automated production line machinery use stock roller followers almost exclusively. The ability to quickly change a worn follower for a new one taken from the stockroom, without losing much production time on the "line," is a strong argument in this environment. Roller followers come in several varieties (see Figure 1-7, p. 6). They are typically based on roller or ball bearings. Plain bearing versions are also available for low-noise requirements. Roller bearing followers are available in two types, "caged" and "full complement." A caged roller bearing has its rollers separated by the cage so that they do not touch. A full complement roller or needle bearing has no cage and its rollers are free to rub on one another. The larger number of rollers in a full complement bearing allow higher loads, but do so at the expense of potentially higher wear. The close spacing of the rollers leaves little volume for grease and the rubbing of roller on roller accelerates their abrasive and adhesive wear. They require frequent relubrication. Another variety of roller follower uses dual-row ball bearings. These are typically caged and have good grease capacity. They also can take some thrust load, which roller bearings cannot.

The outer surface of the roller follower, which rolls against the cam, can be either cylindrical or crowned (i.e., an oblate-spheroid) in shape. The "crown" on this follower is slight—having a large radius, but it allows the follower to ride near the center of a flat cam, despite some nonparallel misalignment of the cam and follower axes. If a cylindrical follower is chosen, and care is not taken to accurately align and stiffly support the axes of cam and roller follower, the follower and cam will ride on one edge and wear rapidly. Crowned rollers are recommended unless the loads are so high that the stresses become excessive. A cylindrical follower will give lower stresses than a crowned follower only if the parallel alignment between cam axis and roller axis is very accurately maintained and dynamic deflections are kept very small by designing in sufficient bending and torsional stiffness in the roller support. If not, the stresses in a cylindrical follower can be higher than for a crowned follower.

Commercial roller followers are typically made of high carbon alloy steel such as AISI 52100 and hardened to Rockwell HRC 60-62. The 52100 alloy is well suited to thin sections that must be heat-treated to a uniform hardness. Because the roller makes many revolutions for each cam rotation, its wear rate is usually higher than that of the cam. Chrome plating the follower can markedly improve its life. Chrome on steel is harder than the base steel at about HRC 70. Tool steel cams are typically hardened to a range of HRC 58 - 62. Hardened ductile iron cams will be somewhat softer, about HRC 55.

Roller followers are available in two types, as shown in Figure 1-7 (p. 6). One has a stud fitted in the inner race and is intended to be cantilever mounted. The other has a hole through the inner race for an axle that will be straddle mounted in a yoke. While packaging geometry and service considerations often dictate the use of a cantilevered stud-mount follower, if the choice is available, it is preferable to use a straddle-mounted follower. This follows from simple mechanics principles. A simply supported beam of given dimension and span will have a smaller deflection than a cantilevered beam of the same dimension and extension ("span"). Consideration needs also be given to the stiffness of the link to which the follower is attached. Ideally, a straddle-mounted roller can have the axis of the link coplanar with the cam, thus eliminating torsional moments on the link. A cantilevered follower will always add a torsional moment to the bending moment in the follower arm, thus potentially increasing deflections at the cam follower and exacerbating misalignment. Note also that the torsional stiffness of an offset follower arm combines in series with the bending stiffness to give a potentially lower overall stiffness for the follower system. This can worsen the system's dynamic response as was shown in Chapter 10. Nevertheless, cantilevered, studded followers are attractive as they are simpler to mount and use simpler follower arm geometry. As in any design, one must balance the trade-offs.

To Dwell or Not to Dwell?

The need for a dwell is usually clear from the problem specifications. If the follower must be held precisely stationary for any time, then a dwell is required. Some cam designers tend to insert dwells in situations where they are not specifically needed for follower stasis, in a belief that this is preferable to providing a rise-return motion when that is what is really needed. They are usually mistaken. We have made significant improvements in machine operating speeds with reduced vibration by eliminating unnecessary dwells and using the added time to gently reverse the direction of the follower with concomitantly lower peak accelerations and smoother jerk. With B-splines especially, the shape of the follower motion can be tailored to create "pseudo-dwells" during which the follower moves very little but has smooth and gentle, low-valued, higher derivatives.

If the designer is attempting to use a double-dwell program in what really needs only to be a single-dwell case, with the motivation to "let the vibrations settle out" by providing a "short dwell" at the end of the motion, he or she is misguided. The discussion in Chapter 11 should put the lie to this "old designer's tale." Instead, the designer probably should be using a different cam program, perhaps a polynomial or a B-spline tailored to the specifications. Taking the follower acceleration to zero, whether for an instant or for a "short dwell," is generally undesirable unless absolutely required for machine function (e.g., at some point you do have to reverse direction). See Chapters 10 and 11 for more information on this topic. A dwell should be used only when the follower is required to be stationary in an accurate location for some measurable time. Moreover, if you do not need any dwell at all, consider using a pure fourbar or sixbar linkage instead of a cam and follower. Linkages are easier to manufacture, have fewer vibration problems, and can be very reliable if well designed.

To Grind or Not to Grind?

Some production machinery cams are used as-milled, and not ground. Automotive valve cams are always ground and are sometimes also polished after grinding. The reasons are largely due to cost and quantity considerations as well as the high speeds of automotive cams. There is no question that a ground cam is superior to a milled cam. The question in each case is whether the advantage gained is worth the cost. In small quantities, as are typical of production machinery, grinding can nearly double the cost of a cam though this depends on a number of factors. The advantages in terms of smoothness and quietness of operation, and of wear, are not always in the same ratio as the cost difference. A well-machined cam can perform nearly as well as a well-ground cam and better than a poorly ground cam in many applications.[1], [3]

As was described in Chapter 14, electrical discharge machining using wire (wire EDM) has been used successfully to make prehardened, full-hard open-faced radial cams "in one pass" with a finish and accuracy good enough to not require any subsequent grinding.[2] This may yet prove to be a viable competitor to the conventional manufacturing process of "machine soft—harden—grind" traditionally used for manufacturing of this type of cam.

Automotive cams are made in large quantities, run at very high speeds, and are expected to last for a very long time with minimal maintenance. This is a very challenging specification. It is a great credit to the engineering of these cams-follower trains that they very seldom fail in 150,000 miles or more of operation. These cams are made on specialized equipment which keeps the cost of their grinding to a minimum. Some manufacturers also subject them to a fine polishing operation after grinding. The finer the finish, the better the lubrication condition can be, as explained in Chapter 15.

Industrial production machine cams also see very long lives, often 10 to 20 years, running into billions of cycles at typical machine speeds. Unlike the typical automotive application, industrial cams often run around the clock, 7 days a week, 50+ weeks a year.

To Lubricate or Not to Lubricate?

Cams ideally need lots of lubrication. Automotive cams are literally drowned in a flow of filtered engine oil. Many production machine cams run immersed in an oil bath. These are reasonably happy cams. Others are not so fortunate. Cams which operate in close proximity to the product on an assembly machine in which oil would cause contamination of the product (food products, personal products) often are run dry. If lightly loaded, these dry cams can last an impressively long time, such as 4-5 years of round-the-clock operation at a few hundred rpm. Camera mechanisms, which are full of linkages and cams, are usually run dry. Lubricant would eventually find its way to the film.

Unless there is some good reason to eschew lubrication, a cam-follower should be provided with a generous supply of clean lubricant, preferably a hypoid-type oil containing additives for boundary lubrication conditions. The geometry of a cam-follower joint (half-joint) is among the worst possible from a lubrication standpoint. Unlike a journal bearing, which tends to trap a film of lubricant within the joint, the half joint is continually trying to squeeze the lubricant out of itself. This can result in a boundary, or mixed boundary/EHD* lubrication state in which some metal-to-metal contact will occur. Lubricant must be continually resupplied to the joint. Another purpose of the liquid lubricant is to remove the heat of friction from the joint. If run dry at high speed, significantly higher material temperatures can result, with accelerated wear and possible early failure.

* Elastohydrodynamic (see Chapter 15)

What Double-Dwell Cam Program to Use?

For a conventional double-dwell motion with no constraints on the intermediate positions of the follower during the rise or fall, a good compromise cam program is the 3-4-5 polynomial. If lower peak velocity is needed because of a high inertia follower train or if pressure angle limitations are an issue, then the modified sine has an advantage over the 3-4-5 polynomial. The modified trapezoid is less desirable than either of the above mentioned functions due to its vibration-induced, higher actual peak acceleration.

If low vibrations are paramount, then smooth jerk is an issue and the 4-5-6-7 polynomial or the cycloid are a better choice despite their higher nominal peak acceleration. The actual peak acceleration may in fact be no higher with these low vibration functions than with ones that boast low theoretical acceleration but create large vibration. Other low vibration double-dwell functions are also available as described in Chapter 11.

As machine speeds increase, exacerbating vibration problems, the use of what were one considered "verboten" double-dwell programs such as cycloidal and 4-5-6-7 motions begins to look very good. A significant enabling factor is recent improvement in the quality of manufacturing equipment that to a large degree renders moot the old caveats about the inability to manufacture these slow-changing functions. If you need high follower accuracy at moderate to high speeds, and you are building machinery that will produce millions or even billions of product per year, then your budget can probably afford the cost of using expensive, high strength steel alloys for the cam and the expense of their manufacture on the best available CNC equipment.

What Cam Program to Use For Difficult or Complicated Motions?

The B-spline is the most versatile cam function available. Its mathematical flexibility in respect to satisfying a large number of boundary conditions with a relatively low order, and thus well-behaved, function is unparalleled. For less critical applications a conventional polynomial may do the trick, but for the really difficult cam design problem, the B-spline is the ultimate answer. For conventional double-dwell motions it offers no advantage, but when a combination of multiple displacement, velocity, or acceleration constraints are demanded, the B-spline is the best solution.

To Polydyne or Not to Polydyne?

For any application where cam speed is expected to be essentially constant, as is the case with most industrial machinery,* there appears to be little or no disadvantage and quite significant advantages to creating polydyne or splinedyne functions for most applications. We have consistently achieved significant reductions in vibrations, lower noise levels (as much as 5 dB reduction) and lower scrap rates by using the splinedyne technique on production machinery in combination with increased operating speeds.

In our experience, for a typical assembly machine follower train, the difference in contour dimensions on a point-by-point basis between the nonpolydyne and polydyne cam seldom exceeds about 0.010 in (0.25 mm). So the cam's radii of curvature and pressure angles are not adversely affected, nor does the cam become too difficult to make after polydyning or splinedyning, as the case may be.

* Constant speed for at least some months or years, until management decides to speed them up for increased production. Luckily, such a project will usually be budgeted to allow replacement of cams with ones optimized for the new machine speed.

Also, despite the theoretical requirement that the displacement function used for a polydyne cam be ping continuous, we have good empirical evidence that one can sometimes break this rule without serious consequence provided that the compliance of the follower train is not excessive. A well-designed follower train should meet this criterion, and if not, it should be redesigned to be stiffer anyway.

It is true that when management decides that they want to take the machine to higher speeds next year, the cams will have to be re-splinedyned for the new speed and remanufactured, but this cost should be readily accommodated in the budget allocated for the speedup program. The cam contour differences required by a speed change will likely be small. In fact, the existing cams can often be reground to the new contour and the follower linkage adjusted for the slight reduction in prime circle radius.

Camshaft Design

Too many machine designers neglect the deflections of components in their designs. A camshaft designed to control stresses against possible fatigue failure, can have significant deflections under stress-safe applied loads. If designed to control deflection, shafts and levers will usually have low stresses and large safety factors against fatigue failure. A camshaft designed for minimum deflection will often appear "overdesigned" to the engineer's eye, and from a stress standpoint, it probably is. Both bending and torsional deflections should be calculated and set to limits that minimize the error in end effector position. The topic of shaft design is addressed in detail in other references such as [4]. The details are too involved to go into here, but issues of bearing type and placement as well as shaft geometry all enter into the design. The reader is directed to the reference for more information.

Follower Train Design

Two words. Light and stiff. Mutually incompatible, you say? Not if link geometry is proper. Good basic engineering design is all that is needed. Use stiff section shapes such as I-beams. Add lightening holes where they do not compromise section stiffness or increase stresses. Minimize end effector mass. Steel has slightly better specific stiffness than cast iron. For the really tough case, consider one of the exotic new composite materials whose specific strengths and specific stiffnesses far exceed the common engineering metals.

One aspect of follower linkage design that is usually ignored is the issue of moment balance. If the center of gravity (CG) of a rotating link is not at its pivot, then there will be a dynamic force at the pivot proportional to the product of the link's moment of inertia and its angular acceleration. This dynamic force increases loads on, and reduces life of, the link's bearings. It is a simple task to design rotating links such that their CG is at the pivot, especially if a solids modeling CAD system is used for their design.

Follower Train Dynamics

Many machine designers seem to be blissfully unaware of the rule that effective mass (and effective spring rate) vary as the square of the lever ratio in a linkage follower train.

We have witnessed too many instances of, say, a 10-lb end effector assembly appearing at the roller follower as a 250-lb effective mass. The cam is then expected to, in effect, move a household appliance back and forth, perhaps 4 times a second. This usually comes about because the designer chose (often arbitrarily) to select a cam rise much smaller than the required end effector motion and then amplified the cam motion with levers. Perhaps a better choice would have been to increase the stroke at the cam and reduce the lever ratio. This can (and should) be done up to the point that pressure angle limitations occur. Since follower acceleration is a linear function of stroke, but effective mass varies as the square of the lever ratio, the dynamic forces (and stresses) on the cam and follower can be significantly lowered by this simple expedient.

Natural Frequencies

Cams typically have significant high harmonic content. Measurable energy is often present out to 20 or 30 harmonics. This harmonic energy is available to couple with structural natural frequencies in the follower train, camshafts, and other machine components with occasionally disastrous results as demonstrated in the previous chapter. The wise machine designer will calculate at least the first few structural natural frequencies of the follower train and drive system, and try to avoid any coincidence between these frequencies and significant cam harmonics. Better yet, a modal analysis of the follower train components can be done fairly easily if a solids model is available along with an FEA package. The cam harmonics are easily determined by a Fourier analysis of the cam acceleration function for the entire rotation of the shaft, including any dwells. (Program DYNACAM provides a Fourier analysis.) As the case studies in the previous chapter point out, small changes in operating speed can have a large effect on follower train vibrations due to coupling between cam harmonics and structural natural frequencies.

Backlash

Any clearances in the follower train, whether between roller and cam track or between pivot pins and bushings in the linkage, create impacts and a nonlinear response of the system. An impact event has a Fourier spectrum that is theoretically uniform in energy at all frequencies from zero to infinity. Thus, it excites all the structural natural frequencies in the system. This can significantly increase vibrations. So impacts are best avoided if possible. Unfortunately, not only is it virtually impossible to completely eliminate clearances in pivots and cam joints, often the function of the follower train requires a deliberate impact between the end effector and its workpiece, e.g., to close a valve on its seat, to clinch a rivet, clamp a part while it is welded, or just to come against a hard stop in order to guarantee accurate positioning of the end effector. In such cases, vibrations can be reduced by tailoring the follower motion to lower its velocity at impact. B-splines are particularly good for this task.

How Important is Theoretical Peak Acceleration?

Cam designers have long held to the belief that a cam function with lower theoretical peak acceleration is, de facto, superior to one with a higher value for this parameter. This author once also believed that, but is no longer so convinced. His perspective about cam design has changed considerably in recent years as a result of exposure to many instances where residual vibrations of the follower train have proved to be the major limitation on machine accuracy, speed, and product quality. The classic rule of thumb that cites the theoretical peak acceleration of the follower as the most important design parameter has proven to be less critical to many cam designs' successes than has the actual dynamic peak acceleration that includes vibration of the follower. Recent experience has repeatedly shown that vibrations in precision machinery are a significant impediment to both product quality and production quantity. This has been unequivocally proven through direct, *in situ* measurement of the acceleration on follower trains, as described in the case studies of the previous chapter. These results were corroborated by simultaneous high-speed video analysis of follower vibrations as well as by reductions in scrap rate concomitant with reduced vibrations.

What, after all, is the downside of higher acceleration from an engineering standpoint? Many machine parts experience extreme levels of acceleration in regular use without compromising their effectiveness or life. A poppet valve in an IC engine will see over 1000g of actual acceleration at redline. The piston and wrist pin see up to 1000g of theoretical acceleration at highway speeds and over 2000g at redline. And they can survive 200 000 miles. Clearly the real issues are dynamic force and stress. If the mass of the follower train is kept small, then it can sustain high accelerations without failure, as long as its dynamic forces due to mass times acceleration are kept reasonable.

Cams in industrial production machinery often have low stress due to their relatively large size and relatively low speeds compared to IC engine applications. Thus, in such circumstances it seems a good trade-off to choose cam functions that are designed to give lower vibration at the expense of higher theoretical peak acceleration. In fact, the actual peak acceleration may end up lower with this approach because of the lower vibrations. Reducing harmonic content of the cam function by eliminating unnecessary dwells and using higher-order functions that control or at least smooth the higher derivatives (jerk and ping) can be beneficial as long as the increased peak acceleration that may result does not overstress the cam-follower train or stall the drive motor. It makes sense to "use the material" and accept higher stresses as long as they are within acceptable safety factors, especially if it results in smoother machine function that translates to lower scrap rates. When millions or billions of product per year are being made, even a fraction of a percent reduction in scrap can translate to significant dollar savings, and they go right to "the bottom line" as pure profit.

18.2 RULES OF THUMB FOR CAM DESIGN

Rules of thumb are meant to be broken! They are only general guidelines and should not be taken as gospel. Nevertheless, they can be of some value in guiding the new cam designer as they tend to capture the results of successful experience. With that caveat in mind, we present the following short list of rules that have proven themselves over time.

1 Do not arbitrarily insert dwells in a cam motion function unless they are really needed. Any stopping of the follower will exact a dynamic penalty in terms of reduced time available for the required motions. It may be better to overtravel the follower in a gently curved path that reverses its direction for the return motion than to stop and restart it.

2 For conventional double-dwell motions, the modified sine or 3-4-5 polynomial functions are good all-around "workhorse" choices, and are better than modified trapezoidal motion when vibration of the follower is considered.

3 For low vibrations in conventional double-dwell motions, the 4-5-6-7 polynomial is superior and the cycloidal curve is an acceptable substitute as long as their somewhat higher accelerations do not overstress the parts. Differences in theoretical peak accelerations between various functions is less important than the actual peak value when follower vibration is factored in.

4 To obtain extremely low vibrations in applications where camshaft speed is essentially constant, a polydyne or splinedyne approach is best.

5 For nonconventional (i.e., not double dwell) motions, especially when intermediate follower positions or velocities are specified, a polynomial or B-spline function will usually give the best design. If many boundary conditions are specified, then a polynomial may not work (or will be suboptimal) and a B-spline will be needed.

6 Pressure angles should be limited to about 30° with conventional translating followers, though if a low-friction ball-slide is used to guide the translating follower, then pressure angles up to about 35° may be tolerable. For oscillating arm followers, a maximum pressure angle of about 35° is generally acceptable.

7 When a roller follower is used, the absolute value of the radius of curvature ρ of the path of the roller centerline (pitch curve) must be kept larger than the radius R_f of the roller at all points to avoid undercutting. A ratio of $| R_f | / \rho > 2$ is a good target, though a ratio of 1.5 has been used successfully in some applications. The Hertzian surface stress should definitely be checked when this ratio is small.

8 When a flat follower is used, there can be no negative radius of curvature allowed in the cam surface contour.

9 When air cylinders are used as follower return springs, an accumulator should be used and the fittings, hoses, valves, etc., connected to the cylinder should be as large in diameter and as short as possible to minimize impedance.

10 "Full-complement" needle roller bearings used as cam followers may be shorter lived than caged roller bearings despite the superior load capacity of the former type. This is due to their poorer grease storage capacity and the fact that the uncaged needles rub on one another.

11 While a cylindrical roller follower has theoretically lower surface stress than a crowned follower in the same application, unless parallel alignment of the axes of the cylindrical roller follower and the cam is accurately and stiffly maintained, a cylindrical roller follower can actually have potentially higher surface stress than a crowned roller. If the alignment between cam and roller axes cannot be made accurate (including dynamic deflections), then a crowned roller may be needed, though its gentle crown radius can accommodate only slight axial misalignment.

12 Track cams with single roller followers will experience crossover shock, and the roller will have to reverse direction at the load reversal points, causing slip and wear. Some designers of track cams provide a follower spring to load the roller follower always against one side of the track. The other side of the track then becomes essentially an "insurance policy" against possible gross follower jump in the event of a follower spring failure or a tooling jam. Another approach to cure crossover shock and roller reversal is to use two rollers in the track, each contacting only one side of the groove and spring-loaded apart to accommodate slight deviations in groove width. A ribbed cam with two rollers pinching across the rib gives the same effect.

13 Topical grease lubrication of open cam surfaces in a dirty environment can be "worse than nothing" as the grease will trap airborne dirt particles and increase abrasive wear rate. Lubrication of a cam-follower joint is best done by enclosing both in an oil-filled box.

14 Yoke mounted roller followers are preferred to stud mounted rollers because of their axle's smaller bending deflections and the possibility of centered loading that can reduce or eliminate torsional moments on the follower arm.

15 Attention should be paid to the design of follower linkages and camshafts to minimize or balance applied bending and torsional moments and their resulting deflections.

16 Stiff, light cam-follower trains will have superior dynamic performance with reduced vibration and improved fidelity to the theoretical motion program specified.

17 Avoid large amplification ratios between the end effector stroke and that of the cam follower motion as the reflected inertia (effective mass) of the end effector and its intervening links is increased by the square of this ratio. It may be better to suffer a larger cam follower motion and its concomitantly larger maximum pressure angle in return for a reduced amplification ratio.

18 Cam-follower systems typically have very low structural damping (are high "Q") having dimensionless damping ratios ζ of between 0.05 and 0.10, and so are prone to vibration problems.

19 Care should be taken to reduce clearances in follower train pivots and joints to minimize within-joint impacts on load reversal, which will exacerbate vibration. It is possible to use springs at the joints between links to take-up the joint clearance in one direction with sufficient force to eliminate impact and improve accuracy. Placing the follower return spring at the end effector as in the automotive valve train, rather than near the cam, can have the same effect.

20 When high accuracy is required of the end effector position in a cam-follower train, it is common practice to design the system with deliberate overtravel (sometimes called lost motion) and provide a "hard stop" on ground that the end effector contacts before the cam motion is complete. Some deliberate compliance in the form of a spring must then be provided within the follower train to avoid jamming. This technique also creates an impact event. To minimize the force and vibration of the impact, the cam can be designed (using polynomials or B-splines) to provide a low magnitude, constant velocity ramp at the end of the stroke to control impact velocity as is done in automotive valve cams.[5]

18.3 REFERENCES

1 **Norton, R. L.** (1988). "Effect of Manufacturing Method on Dynamic Performance of Cams." *Mechanism and Machine Theory*, **23**(3), pp. 191-208.

2 **Hollis, P.** (2001), Tyco Electronics Inc. Personal Communication

3 **Norton, R. L., et al.** (1988). "Analysis of the Effect of Manufacturing Methods and Heat Treatment on the Performance of Double Dwell Cams." *Mechanism and Machine Theory*, **23**(6), pp. 461-473.

4 **Norton, R. L.**, (2000). *Machine Design: An Integrated Approach*, Prentice-Hall, Upper Saddle River, NJ. Chapter 9.

5 **Norton, R. L., D. Eovaldi, J. W. Westbrook III** and **R. L. Stene**, *Effect of Valve-Cam Ramps on Valve Train Dynamics*, SAE 1999-01-0801, SAE International Symposium, Detroit, MI, March 1-4, 1999.

Appendix A

COMPUTER PROGRAMS

A.0 INTRODUCTION

There are limited-time license, demonstration versions of four custom computer programs provided on the CD-ROM with this book: programs DYNACAM, FOURBAR, SIXBAR, and SLIDER. These are professional editions of the programs intended to be used by qualified personnel familiar with the principles of good engineering practice and with the topics discussed in this book and in reference [1]. **The authors and publishers are not responsible for any damages that may result from the use or misuse of these programs.**

A.1 GENERAL INFORMATION

Hardware Requirements

These programs need a Pentium or better processor. The programs, especially DYNACAM, are large and use significant computer resources. A minimum of 64M of RAM is desirable. DYNACAM may require that all other applications be shut off in order to run in a computer with otherwise insufficient memory. A CD-ROM drive is needed, as is a hard disk drive.

Operating System Requirements

These custom programs are written in Microsoft Visual Basic 6.0 and work on any computer that runs Windows 95/98/2000/NT.

Demonstration Versions

The demonstration versions of these programs are supplied with this book at no charge and allow up to 30 runs over a period of 90 days from first installation. Certain features are disabled in the demonstration versions, such as the ability to open or save files,* and to output

* Program DYNACAM can open the example files provided with this book, but cannot open other files created with licensed versions of the program.

cam profile or linkage coordinate data. If you wish to have the fully functional program, you must register it and pay the license fee defined on the *Registration* screen.

Installing the Software

The CD-ROM contains the executable program files plus all necessary Dynamic Link Library (DLL) and other ancillary files needed to run the programs. Run the **Setup.exe** file from the individual program's folder on the CD-ROM to automatically decompress and install all of its files on your hard drive. The program name will appear in the list under the *Start* button's *Program* menu after installation and can be run from there.

How to Use This Manual

This manual is intended to be used while running the programs. To see a screen referred to, bring it up within the program to follow its discussion.

A.2 GENERAL PROGRAM OPERATION

All programs in the set have similar features and operate in a consistent way. For example, all printing and plotting functions are selected from identical screens common to all programs. Opening and saving files are done identically in all programs. These common operations will be discussed in this section independent of the particular program. Later sections will address the unique features and operations of each program.

Running the Programs (All Programs)

At start-up, a splash screen appears which identifies the program name, version, revision number, and revision date. Click the button labeled *Start* or press the *Enter* key to run the program. A *Disclaimer* screen next appears that defines the registered owner and allows the printing of a registration form if the software is as yet unregistered. A registration form can be accessed and printed from this screen.

The next screen, the *Title* screen, allows the input of any user and/or project identification desired. This information must be provided to proceed and is used to identify all plots and printouts from this program session. The second box on the *Title* screen allows any desired file name to be supplied for storing data to disk. This name defaults to **Model1** and may be changed at this screen and/or when later writing the data to disk. The third box allows the typing of a starting design number for the first design. This design number defaults to 1 and is automatically incremented each time you change the basic design during this program session. It is used only to identify plots, data files, and printouts so they can be grouped, if necessary, at a later date. When the *Next* button on the *Title* screen is clicked, the *Home* screen appears.

The *Home Screen* (All Programs)

All program actions start and end at the *Home* screen which has several pulldown menus and buttons, and some of its commands (*File*, *New*, *Open*, *Save*, *Save As*, *Units*, *About*, *Plot*, *Print*, *Quit*) are common to all programs. These will be described below.

General User Actions Possible Within a Program (All Programs)

The programs are constructed to allow operation from the keyboard or the mouse or with any combination of both input devices. Selections can be made either with the mouse or, if a button is highlighted (showing a dotted square within the button), the *Enter* key will activate the button as if it had been clicked with the mouse. Text boxes are provided where you need to type in data. These have a yellow background. Double-clicking on a text box will select its contents. In general, what you type in any text box is not accepted until you hit the *Enter* key or move off that box with the *Tab* key or the mouse. This allows you to retype or erase with no effect until you leave the text box. You can move between available input fields with the *Tab* key (and back up with *Shift-Tab*) on most screens. If you are in doubt as to the order in which to input the data on any screen, try using the *Tab* key, as it will take you to each needed entry field in a sensible order. You can then type or mouse click to input the desired data in that field. Remember that a yellow background means typed input data is expected. Boxes with a cyan background feed information back to you, but will not accept input.

Other information required from you is selected from drop-down menus or lists. These have a white background. Some lists allow you to type in a value different than any provided in the available list of selections. If you type an inappropriate response, it will simply ignore you or choose the closest value to your request. Typing the first few letters of a listed selection will sometimes cause it to be selected. Double clicking on a selectable item in a list will often provide a shortcut and sometimes a help screen.

Units (All Programs)

The *Units* menu defines several units systems to choose from. *Note that all programs work entirely in pure numbers without regard to units.* It is your responsibility to ensure that the data as input are in some consistent units system. **No units conversion is done within the programs**. The *Units* menu selection that you make has only one effect, namely to change the labels displayed on various input and output parameters within the program. You mix units systems at your own peril. (Remember the Mars orbiter.)

Examples (All Programs)

All the programs have an *Examples* pulldown menu on the *Home* screen which provides some number of example mechanisms that will demonstrate the program's capability. Selecting an example from this menu will cause calculation of the mechanism and open a screen to allow viewing the results. In some cases, you may need to hit a button marked *Calculate*, *Run*, or *Animate* on the presented screen to see the results. Some programs also provide access to examples from other screens.

Creating New, Saving, and Opening Files (*File* - All Programs)

The standard *Windows* functions for creating new files, saving your work to disk, and opening a previously saved file are all accessible (in registered installations) from the pulldown menu labeled *File* on each program's *Home* screen. Selecting *New* from this menu will zero any data you may have already created within the program, but before doing so will give warning and prompt you to save the data to disk.

The *Save* and *Save As* selections on the *File* menu prompt you to provide a file name and disk location to save your current model data to disk. The data are saved in a custom format and with a three-character suffix unique to the particular program. You should use the recommended suffix on these files as that will allow the program to see them on the disk when you want to open them later. If you forget to add the suffix when saving a file, you can still recover the file as described below.

Selecting *Open* from the *File* menu prompts you to pick a file from those available in the disk directory that you choose. If you do not see any files with the program's suffix, use the pulldown menu within the *Open File* dialog box to choose *Show All Files* and you will then see them. They will be read into the program properly with or without the suffix in their name as long as they were saved from the same program.

Copying Screens to Clipboard or Printer (*Copy* - All Programs)

Any screen can be copied as a graphic to the clipboard by using the standard *Windows* keyboard combo of *Alt-PrintScrn*. It will then be available for pasting into any compatible *Windows* program such as *Word* or *Powerpoint* that is running concurrently in *Windows*. Most screens also provide a button to dump the screen image to an attached laser printer. However, the quality of that printed image may be less than could be obtained by copying and pasting the image into a program such as *Word* or *Powerpoint* and then printing it from that program. It seems that *Visual Basic* does not print graphics as well as some other *Windows* applications. NOTE: ***In some cases the plotted functions may not print properly from the PrintScreen button. If so, copy the screen to the clipboard, paste it into Word and print from Word.*** **This is the recommended approach in any case.**

Printing to Screen, Printer, and Exporting Disk Files (*Print* Button)

Selecting the *Print* button from the *Home* screen will open the *Print Select* screen (see Figure A-1) containing lists of variables that may be printed. Buttons on the left of this screen can be clicked to direct the printed output to one of *Screen*, *Printer*, or *Disk*. This choice defaults to *Screen* and so must be clicked each time the screen is opened to obtain either of the other options. The output is different with each of these selections.

Selecting *Screen* will result in a scrollable screen window full of the requested data. Scrolling will allow you to view all data requested serially. This data screen can be copied to the clipboard or dumped to a printer as described above, but this clip or dump will typically show only a portion of the available data, i.e., one screen-full.

Selecting *Printer* as the output device will cause the entire selection of data to print to an available printer. Only some of the side-bar information shown on the screen display will be included in this printout.

Selecting *Disk* as the output device will cause your selections to be sent to the file of your choice in an ASCII text format (tab delimited) that can be opened in a spreadsheet program such as *Lotus 123* or *Excel*. You can then do further calculations or plotting of data within the spreadsheet program.

The *Print Select* screen has two modes for data selection, *Preset Formats* and *Mix and Match*. The former provides preselected sets of four variables for printing. Selecting *Mix and Match* allows you to pick any four of the available variables for printing. You must print four variables at a time in either mode. Depending on the program, you may be able to select other ancillary parameters such as the number of decimal places or the frequency of data to be printed.

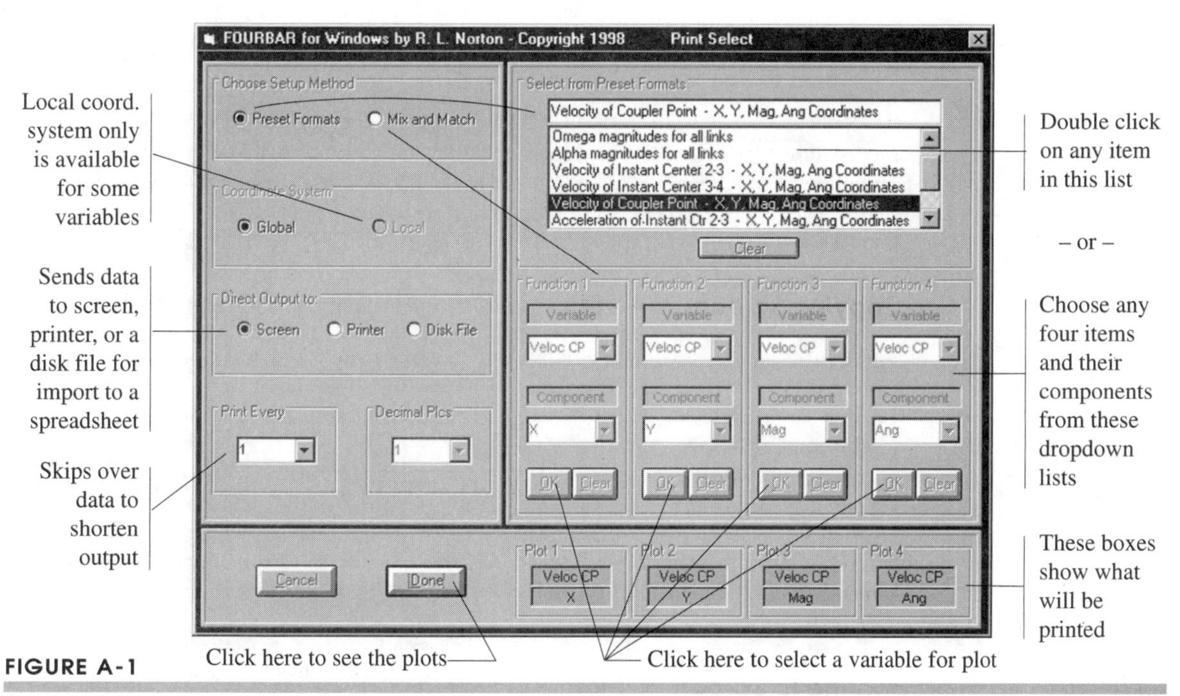

FIGURE A-1

The *Print Select* screen is common to all programs (not all programs allow component selection)

Plotting Data (*Plot* Button)

The *Plot* button on the *Home* screen brings up the *PlotType* screen (see Figure A-2) which is the same in all programs. Variables in these programs can be plotted in one of several formats, three Cartesian (see below) and one polar (see below). This screen allows a choice among these four "flavors" of plots shown as plot-style icons. The first icon (upper left) provides four functions plotted on Cartesian axes in four separate windows. The second icon (upper right) plots two functions on Cartesian axes in separate windows. The third icon (lower left) allows one to four functions to be plotted on common Cartesian axes in a single window. This choice is of value to show a single function full screen or to overlay multiple functions. (Be advised, however, that multiple functions will scale to the largest function of the set, so if there are large differences in magnitude between the members of the set, it may be difficult to see and interpret the smaller ones.) The fourth icon (lower right) provides a polar plot of one selected function. You may select any of these four plot styles by clicking on its icon or on the *Select* button above it, and then clicking *Next*. Double clicking on a plot icon will bring up the next screen immediately.

CARTESIAN PLOTS depict a dependent variable versus an independent variable on Cartesian (x, y) axes. In these programs, the independent variable shown on the x axis may be either time or angle, depending on program and the calculation choice made in a particular program. The variable for the y axis is selected from the plot menu. Angular velocities and torques are vectors, but are directed along the z axis in a two-dimensional system. So their magnitudes can be plotted on Cartesian axes and compared because their directions are constant, known, and the same.

A

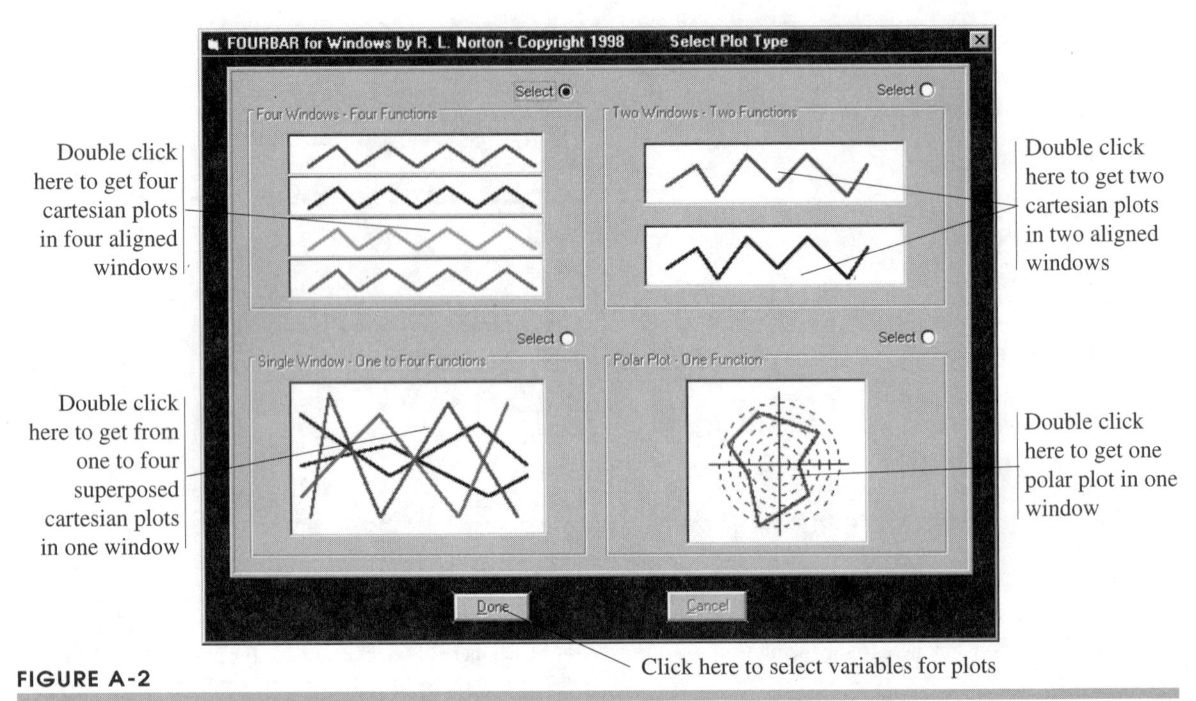

FIGURE A-2

The *PlotType* screen is common to all programs that allow plotting

POLAR PLOTS Plots of linear velocities, linear accelerations, and forces require a different treatment than the Cartesian plots used for the angular vector parameters. Their directions are not the same and vary with time or input angle. One way to represent these linear vectors is to make two Cartesian plots, one for magnitude and one for angle of the vector at each time or angle step. Alternatively, the x and y components of the vector at each time or angle step can be presented as a pair of Cartesian plots. Either of these approaches requires two plots per vector and has the disadvantage of being difficult to interpret. A better method for vectors that act on a moving point (such as a force on a moving pin) can be to make a polar plot with respect to a local, nonrotating axis coordinate system (LNCS) attached at the moving point. This local, nonrotating x, y axis system translates with the point as it moves but remains always parallel to the global axis system X, Y. By plotting the vectors on this moving axis system we can see both their magnitude and direction at each time or angle step, since we are attaching the roots of all the vectors to the moving point at which they act.

In some of the programs, polar plots can be paused between the plotting of each vector. Without a pause, the plot can occur too quickly for the eye to detect the order in which they are drawn. When a mouse click is required between the drawing of each vector, their order is easily seen. With each pause, the current value of the independent variable (time or angle) as well as the magnitude and angle of the vector are displayed.

The programs also allow alternate presentations of polar plots, showing just the vectors, just the envelope of the path of the vector tips, or both. A plot that connects the tips of the vectors with a line (its envelope) is sometimes called a **hodograph**.

SELECTING PLOT VARIABLES Choosing any one of the four plot types from the *Plot Type* screen brings up a *Plot Select* screen that is essentially the same in all programs. (See Figure A-3.) As with the *Print Select* screen, two arrangements for selecting the functions to be plotted are provided, *Preset Formats* and *Mix and Match.* The former provides preselected collections of functions, and the latter allows you to select up to four functions from those available on the pulldown menus. In some cases you will also have to select the component of the function desired, i.e., *x, y, mag,* or *angle*.

PLOT ALIGNMENT Some of the *Plot Select* screens offer a choice of two further plot style variants labeled *Aligned* and *Annotated.* The aligned style places multiple plots in exact phase relationship, one above the other. The annotated style does not align the plots, but allows more variety in their display such as fills and grids. The data displayed are the same in each case.

COORDINATE SYSTEMS For particular variables in some programs, a choice of coordinate system is provided for display of vector information in plots. The *Coordinate System* panel on the *Plot Select* screen will become active when one of these variables is selected. Then either the *Global* or *Local* button can be clicked. (It defaults to *Global.*)

GLOBAL COORDINATES The *Global* choice in the *Coordinate System* panel refers all angles to the *XY* axes of Figure A-14 (p. 553). For polar plots, the vectors shown with the *Global* choice actually are drawn in a local, nonrotating coordinate system (LNCS) that remains parallel to the global system such as x_1, y_1 at point *A* and x_2, y_2 at point *B* in Figure A-13 (p. 552). The LNCS x_2, y_2 at point *B* behaves in the same way as the LNCS x_1, y_1 at point *A*; that is, it travels with point *B,* but remains parallel to the world coordinate system *X, Y* at all times.

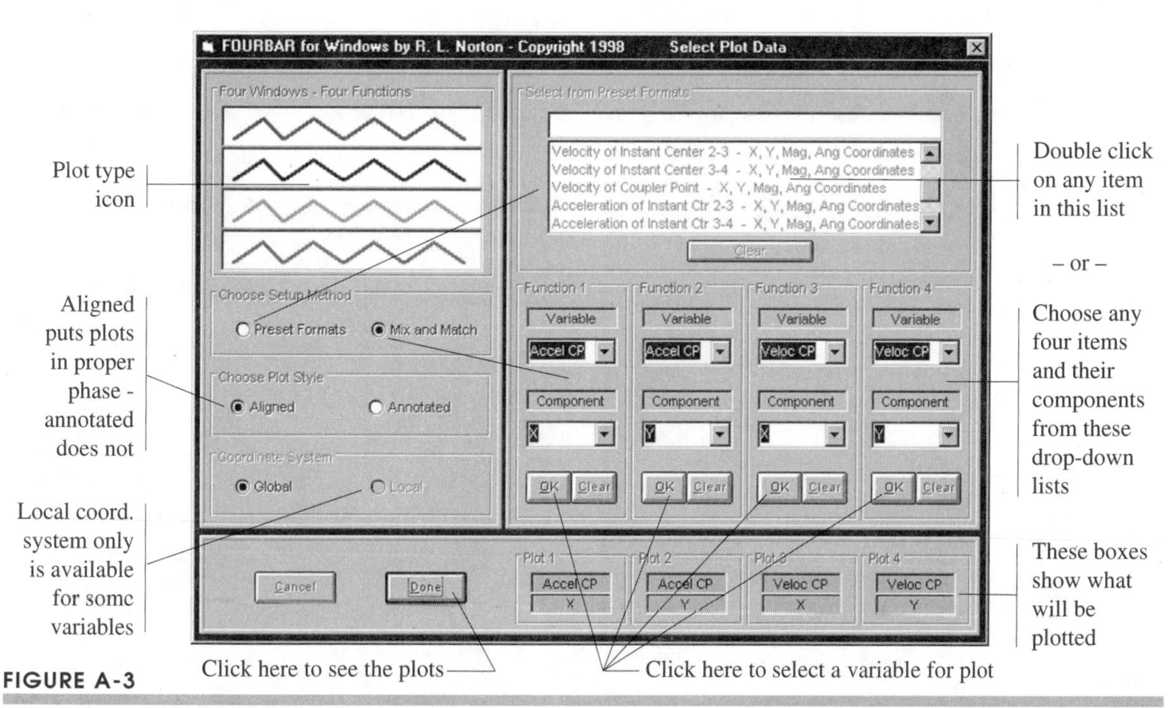

FIGURE A-3

The *PlotSelect* screens are common to all programs; this shows one of four styles of *PlotSelect* screens

LOCAL COORDINATES The coordinate system x', y' also travels with point B as its origin, but is embedded in link 4 and rotates with that link, continuously changing its orientation with respect to the global coordinate system X, Y, making it an LRCS. Each link has such an LRCS but not all are shown in Figure A-13. The *Local* choice in the *Coordinate System* panel uses the LRCS, for each link to allow the plotting and printing of the tangential and radial components of acceleration or force on a link. This is of value if, for example, a bending stress analysis of the link is wanted. The dynamic force components perpendicular to the link due to the product of the link mass and tangential acceleration will create a bending moment in the link. The radial component will create tension or compression.

PLOTTING Once your selections are made and are shown in the cyan boxes at the lower right of the *Plot Select* screen, the *Next* button will become available. Clicking it will bring up the plots that you selected. Figure A-4 shows examples of the four plot types available. From this *Plot* screen, you may copy to the clipboard for pasting into another application or

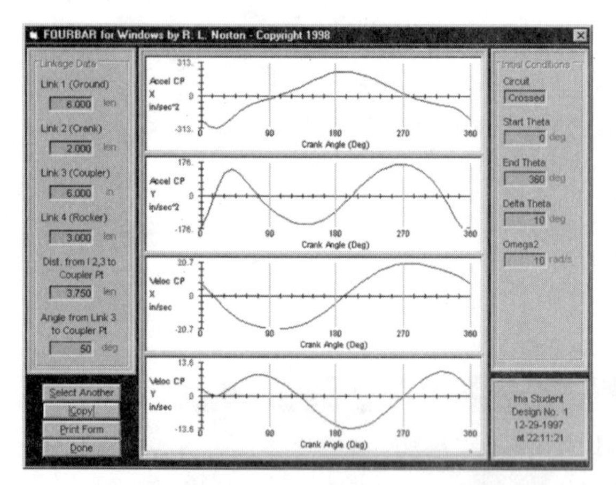

(*a*) Four aligned plots in separate windows

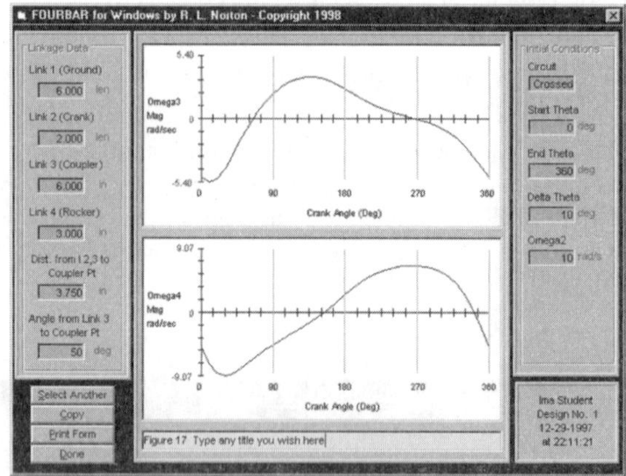

(*b*) Two aligned plots in separate windows

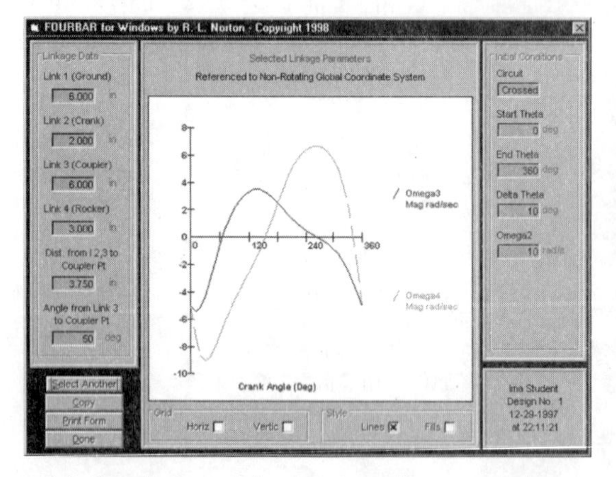

(*c*) One to four plots superposed in one window

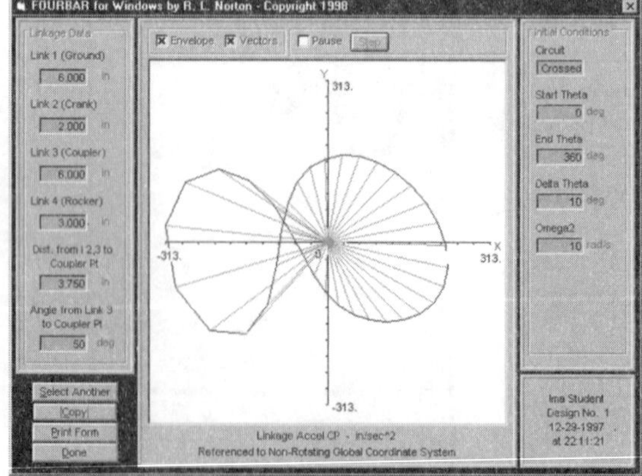

(*d*) Single polar plot

FIGURE A-4

The four styles of plots available in all programs; sidebar information is different in each program

dump the *Plot* screen to a printer. The *Select Another* button returns you to the previous *Plot Select* screen. *Next* returns you to the *Home* screen.

The *About* Menu (All Programs)

The *About* pulldown menu on the *Home* screen will display a splash screen containing information on the edition and revision of your copy of the program. The *Disclaimer* and *Registration* form can also be accessed from this menu.

Exiting a Program (All Programs)

Choosing either the *Quit* button or *Quit* on the *File* pulldown menu on the *Home* screen will exit the program. If the current data has not been saved since it was last changed, it will prompt you to save the model using an appropriate suffix. In all cases, it will ask you to confirm that you want to quit. If you choose yes, the program will terminate and any unsaved data will be gone at that point. Note that the Home Screen lacks a check box at its upper right corner that is sometimes used to exit a program. This omission is deliberate in order to force you to do a normal exit via the *Quit* command, which "cleans up" on the way out and shuts down the graphics server that is running in the background. If the graphics server is left running, it may crash the system when other programs are later run.

Support (All Programs)

Please notify the author of any bugs via email to *norton@designofmachinery.com* or *norton@wpi.edu*.

A.3 PROGRAM DYNACAM

DYNACAM *for Windows* is a cam design and analysis program intended for use by engineers and other professionals who are knowledgeable in the art and science of cam design. It is assumed that the user knows how to determine whether a cam design is good or bad and whether it is suitable for the application for which it is intended. The program will calculate the kinematic and dynamic data associated with any cam design, but cannot substitute for the engineering judgment of the user. The cam theory and mathematics on which this program is based are shown in various chapters of this text. Please consult them for explanations of the theory and mathematics involved.

The DYNACAM *Home* Screen

Initially, only the *SVAJ* and *Quit* buttons are active on the *Home* screen. Typically, you will start a cam design with the *SVAJ* button, but for a quick look at a cam as drawn by the program, one of the examples under the *Example* pulldown menu can be selected and it will draw a cam profile. If you activate one of these examples, when you return from the *Cam Profile* screen you will find all the other buttons on the *Home* screen to be active. We will address each of these buttons in due course below.

Input Data (DYNACAM *Input* Screen)

Much of the basic data for the cam design is defined on the *Input* screen shown in Figure A-5, which is activated by selecting the *SVAJ* button on the *Home* screen. When you open this

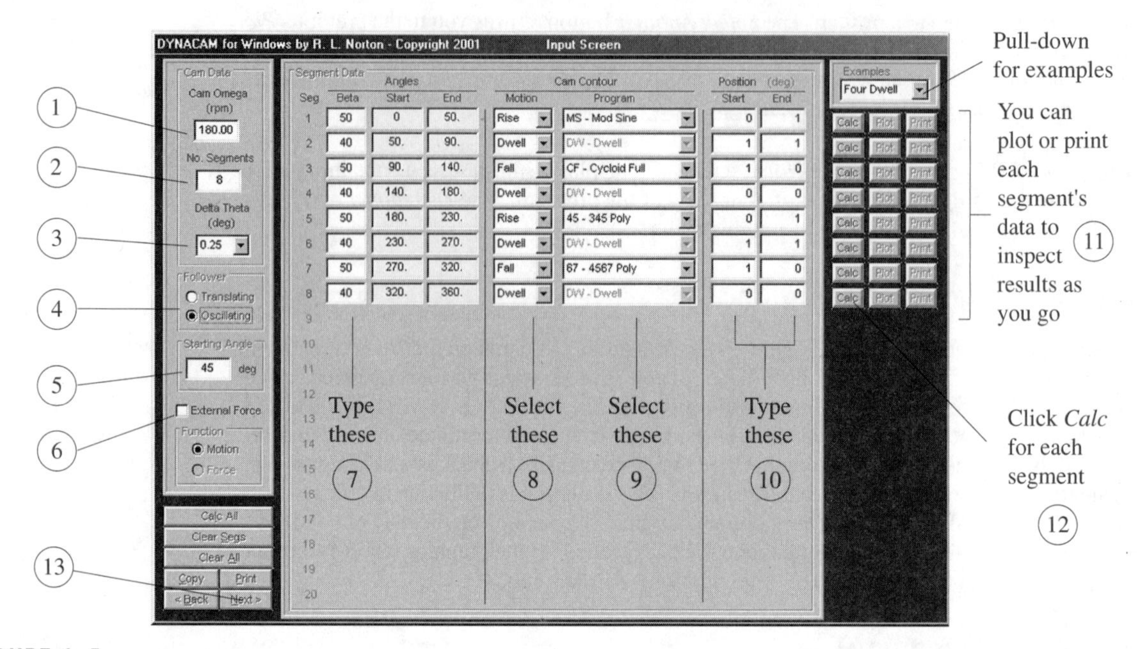

FIGURE A-5

S V A J input screen for program DYNACAM

screen for the first time, it will be nearly blank, with only one segment's row visible. (Note that the built-in examples can also be accessed from this form at its upper right corner.) If you had selected an example cam from the pulldown menu on the *Home* screen, making the *Input* screen nonblank, please now select the *Clear All* button on the *Input* screen to zero all the data and blank the screen, in order to better follow the presentation below. We will proceed with the explanation as if you were typing your data into an initially blank *Input* screen.

If you use the *Tab* button, it will lead you through the steps needed to input all data as indicated by the circled numbers in Figure A-5. On a blank *Input* screen, *Tab* first to the *Cam Omega* box in the upper left corner and type in the speed of the camshaft in rpm. *Tab* again (or mouse click if you prefer) to the *Number of Segments* box and type in any number desired between 1 and 20. That number of rows will immediately become visible on the *Input* screen.

Tab or click to the *Delta Theta* pulldown and select one of the offerings on the menu. No other values than those shown may be used. Next select either *Translating* or *Oscillating* for the follower motion.

The *Starting Angle* box allows any value to be typed to represent the angle on the timing diagram that you choose to begin the first segment of your cam, i.e., cam zero. Unless the timing diagram places machine zero within a motion event, this can be left as zero, making cam zero the same as machine zero (the default).

The *External Force* check box and the *Motion/Force* option buttons are provided for situations in which the cam follower is subjected to a substantial external force during operation, such as in a compactor mechanism. Checking this box temporarily converts the cam design program into a force-time function design program in which the shape of the force-time (or force-angle) function can be defined as if it were a cam displacement function with units of force instead of length. When calculated, the force data is stored for later superposition on the follower's dynamic forces due to motion. When *External Force* is checked, a dialog box pops up with further information on how to use it.

Another *Tab* should put your cursor in the box for the *Beta* (segment duration angle) of segment 1. Type any desired angle (in degrees). Successive *Tabs* will take you to each *Beta* box to type in the desired angles. The *Betas* must, of course, sum to 360 degrees. If they do not, a warning will appear.

As you continue to *Tab* (or click your mouse in the appropriate box, if you prefer), you will arrive at the boxes for *Motion* selection. These boxes offer a pulldown selection of *Rise, Fall, Poly, Dwell*, and *Spline*. You may select from the pulldown menus with the mouse, or you can type the first letter of each word to select it. Rise, fall, and dwell have obvious meanings. The *Poly* choice indicates that you wish to create a customized polynomial function for that segment, and this will later cause a new screen to appear on which you will define the order and boundary conditions of your desired polynomial function. The *Spline* choice indicates that you wish to create a customized *B-Spline* function for that segment, and this will later cause a set of new screens to appear on which you will define the boundary conditions, order, and knot locations of your desired *B-Spline* function.

The next set of choices that you will *Tab* or mouse click to are the *Program* pull-downs. These provide a menu of standard cam functions such as *Modified Trapezoid, Modified Sine,* and *Cycloid,* as well as a large number of specialized functions as described in this text. Also included are portions of functions such as the first and second halves of cycloids and simple harmonics that can be used to assemble piecewise continuous functions for special situations. The standard double-dwell polynomials, 3-4-5 and 4-5-6-7, as well as a symmetrical single-dwell 3-4-5-6 rise-fall function are provided as menu picks though they can also be created with a *Poly* choice and subsequent definition of their boundary conditions.

After you have selected the desired *Program* functions for each segment, you will *Tab* or *Click* to the *Position Start* and *End* boxes. *Start* in this context refers to the beginning displacement position for the follower in the particular segment, and *End* for its final position. You may begin at the "top" or "bottom" of the displacement "hill" as you wish, but be aware that the position values of the follower must be in a range from zero to some positive value over the whole cam. In other words, **you cannot include any anticipated base or prime circle radius in these position data**. These position values represent the excursion of the so-called *S* diagram (displacement) of the cam and cannot include any prime circle information (which will be input later). **Note that if you selected a translating follower, then the displacement values are defined in length units, but if you chose an oscillating follower, they must be in degrees.**

As each row's (segment's) input data are completed, the *Calc* button for that row will become enabled. Clicking on this button will cause that segment's *S V A J* data to be calculated and stored. After the *Calc* button has been clicked for any segment's row, the *Plot* and *Print* buttons for that segment will become available. Clicking on these buttons will bring up a plot or a printed table of data for *S V A J* data for that segment only. More detailed plots and printouts can be obtained later from the *Home* screen. The plots and prints are enabled at this location to allow you to determine the values of boundary conditions as you work your way through a piecewise function.

Polynomial Functions

If any of your segments specified a *Poly* motion, clicking the *Calc* button will bring up the *Boundary Condition* screen shown in Figure A-6. The cursor will be in the box for *Number of Conditions Requested.* Type the number of boundary conditions (BCs) desired, which must be between 2 and 40 inclusive. When you hit *Enter* or *Tab,* or mouse click away from this box, the rest of the screen will activate, allowing you to type in the desired values of BCs. Note that the start and end values of position that you typed on the *Input* screen are already entered in their respective *S* boundary condition boxes at the beginning and end of the segment. Type your additional end of interval conditions on *V, A, J,* and *P* as desired. If you also need some BCs within the interval, click or tab to one of the boxes in the row labeled *Local Theta* at the top of the screen and type in the value of the angle at which you wish to provide a BC. That column will activate and you may type whatever additional BCs you need.

The box labeled *Number of Conditions Selected* monitors the BC count, and when it matches the *Number of Conditions Requested,* the *Next* button becomes available. Note that what you type in any (yellow) text box is not accepted until you hit *Enter* or move off that box with the *Tab* key or the mouse, allowing you to retype or erase with no effect until you leave the text box. (This is generally true throughout the program.)

Selecting the *Next* button from the BC screen calculates the coefficients of the polynomial by a Gauss-Jordan reduction method with partial pivoting. All computations are done in double precision for accuracy. If an inconsistent set of conditions is sent to the solver, an error message will appear. If the solution succeeds, it calculates *s v a j* for the segment. When finished, it brings up a summary screen that shows the BCs you selected and also the coefficients of the polynomial equation that resulted. You may at this point want to print this screen

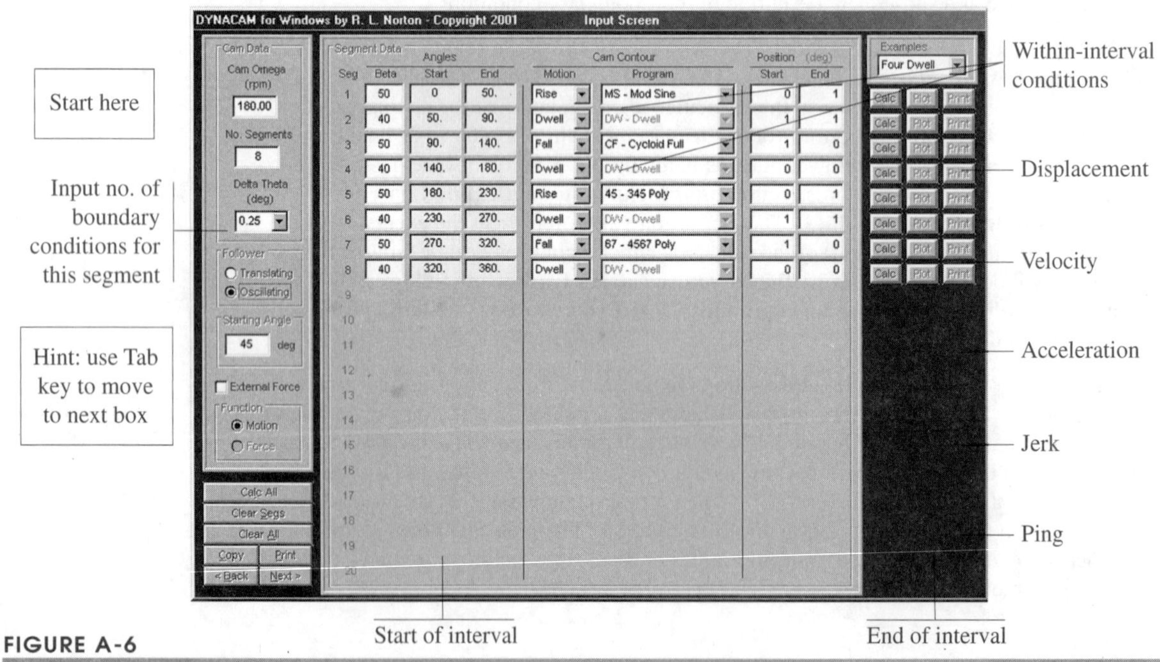

FIGURE A-6

Boundary Condition Input screen for polynomial functions in program DYNACAM—3-4-5-6 single-dwell function shown

to the printer or copy and paste it into another document for your records. You will only be able to reconstruct it later by again defining the BCs and recalculating the polynomial.

Spline Functions

If any of your segments specified a *Spline* motion, clicking the *Calc* button will bring up the same *Boundary Condition* screen shown in Figure A-6 that is used for polynomial functions. These are defined in exactly the same way as described for polynomials in the preceding section.

Once you finish selecting boundary conditions for your spline and hit the Next button, it takes you to the *Spline Function* screen shown in Figure A-7. This screen shows the current segment, the number of boundary conditions that were selected, and requires you to choose a spline order between 4 and the number of boundary conditions previously selected. It then displays the total number of spline knots used and the number of those available for distribution as interior knots. It also calculates the spline functions and displays their *S V A J* functions for the default assumption of evenly spaced interior knots. The current locations of the knots are displayed in the right side-bar.

The shape of the spline can be manipulated as described in Chapter 5 by moving the interior knots around. There are several ways to do this. One is to select a knot with its radio button in the right side-bar and type the angle to move it to in the yellow box. Alternatively, you may select a knot with its radio button and then click the mouse on any one of the *svaj* plots at any location that you want that knot to move to. A shortcut for picking a knot is to *Shift Click* near the knot you want to activate and this will select its radio button for you. Then a *Click* will move that selected knot. Note that because knots must be in ascending order, it

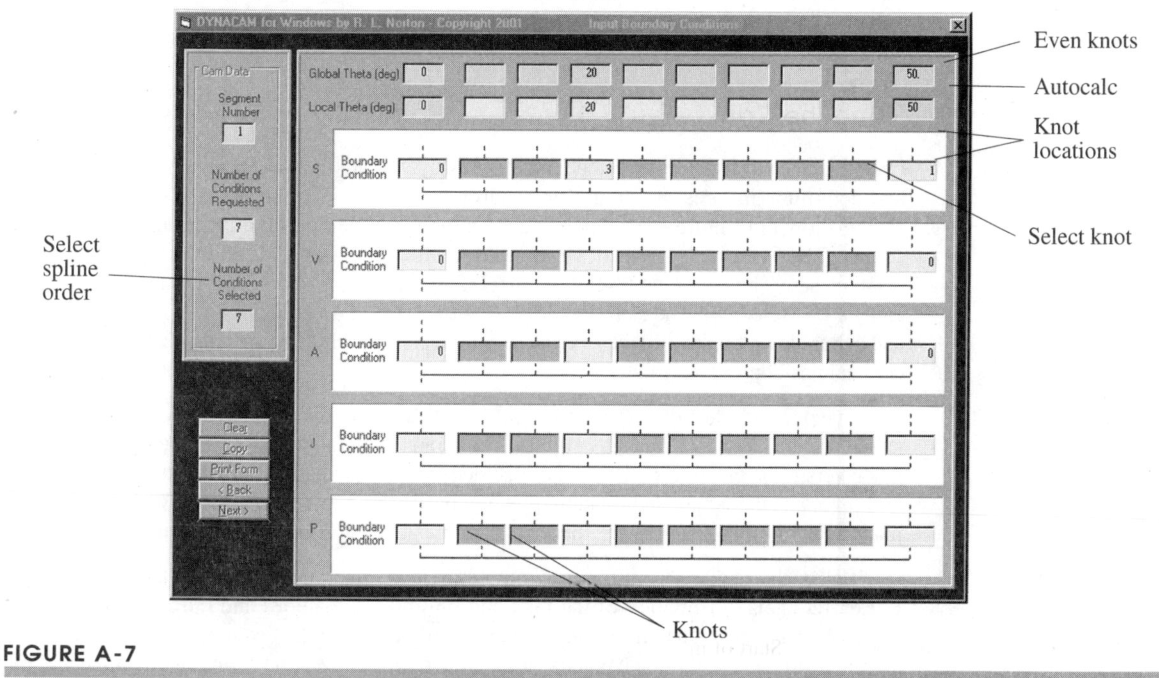

FIGURE A-7

Spline function screen with adjustable interior knot locations will refuse to violate their order if you request an inappropriate knot location. The plots and extreme values will update immediately unless you unchecked the *Autocalc On* box at upper right. If you are making a large number of knot changes, turning off *autocalc* will speed the process by suppressing the screen updates. It may also avoid tripping error messages engendered by a poor initial distribution of multiple knots until you can get them more or less where you want them before allowing it to recalculate the splines.

The back button returns you to the *Boundary Condition* screen if you wish to change them. The *Show Splines* button will display the basis functions that make up the B-Splines. *Plot Functions* returns the B-Spline plots. *Next* returns you to the *Input Screen.*

Back to the *Input* Screen

Completing a polynomial or spline function returns you to the *Input* screen. When all segments have been calculated, select the *Next* button on the *Input* screen (perhaps after copying it to the clipboard or printing it to the printer with the appropriate buttons). This will bring up the *Continuity Check* screen.

Continuity Check Screen

This screen provides a visual check on the continuity of the cam design at the segment interfaces. The values of each function at the beginning and end of each segment are grouped together for easy viewing. The fundamental law of cam design requires that the *S, V,* and *A* functions be continuous. This will be true if the boundary values for those functions shown grouped as pairs are equal. If this is not true, then a warning dialog box will appear when the *Next* button is clicked. It displays any errors between adjacent segments in both absolute and percentage terms. You must decide if any errors indicated are significant or due only to computational roundoff error *Cancel* will return you to the *Input Data* screen to correct the problem and *OK* will return you to the *Home* screen.

Sizing the Cam

Once the *S V A J* functions have been defined to your satisfaction, it remains to size the cam and determine its pressure angles and radii of curvature. This is done from the *Size Cam* screen shown in Figure A-8, which is accessible from the button of that name on the *Home* screen. The *Size Cam* screen allows the cam rotation direction and follower type (flat or roller) to be set. The cam type (radial or barrel) and follower motion (translating or oscillating) can also be selected. The cam start angle (cam zero) as chosen on the *S V A J* screen is shown and a drop down box allows the keyway location versus machine zero to be selected among four cardinal positions.

The prime circle radius, (base circle for a flat follower), roller radius (if any), and cutter radius can be typed into their respective boxes. For a translating follower, its offset and angle of translation versus the positive x-axis can be specified. For an oscillating follower, the location of the arm pivot and the length and rotation direction of the arm must be provided. When a change is made to any of these parameters, the schematic image of cam and follower is updated.

Select the *Calculate* button to compute the cam size parameters and the cam contour. For barrel cams, a summary of the max and min pressure angles and radii of curvature appears. *Next* returns you to the *Home Screen*. Barrel cam contours cannot be displayed as they are three-dimensional. The cam contour is now ready for export. A dialog box appears with instructions on exporting the cam contour data for manufacturing.

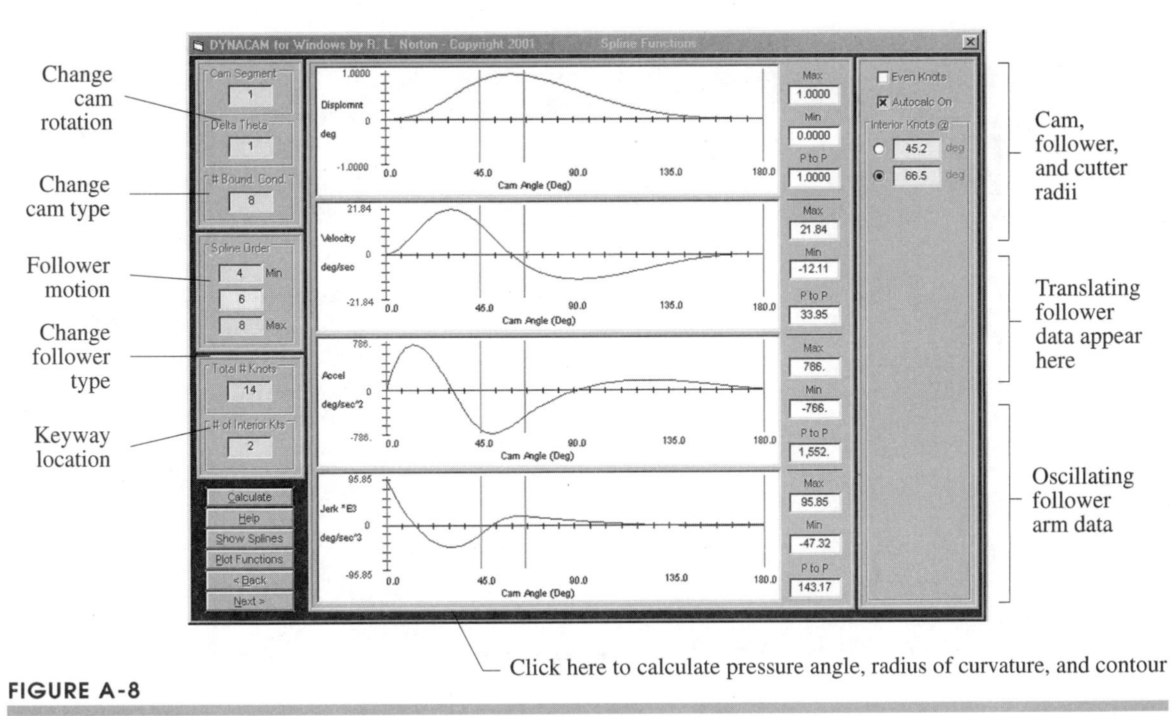

FIGURE A-8

Size Cam screen from DYNACAM for a radial cam with oscillating roller follower

Drawing the Cam

For radial cams, the cam contour is displayed on the *Draw Cam* screen after calculation as shown in Figure A-9. This screen allows the prime (or base) circle and, if translating, the offset, to be changed further. Cam rotation direction can also be changed here and the cam contour will be recalculated and redisplayed. Changes to other parameters can be accomplished via the *Back* button which returns you to the *Size Cam* screen. The maximum and minimum pressure angles and radii of curvature are shown in the right side-bar. For more information on these parameters, select the *Show Summary* radio button at the lower right. This will change the cam image to a summary of pressure angle and radius of curvature information. The *Show Cam* radio button will redraw the cam contour. Selecting the *Show Conjugate* radio button computes and displays the conjugate of the cam previously shown.

In the cam profile drawing, a curved arrow indicates the direction of cam rotation. The initial position of a translating roller follower at cam angle $\theta = 0$ is shown as a filled circle with rectangular stem, and of a flat-faced follower as a filled rectangle. Any eccentricity shows as a shift up or down of the roller follower with respect to the X axis through the cam center. The smallest circle on the cam centerline represents the camshaft, and its keyway is shown as a solid dot. The larger circle is the base circle. The prime circle is not drawn.

The default image shows the inner cam surface. Check boxes above the image allow the outer surface, the track of the follower, or the cutter path around the cam to be displayed instead. When the follower path is shown, the pitch curve is drawn through the locus of the roller follower centers.

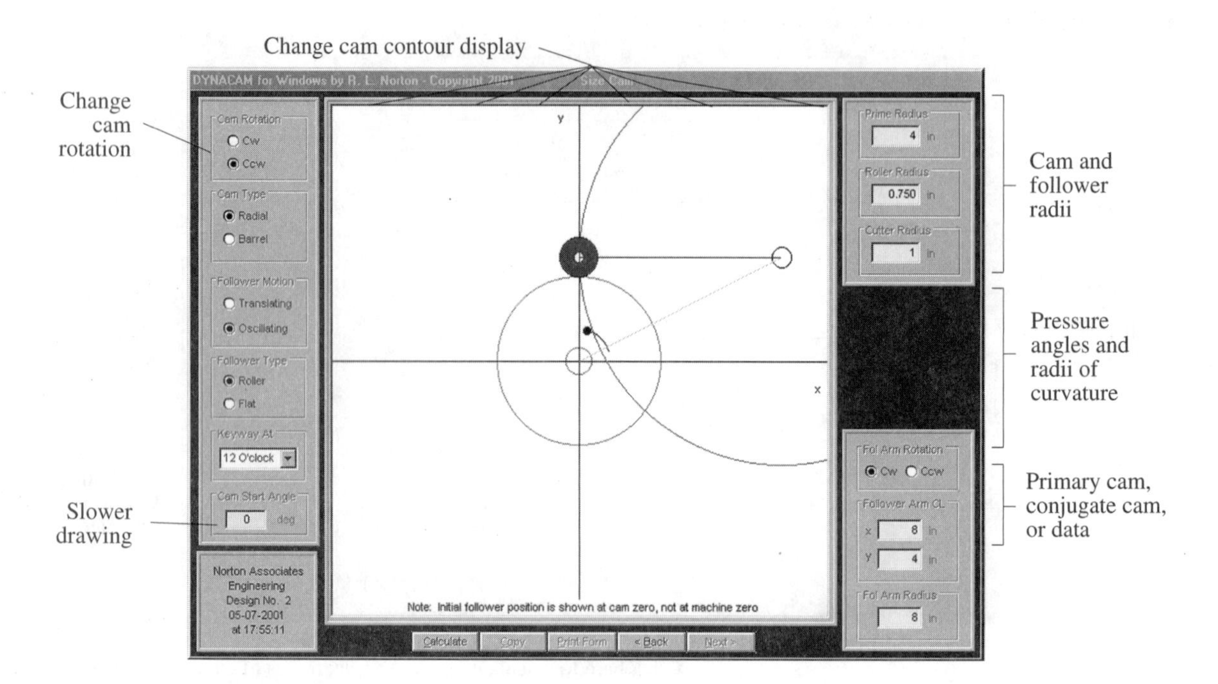

FIGURE A-9

Draw Cam screen from DYNACAM for a radial cam with oscillating roller follower

The check box labeled *Cam vs. Machine Zero* phase-shifts the cam contour to machine zero and redisplays it. In the case of an oscillating arm follower, the cam contour is reoriented to put the keyway at the selected o'clock position. For a translating follower, the follower is kept in a fixed orientation along the x axis and the keyway rotated to the correct relative position. These phase shifts incorporate the start angle, follower angle, and keyway location.

The radial lines that form pieces of pie within the base circle represent the segments of the cam. If the cam turns counterclockwise, the radial lines are numbered clockwise around the circumference and vice versa.

The *Next* button returns you to the Home Screen. The cam profile is now ready for export. Whether the *Show Cam* or *Show Conjugate* radio button was selected when *Next* was clicked determines which of those two cam contours will be the one exported. So to generate cutter data for a pair of conjugate cams requires sequential calculation and export of their respective data to files. The state of the *Inner Surface/Outer Surface* radio buttons at the time *Next* was clicked also dictates which cam surface will be exported. Generating both cam surfaces also requires sequential calculation and export of data.

The cam profile for a radial cam can be displayed at any time from the *Home* screen with the *Draw Cam* button. So, for creation of multiple cam surfaces or conjugates of radial cams, it is only necessary to revisit the *Draw Cam* screen without resizing the cam. Clicking *Next* on the *Draw Cam* screen returns you to the *Home Screen* where a dialog box appears with instructions on exporting the cam contour data for manufacturing.

Exporting Cam Contour Data (DYNACAM Only)

There is more than one way to get the cam contour data out of Dynacam for manufacturing, but only one of these is set up specifically for that purpose, and it is strongly recommended that you use it in order to avoid errors in manufacturing. On the *Home* screen, the *File* pull-down menu has an *Export* selection, within which are selections for *Spreadsheet* and *Profile*.

Profile is the recommended choice. Depending on the type of cam designed (radial or barrel) and the type of follower (roller or flat), the selections under *Profile* will vary. They typically will provide one or several choices that may include surface coordinates, cutter centerline, and roller follower centerline coordinates as appropriate. In all cases, these data are referenced to **cam zero**, not to machine zero, on the premise that if these two angles differ it is probably because machine zero is within a motion segment of the cam. In that case, one should NOT begin cutting the cam contour at machine zero.

The data exported for radial cams is only provided in Cartesian coordinates in order to avoid confusion and possible error by the manufacturer if they were given polar coordinates, especially for oscillating followers or for translating followers with offsets. For compatibility with 3-D CAD/CAM systems, three coordinates are provided, x, y, and z, with z set to zero. The z column is easily deleted in a spreadsheet if not needed.

The exported data for barrel cams is in 3-D cylindrical coordinates, R, θ, z, with R set to a constant value equal to the prime cylinder radius; θ is the cam angle, and z is the axial follower displacement at R, θ.

If you insist on being ornery and want to have your cam cutter data in some other form, choose the *Spreadsheet* option and you will get every piece of data that DYNACAM has calculated for this cam. But don't complain when you pick the wrong columns of data to give to the shop and your cam doesn't work!

Kinetostatic Analysis (DYNACAM Only)

When the cam has been sized, the *Dynamics* button on the *Home* screen will become available. This button brings up the *Dynamics* screen shown in Figure A-10. Text boxes are provided for typing in values of the effective mass of the follower system, the effective spring constant and spring preload for a force-closed follower, and a damping factor. By effective mass is meant the mass of the entire follower system as reflected back to the cam-follower roller centerline or cam contact point as defined in Chapter 8. Any link ratios between the cam-follower and any physical masses must be accounted for in calculating the effective mass. Likewise the effective spring in the system must be reflected back to the follower. The damping is defined by the damping ratio ζ, as defined for second-order vibrating systems. See Chapters 8 and 9 for further information.

The journal diameter and the coefficients of friction are used for calculating the friction torque on the shaft. The *Start New* or *Accumulate* switch allows you to either make a fresh torque calculation or accumulate the torques for several cams running on a common shaft. The energy information in the window can be used to calculate a flywheel needed for any coefficient of fluctuation chosen as described in Chapter 9. The program calculates a smoothed torque function by multiplying the raw camshaft torque by the coefficient of fluctuation specified in the box at lower right of the screen.

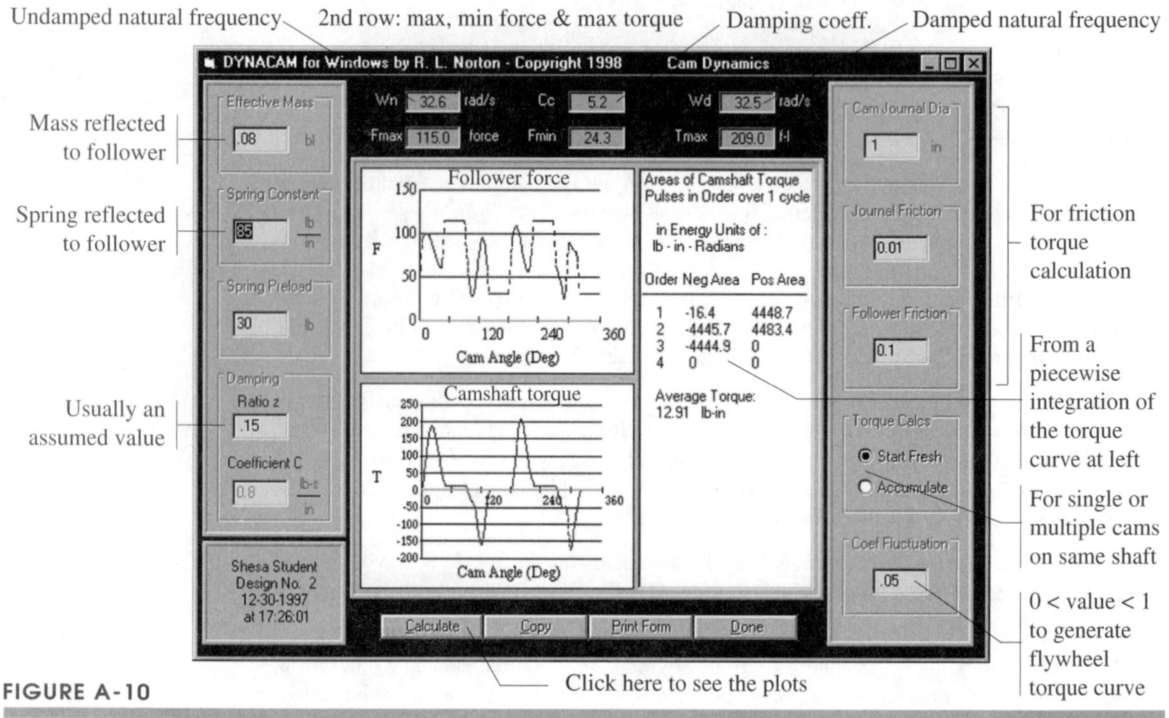

FIGURE A-10

Cam Dynamics screen in DYNACAM

Dynamic Analysis (DYNACAM Only)

True dynamic vibration analysis, as described in Chapter 10, is available from the *Vibration* button on the *Home* screen once the kinetostatic calculation has been done. This brings up the *Select Dynamic Model* screen shown in Figure A-11. Four dynamic models of the types described in Chapter 10 are available from this screen. A fifth model for the case of a form closed follower train is also provided, but is directly invoked without passing through this screen when form closure has been selected on the *Dynamics* screen.

Each model diagram has an *Info* button that will display a message describing its purpose and application. Either selecting the model's radio button and hitting *Next*, or double-clicking on the image of the desired model will take you to the next screen.

Figure A-12 shows the *Dynamic Modeling* screen for the 2-DOF model of an industrial cam-follower system. Text boxes are provided for the input of the relevant effective mass and effective spring data, along with assumed levels of damping.

The box at lower left provides control over the parameters for the 4th-order adaptive Runge-Kutta ODE solver. It is suggested that these be left at their default values. The end time's value defaults to, and cannot be set to less than, two cycles of the camshaft, but can be set longer. Thus, the calculation solves for at least 2 cam revolutions and displays the results in plots of displacement error, $s–x$, x, $\dot{x}$, and $\ddot{x}$ along with their extreme values. The data is saved only for the second revolution in order to discard the effects of any start-up transients. The center bar shows calculated values of the system natural frequencies, natural periods and

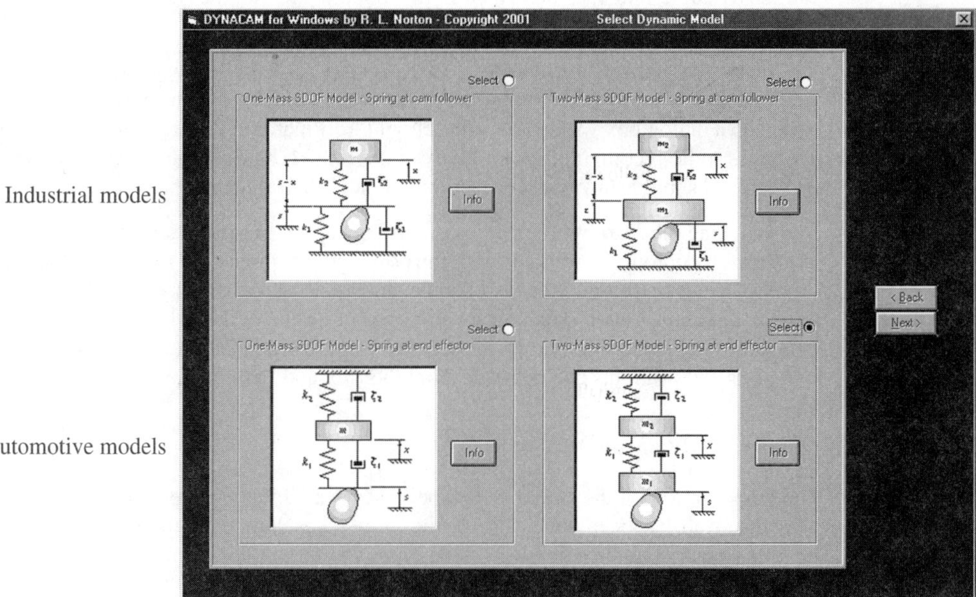

FIGURE A-11

Select Dynamic Model screen in program DYNACAM

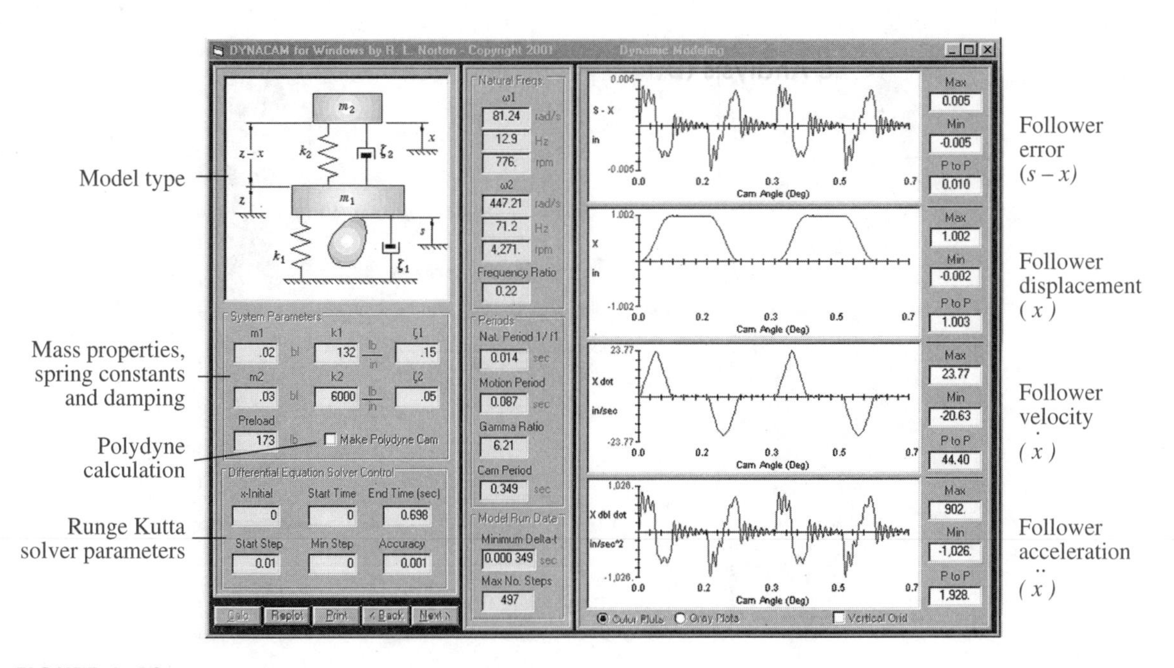

FIGURE A-12

Dynamic Modeling screen in program Dynacam

various dynamic ratios, as described in this book. Clicking the *Next* button at this point returns you to the *Home* screen.

POLYDYNE AND SPLINEDYNE CALCULATION The *Select Dynamic Model* screen provides a check box to make a polydyne or splinedyne cam of the current design. Checking it brings up a dialog box with some instructional information. Proceeding causes the follower dynamics to be recomputed as a polydyne or splinedyne function. The dynamic plots are updated and should show marked improvement in their dynamic behavior.

Clicking the *Next* button at this point brings up the *Size Cam* screen in order to recalculate the cam contour coordinates with the polydyne or splinedyne modifications. Retain the existing values on this and the next screen, and click on *Next* until you are back at the *Home* screen. The cam contour data now awaiting export is that of the polydyne or splinedyne cam, though the original nonpolydyne contour has been saved in another location so that comparison plots and printouts can be made from the *Plot* and *Print* menus if desired.

Stress Analysis (DYNACAM Only)

The *Stress* button on the *Home* screen becomes available only if either a kinetostatic or dynamic analysis has been done. If only the former was done, then the kinetostatic forces will be used to calculate the stresses. If the *Vibration* button has been exercised, then its more accurate dynamic forces are used for the stress calculation instead. The *Stress* screen is shown in Figure A-13. The material parameters for both cam and follower must be supplied but are defaulted to steel for both. The appropriate algorithm for surface stress will be used based on the earlier choice of flat or roller follower, as described in Chapter 12. If a roller

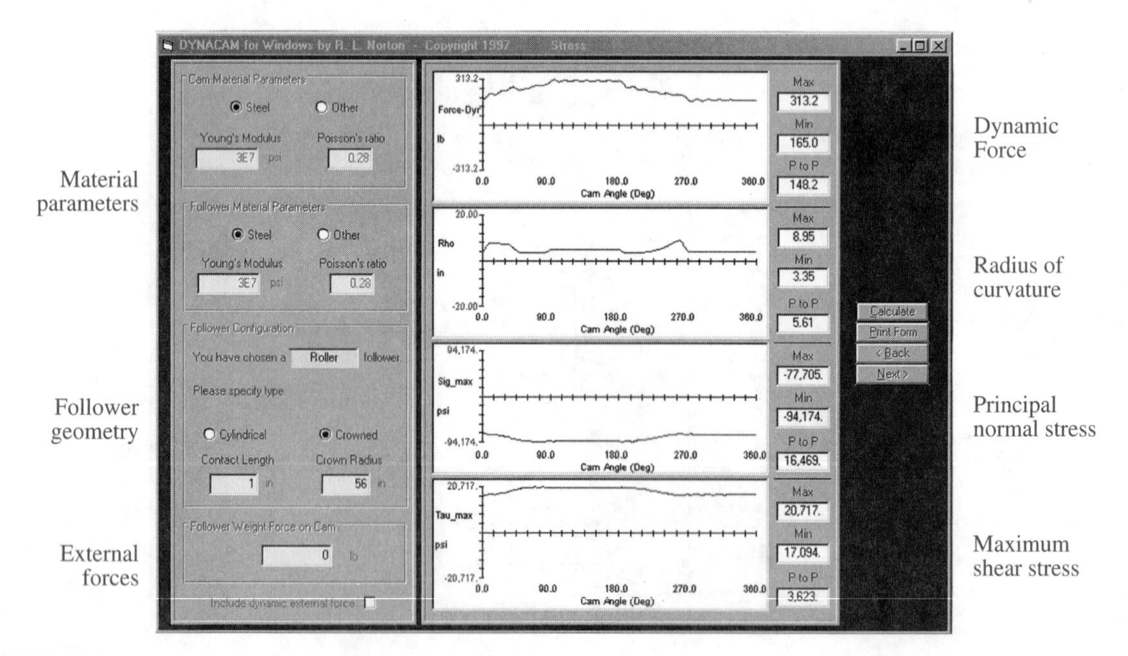

FIGURE A-13

Stress analysis screen in program DYNACAM

follower is used, then you must specify it as cylindrical or crowned and provide the relevant dimensional information. A box is provided for the input of a follower weight force if applicable. If an external force function was calculated on the *S V A J* form, then the check box to include it will become available and it can be included or excluded from the stress calculation as desired.

The calculate button creates and displays the maximum normal and maximum shear stress in cam and follower at each point around the cam and also shows the dynamic force and cam radius of curvature used in the calculation. *Next* returns you to the *Home* screen.

Fourier Transform (FFT)

The Fourier transform of any function calculated within DYNACAM (and the other programs) can be formed by pulling down the menu labeled *FFT* on the *Home* screen. Figure A-14 shows the *Fourier Transform* screen. Any one cam segment or the entire cam can be selected for transformation from the upper dropdown menu. The middle dropdown menu allows any calculated follower function to be chosen for tranformation. The lower dropdown menu chooses the number of harmonics desired in the Fourier spectrum. The *rms sum* and the *spectral power* are displayed in the right side-bar. Once a Fourier calculation is done, the spectral information for the chosen follower function will be available for plotting, printing, and export from the *Home* screen buttons. This FFT data can be used to recreate the functions from the Fourier series harmonics as continuous functions of time in an equation solver for example.

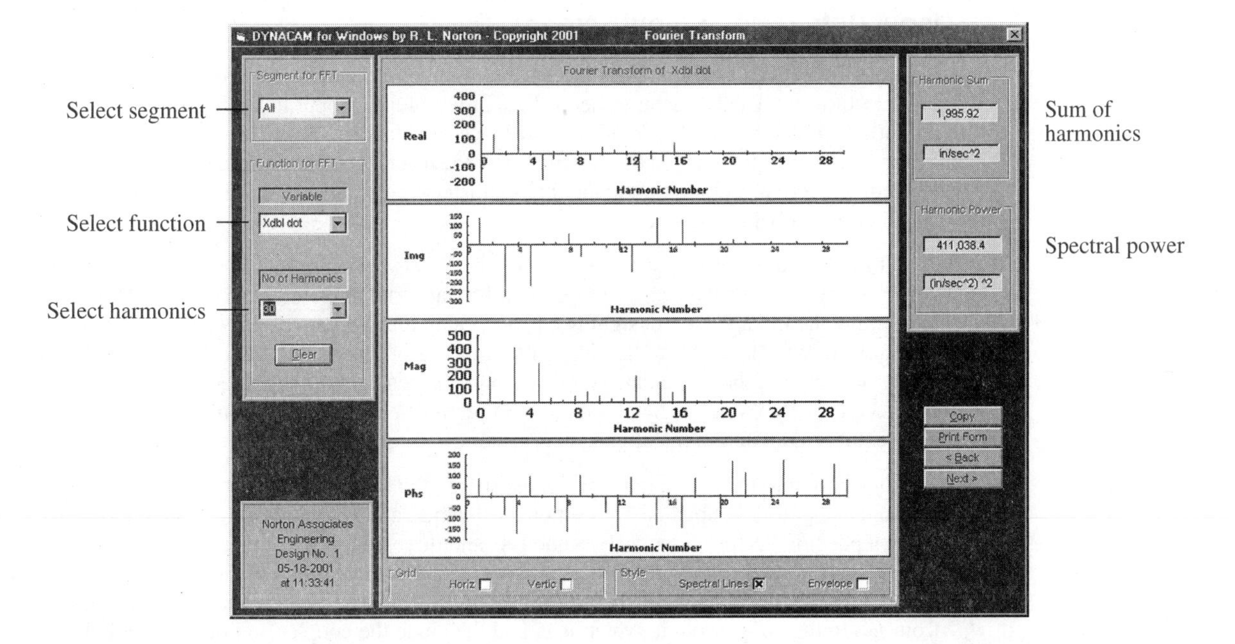

FIGURE A-14

Fourier transform screen in program Dynacam

Other

All the data calculated at each stage is saved and becomes available for plotting, printing, or exporting to disk. See Section A.2, General Program Operation for information on *New, Open, Save, Save As, Plot, Print, Units*, and *Quit* functions.

A.4 PROGRAM FOURBAR

FOURBAR *for Windows* is a linkage design and analysis program intended for use by engineers and other professionals who are knowledgeable in the art and science of linkage design. It is assumed that the user knows how to determine whether a linkage design is good or bad and whether it is suitable for the application for which it is intended. The program will calculate the kinematic and dynamic data associated with any linkage design, but cannot substitute for the engineering judgment of the user. The linkage theory and mathematics on which this program is based are documented in reference [1]. Please consult this reference for explanations of the theory and mathematics involved.

The FOURBAR *Home Screen*

Initially, only the *Input* and *Quit* buttons are active on the *Home* screen. Typically, you will start a linkage design with the *Input* button, but for a quick look at a linkage as drawn by the program, one of the examples under the *Example* pulldown menu can be selected and it will draw a linkage. If you activate one of these examples, when you return from the *Animate* screen you will find all the other buttons on the *Home* screen to be active. We will address each of these buttons in due course below.

Input Data (FOURBAR Input Screen)

Figure A-15 defines the input parameters, link numbering, and the axis system used in program FOURBAR. The link lengths needed are ground link 1, input link 2, coupler link 3, and output link 4, defined by their pin-to-pin distances and labeled a, b, c, d in the figure. The X axis is constrained to lie along link 1, defined by the instant centers $I_{1,2}$ and $I_{1,4}$ which are also labeled, respectively, O_2 and O_4 in the figure. Instant center $I_{1,2}$, the driver crank pivot, is the origin of the global coordinate system.

It might seem overly restrictive to force the X axis to lie on link 1 in this "aligned system." Many linkages will have their ground link at some angle other than zero. However, reorienting the linkage after designing and analyzing it merely involves rotating the piece of paper on which it is drawn to the desired final angle of the ground link. More formally, it means a rotation of the coordinate system through the negative of the angle of the ground link. In effect, the actual final angle of the ground link must be subtracted from all angles of links and vectors calculated in the aligned axis system.

In addition to the link lengths, you must supply the location of one coupler point on link 3 to find that point's coupler curve positions, velocities, and accelerations. This point is located by a position vector rooted at $I_{2,3}$ (point A) and directed to the coupler point P of interest which can be anywhere on link 3. The program requires that you input the polar coordinates of this vector which are labeled p and δ_3 in Figure A-15. The program asks for the distance from $I_{2,3}$ to the coupler point, which is p, and the angle the coupler point makes with link 3 which is δ_3. Note that angle δ_3 is not referenced to either the global coordinate system (GCS) X,Y or to the local nonrotating coordinate system (LNCS) x,y at point A. Rather, it is refer-

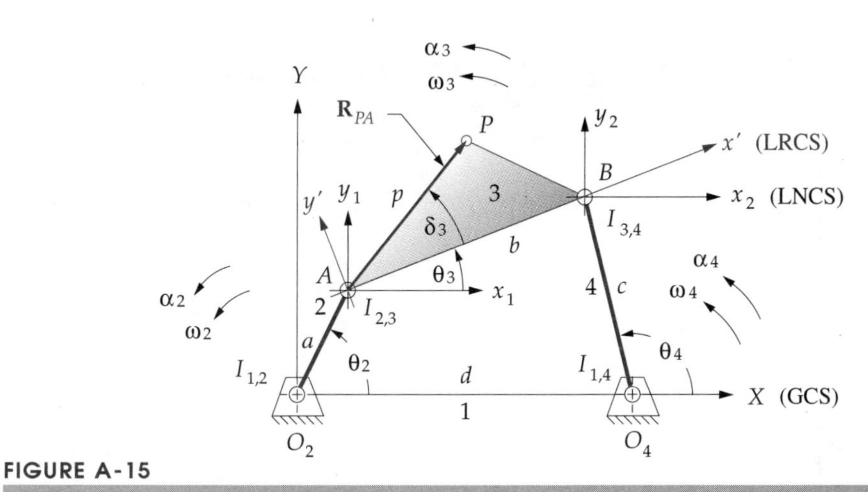

FIGURE A-15

Linkage data for program FOURBAR

enced to the line *AB* which is the pin-to-pin edge of link 3 (LRCS). Angle δ_3 is a property of link 3 and is embedded in it. The angle that locates vector $\mathbf{R}_{CA}$ in the *x,y* coordinate system is the sum of angle δ_3 and angle θ_3. This addition is done in the program, after θ_3 is calculated for each position of the input crank. The coordinate system, dimensions, angles, and nomenclature in Figure A-15 are consistent with those of reference [1] which were used for derivation of the equations used in program FOURBAR.

Calculation (FOURBAR, FIVEBAR, SIXBAR, and SLIDER *Input* Screens)

Basic data for a linkage design is defined on the *Input* screen shown in Figure A-16, which is activated by selecting the *Input* button on the *Home* screen. When you open this screen for the first time, it will have default data for all parameters. The linkage geometry is defined in the *Linkage Data* panel on the left side of the screen. You may change these by typing over the data in the yellow text boxes.

The open or crossed circuit of the linkage is selected in a panel at the lower left of the screen in Figure A-16. Select the type of calculation desired, one of *Angle Steps, Time Steps,* or *One Position* from the *Calculation Mode* in the upper-right panel. The start, finish, and delta step information is different for each of these calculation methods, and the input text box labels in the *Initial Conditions* panel will change based on your choice. Type the desired initial, delta, and final conditions as desired.

ONE POSITION This choice will calculate position, velocity, and acceleration for any one specified input position θ_2, input angular velocity ω_2, and angular acceleration α_2.

ANGLE STEPS This choice assumes that the angular acceleration α_2 of input link 2 is zero, making ω_2 constant. The values of initial and final crank angle θ_2, angle step $\Delta\theta_2$, and the constant input crank velocity ω_2 are requested. The program will compute all linkage parameters for each angle step. This is a steady-state analysis and is suitable for either Grashof or non-Grashof linkages provided that the total linkage excursion is limited in the latter case.

TIME STEPS This choice requires input of a start time, finish time, and a time step, all in seconds. The value for α_2 (which must be either a constant or zero) and the initial position

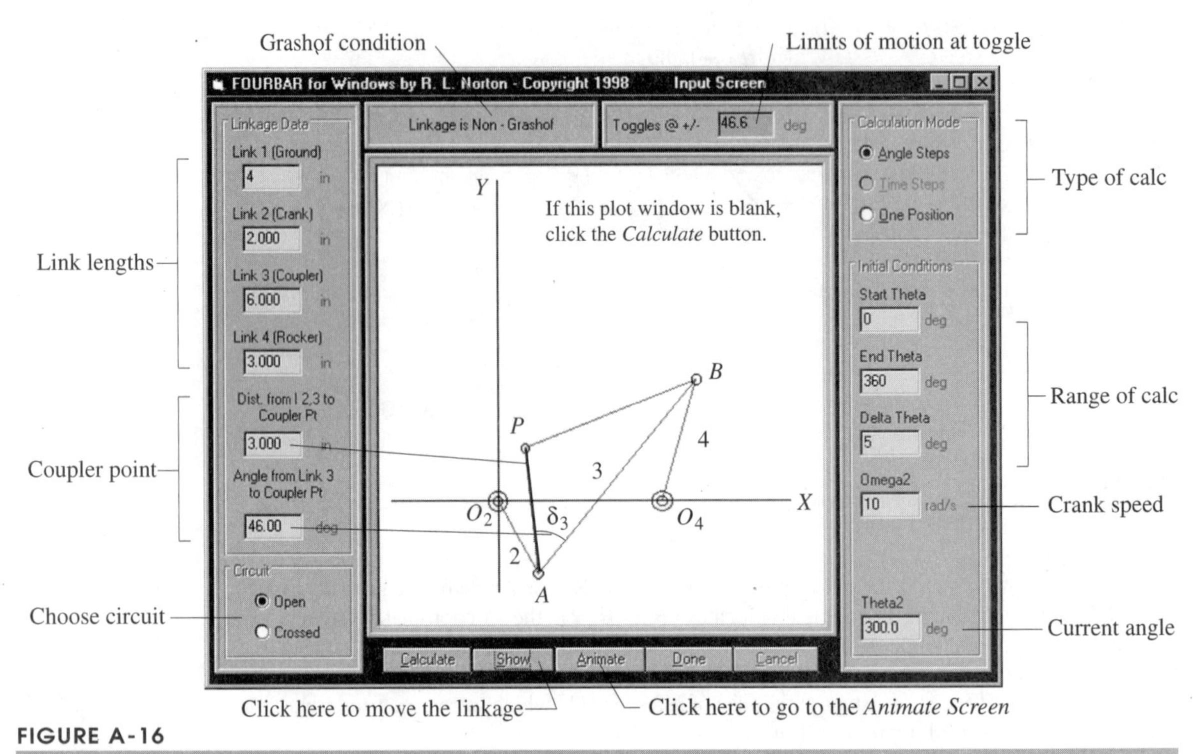

FIGURE A-16

Input Data screen for program FOURBAR; corresponding screens in FIVEBAR, SIXBAR, and SLIDER are similar

θ_2 and initial velocity ω_2 of link 2 at time zero must also be supplied. The program will then calculate all linkage parameters for each time step by applying the specified acceleration, which of course will change the angular velocity of the driver link with time. This is a transient analysis. The linkage will make as many revolutions of the driver crank as is necessary to run for the specified time. This choice is more appropriate for Grashof linkages, unless very short time durations are specified, as a non-Grashof linkage will quickly reach its toggle positions.

Note that a combination of successive **Time Step**, **Crank Angle,** and **Time Step** analyses can be used to simulate the start-up, steady-state, and deceleration phases, respectively, of a system for a complete analysis.

CALCULATE The *Calculate* button will compute all data for your linkage and show it in an arbitrary position in the linkage window on the *Input* screen. If at any time the white linkage window is blank, the *Calculate* button will bring back the image. The *Show* button will move the linkage through its range in "giant steps."

After you have calculated the linkage, the *Animate* and *Next* buttons on the *Input* screen will become available. The *Animate* button takes you to the *Animate* screen where you can run the linkage through any range of motion to observe its behavior. You can also change any of the linkage parameters on the *Animate* screen and then recalculate the results there with the *Recalc* button. The *Next* button on either the *Input* or *Animate* screen returns you to the *Home* screen. The *Plot* and *Print* buttons will now be available as well as the *Animate* button that will send you to the *Animate* screen.

CALCULATION ERRORS If a position is encountered that cannot be reached by the links (in either the angle step or time step calculations), the mathematical result will be an attempt to take the square root of a negative number. The program will then show a dialog box with the message *Links do not connect for Theta2=xx* and present three choices: Abort, Retry, or Ignore. **Abort** will terminate the calculation at this step and return you to the *Input* screen. **Retry** will set the calculated parameters to zero at the current position and attempt to continue the computation at the next step, reporting successive problems as they occur. **Ignore** will continue the calculation for the entire excursion, setting the calculated parameters to zero at any subsequent positions with problems, but will not present any further error messages. If a linkage is non-Grashof and you request calculation for angles that it cannot reach, then you will trip this error sequence. Choosing Ignore will force the calculation to completion, and you can then observe the possible motions of the linkage in the linkage window of the *Input* screen with the *Show* button.

GRASHOF CONDITION Once the calculation is done, the linkage's Grashof condition is displayed in a panel at the top left of the screen. If the linkage is non-Grashof, the angles at which it reaches toggle positions are displayed in a second panel at top right. This information can be used to reset the initial conditions to avoid tripping the "links cannot connect" error.

Animation (FOURBAR)

The *Animate* button on the *Home* screen brings up the *Animate* screen as shown in Figure A-17. Its *Run* button activates the linkage and runs it through the range of motion defined in its most

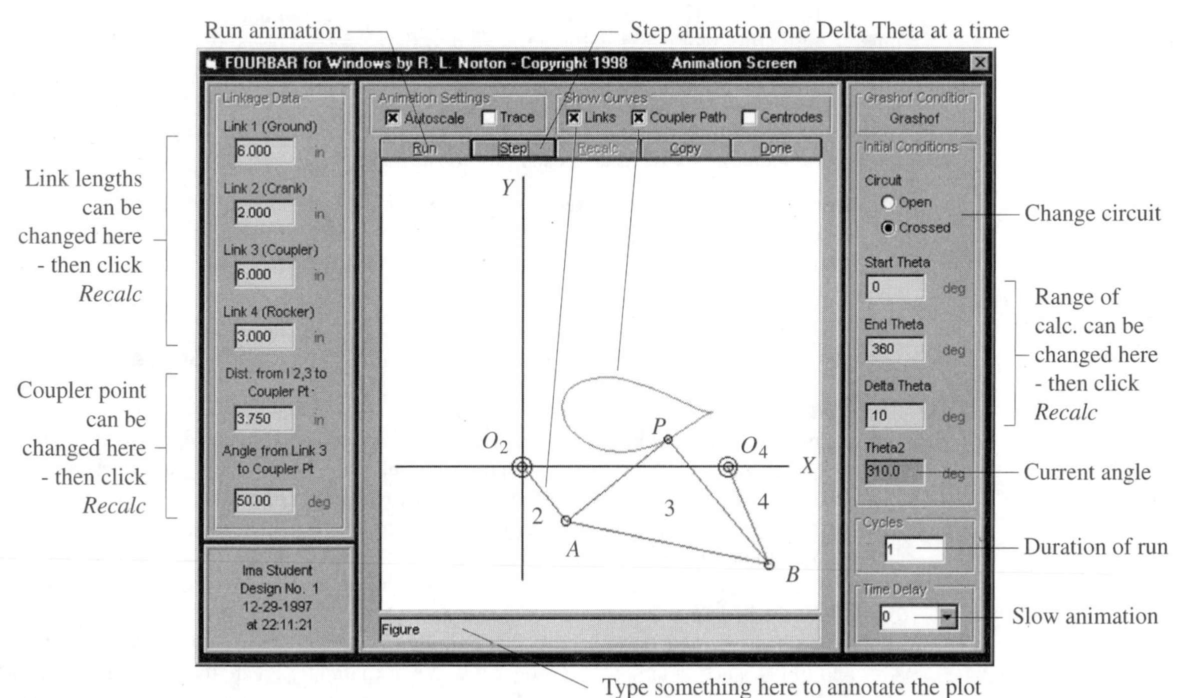

FIGURE A-17

FOURBAR *Animation Screen;* programs FIVEBAR, SIXBAR, and SLIDER have similar animation screens

recent calculation. The Grashof condition is reported at the upper right corner of the screen. The number of cycles for the animation can be typed in the *Cycles* box at the lower right of the screen.

A time delay (defaulted to 0) can be set with a drop-down menu below the *Cycles* box. Any number of cycles not on its list can be typed in the box. This time-delay feature is provided to accommodate variations in speed among computers. If your computer is very fast, the animation may occur too rapidly to be seen. If so, selecting larger positive numbers for the time delay will slow the animation. The *Step* button moves the linkage one increment of the independent variable at a time.

Text boxes in the *Linkage Data* panel allow the linkage geometry to be changed without returning to the *Input* screen. The initial conditions can be redefined in the panel on the left of this screen. The *Open-Crossed* selection can be switched, but the *Calculation Type* can only be changed on the *Input* screen. After any such change, the linkage must be recalculated with the *Recalc* button and then rerun.

Two panels at the top center of the *Animate* screen provide switches to change the animation display. In the *Show Curves* panel, displays of *Links*, *Coupler Path*, and *Centrodes* can be turned on and off in subsequent animations.

CENTRODES Only the FOURBAR program calculates and draws the fixed and moving centrodes (the loci of the instant centers). Different colors are used to distinguish the fixed from the moving centrodes. The centrodes are drawn with their point of common tangency located at the first position calculated. Thus, you can orient them anywhere by your choice of start angle for the calculation.

AUTOSCALE This feature can be turned on or off in the *Animation Settings* panel. The linkage animation plot is normally autoscaled to fit the screen based on the size of the linkage and its coupler curves (but not of the centrodes as they can go to infinity). You may want to turn off autoscaling when you wish to print two plots of different linkages at the same scale for later manual superposition. Turning off autoscaling will retain the most recent scale factor used. When on, it will rescale each plot to fit the screen.

TRACE Turning *Trace* on keeps all positions of the linkage visible on the screen so that the pattern of motion can be seen. Turning *Trace* off erases all prior positions, showing only the current position as it cycles the linkage through all positions giving a dynamic view of linkage behavior.

Dynamics (FOURBAR, FIVEBAR, SIXBAR, and SLIDER *Dynamics* Screen)

The *Dynamics* button on the *Home* screen brings up a screen that allows input of data on link masses, mass moments of inertia, *CG* locations, and any external forces or torques acting on the links. The location of the *CG* of each moving link is defined in the same way as the coupler point, by a position vector whose root is at the low-numbered instant center of each link. That is, for link 2 it is $I_{1,2}$; for link 3, $I_{2,3}$; and for link 4, $I_{1,4}$. These vectors for a fourbar linkage are labeled R_{CGi} in Figure A-18 where i is the link number. Note that the angle of each *CG* vector is measured with respect to a *local rotating coordinate system* (LRCS) whose origin is at the aforementioned instant center and whose x axis lies along the line of centers of the link. For example, in Figure A-18a, link 2's *CG* vector is 3 in at 30°, link 3's is 9 in at 45°, and link 4's is 5 in at 0°. The program will automatically create the necessary position vectors $\mathbf{R}_{12}$, $\mathbf{R}_{32}$, $\mathbf{R}_{23}$, $\mathbf{R}_{43}$, $\mathbf{R}_{34}$, and $\mathbf{R}_{14}$ needed for the dynamic force equations as shown in the free-body diagrams in Figure A-18b. These position vectors must be recalculated in

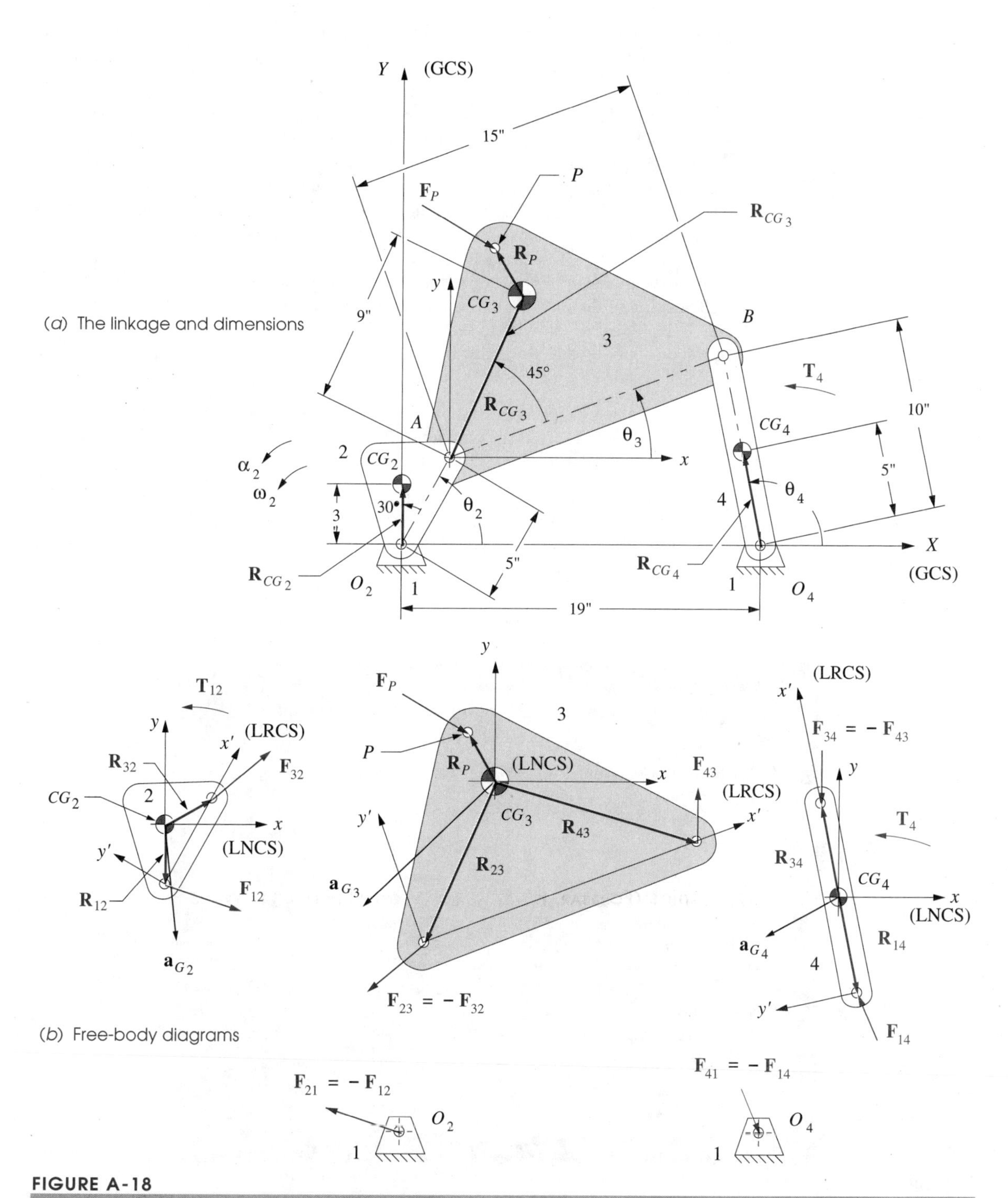

FIGURE A-18

Definition of data needed for dynamics calculations in Program FOURBAR

the *nonrotating local coordinate systems* (LNCS) at the links' *CG*s for each new position of the linkage as the link angles change.

The masses and mass moments of inertia with respect to the CGs of the moving links are also required. Any external forces or torques that are applied to links 3 or 4 are typed in the appropriate boxes on the *Dynamics* screen as shown in Figure A-19. The direction angle of any external force must be measured with respect to the *global coordinate system* (GCS). The program will assume that this angle remains constant for all positions of the linkage analyzed. You must also supply the magnitude and direction of the position vector $\mathbf{R}_P$, which locates any point on the force vector $\mathbf{F}_P$. This $\mathbf{R}_P$ vector is measured in the *rotating, local coordinate system* (LRCS) embedded in the link, as were the *CG*s of the links. $\mathbf{R}_P$ is *not* measured in the global system. The program takes care of the resolution of these $\mathbf{R}_P$ vectors, for each position of the linkage, into coordinates in the nonrotating local coordinate system at the *CG*. Note that if you wish to account for the gravitational force on a heavy link, you may do so by applying that weight as an external force acting through the link's *CG* at 270° in the global system ($\mathbf{R}_P = 0$).

For other linkages such as the sixbar and fourbar slider-crank, the dynamic data input is similar. The only difference is the number of links for which mass property data and possible external forces and torques must be supplied.

After solving (by clicking the *Solve* button), clicking the *Show Matrix* button will display the dynamics matrix for the linkage. The results of the dynamics calculations are automatically stored for later plotting and printing. The menus on the *Plot* and *Print* screens will expand to include forces and torques for all links.

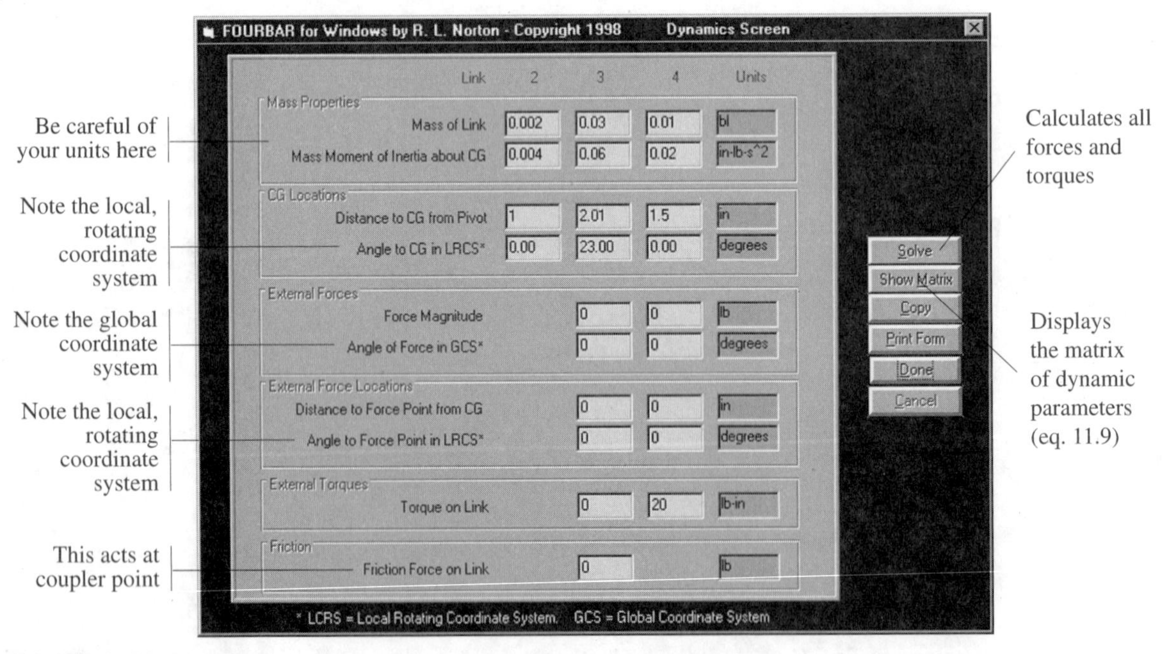

FIGURE A-19

Dynamic Data Input screen for program FOURBAR. Screens for programs FIVEBAR, SIXBAR, and SLIDER are similar

A

Balancing (FOURBAR Only)

The *Balance* button on the FOURBAR *Home* screen brings up the *Balance* screen shown in Figure A-20, which immediately displays the mass-radius products needed on links 2 and 4 to force balance it and reduce the shaking force to zero. If you place the total amounts of the calculated mass-radius products on the rotating links, the shaking force will become zero and the shaking torque will increase. A partial balance condition can be specified by reducing the balance masses and accepting some nonzero shaking force in return for a smaller increase in torque.

The FOURBAR *Balance* screen in Figure A-20 displays two mini-plots that superpose the shaking forces and shaking torques before and after balancing. Effects on the shaking force and torque from changes in the amount of balance mass-radius product placed on each link can be immediately seen in these plots. The energy in each pulse of the torque-time curve is also displayed in a side-bar on the right of this screen for use in a flywheel sizing calculation. See Section 9.8 (p. 251) for a discussion of the meaning and use of these data. The program calculates a smoothed torque function by multiplying the raw torque by the coefficient of fluctuation specified in the box at lower left.

Cognates (FOURBAR Only)

The *Cognates* pulldown menu allows switching among the three cognates that create the same coupler curve. Switching among them requires recalculation of all kinematic and dynamic parameters via the *Input*, *Dynamics*, and *Balance* buttons. The previously used mass property data is retained but can be changed easily by selecting the *Dynamics* button. The

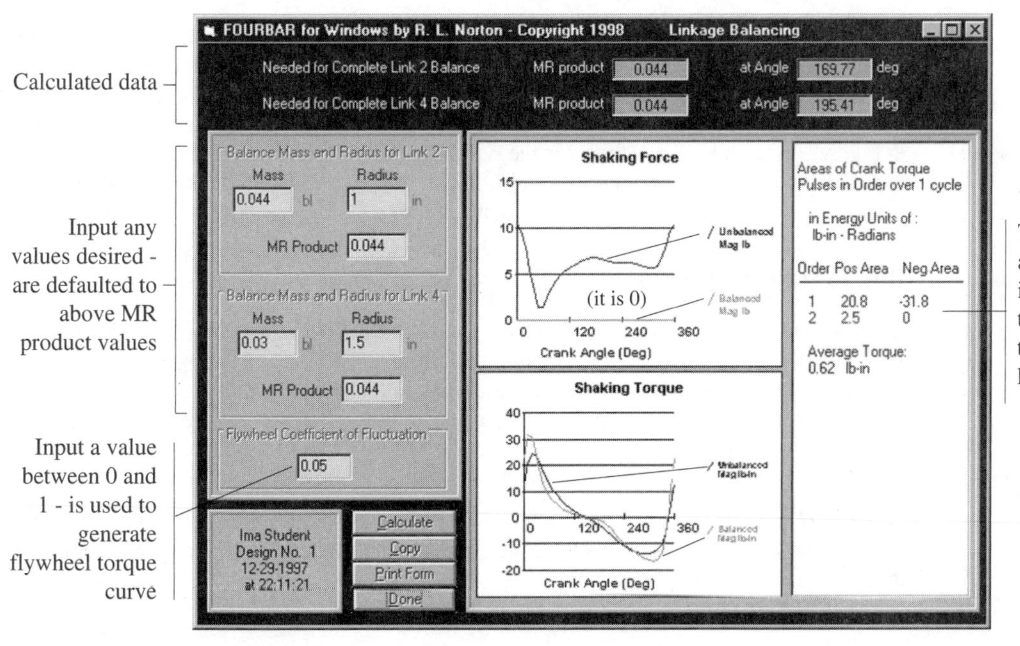

FIGURE A-20

Linkage Balancing screen in program FOURBAR only; other programs allow flywheel calculations without balancing

Cayley Diagram menu pick under *Cognates* displays that diagram of all three cognates. Whenever linkage data are changed on the *Input Screen* and recalculated, the program automatically calculates the dimensions of that linkage's two cognates. These can be switched to, calculated, and investigated at any time.

Synthesis (FOURBAR Only)

This pulldown menu allows selection of two- or three-position synthesis of a linkage, each with a choice of two methods. When the linkage is synthesized, its link geometry is automatically put into the input sheet and recalculation is then required.

Other

See Section A.2, General Program Operation for information on *New, Open, Save, Save As, Plot, Print, Units*, and *Quit* functions.

A.5 PROGRAM SIXBAR

SIXBAR *for Windows* is a linkage design and analysis program intended for use by students, engineers, and other professionals who are knowledgeable in or are learning the art and science of linkage design. It is assumed that the user knows how to determine whether a linkage design is good or bad and whether it is suitable for the application for which it is intended. The program will calculate the kinematic and dynamic data associated with any linkage design but cannot substitute for the engineering judgment of the user. The linkage theory and mathematics on which this program is based are documented in reference [1]. Please consult it for explanations of the theory and mathematics involved.

These are two of the five distinct sixbar isomers. The **Watt's II** mechanism is shown in Figure A-21 with the program's input parameters defined. The **Stephenson's III** mechanism is shown in Figure A-22 with its input parameters defined. Note that the program divides the sixbar into two stages of fourbar linkages. Stage 1 is the left half of the mechanism as shown in Figures A-21 and A-15. Stage 2 is the right half. The X axis of the global coordinate system is defined by instant centers $I_{1,2}$ and $I_{1,4}$ with its origin at $I_{1,2}$. The third fixed pivot $I_{1,6}$ can be anywhere in the plane. Its coordinates must be supplied as input.

The SIXBAR *Home* Screen

Initially, only the *Input* and *Quit* buttons are active on the *Home* screen. Typically, you will start a linkage design with the *Input* button, but for a quick look at a linkage as drawn by the program, one of the examples under the *Example* pulldown menu can be selected and it will draw a linkage. If you activate one of these examples, when you return from the *Animate* screen you will find all the other buttons on the *Home* screen to be active. We will address each of these buttons in due course below.

The *Home* screen's *Examples* pulldown menu includes both Watt's and Roberts' straight-line fourbar linkage stages driven by dyads (making them sixbars), a single-dwell sixbar linkage, and a double-dwell sixbar linkage.

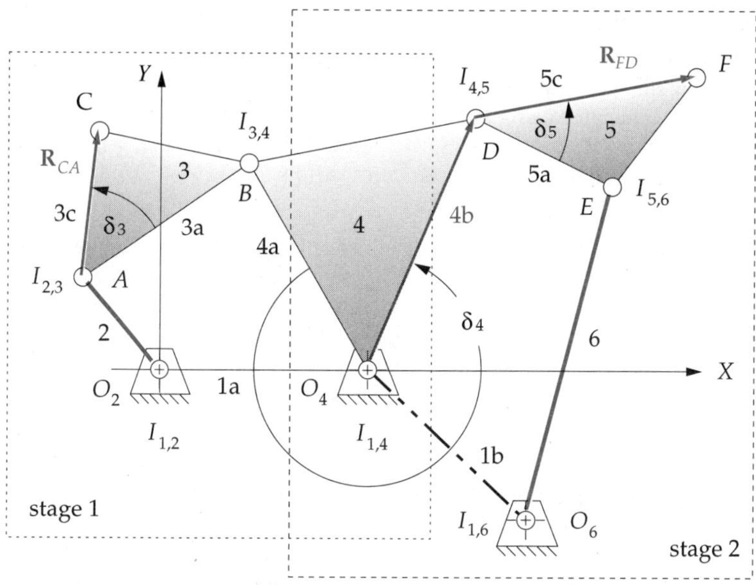

FIGURE A-21

Input data for program SIXBAR—a Watt's II linkage

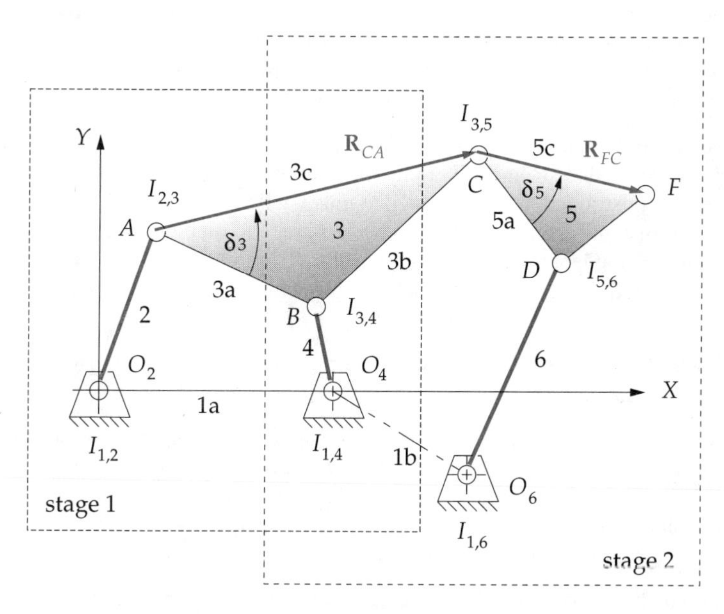

FIGURE A-22

Input data for program SIXBAR—a Stephenson's III linkage

Input Data (SIXBAR *Input* Screen)

Much of the basic data for the linkage design is defined on the *Input* screen which is activated by selecting the *Input* button on the *Home* screen. When you open this screen for the first time, it will have default data for all link parameters. You may change any of these by typing over the data in the yellow text boxes. A choice of Watt's or Stephenson's linkage must be made on the *Input* screen. The link information differs for the Watt's and Stephenson's linkages.

WATT'S II LINKAGE For the Watt's linkage (Figure A-21) the stage 1 data is: crank, first coupler, first rocker, and ground link segment from instant centers $I_{1,2}$ to $I_{1,4}$. These correspond to links 2, 3a, 4a, and 1a, respectively, as labeled in the Figure A-12. The stage 2 data are: second crank, second coupler, second rocker, corresponding respectively to links 4b, 5a, and 6 in Figure A-21. The angle δ_4 that the second crank (4b) makes with the first rocker (4a) is also requested. Note that this angle obeys the right-hand rule as do all angles in these programs.

Two coupler points are allowed to be defined in this linkage, one on link 3 and one on link 5. The method of location is by polar coordinates of a position vector embedded in the link as was done for the fourbar and fivebar linkages. The first coupler point C is on link 3 and is defined in the same way as in FOURBAR. Program SIXBAR requires the length (3c) of its position vector $\mathbf{R}_{ca}$ and the angle δ_3 which that vector makes with line 3a in Figure A-19. The second coupler point F is on link 5 and is defined with a position vector $\mathbf{R}_{fD}$ rooted at instant center $I_{4,5}$. The program requires the length (5c) of this position vector $\mathbf{R}_{fD}$ and the angle δ_5 which that vector makes with line 5a. The X and Y components of the location of the third fixed pivot $I_{1,6}$ are also needed. These are with respect to the global X,Y axis system whose origin is at $I_{1,2}$.

STEPHENSON'S III LINKAGE The stage 1 data for the Stephenson's linkage is similar to that of the Watt's linkage. The first stage's crank, first coupler, first rocker, and ground link segment from instant centers $I_{1,2}$ to $I_{1,4}$ correspond to links 2, 3a, 4, and 1a, respectively, in Figure A-20. The link lengths in stage 2 of the linkage are: second coupler and second rocker corresponding, respectively, to links 5a and 6 in Figure A-20.

Note that, unlike the Watt's linkage, there is no "second crank" in the Stephenson's linkage because the second stage is driven by the coupler (link 3) of the first stage. In this program link 5 is constrained to be connected to link 3 at link 3's defined coupler point C, which then becomes instant center $I_{3,5}$. The data for this are requested in similar format to the Watt's linkage, namely, the length of the position vector $\mathbf{R}_{ca}$ and its angle δ_3. The second coupler point location on link 5 is defined, as before, by position vector $\mathbf{R}_{fc}$ with length 5c and angle δ_5.

The X and Y components of the location of the third fixed pivot $I_{1,6}$ are required. These are with respect to the global X,Y axis system.

Select the type of calculation desired in the upper right corner of the screen, one of *Angle Steps, Time Steps*, or *One Position*. The *Calculate* button will compute all data for your linkage and show it in an arbitrary position on the screen. If at any time the linkage display window is blank, the *Calculate* button will bring back the image. See the description of calculations for program FOURBAR in Section A.4 (p. 553). They are similar in SIXBAR.

After you have calculated the linkage, the *Animate* and *Next* buttons on the *Input* screen will become available. *Animate* takes you to the *Animate* screen where you can run the linkage through any range of motion to observe its behavior. You can also change any of the linkage parameters on the *Animate* screen and then recalculate the results with the *Recalc* button. The *Next* button on either the *Input* or *Animate* screen returns you to the *Home* screen. The *Plot* and *Print* buttons will now be available as well as the *Animate* button that returns you to the *Animate* screen.

Animation (SIXBAR)

The *Animation* screen and its features in SIXBAR are essentially similar to those of Program FOURBAR. The only exception is the lack of a *Centrode* selection in SIXBAR. See the Animation discussion for FOURBAR in Section A.4 for more information.

Dynamics (SIXBAR *Dynamics* Screen)

Input data for dynamics calculation in SIXBAR are similar to that for program FOURBAR with the addition of two links. See the discussion of dynamics calculations for program FOURBAR in Section A.4 for more information.

Other

See Section A.2, General Program Operation (p. 532) for information on *New, Open, Save, Save As, Plot, Print, Units*, and *Quit* functions.

A.6 PROGRAM SLIDER

SLIDER *for Windows* is a linkage design and analysis program intended for use by students, engineers, and other professionals who are knowledgeable in the art and science of linkage design. It is assumed that the user knows how to determine whether a linkage design is good or bad and whether it is suitable for the application for which it is intended. The program will calculate the kinematic and dynamic data associated with any linkage design, but cannot substitute for the engineering judgment of the user. The linkage theory and mathematics on which this program is based are documented in reference [1]. Please consult it for explanations of the theory and mathematics involved.

The Slider Home Screen

Initially, only the *Input* and *Quit* buttons are active on the *Home* screen. Typically, you will start a linkage design with the *Input* button, but for a quick look at a linkage as drawn by the program, one of the examples under the *Example* pulldown menu can be selected and it will draw a linkage. If you activate one of these examples, when you return from the *Animate* screen you will find all the other buttons on the *Home* screen to be active. We will address each of these buttons below.

Input Data (SLIDER *Input* Screen)

Basic data for the slider linkage is defined on the *Input* screen, which is activated by selecting the *Input* button on the *Home* screen. When you open this screen for the first time, it will have default data for all link parameters. You may change any of these by typing over the data in the yellow text boxes.

Figure A-23 defines the input parameters for the fourbar slider-crank linkage. The link lengths needed are input link 2 and coupler link 3, defined by their pin-to-pin distances and labeled a and b in the figure. The X axis lies along the line d, through instant center $I_{1,2}$ (point O_2) and parallel to the direction of motion of slider 4. Instant center $I_{1,2}$, the driver crank pivot, is the origin of the global coordinate system. The slider offset c is the perpendicular distance from the X axis to the sliding axis. Slider position d will be calculated for all specified positions of the linkage.

In addition to the link lengths, you must supply the location of one coupler point on link 3 to find that point's coupler curve positions, velocities, and accelerations. This point is located by a position vector rooted at $I_{2,3}$ (point A) and directed to the coupler point P of interest that can be anywhere on link 3. The program requires that you input the polar coordinates of this vector which are labeled p and δ_3 in Figure A-14. The program needs the distance from $I_{2,3}$ to the coupler point p and the angle δ_3 that the coupler point makes with link 3. Note that angle δ_3 is not referenced to either the global coordinate system X, Y or to the local nonrotating coordinate system x,y at point A. Rather, it is referenced to the line AB, which is the pin-to-pin edge of link 3. Angle δ_3 is a property of link 3 and is embedded in it. The angle that locates vector $\mathbf{R}_{CA}$ in the x,y coordinate system is the sum of angle δ_3 and angle θ_3. This addition is done in the program, after θ_3 is calculated for each position of the input crank.

Calculation (SLIDER *Input* Screen)

See the description of calculations for program FOURBAR in Section A.4 (p. 553) for more information. They are similar in SLIDER. Select the type of calculation desired in the upper right corner of the screen, one of *Angle Steps, Time Steps*, or *One Position*. The *Calculate* button will compute all data for your linkage and show it in an arbitrary position on the screen. If at any time the white linkage window is blank, the *Calculate* button will bring back the image.

After you have calculated the linkage, the *Animate* and *Next* buttons on the *Input* screen will become available. *Animate* takes you to the *Animate* screen where you can run the linkage through any range of motion to observe its behavior. You can also change any of the linkage parameters on the *Animate* screen and then recalculate the results with the *Recalc* button. The *Next* button on either the *Input* or *Animate* screen returns you to the *Home* screen. The *Plot* and *Print* buttons will now be available as well as the *Animate* button that returns you to the *Animate* screen.

Animation (SLIDER *Animation* Screen)

The *Animation* screen and its features in SLIDER are essentially similar to those of program FOURBAR. An exception is the lack of a *Centrode* selection in SLIDER. See the discussion of the *Animation* screen for FOURBAR in Section A.4 for more information.

Dynamics (SLIDER *Dynamics* Screen)

Input data for dynamics calculation in SLIDER are similar to that for program FOURBAR. See the discussion of dynamics calculations for Program FOURBAR in Section A.4 for more information.

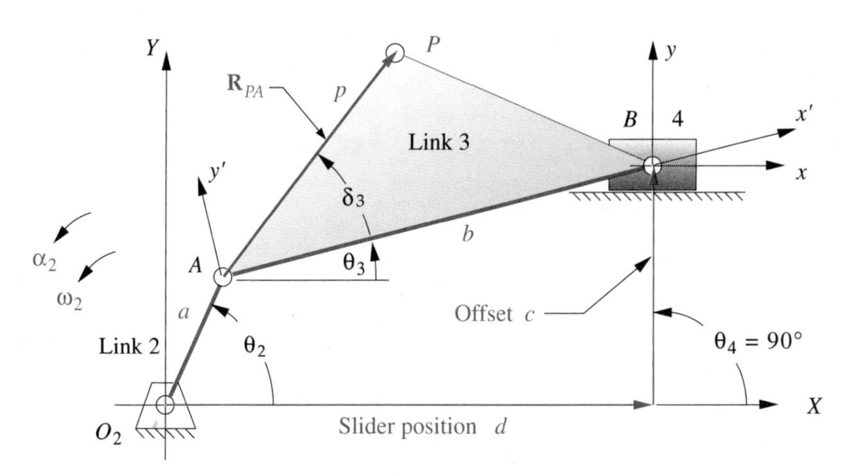

FIGURE A-23

Input data for Program SLIDER

Other

See Section A.2, General Program Operation (p. 534) for information on *New, Open, Save, Save As, Plot, Print, Units*, and *Quit* functions.

REFERENCES

1 **Norton, R. L.** (2001). *Design of Machinery*, 2ed. McGraw-Hill: New York.

A

Appendix B

MATERIAL PROPERTIES

For selected engineering materials. Many other alloys are available.

The following tables contain approximate values for strengths and other specifications of a variety of engineering materials compiled from various sources. In some cases the data are minimum recommended values. In other cases, they are data from a single test specimen. These data are suitable for use in the engineering exercises contained in this text, but should not be considered as statistically valid representations of specifications for any particular alloy or material. The designer should consult the materials' manufacturers for more accurate and up-to-date strength information on materials used in engineering applications or conduct independent tests of the selected materials to determine their ultimate suitability to any application.

Table No.	Description
B-1	Physical Properties of Some Engineering Materials
B-2	Mechanical Properties for Some Wrought-Aluminum Alloys
B-3	Mechanical Properties for Some Carbon Steels
B-4	Mechanical Properties for Some Cast-Iron Alloys
B-5	Mechanical and Physical Properties for Some Engineering Plastics

Table B-1 Physical Properties of Some Engineering Materials

Data from various sources.* These properties are essentially similar for all alloys of the particular material

Material	Modulus of Elasticity E		Modulus of Rigidity G		Poisson's Ratio ν	Weight Density γ	Mass Density ρ	Specific Gravity
	Mpsi	GPa	Mpsi	GPa		lb/in^3	Mg/m^3	
Aluminum Alloys	10.4	71.7	3.9	26.8	0.34	0.10	2.8	2.8
Beryllium Copper	18.5	127.6	7.2	49.4	0.29	0.30	8.3	8.3
Brass, Bronze	16.0	110.3	6.0	41.5	0.33	0.31	8.6	8.6
Copper	17.5	120.7	6.5	44.7	0.35	0.32	8.9	8.9
Iron, Cast, Gray	15.0	103.4	5.9	40.4	0.28	0.26	7.2	7.2
Iron, Cast, Ductile	24.5	168.9	9.4	65.0	0.30	0.25	6.9	6.9
Iron, Cast, Malleable	25.0	172.4	9.6	66.3	0.30	0.26	7.3	7.3
Magnesium Alloys	6.5	44.8	2.4	16.8	0.33	0.07	1.8	1.8
Nickel Alloys	30.0	206.8	11.5	79.6	0.30	0.30	8.3	8.3
Steel, Carbon	30.0	206.8	11.7	80.8	0.28	0.28	7.8	7.8
Steel, Alloys	30.0	206.8	11.7	80.8	0.28	0.28	7.8	7.8
Steel, Stainless	27.5	189.6	10.7	74.1	0.28	0.28	7.8	7.8
Titanium Alloys	16.5	113.8	6.2	42.4	0.34	0.16	4.4	4.4
Zinc Alloys	12.0	82.7	4.5	31.1	0.33	0.24	6.6	6.6

* *Properties of Some Metals and Alloys,* International Nickel Co., Inc., NY; *Metals Handbook,* American Society for Metals, Materials Park, OH

Table B-2 Mechanical Properties for Some Wrought-Aluminum Alloys

Data from various sources.* Approximate values. Consult manufacturers for more accurate information

Wrought-Aluminum Alloy	Condition	Tensile Yield Strength (2% offset)		Ultimate Tensile Strength		Fatigue Strength at 5E8 cycles		Elongation over 2 in	Brinell Hardness
		kpsi	MPa	kpsi	MPa	kpsi	MPa	%	-HB
1100	Sheet annealed	5	34	13	90			35	23
	Cold rolled	22	152	24	165			5	44
2024	Sheet annealed	11	76	26	179			20	-
	Heat treated	42	290	64	441	20	138	19	-
3003	Sheet annealed	6	41	16	110			30	28
	Cold rolled	27	186	29	200			4	55
5052	Sheet annealed	13	90	28	193			25	47
	Cold rolled	37	255	42	290			7	77
6061	Sheet annealed	8	55	18	124			25	30
	Heat treated	40	276	45	310	14	97	12	95
7075	Bar annealed	15	103	33	228			16	60
	Heat treated	73	503	83	572	14	97	11	150

* *Properties of Some Metals and Alloys,* International Nickel Co., Inc., NY; *Metals Handbook,* American Society for Metals, Materials Park, OH

A

Table B-3 Mechanical Properties for Some Carbon Steels

Data from various sources.* Approximate values. Consult manufacturers for more accurate information

SAE / AISI Number	Condition	Tensile Yield Strength (2% offset)		Ultimate Tensile Strength		Elongation over 2 in	Brinell Hardness
		kpsi	MPa	kpsi	MPa	%	-HB
1010	Hot rolled	26	179	47	324	28	95
	Cold rolled	44	303	53	365	20	105
1020	Hot rolled	30	207	55	379	25	111
	Cold rolled	57	393	68	469	15	131
1030	Hot rolled	38	259	68	469	20	137
	Normalized @ 1650°F	50	345	75	517	32	149
	Cold rolled	64	441	76	524	12	149
	Q&T @ 1000°F	75	517	97	669	28	255
	Q&T @ 800°F	84	579	106	731	23	302
	Q&T @ 400°F	94	648	123	848	17	495
1035	Hot rolled	40	276	72	496	18	143
	Cold rolled	67	462	80	552	12	163
1040	Hot rolled	42	290	76	524	18	149
	Normalized @ 1650°F	54	372	86	593	28	170
	Cold rolled	71	490	85	586	12	170
	Q&T @ 1200°F	63	434	92	634	29	192
	Q&T @ 800°F	80	552	110	758	21	241
	Q&T @ 400°F	86	593	113	779	19	262
1045	Hot rolled	45	310	82	565	16	163
	Cold rolled	77	531	91	627	12	179
1050	Hot rolled	50	345	90	621	15	179
	Normalized @ 1650°F	62	427	108	745	20	217
	Cold rolled	84	579	100	689	10	197
	Q&T @ 1200°F	78	538	104	717	28	235
	Q&T @ 800°F	115	793	158	1 089	13	444
	Q&T @ 400°F	117	807	163	1 124	9	514
1060	Hot rolled	54	372	98	676	12	200
	Normalized @ 1650°F	61	421	112	772	18	229
	Q&T @ 1200°F	76	524	116	800	23	229
	Q&T @ 1000°F	97	669	140	965	17	277
	Q&T @ 800°F	111	765	156	1 076	14	311
1095	Hot rolled	66	455	120	827	10	248
	Normalized @ 1650°F	72	496	147	1 014	9	13
	Q&T @ 1200°F	80	552	130	896	21	269
	Q&T @ 800°F	112	772	176	1 213	12	363
	Q&T @ 600°F	118	814	183	1 262	10	375

* *SAE Handbook*, Society of Automotive Engineers, Warrendale PA; *Metals Handbook*, American Society for Metals, Materials Park, OH

Table B-4 Mechanical Properties for Some Cast-Iron Alloys

Data from various sources.* Approximate values. Consult manufacturers for more accurate information

Cast-Iron Alloy	Condition	Tensile Yield Strength (2% offset)		Ultimate Tensile Strength		Compressive Strength		Brinell Hardness
		kpsi	MPa	kpsi	MPa	kpsi	MPa	HB
Gray Cast Iron—Class 20	As cast	–	–	22	152	83	572	156
Gray Cast Iron—Class 30	As cast	–	–	32	221	109	752	210
Gray Cast Iron—Class 40	As cast	–	–	42	290	140	965	235
Gray Cast Iron—Class 50	As cast	–	–	52	359	164	1 131	262
Gray Cast Iron—Class 60	As cast	–	–	62	427	187	1 289	302
Ductile Iron 60-40-18	Annealed	47	324	65	448	52	359	160
Ductile Iron 65-45-12	Annealed	48	331	67	462	53	365	174
Ductile Iron 80-55-06	Annealed	53	365	82	565	56	386	228
Ductile Iron 120-90-02	Q & T	120	827	140	965	134	924	325

* *Properties of Some Metals and Alloys*, International Nickel Co., Inc., NY; *Metals Handbook*, American Society for Metals, Materials Park, OH

Table B-5 Mechanical and Physical Properties for Some Engineering Plastics

Data from various sources.* Approximate values. Consult manufacturers for more accurate information

Material	Approximate Modulus of Elasticity E †		Ultimate Tensile Strength		Ultimate Compressive Strength		Elongation over 2 in	Max Temp	Specific Gravity
	Mpsi	GPa	kpsi	MPa	kpsi	MPa	%	°F	
ABS	0.3	2.1	6.0	41.4	10.0	68.9	5-25	160-200	1.05
20-40% glass filled	0.6	4.1	10.0	68.9	12.0	82.7	3	200-230	1.30
Acetal	0.5	3.4	8.8	60.7	18.0	124.1	60	220	1.41
20-30% glass filled	1.0	6.9	10.0	68.9	18.0	124.1	7	185-220	1.56
Acrylic	0.4	2.8	10.0	68.9	15.0	103.4	5	140-190	1.18
Fluoroplastic (PTFE)	0.2	1.4	5.0	34.5	6.0	41.4	100	350-330	2.10
Nylon 6/6	0.2	1.4	10.0	68.9	10.0	68.9	60	180-300	1.14
Nylon 11	0.2	1.3	8.0	55.2	8.0	55.2	300	180-300	1.04
20-30% glass filled	0.4	2.5	12.8	88.3	12.8	88.3	4	250-340	1.26
Polycarbonate	0.4	2.4	9.0	62.1	12.0	82.7	100	250	1.20
10-40% glass filled	1.0	6.9	17.0	117.2	17.0	117.2	2	275	1.35
HMW Polyethylene	0.1	0.7	2.5	17.2	–	–	525	–	0.94
Polyphenylene Oxide	0.4	2.4	9.6	66.2	16.4	113.1	20	212	1.06
20-30% glass filled	1.1	7.8	15.5	106.9	17.5	120.7	5	260	1.23
Polypropylene	0.2	1.4	5.0	34.5	7.0	48.3	500	250-320	0.90
20-30% glass filled	0.7	4.8	7.5	51.7	6.2	42.7	2	300-320	1.10
Impact Polystryrene	0.3	2.1	4.0	27.6	6.0	41.4	2-80	140-175	1.07
20-30% glass filled	0.1	0.7	12.0	82.7	16.0	110.3	1	180-200	1.25
Polysulfone	0.4	2.5	10.2	70.3	13.9	95.8	50	300-345	1.24

* *Modern Plastics Encyclopedia*, McGraw-Hill, New York; *Machine Design Materials Reference Issue*, Penton Publishing, Cleveland, OH

† Most plastics do not obey Hooke's Law. These apparent moduli of elasticity vary with time and temperature.

Appendix C

GEOMETRIC PROPERTIES

DIAGRAMS AND FORMULAS TO CALCULATE THE FOLLOWING PARAMETERS FOR SEVERAL COMMON GEOMETRIC SOLIDS

V = volume

m = mass

C_g = location of center of mass

I_x = second moment of mass about x axis $= \int \left(y^2 + z^2\right) dm$

I_y = second moment of mass about y axis $= \int \left(x^2 + z^2\right) dm$

I_z = second moment of mass about z axis $= \int \left(x^2 + y^2\right) dm$

k_x = radius of gyration about x axis

k_y = radius of gyration about y axis

k_z = radius of gyration about z axis

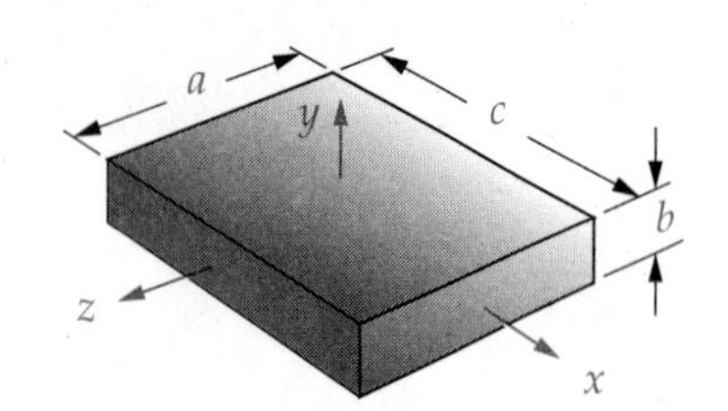

(*a*) Rectangular prism

$V = abc$ $\quad m = V \cdot$ mass density

$x_{Cg} @ \frac{c}{2}$ $\quad y_{Cg} @ \frac{b}{2}$ $\quad z_{Cg} @ \frac{a}{2}$

$$I_x = \frac{m(a^2 + b^2)}{12} \quad I_y = \frac{m(a^2 + c^2)}{12} \quad I_z = \frac{m(b^2 + c^2)}{12}$$

$$k_x = \sqrt{\frac{I_x}{m}} \quad k_y = \sqrt{\frac{I_y}{m}} \quad k_z = \sqrt{\frac{I_z}{m}}$$

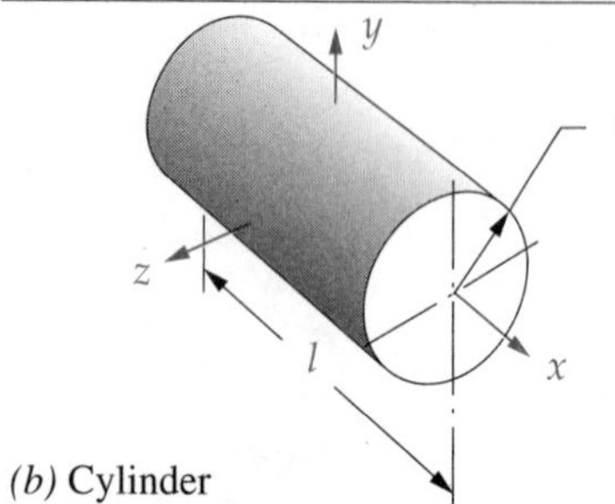

(*b*) Cylinder

$V = \pi r^2 l$ $\quad m = V \cdot$ mass density

$x_{Cg} @ \frac{l}{2}$ $\quad y_{Cg}$ on axis $\quad z_{Cg}$ on axis

$$I_x = \frac{mr^2}{2} \quad I_y = I_z = \frac{m(3r^2 + l^2)}{12}$$

$$k_x = \sqrt{\frac{I_x}{m}} \quad k_y = k_z = \sqrt{\frac{I_y}{m}}$$

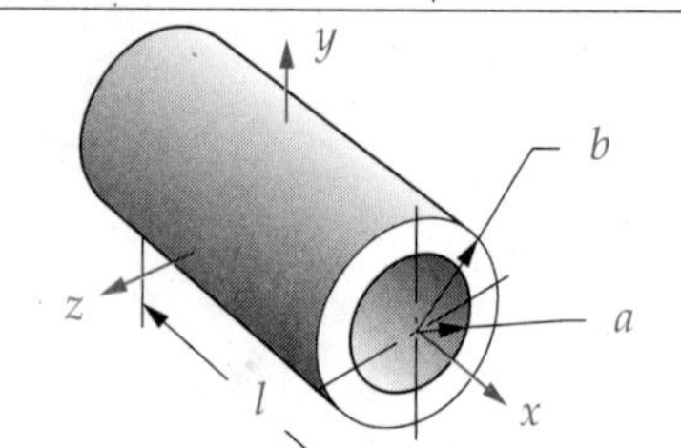

(*c*) Hollow cylinder

$V = \pi(b^2 - a^2)l$ $\quad m = V \cdot$ mass density

$x_{Cg} @ \frac{l}{2}$ $\quad y_{Cg}$ on axis $\quad z_{Cg}$ on axis

$$I_x = \frac{m(a^2 + b^2)}{2} \quad I_y = I_z = \frac{m(3a^2 + 3b^2 + l^2)}{12}$$

$$k_x = \sqrt{\frac{I_x}{m}} \quad k_y = k_z = \sqrt{\frac{I_y}{m}}$$

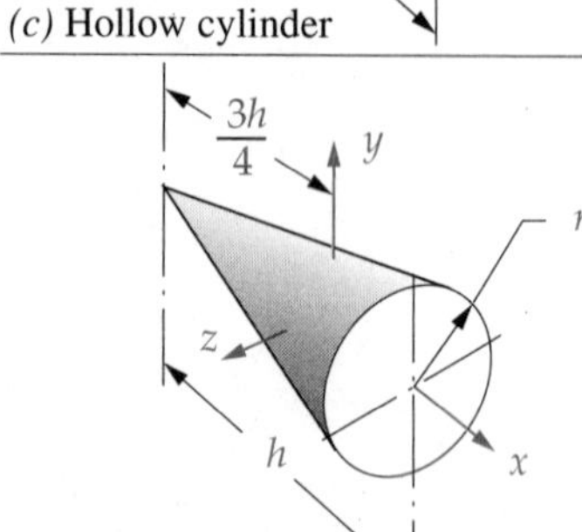

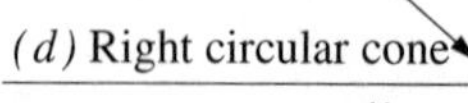

(*d*) Right circular cone

$V = \pi \frac{r^2 h}{3}$ $\quad m = V \cdot$ mass density

$x_{Cg} @ \frac{3h}{4}$ $\quad y_{Cg}$ on axis $\quad z_{Cg}$ on axis

$$I_x = \frac{3}{10} mr^2 \quad I_y = I_z = \frac{m(12r^2 + 3h^2)}{80}$$

$$k_x = \sqrt{\frac{I_x}{m}} \quad k_y = k_z = \sqrt{\frac{I_y}{m}}$$

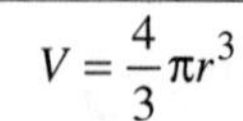

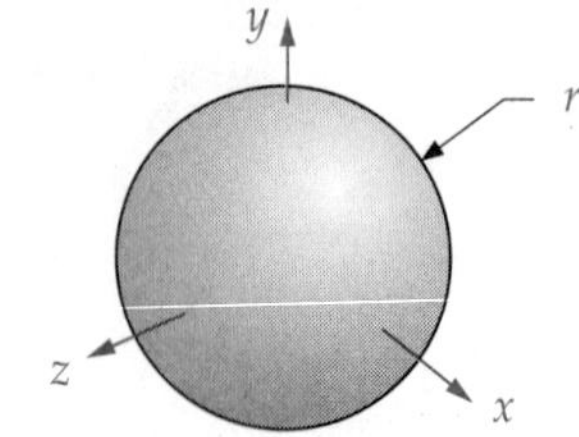

(*e*) Sphere

$V = \frac{4}{3}\pi r^3$ $\quad m = V \cdot$ mass density

x_{Cg} at center $\quad y_{Cg}$ at center $\quad z_{Cg}$ at center

$$I_x = I_y = I_z = \frac{2}{5} mr^2$$

$$k_x = k_y = k_z = \sqrt{\frac{I_y}{m}}$$

Appendix D

SPRING DATA

The following catalog pages of helical compression and extension spring data provided courtesy of *Hardware Products Co., Chelsea, Massachusetts.*

COMPRESSION SPRINGS

Will go in hole In.		7/16			1/2				5/8				3/4				7/8			
Wire Dia. In.		.031	.047	.062	.047	.062	.078	.094	.047	.062	.078	.094	.062	.078	.094	.125	.062	.078	.094	.125
FREE LENGTHS 7/16	Catalog No.	247	248	249																
	Price Code	HB	HB	HC																
	lbs./in.	12	55	180																
	Max. Defl.	.32	.23	.16																
1/2	Catalog No.	250	251	252	283	284	285	286												
	Price Code	HB	HB	HC	HB	HB	HE	HE												
	lbs./in.	10	47	150	37	110	320	840												
	Max. Defl.	.37	.27	.19	.29	.22	.15	.10												
5/8	Catalog No.	253	254	255	287	288	289	290	331	332	333	334								
	Price Code	HB	HB	HC	HB	HB	HE	HE	HB	HB	HD	HE								
	lbs./in.	7.9	36	175	29	85	240	610	20	54	140	320								
	Max. Defl.	.47	.35	.25	.38	.29	.20	.14	.43	.35	.27	.20								
3/4	Catalog No.	256	257	258	291	292	293	294	335	336	337	338	375	376	377	378				
	Price Code	HB	HB	HC	HB	HB	HE	HE	HB	HB	HD	HE	HD	HD	HF	HG				
	lbs./in.	6.4	29	90	23	68	185	470	16	43	105	250	32	78	170	650				
	Max. Defl.	.58	.44	.32	.48	.37	.26	.18	.53	.44	.34	.26	.59	.40	.32	.19				
7/8	Catalog No.	259	260	261	295	296	297	298	339	340	341	342	379	380	381	382	419	420	421	422
	Price Code	HB	HB	HC	HB	HB	HE	HE	HB	HB	HD	HE	HD	HD	HF	HG	HF	HF	HJ	HK
	lbs./in.	5.4	24	75	19	56	155	384	13	36	90	204	27	65	140	520	21	49	100	350
	Max. Defl.	.68	.52	.39	.57	.44	.32	.23	.64	.53	.42	.32	.58	.48	.39	.24	.62	.53	.44	.30
1	Catalog No.	262	263	264	299	300	301	302	343	344	345	346	383	384	385	386	423	424	425	426
	Price Code	HB	HB	HC	HB	HC	HE	HE	HB	HC	HD	HE	HD	HD	HF	HG	HF	HF	HJ	HK
	lbs./in.	4.7	21	65	17	48	130	320	11	31	77	170	23	55	115	430	18	42	86	290
	Max. Defl.	.79	.60	.45	.66	.51	.37	.27	.74	.62	.49	.38	.68	.57	.46	.29	.73	.63	.53	.36
1¼	Catalog No.	265	266	267	303	304	305	306	347	348	349	350	387	388	389	390	427	428	429	430
	Price Code	HB	HC	HC	HC	HC	HF	HF	HC	HC	HD	HE	HE	HE	HF	HG	HG	HG	HJ	HK
	lbs./in.	3.7	16	50	13	38	100	245	9.0	24	59	130	18	42	89	320	14	32	66	220
	Max. Defl.	1.0	.77	.58	.84	.66	.49	.35	.94	.80	.64	.49	.88	.74	.60	.39	.94	.81	.69	.47
1½	Catalog No.	268	269	270	307	308	309	310	351	352	353	354	391	392	393	394	431	432	433	434
	Price Code	HB	HC	HC	HC	HC	HF	HF	HD	HD	HE	HF	HE	HE	HF	HG	HG	HG	HJ	HK
	lbs./in.	3.1	14	41	11	31	83	200	7.4	20	48	105	15	34	72	260	12	26	53	175
	Max. Defl.	1.2	.94	.70	1.0	.81	.60	.43	1.1	.98	.78	.61	1.1	.91	.74	.48	1.1	1.0	.85	.59
1¾	Catalog No.	271	272	273	311	312	313	314	355	356	357	358	395	396	397	398	435	436	437	438
	Price Code	HC	HD	HE	HC	HD	HF	HF	HD	HD	HC	HF	HE	HE	HF	HG	HG	HG	HJ	HK
	lbs./in.	2.6	11	35	9.1	26	70	170	6.2	17	41	90	12.4	29	61	216	9.9	22	45	147
	Max. Defl.	1.4	1.1	.84	1.2	.96	.71	.52	1.3	1.1	.93	.73	1.3	1.1	.89	.58	1.35	1.2	1.0	.71
2	Catalog No.	274	275	276	315	316	317	318	359	360	361	362	399	400	401	402	439	440	441	442
	Price Code	HD	HD	HE	HC	HD	HF	HF	HD	HD	HE	HF	HE	HE	HF	HG	HG	HG	HJ	HL
	lbs./in.	2.3	10	30	7.9	23	60	145	5.4	14	35	77	11	25	52	185	8.6	19	38	125
	Max. Defl.	1.6	1.3	.96	1.4	1.1	.82	.60	1.5	1.3	1.1	.85	1.4	1.2	1.0	.68	1.5	1.4	1.2	.83
3	Catalog No.	277	278	279	319	320	321	322	363	364	365	366	403	404	405	406	443	444	445	446
	Price Code	HD	HE	HE	HF	HF	HG	HG	HD	HD	HF	HG	HF	HF	HG	HK	HJ	HJ	HK	HL
	lbs./in.	1.5	6.6	20	5.2	15	39	94	3.6	9.4	23	50	7	16	34	115	5.6	12	25	80
	Max. Defl.	2.4	1.9	1.4	2.1	1.7	1.2	.93	2.4	2.0	1.6	1.3	2.2	1.9	1.6	1.0	2.4	2.1	1.8	1.3
4	Catalog No.	280	281	282	323	324	325	326	367	368	369	370	407	408	409	410	447	448	449	450
	Price Code	HE	HE	HE	HE	HF	HG	HG	HE	HE	HF	HG	HF	HF	HG	HL	HJ	HJ	HL	HM
	lbs./in.	1.1	4.9	15	3.9	11	29	69	2.6	6.9	17	37	5.2	12	25	86	4.2	9.2	18	59
	Max. Defl.	3.3	2.6	2.0	2.8	2.3	1.7	1.2	3.2	2.7	2.2	1.8	3.0	2.6	2.1	1.4	3.2	2.8	2.5	1.8
6	Catalog No.				327	328	329	330	371	372	373	374	411	412	413	414	451	452	453	454
	Price Code				HF	HF	HG	HG	HF	HF	HG	HG	HG	HG	HJ	HN	HK	HK	HL	HO
	lbs./in.				2.5	7.0	17	45	1.8	4.6	11	24	3.4	7.9	16	56	2.7	6.0	12	38
	Max. Defl.				4.4	3.5	2.5	2.	4.8	4.2	3.4	2.7	4.6	3.9	3.3	2.2	4.9	4.3	3.8	2.7
8	Catalog No.												415	416	417	418	455	456	457	458
	Price Code.												HJ	HJ	HK	HO	HL	HL	HM	HP
	lbs./in.												2.6	6	11	40	2.0	4.5	8.9	28
	Max. Defl.												6.1	5.2	4.5	3.0	6.5	5.8	5.1	3.7
Maximum Load		3.7	12.7	29	11	25	45	88	8.3	19	38	66	15.8	31.2	54	125	13.4	26.3	45	105
Will work free over		.347	.315	.285	.375	.345	.313	.281	.505	.475	.443	.411	.585	.554	.522	.460	.700	.670	.638	.576
Pitch		.195	.141	.128	.173	.151	.141	.141	.259	.214	.188	.177	.284	.240	.217	.204	.371	.306	.268	.239
Solid Stress (000 omitted)		125	118	113	118	113	109	105	118	113	109	105	113	109	105	99	113	109	105	99

Pounds per inch figure is a constant for each spring, and represents the number of pounds required to compress the spring 1". To compress the spring ½" or ¼" requires ½ or ¼ of this value.

Maximum Deflection is the amount spring deflects to give the maximum load. This value subtracted from the free length gives the solid or compressed length.

NOTE: Stock springs can be ordered in stainless steel or plated. Prices quoted upon request.

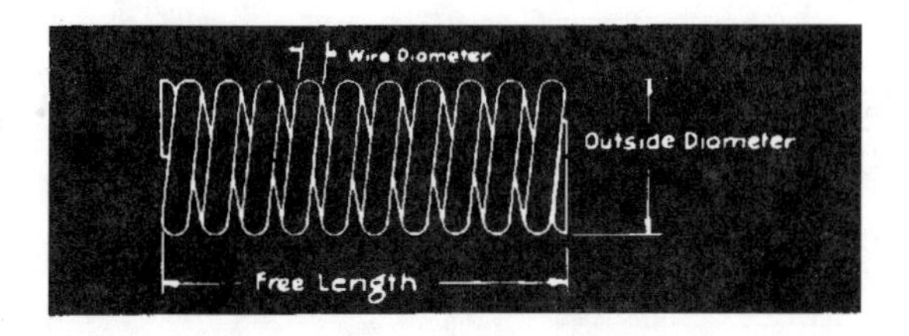

Free Lengths	Will go in hole In.	1				1 1/4			1 1/2			2			3			4			6	
	Wire Dia. In.	.078	.094	.125	.187	.094	.125	.187	.125	.187	.250	.187	.250	.375	.250	.375	.500	.375	.500	.750	.750	1.000
1	Catalog No.	459	460	461	462																	
1	Price Code	HK	HL	HL	HR																	
1	lbs./in.	34	67	210	1500																	
1	Max. Defl.	.67	.58	.41	.19																	
1¼	Catalog No.	463	464	465	466	499	500	501														
1¼	Price Code	HL	HM	HM	HS	HN	HN	HM														
1¼	lbs./in.	26	52	160	1100	35	100	600														
1¼	Max. Defl.	.87	.76	.55	.26	.85	.67	.37														
1½	Catalog No.	467	468	469	470	502	503	504	526	527	528											
1½	Price Code	HL	HM	HN	HS	HN	HO	HT	HR	HX	HAC											
1½	lbs./in.	21	42	130	870	29	82	460	60	300	1200											
1½	Max. Defl.	1.0	.93	.69	.34	1.1	.84	.48	.95	.60	.35											
1¾	Catalog No.	471	472	473	474	505	506	507	529	530	531											
1¾	Price Code	HL	HM	HN	HS	HN	HO	HT	HR	HX	HAC											
1¾	lbs./in.	18	35	108	712	24	68	379	50	244	960											
1¾	Max. Defl.	1.3	1.1	.83	.41	1.3	1.0	.59	1.1	.74	.44											
2	Catalog No.	475	476	477	478	508	509	510	532	533	534	553	554	555								
2	Price Code	HM	HN	HO	HT	HO	HP	HU	HS	HZ	HAE	HAA	HAG	HZZ								
2	lbs./in.	16	30	93	600	21	59	320	43	200	800	115	390	3000								
2	Max. Defl.	1.4	1.3	.97	.49	1.4	1.2	.70	1.3	.87	.53	1.1	.77	.34								
3	Catalog No.	479	480	481	482	511	512	513	535	536	537	556	557	558	577	578	579					
3	Price Code	HN	HO	HP	HZ	HP	HR	HAA	HT	HAD	HAL	HAE	HAN	HZZ	HAR	HZZ	HZZ					
3	lbs./in.	10	19	59	370	13	37	200	27	130	480	73	230	1650	105	560	2300					
3	Max. Defl.	2.2	2.0	1.5	.79	2.2	1.8	1.1	2.1	1.4	.89	1.8	1.3	.61	1.8	1.1	.64					
4	Catalog No.	483	484	485	486	514	515	516	538	539	540	559	560	561	580	581	582	598	599	610		
4	Price Code	HP	HR	HS	HAC	HS	HT	HAD	HW	HAG	HAO	HAJ	HAR	HZZ	HAT	HZZ	HZZ	HZZ	HZZ	HZZ		
4	lbs./in.	7.4	14	43	270	9.9	27	144	20	93	340	53	170	1150	76	390	1500	210	720	4600		
4	Max. Defl.	3.0	2.7	2.1	1.1	3.0	2.5	1.5	2.8	1.9	1.2	2.5	1.8	.88	2.5	1.6	.96	2.1	1.4	.8		
6	Catalog No.	487	488	489	490	517	518	519	541	542	543	562	563	564	583	584	585	600	601	611	616	621
6	Price Code	HR	HT	HU	HAD	HU	HW	HAE	HX	HAJ	HAT	HAM	HAW	HZZ	HAZ	HZZ	HZZ	HZZ	HZZ	HZZ	HZZ	HZZ
6	lbs./in.	4.9	9.4	28	175	6.5	18	93	13	60	220	34	105	710	49	240	920	130	430	2840	850	3500
6	Max. Defl.	4.6	4.1	3.2	1.7	4.7	3.9	2.4	4.3	3.0	1.9	3.8	2.8	1.4	4.0	2.6	1.6	3.4	2.4	1.3	1.9	1.4
8	Catalog No.	491	492	493	494	520	521	522	544	545	546	565	566	567	586	587	588	602	603	612	617	622
8	Price Code	HS	HU	HW	HAL	HW	HX	HAM	HAA	HAP	HAW	HAR	HAZ	HZZ	HBD	HZZ	HZZ	HZZ	HZZ	HZZ	HZZ	HZZ
8	lbs./in.	3.6	7.0	21	125	4.8	13	68	9.6	44	160	25	79	510	36	175	660	95	310	2050	630	2500
8	Max. Defl.	6.2	5.6	4.3	2.3	6.3	5.2	3.2	5.9	4.1	2.7	5.2	3.9	1.9	5.4	3.6	2.2	4.5	3.4	1.8	2.7	2.0
12	Catalog No.	495	496	497	498	523	524	525	547	548	549	568	569	570	589	590	591	604	605	613	618	623
12	Price Code	HT	HW	HZ	HAP	HX	HAA	HAR	HAC	HAU	HBA	HAZ	HBE	HZZ	HBK	HZZ	HZZ	HZZ	HZZ	HZZ	HZZ	HZZ
12	lbs./in.	2.4	4.6	14	84	3.2	8.7	45	6.3	29	105	16	52	330	23	110	420	61	195	1325	400	1580
12	Max. Defl.	9.4	8.4	6.5	3.5	9.5	7.9	5.0	8.9	6.2	4.1	8.0	5.9	3.0	8.3	5.5	3.5	7.3	5.3	2.6	4.3	3.1
16	Catalog No.								550	551	552	571	572	573	592	593	594	606	607	614	619	624
16	Price Code								HAE	HAW	HBD	HAZ	HBG	HZZ	HBL	HZZ	HZZ	HZZ	HZZ	HZZ	HZZ	HZZ
16	lbs./in.								4.7	21	74	12	38	240	17	83	310	45	145	975	300	1170
16	Max. Defl.								11.9	8.5	5.6	10.7	8.0	4.1	11.3	7.5	4.0	10	7.3	3.8	6.1	4.3
24	Catalog No.											574	575	576	595	596	597	608	609	615	620	625
24	Price Code											HBA	HBL	HZZ	HBP	HZZ	HZZ	HZZ	HZZ	HZZ	HZZ	HZZ
24	lbs./in.											7.8	23.4	150	11.4	54	200	29	94	640	175	760
24	Max. Defl.											16.3	12.1	7.0	17.1	11.5	7.3	15.2	11.1	5.8	9.4	6.5
	Maximum Load	23	39	90	295	30	69	224	57	180	428	131	307	1000	195	624	1470	449	1040	3700	2000	4800
	Will work free over	.784	.752	.690	.565	1.00	.940	.815	1.19	1.06	.940	1.52	1.39	1.14	2.33	2.08	1.83	3.08	2.83	2.25	4.25	3.75
	Pitch	.382	.328	.279	.268	.481	.384	.327	.516	.403	.388	.596	.518	.514	.917	.741	.736	1.08	.969	1.00	1.3	1.4
	Solid Stress (000 omitted)	109	105	99	90	105	99	90	99	90	85	90	85	77	85	77	73	77	73	70	70	65

Pounds per inch figure is a constant for each spring, and represents the number of pounds required to compress the spring 1". To compress the spring ½" or ¼" requires ½ or ¼ of this value.

Maximum Deflection is the amount spring deflects to give the maximum load. This value subtracted from the free length gives the solid or compressed length.

NOTE: Stock springs can be ordered in stainless steel or plated. Prices quoted upon request.

Hardware Products Company, Inc.

A

EXTENSION SPRINGS

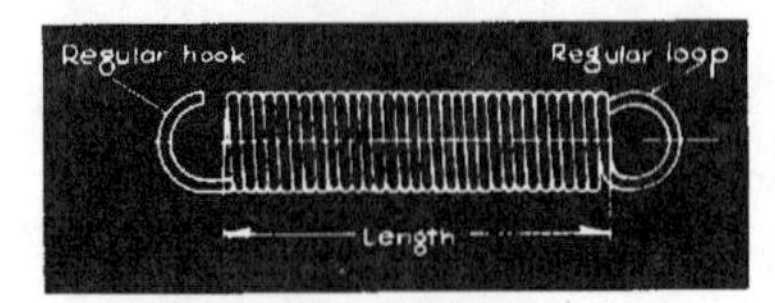

ORDER BY:
SE LENGTH x O.D. x WIRE DIA.

SPECIFY HOOKS OR LOOPS

The figures given for "Maximum Extension" and "lbs. per inch" are for a spring 1" long. For other lengths multiply the "Maximum Extension" and divide the "lbs. per inch" by the length in inches. The "Maximum Load" and "Initial Tension" remain constant for any length.

Example: A spring ½" diam. .062" wire and 4" long will have a safe maximum extension of 3.2" and it will require 4 lbs. to deflect it 1 in. The spring will hold approximately 3.3 lbs. before it starts to extend, and will hold a maximum of 16.1 lbs. without permanent stretch. If 8.5 lbs. is hung on the spring it will deflect 1.3". 8.5 lbs. minus 3.3 lbs. divided by 4 lbs. per inch equals 1.3".

FOR QUICK DELIVERY
Call 617-884-9410

NOTE: Stock springs can be ordered in stainless steel or plated. Prices quoted upon request.

Outside dia.	Wire dia.	Catalog No.	Price code	Safe maximum load in pounds	Safe maximum extension - in.	Approx. initial tension in pounds	Pound per inch extension	Stress at max. load (000 omitted)	Weight per foot (lbs.)
1/8	.012	01	EHE	.6	1.9	.07	.27	100	.012
	.016	02	EHD	1.3	.9	.2	1.2	93	.015
	.023	03	EHD	4.2	.35	.9	9.0	90	.02
5/32	.012	04	EHE	.47	3.5	.01	.12	100	.015
	.016	05	EHD	1.1	1.7	.15	.55	93	.019
	.023	06	EHD	3.2	.7	.5	3.9	90	.027
3/16	.016	07	EHD	.87	2.5	1	3	93	.024
	.023	08	EHD	2.6	1.0	.4	2.2	90	.032
	.031	09	EHD	6.5	.45	1.5	10.7	88	.04
7/32	.016	10	EHD	.75	4.0	.01	.18	93	.028
	.023	11	EHD	2.3	1.6	.32	1.2	90	.039
	.031	12	EHD	5.5	.7	1.0	6.5	88	.048
1/4	.023	13	EHD	1.8	1.9	.26	8	90	.044
	.031	14	EHD	4.7	1.0	.75	3.8	88	.055
	.047	15	EHE	16.0	.3	3.5	40.0	83	.082
5/16	.023	16	EHE	1.5	3.5	.16	.38	90	.058
	.031	17	EHE	3.6	1.6	.55	1.9	88	.072
	.047	18	EHE	12.5	.9	2.2	10.8	83	.108
3/8	.031	19	EHE	2.9	2.5	.37	1.0	88	.084
	.047	20	EHE	10.5	9	1.7	9.5	83	.13
	.062	21	EHF	23.0	.39	5.3	45.0	79	.16
7/16	.031	22	EHF	2.5	3.5	.26	.63	88	.105
	.047	23	EHF	8.5	1.2	1.4	5.7	83	.163
	.062	24	EHF	20.0	.6	4.3	26.0	79	.2
1/2	.047	25	EHF	7.3	1.6	1.1	3.7	83	.18
	.062	26	EHF	17.0	8	3.3	16.0	79	.23
	.078	27	EHG	34.0	45	8.0	57.0	77	.28
	.094	28	EHJ	57.0	.25	16.0	160.0	74	.32
5/8	.047	29	EHG	6.0	3.0	.7	1.7	83	24
	.062	30	EHG	13.3	1.4	2.1	7.6	79	.3
	.078	31	EHG	27.0	.9	5.2	23.0	77	.37
	.094	32	EHJ	45.0	.4	11.0	73.0	74	.44
3/4	.062	33	EHG	10.5	2.2	1.5	4.1	79	.36
	.078	34	EHJ	22.0	1.3	3.5	14.0	77	.46
	.094	35	EHJ	36.0	7	8.0	38.0	74	.51
	.125	36	EHK	85.0	.3	22.0	180.0	69	.64
7/8	.062	37	EHK	9.2	3.3	1.1	2.4	79	.4
	.078	38	EHK	18.0	1.7	2.6	8.7	77	.59
	.094	39	EHL	31.0	1.0	6.0	25.0	74	.64
	.125	40	EHM	72.0	.5	17.0	107.0	69	.8
1	.078	41	EHL	16.0	2.5	2.0	5.5	77	.67
	.094	42	EHL	26.0	1.5	4.5	13.7	74	.70
	.125	43	EHN	65.0	.75	14.0	68.0	69	.90
	.187	44	EHW	200.0	.23	60.0	600.0	63	1.4
1¼	.094	45	EHM	71.0	2.6	2.8	6.8	74	.94
	.125	46	EHO	47.0	1.2	9.0	31.0	69	1.3
	.187	47	EHZ	148.0	.3	40.0	290.0	63	1.8
1½	.125	48	EHS	39.0	1.9	6.0	17.0	69	1.4
	.187	49	EHAA	122.0	.6	33.0	150.0	63	2.2
	.250	50	EHAC	290.0	.27	90.0	720.0	60	2.6
2	.187	51	EHAD	90.0	1.3	20.0	54.0	63	3.1
	.250	52	EHAG	210.0	.6	55.0	260.0	60	3.7

Carried in-stock in 3-foot lengths - cut to length and looped to order.

Hardware Products Company, Inc.

191 WILLIAMS STREET • CHELSEA, MA 02150

BIBLIOGRAPHY

Abell, R. F. (1977). "I. C. Engine Cam and Tappet Wear Experience." SAE 770019.

Akiba, K., and T. Kakiuchi. (1988). "A Dynamic Study of Engine Valving Mechanisms: Determination of the Impulse Force Acting on the Valve." SAE 880389.

Akiba, K., and H. Sakai. (1981). "A Comprehensive Simulation of High Speed Driven Valve Trains." SAE 810865.

Allais, D. C. (1963). "Cycloidal vs. Modified Trapezoid Cams." *Machine Design*, January, 1963, pp. 92-96.

Allen, C. H., et al. (1953). "Discussion of Valve Symposium Papers." *61*, pp. 714-716.

Amarnath, C., and K. C. Gupta, eds. (1978). "Novel Cam-Linkage Mechanisms for Multiple Dwell Generation." *Cams and Cam Mechanisms*, Jones, J., ed., I. Mech. E.: London.

Angeles, J., and C. S. Lopez-Cajun. (1991). *Optimization of Cam Mechanisms*. Kluwer Academic Publishers: Dordrecht.

Angeles, J., et al. (1994). "The Design Optimization of Cam Mechanisms with Oscillating Flat-Face Followers under Curvature Constraints." *Journal of Mechanical Design*, **116**(1).

Angeles, J., et al. (1994). "The Design of Cam Mechanisms with Translating Flat-Face Followers under Curvature Constraints." *Journal of Mechanical Design*, **116**(1).

Anonymous. (1998). "Valvetrain Design" http://www.fev.de/design/valvdesgn.htm, FEV.

Arnold, R. N. (1952). "The Limiting Speeds of Operation of Simple Cam Systems." *Proc. of 8th Intl Conf of Theoretical and Applied Mechanics*, pp. 403-404.

Asmus, T. W. (1982). "Valve Events and Engine Operation." SAE 82079.

Asmus, T. W. (1991). "Perspectives on Applications of Variable Valve Timing." SAE 910445.

Ault, H. K., and R. L. Norton. (1993). "Spline Based Cam Functions for Minimum Kinetic Energy Follower Motion." *Proc. of 3rd Conference on Applied Mechanisms and Robotics*, Cincinnati OH, pp. 2.1-2.6.

Aviza, G. D. (1997). "An Experimental Investigation of Torque Balancing to Reduce the Torsional Vibration of Camshafts." M.S. Thesis, Worcester Polytechnic Institute.

Bagci, C. (1994). "Exact Response Analysis and Dynamic Design of Cam-Follower Systems Using Laplace Transforms." *Proc. of ASME Design Technical Conference*, Minneapolis, MN, pp. 613-629.

Bagepalli, B. S., et al. (1991). "Generalized Modeling of Dynamic Cam-Follower Pairs in Mechanisms." *Journal of Mechanical Design*, **113**(2), p. 192.

Bakonyi, S. M. (1968). "The Advantages of Overhead Camshafts-What Are They?" SAE 680028.

Baniasad, S. M., and M. R. Emes. (1997). "Design and Development of a Method of Valve Train Friction Measurement." SAE 980572.

Baranyi, S. J. (1969). "Cams, Dynamics, and Design." *Design News*, v. 24.

Baranyi, S. J. (1970). "Multiple Harmonic Cam Profiles." ASME 70-Mech-59.

Baratta, F. I. (1954). "When Will a Cam Follower Jump." *Product Engineerinig*, July, 1954.

Barkan, P. (1948). "Comments on Paper "Tests on Dynamic Response of Cam Follower Systems." by D. B. Mitchell." *Mechanical Engineering*, pp. 467-471.

Barkan, P. (1953). "Calculation of High Speed Valve Motion with a Flexible Overhead Linkage." *SAE Transactions*, **61**, pp. 687-700.

Barkan, P. (1958). "Spring Driven Cam Systems." *Proc. of Fifth Mechanisms Conference*, Purdue Univ.

Barkan, P., and R. V. McGarrity. (1965). "A Spring Actuated, Cam-Follower System: Design Theory and Experimental Results." *ASME J. Engineering for Industry*(August), pp. 279-286.

Barwell, F. T., and B. J. Roylance, eds. (1978). "Tribological Considerations in the Design and Operation of Cams: A Review of the Situation." *Cams and Cam Mechanisms*, Jones, J. R., ed., Institution of Mechanical Engineers: London, pp. 100-108.

Baxter, M. L. (1948). "Curvature-Acceleration Relations for Plane Cams." *Trans ASME*, pp. 483-489.

Beard, C. A., ed. (1978). "Mechanism Design Problems: An Engine Designer's Viewpoint." *Cams and Cam Mechanisms*, Jones, J. R., ed., Institution of Mechanical Engineers: London, pp. 49-61.

Beese, J. G., and H. Clarke, eds. (1978). "The Performance of Materials Associated with Cams." *Cams and Cam Mechanisms*, Jones, J. R., ed., Institution of Mechanical Engineers: London, pp. 95-98.

Benedict, C. E. (1978a). "Model Formulation of Complex Mechanisms with Multiple Inputs: Part 1 Geometry." *Journal of Mechanical Design*, **100**(4), pp. 747-754.

Benedict, C. E. (1978b). "Model Formulation of Complex Mechanisms with Multiple Inputs: Part 2 the Dynamic Model." *Journal of Mechanical Design*, **100**(4), pp. 755-761.

Berg, M., et al. (1997). "Mechanical Fully Flexible Valve Control with Delta-St." SAE 970251.

Berzak, N. (1979). "Optimization Criteria in Polydyne Cam Design." *Proc. of Fifth World Congress on Theory of Machines and Mechanisms*, pp. 1303-1310.

Berzak, N. (1982). "Optimization of Cam-Follower Systems with Kinematic and Dynamic Constraints." *Journal of Mechanical Design*, **104**(1), pp. 29-33.

Bialkowicz, B., et al. (1979). "Changes in the Dynamic Properties of the Real Cam Profile During Its Wear." *Proc. of 5th World Congress on Theory of Mechanisms and Machines*, Montreal Canada, pp. 984-987.

Black, T. (1979). "New Indexing Cam Mechanisms." *Product Engineering*, pp. 55-56.

Bloom, D. (1970). "The Dual (or X, Y) Cam Mechanism." *Mechanism and Machine Theory*, **5**, pp. 11-27.

Brittain, J. H. C., and R. Horsnell. (1968). "A Prediction of Some Causes and Effects of Cam Profile Errors." *Proc Instn Mech Engrs*, **182**(Pt 3L), pp. 145-151.

Brüstle, C., and D. Schwarzenthal. (1998). "The "Two-in-One" Engine - Porsche's Variable Valve System (VVS)." SAE 980766.

Buhayar, E. S. (1966). "Computerized Cam Design and Plate Cam Manufacture." ASME 66-MECH-2.

Buuck, B. (1982). "Elementary Design Considerations for Valve Gears." SAE 821574.

Candee, A. H. (1947). "Kinematics of Disk Cam and Follower." *Trans ASME*, **69**, pp. 709-724.

Cardona, A., and M. Geradin. (1993). "Kinematic and Dynamic Analysis of Mechanisms with Cams." *Computer Methods in Applied Mechanics and Engineering*, **103**(1-2), pp. 115-134.

Cartwright, D., et al. (1995). "Look-Clean Hands: Ricardo's Approach to Race Engine Development." *Racecar Engineering*.

Carver, W. B., and B. E. Quinn. (1945). "An Analytical Method of Cam Design." *Mechanical Engineering*, **67**, p. 523.

Chan, C., and A. P. Pisano. (1986). "Dynamic Model of a Fluctuating Rocker-Arm Ratio Cam System." ASME 86-DET-79.

Chan, C., and A. P. Pisano. (1990). "On the Synthesis of Cams with Irregular Followers." *Journal of Mechanical Design*, **112**, pp. 36-41.

Chen, F. Y. (1969). "An Algorithm for Computing the Contour of a Slow Speed Cam." *Journal of Mechanisms*, **4**, pp. 171-175.

Chen, F. Y. (1972). "A Refined Algorithm for Finite-Difference Synthesis of Cam Profiles." *Mechanism and Machine Theory*, **7**, pp. 453 - 460.

Chen, F. Y. (1973a). "Analysis and Design of Cam-Driven Mechanisms with Nonlinearities." *Journal of Engineering for Industry*, pp. 685-694.

Chen, F. Y. (1973b). "Kinematic Synthesis of Cam Profiles for Prescribed Acceleration by a Finite Integration Method." *J. Engineering for Industry, Trans ASME*(May), pp. 519-524.

Chen, F. Y. (1975). "A Survey of the State of the Art of Cam System Dynamics." *Mechanism and Machine Theory*, **12**, pp. 210 - 224.

Chen, F. Y. (1977). "Dynamic Response of a Cam-Actuated Mechanism with Pneumatic Coupling." *Journal of Engineering for Industry*, p. 598.

Chen, F. Y. (1981). "Assessment of the Dynamic Quality of a Class of Dwell-Rise-Dwell Cams." *Journal of Mechanical Design*, **103**, pp. 793 - 802.

Chen, F. Y. (1982). *Mechanics and Design of Cam Mechanisms*. Pergamon Press: New York, 520 pp.

Chen, F. Y., and N. Polvanich. (1975). "Dynamics of High-Speed Cam Driven Mechanisms: Part 1 Linear System Models, Part 2 Nonlinear System Models." *Journal of Engineering for Industry*, **97 Series B**, pp. 769-784.

Chen, F. Y., and A. M. Shah. (1972). "Optimal Design of the Geometric Parameters of a Cam Mechanism Using a Sequential Random Vectors Technique." ASME 72-MECH-73.

Chenard, J. W., and T. J. Chlupsa. (1990). "Computer Implementation of Spline-Based Cam Design." Major Qualifying Project, Worcester Polytechnic Institute.

Cheney, R. E. (1961). "High-Speed Master Cams Generated Mechanically." *Machinery*, v. 68,, pp. 93-99.

Cheng, C. Y. (1999). "Determination of Allowable Pushrod Angle Using a Three-Dimensional Valve Train Model." *J. Dyn. Sys. Meas. and Ctrl.*

Chew, M., and C. H. Chuang. (1995). "Minimizing Residual Vibrations in High Speed Cam-Follower Systems over a Range of Speeds." *Journal of Mechanical Design*, **117**(1), p. 166.

Chew, M., et al. (1982a). "Application of Optimal Control Theory to the Synthesis of High-Speed Cam-Follower Systems. Part 2-System Optimization." ASME 82-DET-101.

Chew, M., et al. (1982b). "Application of Optimal Control Theory to the Synthesis of High-Speed Cam-Follower Systems. Part 1." ASME 82-DET-100.

Choi, J. K., and S. C. Kim. (1994). "An Experimental Study on the Frictional Characteristics in the Valve Train System." SAE 945046.

Choubey, M., and A. C. Rao. (1980). "Optimum Sensitivity Synthesis of a Cam System Incorporating Manufacturing Tolerances." ASME 80-DET.

Clifton, C. J. (1979). "Design of Disc Cams with Translating Roller Followers: Calculation Methods." ESDU: London.

Cokonis, T. J. (1981). "Minicomputers Tackle Cad/Cam." *Machine Design*, Jan. 8, 1981, pp. 121-125.

Colechin, M., et al. (1993). "Analysis of Roller Follower Valve Gear." SAE 930692.

Cooke, P., and D. R. Perkins, eds. (1978). "Computer Controlled Cam Grinding Machine." *Cams and Cam Mechanisms*, Jones, J. R., ed., Institution of Mechanical Engineers: London.

Cox, M. G., ed. (1981). "Practical Spline Approximation." *Lecture Notes in Mathematics 965*, Dold, A. and B. Eckmann, eds., Springer-Verlag: Berlin, pp. 79-112.

Crane, M. E., and R. C. Meyer. (1990). "A Process to Predict Friction in an Automotive Valve Train." SAE 901728.

Cronin, D. L., and G. A. LaBouff. (1981). "Resonances and Instabilities in Dynamic Systems Incorporating a Cam." *Journal of Mechanical Design*, **103**, pp. 914-921.

Crossley, F. R. E., et al. (1979). "On the Modeling of Impacts of Two Elastic Bodies Having Flat and Cylindric Surfaces with Application to Cam Mechanisms." *Proc. of ASME Fifth World Conference on Theory of Machines and Mechanisms*, pp. 1090 - 1092.

Dalpiaz, G., and A. Rivola. (2000). "A Model for the Elastodynamic Analysis of a Desmodromic Valve Train." *Mechanism and Machine Theory*.

Davaine, J. (1995). "Engine Order Analysis: Improving Accuracy and Productivity." Application Note, Syminex: Rockaway NJ.

David, J. W., et al. (1994). "Optimal Design of High Speed Valve Train Systems." SAE 942502.

Davies, R. (1955). "Hydrodynamic Lubrication of a Cam and a Cam Follower." *Lubrication Engineering*, Jan-Feb 1955, pp. 37-39.

deBoor, C. (1978). *A Practical Guide to Splines*. Springer-Verlag: Berlin.

deBoor, C., ed. (1981). "Topics in Multivariate Approximation Theory." *Lecture Notes in Mathematics 965*, Dold, A. and B. Eckmann, eds., Springer-Verlag: Berlin, pp. 39-78.

Dent, W. T., and C. R. Chen. (1989). "A New Approach for the Analysis of OHC Engine Cam Shape." SAE 891768.

Dhanbe, S. G., and J. Chakraborty. (1975). "Mechanical Error Analysis of Cam-Follower Systems - a Stochastic Approach." *Proc. of IFToMM Fourth World Congress on the Theory of Machines and Mechanisms*, pp. 957-962.

Dhande, S. G., et al. (1975). "A Unified Approach to the Analytical Design of Three Dimensional Cam Mechanisms." *Journal of Engineering for Industry*, pp. 327-333.

Dhande, S. G., and J. Chakroborty. (1975). "Mechanical Error Analysis of Cam-Follower Systems - a Stochastic Approach." *I Mech E.*

DiBenedetto, A., and A. Vinciguerra. (1982). "Kinematic Analysis of Plate Cams Not Analytically Defined." *Journal of Mechanical Design*, **104**, pp. 34-38.

Dooley, D., et al. (1997). "Materials and Design Aspects of Modern Valve Seat Inserts." *Proc. of Int. Symposium on Valvetrain System Design and Materials*, Dearborn, MI, p. 55.

Dresner, T., and P. Barkan. (1989a). "The Application of a Two-Input Cam-Actuated Mechanism to Variable Valve Timing." SAE 890676.

Dresner, T., and P. Barkan. (1989b). "A Review and Classification of Variable Valve Timing Mechanisms." SAE 890674.

Dresner, T. L., and P. Barkan. (1995). "New Methods for the Dynamic Analysis of Flexible Single-Input and Multi-Input Cam-Follower Systems." *Journal of Mechanical Design*, **117**(1), p. 151.

Druce, G., et al., eds. (1978). "The Rotary Motion of Cam Followers." *Cams and Cam Mechanisms*, Jones, J. R., ed., I. Mech. E.

Druce, G., and F. Stride, eds. (1978). "Survey of Devices for the Generation of Cam Profiles." *Cams and Cam Mechanisms*, Jones, J. R., ed., Institution of Mechanical Engineers: London.

Du, H. Y. I., and J. S. Chen. (2000). "Dynamic Analysis of a Finger Follower Valve Train System Coupled with Flexible Camshafts." SAE 00P-345.

Dudley, W. M. (1948). "New Methods in Valve Cam Design." *SAE Quarterly Transactions*, **2**(1), pp. 19-33.

Duffy, P. E. (1993). "An Experimental Investigation of Sliding at Cam to Roller Tappet Contacts." SAE paper.

Dyson, A. (1977). "Elastohydrodynamic Lubrication and Wear of Cams Bearing against Cylindrical Tappets." SAE 770018.

Dyson, A. (1980). "Kinematics and Wear Patterns of Cam and Finger Follower Automotive Valve Gear." *Tribology International*(June), pp. 121-132.

Eaton. (1994). "Automotive Valve Train Terminology and Applications." Eaton Corp.: Marshall MI.

Eaton. (1997). "Standard Valvetrain Analysis." Eaton Corp.: Marshall MI.

Elms, D. (1981). "Replacing Cams with Software." *Machine Design*, Feb. 26, 1981, pp. 95 - 98.

Eyre, T. S., and B. Crawley. (1980). "Camshaft and Cam Follower Materials." *Tribology International* (August), pp. 147-152.

Fabien, B. C. (1995). "The Design of Dwell-Rise-Dwell Cams with Reduced Sensitivity to Parameter Variation." *J. of the Franklin Institute*, **332B**(2), pp. 195-209.

Fabien, B. C., et al. (1994). "The Design of High-Speed Dwell-Rise-Dwell Cams Using Linear Quadratic Optimal Control Theory." *Journal of Mechanical Design*, **116**, p. 867.

Fanella, R. J. (1959). "Dynamic Analysis of a Cylindrical Cam." ASME 59-SA-3.

Farin, G. (1993). *Curves and Surfaces for Computer Aided Geometric Design*. Academic Press.

Farin, G. (1995). *Nurb Curves and Surfaces: From Projective Geometry to Practical Use*. A. K. Peters, Wellesley, MA.

Farouki, R. T., et al. (1998). "Design of Rational Cam Profiles with Pythagorean Hodograph Curves." *Mechanism and Machine Theory*, **33**(6), pp. 669-682.

Farouki, R. T., and C. A. Neff. (1990a). "Algebraic Properties of Plane Offset Curves." *Computer Aided Geometric Design*, **7**, pp. 101-127.

Farouki, R. T., and C. A. Neff. (1990b). "Analytic Properties of Plane Offset Curves." *Computer Aided Geometric Design*, **7**, pp. 83-99.

Farouki, R. T., and C. A. Neff. (1995). "Hermite Interpolation by Pythagorean Hodographic Quintics." *Mathematics of Computation*, **64**(212), pp. 1589-1609.

Farouki, R. T., and T. Sakkalis. (1990). "Pythagorean Hodographs." *IBM J. Res. Develop.*, **34**(5), pp. 736-752.

Farouki, R. T., and S. Shah. (1996). "Real Time CNC Interpolators for Pythagorean-Hodograph Curves." *Computer Aided Geometric Design*, **13**, pp. 583-600.

Fawcett, G. F., and J. N. Fawcett, eds. (1978). "Comparison of Polydyne and Non-Polydyne Cams." *Cams and Cam Mechanisms*, Jones, J. R., ed., Institution of Mechanical Engineers: London.

Felszeghy, S. F. (1986). "Steady-State Residual Vibrations in High-Speed, Dwell-Type, Rotating Disk, Cam-Follower Systems." ASME 86-DET-152.

Ferguson, R. C. "NC Jig Grinding Joins the Ranks." Moore Special Tool Co.

Flierl, R., and M. Kluting. (2000). "The Third Generation of Valvetrains - New Fully Variable Valvetrains for Throttle-Free Load Control." SAE 2000-01-1227.

Flugrad, D. R. (1983). "Cam Synthesis by an Iterative Procedure." *Proc. of 8th Applied Mechanisms Conference*, St Louis MO.

Freudenstein, F. (1960). "On the Dynamics of High-Speed Cam Profiles." *Int. J. Mech. Sci.*, **1**, pp. 342 - 349.

Freudenstein, F., et al. (1988). "The Synthesis and Analysis of Variable-Valve-Timing Mechanisms for Internal Combustion Engines." SAE 880387.

Gandhi, A., and A. Myklebust. (1986). "Cam Mechanisms from Concept to Prototype - a System for Rapid Design and 3-D Modeling of Cams for Cad/Cam Databases." ASME, Oct 5, 1986.

Ganter, M. A., and J. J. Uicker. (1979). "Design Charts for Disk Cams with Reciprocating Roller Followers." *Journal of Mechanical Design*, **101 Series B.**(3), pp. 465-470.

Garcez, N. A. N. (1992). "Variable Valve Timing - the "Mirabilis" Camshaft - Its Impact on the Environment and Fuel Consumption." SAE 921476.

Ge, Q. J., and B. Ravani. (1994). "Geometric Construction of Bézier Motions." *Journal of Mechanical Design*, **116**, pp. 749-755.

Ge, Q. J., and L. Srinivasan. (1996). "C2 Piecewise Bezier Harmonics for Motion Specifications of High Speed Cam Mechanisms." 96-DETC/MECH-1173, ASME Design Engineering Technical Conference: Irvine, CA.

Gecim, B. (1993). "Analysis of a Lost-Motion-Type Hydraulic System for Variable Valve Actuation." SAE 930822.

Gecim, B. A. (1997). "A Low-Friction Variable-Valve Actuation Device, Part 2: Analysis and Simulation." SAE 970339.

Giles, W. S. (1966). "Fundamentals of Valve Design and Material Selection." SAE paper.

Giordana, F., and V. Rognoni. (1980). "The Influence of Construction Errors in the Law of Motion of Cam Mechanisms." *Mechanism and Machine Theory*, **15**, pp. 29-45.

Gonzales-Palacios, M. A., and J. Angeles. (1993). *Cam Synthesis*. Kluwer Academic Publishers: Dordrecht, 250 pp.

Goodman, T. P. (1957). "Linkages Vs. Cams." *Proc. of Fourth Conference on Mechanisms*, Purdue Univ, pp. 76-83.

Goodman, T. P. (1962). "Dynamic Effects of Backlash." *Proc. of Seventh Conference on Mechanisms*, Purdue University, pp. 128-138.

Goodman, T. P. (1963). "Dynamic Effects of Backlash." *Machine Design*, May 23, pp. 150-150.

Grant, B., and A. H. Soni. (1979). "A Survey of Cam Manufacturing Methods." *Journal of Mechanical Design*, **101**, pp. 455-464.

Grewal, P. S., and H. R. Newcombe. "Dynamic Performance of High Speed Semi-Rigid Follower Cam Systems - Effects of Cam Profile Errors." *Mechanism and Machine Theory*.

Gupta, K. C., and J. L. Wiederrich. (1982). "Development of Cam Profiles Using the Convolution Operator." ASME 82-DET-87.

Gupta, K. C., and J. L. Wiederrich. (1986). "On the Modification of Cam-Type Profiles." *Mechanism and Machine Theory*, **21**(5), pp. 439-444.

Gutman, A. S. (1961). "To Avoid Vibration, Try This New Cam Profile." *Product Engineering*, December 25.

Hain, K. (1969). "Challenge: To Design Better Cams." *Journal of Mechanisms*, **5**, pp. 283-286.

Hain, K. (1971). "Optimization of a Cam Mechanism to Give Good Transmissibility, Maximal Output Angle of Swing, and Minimal Acceleration." *Journal of Mechanisms*, **6**, pp. 419-434.

Hain, K., ed. (1978). "Optimization of an Intermittent Motion Mechanism with Fixed Cam." *Cams and Cam Mechanisms*, Jones, J. R., ed., I. Mech. E., pp. 111-115.

Hamilton, G. M. (1980). "The Hydrodynamics of a Cam Follower." *Tribology International* (June), pp. 113-119.

Hanachi, S., and F. Freudenstein. (1986). "The Development of a Predictive Model for the Optimization of High-Speed Cam-Follower Systems with Coulomb Damping Internal Friction and Elastic and Fluidic Elements." ASME, Oct 5, 1986.

Hanaoka, M., and S. Fukumura. (1973). "A Study of Valve Train Noises and a Method of Cam Design to Reduce the Noises." SAE 730247.

Hannibal, W., and A. Bertsch. (1998). "Vast: A New Variable Valve Timing System for Vehicle Engines." SAE 980769.

Hanson, R. S., and F. T. Churchill. (1962). "Theory of Envelopes Provides New Cam Design Equations." *Product Engineering*, pp. 45-55.

Hatano, K., et al. (1993). "Development of a New Multi-Mode Variable Valve Timing Engine." SAE 930878.

Helden, A. K. v., et al. (1985). "Dynamic Friction in Cam/Tappet Lubrication." SAE 850441.

Henry, R. R., and B. Lequesne. (1997). "A Novel, Fully Flexible, Electro-Mechanical Engine Valve Actuation System." SAE 970249.

Herrin, R. J. (1982). "Measurement of Engine Valve Train Compliance under Dynamic Conditions." SAE 820768.

Hirschhorn, J. (1958). "Disc Cam Curvature." *Proc. of Fifth Conference on Mechanisms*, Purdue University, pp. 50-63.

Hollingworth, P., and R. A. Hodges. (1991). "The History and Mathematical Development of Cam Profile Design in Rover." SAE 914172.

Holowenko, A. R., and A. S. Hall. (1953). "Cam Curvature." *Machine Design*, August, pp 170-177, Sept pp 162-170, Nov pp 148-156.

Horan, R. P. (1953). "Overhead Valve Gear Problems." *SAE Transactions*, **61**, pp. 678-686.

Horeni, B. (1992). "Double-Mass Model of an Elastic Cam Mechanism." *Mechanism and Machine Theory*, **27**(4), pp. 443-449.

Hoskins, W. D. (1970). "Interpolating Quintic Splines on Equidistant Knots. Algorithm 62." *The Computer J.*, **13**(4), pp. 437-438.

Hrones, J. A. (1948). "An Analysis of the Dynamic Forces in a Cam-Driven System." *Trans ASME*, pp. 473 - 482.

Hsu, W., and A. P. Pisano. (1996). "Modeling of a Finger-Follower Cam System with Verification in Contact Forces." *Journal of Mechanical Design*, **118**, p. 132.

Hundal, M. S. (1963). "Aid of Digital Computer in the Analysis of Rigid Spring-Loaded Valve Mechanisms." *SAE Progress in Technology*, **5**, pp. 4-9, 57.

Jackowski, C. S., and J. F. Dubil. (1967). "Single Disk Cams with Positively Controlled Oscillating Followers." *Journal of Mechanisms*, **2**, pp. 157-184.

Jacobs, F. B. (1921). *Cam Design and Manufacture*. Van Nostrand: New York.

Järvi, I. (1998). "Variable Valve Timing Mechanism with Control Ramp." SAE 980768.

Jeans, H. (1957). "Designing Cam Profiles with Digital Computers." *Machine Design*, October 31, 1957, pp. 103-106.

Jensen, P. B. (1987). *Cam Design and Manufacture*. Marcel Dekker: New York, p. 428.

Jensen, P. W. (1965). "Mechanisms for Generating Cam Curves." *Product Engineering*, March 1, 1965, pp. 41-47.

Jepson, G. (1905). *Cams and the Principles of Their Construction*. Van Nostrand: New York.

Johnson, A. R. (1965). "Motion Control for a Series System of N-Degrees of Freedom Using Numerically Derived and Evaluated Equations." *Journal of Engineering for Industry*, **87 Series B**, pp. 191-204.

Johnson, G. I. (1963). "Studying Valve Dynamics with Electronic Computers." *SAE Progress in Technology*, **5**, pp. 10-28.

Johnson, R. C. (1955). "Method of Finite Differences Provides Simple but Flexible Arithmetical Techniques for Cam Design." *Machine Design*, November, pp. 195-204.

Johnson, R. C. (1956a). "How to Design Cam Mechanisms for Optimum Cam and Follower Proportions." *Machine Design*, January 26, 1956, pp. 85-89.

Johnson, R. C. (1956b). "A New Point of View on Minimizing Cam Vibration." *Machine Design*, December 13, 1956, pp. 103-104.

Johnson, R. C. (1956c). "A Rapid Method for Developing Cam Profiles Having Desired Acceleration Characteristics." *Machine Design*, December 13, 1956, pp. 129-132.

Johnson, R. C. (1957a). "How Profile Errors Affect Cam Dynamics." *Machine Design*, February 7, 1957, pp. 105-108.

Johnson, R. C. (1957b). "Method of Finite Differences in Cam Design." *Machine Design*, November 14, 1957, pp. 159-161.

Johnson, R. C. (1958). "Dynamic Analysis of Cam Mechanisms." *Proc. of Fifth Conference on Mechanisms*, Purdue University, pp. 21-35.

Johnson, R. C. (1984). "Force Reduction by Motion Design in Spring-Loaded Cam Mechanisms." *Journal of Mechanisms, Transmissions, and Automation in Design*, **106**, pp. 278-284.

Jones, J. R. (1978). *Cams and Cam Mechanisms*. Institution of Mechanical Engineers: London, 153 pp.

Jones, J. R., and J. E. Reeve, eds. (1978). "Dynamic Response of Cam Curves Based on Sinusoidal Segments." *Cams and Cam Mechanisms*, Jones, J. R., ed., Institution of Mechanical Engineers: London, pp. 14-24.

Jordan, R. L. A. (1986). "Cams." Delta Engineering Corporation, December 1986.

Jordan, R. L. A. (1993). "A General Approach to Cam Surface Definition." *Proc.*,, pp. 159-162.

Kanango, R. N., and N. Patnaik. (1979). "Improving Dynamic Characteristics of a Cam-Follower Mechanism through Finite Difference Techniques." *Proc. of 5th World Congress on Theory of Machines and Mechanisms*, p. 591.

Kanesaka, H., and K. Akiba. (1977). "A New Method of Valve Cam Design - Hysdyne Cam." SAE 770777.

Kanzaki, K., and K. Itao. (1972). "Polydyne Cam Mechanisms for Typehead Positioning." *ASME Journal of Engineering for Industry*, **94**(1), pp. 250 - 254.

Kaplan, J. (1960). "Cam Control Gets More out of a Planetary Gear." *Product Engineering*, pp. 38-41.

Keribar, R. (2000). "A Valvetrain Design Analysis Tool with Multiple Functionality." SAE 2000-01-0562.

Kerle, H., ed. (1978). "How Effective Is the Method of Finite Differences as Regards Cam Mechanisms." *Cams and Cam Mechanisms*, Jones, J. R., ed., Institution of Mechanical Engineers: London, pp. 131-135.

Khader, K., et al. (1996). "The Synthesis of a Multi-Step Cam Mechanism to Drive a Shaking Belt Conveyor." *Mechanism and Machine Theory*, **31**(7), pp. 913-924.

Kim, D., and J. W. David. (1990). "A Combined Model for High Speed Valve Train Dynamics (Partly Linear and Partly Nonlinear)." SAE 901726.

Kim, H. R., and H. R. Newcombe. (1980). "The Effect of Cam Profile Errors and System Flexibility on Cam Mechanism Output." *Machine and Mechanism Theory*, **17**(1), pp. 57 - 72.

Kim, H. R., and W. R. Newcombe. (1978). "Stochastic Error Analysis in Cam Mechanisms." *Mechanism and Machine Theory*, **13**, pp. 631-641.

Kloomok, M., and R. V. Muffley. (1955a). "Plate Cam Design-Pressure Angle Analysis." *Product Engineering*, v. 26, May, 1955, pp. 155-171.

Kloomok, M., and R. V. Muffley. (1955b). "Plate Cam Design-Radius of Curvature." *Product Engineering*, v. 26, September, 1955, pp. 186-201.

Kloomok, M., and R. V. Muffley. (1955c). "Plate Cam Design-with Emphasis on Dynamic Effects." *Product Engineering*, v. 26, February, 1955, pp. 156-162.

Kloomok, M., and R. V. Muffley. (1956a). "Determination of Pressure Angles for Swinging-Follower Cam Systems." *Transactions of the ASME*, **May 1956**, pp. 803 - 806.

Kloomok, M., and R. V. Muffley. (1956b). "Determination of Radius of Curvature for Radial and Swinging-Follower Cam Systems." *Transactions of the ASME*, **May 1956**, pp. 795 - 802.

Kloomok, M., and R. V. Muffley. (1956c). "Evaluating Dynamic Loads." *Product Engineering*, January, 1956, pp. 178-182.

Klotter, K. (1953) "The Attenuation of Damped Free Vibrations and the Derivation of the Damping Law from Recorded Data." Tech. Report #23, Contract N6-ONR-251-T.O., Stanford Univ., Nov. 1, 1953.

Koster, M. P. (1974). *Vibrations of Cam Mechanisms*. Phillips Technical Library Series, Macmillan Press Ltd.: London.

Koster, M. P. (1975a). "Digital Simulation of the Dynamics of Cam Followers." *I. Mech. E.*, pp. 969-974.

Koster, M. P. (1975b). "Effect of Flexibility of Driving Shaft on the Dynamic Behavior of a Cam Mechanism." *Journal of Engineering for Industry*(2).

Koster, M. P., ed. (1978). "The Effects of Backlash and Shaft Flexibility on the Dynamic Behavior of a Cam Mechanism." *Cams and Cam Mechanisms*, Jones, J. R., ed., I. Mech. E.: London, pp. 141-147.

Kosugi, T., and T. Seino. (1985). "Valve Motion Simulation Method for High-Speed Internal Combustion Engines." SAE 850179.

Koumans, P. W., ed. (1978). "A Special Cam Milling Machine." *Cams and Cam Mechanisms*, Jones, J. R., ed., Institution of Mechanical Engineers: London.

Kozhevnikov, S. N., et al. (1974). "Synthesis of a Cam-Differential Mechanism with Periodic Dwell of the Output Link." *Mechanism and Machine Theory*, **9**, pp. 219-229.

Kreuter, P., et al. (1998). "The Meta VVH System - A Continuously Variable Valve Timing System." SAE 980765.

Kreuter, P., and G. Mass. (1987). "Influence of Hydraulic Valve Lash Adjusters on the Dynamic Behavior of Valve Trains." SAE 870086.

Kreuter, P., and F. Pischinger. (1985). "Valve Train Calculation Model with Regard to Oil Film Effects." SAE 850399.

Krouse, J. K. (1980). "Cad/Cam - Bridging the Gap from Design to Production." *Machine Design*, June 12, 1980, pp. 117 - 125.

Krouse, J. K. (1981a). "Computer Time-Sharing for Cad/Cam." *Machine Design*, April 23, 1981, pp. 57 - 64.

Krouse, J. K. (1981b). "Sculptured Surfaces for Cad/Cam." *Machine Design*, March 12, 1981, pp. 115 - 120.

Kumagai, A., et al. (1991). "Influence of Pressure Angle on Dynamic Stability of Cam-Modulated Linkages." *Proc. of 2nd Applied Mechanisms and Robotics Conference*, Cincinnati OH, pp. VIA.4-1.

Kwakernaak, H., and J. Smit. (1968). "Minimum Vibration Cam Profiles." *J. Mechanical Engineering Science*, **10**(3), pp. 219-227.

Lancefield, T., et al. (2000). "The Application of Variable Event Valve Timing to a Modern Diesel Engine." SAE 2000-01-1229.

Lancefield, T. M., et al. (1993). "The Practical Application and Effects of a Variable Event Valve Timing System." SAE 930825.

Lee, J., and D. J. Patterson. (1995). "Analysis of Cam/Roller Follower Friction and Slippage in Valve Train Systems." SAE 951039.

Lee, J., et al. (1994). "Friction Measurement in the Valve Train with a Roller Follower." SAE 940589.

Lenz, H. P., et al. (1988). "Variable Valve Timing - A Possibility to Control Engine Load without Throttle." SAE 880388.

Li, Y. M., and V. Y. Hsu. (1998). "Curve Offsetting Based on Legendre Series." *Computer Aided Geometric Design*, **15**, pp. 711-720.

Lin, A. C., et al. (1988). "Computerized Design and Manufacturing of Plate Cams." *Int. J. of Production Research*, **26**(8), pp. 1395-1430.

Lin, A. C., and H. Wang. (1988). "Development of a Computer-Integrated Design and Manufacture System for Plate Cams." *Manufacturing Review*, **1**(3), pp. 198-213.

Liniecki, A. G. (1975). "Optimum Design of Disk Cams by Nonlinear Programming." *Proc. of 4th OSU Applied Mechanisms Conference*, Chicago, IL.

Loeser, E. H., et al. (1958). "Cam and Tappett Lubrication III—Radioactive Study of Phosphorous in the Ep Film." *Proc. of ASLE Annual Meeting*, Cleveland OH.

Love, R. J., and F. C. Wykes. (1975). "European Practice in Respect of Automotive Cams and Followers." SAE 750865.

Lucas, R. A., and Y. Li. "Design Charts for Plate Cams with Oscillating Roller Followers." Lehigh University.

MacCarthy, B. L. (1988). "Quintic Splines for Kinematic Design." *Computer-Aided Design*, **20**(7), pp. 406-415.

MacCarthy, B. L., and N. D. Burns. (1985). "An Evaluation of Spline Functions for Use in Cam Design." *IMechE*, **199**(C3), pp. 239-248.

Mahkijani, V. S. (1984). "Study of Contact Stresses as Developed on a Radial Cam Using Photoelastic Model and Finite Element Analysis." M.S. Thesis, Worcester Polytechnic Institute.

Mahyuddin, A. I., and A. Midha. (1991). "Effects of Pressure Angle Variation on Parametric Instability in Flexible Cam Follower Mechanisms." *Proc. of 2nd Applied Mechanisms and Robotics Conference*, Cincinnati OH, pp. VIA.3-1.

Mahyuddin, A. I., and A. Midha. (1994). "Influence of Varying Cam Profile and Follower Motion Event Type on Parametric Vibration and Stability of Flexible Cam-Follower Systems." *Journal of Mechanical Design*, **116**(1), p. 298.

Mahyuddin, A. I., et al. (1994). "Evaluation of Parametric Vibration and Stability of Flexible Cam-Follower Systems." *Journal of Mechanical Design*, **116**(1), p. 291.

Maier, K. (1957). "Surge Waves in Compression Springs." *Product Engineering*, August, 1957, pp. 167-174.

Mathieson, M. F., et al. (1993). "A Hydraulic Tappet with Variable Timing Properties." SAE 930823.

Matsuda, T., and M. Sato. (1989). "Dynamic Modeling of Cam and Follower System. Evaluation of One Degree of Freedom Model." *Proc. of ASME Vibration Analysis - Techniques and Applications*, Montreal, Canada.

Matthew, G. K. (1979). "The Modified Polynomial Specification for Cams." *Proc. of 5th World Congress on Theory of Machines and Mechanisms*, pp. 1299-1302.

Matthew, G. K., and D. Tesar. (1975a). "Cam System Design: The Dynamic Synthesis and Analysis of the One Degree of Freedom Model." *Mechanisms and Machine Theory*, v. 11, August 24, 1975, pp. 247 - 257.

Matthew, G. K., and D. Tesar. (1975b). "The Design of Modeled Cam Systems Part I: Dynamic Synthesis and Design Chart for the Two-Degree-of-Freedom Model, Part Ii: Minimization of Motion Distortion Due to Modeling Errors." *Journal of Engineering for Industry*(November 1975), pp. 1175-1189.

McAllister, D. F., and J. A. Roulier. (1981). "An Algorithm for Computing a Shape-Preserving Osculatory Quadratic Spline." *ACM Trans Mathematical Software*, **7**(3), pp. 331-347.

McKellar, M. R. (1966). "Overhead Camshaft Stirs New Tempest." *SAE Journal*, **74**(11), pp. 69-73.

McPhate, A. J., and L. R. Daniel. (1962). "A Kinematic Analysis of Fourbar Equivalent Mechanisms for Plane Motion Direct Contact Mechanisms." *Proc. of Seventh Conference on Mechanisms*, Purdue University, pp. 61-65.

Meeusen, H. J. (1966). "Overhead Cam Valve Train Design Analysis with a Digital Computer." SAE 660350.

Mendez-Adriani, J. A. (1983). "Variation of the Range of Jump Phenomenon with the Harmonic Cam Stiffness." ASME 83-DET-5.

Mendez-Adriani, J. A. (1985). "Design of a General Cam-Follower Mechanical System Independent of the Effect of Jump Resonance." ASME 85-DET-56.

Meneghetti, U., and A. O. Andrisano. (1979). "On the Geometry of Cylindrical Cams." *Proc. of 5th World Congress on Theory of Machines and Mechanisms*, p. 595.

Mercer, S., and A. R. Holowenko. (1958). "Dynamic Characteristics of Cam Forms Calculated by the Digital Computer." *Trans ASME*, **80**(8), pp. 1695-1705.

Midha, A., et al. (1979). "Periodic Response of High Speed Cam Mechanism with Flexible Follower and Camshaft Using a Closed-Form Numerical Algorithm." *Proc. of 5th World Congress on Theory of Machines and Mechanisms*, pp. 1311-1314.

Midha, A., and D. A. Turcic. (1980). "On the Periodic Response of Cam Mechanism with Flexible Follower and Camshaft." *Journal of Dynamic Systems, Measurement and Control*, **102**, pp. 255-264.

Mills, J. K., et al. (1993). "Optimal Design and Sensitivity Analysis of Flexible Cam Mechanisms." *Mechanism and Machine Theory*, **28**(4), pp. 563-581.

Mitchell, D. B. (1948). "Tests on Dynamic Response of Cam Follower Systems." *Mechanical Engineering*, pp. 467-471.

Mizuno, N., et al. (1988). "Analysis of Synchronous Belt Vibration in Automotive Valve Train." SAE 880077.

Molian, S. (1968). *The Design of Cam Mechanisms and Linkages*. American Elselvier: New York.

Moon, C. H. (1962). *Cam Design*. AMCAM Corporation: Farmington, Conneticut.

Morehead III, J. C. "Intepretation of Cam-Follower Acceleration Data.", Miller Printing Equipment Corporation.

Moriya, Y., et al. (1996). "A Newly Developed Intelligent Variable Valve Timing System - Continuously Controlled Cam Phasing as Applied to a New 3 Liter Inline 6 Engine." SAE 960579.

Morrison, R. A. (1968). "Test Data Let You Develop Your Own Load/Life for Gear and Cam Materials.", USM Machinery Co, 1968.

Morrison, R. A., and W. Cram. (1968). "Load/Life Curves for Gear and Cam Materials." *Machine Design*, v. 40, Aug 1, 1968, pp. 102-108.

Mosier, R. G. (2000). "Modern Cam Design." *Int. J. Vehicle Design*, **23**(1/2), pp. 38-55.

Nagaya, K., et al. (1993). "Vibration Analysis of High Rigidity Driven Valve System of Internal Combustion Engine." *J. Sound and Vibration*, **165**(1), pp. 31-43.

Nagle, P. D. (1992). "Making the Most of Valve Train Lift Data." SAE 921665.

Neamtu, M., et al. (1998). "Designing Nurbs Cam Profiles Using Trigonometric Splines." *Journal of Mechanical Design*, **120**(2), pp. 175-180.

Neklutin. (1952). "Designing Cams for Controlled Inertia and Vibration." *Machine Design*, June, 1952, pp. 143-160.

Neklutin, C. N. (1954). "Vibration Analysis of Cams." *Proc. of The Second Conference on Mechanisms*, Purdue University, pp. 6-14.

Neklutin, C. N. (1959). "Trig-Type Cam Profiles." *Machine Design*, v. 31, October 15, 1959, pp. 175-187.

Neklutin, C. N. (1969). *Mechanisms and Cams for Automatic Machinery*. American Elsevier: New York.

Newcombe, H. R., and H. R. Kim. (1983). "Some Results of the Effect of Waviness in Cam Profiles on the Output Motion." *Proc. of 8th Applied Mechanisms Conference*, St. Louis MO, pp. 22-1 to 22-7.

Nishioka, M. "Locus Compensation of Tool for Contouring of Cams by Specific Machine Tools." Wissenschaftliche Zeitschrift Der Wilhem-Pieck-Universitat Rostock, 36. Jahrgang 1987.

Nishioka, M. (1984). "Development of Data Processing System for Cam and Linkage Mechanisms." *International Symposium on Design and Synthesis*, pp. 788-793.

Nishioka, M. (1984). "Study of Cam Mechanism." *Proc. of Iftomm-Symposium*, Karl-Marx Stadt Deutsche Demokratische Republik, pp. 113-118.

Nishioka, M. (1996). "Modular Structure of Spatial Cam Linkage Mechanism." *Mechanism and Machine Theory*, **31**(6), pp. 813-819.

State of the Art of Torque Compensation Cam Mechanisms.", Sankyo America, Report: 11/24/97.

Nitao, J. J., and J. L. Weiderrich. (1986). "Estimation of an Effective Dissipation Function in Machines." ASME 86-DET-11.

Noortgate, L. V. D., and J. DeFraine. (1977). "A General Computer Aided Method for Designing High Speed Cams Avoiding the Dangerous Excitation of the Machine Structure." *Mechanism and Machine Theory*, **12**(3), pp. 237-245.

Norton, R. L. (1986). "Accuracy, Dynamic Performance, and Audible Sound of Plate Cams Made by Various Methods." *Proc. of Design Engineering Technical Conference*, Columbus, Ohio, pp. 1-8.

Norton, R. L. (1988a). "Effect of Manufacturing Method on Dynamic Performance of Eccentric Cams - Part I." *Mechanism and Machine Theory*, **23**(3), pp. 191-200.

Norton, R. L. (1988b). "Effect of Manufacturing Method on Dynamic Performance of Double Dwell Cams - Part II." *Mechanism and Machine Theory*, **23**(3), pp. 201-208.

Norton, R. L., et al. (1988c). "Analysis of the Effect of Manufacturing Methods and Heat Treatment on the Performance of Double Dwell Cams." *Mechanism and Machine Theory*, **23**(6), pp. 461-473.

Norton, R. L. (1992). "*Machinery* - a Computer Aided Design Package for the Synthesis and Analysis of Linkages and Cams." *The COeD Journal, ASEE*, **11**(2), pp. 1-11.

Norton, R. L., ed. (1993). "Cams and Cam Followers." *Modern Kinematics*, Erdman, A. E., ed., Wiley: New York, pp. 271-332.

Norton, R. L., et al. (1998). "Analyzing Vibrations in an IC Engine Valve Train." SAE 980570.

Norton, R. L., et al. (1999). "Effect of Valve-Cam Ramps on Valve Train Dynamics." SAE 1999-01-0801.

Nourse, J. H. (1965). "Recent Developments in Cam Profile Measurement and Evaluation." SAE 650259.

Nourse, J. H., et al. (1960). "Recent Developments in Cam Design."

Nuccio, P., et al. (1992). "Variable Intake Valve Closing as a Means to Improve the SI Engine Torque Characteristic." SAE 92A056.

Nutbourne, A. W., et al. (1972). "A Cubic Spline Package: Part 1 - the User's Guide, Part 2 the Mathematics." *Computer Aided Design*, **4, 5**, pp. Part 1: .228-238, Part 2: 7-13.

Ogino, S. (1967a). "Design Characteristics of Constant Diameter Cam." *Bukketin of Japan Soc. of Mech Eng.*, **10**(42), pp. 1032-1038.

Ogino, S. (1967b). "Design Characteristics of Constant Diameter Cam (Contribution to Brunell's Paper)." *Bulletin of The JSME*, **10**(42), pp. 1032-1038.

Ogozalek, F. J. (1966). "Theory of Catenoidal-Pulse Motion and Its Application to High-Speed Cams." ASME 66-Mech-45.

Okcuoglu, S. A. (1960). "An Application of Polydyne Cam Design." *Proc. of Sixth Conference on Mechanisms*, Purdue University, pp. 44-47.

Pagel, P. A. (1978). "Sizing Cams for Long Life." *Machine Design*, Sept. 7, 1978, pp. 104-109.

Paranjpe, R. S. (1990). "Dynamic Analysis of a Valve Spring with a Coulomb-Friction Damper." *Journal of Mechanical Design*, **112**(4), p. 509.

Parmater, J. Q. (1980). "Computer Aided Engineering: The Step Beyond Cad/Cam." *Machine Design*, October 23, 1980, pp. 55 - 59.

Peisakh, E. E. (1966). "Improving the Polydyne Cam Design Method." *Russian Engineering Journal*, **XLVI**(12), pp. 25-27.

Pham, B. (1992). "Offset Curves and Surfaces: A Brief Survey." *Computer Aided Design*, **24**, pp. 223-229.

Phlips, P. J., et al. (1989). "An Efficient Model for Valvetrain and Spring Dynamics." SAE 890619.

Pierce, M., and R. Mollica. (1984). "FEA Model of Cam Follower Arm." Major Qualifying Project, Worcester Polytechnic Institute.

Pierik, R. J., and J. F. Burkhard. (2000). "Design and Development of a Mechanical Variable Valve Actuation System." SAE 2000-01-1221.

Pierik, R. J., and B. A. Gecim. (1997). "A Low-Friction Variable-Valve-Actuation Device, Part 1: Mechanism Description and Friction Measurements." SAE 970338.

Pisano, A. P. (1981). "The Analytical Development and Experimental Verification of a Predictive Model of a High-Speed Cam-Follower System." Ph.D. Thesis, Columbia University.

Pisano, A. P. (1984). "Coulomb Friction in High-Speed Cam Systems." ASME 84-DET-19.

Pisano, A. P. (1986). "Cam Systems with Coulomb Friction: Comparison of Two Models and Experiment." ASME 86-DET-96.

Pisano, A. P., and H. T. Chen. (1985). "Coulomb Friction and Optimal Rocker Arm Ratio for High-Speed Cam Systems." ASME 85-DET-61.

Pisano, A. P., and F. Freudenstein. (1982). "An Experimental and Analytical Investigation of the Dynamic Response of High-Speed Cam-Follower Systems Part I: Experimental Investigation." ASME 82-DET-135.

Pisano, A. P., and F. Freudenstein. (1983a). "An Experimental and Analytical Investigation of the Dynamic Response of a High-Speed Cam-Follower System Part 2: A Combined, Lumped/Distributed Parameter Dynamic Model." *Journal of Mechanisms, Transmissions, and Automation in Design*, **105**, pp. 699-704.

Pisano, A. P., and F. Freudenstein. (1983b). "An Experimental and Analytical Investigation of the Dynamic Response of a High-Speed Cam-Follower System. Part I: Experimental Investigation, Part 2: A Combined, Lumped/Distributed Parameter Dynamic Model." *Journal of Mechanisms, Transmissions, and Automation in Design*, **105**(4), pp. 692-704.

Pisano, A. P., and F. Freudenstein. (1983c). "An Experimental and Analytical Verification of the Dynamic Response of a High-Speed Cam-Follower System." *Journal of Mechanisms, Transmissions, and Automation in Design*, **105**, pp. 692-704.

Plesse, L. G., and J. J. Rogers. (1977). "Cam and Lifter Wear as Affected by Engine Oil ZDP Concentration and Type." SAE 77087.

Priebsch, H. H., et al. (1992). "Valve Train Dynamics and Their Contribution to Engine Performance." SAE paper.

Pryor, R. F., et al. (1979). "On the Classification and Enumeration of Six-Link and Eight-Link Cam-Modulated Linkages." *Proc. of 5th World Congress on Theory of Machines and Mechanisms*, p. 1315.

Qimi, J. (1992). "The Design of a Cam, Steel-Strip Function Mechanism." JMD478, WPI, June 1, 1992.

Raghavacharyulu, E., and J. S. Rao. (1976). "Jump Phenomena in Cam-Follower Systems a Continuous-Mass-Model Approach." ASME 76-WA/DE-26.

Raghavacharyulu, E., and J. S. Rao. (1983). "Nonlinear Vibration Analysis of Cam-Follower Systems with Pneumatic Coupling." *Proc. of Proceedings of the Sixth World Congress on Theory of Machines and Mechanisms-1983*, pp. 1213 - 1216.

Raghavacharyulu, E., and B. L. Sachdeva. (1984). "Dynamic Characteristics of Cam-Follower Systems." *Journal of Engineering Design*, **2**(1), pp. 20-28.

Rahman, Z. U., and W. H. Bussell. (1971). "An Iterative Method for Analyzing Oscillating Cam Follower Motion." *Journal of Engineering for Industry*, pp. 149-156.

Rao, A. C., and M. Choubey. (1980). "Optimum Sensitivity Synthesis of a Cam System Incorporating Manufacturing Tolerances." ASME 80-DET-01.

Rao, S. S., and S. S. Gavane. (1982). "Analysis and Synthesis of Mechanical Error in Cam-Follower Systems." *J. Mechanical Design*, **104**(1), pp. 52 - 62.

Raven, F. H. (1959). "Analytical Design of Disk Cams and Three-Dimensional Cams by Independent Position Equations." *Trans ASME*, **26 Series E**(1), pp. 18-24.

Reeve, J. (1995). *Cams for Industry*. Mechanical Engineering Publications Ltd.: Suffolk G.B., p. 364.

Renstrom, R. C. (1976). "Desmodromics: Ultimate Valve Gear?" *Motorcycle World*, May, 1976.

Ricardo. (1996). "Software for Valve Train Design." Ricardo Consulting Engineers Ltd.: Shoreham-by-Sea, England.

Riley, M. B., et al. (1997). "A Mechanical Valve System with Variable Lift, Duration, and Phase Using a Moving Pivot." SAE 970334.

Rob, J., and M. Arnold. (1993). "Analysis of Dynamic Interactions in Valve Train Systems of Ic Engines by Using a Simulation Model." SAE 930616.

Rodrigues, H. (1997). "Sintered Valve Seat Inserts and Valve Guides: Factors Affecting Design, Performance & Machinability." *Proc. of Int. Symp. on Valvetrain System Design and Materials*, Dearborn, MI, p. 107.

Rogers, M. D., and R. R. Schaffer. (1963). "Kinematic Effects of Cam Profile Errors." ASME 63-WA-269.

Roggenbuck, R. A. (1953). "Designing the Cam Profile for Low Vibration at High Speeds." *SAE Transactions*, **61**, pp. 701 - 705.

Rooney, G. T., and P. Deravi. (1982). "Coulomb Friction in Mechanism Sliding Joints." *Mechanism and Machine Theory*, **17**(3), pp. 207-211.

Roskilly, M., et al. (1986). "Valve Gear Design Analysis." SAE 865027.

Rothbart, H. A. (1956). *Cams: Design, Dynamics, and Accuracy*. John Wiley & Sons: New York, 350 pp.

Rothbart, H. A. (1957a). "Catalog of Equivalent Mechanisms for Cams for Simplified Graphical Analysis of Complex Cam-Follower Motions." *Proc. of Fourth Conference on Mechanisms*, Purdue University, pp. 25-30.

Rothbart, H. A. (1957b). "Limitations on Cam Pressure Angles." *Product Engineering*, Jan., 1957.

Rothbart, H. A. (1958a). "Cam Torque Curves." *Proc. of Fifth Conference on Mechanisms*, Purdue University, pp. 36-41.

Rothbart, H. A. (1958b). "Which Way to Make a Cam?" *Product Engineering*, March 3. 1958.

Sacks, E., and L. Joskowicz. (1998). "Dynamical Simulation of Planar Systems with Changing Contacts Using Configuration Spaces." *Journal of Mechanical Design*, **120**(2), pp. 181-187.

Sadek, K. S. H., and A. Daadbin. (1993). "Modeling of a Torsion Bar Cam Mechanism." *Mechanism and Machine Theory*, **28**(5), pp. 631-640.

Sadek, K. S. H., et al. (1990). "Natural Frequencies of a Cam Mechanism." *Proceedings of the Institution of Mechanical Engineers*, **204**(4), pp. 255-261.

Sadler, J. P., and Z. Yang. (1990). "Optimal Design of Cam-Linkage Mechanisms for Dynamic Force Characteristics." *Mechanism and Machine Theory*, **25**(1), pp. 41-57.

Sanchez, M. N., and J. G. deJalon. (1980). "Application of B-Spline Functions to the Motion Specifications of Cams." *Proc. of ASME Design Engineering Technical Conference*, Beverly Hills, CA.

Sandgren, E., and R. Wainman. (1985). "Cam Profile Optimization for Dynamic Systems." *Proc. of Ninth Applied Mechanisms Conference*, Kansas City, MO, pp. V1-V7.

Sandgren, E., and R. L. West. (1990). "Shape Optimization of Cam Profiles Using B-Spline Representation." *ASME Journal of Mechanisms, Transmissions, and Automation in Design*, **111**, pp. 195-201.

Sandler, B. Z. (1977). "Cam Mechanism Accuracy." ASME 77-WA/DE-19.

Sankar, S. (1977). "Dynamic Accuracy of Hybrid Profiling Mechanisms in Cam Manufacturing." ASME 77-WA/DE-3.

Sarring, E. J. (1982). "Torque Compensation for Cam Systems." *Proc. of 7th Mechanism Conference*, Purdue Univ., pp. 179-185.

Sato, K., et al. (1997). "The Progress of Valvetrain Design and Exhaust Valve Material Research for Automobiles." *Proc. of Int. Symp. on Valvetrain System Design and Materials*, Dearborn, MI, p. 87.

Saunders, C. G., and G. W. Vickers. (1983). "A Generalized Approach to the Replication of Cylindrical Bodies with Compound Curvature." ASME 83-DET-2.

Savage, J. C., and J. P. Matterazzo. (1993). "Application of Design of Experiments to Determine the Leading Contributors to Engine Valve Noise." SAE 930884.

Schamel, A. R. (1991). "Development of a Mathematical Model for the Dynamics of Nonlinear Valve Springs." M.S. Thesis, Loughborough U. of Technology.

Schamel, A. R., et al. (1993). "Modeling and Measurement Techniques for Valve Spring Dynamics in High Revving Internal Combustion Engines." SAE 930615.

Schmidt, E. (1960). "Continuous Cam Curves." *Machine Design*, **32**(1), pp. 127-132.

Schumaker, L. (1983). "On Shape Preserving Quadratic Spline Interpolation." *SIAM J. Numer. Anal.*, **20**(4), pp. 854-864.

Schumaker, L. L. (1981). *Spline Functions: Basic Theory*. John Wiley & Sons: New York.

Sefler, J. F., and A. P. Pisano. (1993). "The Design, Experimentation, and Simulation of a Novel Coulomb Friction Device for Automotive Valve Spring Damping." *Journal of Mechanical Design*, **115**(4), pp. 871-876.

Seidlitz, S. (1989). "Valve Train Dynamics - a Computer Study." SAE 890620.

Seidlitz, S. (1990). "An Optimization Approach to Valve Train Design." SAE 901638.

Seller, J. F., and A. P. Pisano. (1993). "The Design, Experimentation, and Simulation of a Novel Coulomb Friction Device for Automotive Valve Spring Damping." *Journal of Mechanical Design*, **115**(4), p. 871.

Sendyka, B. (1993). "The Modifying of a Camshaft Profile to Decrease Acceleration and Jump of a Valve of a High Speed Combustion Engine." SAE 930059.

Sendyka, B. (1994). "The Analysis of the Dynamic Contact Stresses Occurring in the Valve-Camshaft System of the Internal Combustion Engine." SAE 940214.

Sidlof, P., et al. "The Influence of Dimensional Accuracy of the Function Contact Face of a Cam on the Follower Motion in Cam Mechanisms." *Proc.*

Skreiner, M., and P. Barkan. (1971). "On a Model of a Pneumatically Actuated Mechanical System." *Journal of Engineering for Industry*, **93**, p. 211.

Smoluk, G. R. "Toggle Linkage Found Efficient for Valve Actuator." *New Design Ideas*.

Speckhart. (1984). "Design of Maximum Acceleration Plate Cam Profiles Subject to Stress and Curvature Constraints." ASME 84-DET-131.

Srinivasan, L. N., and Q. J. Ge. (1996). "Parametric Continuous and Smooth Motion Interpolation." *Journal of Mechanical Design*, **116**(4), pp. 494-498.

Srinivasan, L. N., and Q. J. Ge. (1998). "Designing Dynamically Compensated and Robust Cam Profiles with Bernstein-Bezier Harmonic Curves." *Journal of Mechanical Design*, **120**(1), pp. 40-45.

Steinberg, R., et al. (1998). "A Fully Continuous Variable Cam Timing Concept for Intake and Exhaust Phasing." SAE 980767.

Stene, R. L., et al. (1998). "Analyzing Vibrations in an Ic Engine Valve Train." SAE 980570.

Stoddart, D. A. (1953a). "Polydyne Cam Design-I." *Machine Design*, January, 1953, pp. 121-135.

Stoddart, D. A. (1953b). "Polydyne Cam Design-II." *Machine Design*, February, 1953, pp. 146-154.

Stoddart, D. A. (1953c). "Polydyne Cam Design-III." *Machine Design*, March, 1953, pp. 149-164.

Strong, G., and X. Liang. (1997). "A Review of Valve Seat Insert Materials Required for Success." *Proc. of Int. Symp. on Valvetrain System Design and Materials*, Dearborn, MI, p. 121.

Subramanian, A. K. (1978). "Evaluation of Internal Combustion Engine Valve Trains by an Empirically Tuned Simulation Model." ASME 78-DGP-9.

Sugiura, A., et al. (1983). "Development of a Self-Contained Hydraulic Valve Lifter." SAE 831221.

Tanaka, H. (1976). "Kinematic Design of Cam-Follower Systems." Ph.D. Thesis, Columbia Univ.

Taylor, G. T., and T. Campbell. (1989). "Design Analysis of Cam and Tappet Interaction for Wear Reduction in Marine Diesel Engines." SAE 894309.

Terauchi, Y., and S. A. El-Shakery. (1983). "A Computer-Aided Method for Optimum Design of Plate Cam Size Avoiding Undercutting and Separation Phenomena - I." *Mechanism and Machine Theory*, **18**(No 2.), pp. 157 - 163.

Tesar, D. (1978). "Mission-Oriented Research for Light Machinery." *Science*, **201**(September 8, 1978), pp. 880 - 887.

Tesar, D., and G. K. Matthew. (1976). *The Dynamic Synthesis, Analysis, and Design of Modeled Cam Systems*. Lexington Books, DC Heath Co.: Lexington MA, p. 350.

Tesar, D., and G. K. Matthew, eds. (1978). "The Design of Modelled Cam Systems." *Cams and Cam Mechanisms*, Jones, J. R., ed., Institution of Mechanical Engineers: London, pp. 30-42.

Thoren, T. R., et al. (1952). "Cam Design as Related to Valve Train Dynamics." *SAE Quarterly Transactions*, **6**(1), pp. 1-14.

Tidwell, P. H., et al. (1994). "Synthesis of Wrapping Cams." *Journal of Mechanical Design*, **116**(2), p. 634.

Ting, K. L., et al. (1990). "Bezier Motion Programs in Cam Design." *Proc. of ASME 21st Biennial Mechanisms Conference*, Chicago, pp. 141-148.

Ting, K. L., et al. (1994). "Synthesis of Polynomial and Other Curves with the Bezier Technique." *Mechanism and Machine Theory*, **29**(6), pp. 887-903.

Torazza, G. (1972). "A Variable Lift and Event Control Device for Piston Engine Valve Operation." SAE 725025.

Tsay, D. M., and C. O. Huey. (1986). "Cam Profile Synthesis Using Spline Functions to Approximately Satisfy Displacement, Velocity, and Acceleration Constraints." 86-DET-109, ASME: Columbus OH, Oct 5, 1986.

Tsay, D. M., and C. O. Huey. (1987a). "Cam Motion Using Spline Functions: Part I - Basic Theory and Elementary Applications." *ASME Advances in Design Automation*, **2**, pp. 143-150.

Tsay, D. M., and C. O. Huey. (1987b). "Cam Motion Using Spline Functions: Part II - Applications." *ASME Advances in Design Automation*, **2**, pp. 151-157.

Tsay, D. M., and C. O. Huey. (1988). "Cam Motion Synthesis Using Spline Functions." *Journal of Mechanisms, Transmissions, and Automation in Design*, **110**, pp. 161-165.

Tsay, D. M., and C. O. Huey. (1989). "Synthesis and Analysis of Cam-Follower Systems with Two Degrees of Freedom." *Proc. of ASME Advances in Design Automation*, Montreal, Canada, pp. 281-288.

Tsay, D. M., and C. O. Huey. (1990). "Spline Functions Applied to the Synthesis and Analysis of Non-Rigid Cam-Folllower Systems." *Journal of Mechanisms, Transmissions, and Automation in Design*, **111**, pp. 561-569.

Tsay, D. M., and C. O. Huey. (1993). "Application of Rational B-Splines to the Synthesis of Cam-Follower Motion Programs." *Journal of Mechanical Design*, **115**(3), pp. 621-626.

Tsay, D. M., and G. S. Hwang. (1992). "Application of the Theory of Envelopes to to the Determination of Camoid Profiles with Translating Followers." *Proc. of ASME Mechanism Conference, Mechanism Design and Synthesis, DE-Vol. 46*, pp. 345-352.

Tsay, D. M., and H. M. Wei. (1993). "Profile Determination and Analysis of Cylindrical Cams with Oscillating Roller Followers." *Proc. of ASME Design Automation Conference, DE-Vol. 65-1*, pp. 711-718.

Tsay, D. M., and H. M. Wei. (1994). "A General Approach to the Determination of Planar and Spatial Cam Profiles." *Proc.* pp. 261-268.

Turqut, T. S., and U. Y. Samim. (1991). "Nondimensional Analysis of Jump Phenomenon in Force-Closed Cam Mechanisms." *Mechanism and Machine Theory*, **26**(4), pp. 421-432.

Turkish, M. C. (1953). "Relationship of Valve-Spring Design to Valve Gear Dynamics and Hydraulic Lifter Pump-Up." *SAE Transactions*, **61**, pp. 706-714.

Uhtenwoldt, H. R. (1974). "Cam Controlled Machine for Grinding a Noncircular Surface.": U.S. Patent 3,828,481.

Unlusoy, Y. S., and S. T. Tumer. (1993). "Analytical Dynamic Response of Elastic Cam-Follower Systems with Distributed Parameter Return Spring." *Journal of Mechanical Design*, **115**(3), p. 612.

Vogel, O., et al. (1997). "Variable Valve Timing Implemented with a Secondary Valve on a Four Cylinder SI Engine." SAE 970335.

Wampler, C. W. (1993). "Type Synthesis of Mechanisms for Variable Valve Actuation." SAE 930818.

Watson, H. C., and E. E. Milkins, eds. (1978). "The Design of Camshafts for Racing Engines." *Cams and Cam Mechanisms*, Jones, J. R., ed., Institution of Mechanical Engineers: London, pp. 54-59.

Weber, T. (1960a). "Cam Dynamics Via Filter Theory." *Machine Design*, v. 32, October 13, p. 160.

Weber, T. (1960b). "Filter Theory Applied to Cam Dynamics." *Proc. of Sixth Conference on Mechanisms*, Purdue University, pp. 48-54.

Weber, T. (1975a). "Cam Dynamics by Energy Economy Method." Report #1-9, Cam Technology, Inc.: Hawthorne, NY.

Weber, T. (1975b). "Cam Parameters." Report #1-2, CAM Technology, Inc.: Hawthorne, NY.

Weber, T. (1975c). "Generalization of Cam Transitions under Various Terminal Conditions." Report #1-4, Cam Technology, Inc.: Hawthorne, NY.

Weber, T. (1975d). "Parallel Curves and Cutoff." Report #1-5, Cam Technology, Inc.: Hawthorne, NY.

Weber, T. (1975e). "The Pivoted Arm Follower Cam." Report #1-3, CAM Technology, Inc..: Hawthorne, NY.

Weber, T. (1975f). "Precise Measurement of Contours by the Sweep Method." Report #1-6, Cam Technology, Inc.: Hawthorne, NY.

Weber, T. (1975g). "Self Conjugate Cams." Report #1-7, CAM Technology, Inc.: Hawthorne, NY.

Weber, T. (1976). "Facc Cams." Report #1-10, Cam Technology, Inc.: Hawthorne, NY.

Weber, T. (1977). "Cam Dynamics by Acceleration Vector Method." Report #1-8, Cam Technology, Inc.: Hawthorne, NY.

Weber, T. (1979). "Cam Curves by Fourier Synthesis." Report #1-11, Cam Technology, Inc.: Hawthorne, NY.

Weber, T. (1979). "Simplifying Complex Cam Design." *Machine Design*, March 22, pp. 115-119.

Weber, T. (1981). "Curvilinear Interpolation." Report #1-1, Cam Technology, Inc.: Hawthorne, NY.

Weber, T. (1983). "Useful Relationships between Cam Curvature and Acceleration Force." Report #1-12, Cam Technology, Inc.: Hawthorne, NY.

Wiederrich, J. L. (1973). "Design of Cam Profiles for Systems with High Inertial Loading." Ph.D. Thesis, Stanford University.

Wiederrich, J. L. (1981). "Residual Vibration Criteria Applied to Multiple Degree of Freedom Cam Followers." *Journal of Mechanical Design*, **103**, pp. 702-705.

Wiederrich, J. L. (1982). "A Theory for the Identification of a Single Degree of Freedom Machine from Its Periodic Operating Characteristics." ASME 82-DET-1.

Wiederrich, J. L., and B. Roth. (1975). "Dynamic Synthesis of Cams Using Finite Trigonometric Series." *Journal of Engineering for Industry*, pp. 287-293.

Wiederrich, J. L., and B. Roth, eds. (1978). "Design of Low Vibration Cam Profiles." *Cams and Cam Mechanisms*, Jones, J. R., ed., Institution of Mechanical Engineers: London, pp. 3-8.

Wilson, N. O., et al. (1992). "Future Valve Train Systems to Reduce Engine Performance Compromises." SAE 925102.

Winfrey, R. C., et al. (1973). "Analysis of Elastic Machinery with Clearances." *Journal of Engineering for Industry*, pp. 695-703.

Wunderlich, W. (1971). "Contributions to the Geometry of Cam Mechanisms with Oscillating Followers." *Journal of Mechanisms*, **6**, pp. 1-20.

Yagi, S., et al. (1992). "A New Variable Valve Mechanism; 'Shuttle Cam'." SAE 92A048.

Yan, H. S., and H. M. Ting. (1996). "The Effects of Manufacturing Errors on Planar Conjugate Cams." *J. Chinese Soc. of Mech. Eng.*, **17**(2), pp. 145-153.

Yan, H. S., et al. (1996). "An Experimental Study of the Effects of Cam Speeds on Cam Follower Systems." *Mechanism and Machine Theory*, **31**(4), pp. 397-412.

Yang, D. C. H., and F. C. Wang. (1994). "A Quintic Spline Interpolator for Motion Command Generation of Computer-Controlled Machines." *Journal of Mechanical Design*, **116**(1), pp. 226-231.

Yilmaz, Y., and H. Kocabas. (1995). "The Vibration of Disc Cam Mechanism." *Mechanism and Machine Theory*, **30**(5), pp. 695-703.

Yoon, K., and S. S. Rao. (1993). "Cam Motion Synthesis Using Cubic Splines." *Journal of Mechanical Design*, **115**(3), pp. 441-446.

Young, S. D., and T. E. Shoup. (1982). "The Sensitivity Analysis of Cam Mechanism Dynamics." *Journal of Mechanical Design*, **104**, pp. 476-481.

Yu, Q., and H. P. Lee. (1995). "A New Family of Parameterized Polynomials for Cam Motion Synthesis." *Journal of Mechanical Design*, **117**(4), p. 653.

Yu, Q., and H. P. Lee. (1996). "Curvature Optimization of a Cam Mechanism with a Translating Follower." *Journal of Mechanical Design*, **118**, p. 446.

Zhang, W., and M. Ye. (1994). "Local and Global Bifurcations of Valve Mechanism." *Nonlinear Dynamics*, **6**, pp. 301-316.

GLOSSARY OF TERMS

AC	*alternating current*
axial cam	*see barrel cam*
B-spline	*basis spline, a type of mathematical function useful for cam-follower motion*
barrel cam	*one in which follower motion is in the axial direction*
BC	*boundary condition*
blob	*mass unit in the ips system. 1 blob = 1 lb-sec2/in = 12 slugs*
CAD	*computer aided design*
CAM	*computer aided manufacturing, not to be confused with a real "cam" (in lower case)*
camoid	*a three-dimensional cam*
CCW	*counterclockwise*
CDTF	*cam dynamic test fixture*
CEP	*critical extreme position, typically for double-dwell motions*
CG	*center of gravity*
CI	*circular interpolation*
CMM	*coordinate measuring machine*
CNC	*continuous numerical control*
CPM	*critical path motion, a cam program with additional constraints*
CW	*clockwise*
cylindrical cam	*see barrel cam*
damper	*a mechanical energy dissipator. e.g., friction, viscous drag*
DC	*direct current*
DFT	*discrete fourier transform*
DOF	*degree of freedom*
DOHC	*dual overhead camshaft engine (pronounced "dock")*
DSA	*dynamic signal analyzer*
DYNACAM	*computer program for design and analysis of cam-follower systems*
EDM	*electrical discharge machining*
FEA	*finite element analysis (used interchangebly with FEM)*
FEM	*finite element method (used interchangebly with FEA)*
FFT	*fast fourier transform*
FRF	*frequency response function of a system in the Fourier domain*
GCS	*global coordinate system*
GSC	*geometric stress concentration*
HRC	*Rockwell hardness on the C scale*
Hz	*a unit, hertz or cycles per second*
IC	*internal combustion, as in engine*
ICP	*integrated circuit piezoelectric*
LI	*linear interpolation*
LIVM	*low impedance voltage mode*

LNCS	*local nonrotating coordinate system*
LRCS	*local rotating coordinate system*
LVDT	*linear variable differential transformer*
LVT	*linear velocity transducer*
MDOF	*multi-degree of freedom*
mod sine	*colloquial shortening of modified sine*
mod trap	*colloquial shortening of modified trapezoid*
ms	*millisecond*
NC	*numerical control*
NURB	*nonuniform, rational, B-spline*
ODE	*ordinary differential equation*
OHC	*overhead camshaft engine*
OSHA	*Occupational Safety and Health Administration, a federal government agency*
ping	*the 4th time derivative of displacement and the time derivative of jerk*
PM	*permanent magnet, a type of electric motor, also powdered metal*
polydyne	*polynomial + dynamic, refers to a type of cam design for reduced vibration using polynomials*
psi	*pounds per square inch, a measure of pressure in the ips system*
psia	*absolute pressure referenced to zero pressure*
psig	*gage pressure, referenced to local atmospheric pressure*
PSO	*point surface origin*
puff	*the 5th time derivative of displacement and the time derivative of ping*
Q	*quality - a measure of the efficiency or lack of damping in a system*
rad/sec	*radians per second*
radial cam	*one in which follower motion is in the radial direction*
RDFD	*rise dwell fall dwell, a double or multiple dwell cam motion specification*
RF	*rise fall, a type of cam motion specification with no dwell*
RFD	*rise fall dwell, a type of cam motion specification with single dwell*
rms	*root mean square*
rpm	*revolutions per minute*
rps	*revolutions per second*
SCCA	*sine-cosine-constant-acceleration - a family of double-dwell cam functions*
SDOF	*single degree of freedom*
SEM	*scanning electron microscope*
SHM	*simple harmonic motion*
SOHC	*single overhead camshaft engine (pronounced "sock")*
splinedyne	*spline + dynamic; refers to a type of cam design for reduced vibration using B-splines*
SSU	*standard Saybolt units (viscosity)*
$s\ v\ a\ j$	*displacement (s), velocity (v), acceleration (a), and jerk (j) cam motion funtions with units versus radians*
$S\ V\ A\ J$	*displacement (s), velocity (v), acceleration (a), and jerk (j) cam motion funtions with units versus time in seconds*
TDC	*top dead center, as in an IC engine*
TF	*transfer function of a system in the Laplace domain*
TPB	*truncated power basis*
VF	*variable frequency, a type of electrical servomotor*
WPI	*Worcester Polytechnic Institute*

INDEX

A

B

C

G

H

Q

R

S

Index to CD-ROM

I Custom Programs

II Example Files

III TKSolver Files

COMPUTER PROGRAMS

The attached CD-ROM contains four computer programs in demonstration versions with limited time licenses for their use. Each program may be installed on a computer and used for a period of ninety (90) days after installation. During that period the program may be run a maximum of 30 times. After either of these limits is reached, the program will display a dialog box containing a multi-digit code number. If you wish to license the program, you can send the license fee (defined on the registration screen within each program) and the displayed code number to the address shown and an unlock code will be supplied to enable the program in its full, professional version. The demonstration versions supplied will perform all program functions except for *Saving* or *Opening* a data file (except for the example files supplied), or *Exporting* resulting data to files for manufacturing purposes. All output values and plotted functions can nevertheless be viewed on the screen. The example files supplied on the CD-ROM can be opened in their parent program.

The programs supplied are:

DYNACAM - for cam design and analysis

FOURBAR - for design and analysis of fourbar linkages

SIXBAR - for design and analysis of sixbar linkages

SLIDER - for design and analysis of crank-slide linkages

SYSTEM REQUIREMENTS FOR THE PROGRAMS

- Pentium processor or better
- Windows® 95/98/2000 or Windows NT™ 4.0 or higher.
- 32 megabytes of RAM (64 MB strongly recommended).
- CD-ROM drive.

INSTALLATION INSTRUCTIONS FOR PROGRAMS AND FILES

1. Quit all applications currently running on your computer.
2. Insert the CD-ROM into your computer's CD-ROM drive.
3. Use Explorer or other method to see the contents of the CD-ROM.
4. Open the folder of the program that you wish to install, e.g., DYNACAM, FOURBAR, SIXBAR, or SLIDER and locate the file **Setup.exe**.
5. Double click on **Setup.exe**, or *Run* **Setup.exe** from Explorer.
6. Follow the on-screen instructions.
7. You may see a dialog box telling you that it is attempting to replace a file already on your computer with an older version. You should always keep your newer version rather than replace it with an older one. The correct response to accomplsh this is "Yes" (the default). You also may see a dialog box telling you that a file cannot be installed because it is protected or in use. You should choose "Ignore" and proceed in this case.
8. The data files provided can be copied from the CD-ROM to a hard drive with standard Windows copying techniques. Some of these require the program *TKSolver*, a free demo of which can be downloaded from http://www.uts.com.